T0187778

Nanoengineering, Quantum Science, and, Nanotechnology Handbook

Nanoengineering, Quantum Science, and, Nanotechnology Handbook

Edited By

Sergey Edward Lyshevski

CRC Press is an imprint of the
Taylor & Francis Group, an **informa** business

CRC Press
Taylor & Francis Group
6000 Broken Sound Parkway NW, Suite 300
Boca Raton, FL 33487-2742

Printed on acid-free paper

International Standard Book Number-13: 978-0-367-19751-3 (Hardback)

Visit the Taylor & Francis Web site at
http://www.taylorandfrancis.com

and the CRC Press Web site at
http://www.crcpress.com

Contents

Preface

It was a great pleasure to edit the *Nanoengineering, Quantum Science, and, Nanotechnology Handbook,* which consists of outstanding chapters written by acclaimed experts in the fields on nanotechnology, nanobiotechnology, quantum mechanics, and information sciences. The overall objective was to provide coherent coverage of a broad spectrum of issues in nanoscale electronics, energy sources, and materials. Basic fundamentals, recent innovations, novel solutions, emerging technologies, far-reaching developments, and new paradigms are covered. Molecular electronics and material sciences are transformative developments. The handbook's chapters document practical solutions with applications. The contributed chapters focus on balanced coverage on envisioned innovations and discoveries with substantiation and justifications toward technology transfers.

Due to high-risk but high-payoff developments and the abrupt nature of fundamental discoveries, it is difficult to accurately predict when new discoveries mature to a commercial product. Matured microelectronics and biotechnology have profoundly contributed to technological progress and social welfare. Further progress is envisioned in life and physical sciences. The evolutionary and revolutionary developments encountered significant fundamental and technological challenges. In attempts to find new practical solutions and define novel inroads, innovative paradigms and technologies are under development. The progress accomplished, promising solutions, and innovative paradigms are reported in this handbook.

The handbook's chapters report the authors' results. Reading chapters, the reader may observe variations of views, as well as differences in outlooks. This may be not a weakness, but rather a strength. The reader should be aware of the differences in opinions, different approaches, and alternative solutions. I truly enjoyed collaborating with all contributors and appreciate their valuable contributions. All chapters emphasize the need for research and further developments in nano and quantum engineering, science, and technology, which are the forefront frontier.

The views, findings, recommendations, and conclusions documented in the handbook's chapters are those of authors and do not necessarily reflect the editor's opinion. No matter how many times the material has been reviewed and efforts are spent to guarantee the highest quality, there is no guarantee that the handbook is free from minor errors, deficiencies and shortcomings. If you find something that you feel needs correction, adjustment, clarification, and/or modification, please notify me or the authors. Your help and assistance are greatly appreciated and deeply acknowledged.

Acknowledgments

Many people contributed to this book. First, thanks go to all contributors, to whom I would like to express my sincere acknowledgments and gratitude. It gives me a great pleasure to acknowledge the help I received from many people in the preparation of this handbook. The outstanding CRC Press team, especially Nora Konopka (Acquisitions Editor, Electrical Engineering) and Glenon Butler (Project Editor), tremendously helped and assisted me by providing valuable and deeply treasured feedback. Many thanks to all of you.

March 2019, Rochester, NY

Sergey Edward Lyshevski, PhD, Professor of Electrical Engineering
Department of Electrical and Microelectronic Engineering
Rochester Institute of Technology, Rochester, NY 14623, USA
E-mail: Sergey.Lyshevski@mail.rit.edu
URLs: https://people.rit.edu/seleee

Editor

Dr. Sergey Edward Lyshevski received an MS and a PhD in Electrical Engineering from Kiev Polytechnic Institute in 1980 and 1987, respectively. From 1980 to 1993, he held research and faculty positions at the Department of Electrical Engineering, Kiev Polytechnic Institute, and the Academy of Sciences of Ukraine. From 1989 to 1993, he was the Microelectronic and Electromechanical Systems division head at the Academy of Sciences of Ukraine. From 1993 to 2002, he was associate professor of Electrical and Computer Engineering, Purdue School of Engineering IUPUI. In 2002, he joined Rochester Institute of Technology as professor of Electrical Engineering. Dr. Lyshevski served as professor of Electrical and Computer Engineering in the US Department of State Fulbright program. He is a full professor faculty fellow at the Air Force Research Laboratories, US Naval Surface Warfare Center, and US Naval Undersea Warfare Center.

Dr. Lyshevski is the author and co-author of 13 books, 14 handbook chapters, more than 80 journal articles, and over 300 refereed conference papers. He serves as an editor of encyclopedia and handbooks and has conducted more than 100 invited tutorials, workshops, and keynote talks.

As a principal investigator (project director), he performed contracts and grants for high-technology industry (Allison Transmission, Cummins, Delco, Delphi, Harris, Lockheed Martin, Raytheon, General Dynamics, General Motors, and others), the US Department of Defense (AFRL, AFOSR, DARPA, ONR, and Air Force), and government agencies (DoE, DoT, and NSF). Dr. Lyshevski conducts research and technology developments in microsystems, cyber-physical systems, information sciences, and electromechanical systems. He has made a significant contribution in the design, deployment, and commercialization of advanced aerospace, automotive, and naval systems.

Contributors

Rita Aghjayan
Pace University
New York, New York

Lucas Fernandes Aguiar
Chemistry Institute
Federal University of Goiás
Goiânia, Brazil

Ellen T. Chen
Advanced Biomimetic Sensors, Inc.
Rockville, Maryland

Diéricon Sousa Cordeiro
Chemistry Institute
Federal University of Goiás
Goiânia, Brazil

Adão Marcos Ferreira Costa
Chemistry Institute
Federal University of Goiás
Goiânia, Brazil

G.C. Dannangoda
Department of Physics and Astronomy
University of Texas at Rio Grande Valley
Brownsville, Texas

Sahar Daraeizadeh
Department of Electrical and Computer Engineering
Portland State University
Portland, Oregon
and
Intel Labs, Intel Corporation
Hillsboro, Oregon

Brad C. Dodrill
Lake Shore Cryotronics, Inc.
Westerville, Ohio

Lixin Dong
Department of Biomedical Engineering
City University of Hong Kong
Kowloon Tong, Hong Kong, China

Rodolfo Fabián Estrada-Guerrero
Department of Physics and Mathematics
Iberoamericana University
Mexico City, Mexico

Edward P. Furlani
Department of Electrical Engineering
Department of Chemical and Biological Engineering
The State University of New York at Buffalo
Buffalo, New York, USA

Danko D. Georgiev
Eastern Illinois University
Charleston, Illinois
and
Institute for Advanced Study
Varna, Bulgaria

James F. Glazebrook
Department of Mathematics and Computer Science
Eastern Illinois University
Charleston, Illinois

Jennifer L. Gray
Materials Research Institute
Penn State University
University Park, Pennsylvania

Parameswar Hari
Department of Physics
The Oklahoma Photovoltaic Research Institute
The University of Tulsa
Tulsa, Oklahoma

Justin D. Holmes
Materials Chemistry and Analysis Group
School of Chemistry and the Tyndall National Institute
University College Cork
Cork, Ireland
and
Centre for Research on Adaptive Nanostructures and Nanodevices (CRANN)
Trinity College Dublin
Dublin, Ireland

Chaojian Hou
Key Laboratory of Microsystems and Microstructures Manufacturing
Ministry of Education and School of Mechatronics Engineering
Harbin Institute of Technology
Harbin, Heilongjiang, China

Naoki Inomata
Graduate School of Engineering
Tohoku University
Sendai, Japan

Amrit Kaphle
Department of Physics
The University of Tulsa
Tulsa, Oklahoma

Sungho Kim
Department of Electrical & Computer Engineering
Carnegie Mellon University
Pittsburgh, Pennsylvania

A.N. Kocharian
California State University
Los Angeles, California

G.A. Kumar
Department of Physics and Astronomy
University of Texas at San Antonio
San Antonio, Texas

Alexander S. Liberson
Mechanical Engineering Department
Rochester Institute of Technology
Rochester, New York

Kai Liu
Department of Electrical Engineering
The State University of New York at Buffalo
Buffalo, New York, USA

Arthur Luniewski
Pace University
New York, New York

Sergey Edward Lyshevski
Department of Electrical and Microelectronic Engineering
Rochester Institute of Technology
Rochester, New York

Tatiana Duque Martins
Chemistry Institute
Federal University of Goiás
Goiânia, Brazil

K.S. Martirosyan
Department of Physics and Astronomy
University of Texas at Rio Grande Valley
Brownsville, Texas

Anne Y. Matsuura
Intel Labs, Intel Corporation
Portland, Oregon

Efstathios I. Meletis
Department of Materials Science and Engineering
University of Texas at Arlington
Arlington, Texas

María Eugenia Mena-Navarro
Department of Physics and Mathematics
Iberoamericana University
Mexico City, Mexico

Ramon Silva Miranda
Chemistry Institute
Federal University of Goiás
Goiânia, Brazil

Amir Mokhtare
Department of Chemical and Biological Engineering
The State University of New York at Buffalo
Buffalo, New York, USA

R.M. Movsesyan
The State Engineering University of Armenia
Yerevan, Armenia

Ganga R. Neupane
Department of Physics
The University of Tulsa
Tulsa, Oklahoma

Takahito Ono
Graduate School of Engineering
Tohoku University
Sendai, Japan

Ece Isenbike Ozalp
Department of Electrical & Computer Engineering
Carnegie Mellon University
Pittsburgh, Pennsylvania

Anthony Palumbo
Department of Mechanical Engineering
Stevens Institute of Technology
Hoboken, New Jersey

Tereza M. Paronyan
Hexalayer, LLC
Louisville, Kentucky

Marek Perkowski
Department of Electrical and Computer Engineering
Portland State University
Portland, Oregon

A.V. Poghosyan
The State Engineering University of Armenia
Yerevan, Armenia

Shavindra P. Premaratne
Intel Labs, Intel Corporation
Portland, Oregon

Ivan Puchades
Department of Electrical and Microelectronic Engineering
Rochester Institute of Technology
Rochester, New York

Kamil Rahme
Department of Sciences
Faculty of Natural and Applied Science
Notre Dame University-Louaize
Zouk Mosbeh, Lebanon
and
Materials Chemistry and Analysis Group
School of Chemistry and the Tyndall National Institute
University College Cork
Cork, Ireland

Manuel Reali
Department of Engineering Physics
Polytechnique Montréal
Montréal, Québec, Canada

Innem V.A.K. Reddy
Department of Electrical Engineering
The State University of New York at Buffalo
Buffalo, New York, USA

Antonio Carlos Chaves Ribeiro
Federal Institute of Education, Science and Technology of Goiano
Morrinhos, Goiás, Brazil

A.S. Sahakyan
The State Engineering University of Armenia
Yerevan, Armenia

Clara Santato
Department of Engineering Physics
Polytechnique Montréal
Montréal, Québec, Canada

Farad Sarker
Department of Materials Science and Engineering
The University of Texas at Arlington
Arlington, Texas

Meet Shah
Department of Materials Science and Engineering
The University of Texas at Arlington
Arlington, Texas

Jeffrey Shallenberger
Materials Research Institute
Penn State University
University Park, Pennsylvania

Madhu Singh
John and Willie Leone Family Department of Energy and Mineral Engineering
EMS Energy Institute
Penn State University
University Park, Pennsylvania

Geovany Albino de Souza
Chemistry Institute
Federal University of Goiás
Goiânia, Brazil

Viktor Sukhotskiy
Department of Electrical Engineering
The State University of New York at Buffalo
Buffalo, New York, USA

Vignesh Sundar
Department of Electrical & Computer Engineering
Carnegie Mellon University
Pittsburgh, Pennsylvania

Harry F. Tibbals
Department of Materials Science and Engineering
The University of Texas at Arlington
Arlington, Texas

Nguyen Van Toan
Graduate School of Engineering
Tohoku University
Sendai, Japan

Masaya Toda
Graduate School of Engineering
Tohoku University
Sendai, Japan

Luis Adolfo Torres-González
Department of Science and Engineering
Iberoamericana University
León, Mexico

Randy L. Vander Wal
John and Willie Leone Family Department of Energy and Mineral Engineering
EMS Energy Institute
Penn State University
University Park, Pennsylvania

Kamil Walczak
Pace University
New York, New York

Genwang Wang
Key Laboratory of Microsystems and Microstructures Manufacturing
Ministry of Education and School of Mechatronics Engineering
Harbin Institute of Technology
Harbin, Heilongjiang, China

Yang Wang
Key Laboratory of Microsystems and Microstructures Manufacturing
Ministry of Education and School of Mechatronics Engineering
Harbin Institute of Technology
Harbin, Heilongjiang, China

Jeffrey A. Weldon
Department of Electrical Engineering
University of Hawaii at Manoa
Honolulu, Hawaii

Jian Xu
Department of Mechanical Engineering
Stevens Institute of Technology
Hoboken, New Jersey

Xiaozheng Xue
Department of Chemical and Biological Engineering
The State University of New York at Buffalo
Buffalo, New York, USA

Suni Yadev
Department of Materials Science and Engineering
The University of Texas at Arlington
Arlington, Texas

Eui-Hyeok Yang
Department of Mechanical Engineering
Stevens Institute of Technology
Hoboken, New Jersey

Lijun Yang
Key Laboratory of Microsystems and Microstructures Manufacturing
Ministry of Education and School of Mechatronics Engineering
Harbin Institute of Technology
Harbin, Heilongjiang, China

Zhan Yang
School of Mechatronics Engineering
Soochow University
Soochow, Jiangsu, China

Runzhi Zhang
Department of Mechanical Engineering
Stevens Institute of Technology
Hoboken, New Jersey

Jian-Gang (Jimmy) Zhu
Department of Electrical & Computer Engineering
Carnegie Mellon University
Pittsburgh, Pennsylvania

Tawanda J. Zimudzi
Materials Research Institute
Penn State University
University Park, Pennsylvania

1 Capacitative Silicon Resonator Technologies

Nguyen Van Toan, Naoki Inomata, Masaya Toda, and Takahito Ono
Graduate School of Engineering, Tohoku University, Sendai, Japan

CONTENTS

1.1 INTRODUCTION

Microfabricated resonators play an essential role in a variety of applications [1–9]. Resonating ultra-thin cantilevers for the mass sensing was presented by Ono and Kim [1–2], where the resonant frequency of the cantilever monitors an adsorbed mass on its surface. Timing reference applications that use the resonator structures to generate clock signals in electronic systems were demonstrated in many studies [3–6]. Microfabricated resonators were also employed for filtering applications [7–9], which can be of use in the radio frequency transmitter and receiver modules nowadays. Many transduction mechanisms—consisting of piezoelectric [7–10], piezoresistive [11–12], and capacitive [5–6, 13] resonators—have been studied to excite and detect the motion of resonators. All of the above methods possess both advantages and disadvantages.

Piezoelectric resonators, usually used for filtering applications [9, 14], offer low insertion losses and small motional resistance. But the quality factors (Q) of resonators are small and good piezoelectric materials are vital. Piezoelectric micromechanical resonators have similar structures to quartz structures that consist of piezoelectric material sandwiched between two metal electrodes. When the electric polarization applies to these electrodes, the mechanical stress or vice versa occurs. Many piezoelectric materials, such as lead zirconate titanate (PZT), aluminum nitride (AlN), or zinc oxide (ZnO), have been investigated; however, further improvements are necessary to achieve high piezoelectric coefficients and low residual stress.

Piezoresistive resonators were presented by Beek and Rahafrooz [11–12] for the applications of oscillators and thermal-piezoresistive energy pumps. Piezoresistive sensing with the thermal actuation shows a high-quality factor and low insertion loss; however, it faces significant power consumption. This technique is not appropriate for the construction of a high-sensitivity products due to strong effects from environmental conditions, such as temperature dependence. There have been some efforts to improve the piezoresistive transduction sensitivity recently [15–16]. Self-sustained oscillators without any amplifying circuitry can be achieved by piezoresistive heat engines, which are based on thermodynamic cycles.

Their operation principles are as follows:

> Due to the higher resistance of narrow piezoresistive beams over other parts of the resonator structure, the piezoresistive beams are heated up when a DC voltage is applied to the piezoresistive elements. This results in the expansion of the piezoresistive beams, which causes the resistance to increase due to the piezoresistive effect. As a consequence of this resistance increase, the current passing through the piezoresistive beams decreases, which also decreases the temperature of the piezoresistive beams. Thus, the piezoresistive beams are compressed, and the resistance decreases due to the piezoresistive effect, resulting in an increase in the current. From this, a thermomechanical actuation power is generated by the above cycle in the piezoresistive elements. Therefore, the self-oscillators can be achieved [15].

Capacitive resonators, typically employed for sensing and timing applications, are based on the measurement of the change in the capacitance between a sensing electrode and the resonant body. Capacitive resonators exhibit a high quality factor Q, which results in the stability and low internal friction. However, their drawbacks are large motional resistances and high insertion losses, which make it difficult to satisfy oscillation conditions. Also, the large motional resistance results in a significant phase noise in the oscillator; therefore, the motional resistance should be as small as possible. Methods for lowering the motional resistance presented in [17–24] utilize the reduction of the capacitive gap width [17–19], and the increase of the overlap area of the capacitance, the quality factor Q and the polarization voltage V_{DC}. Unfortunately, each of the above methods has some drawbacks. Decreasing the capacitive gap width is very efficient in lowering the motional resistance; nevertheless, the fabrication of a nanogap is very difficult, and the applicable maximum polarization voltage decreases due to the pull-in phenomenon. Smaller capacitive gaps result in lower pull-in voltages, which easily cause short circuit situations. An increase in the electrode areas being utilized—such as resonator arrays [4, 20] and mechanical coupling [21]—can reduce the motional resistance, but the frequency response and the Q factor of the resonators suffer from the mismatches in the individual resonant frequencies. To increase the Q factor, some of the methods to reduce the energy losses, including external and internal losses, have been reported, but there are material and structural limitations [5, 22–24]. Increasing the polarization voltage has a limit because a high voltage is not available on a complementary metal-oxide-semiconductor (CMOS) chip. It means that more challenges and difficulties face the large-scale integration (LSI) process for the silicon resonator.

In this chapter, we present the recent progress on the capacitive silicon resonators, including the different fabrication technologies and design considerations to achieve small motional resistance, low insertion loss, and high-quality factor, which are mainly based on the works of the authors. The paper is organized as follows: In section 1.2, basic concepts of the capacitive silicon resonator structures, including the device structure, working principles an equivalent circuit model, are introduced. Also, the vibration modes are demonstrated by a finite element method (FEM). In section 1.3, we present the performance enhancement methods of the capacitive silicon resonators, which consisted of:

- A hermetically packaged silicon resonator on a low temperature co-fired ceramic (LTCC) substrate.
- A long bar-type silicon resonator with a high Q factor.
- Capacitive silicon resonators with a low-damage process using neutral beam-etching technology.
- Mechanically coupled capacitive silicon nanomechanical resonators.

- Capacitive silicon resonator with movable electrode structures.
- Capacitive silicon resonators with piezoresistive heat engines.

Finally, we summarize the topic in section 1.4.

1.2 DEVICE STRUCTURE AND WORKING PRINCIPLE

Several kinds of capacitive silicon resonators, including bar type [6, 17–19, 23], square type [22, 25–26], two-arm type [6, 11], disk-type [27–28], etc., have been studied. Basic components of capacitive silicon resonators are the resonant body, driving/sensing electrodes, supporting beams, and capacitive gaps. Figures 1.1(a) and (c) show the two-arm type resonator and bar-type resonator, respectively, where the resonant bodies are placed between driving/sensing electrodes, supported by two thin beams on the sides, and separated by the narrow capacitive gaps.

Resonators are excited in their horizontal extensional mode at the fundamental resonant frequency. This is demonstrated via the finite element method (FEM) simulation, as shown in Figures 1.1(b) and (d).

The capacitive resonator's operation is described as follows:

An AC voltage V_{AC} is supplied to the driving electrode. The resulting electrostatic force induces a bulk acoustic wave in the resonant body. Additional DC voltage V_{DC} is applied to the driving/sensing electrodes to amplify the electrostatic force. This electrostatic force makes

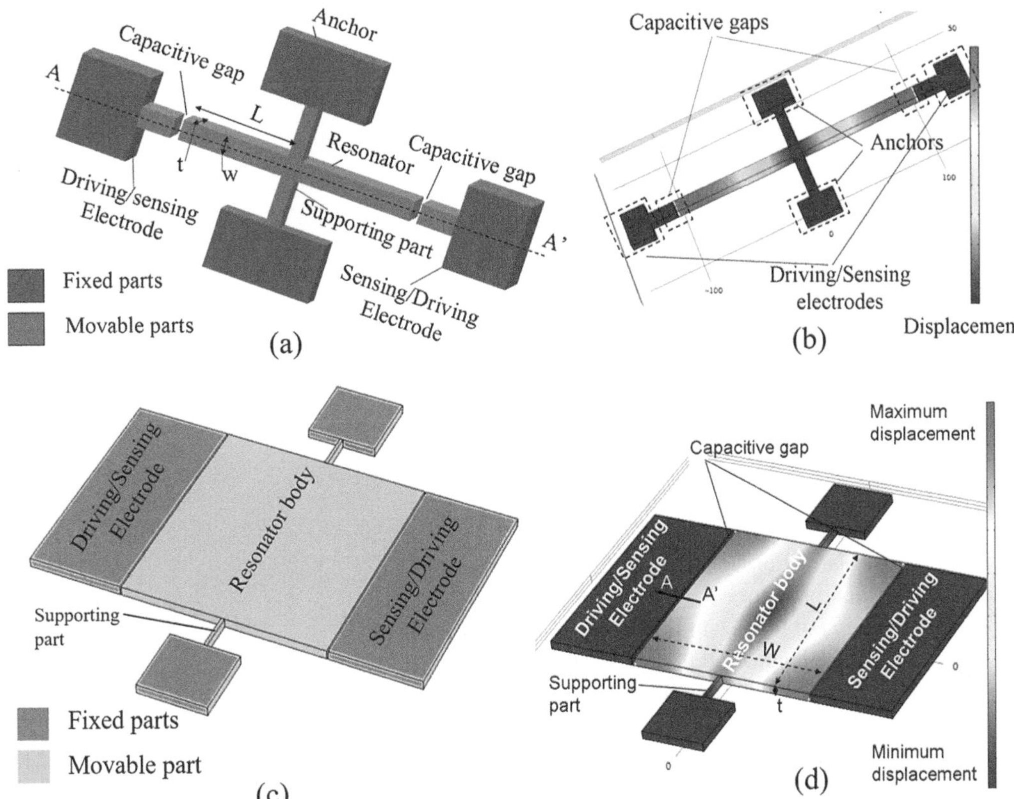

FIGURE 1.1 (a) Two-arm type resonator structure. (b) FEM simulation result on vibration mode of two-arm type resonator. (c) Bar-type resonator structure. (d) FEM simulation result on vibration mode of bar-type resonator.

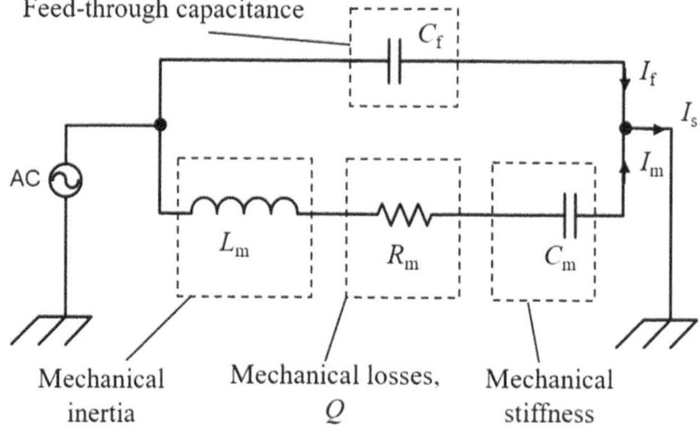

FIGURE 1.2 Equivalent circuit model.

the resonant body actuated. Small changes in the capacitive gap width between the sensing electrode and resonant body generate a voltage on the sensing electrode.

Figure 1.2 illustrates the equivalent circuit model of the capacitive resonator. It consists of the motional resistance R_m, the motional inductance L_m, the motional capacitance C_m, and the feed-through capacitance C_f, as presented in [11, 29–30]. The motional resistance R_m represents mechanical losses of vibration. The motional inductance L_m reveals mechanical inertia. The motional capacitance C_m corresponds to mechanical stiffness.

$$R_m = \frac{\sqrt{k_{eff} m_{eff}}}{Q\eta^2} = \frac{\sqrt{k_{eff} m_{eff}}}{QV_{DC}^2 \varepsilon_0^2 L^2 t^2} g^4, \tag{1.1}$$

$$L_m = \frac{m_{eff}}{\eta^2}, \tag{1.2}$$

$$C_m = \frac{\eta^2}{k_{eff}}, \tag{1.3}$$

$$C_f = \varepsilon_0 \frac{A_{el}}{g}, \tag{1.4}$$

$$\eta = V_{DC} \frac{\partial C_f}{\partial g} = -V_{DC} \frac{C_f}{g}, \tag{1.5}$$

where,

- k_{eff} is an effective spring constant.
- m_{eff} is an effective mass.
- V_{DC} is a polarization voltage.
- Q is the quality factor of the resonator.
- A_{el} is the area of the electrode plate.
- L and t are the length and thickness of resonator, respectively.
- g is the capacitive gap between resonant body.
- ε_0 is the electric constant ($\varepsilon_0 = 8.854 \times 10^{-12}$ Fm^{-1}).
- η is an electromechanical transduction factor.

The resonant frequency f_0 of the fundamental extensional mode is given by:

$$f_0 = \frac{1}{2\pi} \sqrt{\frac{k_{eff}}{m_{eff}}} = \frac{1}{2W} \sqrt{\frac{E}{\rho}} \tag{1.6}$$

where W is the width of the resonant body. E and ρ are Young's modulus (170 GPa) and density (2.32 g/cm^3) of the structural material, respectively.

1.3 PERFORMANCE ENHANCEMENT METHODS

1.3.1 Hermetically Packaged Silicon Resonator on the LTCC Substrate

Packaging process [31] for encapsulating and electrical interconnections is a critical technology for practical applications of the microfabricated resonators. It is to avoid not only the viscous damping for obtaining a high-quality factor and high stability of the resonance, but also by the ability to protect the device from the external environment and to avoid issues with moisture and particles for the long-term operation. The successful integration of a resonator with LSI can easily compensate the temperature drift of the resonator by using a temperature compensation circuit.

Encapsulation using a thin-film package was presented in [32]. Using a high-temperature process for the deposition of the thin film and releasing a significant amount of hydrogen are the disadvantages of this method. Thin-film packaging process based on a metal-organic thin-film [33] is available at low temperature (<110°C), but it also suffers from process incompatibilities. Although getters are typically employed for the vacuum packaging to absorb the trapped and desorbed gases during the sealing process, to become active, getters need to be heated in a vacuum for times and temperatures from 300°C to 500°C. The packaging method using a conventional borosilicate glass substrate with electrical feed-through was demonstrated in [34–35]. Nevertheless, the quality of via-holes with metal is poor. The packaging process using LTCC substrate offers great potential to reduce cost and improve reliability. This substrate is a multilayer glass ceramic substrate, which contains the metal feed-throughs and can be anodically bonded to silicon. The sealed packaging of radio-frequency microelectrical systems (RFMEMS) produced by the anodic bonding process using LTCC was reported in [36–38].

Silicon resonators with the hermetic packaging based on the utilization of the LTCC substrate were proposed for the integration of the resonator with LSI, as presented in our works [6, 19, 23–24]. The resonator structure is transferred onto the LTCC substrate using the anodic bonding between silicon and LTCC for electrical interconnections. The resonator structure is packaged hermetically by the second anodic bonding of the silicon and Tempax glass. These works bring out a new approach to combine the vacuum-packaging technology with the silicon resonator on the LTCC substrate via the anodic bonding method. Figures 1.3(a) and (b) describe the titled and cross section views of the proposed capacitive silicon resonators, respectively.

Figure 1.4 describes the fabrication process of the hermetically packaged silicon resonator on LTCC substrate. Resonator structures are produced by a combination of electron beam lithography and photolithography, and the deep reactive ion etching (RIE) process on a silicon on insulator (SOI) wafer [Figure 1.4(b)], and then transfer to the LTCC substrate by using an anodic bonding

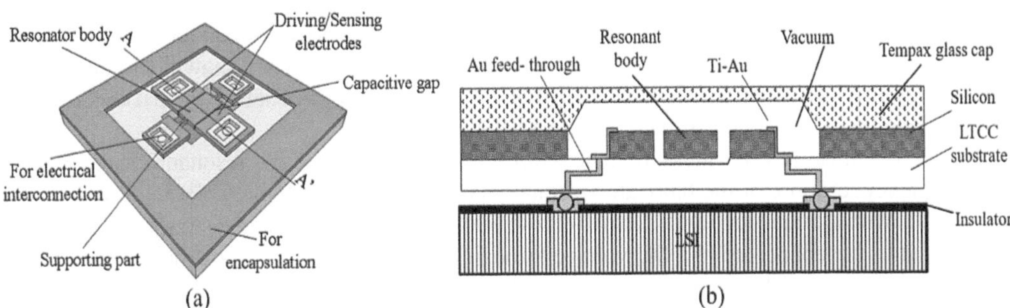

FIGURE 1.3 (a) Titled view of the capacitive silicon resonator forming on LTCC via anodic bonding (before packaging process). (b) Cross section view of the proposed structure (after packaging process).

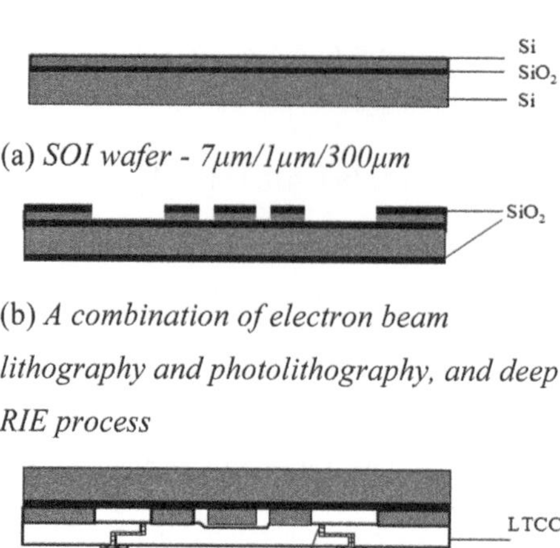

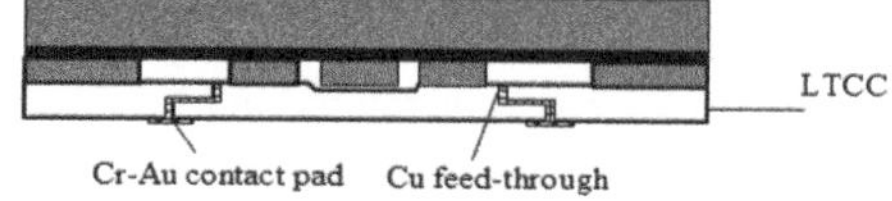

(c) 1st *anodic bonding: Silicon and LTCC*

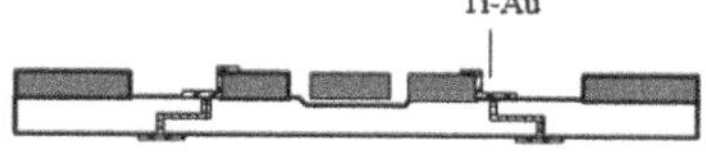

(d) Buried SiO_2 layer and electrical connections

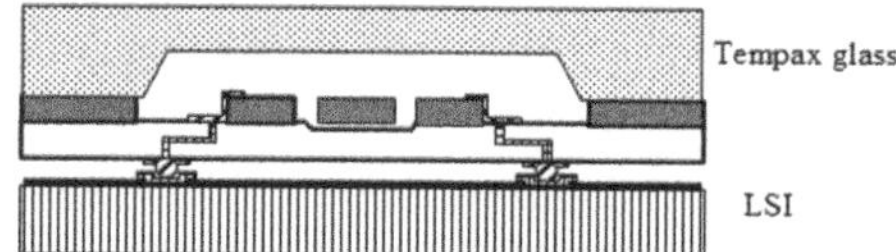

(e) 2nd anodic bonding and Integration of LSI

FIGURE 1.4 Fabrication process.

method. The handling silicon layer of SOI wafer is etched out by plasma etching [using only sulfur hexafluoride (SF_6)gas] and the buried SiO_2 layer is removed by the wet etching method (buffered hydrofluoric acid solution). Figures 1.5(a) and (b) show the fabricated results of two-arm type and bar-type resonators, respectively. An electrical interconnection is by titanium gold (Ti-Au) wire patterns to connect silicon electrodes to LSI via the metal feed-through of the LTCC [Figure 1.4 (d)]. The second anodic bonding process is performed for hermetic sealing between Tempax glass cap and silicon [Figure 1.4(e)]. Figure 1.5(c) shows the image of the capacitive resonators with LTCC integration for the encapsulation. Front and back sides of the isolated device are shown in Figures 1.5(d) and (e), respectively. The packaged device is mounted on an eight-pin dual in-line package, and the Au wire bonding is finally performed [Figure 1.5(f)].

Figure 1.6 illustrates the measurement setup for an evaluation of the resonant characteristics. The frequency response of the fabricated resonators is measured using a network analyzer (Anritsu MS6430B) with a frequency range from 10 Hz to 300 MHz. Both driving and sensing electrodes of the resonators are connected to a DC voltage through 100 kΩ resistors to block the AC signal from the network analyzer. The AC signal from the RF output of the network analyzer is supplied to the driving electrode; its input is connected to the sensing electrode via the 100 nF capacitors.

The frequency characteristics of the resonators forming on the LTCC substrate are evaluated in cases with and without the packaging process. Before the packaging, the resonant characteristics are

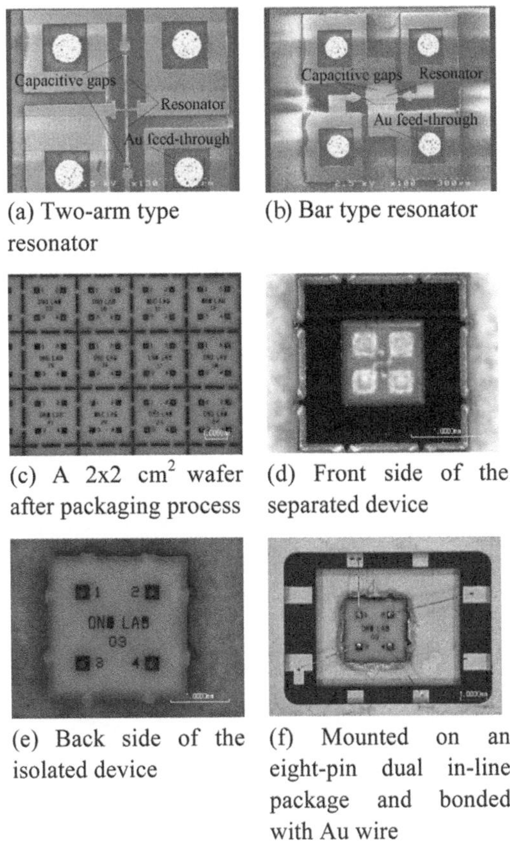

(a) Two-arm type resonator
(b) Bar type resonator
(c) A 2x2 cm^2 wafer after packaging process
(d) Front side of the separated device
(e) Back side of the isolated device
(f) Mounted on an eight-pin dual in-line package and bonded with Au wire

FIGURE 1.5 Fabricated results.

measured by using a vacuum chamber with the coaxial feed-through. The Q factor of the two-arm–type resonator is of 4,400 at 0.5 Pa in the vacuum environment, which is larger than that of 1,200 in an ambient atmosphere because of the absence of the air viscous damping, as shown in Figure 1.7. Transmission (S_{21}) and phase responses of the bar-type resonator are indicated in Figure 1.8, and a resonant peak is found at 44.66 MHz with the Q factor of 22,300 at 0.5 Pa in the vacuum chamber. The packaging process is performed in the low-pressure chamber of 0.02 Pa. The two-arm–type silicon capacitive resonator is successfully fabricated, and its resonant characteristic is shown in Figure 1.9.

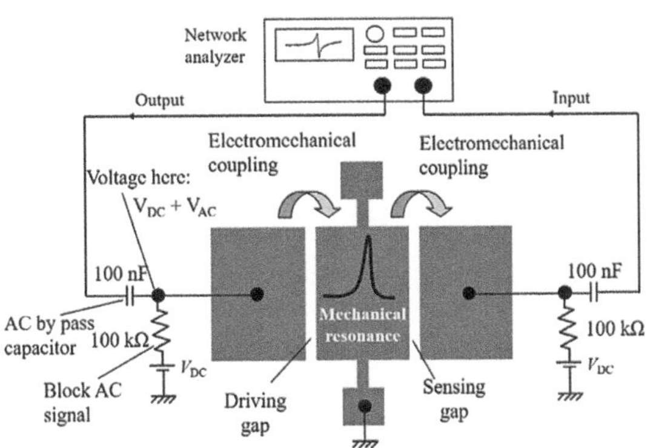

FIGURE 1.6 Measurement setup.

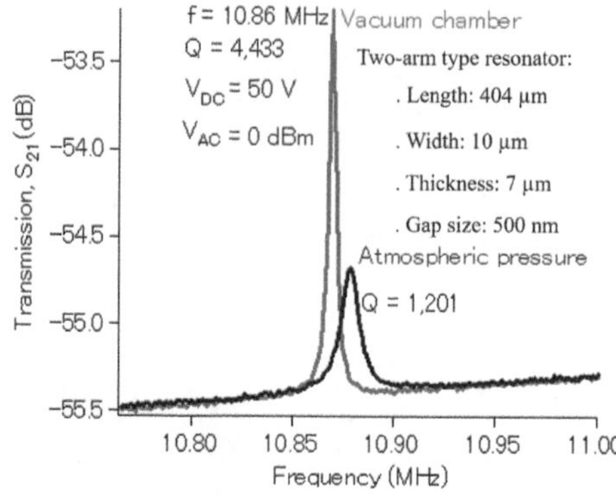

FIGURE 1.7 Two-arm type resonator: vacuum chamber and atmospheric pressure.

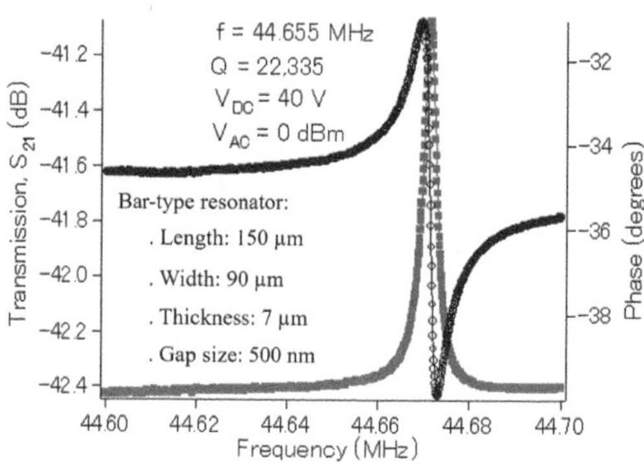

FIGURE 1.8 Bar-type resonator: frequency response and phase response before packaging.

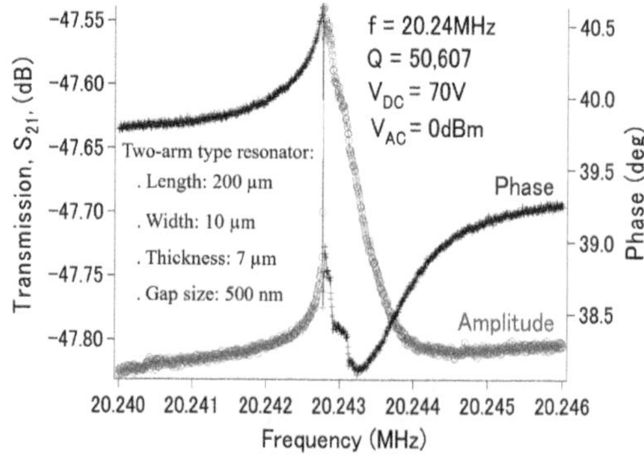

FIGURE 1.9 Two-arm–type resonator after packaging at 0.02 Pa.

A resonant peak is observed at 20.24 MHz with the Q factor of 50,600 under measurement conditions of $V_{DC} = 70$ V and $V_{AC} = 0$ dBm.

In summary, the new approach for the hermetically packaged silicon resonators was developed and demonstrated. The packaging process using the LTCC was performed successfully, and the resonance with the high Q factor was observed. We believe that the proposed hermetically sealed resonators based on LTCC substrate can simplify the integration of LSI for the application of timing device.

1.3.2 A Long Bar-type Capacitive Silicon Resonator with a High Q Factor

Single crystal silicon still keeps a major role in the micromechanical resonator's fabrication because of its small internal frictional loss, which consequently results in the high mechanical Q factor, as reported in [3–5]. The high Q factor helps not only to reduce the motional resistance, but also to decrease the phase noise in the oscillators. Besides, the selection of the structural material, the Q factor is also affected by the topology of the device structure. A study of the geometries of the supporting beams of a square-type resonator for reducing the anchor loss, and accordingly increasing the Q factor, was demonstrated in [22]. A bulk mode resonance of the single crystal silicon resonators with a deep cavity under the resonator to avoid the parasitic capacitance between the device and the substrate exhibited the high Q factor [38].

A simple method to obtain the high Q factor and low motional resistance by changing the dimensions of the resonant structure is presented in this section. The Q factor, as a function of resonant frequency f_0 and a characteristic resonant dimension D, was introduced by [5], as given by:

$$Q = 2\pi \frac{f_0 m}{\gamma} = \sim 2\pi f_0 D, \tag{1.7}$$

where m is the effective mass and γ is the damping coefficient.

Equation 1.7 above exhibits that the Q factor is a function of the resonant dimension D. Even if the resonators of the different structures have the same resonant frequency, the Q factor is variable by their various aspects. As the resonant frequency constantly keeps and the D value increases, the Q factor gets larger value. The bar-type silicon resonator is one of the promising designs because the dimensions can be changed by the length or thickness of the resonant body without degradation of its resonant frequency. However, if the thickness of the resonator structure increases, the high aspect ratio etching silicon must be considered to achieve the narrow capacitive gap width. Therefore, changing resonant length is a preferred choice to improve the Q factor. Also, increasing the resonant length increases a transduction area. Therefore, not only can the Q factor of the resonator structure be enhanced, but the motional resistance can also be decreased.

Figures 1.10(a) and (b) show the fabricated bar-type resonator and the cross section of the capacitive gap, respectively. The fabrication process is the same as above. First, the resonator structures are transferred onto the LTCC substrate for electrical interconnections via an anodic bonding technique. Then, the resonator structures are packaged hermetically through another step of the anodic bonding between silicon and Tempax glass. Figure 1.10 (c) shows the hermetically packaged device. The measurement setup for the characterization of this device was presented previously in this paper. There exist many parasitic capacitances in the measurement setup, which come from several sources, including feed-through capacitance C_f, pad capacitances C_p, and other parasitic capacitances associated with the printed circuit board. Although the estimated parasitic capacitances are quite small, for the high-frequency applications, the electrical paths through these capacitances are significant and could potentially obscure the resonant signal. The small resonant signal is a big challenge for the practical applications. The amplifier circuit is commonly employed to enhancing the resonant signal [39–42]. Using the capacitive compensation to suppress the parasitic elements is also the suitable method, as presented by Lee [43]. Here, we present the simple method to observe the resonant signal only for precisely evaluating the Q factor and the motional resistance of the capacitive silicon

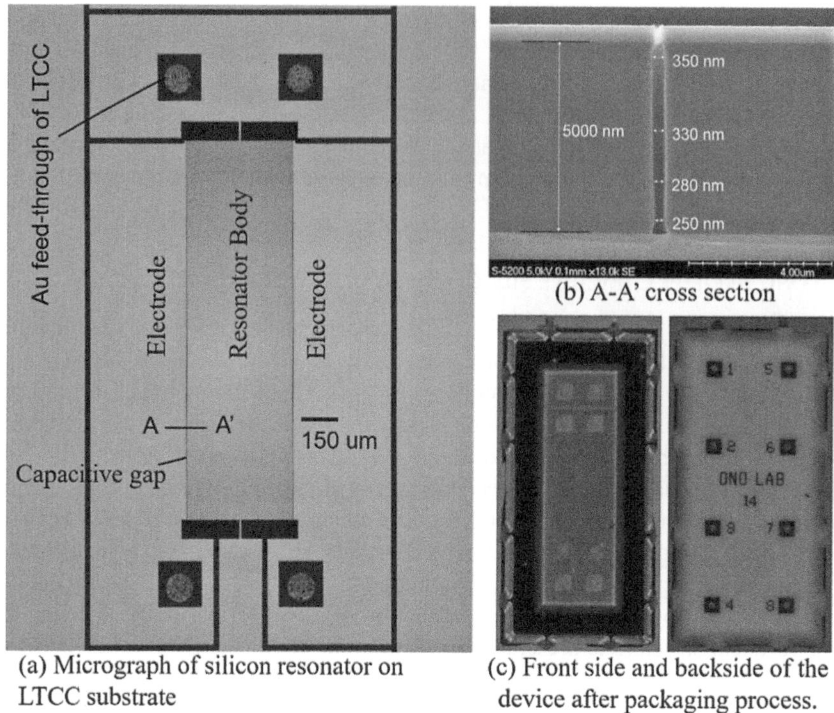

FIGURE 1.10 Fabricated results. (a) Bar-type silicon resonator forming on LTCC substrate. (b) A-A' cross section of the capacitive gap. (c) Completely fabricated device showing the front side and backside.

resonators. The transmission characteristic of the resonators without applied bias is dominated by the parasitic elements. The transmission characteristic, measured with applied DC bias, consists of both the motional and parasitic components. In this manner, it is possible to extract the motional components by subtracting the measured results under the two different conditions above (with and without the DC bias). The accurate measurement of the motional components can be achieved using this method. The cancellation of the effects of the parasitic elements is performed on the network analyzer (Agilent E5071B, frequency range from 300 kHz to 8.5 GHz). A clear resonant peak can be observed after performing the process above.

The resonant characteristics of the unpackaged and packaged devices are summarized in Table 1.1. The measurement results show that the long bar-type resonant body has a higher Q factor and a lower motional resistance than those of the shorter bar-type resonant body. The obtained experimental results are in agreement with Equation 1.7. The resonant structure with the long length of the resonant body of 1500 μm shows a Q factor of 368,000 and a motional resistance of 11.7 kΩ. Those of the resonant structure with the short length of the resonant body of 500 μm are 64,000 and 100 kΩ, respectively.

Figure 1.11 shows the relationship between the Q factor and ambient pressure. The Q factor of the fabricated device is measured under the different pressure levels from the atmosphere to the vacuum pressure of 0.01 Pa. The effects of viscous damping on the Q factor are observed when the pressure changes. The Q factor of the resonator is 2,680 in the air and 368,000 at 0.01 Pa in the vacuum chamber. A comparison of the Q factor in the air, in the vacuum chamber, and in hermetic packaging is shown in Figure 1.12. The packaging process is performed at the low-vacuum pressure level of 0.008 Pa; however, the Q factor of 341,000 of the hermetically packaged device is smaller than that of the capacitive silicon resonator, evaluated at 0.01 Pa in the vacuum chamber pressure. One of the main reasons is due to the outgassing in the resonator's cavity, which generated during

TABLE 1.1
Summary of Parameters of the Fabricated Resonators

	Unpackaged Device			Packaged Device
		Resonator Parameters:		
Length of resonator body	$L = 500\ \mu m$	$L = 1000\ \mu m$	$L = 1500\ \mu m$	$L = 1500\ \mu m$
Width of resonator body	$W = 440\ \mu m$	$W = 440\ \mu m$	$W = 440\ \mu m$	$W = 440\ \mu m$
Thickness of resonator body	$t = 5\ \mu m$	$t = 5\ \mu m$	$t = 5\ \mu m$	$t = 5\ \mu m$
Capacitive gap	$g = 440$ nm	$g = 440$ nm	$g = 440$ nm	$g = 440$ nm
		Applied Conditions:		
V_{DC}	$V_{DC} = 25$ V	$V_{DC} = 10$ V	$V_{DC} = 10$ V	$V_{DC} = 10$ V
V_{AC}	$V_{AC} = 0$ dBm	$V_{AC} = 0$ dBm	$V_{AC} = 0$ dBm	$V_{AC} = 0$ dBm
		Theoretical Calculation:		
Effective mass	$m_{eff} = 1.28 \times 10^{-9}$ kg	$m_{eff} = 2.56 \times 10^{-9}$ kg	$m_{eff} = 3.84 \times 10^{-9}$ kg	$m_{eff} = 3.84 \times 10^{-9}$ kg
Effective stiffness	$k_{eff} = 4.71 \times 10^{6}$ NM^{-1}	$k_{eff} = 9.51 \times 10^{6}$ NM^{-1}	$k_{eff} = 1.43 \times 10^{7}$ NM^{-1}	$k_{eff} = 1.43 \times 10^{7}$ NM^{-1}
Resonant frequency	$f_0 = 9.67$ MHz	$f_0 = 9.67$ MHz	$f_0 = 9.67$ MHz	$f_0 = 9.67$ MHz
Electromechanical transduction	$\eta = 3.46 \times 10^{-6}$	$\eta = 4.92 \times 10^{-6}$	$\eta = 7.38 \times 10^{-6}$	$\eta = 7.38 \times 10^{-6}$
		Results:		
Resonance frequency	$f_0 = 9.65$ MHz	$f_0 = 9.68$ MHz	$f_0 = 9.69$ MHz	$f_0 = 9.69$ MHz
Quality factor	$Q = 64{,}000$	$Q = 148{,}000$	$Q = 368{,}000$	$Q = 341{,}000$
Insertion loss	$IL = -71$ dB	$IL = -62$ dB	$IL = -56$ dB	$IL = -56$ dB
Motional resistance	$R_m = 100$ kΩ	$R_m = 43.3$ kΩ	$R_m = 11.7$ kΩ	$R_m = 12.5$ kΩ
Motional capacitance	$C_m = 2.54$ aF	$C_m = 2.6$ aF	$C_m = 3.8$ aF	$C_m = 3.8$ aF
Motional inductance	$L_m = 107$ H	$L_m = 10.6$ H	$L_m = 70$ H	$L_m = 70$ H
Feed-through capacitance	$C_f = 0.06$ pF	$C_f = 0.15$ pF	$C_f = 0.22$ pF	$C_f = 0.22$ pF

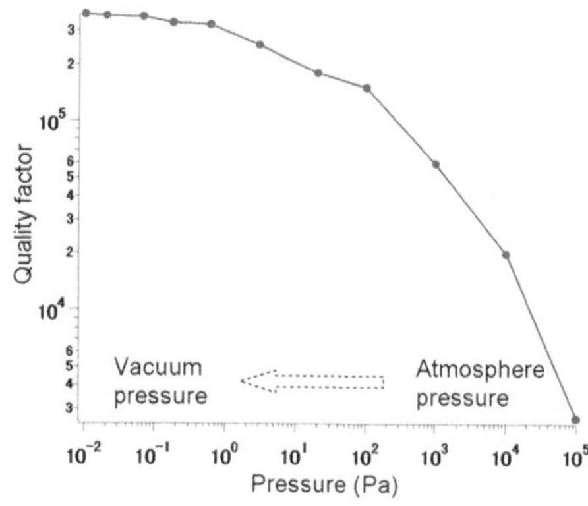

FIGURE 1.11 Pressure versus quality factor.

the packaging process. Evaluation, based on the Q factor, shows that the vacuum level of the sealed cavity is around 10 Pa.

In summary, the long bar–type silicon resonators with the high Q factor and low motional resistance were demonstrated. The effects of the resonant dimension (length of the resonant body) on the Q factor were examined. Moreover, the device was hermetically packaged using an LTCC substrate based on anodic bonding technology and the resonant characteristics before and after packaging processes were evaluated and compared.

1.3.3 Capacitive Silicon Resonator Using Neutral Beam Etching Technology

As presented in the previous sections, the high Q factor value is one of the necessities in a resonator's performance. The Q factor is affected by many other factors [5], such as anchor losses, electrical loading, air damping, etc. Surface defects are one of those factors that make the energy dissipation larger. Thus, the Q factor consequently results in the smaller value. To pattern silicon device layer, deep reactive ion etching (RIE) using the Bosch process are typically employed. Although deep RIE has many benefits, such simple and well-known technology, charge-induced damage or damages due to ultraviolet/vacuum ultraviolet (UV/VUV) light from the high-density plasma during etching process [44–47] are its disadvantages. Also, deep RIE creates the rough etched surfaces, which are called scallops or ripples.

The neutral beam etching (NBE) technology was introduced in [48–52]. NBE does not possess charge, and UV/VUV light is blocked by an aperture during the etching process. Thus, the etched

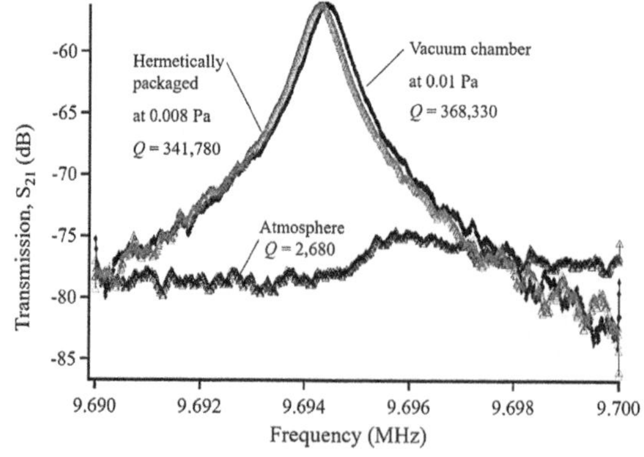

FIGURE 1.12 Atmosphere, vacuum chamber, and hermetically packaged.

TABLE 1.2
Summary of Parameters of the Silicon Resonators Fabricated by Deep RIE and NBE

	NBE (device #1)		Deep RIE (device #2)	
Parameters:				
Length of resonant body	$L = 500\ \mu m$		$L = 500\ \mu m$	
Width of resonant body	$W = 440\ \mu m$		$W = 440\ \mu m$	
Thickness of device layer	$t = 5\ \mu m$		$t = 5\ \mu m$	
Capacitive gap	$g = 230$ nm		$g = 400$ nm	
Applied Conditions:				
V_{DC}	5 V	10 V	5 V	10 V
V_{AC}	0 dBm	0 dBm	0 dBm	0 dBm
Theoretical Calculation:				
Effective mass—m_{eff}	1.3×10^{-9} Kg	1.3×10^{-9} Kg	1.3×10^{-9} Kg	1.3×10^{-9} Kg
Effective stiffness—k_{eff}	4.7×10^{6} Nm^{-1}	4.7×10^{6} Nm^{-1}	4.7×10^{6} Nm^{-1}	4.7×10^{6} Nm^{-1}
Electromechanical transduction—η	2.1×10^{-6}	4.2×10^{-6}	6.9×10^{-7}	1.4×10^{-6}
Resonant frequency—f_0	9.67 MHz	9.67 MHz	9.67 MHz	9.67 MHz
Simulation Result:				
Resonant frequency—f_0	9.75 MHz	9.75 MHz	9.75 MHz	9.75 MHz
Experiment Results:				
Measured frequency—f_0	9.66 MHz	9.66 MHz	9.66 MHz	9.66 MHz
Quality factor—Q	77,000	75,000	None	63,000
Insertion loss—IL	−72.2 dB	−63.8 dB	−92.6 dB	−78.7 dB
Motional resistance—R_m	230 kΩ	59 kΩ	None	645 kΩ
Motional capacitance—C_m	0.9 aF	3.7 aF	None	0.4 aF
Motional inductance—L_m	293 H	73 H	None	670 H
Feed-through capacitance—C_f	0.1 pF	0.1 pF	None	0.06 pF

surface with almost no damage and atomically flat silicon surfaces can be achieved. The NBE process is appropriate for the micro/nanodevice fabrication, instead of the deep RIE.

In this section, highly anisotropic etching shapes are demonstrated using NBE technology. Moreover, the silicon resonator's fabrication using this technology are proposed to obtain narrow capacitive gaps and smooth etching surfaces for the small motional resistance, low insertion loss, and high Q factor.

The capacitive silicon resonators are fabricated by both the deep RIE and NBE technologies. Table 1.2 shows the summary of the parameters of the fabricated devices. Figure 1.13(a) demonstrates the fabricated resonator formed on the LTCC substrate. Capacitive gap widths of the 5 μm-thick devices produced by the deep RIE and NBE are around 400 nm and 230 nm (average value of the capacitive gap width), as shown in Figures 1.13(b) and (c), respectively. The gap size produced by NBE is smaller than that by deep RIE. Besides, the trench formed by deep RIE exhibits scallops of around 30 nm (low-scallop recipe: etching time of 4 s and passivation time of 3 s) while NBE has a smooth surface.

Figure 1.14 gives a comparison on the transmission S_{21} of the fabricated devices by NBE (device #1) and deep RIE (device #2). Although devices #1 and #2 are designed in the same parameters of length L, width W, and thickness t, their resonant frequencies are a little different. The fabrication errors in the size—making the effective mass and spring constant slightly changed—would be a reasonable reason for such the behavior. The measurement results show the motional resistance of device #1 is reduced by almost 11 times from 645 kΩ to 59 kΩ and its insertion loss is increased by

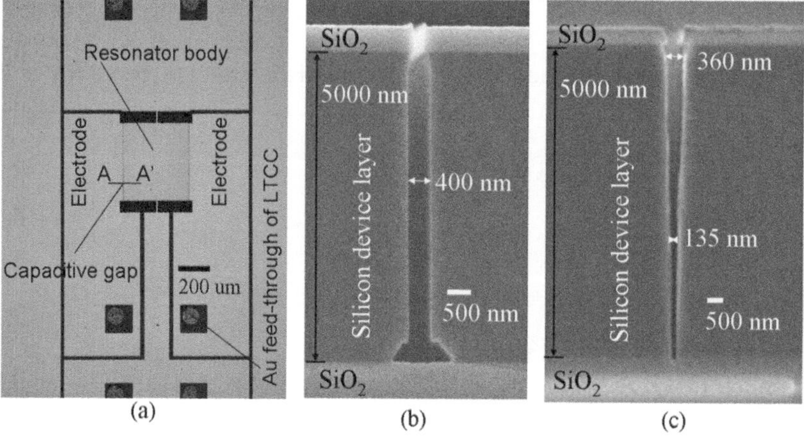

FIGURE 1.13 (a) Silicon resonator formed on the LTCC substrate. (b) A-A' cross section of the capacitive gap of the device fabricated by deep RIE. (c) A-A' cross section of the capacitive gap of the device produced by NBE.

approximately 15 dB, in comparison with those of device #2. Also, device #1 indicates its Q factor (Q = 75,000) is higher than that of device #2 (Q = 63,000).

Figure 1.15 confirms that the Q factor of the devices produced by NBE is greater than that of the devices produced by deep RIE. Devices fabricated by both technologies are measured and compared under the same measurement conditions V_{DC} of 10 V, V_{AC} of 0 dBm, and the pressure vacuum of 0.01 Pa. The Q factors of resonators fabricated by NBE are from 75,000 to 82,000; those made by deep RIE are from 57,000 to 66,000. The motional resistance of resonators decreases when reducing the capacitive gap width; however, this causes the drop in the Q factor, as presented in [19, 53–54]. Although device #1 has a smaller gap width than that of device #2, a higher Q factor is observed. The possible reasons for such behavior are due to the roughness and lattice defects on the device surfaces that lead to the mechanical deterioration. The etched surfaces of the devices produced by NBE are smooth while those of the devices fabricated by deep RIE are rough due to scallops, as mentioned previously. In addition, the plasma irradiations (UV/VUV) of deep RIE possibly cause defects on the device surfaces, as given in [55–56]. However, no damages are generated in turn on the neutral beam irradiation due to the high aspect ratio carbon aperture [48–52].

In summary, devices produced by both deep RIE and NBE were evaluated and compared with each other. The fabricated devices using NBE provided the higher Q factor, lower insertion loss, and

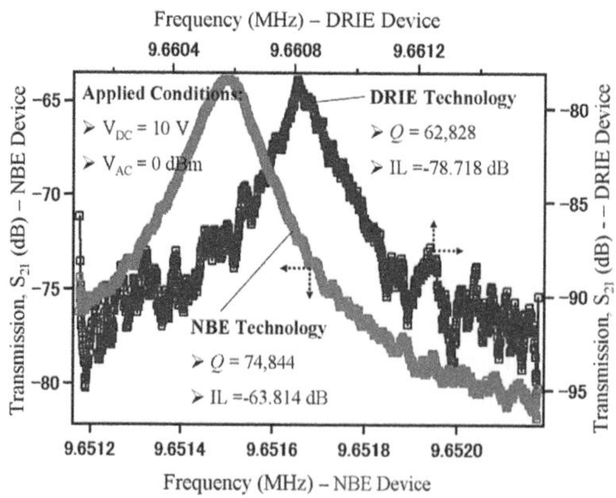

FIGURE 1.14 Frequency response of NBE and deep RIE devices at V_{DC} of 10 V.

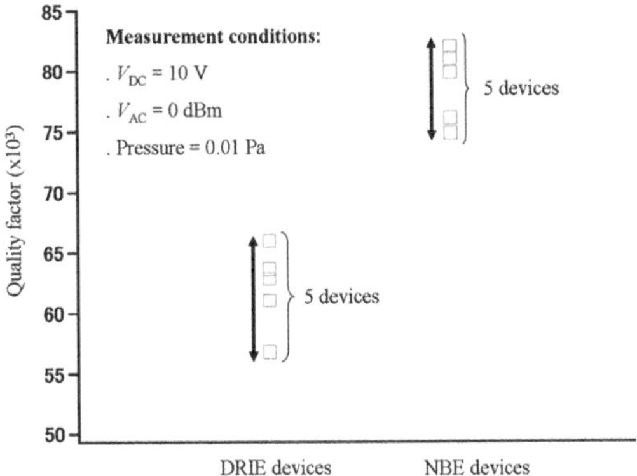

FIGURE 1.15 Comparison of quality factors of NBE device with that of deep RIE devices.

smaller motional resistance over those of the fabricated devices using deep RIE. This work brings out a new approach for fabricating silicon micromechanical resonator using NBE.

1.3.4 Mechanically Coupled Capacitive Nanomechanical Silicon Resonator

Capacitive micromechanical silicon resonators are used for timing [3–5] and filtering [9] applications; however, other applications, including high-sensitive sensors [57–59], quantum information processing [60], etc., also have considerable potential. As such, downscaling of resonator structures can enhance the sensitivity characteristic of devices due to their large surface area relative to volume. To detect the motion of the micromechanical resonators, monitoring the changes in the capacitance were presented in the previous sections. However, this may not be effective for nanomechanical resonators because of their minuscule change of the motional capacitance. In addition, the nanomechanical resonator exhibits a large motional resistance, and hence, high insertion loss that makes it difficult for practical applications.

Development of arrays of the nanomechanical resonator is one of the solutions for increasing the motional capacitance and lowering motional resistance. An array of 20, 49, and 100 resonators was designed and fabricated in [61]. Then all individual resonators were electrically interconnected. The small motional resistances were achieved; however, it may be faced with a large parasitic capacitance and mismatch of the resonant frequencies that result in difficulty in capacitive detection.

In this section, mechanically coupled capacitive nanomechanical silicon resonator is investigated for lowering the motional resistance of nanomechanical silicon resonator. Figure 1.16(a) sketches the single beam nanomechanical resonator (device #1). In turn, designed mechanically coupled nanomechanical resonator (device #2) is shown in Figure 1.16(b), which consists of 100 single nanomechanical resonators connected by mechanically coupled elements. Table 1.3 presents the summarized design parameters of devices #1 and #2. Vibration modes of devices #1 and #2 are demonstrated via the FEM simulation, as shown in Figures 1.16(c) and (d), respectively. Device #1 vibrates at a flexural mode, while a synchronization of the mechanical elements indicates for device #2. Its vibration synchronizes because of mechanical interaction via the vibration of the coupling element connecting those resonators. The individual resonators are designed with the same parameters and fabricated at the same time and with the same process. It is expected to have the same resonant frequency. The effective lowering of the motional resistance by the numbers of the resonator can be achieved. The total motional resistance R_{total} is given by [62]:

$$R_{total} = \frac{R_m}{n} \tag{1.8}$$

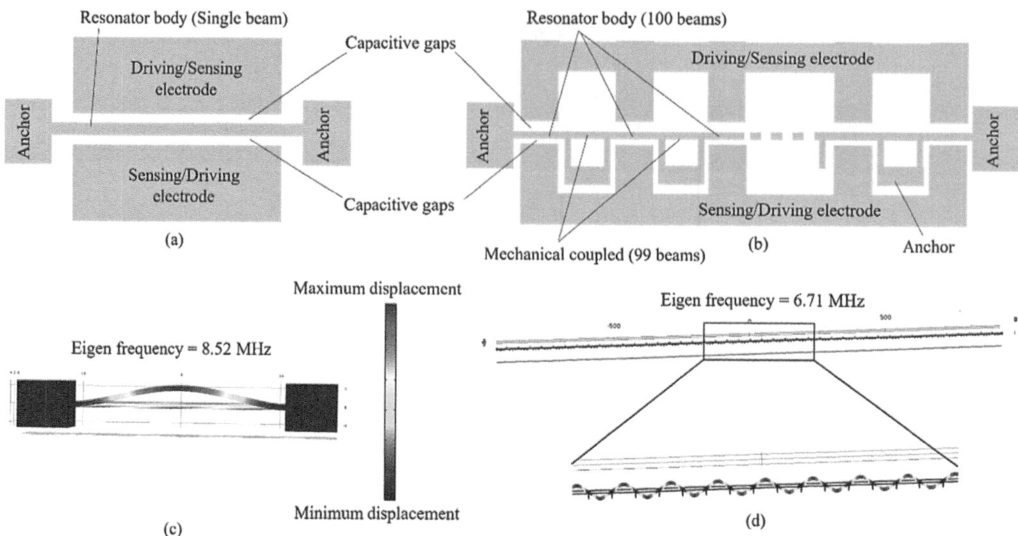

FIGURE 1.16 Proposed device structures and FEM simulation results. (a) Single capacitive nanomechanical silicon resonator (device #1). (b) Mechanical coupled capacitive nanomechanical silicon resonator (device #2). (c) FEM simulation of the single capacitive resonator. (d) FEM simulation of the mechanical coupled capacitive resonator.

Figure 1.17(a) shows the cross-section SEM image of the resonant body of 500 nm and a capacitive gap width of 300 nm. Single (device #1) and mechanically coupled (device #2) nanomechanical resonators are successfully fabricated, as shown in Figures 1.17(b) and (c).

The transmission S_{21} of device #1 is shown in Figure 1.18(a). A resonant peak of device #1, which is observed under measurement conditions $V_{DC} = 10$ V and $V_{AC} = 0$ dBm, is found at 9.9 MHz. Jump and drop phenomenon also arose in device #1. The amplitude of the transmission increases as the

TABLE 1.3
Summary of Parameters of Single and Mechanically Coupled Capacitive Nanomechanical Silicon Resonators

Parameters	Single (device #1)	Coupled (device #2)
Width of resonant body (W)	500 nm	500 nm
Length of resonant body (L)	21.3 μm	21.3 μm
SOI wafer:		
• Device layer (t) (resonator height)	5 μm	5 μm
• Buried SiO_2 layer	0.4 μm	0.4 μm
• Silicon handling layer	546 μm	546 μm
Capacitive gap (g)	300 nm	300 nm
Number of resonators	1	100
Number of coupling	0	99
V_{DC}	10 V	15 V
V_{AC}	0 dBm	0 dBm
Measured frequency (f_0)	9.9 MHz	7.2 MHz
Amplitude peak (A)	0.5 dB	20 dB
Quality factor (Q)	Nonlinear response	1000
Motional resistance (R_m)		1.2 kΩ

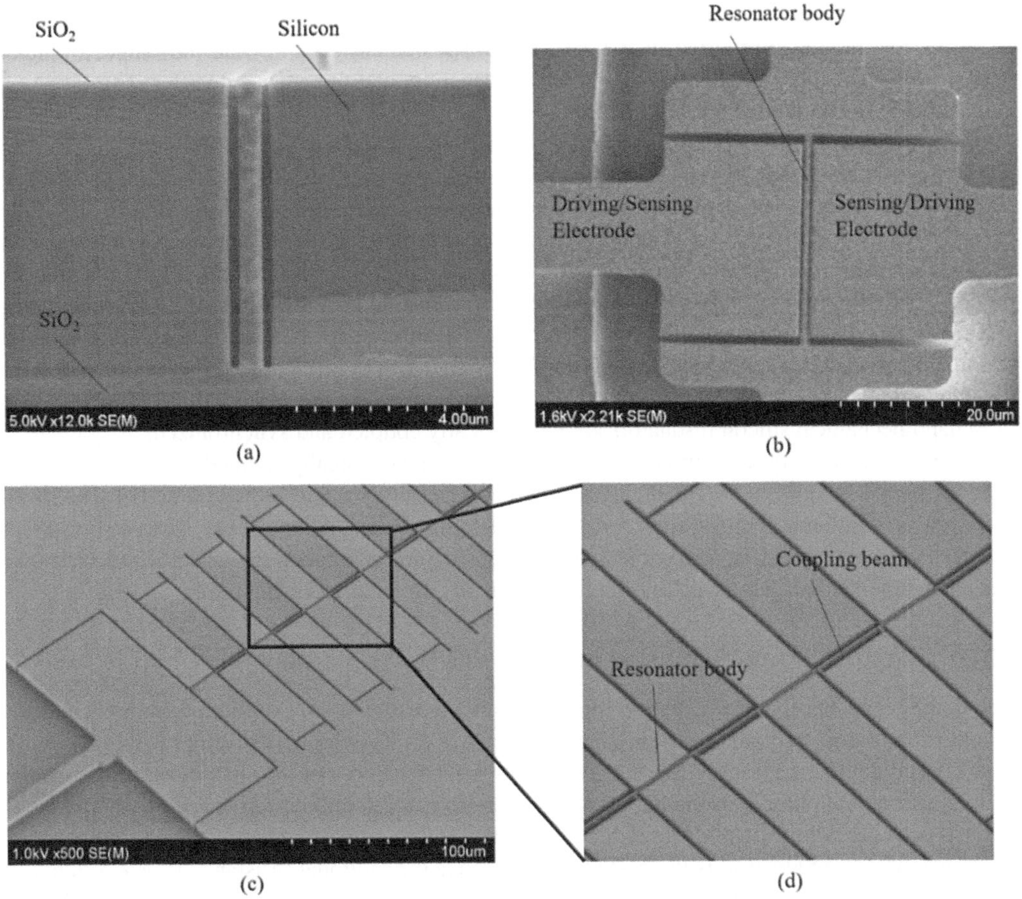

FIGURE 1.17 Fabricated results. (a) Cross-sectional image of resonant body and capacitive gap width. (b) Single capacitive nanomechanical silicon resonator. (c) Mechanically coupled capacitive nanomechanical silicon resonator. (d) Close-up image of mechanical coupled capacitive silicon resonator.

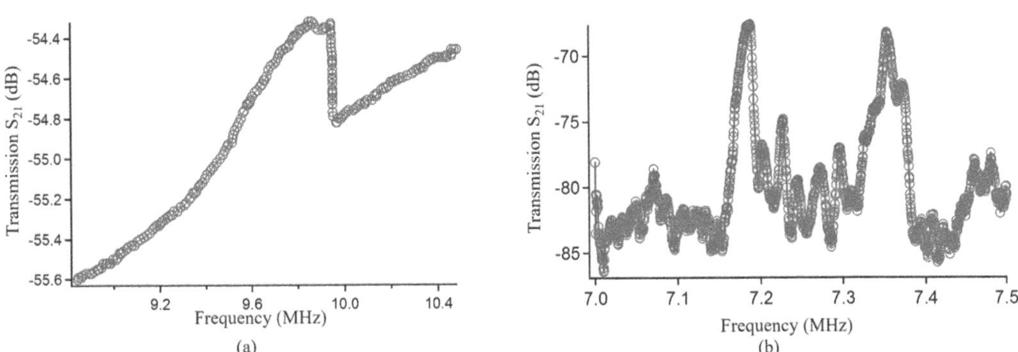

FIGURE 1.18 Frequency response of the fabricated devices. (a) Single capacitive nanomechanical silicon resonator. (b) Mechanically coupled capacitive nanomechanical silicon resonator.

frequency is swept upward, and then suddenly drops to a lower value. This hard spring effect of beam resonators is also reported in [63–64]. The displacement goes beyond the ranges of small perturbations, tension is applied to the resonating structure, and this makes it stiffer. This situation is identical to the hardening spring.

The resonant peaks of device #2 are observed at around 7.2 MHz, as shown in Figure 1.18(b). The resonant peaks are clearly observed with the resonant amplitude of 20 dB, and the high Q factor of 1000 is achieved for the mechanically coupled nanomechanical resonators with a number of 100. The observed resonant frequency of device #2 is lower than that of device #1. It is in good agreement with the FEM simulation (Comsol Multiphysics, Inc.), as shown in Figures 1.16(c) and (d). To evaluate the motional resistance of the coupling capacitive resonator, we assume that the quality factor of coupling and single resonators are same. The motional resistance of the single capacitive resonator without coupling is about 120 kΩ. In turn, the motional resistance of the mechanical coupling resonator of the numbers of 100 exhibits about 1.2 kΩ. Some resonant peaks are observed, which shows that most nanomechanical silicon resonators are mechanically coupled and synchronized.

In summary, we designed, fabricated, and evaluated the mechanically coupled capacitive nanomechanical silicon resonators for enhancing the motional capacitance and for lowering motional resistance towards high emerging sensing, image, and data processing technologies. Resonant peaks of the mechanically coupled capacitive nanomechanical silicon resonators were observed, which shows that most nanomechanical resonators were mechanically coupled and synchronized.

1.3.5 Capacitive Silicon Resonator with Movable Electrode Structures

One of the biggest problems facing the capacitive silicon resonators for oscillators and other applications is their large motional resistance R_m, and hence, high insertion loss that makes them difficult to meet oscillation conditions. Also, the greater motional resistance of the silicon resonators results in more substantial phase noise in the oscillator; therefore, the motional resistance R_m should be as small as possible. Many methods for reducing the motional resistance were mentioned previously in this paper. The gap reduction is the best method because the motional resistance is proportional to the fourth order of the gap width, as delineated in Equation 1.1.

Many works have been investigated to reduce the capacitive gap width. For instance, the resonant structures with a solid filled in the gap between electrodes and the resonant body were presented [65–66]. A high dielectric constant thin film, such as silicon nitride, is sandwiched between the electrode and the resonant body instead of the air gap. The motional resistance of these structures is of small value because of the high dielectric constant and the very narrow solid gap (15 nm-thick silicon nitride in [65] and 20 nm-thick silicon nitride in [66]). However, the resonator structure suffers from a material mismatch between silicon and silicon nitride material, which decreases the mechanical Q factor due to the high internal loss of the interface losses induced by the deposited silicon nitride film [5]. The fabrication method of the capacitive silicon resonator with the small capacitive gaps, using a thin oxide film as a sacrificial layer, was presented in [17–18, 67–69]. However, this approach is slightly complicated, and the high-temperature process is required.

In this section, we present the design and fabrication of the capacitive silicon resonator with movable electrodes to decrease capacitive gap widths, which results in smaller motional resistance and lower insertion loss. The proposed device also helps to increase tuning frequency range for compensation of temperature drift of silicon resonator.

The schematic of the silicon resonator structure with movable electrodes, together with the FEM simulation, is shown in Figure 1.19. The operation of this device is as follows: The driving/sensing movable electrodes are connected to a DC source V_{DC} while the resonant body is linked to ground. Thus, electrostatic forces on the both sides of the resonant body are generated. When V_{DC} increases, the movable electrodes move toward the body due to the electrostatic parallel actuation. The electrodes attract to the body until they touch the stoppers, when V_{DC} is raised to a certain voltage called *pull-in voltage*. Therefore, the capacitive gaps can be reduced. A V_{AC} is applied to the driving

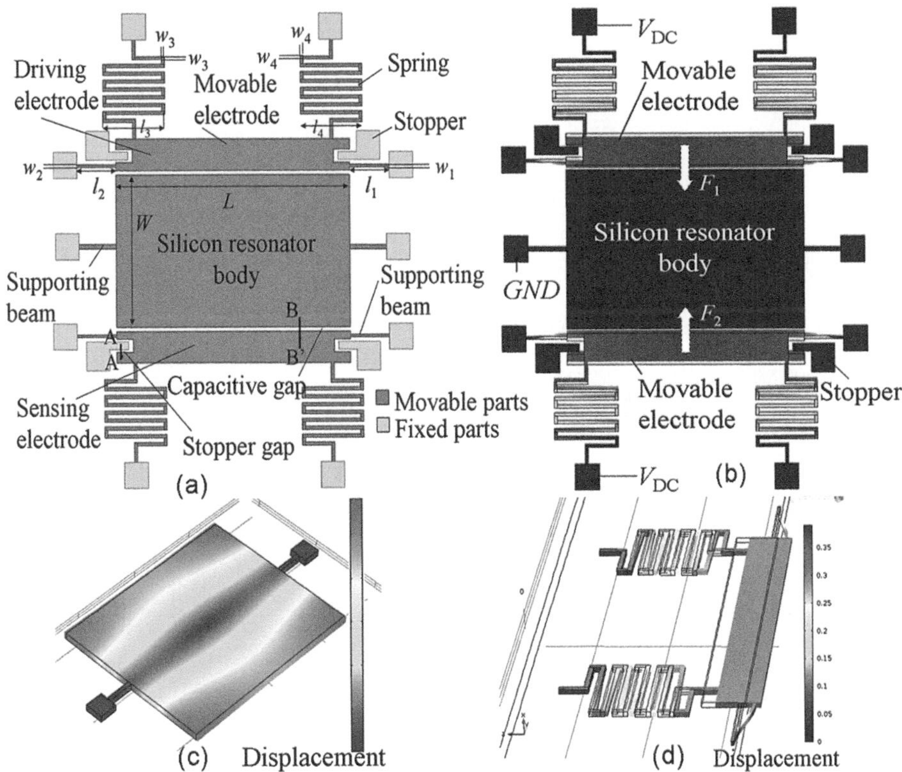

FIGURE 1.19 (a) Silicon resonator structure with the movable electrodes (device #1). (b) FEM simulation of device #1. (c) Extensional mode of the resonant body. (d) FEM simulation of the movable electrode structure.

electrode, which results in electrostatic force that causes a bulk acoustic wave in the resonant body. The resonant body is electrostatically excited and vibrated at extensional modes.

The resonator structures with (device #1) and without (device #2) movable electrodes are successfully fabricated [Figures 1.20(a) and 1.20(c)]. The summarized parameters of devices #1 and #2 are shown in Table 1.4. The stopper gap widths with A-A' cross section and the capacitive gap widths with B-B' cross section of device #1 are 400 nm and 500 nm (Figure 1.20(d)), respectively. Therefore, the final gap widths can reach at 100 nm in this design, as discussed above. The capacitive gap width of device #2 is 400 nm (Table 1.4). The resonant characteristics of devices #1 and #2 are compared to each other under the same conditions (Table 1.4 and Figure 1.21). It shows that motional resistance of device #1 reduces by 200 times, its insertion loss increases 21 dB, and its tuning characteristic of frequency increases seven times over those of device #2.

In summary, we designed, fabricated, and evaluated the capacitive silicon resonators with movable electrode structures for smaller motional resistance and lower insertion loss and wider tuning range frequency, Also we evaluated the compensation of the temperature drifts of a silicon resonator over those of the conventional capacitive silicon resonator.

1.3.6 Capacitive Silicon Resonators with Piezoresistive Heat Engines

As presented in the Introduction, there are several transduction mechanisms—including piezoelectric, capacitive and piezoresistive—to actuate and sense the motion of resonators. All of the aforementioned methods have both advantages and disadvantages. Achieving all of the benefits in the different transduction mechanisms on a single resonator are highly desired. Combined capacitive

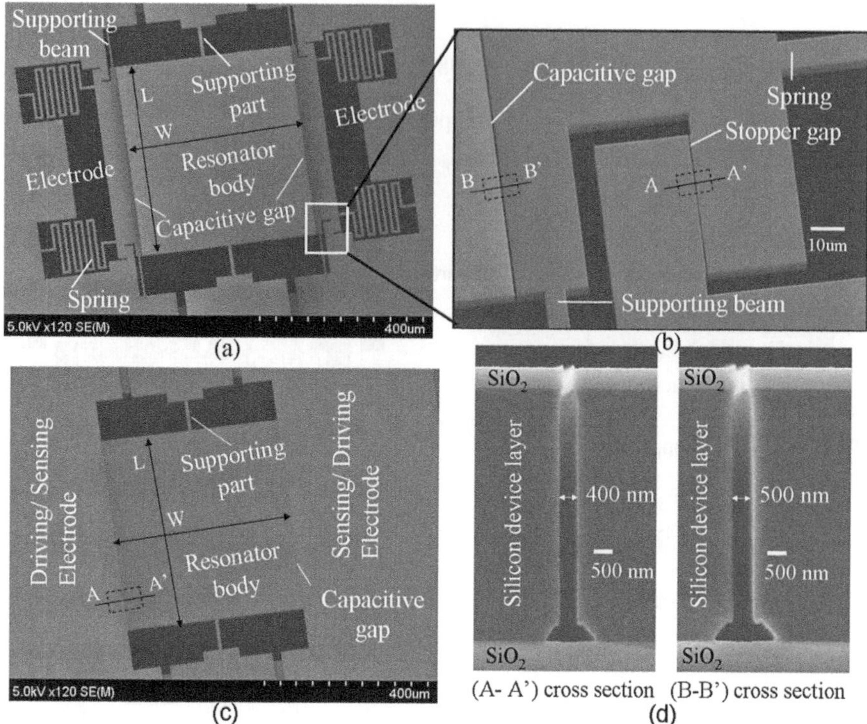

FIGURE 1.20 (a) Fabricated silicon resonator structure with movable electrodes (device #1). (b) Magnification image of a corner of the resonator structure. (c) Fabricated silicon resonator without movable electrode (device #2). (d) Profile shape of (A-A') and (B-B') cross-section.

and piezoelectric transductions for the high-performance micromechanical silicon resonators were demonstrated in [70]. An integration of a capacitive actuation and piezoresistive sensing in micromechanical resonators was presented in [71–72].

In this section, capacitive silicon resonators with piezoresistive heat engines are proposed and examined. The piezoresistive thermal actuators are used for the excitation of the vibration to enhance the driving force. Two types of devices, including single and multiple piezoresistive beams, are demonstrated. The fabricated devices are evaluated and compared with each other in cases with and without the piezoresistive effect.

The proposed device is shown in Figure 1.22(a). It consists of the resonant body, supporting beams, electrodes, piezoresistive beams, and capacitive gaps. The *resonant body* is a square frame structure that is fixed at four corners of the square plate via the supporting beams. The resonant body is divided into many sections that are connected to others using the small piezoresistive beams. Four electrodes create narrow capacitive gaps with the resonant body.

Two types of resonator design are investigated. A single piezoresistive beam at the connection areas (connecting two parts of the resonant body) is used in device #1. The resonant body of device #1 is split into four sections, which are connected to each other by the single piezoresistive beams. The beam is placed at the center of the resonant body. In device #2, the multiple piezoresistive beams are employed, which are located near the edges of the resonant body to enhance the vibration amplitude. Its resonant body is divided into the twelve elements. The design parameters of devices #1 and #2 are shown in Table 1.5.

The working principle of these devices is as follows: The resonator structure works in a capacitive mode, in which the output voltage results from the changes in the capacitive gap on the sensing

TABLE 1.4
Summary of Parameters of the Silicon Resonators with and without Movable Electrode Structures

	With Movable Electrodes (resonator #1)	Without Movable Electrodes (resonator #2)
	Parameters:	
Length of resonant body—L	500 μm	500 μm
Width of resonant body—W	440 μm	440 μm
Thickness of SOI device layer—t	5 μm	5 μm
Capacitive gap—g	100 nm	400 nm
Length of the supporting beams—l_1, l_2	100 μm	none
Width of the supporting beams—w_1, w_2	3 μm	none
Length of the springs—l_3, l_4	120 μm	none
Width of the springs—w_3, w_4	6 μm	none
	Applied Conditions:	
V_{DC}	25 V	25 V
V_{AC}	0 dBm	0 dBm
Pressure level of vacuum chamber	0.01 Pa	0.01 Pa
	Theoretical Calculation:	
Effective mass of resonant body—m_{eff}	1.28×10^{-9} Kg	1.28×10^{-9} Kg
Effective stiffness of resonant body—k_{eff}	4.71×10^{6} Nm^{-1}	4.71×10^{6} Nm^{-1}
Movable electrode displacement at 25V—D	385 nm	none
Resonance frequency—f_0	9.67 MHz	9.67 MHz
	Simulation Results:	
Resonance frequency—f_0	9.75 MHz	9.75 MHz
Movable electrode displacement at 25V	402 nm	none
	Measurement Results:	
Measured frequency—f_0	9.652260 MHz	9.654576 MHz
Quality factor—Q	48, 607	63,952
Insertion loss—IL	−50.908 (dB)	−71.938 (dB)
Motional resistance—R_m	0.5 kΩ	100 kΩ
Motional capacitance—C_m	6.50×10^{-16} F	2.54×10^{-18} F
Motional inductance—L_m	4.18×10^{-1} H	1.07×10^{2} H
Feed-through capacitance—C_f	2.21×10^{-13} F	5.53×10^{-14} F
Tuning slope frequency	−253.5 Hz/V	−37.3 Hz/V

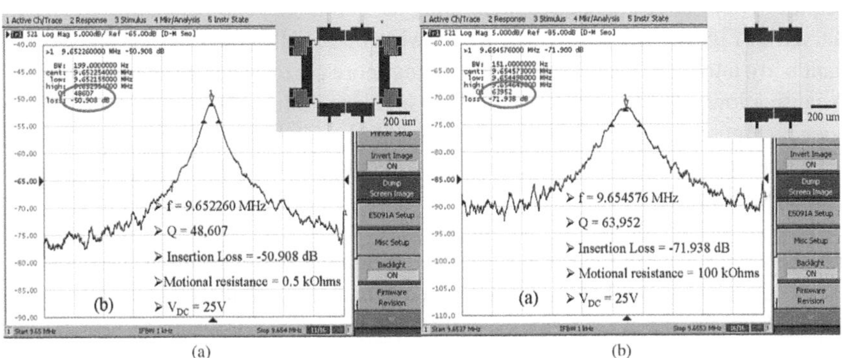

FIGURE 1.21 (a) Frequency response of device #1. (b) Frequency response of device #2.

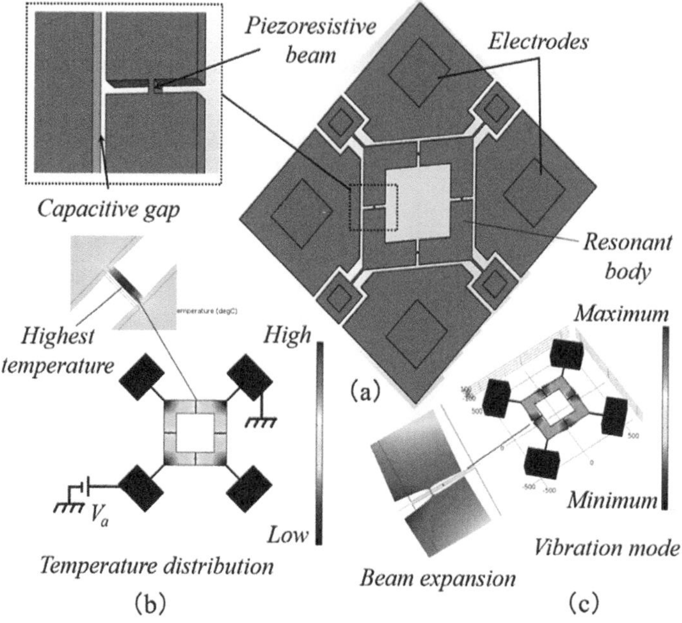

FIGURE 1.22 Capacitive silicon resonator with piezo-resistive heat engines. (a) Proposed structure. (b) Temperature distribution. (c) Vibration mode.

electrode. A current controlled by a voltage source V_b is applied to the resonant body. Due to the higher electrical resistance of the narrow actuator beams than that of other parts of the body, Joule heating mostly occurs at the actuator beams [Figure 1.22(b)—temperature distribution]. The vibration of the resonant body is mainly caused by the capacitive transduction in addition to the piezoresistive engine using the piezoresistive beams. As described in the Introduction, the temperature modulation—thermodynamic cycles consist of heating and cooling cycles—in the piezoresistive beam causes thermomechanical force. Thus, the thermodynamic cycles (thermal actuators) are possibly used for an efficient excitation of vibration with the driving force enhancement.

Figures 1.23(a) and (b) show the successfully fabricated devices #1 and #2, respectively. The resonant peak of device #1 is found at 3.166 MHz, with a quality factor of 3000. Figure 1.24(a) shows the frequency response of device #1 without connecting bias voltage (capacitive effect only). Figure 1.24(b) shows that its output signal (insertion loss) is enhanced by 10 dB in the case that 7V bias voltage is applied to the piezoresistive engine.

To increase the piezoresistive electrothermal force, device #2 with multiple piezoresistive beams, which are placed near edges of the resonant body, is examined. The resonant frequency of device #2 is observed in Figure 1.25(a) at 1.333 MHz, which is smaller than that of device #1. Dividing the resonant body into more sections and embedding more piezoresistive beams within the resonant body can cause a lower stiffness because of their smaller widths. Thus, it results in the lower resonant frequency. Figure 1.25(b) shows the achievement of a 20 dB improvement in the insertion loss when the 7V bias voltage is supplied. Nevertheless, another peak has been observed. There are some possible reasons for this behavior:

- First, the vibration amplitude of piezoresistive beams increases when the bias voltage V_b increases. However, it would be confined by a mechanical saturation that leads to an unstable region [16].

TABLE 1.5
Summary of Parameters of Devices #1 and #2

	Single Piezoresistive Beam (device #1)	Multiple Piezoresistive Beams (device #2)
Device Parameters:		
Type of semiconductor	p-type	p-type
Electrical resistivity	0.02 Ω.cm	0.02 Ω.cm
Outer length/width of resonator body—L_1 (μm)	500	500
Inner length/width of resonator body—L_2 (μm)	400	400
Thickness of device layer—t (μm) (SOI (Silicon on Insulator) wafer of 7 μm /1 μm /300 μm)	7	7
Capacitive gap—g (nm)	250	250
Length of supporting beam—x (μm)	100	100
Width of supporting beam—y (μm)	10	10
Length of piezoresistive beam—l (μm)	3	3
Width of piezoresistive beam—w (μm)	0.8	0.8
Number of piezoresistive beams	Single	Multiple
Resonant body is divided into	4	12
Evaluated Results:		
Resonant frequency	3.166 MHz	1.333 MHz
Quality factor (Q)	3000	2000
Insertion Loss:		
$V_{\text{bias}} = 0$ V	−50 dB	−68 dB
$V_{\text{bias}} = 7$ V	−40 dB	−48 dB
Enhanced	**10 dB**	**20 dB**
Motional Resistance:		
$V_{\text{bias}} = 0$ V	15.7 kΩ	125.5 kΩ
$V_{\text{bias}} = 7$ V	4.9 kΩ	12.5 kΩ
Reduced	**70 %**	**90 %**

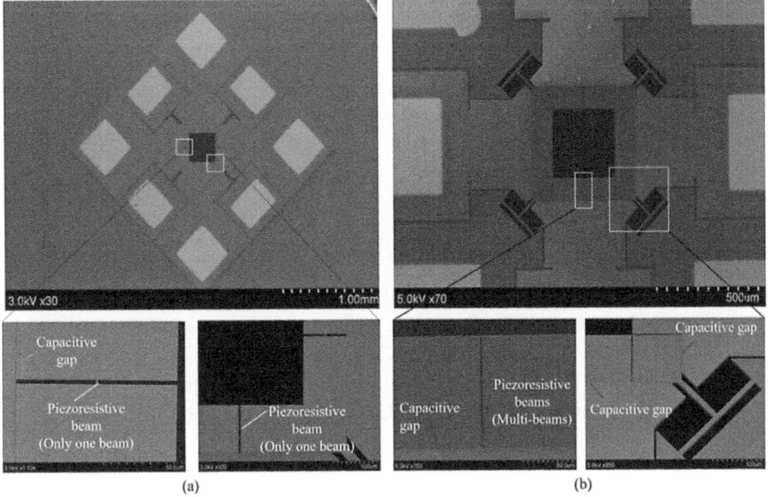

FIGURE 1.23 Fabricated devices. (a) Capacitive silicon resonator with the single piezoresistive beam. (b) Capacitive silicon resonator with multiple piezoresistive beams.

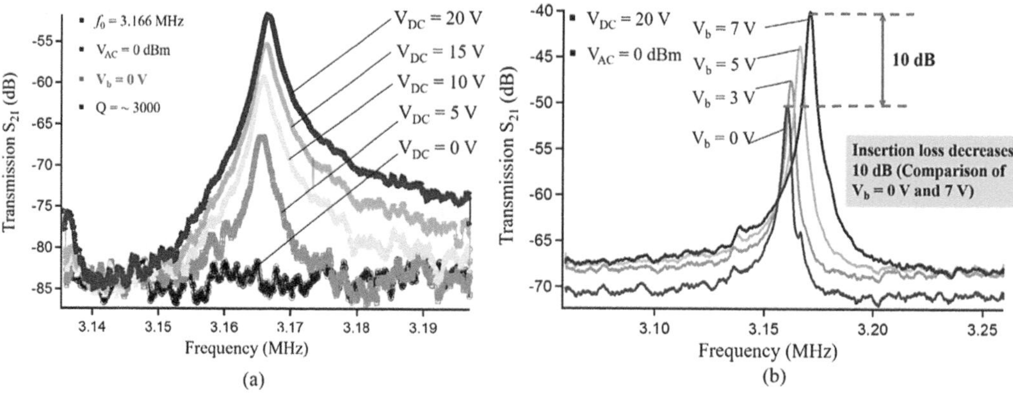

FIGURE 1.24 Frequency response of device #1. (a) Capacitive effect. (b) Piezoresistive heat engine effect.

- Second, this behavior may come from a thermoelastic dissipation [73], which causes a high energy loss.
- Last, a thermomechanical coupling with a shifted phase may also be one of the possible reasons.

By tuning the bias voltage V_b from 0 V to 7 V, the motional resistance of devices #1 and #2 is tuned from 15.7 kΩ to 4.9 kΩ and 125.5 kΩ to 12.5 kΩ, respectively. Thus, a reduction of 70 percent and 90 percent can be achieved for devices #1 and #2, respectively.

In summary, we proposed and demonstrated that piezoresistive heat engines could enhance the performance of the capacitive silicon resonators. Capacitive silicon resonators with single and multiple piezoresistive beams were fabricated and evaluated. The improvement of the insertion loss and reduction of the motional resistance were achieved.

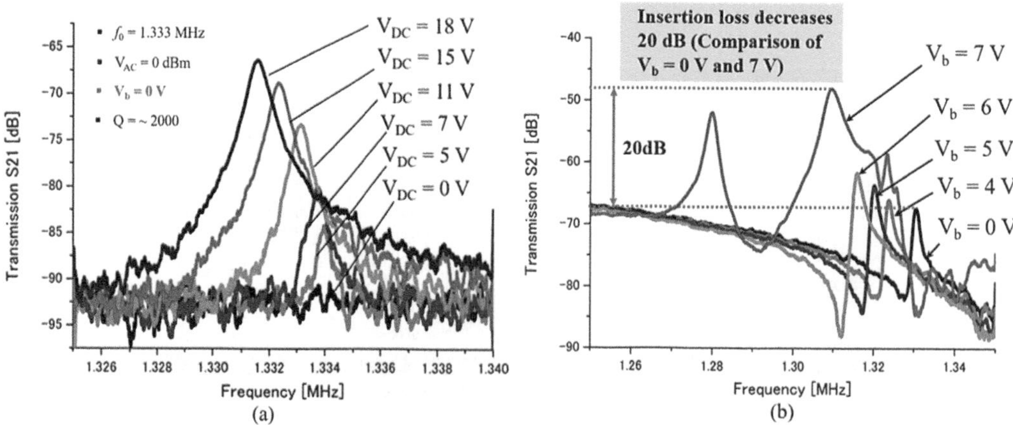

FIGURE 1.25 Frequency response of device #2. (a) Capacitive effect. (b) Piezoresistive heat engine effect.

1.4 CONCLUSIONS

In this chapter, various methods for enhancing performance capacitive silicon resonators were demonstrated. The novel different fabrication technologies, including hermetic packaging based on LTCC substrate, deep RIE and NBE, as well as design considerations on mechanically coupled, movable electrode structures and piezoresistive heat engines were investigated to enhance the Q factor, lower motion resistance and insertion loss, and wider tuning frequency range. The contents of this paper are based on the experimental research of the authors. It is the authors hope that this work may be a useful reference to those working in the field of micro- and nanotechnology.

1.5 ACKNOWLEDGMENTS

Part of this work was performed in the Micro/Nanomachining Research Education Center (MNC), and the Micro System Integration Center (μSIC) of Tohoku University. This work was supported in part by JSPS KAKENHI for Young Scientists B (Grant number: 17K14095), and also supported in part by Grants-in-Aid for Scientific Research -KAKENHI- (Grant number: 16K14189).

REFERENCES

1. Ono, T., and M. Esashi, "Mass Sensing with Resonating Ultra-thin Silicon Beams Detected by a Double-Beam Laser Doppler Vibrometer," *Measurement Science and Technology*, **15**, 1977–1981, 2004.
2. Kim, S. J., T. Ono, and M. Esashi, "Mass Detection using Capacitive Resonant Silicon Resonator Employing LC Resonant Circuit Technique," *Review of Science Instruments*, **78**, 085103, 2007.
3. Nguyen, C. T. C. "MEMS Technology for Timing and Frequency Control," *IEEE Transactions On Ultrasonics, Ferroelectrics, and Frequency Control*, **54**, 251–270, 2007.
4. Ayazi, F., "MEMS for Integrated Timing and Spectral Processing," *IEEE Custom Integrated Circuits Conference*, California, USA, 65–72, 2009.
5. Van Beek, J. T. M., and R. Puers, "A Review of MEMS Oscillations for Frequency Reference and Timing Applications," *Journal of Micromechanics and Microengineering*, **22**, 013001, 2012.
6. Toan, N. V., H. Miyashita, M. Toda, Y. Kawai, and T. Ono, "Fabrication of an Hermetically Packaged Silicon Resonator on LTCC Substrate," *Microsystem Technology*, **19**, 1165–1175, 2013.
7. Piazza, G., P. J. Stephanou, J. M. Porter, M. B. J Wijesundara, and A. P. Pisano, "Low Motional Resistance Ring-shaped Countour-mode Aluminum Nitride Piezoelectric Micromechanical Resonators for UHF Application," *The 18th IEEE International Conference on Micro Electro Mechanical Systems*, Florida, USA, 20–23, 2005.
8. Sorenson, L., J. L. Fu, and F. Ayazi, "One-Dimensional Linear Acoustic Bandgap Structures for Performance Enhancement of AlN-on-Silicon Micromechanical Resonators," *The 16th International Conference on Solid State Sensors, Actuators and Microsystems*, Beijing, China, 918–921, 2011.
9. Nguyen, N., A. Johannesen, and U. Hanke, "Design of High Q Thin Film Bulk Acoustic Resonator using Dual-mode Reflection, *IEEE International Ultrasonics Symposium*, Illinois, USA, 487–490, 2014.
10. Ali, A., and J. E. Y. Lee, "Novel Platform for Resonant Sensing in Liquid with Fully Electrical Interface based on an In-plane-mode Piezoelectric-on-Silicon Resonator," *Procedia Engineering*, **120**, 1217–1220, 2015.
11. Van Beek, J. T. M., P. G. Steeneken, and B. Giesbers, "A 10 MHz Piezoresistive MEMS Resonator with High Q," *IEEE International Frequency Control Symposium and Exposition*, Florida, USA, 475–480, 2006.
12. Rahafrooz, A., and S. Pourkamali, "Thermal-Piezoresistive Energy Pumps in Micromechanical Resonator Structures," *IEEE Transactions on Electron Devices*, **59**, 3587–3593, 2012.
13. Ho, G. K., K. Sundaresan, S. Pourkamali, and F. Ayazi, "Low Motional Impedance Highly Tunable I^2 Resonators for Temperature Compensated Reference Oscillators," *18th IEEE International Conference on Micro Electro Mechanical Systems,* Florida, USA, 116–120, 2005.

14. Hashimoto, K. Y., T. Kimura, T. Matsumura, H. Hirano, M. Kadota, M. Esashi, and S. Tanaka, "Moving Tunable Filters Forward," *IEEE Microwave Magazine*, **16**, 89–97, 2015.
15. Steeneken, P. G., K. L. Phan, M. J. Goossens, G. E. J. Koops, G. J. A. M. Bronm, C. V. D. Avoort, and J. T. M. Van Beek, "Piezoresistive Heat Engine and Refrigerator," *Nature Physics*, **7**, 354–359, 2011.
16. Ramezany, A., M. Mahdavi, and S. Pourkamali, "Nanoelectromechanical Resonant Narrow Band Amplifiers," *Microsystems & Nanoengineering*, **2**, 16004, 2016.
17. Pourkamali, S., G. K. Ho, and F. Ayazi, "Low Impedance VHF and UHF Capacitive Silicon Bulk Acoustic Wave Resonators—Part I: Concept and Fabrication," *IEEE Transactions on Electron Devices*, **54**, 2017–2023, 2007.
18. Pourkamali, S., G. K. Ho, and F. Ayazi, "Low Impedance VHF and UHF Capacitive Silicon Bulk Acoustic Wave Resonators—Part II: Measurement and Characterization, *IEEE Transactions on Electron Devices*, **54**, 2024–2030, 2007.
19. Toan, N. V., M. Toda, Y. Kawai, and T. Ono, "A Capacitive Silicon Resonator with a Movable Electrode Structure for Gap Width Reduction," *Journal of Micromechanics and Microengineering*, **24**, 025006, 2014.
20. Qishu, Q., S. Pourkamali, and F. Ayazi, "Capacitively Coupled VHF Silicon Bulk Acoustic Wave Filters," *IEEE Ultrasonics Symposium*, 1649–1652, 2007.
21. Toan, N. V., T. Shimazaki, and T. Ono, "Single and Mechanically Coupled Capacitive Silicon Nanomechanical Resonator," *Micro & Nano Letters*, **11**, 591–594, 2016.
22. Lee, J. E. Y., J. Yan, and A. A. Seshia, "Study of Lateral Mode SOI-MEMS Resonators for Reduced Anchor Loss," *Journal of Micromechanics and Microengineering*, **21**, 045011, 2011.
23. Toan, N. V., T. Kubota, H. Sekhar, S. Samukawa, and T. Ono, "Mechanical Quality Factor Enhancement in Silicon Micromechanical Resonator by Low-Damage Process Using Neutral Beam Etching Technology," *Journal of Micromechanics and Microengineering*, **24**, 085005, 2014.
24. Toan, N. V., M. Toda, Y. Kawai, and T. Ono, "A Long Bar Type Silicon Resonator with a High Quality Factor," *IEEE Transactions on Sensors and Micromachines*, **134**, 26–31, 2014.
25. Palaniapan, M., and L. Khine, "Micromechanical Resonator with Ultra-High Quality Factor," *Electronics Letters*, **43**, 1090–1092, 2007.
26. Lee, L. E. Y., Y. Zhu, and A. A. Seshia, "A Bulk Acoustic Mode Single Crystal Silicon Microresonator with a High-Quality Factor," *Journal of Micromechanics Microengineering*, **18**, 064001, 2008.
27. Clark, J. R, W. T. Hsu, and C. T. C. Nguyen, "Measurement Techniques for Capacitively-Transduced VHF-to-UHF Micromechanical Resonators," In *Proceedings of Conference on Solid-State Sensors & Actuators (Transducers'01)*, Munich, Germany, 1118–1121, 2001.
28. Lin, Y. W., S. S. Li, Y. Xie, Z. Ren, and C. T. C. Nguyen, "Vibrating Micromechanical Resonators with Solid Dielectric Capacitive Transducer Gaps," In *Proceedings of the IEEE International Conference on Frequency Control Symposium and Exposition*, 128–134, 2005.
29. Mattila, T., J. Kihamaki, T. Lamminmaki, O. Ranatakari, A. Oja, H. Seppa, H. Karelus, and I. Tittonen, "A 12 MHz Microchemical Bulk Acoustic Mode Oscillator," *Sensors and Actuators A: Physical,* **101**, 1–9, 2002.
30. Lee, J. E. Y., and A. A. Seshia, "A 5.4 MHz single crystal silicon wine glass mode disk resonator with quality factor of 2 million," *Sensors and Actuators A: Physical*, **156**, 28–35, 2009.
31. Esashi, M. "Wafer Level Packaging of MEMS," *Journal of Micromechanics and Microengineering*, **18**. 073001, 2008.
32. Pourkamali, S., and F. Ayazi, "Wafer Level Encapsulation and Sealing of Electrostatic HARPSS Transducers," In *Proceedings of IEEE Sensors*, 49–52, 2007.
33. Fang, J., J. Fu, and F. Ayazi, "Meal Organic Thin Film Encapsulation for MEMS," *Journal of Micromechanics and Microengineering*, **18**, 105002, 2008.
34. Li, X, T. Abe, Y. Liu, and M. Esashi, "Fabrication of High-Density Electrical Feed through by Deep Reactive Ion Etching of Pyrex Glass," *Journal of Microelectromechanical Systems*, **11**, 625–629, 2002.
35. Nakamura, K., F. Takayanagi, Y. Moro, H. Sanpei, M. Onozawa, and M. Esashi, "Development of RF MEMS Switch," In *Proceedings of Advantest Technical Report*, **22**, 9–15, 2004.
36. Tanaka, S., S. Matruzaki, M. Mohri, A. Okada, H. Fukushi, and M. Esashi, "Wafer Level Hermetic Packaging Technology for MEMS Using Anodically Bondable LTCC Wafer," In *Proceedings of Micro-Electro-Mechanical Systems*, 376–379, 2011.

37. Tanaka, S., M. Mohri, A. Okada, H. Fukushi, and M. Esashi, "Versatile Wafer Level Hermetic Packaging Technology Using Anodically Bondable LTCC Wafer with Compliant Porous Gold Bumps Spontaneously Formed in Wet Etched Cavities," In *Proceedings of Micro-Electro-Mechanical Systems*, 369–372, 2012.
38. Lin, Y. C., W. S. Wang, L. Y. Chen, M. W. Chen, T. Gessner, and M. Esashi, "Anodically Bondable LTCC Substrate with Novel Nano-structured Electrical Interconnection for MEMS Packaging. In *Proceedings of Solid-State Sensors, Actuators and Microsystems Conference*, 2351–2354, 2011.
39. Wu, G., D. Xu, B. Xiong, and Y. Wang, "A High Performance Bulk Mode Single Crystal Silicon Microresonator based on a Cavity—SOI Wafer," *Journal of Micromechanics and Microengineering*, **22**, 025020, 2012.
40. Mattila, T., J. Kiihamaki, T. Lamminmaki, O. Jaakkola, P. Rantakari, A. Oja, H. Seppa, H. Kattelus, and I. Titonen, "A 12 MHz Micromechanical Bulk Acoustic Mode Oscillator," *Sensors and Actuators A*, **101**, 1–9, 2012.
41. Tanaka, S., S. Matsuzaki, M. Mohri, A. Okada, H. Fukushi, and M. Esashi, "Wafer Level Hermetic Packaging Technology for MEMS Using Anodic-Bondable LTCC Wafer," *The 24th IEEE International Conference on MEMS*, 376–379, 2011.
42. Kim, B., M. A. Hopcroft, R. N. Candler, C. M. Jha, M. Agarwal, R. Melamud, S. A. Chandorkar, G. Yama, and T. W. Kenny, "Temperature Dependence of Quality Factor in MEMS Resonators," *Journal of Microelectromechanical Systems*, **17**, 755–766, 2008.
43. Lee, J. E. Y., and A. A. Seshia, "Parasitic Feed-through Cancellation Techniques for Enhanced Electrical Characterization of Electrostatic Microresonators," *Sensors and Actuators A*, **156**, 36–42, 2009.
44. Tomura, M., C.H. Huang, Y. Yoshida, T. Ono, S. Yamasaki, and S. Samukawa, "Plasma Induced Deterioration of Mechanical Characteristic of Microcantilever," *Japanese Journal of Applied Physics*, **49**, 04DL20, 2010.
45. Oehrlein, G. S., R. M. Tromp, Y. H. Lee, and E. J. Petrillo, "Study of Silicon Contamination and Near Surface Damage Caused by CF_4/H_2 Reactive Ion Etching," *Applied Physics Letters* **45**, 420–22, 1984.
46. Petti, C. J., J. P. Mcvitte, and J. D. Plummer, "Characterization of Surface Mobility on the Sidewalls of Dry Etched Trenches," *Electron Devices Meeting*, 104–107, 1988.
47. Wada, A., Y. Yanagisawa, B. Altansukh, T. Kubota, T. Ono, S. Yamasaki, and S. Samukawa, "Energy-loss of Single Crystal Silicon Microcantilever due to Surface Defects Generated during Plasma Process," *Journal of Micromechanics and Microengineering*, **23,** 065020, 2013.
48. Samukawa, S., "Novel Neutral Beam Etching Processes for Future Nanoscale Devices," *The International Microprocesses and Nanotechnology Conference*, 472–473, 2007.
49. Samukawa, S., K. Sakamoto, and K. Ichiki, "Generating High-Efficiency Neutral Beams by Using Negative Ions in an Inductively Coupled Plasma Source," *Journal of Vacuum Science and Technology A*, **20**, 1566–1573, 2002.
50. Endo, K., S. Noda, M. Masahara, T. Kubota, T. Ozaki, S. Samukawa, Y. Liu, K. Ishii, Y. Ishikawa, E. Sugimata, T. Matsukawa, H. Takashima, H. Yamauchi, and E. Suzuki, "Damage Free Neutral Beam Etching Technology for High Mobility FinFETs," *Electron Devices Meeting,* 104–107, 2005.
51. Kubota, T., O. Nukaga, S. Ueki, M. Sugiyama, Y. Inamato, H. Ohtake, and S. Samukawa, "200 mm Diameter Neutral Beam Source based on Inductively Coupled Plasma Etcher and Silicon Etching," *Journal of Vacumn Science and Technology A,* **28**, 1169–1174, 2010.
52. Kubota, T., N. Watanabe, S. Ohtsuka, T. Iwasaki, K. Ono, Y. Iriye, and S. Samukawa. "Numerical Study on Electron Transfer Mechanism by Collision of Ions at Graphite Surface in Highly Efficient Neutral Beam Generation," *Journal of Physics D: Applied Physics*, **45**, 095202, 2012.
53. Oka, T., T. Ishino, H. Tanigawa, and K. Suzuki, "Silicon Beam Micromechanical Systems Resonator with a Sliding Electrode," *Japanese Journal of Applied Physics*, **50**, 06GH02, 2011.
54. Shao, L. C., M. Palaniapan, L. Khine, and W. W. Tan, "Micromechanical Resonators with Submicron Capacitive Gaps in 2 um Process," *Electronics Letters*, **43**,1427–1428, 2007.
55. Samukawa, S., Y. Ishikawa, S. Kumagai, and M. Okigawa, "On-Wafer Monitoring of Vacuum-Ultraviolet Radiation Damage in High Density Plasma Process," *Japanese Journal of Applied Physics*, **40**, L1346, 2001.
56. Yunogami, T., T. Mizutani, K. Suzuki, and S. Nishimatsu, "Radiation Damage in SiO2/Si Induced by VUV Photons," *Japanese Journal of Applied Physics*, **28**, 2172, 1989.

57. Chaste, J., A. Eichler, J. Moser, G. Ceballos, R. Rurali, and A. Bachtold, "A Nanomechanical Mass Sensor with Yoctogram Resolution," *Nature Nanotechnology*, **7**, 301–304, 2012.
58. Seo, Y. J., M. Toda, and T. Ono, "Si Nanowire Probe with Nd-Fe-B Magnet for Attonewton Scale Force Detection," *Journal of Micromechanics and Microengineering*, **25**, 045015, 2015.
59. Inomata, N., M. Toda, M. Sato, A. Ishijima, and T. Ono, "Pico Calorimeter for Detection of Heat Produced in an Individual Brown Fat Cell," *Applied Physics Letter*, **100**, 154104, 2012.
60. Rips, S., and M. J. Hartmann, "Quantum Information Processing with Nanomechanical Qubits," *Physical Review Letters*, **111**, 049905, 2013.
61. Sage, E., O. Martin, C. Dupre, T. Ernst, G. Billiot, L. Duraffourg, E. Colinet, and S. Hentz, "Frequency Addressed NEMS Arrays for Mass and Gas Sensing Applications," *Proceedings of Transducers Conference*, 665–668, 2013.
62. Demirci, M. U., M. A. Abdelmoneum, and C. T. C. Nguyen, "Mechanically Corner-coupled Square Microresonator Array for Reduced Series Motional Resistance," *Journal of Microelectromechanical Systems*, **15**, 1419–1936, 2006.
63. Mestron, R. M. C., R. H. B. Fey, K. L. Phan, and H. Nijmeijer, "Experimental Validation of Hardening and Softening Resonances in a Clamped-Clamped Beam MEMS Resonator," *Proceedings of the Eurosensors XXIII Conference*, 812–815, 2009.
64. Husain, A., J. Hone, H. C. Ch. Postma, M. H. Huang, T. Drake, and M. Barbic, A. Scherer, M. L. Roukes, "Nanowire-based Very High Frequency Electromechanical Resonator," *Applied Physics Letter*, **83**, 1240–1242, 2003.
65. Weinstein, D., and S. A. Bhave, "Piezoresistive Sensing of a Dielectrically Actuated Silicon Bar Resonator," In *Solid-state Sensors, Actuators, and Microsystems Workshop,* 368–71, 2008.
66. Lin, Y. W., S. S. Li, Y. Xie, Z. Ren, and C. T. C. Nguyen, "Vibrating Micromechanical Resonators with Solid Dielectric Capacitive Transducer Gaps," In *Frequency Control Symposium and Exposition*, 128–34, 2005.
67. Qishu, Q., S. Pourkamali, and F. Ayazi, "Capacitively Coupled VHF Silicon Bulk Acoustic Wave Filters," In *IEEE Ultrasonics Symposium*, 1649–1652, 2007.
68. Ho, G. K., K. Sundaresan, S. Pourkamali, and F. Ayazi, "Micromechanical IBARs: Tunable High Q Resonator for Temperature Compensated Reference Oscillators," *Journal of Microelectromechanical Systems*, **19**, 503–515, 2010.
69. Hsu, W. T., J. R. Clark, and C. T. C. Nguyen, "A Sub-micron Capacitive Gap Process for Multiple Metal Electrode Lateral Micromechanical Resonators," In *Micro Mechanical Systems Conference*, 349–352, 2001.
70. Samarao, A. K., and F. Ayazi, "Combined Capacitive and Piezoelectric Transduction for High Performance Silicon Microresonators," *24th IEEE International Conference on Micro Electro Mechanical Systems*, Cancun, Mexico, 169–172, 2011.
71. Van Beek, J. T. M., G. J. A. M. Verheijden, G. E. Koops, K. L. Phan, and C. V. D. Avoort, "Scalable 1.1 GHz Fundamental Mode Piezo-Resistive Silicon MEMS Resonator," *IEEE International Electron Devices Meeting,* Washington, DC, USA, 411–414, 2007.
72. Weinstein, D., and S. A. Bhave, "Piezoresistive Sensing of a Dielectrically Actuated Silicon Bar Resonator," *Solid-State Sensors, Actuators, and Microsystems Workshop*, South Carolina, USA, 368–371, 2008.
73. Chandorkar, S. A., R. V. Candler, A. Duwel, R. Melamud, M. Agarwal, K. E. Goodson, T. W. Kenny, "Multimode Thermoplastic Dissipation," *Journal of Applied Physics*, **105**, 043505, 2009.

2 Photodetection Based On Two-Dimensional Layered Semiconductor Materials and Their Heterojunction

Chaojian Hou[a], Lijun Yang[a], Genwang Wang[a], Yang Wang[a], Zhan Yang[b], and Lixin Dong[c]

[a]Key Laboratory of Microsystems and Microstructures Manufacturing, Ministry of Education, and School of Mechatronics Engineering, Harbin Institute of Technology, Harbin, Heilongjiang, China

[b]School of Mechatronics Engineering, Soochow University, Soochow, Jiangsu, China

[c]Department of Biomedical Engineering, City University of Hong Kong, Kowloon Tong, Hong Kong, China

CONTENTS

2.1 INTRODUCTION

Photodetector, as an important functional sensing module in optoelectronic devices, provides an effective way to convert light or other electromagnetic radiation into electrical signals, allowing one to use it to realize photoelectric signal detection, energy harvest, and so forth [1]. Based on their unique functional features, photodetectors hold a multitude of significant applications in modern society. These applications include video imaging, environment monitoring, night vision, military early warning, gas sensing, remote sensing, optical communications, security checks, motion detection, scientific research, industrial processing control [2], as well as affecting many aspects of our daily lives and national strategic development. In the past decade, various types of photodetectors had been created based on elemental or compound semiconductors. According to different working mechanisms, photodetectors can be divided into photoemission devices [3], photovoltaic devices [4], thermal devices [5], photochemical devices [6], and polarization-sensitive devices [7], resulting in the achievement of a high level of maturity due to the development of high-performance material synthetic technology [8], large-scale production [9], and integration technologies [10]. Generally, the current configurations of commercial photodetectors are composed of traditional bulk semiconductor materials, including monocrystalline or polycrystalline silicon, HgCdTe, GaSe, In_2Se_3, SnS_2, and so on [11]. The most representative high-performance photodetectors dominating the market are made from crystalline silicon (Si), which mainly offer a visible and near-infrared (NIR) photodetection [12]. Moreover, quantum well infrared photodetectors—made of multilayer compound semiconductors (such as AlGaAs-GaAs-AlGaAs or other multi-well heterostructures) via molecular beam epitaxy (MBE) technology—exhibit a broader wavelength range into mid-infrared wavelength [13].

However, there exist many fatal drawbacks among photodetectors based on traditional semiconductor materials in bulk form, offsetting their advantages in photodetection and energy harvest. These fatal shortcomings are as follow:

- First, photodetectors comprised of 3D (bulk) materials suffer from dangling bonds [14], which lead to intense scattering of the charge carriers. As a result, the expected transfer characteristic of photogenerated carriers (electron-hole pairs) in photodetectors will be inhibitory.
- Second, the low maximum strain of conventional bulk semiconductor materials is the key hurdle for the extent application in flexible, bendable, wearable optoelectronic devices. To our best knowledge, bulk silicon crystalline resulting from the defects and surface imperfections can sustain only ~1.2 percent breaking strains, hindering Si-based photodetector in the application of flexible devices in the future [15].
- Finally, immutable energy band gap of bulk semiconductor materials is not conducive to high light-absorption efficiencies with broadband wavelengths.

For instance, a Si-based photodetector can typically operate from 400 nm to 1100 nm owing to its larger energy band gap at ~1.14eV, leading to a weak photo response and energy harvest efficiency under longer wavelength range [16]. Thus, the above disadvantages of bulk traditional semiconductor materials in thick material usage, brittle material properties, and rigorous operation demands compel conventional bulk-based optoelectronic devices to reach their physical limits. To overcome the limitation of bulk semiconductor materials for photodetection, there is an urgent demand for alternative photodetection materials that match the high-performance photodetector's requirements, including a device structure that is lighter, smaller, more flexible, and portable, and a device performance that has faster photo response, higher efficiency, longer wavelength range, more transparency, and more flexible integrability with traditional complementary metal-oxide-semiconductor (CMOS). In the past few years, two-dimensional (2D) semiconductor layered material-based photodetectors exhibit their tremendous development prospects in high-performance photoelectric detection and energy collection regions, providing a vital complement of shortages of current Si technologies. The category of 2D materials are as shown in Figure 2.1[17].

Compared with traditional bulk semiconductor materials, many remarkable advantages of 2D semiconductor-layered materials make them the most suitable platform for the realization of future state-of-the-art multifunctional photodetectors.

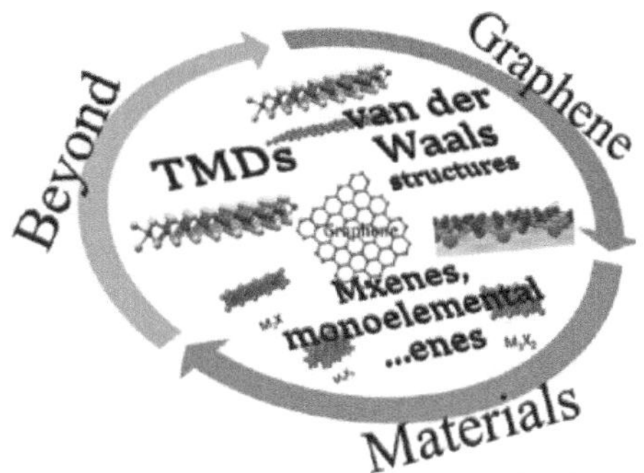

FIGURE 2.1 The category of 2D materials [17].

- First, due to their atomic-scale thickness, 2D semiconductor-layered materials not only can be active channels for photodetectors, but they also can offer a thinner photo-generated carriers' (Electrons and holes, EHs) transport channel than traditional bulk semiconductor materials. Specifically, benefiting from the proportional relationship ($\lambda = \sqrt{t_s t_{ox} \varepsilon_s / \varepsilon_{ox}}$ [18], where λ is the characteristic length, t_s and ε_s are the thickness and dielectric constants of semiconducting channel, and t_{ox} and ε_{ox} are the thickness and dielectric constants of oxidation dielectric layer) between characteristic length λ of a conventional field-effect-transistors and the thickness t_s of channel material, the thinner thickness of 2D semiconductor-layered materials provide an important means of implementation for photodetectors with shorter transport channel, which is favorable for further device miniaturization. Moreover, different from the coupling effect between photo-generated carriers and phonons or interface states in bulk materials, all photogenerated carriers in 2D semiconductor-layered materials are confined in an atomic-scale transport channel, enabling an impactful suppression of current leakage. This special performance benefits by reducing current leakage, enabling a smaller dark current in 2D-based photodetectors. In addition, the naturally atomic-scale thickness of 2D semiconductor-layered materials motivate the appealing flexible, wearable, or portable electronics application, resulting from their outstanding mechanical flexibility. To our knowledge, the excellent strain tolerance of 2D semiconductor-layered materials is larger ~10 percent before rupture [19], enabling that strain engineering can be an efficacious technology for the implementation of novel device architectures with diverse functionalities, which greatly promote the development of wearable, bendable devices and devices.
- Second, another inherent physical property of 2D semiconductor-layered materials is their layer-dependent electronic band structures [20], resulting in a significant impact on the photon absorption, photocarriers generation, transport, separation, recombination, and even trapping processes. This allows them to be utilized in versatile photodetectors. Thus, 2D semiconductor-layered materials exhibit more varied photodetection ability over a wide range of electromagnetic spectrum, which is consistent with their different band energy gap and structural characteristics. For instance, layer-dependent different modulation regions of energy bandgap lead to most suitable wavelength response region of molybdenum disulfide (MoS_2) and few layered black phosphorus (BP) located at visible [21] and infrared region [22], respectively. This paves the way to provide a remedy of the narrow wavelength response of Si-based photodetector. Meanwhile, the existence of Van Hove singularities in electronic density of states in 2D semiconductor-layered materials contribute to strong light-matter interactions, enabling an enhancement of photon absorption and photo-carriers generation [23]. Moreover, the electronic band structure can be easily modulated via chemical doping [24], electrostatic doping [25], and so forth, which not only allow the expansion

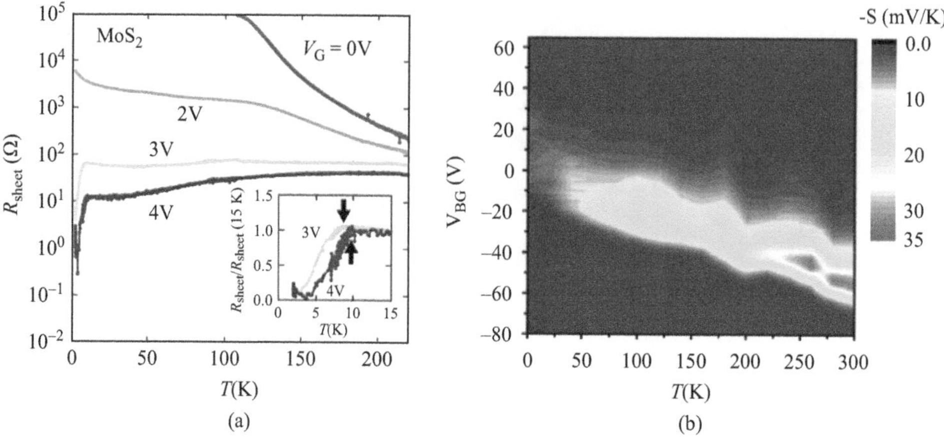

FIGURE 2.2 (a) Superconductivity effect [27]. (b) Thermoelectric effect [30].

of imminent attributes of 2D semiconductor-layered materials, but also provide more practical means for the development of 2D functional devices with excellent performance and diverse functions. For example, the transition from indirect band gap to direct band gap, with the layers decrease of 2D semiconductor-layered materials, allows the reduction of the influence of phonons under the photoexcitation process, which can then be used for future ultra-fast photodetection [26].

- Third, structural characteristics without surface dangling bonds offer us numerous plentiful lattice-match platforms for construct functional heterostructures. Especially, it has also been reported that the heterostructures integration categories of different 2D semiconductor-layered materials is achievable via either mechanical peeling and all-dry transferring or chemical vapor deposition technology, which render 2D semiconductor-layered materials promising competitors for future ultra-fast and ultra-sensitive photoelectric detection and high-efficiency energy collection.
- Last, under special conditions, colorful physical phenomena (such as superconductivity [27], as shown in Figure 2.2(a), piezoelectric effect [28], valley spin effect [29], thermoelectric effect [30], as shown in Figure 2.2(b), and so on are also exhibited in 2D semiconductor-layered materials, which can provide a research platform for scientists to explore novel physical phenomena. The anisotropic nature within the plane of the layers in few-layered BP nanoflakes brings the in-plane anisotropic electron transport and excitation of photogenerated carriers, which can be used for future polarization-sensitive optoelectronic devices [31]. All in all, unique properties and advantages of 2D semiconductor-layered materials, including high transparency, strong light-matter interactions, good flexibility, ease of processing and so on, can make up for the shortcomings of traditional bulk semiconductors—ranging from huge fabrication cost, meticulous fabrication condition, and rigorous operation demands. Therefore, it has great academic value and application prospects to develop 2D-semiconductor optoelectronic devices with excellent performance and diverse functions.

In this chapter, we will review the development of 2D-semiconductor photodetector based on both layered semiconductors and their heterostructures:

- First, we will give a brief introduction of the most commonly utilized 2D semiconductor materials and their heterojunction systems. We will then discuss basic physical mechanisms for photodetection and important performance metrics of photodetectors.

- Next, we will present an overview of typically demonstrated 2D-semiconductor photodetectors in the past decade.
- Finally, we will provide outlooks to shed light on the future development of this research area.

2.2 2D SEMICONDUCTOR-LAYERED MATERIALS AND THEIR HETEROJUNCTION SYSTEMS

The emergence of single-layer graphene by Konstantin Novoselov and Andre Geim in 2004 initiated the development of 2D layered material in condensed matter physics. Thousands of scientific publications [32] contained articles about this material. However, there are many limitations hindering the application of graphene in future optoelectronic devices.

- First, the invariable absorption coefficient of single-layer graphene is only ~2.3 percent—under visible and infrared (IR) light [33]—enabling to a limited application and requiring strong light absorption.
- Second, the gap-lessness electronic band structure of graphene [34] in its pristine form not only makes the photocarriers' lifetime much shorter, but it also results in a huge dark current, which is not conducive to high-performance photodetectors. .

Thus, recent scientific interest focuses on the development of high-performance photodetectors based on several 2D semiconductor materials with inherent electronic bandgap. Until now, many kinds of 2D semiconductor-layered materials have been successfully synthesized, including transition metal dichalcogenides (TMDCs), black phosphorous (BP), and derivative 2D Van der Waals heterostructures. In this part, we will elaborate on the features of atomic structure and energy band structure of these 2D semiconductor layer materials.

2.2.1 TRANSITION METAL DICHALCOGENIDES

The general formula of TMDCs is MX_2, where M represents transition metal element (including group IV, V, or VI) and X represents the chalcogen (including S, Se, Te). The single-layer TMDCs consist of transition metal elements sandwiched between upper and lower chalcogenide elements. Multilayer TMDCs form by strong ionic-covalent bonds in the inner layer to ensure the in-plane stability of the layer, whereas interlayer TMDCs maintain their stability through van der Waals force interactions [35]. The atomic structure of TMDCs mainly includes three types: 2H state (hexagonal (H) symmetry), 3R state (rhomboidal (R) symmetry), and 1T state (tetragonal (T) symmetry) [36]. Commonly, the first two states of TMDCs exhibit semiconducting, such as $MoSe_2$, WS_2, and WSe_2. By contrast, the last state with octahedral coordination of the Mo atoms is metallic, including TaS_2, $TaSe_2$, NbS_2, and $NbSe_2$. Among all TMDCs, the most stable atomic structure is the 2H state. Thus, most of researchers focus on the semiconducting 2H-TMDCs. In the TMDC family, molybdenum disulfide (MoS_2) has been widely studied as a notable representative of TMDCs. The thickness of monolayer MoS_2 is 6.5Å [37]. The lattice parameters of 2H-MoS_2 are $a = b = 3.16$Å and $c = 12.29$Å.

The characteristics of MoS_2 band structure determine the potential application in many aspects:

- First, different from graphene band structure, MoS_2 has a large band gap range (1.2eV-1.8eV) [38], so it has a wide application potentiality in high-performance electronic devices and optoelectronic devices in visible light band. For example, single-layer MoS_2 material can absorb 10 percent of the incident light at 615 or 650 nm [39].

- Second, bulk MoS_2 material shows indirect band gap. By contrast, single-layer MoS_2 transforms into direct band gap. Therefore, the transition from indirect band gap to direct band gap makes MoS_2 hold excellent photoluminescence characteristics and fast photoelectric response characteristics.

In addition, MoS_2 also has other excellent properties, such as a strong ion-covalent bond in MoS_2 surface, which makes it withstand mechanical deformation of ~25 percent [40], so MoS_2 can bend and fold like paper, forming folds, ripples, and other structures. MoS_2 also has excellent thermal stability, mainly reflected in its thermal stability temperature of 1100 degrees [41], enabling MoS_2 to maintain performance in a harsh working environment.

2.2.2 Black Phosphorous

According to the variation range of MoS_2's band gap, the minimum band gap is still larger than 1eV, hindering the application of 2D semiconductor optoelectronic devices into near-infrared and infrared wavelength range. Therefore, exploring 2D semiconductor materials with smaller bandgap is the future trend in the development of materials science. In 2014, black phosphorus (BP) entered the field of material researchers. For the single-layer BP material, each phosphorus atom forms covalent bonds with three adjacent phosphorus atoms, forming a folded honeycomb structure with a single-layer thickness of about ~5Å [42]. The lattice parameters of BP are a = 4.374Å, b = 3.31 Å, and c = 10.473Å. For the characteristics of energy band structure, BP also holds a thickness-dependent energy band gap property, ranging from 0.3eV to 2eV [43], with the decrease of thickness. Such a large bandgap change range makes BP possible not only as channel material of high-performance field effect transistor, but also as photosensitive material of optoelectronic device with photoelectric detection capabilities ranging from ultraviolet to microwave. Moreover, compared with other 2D semiconductor-layered materials, BP exhibits unique anisotropic property [44], resulting from its asymmetry lattice structure. From BP's atomic structure, it can be seen that the lattice structure of BP in the armchair direction and the zigzag direction has obvious asymmetry, leading to some differences in its electron transport and heat transport properties along the two directions. These unique properties make black phosphorus widely used in future nano-electronic devices and nano-optoelectronic devices.

2.3 2D VAN DER WAALS HETEROSTRUCTURES

With the emergence of above 2D semiconductor layer materials, researchers have stacked two or more 2D layered materials together, enabling the formation of 2D Van der Waals heterostructures (vdW) [45]. Due to similar lattice structure and the absence of surface dangling bonds in 2D semiconductor layer materials, there is no lattice mismatch in the 2D Van der Waals heterostructure in contrast with the traditional semiconductor heterojunction structure; this avoids the effect of lattice mismatch on the electron transport and photocarrier's harvest at the interface. This advantage provides opportunities for further development of high-performance functional nano-electronic devices and nano-optoelectronic devices. So far, a variety of two-dimensional Van der Waals heterojunctions have been achieved, including semimetal-semiconductor, semimetal-insulator, semiconductor-semiconductor, and semiconductor-insulator heterojunctions [46]. In particular, the successful utilization of chemical vapor deposition [47] pave a way to form stable bonds among different interface atoms, thus enhancing the interface transport properties of materials.

An exploratory scheme for infrared light emitting diodes based on 2D Van der Walls heterojunction is discussed by Fengnian Xia [48]. Among them, black phosphorus material is used as the activation layer material for infrared light emission, and its band gap can be controlled by the thickness of black phosphorus. P-type WSe_2 can provide holes for the activation layer, where n-type MoS_2 can provide electrons for the activation layer. Then the injected carriers are captured by black phosphorus, and finally the recombination of electron-hole pairs give rise to emit infrared light. Additionally, graphene can also be utilized as the electrode material to improve the device contact quality, enabling a more effective collection of photocarriers. Therefore, the integrated 2D Van der Waals heterojunction can provide new research ideas for novel functional devices.

2.4 PHYSICAL MECHANISMS FOR PHOTODETECTION

The interaction between incident photons and semiconductor materials mainly involves three processes: *photon absorption process*, *photocarrier's transport process*, and *photocarrier's recombination* or *capture process*. Upon photodetector irradiated under light, the interaction between incident photons with sufficient energy and semiconductor materials produces electron-hole pairs in semiconductor materials along with the process of photon absorption, enabling electrons to excite in the valence band of semiconductor materials into the conduction band, thus creating a corresponding hole in the valence band at the same time. Under a certain bias voltage or potential difference, the electrons and holes propagate in the opposite direction to form photocurrent. Based on the above process of generation of photocurrent, in the following section the physical mechanisms of photodetectors are introduced, including *photoconductive effect*, *photovoltaic effect*, *photo-thermoelectric effect*, *bolometric effect*, and *photogating effect*.

2.4.1 Photoconductive Effect

Photoconductive effect utilizes excess carriers of semiconductor material generated in the case of incident photon energy larger than material bandgap. These generated electron-hole pairs can participate in transmission between source and drain terminal, which is equivalent to the increase in conductivity of semiconductor material. The phenomenon that the conductivity of this semiconductor material increases under light excitation is called *photoconductivity effect*, which belongs to the internal photoelectric effect [49]. To achieve effective separation of electron-hole pairs, the typical structure of photoconductive devices is semiconductor material as a channel with a perfect ohmic contact terminal. Therefore, photogenerated carriers can easily collect electrons and holes in the channel under external bias. Moreover, the larger the applied bias voltage, the more effective separation and transmission of electron and hole pairs will be, leading to a larger photocurrent. More important, the main source of photoconductive effect depends on inter-band carrier transition and impurity excitation. Thus, the photoconductivity can be divided into *intrinsic photoconductivity* and *impurity photoconductivity*. In addition, the photoconductive gain of photoconductive devices can be greater than 1. The calculation formula of photoconductive gain G is as follows [50]:

$$G = \frac{\tau_{\text{lifetime}}}{\tau_{\text{transit}}} \tag{2.1}$$

where τ_{lifetime} is lifetime of holes and τ_{transit} is electron transit time. To obtain higher photoconductive gain, it is necessary to reduce the transit time of electrons and increase the lifetime of holes. However, the response time of photoconductive devices is related to the recombination process of carriers, i.e., the lifetime of holes. In other words, higher photoconductive gains lead to slower response time. Therefore, for photoconductive devices, it is necessary to take a comprehensive consideration between photoconductive gains and response time for the concrete device application. The photocurrent I_{PC} under photoconductive effect can be expressed as follows [51]:

$$I_{\text{PC}} = \left(\frac{W}{L}\right) V_D \Delta\sigma \tag{2.2}$$

where W is channel width, L is channel length, V_D is the drain voltage, and $\Delta\sigma$ is the difference of conductivity before and after light irradiation.

2.4.2 Photovoltaic Effect

Photovoltaic effect is another phenomenon of internal photovoltaic effect, which is a phenomenon of voltage difference generated under light illumination. The photovoltaic effect mainly uses the built-in electric field to separate photogenerated electron-hole pairs. The built-in electric field is usually generated in the space charge region formed either by the contact with different work functions or by different doped semiconductor materials [52]. Generally, optoelectronic devices based on photovoltaic effect are defined as *photodiodes*. According to different separation mechanisms of photogenerated carriers, photodiodes can be divided into two kind, including the p-n *photodiode* formed by different doped semiconductor materials and the *Schottky photodiode* formed at the interface between metal and semiconductor materials. Photodiodes usually exhibit asymmetric rectification characteristics in the darkness. Under illumination conditions, the device can operate in two modes based on external bias voltage or not, including the photovoltaic mode when the applied bias voltage is zero and the photoconductive mode under reverse bias voltage. In the photovoltaic mode, the photogenerated electron-hole pairs are collected in different regions via built-in electric field at the metal-semiconductor interface, thus forming a stable photocurrent named *short-circuit current*. In addition, electrical signals can also be output in the form of photovoltaic. In other words, when the device keeps open, the output voltage is called *open-circuit voltage*. In the photovoltaic mode, the dark current of the device is the lowest, so the sensitivity of the device can be improved. In the photoconductive mode, the external bias voltage is the same as the direction of the built-in electric field, enabling it to be equivalent to the continuous enhancement of the built-in electric field. This enhanced built-in electric field is conducive to the collection of photogenerated carriers and the reduction of carrier transit time. Therefore, according to the above principles, photodiodes are suitable for sensitivity testing in photovoltaic mode and high-speed detection in photoconductive mode. However, it is noteworthy that the higher reverse bias voltage, the more avalanche amplification of device will be, resulting in avalanche photodiodes, thus achieving higher current gain. The photovoltaic effect can be described by the offset of the threshold gate voltage of the transistor. Therefore, the photocurrent I_{PV} under the photovoltaic effect can be calculated in two ways as follows [51]:

$$
\begin{aligned}
I_{\text{PV}} &= I_D(V_G - V_T + \Delta V_T) - I_D(V_G - V_T) \\
I_{\text{PV}} &\approx g_m \Delta V_T \\
g_m &= dI_D / dV_G
\end{aligned}
\tag{2.3}
$$

where I_D is drain current, V_G is gate voltage, V_T is threshold gate voltage, ΔV_T is offset of threshold gate voltage, and g_m is device transconductance.

2.4.3 Photo-Thermoelectric Effect

Photo-thermoelectric effect mainly occurs between two different metal or semiconductor materials with temperature difference. The specific principle is that the temperature gradient at both ends of the material will be generated by different material thermal characteristics when light irradiates two different conductors or semiconductor surfaces, leading to a photoinduced thermoelectric potential. This thermoelectric potential will make the current generated even under zero bias voltage. In order to achieve photothermal effect, local irradiation of devices or global irradiation of materials with different absorption coefficients are usually used. Thermoelectric potential V_{PTE} mainly depends on the difference of the Seebeck coefficient and the temperature difference of materials. The formulas for calculating V_{PTE} are as follows [53]:

$$
V_{\text{PTE}} = (S_1 - S_2)\Delta T \tag{2.4}
$$

where S_1 is the Seebeck coefficient at region 1, S_2 is the Seebeck coefficient at region 2, and ΔT is the temperature difference between region 1 and region 2.

In addition, V_{PTE} can also be obtained by integrating the local electric field caused by the spatial variation of light-induced temperature gradient and the *Seebeck coefficient*. The calculation formula is as follows [1]:

$$V_{PTE} = \int S \bullet \nabla T dx \tag{2.5}$$

where ∇T is temperature gradient between region 1 and region 2. For the Seebeck coefficient, it is also commonly referred to as *thermal power*. The coefficient is related to the conductivity of the material and can be calculated by *Mott formula*, which is as follows [1]:

$$S = -\frac{\pi^2 k_B^2 T}{3q} \frac{1}{\sigma} \frac{\partial \sigma}{\partial \varepsilon} \tag{2.6}$$

where σ is conductivity, ε is Fermi level, and k_B is *Boltzmann's constant*. Note that the Seebeck coefficient can be obtained by the transmission characteristic curve of the device.

2.4.4 Bolometric Effect

The *bolometric effect* is a phenomenon different from photo-thermalelectric effect, which is the effect of light-induced thermal field on the conductivity of materials under uniform light irradiation, thus achieving the modulation of photocurrent [1]. There are two key factors determining the bolometric effect, including conductivity change rate with the change of temperature and the temperature difference of materials. *Radiotherapy* is mainly made of semiconductors or superconducting absorbing materials, which are used in terahertz wavelength range. At present, the key coefficients of the radio calorimeter are *thermal resistance* and *thermal capacity*. The thermal resistance determines the sensitivity of the device, and the thermal capacity determines the response time of the device. At present, the carrier in graphene is easily heated; thus the bolometric effect has been verified in graphene devices. At the same time, the bolometric effect has also been observed in black phosphorus materials with a thickness of 100 nm [54].

2.4.5 Photogating Effect

Photogating effect is a special case of photoconductive effect. Under light irradiation, either photogenerated electrons or photogenerated holes are trapped by a defect in the material or by an adsorbed state on the surface [51], as shown in Figure 2.3. In other words, when an electron-hole pair is formed on the surface of a material, one of the carriers transmits between the source and the drain terminal, whereas the other carriers are bound to the surface state or trapped state. These trapped carriers become a local gate voltage due to electrostatic interaction, giving rise to the modulation of the

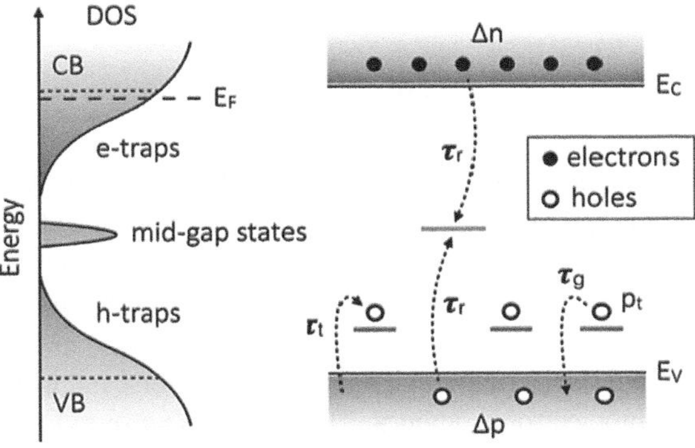

FIGURE 2.3 The diagram of photo-thermoelectric effect [51].

conductance of semiconductor materials. Therefore, the free carriers can travel many times in semiconductors, resulting in high gain. This phenomenon particularly occurs in nanomaterials due to the large specific surface area. The reason is that surface states or trapping states can easily occur in quantum dots, nanowires, and 2D nanomaterials. Similar to the photoconductivity effect, the higher the photoelectric gain, the lower the device response time. Therefore, it is necessary to balance the photoelectric gain and the response speed of the device.

2.5 PERFORMANCE METRICS FOR PHOTODETECTORS

To compare the performance of different devices with different materials and geometries under different work mechanisms, it is necessary to establish a unified evaluation standard for the performance metrics of optoelectronic devices. In general, the performance of photodetectors focuses on several important aspects, such as photoelectric conversion ability, photoelectric response time, photoelectric response wavelength range, noise characteristics, and so forth. The performance metrics of these characteristics are described in the following section.

Photoresponsivity (R) of photodetectors is the ratio of achieved output photocurrent I_{ph} to the actual absorbed optical power P_{ab} irradiated on the effective area of the device $A_{effective}$. The calculation formula of R and actual P_{ab} are as follows:

$$\begin{aligned} R &= \frac{I_{ph}}{P_{ab}} \\ P_{ab} &= P_{in}\frac{A_{effective}}{A_{incident\ light}} \end{aligned} \tag{2.7}$$

where P_{in} is incident optical power, and $A_{incident\ light}$ is the area of incident light. This parameter can be used to evaluate the photoelectric conversion capability of photodetector at a given wavelength.

External quantum efficiency (EQE) refers to the ratio of the number of electron-hole pairs contributing the generation of photocurrent to the number of incident photons in unit time. The calculation formula of *EQE* is as follows:

$$EQE = \frac{I_{ph}/q}{P_{in}/h\nu} = R\frac{hc}{q\lambda} \tag{2.8}$$

where q is the electron charge, h is Planck's constant, ν is frequency of incident light, c is light speed, and λ is wavelength of incident light. *EQE* is another important parameter for characterization of photoelectric conversion capability of photodetector. To attain high *EQE*, it is necessary to utilize the active photosensitive layer with high optical absorption.

Internal quantum efficiency (IQE) is equal to the number of electron-hole pairs collected to generate photocurrent in unit time divided by the number of absorbed photons in the same time. The absorbed photons should take into account the remaining photons after deducting the photon losses due to transmission and reflection. The calculation formula of *IQE* is as follows:

$$IQE = \frac{I_{ph}/q}{P_{ab}/h\nu} \tag{2.9}$$

Response time determines the photoelectronic response ability of optoelectronic devices. At present, rising time t_r and falling time t_f are used to characterize the response speed of optoelectronic devices. *Rising time (falling time)* refers to the corresponding time interval between the rise (falling) of photocurrent from 10 percent (90 percent) to 90 percent (10 percent) during the opening (closing) process of irradiation.

Signal-to-noise ratio (SNR) is the ratio of signal intensity to noise intensity. The lower noise is another important index of optoelectronic devices. The signal-to-noise ratio determines the minimum signal intensity that can be detected in photodetector. Noise usually occurs when the external radiation disturbs the output of the device. Noise belongs to physical process and is inherent in device operation. Its instantaneous value is random fluctuation, so it is uncertain. In addition, when multiple noises coexist, the relationship of different noises is irrelevant, thus the noise power can be superimposed. Noise of optoelectronic devices mainly includes thermal noise, generation-recombination noise, temperature noise, and so on. Usually, the signal-to-noise ratio of photoelectric devices is larger than 1 to ensure that the photoelectric devices can effectively distinguish the detected signal from the noise. The calculation formula of *SNR* is as follows:

$$SNR = \frac{Signal\ power}{Noise\ power} \tag{2.10}$$

Noise equivalent power (NEP) refers to the minimum required incident light power to obtain a signal-to-noise ratio of 1 in a 1 Hz bandwidth. In other words, when the external irradiation signal is weak, it is likely to make the detection signal possess the same amplitude as the noise signal. In this case, the detection signal will be submerged in the noise signal. Therefore, the minimum required incident light power is defined as the noise equivalent power. The calculation formula of *NEP* is as follows:

$$\begin{aligned} NEP &= \frac{\sqrt{2qJ_d}}{R} \\ J_d &= \frac{I_d}{W} \end{aligned} \tag{2.11}$$

where J_d is dark current density, I_d is dark current, and W is channel width.

In addition, the detection capability of photoelectric devices with different materials and geometries is also characterized by the detectivity (D^*). The formula of D^* is as follows:

$$\begin{aligned} D^* &= \frac{\sqrt{A_{\text{effective}}\Delta f}}{NEP} \\ \Delta f &= f_2 - f_1 \end{aligned} \tag{2.12}$$

where Δf is bandwidth of photodetector, f_1 is lower light modulation frequency, and f_2 is upper light modulation frequency. Based on the above formula, specific detectivity is related to the square root of the product of the effective area of the device and the working frequency bandwidth. The larger the specific detectivity, the better the device detection performance. Usually, specific detectivity is related to the sensitivity of the device, the wavelength range of the response, and the noise. When the total noise of the device is dominated by the noise in the dark field, the formula of specific detectivity must be related to the dark current, the effective area of the device, and photoresponsivity. The calculation formula of specific detectivity is as follows:

$$D^* = \frac{\sqrt{A_{\text{effective}}} \bullet R}{\sqrt{2qI_d}} \tag{2.13}$$

2.6 PHOTODETECTORS BASED ON 2D SEMICONDUCTOR-LAYERED MATERIALS AND THEIR HETEROJUNCTION SYSTEMS

At present, 2D semiconductor layered materials have become an important functional component of nano-optoelectronic devices due to their special energy band structure. They are widely used in photoelectric detection, photoelectric imaging, and energy collection. Specifically, the advantages of 2D semiconductor-layered materials in optoelectronic devices mainly include the following four aspects:

- First, 2D semiconductor-layered materials exhibit strong optical absorption ability and strong binding ability to excitons. For example, molybdenum disulfide can absorb 5–10 percent of the sunlight, which is twice as much as silicon and gallium arsenide [55].

- Second, due to the quantum confinement effects and atomic symmetry change in the out-of-plane direction, the band gap of 2D semiconductor-layered materials can be tuned via the change of the number of layers, thus enabling 2D semiconductor-layered materials to present a wide wavelength response range. For instance, compared with silicon material (band gap is 1.12eV), molybdenum disulfide's band gap can change from 1.2eV to 1.9eV with the thickness decrease, and the variation range of black phosphorus's band gap is 0.3eV-2.0eV. Thus, molybdenum disulfide can completely replace the application advantages of silicon material in visible wavelength, whereas black phosphorus can make up for the shortcomings of silicon in near-infrared and far infrared wavelength.
- Third, compared with indirect bandgap characteristics of silicon, single-layer molybdenum disulfide and all black phosphorus belong to the direct bandgap. Therefore, in the process of interaction between photons and 2D semiconductor-layered materials, phonon scattering will be weakened, leading to great advantages in fast photo-switch property. Finally, 2D semiconductor-layered materials also have great advantages in low-cost and simple manufacturing process.

Therefore, the study of high-performance photodetector based on 2D semiconductor-layered materials is one of the hot issues in the future development of nano-optoelectronic devices. Based on the above analysis, this part mainly analyzes and summarizes the related research of nano-optoelectronic devices based on novel 2D semiconductor-layered materials (MoS_2 and BP) and 2D van der Waals heterostructure.

2.6.1 Photodetectors Based on Molybdenum Disulfide

In 2011, Zongyu Yin et al. [56], investigated the performance of single-layer MoS_2-based photodetector fabricated via mechanically exfoliation for the first time This is shown in Figure 2.4(a). The results showed that photocurrent generated from the photodetector is solely determined by the illuminated optical power at a constant drain or gate voltage. The cut-off response wavelength of the single-layer MoS_2-based photodetector was ~670 nm, demonstrating its strong photoelectronic ability in visible spectrum. The photoresponsivity of the device was ~7.5mA/W at the gate voltage of 50V, which is larger than a graphene-based photodetector (~1mA/W) at a gate voltage of 60V; this proved the potential application of 2D semiconductor layered materials in optoelectronic devices. In addition, the switching behavior of photocurrent generation and annihilation could be completely finished within ca. 50ms, and it showed good stability. Thus, above unique characteristics of incident-light control, prompt photo-switching, and good photoresponsivity from the MoS_2

(a)

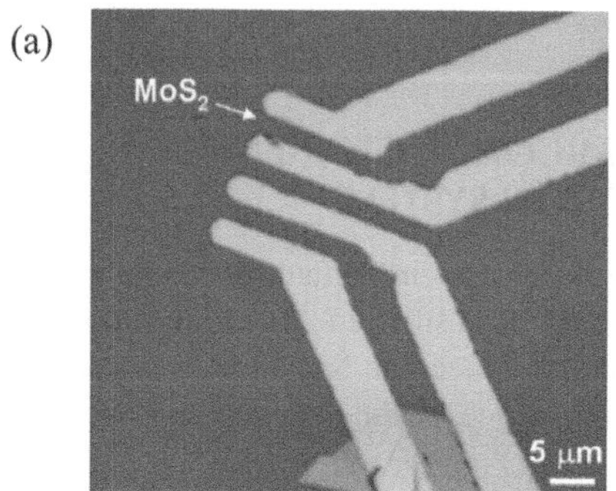

(b)

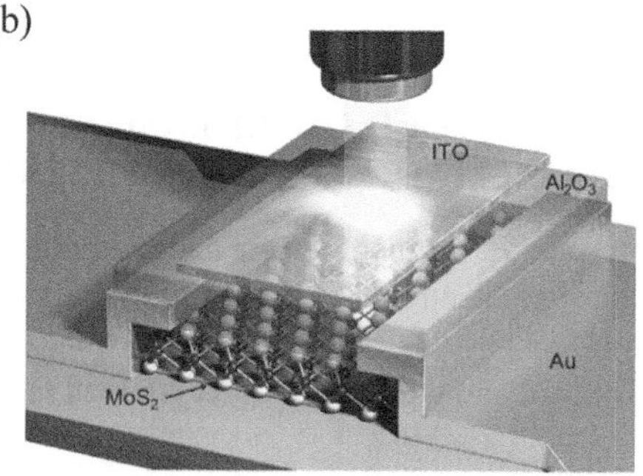

FIGURE 2.4 MoS_2-based photodetector. (a) Monolayer MoS_2 photodetector [56]. (b) Thickness dependent photodetector [57].

photodetector pave an avenue to develop the single-layer semiconducting materials for multifunctional optoelectronic device applications in the future. Afterwards, to obtain high photoresponsivity and wide wavelength response range, many research groups utilized multi-layer MoS_2 photodetector. In 2012, Hee Sung Lee et al. [57], studied the effects of three different thicknesses of MoS_2 on the performance of photodetector, as shown in Figure 2.4(b). The results showed that devices with triple MoS_2 layers exhibited excellent photodetection capabilities for red light, while those with single- and double-layers turned out to be quite useful for green light detection. The varied functionalities were attributed to energy gap modulation by the number of MoS_2 layers. In 2012, Woong Choi et al. [58], explored the response of multilayer MoS_2 photodetector with spectral response from ultraviolet to infrared wavelengths. The results showed that the photoresponsivity of multilayer MoS_2 photodetector was more than 100 mA/W, and the response wavelength range could extend into near-infrared wavelength, which could completely provide a strong remedy for the narrow detection range of GaN, Si, GaAs photodetectors. In 2013, Oriol Lopez-Sanchez et al. [59], demonstrated ultrasensitive monolayer MoS_2 photodetector with improved device mobility and ON current. More important, this photodetector was treated via KOH solution and plasma cleaning of the device substrate, enabling a maximum external photoresponsivity of 880 A/W at a wavelength of 561 nm and a photoresponse in the 400–680 nm range. Wenjing Zhang et al. [60], realized that the photoresponsivity and photogain were up to 2200 A/W and 5000, respectively, for a CVD MoS_2 monolayer at room temperature in a high vacuum. The shallow trapped long-range Coulomb potentials, which could be from the charged impurities at the MoS_2/substrate interface, drastically affected the recombination of the photogenerated carriers, accounting for the observation of the persistent photocurrent at room temperature. In addition, the adsorbates from ambient air interact strongly with MoS_2, resulted in the change of the out-of-plane Raman A_{1g} mode, the enhancement of carrier scattering, and the decrease in mobility, photoresponsivity, and photogain. The study of the photocurrent decay time revealed that the adsorption/desorption processes of the adsorbates assist the recombination of the photogenerated carriers, causing the decrease of the decay time.

In recent years, there are numerous efforts towards the integration of novel materials, fabrication processes and novel nanostructures, and working environment with MoS_2 for achieving outstanding performance of photodetector. In 2014, Junpeng Lu et al. [61], employed a facile, effective, and well-controlled technique to achieve micropatterning of MoS_2 films with a focused laser beam, as shown in Figure 2.5. The results demonstrated that a direct focused laser beam irradiation was able to achieve localized modification and thinning of as-synthesized MoS_2 films. With a scanning laser beam, microdomains with well-defined structures and controllable thickness were created on the same film. In addition, it was found that laser modification altered the photoelectrical property of the MoS_2 films, and subsequently, photodetectors with improved performance have been fabricated

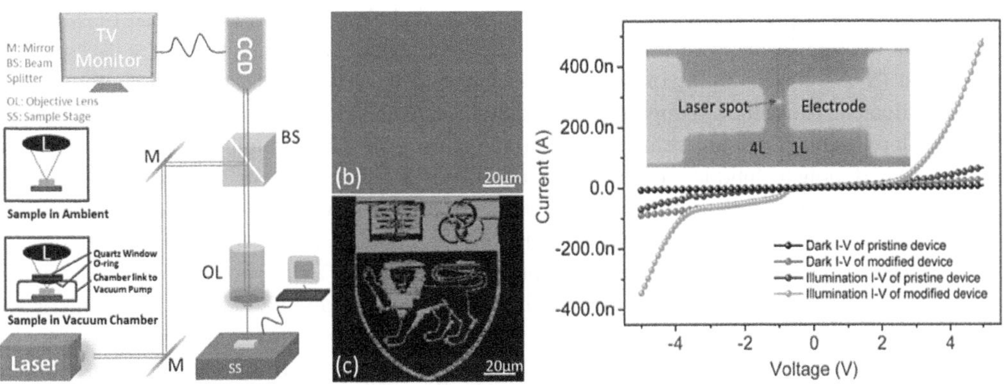

FIGURE 2.5 The 2D semiconductor photodetector based on laser-induced surface modification [61].

and demonstrated using laser modified films. This work also demonstrated that direct patterning of ultrathin MoS_2 films—with well-defined structures and controllable thickness—are appealing since the properties of MoS_2 sheets are sensitive to the number of layer and surface properties.

In 2015, Junyeon Kwon et al. [62], focused on the optoelectronic design of multilayer MoS_2 phototransistors to enhance photocurrent and demonstrate an alternative approach to obtain high photoresponsivity using indirect-bandgap multilayer MoS_2. Unlike the previous multilayer MoS_2 phototransistors with global bottom-gate structures, these multilayer MoS_2 phototransistors were fabricated in a patterned local bottom-gate thin-film transistor configuration to achieve a giant improvement in photoresponse. For the light with the wavelength of 532 nm, phototransistors based on mechanically exfoliated MoS_2 flakes exhibited high responsivity (up to 342.6 A/W at 2 mW/cm) and linear relationship between photocurrent and incident power density over a wide range of optical power. In particular, the inclusion of ungated region in the MoS_2 channel increases responsivity by 3 orders of magnitude as compared to that of global-gate multilayer MoS_2 phototransistors.

In 2015, Dominik Kufer et al. [63], focused on the effect of the detrimental and uncontrollable environmental adsorbates on devices performance. The authors reported on highly stable and high-performance monolayer and bilayer MoS_2 photodetectors encapsulated with atomic layer deposited hafnium oxide. This protected device showed enhanced electronic properties by isolating them from the ambience as strong n-type doping, vanishing hysteresis, and reduced device resistance. By controlling the gate voltage, the responsivity and temporal response could be tuned by several orders of magnitude with $R\sim 10\text{-}10^4$A/W and $t\sim$10ms to 10 s. At strong negative gate voltage, the detector was operated at a higher speed and simultaneously exhibited a low-bound, record sensitivity of $D^* \geq 7.7 \times 10^{11}$ Jones. This work led the way for future application of ultrathin, flexible, and high-performance MoS_2 detectors and prompted further investigation in encapsulated transition metal dichalcogenide optoelectronics.

In 2015, Xudong Wang et al. [64], reported about the MoS_2 transistor with a ferroelectric gate used as a photodetector, wherein the few-layer MoS_2 served as the photosensitive semiconducting channel while the remnant polarization of P(VDF-TrFE) was employed to depress the dark current of the MoS_2 semiconducting channel. The stable remnant polarization provided an ultrahigh local electrostatic field ($\approx$109 V/m within a several nanometer scale) in the semiconductor channel, which is larger than that produced by gate bias in traditional field effect transistors. With such an ultrahigh electrostatic field, the few-layer MoS_2 channel was maintained in a fully depleted state, significantly increasing the sensitivity of the detector even at ZERO gate voltage. Based on these special properties, a photodetector with high detectivity $\approx 2.2 \times 10^{12}$ Jones and photoresponsivity up to 2570 A/W was achieved. In addition, for the first time the photoresponse wavelengths of the ferroelectric polarization gating MoS_2 photodetector were extended from the visible to the near-infrared (0.85-1.55 μm).

In 2015, most doping research into transition metal dichalcogenides (TMDs) had been mainly focused on the improvement of electronic device performance. Here, the effect of self-assembled monolayer (SAM)-based doping on the performance of MoS_2-based transistors and photodetectors was investigated. The achieved doping concentrations were $\approx 1.4 \times 10^{11}$ for octadecyltrichlorosilane (OTS) p-doping and $\approx 10^{11}$ for aminopropyltriethoxysilane (APTES) n-doping (nondegenerate). Using this SAM doping technique, the field-effect mobility was increased from 28.75 to 142.2 $cm^2V^{-1}s$ in APTES/MoS_2 transistors. For the photodetectors, the responsivity was improved by a factor of $\approx$26.4 (from 219 to 5.75×10^3 A/W) in the APTES/MoS_2 devices. The enhanced photoresponsivity values were much higher than that of the previously reported TMD photodetectors. The detectivity enhancement was $\approx$24.5-fold in the APTES/MoS_2 devices and was caused by the increased photocurrent and maintained dark current after doping. The optoelectronic performance was also investigated with different optical powers and the air-exposure times. This doping study, performed on TMD devices, will play a significant role for optimizing the performance of future TMD-based electronic/optoelectronic applications [65].

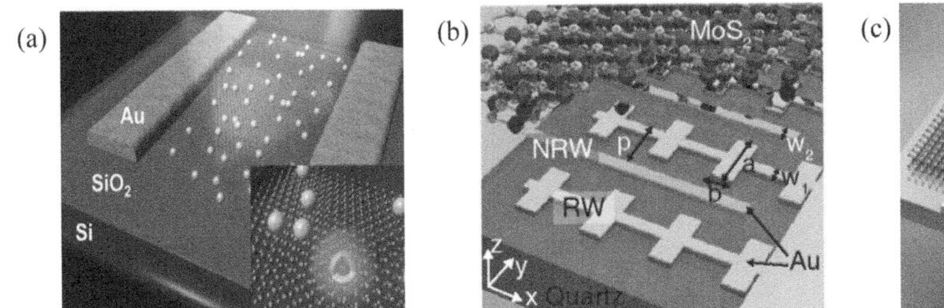

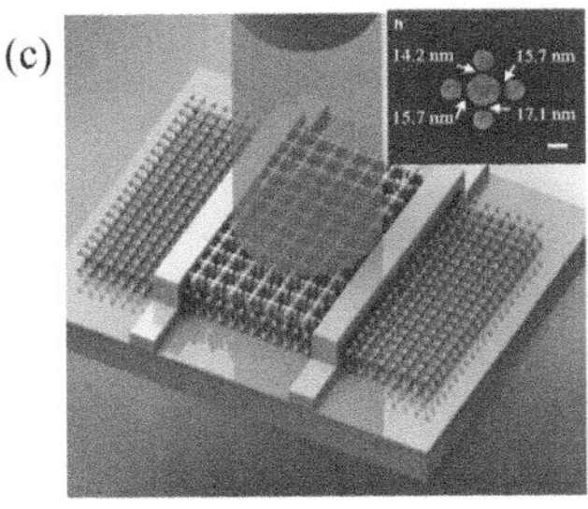

FIGURE 2.6 The 2D semiconductor photodetector based on the plasmonic enhancement. (a) Au nanoparticles-enhanced photodetection [67]. (b) Au plasmon electrode-enhanced photodetection [69]. (c) Near-infrared pentamer optical nano-antennas-enhanced photodetection [70].

In addition, Huamin Li et al. [66], fabricated vertical *p-n* junctions based on ultrathin MoS_2 by introducing $AuCl_3$ and benzyl viologen dopants. Unlike usual unipolar MoS_2, the MoS_2 *p-n* junctions show ambipolar carrier transport,,current rectification via modulation of potential barrier in films thicker than 8 nm, and reversed current rectification via tunneling in films thinner than 8 nm. The ultimate thinness of the vertical p-n homogeneous junctions in MoS_2 was experimentally found to be 3 nm, and the chemical doping depth was found to be 1.5 nm. The ultrathin MoS_2 *p-n* junctions present a significant potential of the two-dimensional crystals for flexible, transparent, high-efficiency electronic, and optoelectronic applications.

The combination of plasma structure and two-dimensional semiconductor optoelectronic devices has become an important means to enhance the photoelectric detection ability in recent years. In 2014, Ali Sobhani [67] used Au nanoparticles to enhance MoS_2 optoelectronic devices, as shown in Figure 2.6(a). By tuning plasmonic core-shell nanoparticles to the direct bandgap of monolayer MoS_2 and depositing them sparsely (<1% coverage) onto the material's surface, a threefold increase in photocurrent and a doubling of photoluminescence signal for both excitonic transitions was observed, amplifying but not altering the intrinsic spectral response.

In 2015, Jinshui Miao et al. [68], used rectangular gold nanoarrays to realize the secondary enhancement of photocurrent in MoS_2 optoelectronic devices. In 2015, Wenyi Wang et al. [69], fabricated plasma resonance electrode structure, which increased the photoelectric detection capability of MoS_2 photoelectric devices in near-infrared bands, as shown in Figure 2.6(b). By investigating hot electron-based photodetection in a device consisting of bilayer MoS_2 integrated with a plasmonic antenna array, the authors demonstrated sub-bandgap photocurrent originating from the injection of hot electrons into MoS_2 as well as photoamplification that yielded a photogain of 10^5. The large photogain resulted in a photoresponsivity of 5.2 A/W at 1070 nm, which is far above similar silicon-based hot electron photodetectors in which no photo-amplification is present. Chaojian Hou et al. [70], exploited a near-infrared pentamer optical nano-antennas taking few-layer MoS_2 as photodetectors to investigate the effect of common locations of optical nano-antennas on photocurrent amplification, as shown in Figure 2.6(c). The results showed that additional photocurrent amplification (larger than ~60 percent in our devices) can be achieved by placing optical nano-antennas close to the depletion regions, i.e., the interface between functional elements and one of the metal electrodes, demonstrating the strong position sensitivity of optical nano-antennas on optoelectronic devices. Additionally, the maximum photoresponsivity of few-layer MoS_2 photodetectors under the wavelength of 830 nm had been achieved up to ~250 mA/W based on the optical fled enhancement and absorption modulation of MoS_2 nanosheets with our near-infrared pentamer optical nano-antennas, indicating its application for near-infrared photodetection. Finally, the maximum responsivity of MoS_2-based nanodevice with pentamer optical nano-antennas (~0.111 mA/W) under self-powered conditions was enhanced by a factor of ~7 as against MoS_2-based nanodevice without pentamer optical nano-antennas.

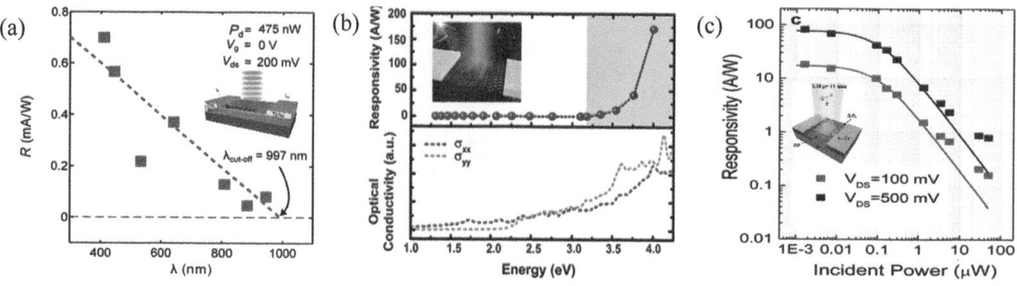

FIGURE 2.7 BP-based photodetector (a) Monolayer BP photodetector [71]. (b) Wavelength dependent photocurrent [72]. (c) Mid-Infrared BP photodetector [74].

2.6.2 Photodetectors Based on Black Phosphorus

Due to the wide energy band gap, black phosphorus is the most suitable channel material for optoelectronic devices with detection wavelength of near-infrared and infrared. In 2014, Michele Buscema et al. [71], studied the photo-response of field-effect transistors (FETs) made of few-layer black phosphorus with the thickness of 3 to 8 nm, as shown in Figure 2.7(a), as a function of excitation wavelength, power, and frequency. In the dark state, the black phosphorus field-effect transistors (FETs) can be tuned both in hole and electron doping regimes allowing for ambipolar operation. The author measured mobilities in the order of 100 cm^2/V s and a current ON/OFF ratio larger than 103. Upon illumination, the black phosphorus transistors showed a response to excitation wavelengths from the visible region up to 940 nm and a rise time of about 1 ms, demonstrating broadband and fast detection. The responsivity reached 4.8 mA/W, and it could be drastically enhanced by engineering a detector based on a PN junction. The ambipolar behavior coupled to the fast and broadband photodetection make few-layer black phosphorus a promising 2D material for photodetection across the visible and near-infrared part of the electromagnetic spectrum.

In order to further explore the photoelectric response of black phosphorus at different wavelengths, as shown in Figure 2.7(b), Jing Wu et al. [72], studied the optoelectronics characteristics of high-quality, few-layer black phosphorus-based photodetectors over a wide spectrum ranging from near-ultraviolet (UV) to near-infrared (NIR). The study demonstrated for the first time that black phosphorus could be configured as an excellent UV photodetector with a specific detectivity $\sim 3 \times 10^{13}$ Jones. More critically, it was found that the UV photoresponsivity could be significantly enhanced to $\sim 9 \times 10^4$ A/W by applying a source-drain bias (V_{SD}) of 3 V, which is the highest ever measured in any 2D material and 10^7 times higher than the previously reported value for black phosphorus. Such a colossal UV photoresponsivity is attributed to the resonant-interband transition between two specially nested valence and conduction bands. In 2016, Mingqiang Huang et al. [73], adopted the methods of lowering the temperature and shortening the length of the channel to enhance photoresponsivity. The results demonstrated a high-performance broadband BP-based photodetector in the wavelength range from 400 to 900 nm. The record-high photoresponsivity reached up to about 7×10^6 A/W at 20 K and 4.3×10^6 AW at 300 K for the 100 nm device in a broadband spectrum. A systematic study of the photo response dependence on temperature, incident laser power density, photon energy, and channel length for BP-based photodetectors was carried out for the first time. These performance parameters were among the best in the 2D materials-based photodetectors reported so far, showing great potential of black phosphorus in the application of broadband photodetection at a wide temperature range.

In addition to exploring the photoelectric detection capability of black phosphorus in the near-infrared and mid-infrared region, Qiushi Guo et al. [74], demonstrated BP mid-infrared detectors at 3.39 μm with high internal gain, resulting in an external responsivity of 82 A/W. Noise measurements showed that such BP photodetectors were capable of sensing mid-infrared light in the picowatt

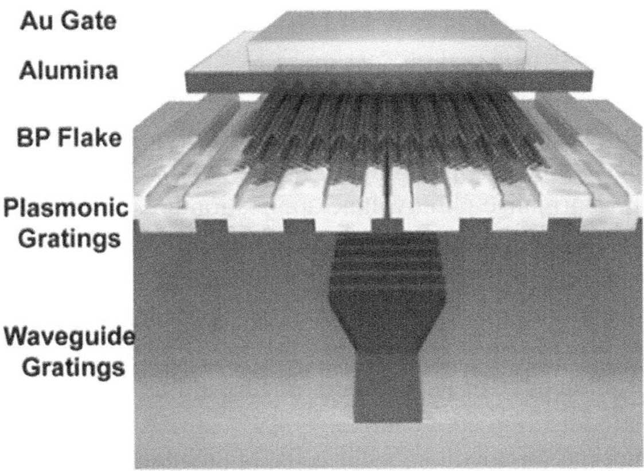

FIGURE 2.8 Three-dimensional integration of black phosphorus photodetector with silicon photonics and nanoplasmonics [77].

range, as shown in Figure 2.7(c). Moreover, the high photo response remained effective at kilohertz modulation frequencies because of the fast carrier dynamics arising from BP's moderate bandgap. The high photo response at mid-infrared wavelengths and the large dynamic bandwidth, together with its unique polarization dependent response induced by low crystalline symmetry, could be coalesced to promise photonic applications such as chip-scale mid-infrared sensing and imaging at low light levels.

In 2018, Chaojian Hou et al. [75], investigated the thickness-dependence of the intrinsic property of BP photodetectors in the dark. Then the photo-response performance (including responsivity, photogain, photo-switching time, noise equivalent power, and specific detectivity) of BP photodetectors with relative thicker thickness was explored under a near-infrared laser beam (λ_{IR} = 830 nm). The experimental results revealed the impact of BP's thickness on the current intensity of the channel and showed that degenerated p-type BP is beneficial for larger current intensity. More importantly, the photo-response of thicker BP photodetectors exhibited a larger responsivity up to 2.42 A/W than the few-layer ones and a fast response photo-switching speed (response time is ~2.5ms) comparable to thinner BP nanoflakes was obtained, indicating that BP nanoflakes with larger layer thickness are also promising for application for ultra-fast and ultra-high near-infrared photodetectors.

Similar to MoS_2, in order to enhance the interaction between black phosphorus and incident light, researchers have also adopted special processes to improve the performance of devices. In 2015, Nathan Youngblood et al. [76], demonstrated a gated multilayer black phosphorus photodetector integrated on a silicon photonic waveguide operating in the near-infrared telecom band. In a significant advantage over graphene devices, black phosphorus photodetectors can operate under bias with very low dark current and attain an intrinsic responsivity up to 135 mA/W and 657 mA/W in 11.5-nm-thick and 100-nm-thick devices, respectively, at room temperature. The photocurrent is dominated by the photovoltaic effect with a high response bandwidth exceeding 3 GHz.

In 2017, Che Chen et al. [77], demonstrated the integration of a black phosphorus photodetector in a hybrid, three-dimensional architecture of silicon photonics and metallic nanoplasmonics structures, as shown in Figure 2.8. This integration approach combined the advantages of the low propagation loss of silicon waveguides, high-field confinement of a plasmonic nanogap, and the narrow bandgap of black phosphorus to achieve high responsivity for detection of telecom-band, near-infrared light. Benefiting from an ultrashort channel (~60 nm) and near-field enhancement enabled by the nanogap structure, the photodetector showed an intrinsic responsivity as high as 10 A/W afforded by internal gain mechanisms, and a 3dB roll-off frequency of 150 MHz. This device demonstrated a promising approach for on-chip integration of three distinctive photonic systems, which, as a generic platform, may lead to future nanophotonic applications for biosensing, nonlinear optics, and optical signal processing.

In 2017, chemical doping was also utilized in black phosphorus, enabling the device to exhibit photovoltaic effect. The authors reported the demonstration of a spatially controlled aluminum doping technique that enabled a p-n homojunction diode to be realized within a single 2D blak phosphorus nanosheet for high performance photovoltaic application [78]. The diode achieved a near unity ideality factor of 1.001 along with an on/off ratio of $\approx 5.6 \times 10^3$ at a low bias of 2 V, which allowed for low-power dynamic current rectification without signal decay or overshoot. When operated under a photovoltaic regime, the diode's dark current could be significantly suppressed. The presence of a built-in electric field additionally gives rise to temporal short-circuit current and open-circuit voltage under zero external bias, indicative of its enriched functionalities for self-powered photovoltaic and high signal-to-noise photodetection applications.

Moreover, in conventional photovoltaic solar cells, photogenerated carriers are extracted by the built-in electric field of a semiconductor PN junction, defined by ionic dopants. In atomically thin semiconductors, the doping level could be controlled by the field effect, enabling the implementation of electrically tunable PN junctions. However, most two-dimensional (2D) semiconductors do not show ambipolar transport, which is necessary to realize PN junctions. Few-layer black phosphorus (b-P) is a recently isolated 2D semiconductor with direct bandgap, high mobility, large current on/off ratios, and ambipolar operation.

In 2014, Michele Buscema et al. [79], fabricated few-layer b-P field-effect transistors with split gates and hexagonal boron nitride dielectric and demonstrated electrostatic control of the local charge carrier type and density in the device. Illuminating a gate-defined PN junction, we observed zero-bias photocurrents and significant open-circuit voltages due to the photovoltaic effect. The small bandgap of the material allowed power generation for illumination wavelengths up to 940 nm, attractive for energy harvesting in the near-infrared.

2.6.3 Photodetectors Based on 2D van der Waals Heterostructure

With the development of 2D semiconductor optoelectronic devices, 2D van der Waals heterojunction optoelectronic devices became the focus of attention in material science. The first structure of 2D van der Waals heterojunction was composed of graphene-2D semiconductor material-graphene structure. In this configuration, 2D semiconductor material was used as optical absorption layer, whereas graphene was used as an electrode to separate and collect the photogenerated carriers, thus realizing photoelectric detection.

In 2013, L. Britnell et al. [80], demonstrated that two-dimensional van der Waals heterojunctions have strong light-matter interaction for the first time, as shown in Figure 2.9(a). Van Hove singularities in the electronic density of states of TMDC guarantees enhanced light-matter interactions, leading to enhanced photon absorption and electron-hole creation (which are collected in transparent graphene electrodes). This allowed development of extremely efficient flexible photovoltaic devices with photoresponsivity above 0.1 ampere per watt (corresponding to an external quantum efficiency of above 30 percent).

In 2013, Kallol Roy et al. [81], proposed the graphene-MoS_2-graphene structure and characterized its photoelectric properties. The results showed that graphene-on-MoS_2 binary heterostructures display remarkable dual optoelectronic functionality, including highly sensitive photodetection and gate-tunable persistent photoconductivity. The responsivity of the hybrids was found to be nearly 1×10^{10} A/W at 130 K and 5×10^8 A/W at room temperature, making them the most sensitive graphene-based photodetectors. When subjected to time-dependent photo illumination, the hybrids could also function as a rewritable optoelectronic switch or memory, where the persistent state showed almost no relaxation or decay within experimental timescales, indicating near-perfect charge retention. These effects could be quantitatively explained by the gate-tunable charge exchange between the graphene and MoS_2 layers, and may lead to new graphene-based optoelectronic devices that are

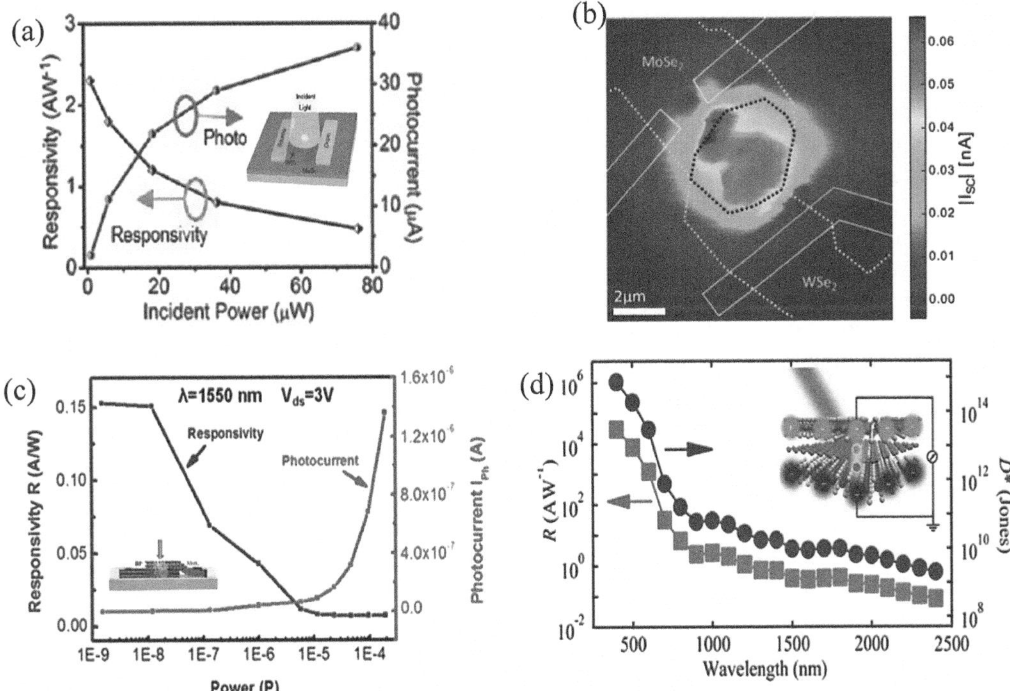

FIGURE 2.9 Vertical heterojunction photodetector (a) WS_2/MoS_2 [86]. (b) $WSe_2/MoSe_2$ [88]. (c) BP/MoS_2 [90]. (d) MoS_2-graphene-WSe_2[92].

naturally scalable for large-area applications at room temperature. At the same time, Woo JongYu et al. [82], also demonstrated that vertically stacked graphene-MoS_2-graphene and graphene-MoS_2-metal junctions could be created with a broad junction area for efficient photon harvesting. The weak electrostatic screening effect of graphene allowed the integration of single or dual gates under and/or above the vertical heterostructure to tune the band slope and the photocurrent generation. It was also demonstrated that the amplitude and polarity of the photocurrent in the gated vertical heterostructures could be readily modulated by the electric field of an external gate to achieve a maximum external quantum efficiency of 55 percent and internal quantum efficiency up to 85 percent. Their study established a method to control photocarrier generation, separation, and transport processes using an external electric field.

In 2014, Wenjing Zhang et al. [83], placed graphene on MoS_2 to achieve ultra-high gain ($>10^8$) optoelectronic devices. The results showed that the electron-hole pairs are produced in the MoS_2 layer after light absorption and subsequently separated across the layers. Contradictory to the expectation based on the conventional built-in electric field model for metal-semiconductor contacts, photoelectrons are injected into the graphene layer rather than trapped in MoS_2 due to the presence of a perpendicular effective electric field caused by the combination of the built-in electric field, the applied electrostatic field, and charged impurities or adsorbates, resulting in a tunable photoresponsivity.

In 2016, M. Masicotte et al. [84], demonstrated that that graphene/WSe_2/graphene heterostructures ally the high photodetection efficiency of transition-metal dichalcogenides with a picosecond photo response comparable to that of graphene, thereby optimizing both speed and efficiency in a single photodetector. This configuration follows the extraction of photoexcited carriers in these devices using time-resolved photocurrent measurements and demonstrate a photo-response time as short as 5.5 ps, which tuned by applying a bias and by varying the transition-metal dichalcogenide layer thickness. Their study provided direct insight into the physical processes governing the detection speed and quantum efficiency of these van der Waals heterostructures, such as out-of-plane carrier drift and recombination. The observation and understanding of ultrafast and efficient photodetection demonstrated the potential of hybrid transition-metal dichalcogenide-based heterostructures as a platform for future optoelectronic devices.

The second structure of 2D van der Waals heterojunction is composed of 2D semiconductor material and 2D semiconductor material. Owing to different doping types from two semiconductor materials, this combination could also be used as the basic unit of photovoltaic devices. According to the different carrier transport paths, these configurations could be divided into vertical heterojunctions and lateral heterojunctions. For vertical heterojunction optoelectronic devices, the dry transfer method is usually utilized to form type II band arrangement. For example, taking *n*-type MoS_2 and *p*-type WSe_2 as examples, smaller band gap can be formed in the overlapping region via bottom conduction band of MoS_2 and top valence band of WSe_2, enabling the recombination of electron-hole pairs formed in the overlapping region. Therefore, the advantages of type II band structures lie in strong interlayer carrier coupling and ultra-fast carrier transport.

Chu-Ho Lee et al. [85], proposed optoelectronic devices based on 2D van der Waals heterojunction at atomic level in 2014. In this study, in order to facilitate the injection of electrons and holes, Pd and Al metals were used to form metal semiconductor contact junctions with WSe_2 and MoS_2, respectively. The results showed that the device exhibits remarkable grid-adjustable photovoltaic characteristics when irradiated by laser. The photocurrent imaging of the device showed that the region of photocurrent generation is mainly concentrated in the overlapping part, and further revealed that the separation of photocurrent carriers at the interface is the main reason for the photovoltaic characteristics.

Since then, a variety of heterojunction mainly included WS_2/MoS_2[86], as shown in Figure 2.9(a), WSe_2/WS_2[87], and $WSe_2/MoSe_2$[88], as shown in Figure 2.9(b). Especially in 2016, Atiye Pezeshki [89] detected the photoelectric response from the visible to the near-infrared region (800 nm) based on α-$MoTe_2/MoS_2$. The results showed that the response of heterojunction is up to 322 mA/W in the visible region and 38 mA/W in the near-infrared region. In 2016, Lei Ye et al. [90], studied the near-infrared response characteristics of MoS_2/BP heterojunctions, as shown in Figure 2.9(c). The results showed that the photo responsivity of the device reaches 22.3 A/W under 532 nm light irradiation and 153.4 mA/W under 1550 nm light irradiation.

In 2018, Do-Hyun Kwak et al. [91], carried out performance characterization at 405 nm and solar spectrum. The results showed that the external quantum efficiency of the device is up to 4.4 percent and the photovoltaic efficiency is up to 4.6 percent under 405 nm illumination. In addition, sandwich heterojunctions have also been reported. In 2016, Mingsheng Long et al. [92], proposed the sandwich heterostructure MoS_2-graphene-WSe_2, as shown in Figure 2.9(d). Because graphene has a wide absorption spectrum, the structure successfully extends the detection range of the device to the infrared band, which greatly improves the response wavelength range of the device. In 2017, Hao Li et al. [93], implemented WSe_2/BP/MoS_2 sandwich heterojunction to achieve a photoresponsivity of ~1.12A/W under 1550 nm light.

The rapid development of lateral heterojunction optoelectronic devices is attributed to the maturity of chemical vapor deposition technology. At present, one-step chemical vapor deposition and two-step chemical vapor deposition have been able to achieve reliable bonding between two different layered semiconductor materials. Compared with all-dry transfer technology, the interface of heterojunction obtained by this method improves transport performance, and greatly reduces the effect of interface on the separation and collection of electron-hole pairs. See the chemical vapor deposition schematic diagram of WSe_2/MoS_2 heterojunction [94]. In the process of growth, monolayer WSe_2 was obtained by epitaxy growth on the substrate at 925°C, and then MoS_2 was grown along the front of W atom at 755°C, leading to obtain heterostructure. By controlling the growth time, another group obtained $WSe_2/MoSe_2$ lateral heterojunction and vertical [95] as shown in Figure 2.10. After the stable heterojunction was synthetized, the electrical and optoelectronic properties of the heterojunction

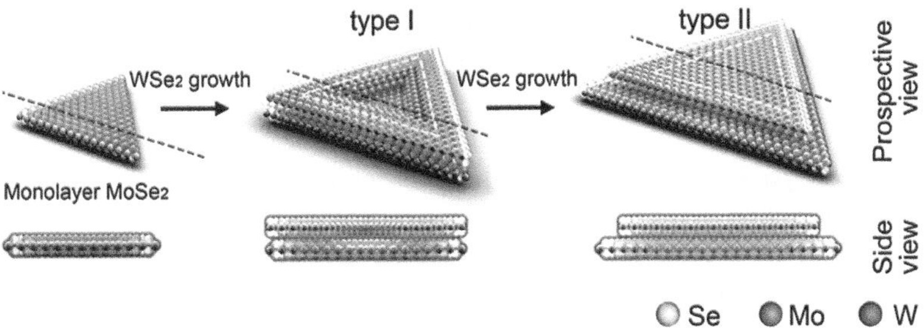

FIGURE 2.10 Schematic illustration of the process of $WSe_2/MoSe_2$ [95].

were characterized. In 2014, Chunming Huang et al. [96], first characterized $WSe_2/MoSe_2$ heterojunctions by fluorescence spectra. The results show that there are obvious photoluminescence characteristics at the interface. In 2014, Xidong Duan et al. [97], characterized the performance of $MoS_2/MoSe_2$ and WS_2/WSe_2 heterojunctions. The results showed that the heterojunction has adjustable gate rectification characteristics and can be used in complementary inverters. At the same time, the structure has obvious photovoltaic characteristics. Thereafter, lateral heterojunctions, such as WSe_2/MoS_2 [98], have been also successfully fabricated, which is not described here.

2.7 CONCLUSIONS AND OUTLOOK

By investigation of the research status of optoelectronic devices formed by different kinds of 2D semiconductor materials in the last section, it showed that the pursuit of high-performance 2D semiconductor nano-optoelectronic devices is not only the development direction of nanodevices, but also the key to the application of novel functional nanodevices in the future. According to the different application fields, the focus of attention on the performance of nanodevices is also different. For nano-optoelectronic devices, the main research focus is on the photoresponsivity, sensitivity, response speed, spectral response range, noise, and photovoltaic response characteristics, and so on. It can be concluded that how to further improve 2D photodetector performance is still a hot issue of current research at home and abroad.

At present, several problems in current research still exist, which are analyzed as follows. From the point of view of the existing methods for performance control of nanodevices, we can see that further exploration and extraction of effective performance enhancement and optimization strategies are of great significance for the practical and commercial applications of 2D nanodevices in the future. Here, depending on the intrinsic attributes of materials or not, the performance control strategies of nanodevices are divided into two types: the performance control strategy based on the intrinsic attributes of 2D semiconductor materials and the performance control strategy based on the non-intrinsic attributes of 2D semiconductor materials.

The performance control strategy based on the intrinsic properties of 2D semiconductor materials is to improve the carrier transport, separation, and collection capabilities of nanodevices by designing and optimizing the geometrical structure of devices on the basis of maintaining the intrinsic properties of 2D semiconductor materials. These performance control methods include:

- Optimizing device structure
- Reducing contact resistance
- Oxide film packaging
- Graphene sandwich packaging

The method of *optimizing device structure* is mainly to optimize the geometrical structure of electrodes and transport channels in nanodevices, aimed at reducing the scattering effect of carriers.

The method of *reducing contact resistance* is mainly to optimize the material of device electrodes or annealing the device, aimed at obtaining excellent contacts between electrodes and 2D semiconductor materials. The method of *oxide film encapsulation* encapsulates the nanodevices with insulating layer to isolate impurity trapping in the air and reduce phonon scattering in the channel. The method of *graphene sandwich packaging* uses the ultra-high carrier mobility of graphene material to enhance the carrier collection ability in the devices.

Despite the improvement of device performance achieved by the above-mentioned methods, high requirements in both cost and fabrication difficulty limit the application of devices. For example, the optimization of device channel with the channel length lower than 20 nm can't be achieved by conventional ultraviolet lithography. Moreover, special metal materials, such as scandium and niobium, not only belong to trace elements, but also have higher material and manufacturing costs. In addition, the methods of oxide film packaging and graphene sandwich packaging both need to rely on complex growth and transfer processes.

The performance control strategy based on the non-intrinsic properties of 2-dimensional semiconductor materials uses physical or chemical methods to control the carrier transport type and concentration of 2D semiconductor materials. These methods can expand the intrinsic attributes of materials by adjusting the band structure of 2D semiconductor materials with the help of new materials and structures. Such performance control methods include:

- Electrostatic control
- Chemical doping
- Plasma enhancement
- Waveguide integration

The method of *electrostatic control* must be based on bipolar 2D semiconductor materials. The method of local electrostatic control is used to improve the carrier transmission polarity in different regions between channels, realizing the conversion of different operating modes. The method of *chemical doping* uses different doping materials to control the carrier transport concentrations and type of 2D semiconductor materials. The method of *plasma enhancement* is to improve the concentration of electrons in nanodevices by means of plasmon-induced 'hot' electrons generated by plasmomic structure, and then enhance the performance of devices. The method of *waveguide integration* combines optical waveguide and the transmission channel of nanodevices to enhance the response of devices. The method of *electrostatic control* is limited to bipolar materials. The method of *chemical doping* needs to select suitable doped materials, whereas the method of *mismatched doping* easily leads to the actual effect of material properties. The methods of *plasma enhancement* and *waveguide integration* are in the initial stage of research. Therefore, it is of great significance to develop new device structures and fabrication processes for the performance control of nanodevices based on different manufacturing methods and novel 2D semiconductor materials.

Although the current 2D nanodevices exhibit strong photoelectric response ability, most of the research focuses on the detection and response of devices in the visible band. For example, the photoelectric response of molybdenum disulfide in visible band is very significant because of its strong absorption peak in visible band. However, the research on detection and response of 2D nanodevices in near-infrared band is very few. The reported 2D nanodevices show weak photoelectric response characteristics in near-infrared band, which greatly limits the application prospect of 2D nanodevices in optical fiber communication, unmanned driving, pattern recognition, and photoelectric coupling. Therefore, it is of great research value and academic significance to develop 2D near-infrared semiconductor nanodevices by combining new manufacturing processes and new nanostructures.

REFERENCES

1. Koppens, F. H. L, T. Mueller, P. Avouris, et al., "Photodetectors Based on Graphene, Other Two-dimensional Materials and Hybrid Systems," *Nature Nanotechnology*, 2014, **9**(10): 780–793.

2. Yan, F., Z. Wei, X. Wei, et al., "Toward High-Performance Photodetectors Based on 2D Materials: Strategy on Methods," *Small Methods*, 2018, **2**(5): 1700349.
3. Casalino M., G. Coppola, M. Iodice, et al., "Critically Coupled Silicon Fabry-Perot Photodetectors Based on the Internal Photoemission Effect at 1550 nm," *Optics Express*, 2012, **20**(11): 12599–12609.
4. Bahrami-Yekta, V., et al., "Limiting Efficiency of Indoor Silicon Photovoltaic Devices," *Optics Express*, 20148, **26**(22): 28238–28248.
5. Keneth, O., P. Arora, N. Mazurski, and U. Levy, "Demonstration of On-Chip Thermocouple Photodetector in Infrared Regime through Field Enhancement by Plasmonic Nano Focusing." (May 2018). In CLEO: *QELS_Fundamental Science* (pp. FTh1P-2), Optical Society of America.
6. Zhong, N., M. Chen, Z. Wang, X. Xin, et al., "Photochemical Device for Selective Detection of Phenol in Aqueous Solutions," *Lab on a Chip*, 2018, **18**(11):1621–1632.
7. Mukherjee, S., K. Das, S. Das, et al., "Highly Responsive, Polarization Sensitive, Self-Biased Single GeO_2-Ge Nanowire Device for Broadband and Low Power Photodetectors," *ACS Photonics*, 2018, **5**(10): 4170–4178.
8. Zheng, Z., R. Grünker, et al., "Synthetic Two-Dimensional Materials: A New Paradigm of Membranes for Ultimate Separation," *Advanced Materials*, 2016, **28**(31): 6529–6545.
9. Varrla, E., C. Backes, K. R. Paton, A. Harvey, Z. Gholamvand, J. McCauley, et al., "Large-Scale Production of Size-Controlled MoS_2 Nanosheets by Shear Exfoliation," *Chemistry of Materials*, 2015, **27**(3): 1129–1139.
10. Zadeh, I. E., A. W. Elshaari, K. D. Jöns, A. Fognini, D. Dalacu, P. J. Poole, et al., "Deterministic Integration of Single Photon Sources in Silicon Based Photonic Circuits," *Nano Letters*, 2016, **16**(4): 2289–2294.
11. de Arquer, F. P. G., A. Armin, P. Meredith, et al., "Solution-Processed Semiconductors for Next-generation Photodetectors," *Nature Reviews Materials*, 2017, **2**(3), 16100.
12. Ludin, N. A., N. I. Mustafa, M. M. Hanafiah, M. A. Ibrahim, M. A. M. Teridi, S. Sepeai, et al., "Prospects of Life Cycle Assessment of Renewable Energy from Solar Photovoltaic Technologies: A Review," *Renewable and Sustainable Energy Reviews*, 2018, **96**: 11–28.
13. Goldan, A. H., J. A. Rowlands, M. Lu, et al., "Nanopattern Multi-Well Avalanche Selenium Detector with Picosecond Time Resolution. In *Nuclear Science Symposium and Medical Imaging Conference (NSS/MIC)*, 2014 IEEE (pp. 1–3). IEEE.
14. Zhang, H., J. Wang, T. Hasan, et al., "Photonics of 2D Materials," *Optics Communications,* 2018, **406**: 1–2.
15. Van Der Zande, A., and J. Hone, "Optical Materials: Inspired by Strain," *Nature Photonics*, 2012, **6**(12), 804.
16. Duong, T., Y. Wu, H. Shen, J. Peng, X. Fu, D. Jacobs, et al., "Rubidium Multication Perovskite with Optimized Bandgap for Perovskite-Silicon Tandem with over 26% Efficiency," *Advanced Energy Materials*, 2017, **7**(14), 1700228.
17. Ganesh R. B., et al. "Recent Advances in Two-Dimensional Materials beyond Graphene," *ACS Nano*, 2015, **9**(12):11509–11539.
18. Ferrain, I., C. A. Colinge, and J. P. Colinge, "Multi-gate Transistors as the Future of the Classical Metal-oxide Semiconductors Field-effect Transistors," *Nature*, 2011, **479**:310–316.
19. Lee, C., X. Wei, and J. W. Kysar, et al., "Measurement of the Elastic Properties and Intrinsic Strength of Monolayer Graphene," *Science*, 2008, **321**(5887):385–388.
20. Yun, W. S., S. W. Han, S. C. Hong, et al., "Thickness and Strain Effects on Electronic Structures of Transition Metal Dichalcogenides: 2H-MX_2 Semiconductors (M= Mo, W; X= S, Se, Te)," *Physical Review B*, 2012, **85** (4): 033305.
21. Splendiani, A., L. Sun , Y. Zhang, et al., "Emerging Photoluminescence in Monolayer MoS_2," *Nano Letters*, 2010, **10** (4):1271–1275.
22. Tran, V., R. Soklaski, Y. Liang, et al. "Layer-Controlled Band Gap and Anisotropic Excitons in Few-Layer Black Phosphorus." *Physical Review B*, 2014, **89** (23): 235319.
23. Novoselov, K. S., A. Mishchenko, A. Carvalho, et al., "2D Materials and van der Waals Heterostructures," *Science*, 2016, **353**(6298): 9439.
24. Mouri, S, Y. Miyauchi, K. Matsuda, "Tunable Photoluminescence of Monolayer MoS_2 Via Chemical Doping," *Nano Letters*, 2013, **13**(12): 5944–5948.
25. Ross, J. S., P. Klement, A. M. Jones, et al., "Electrically Tunable Excitonic Light-emitting Diodes Based on Monolayer WSe_2 p-n Junctions," *Nature Nanotechnology*, 2014, **9**(4):268–272.

26. Tongay, S., J. Zhou, C. Ataca, et al., "Thermally Driven Crossover from Indirect Toward Direct Bandgap in 2D Semiconductors: $MoSe_2$ Versus MoS_2," *Nano Letters*, 2012, **12**(11): 5576–5580.
27. Taniguchi, K., A. Matsumoto, H. Shimotani, et al., "Electric-field-induced Superconductivity at 9.4 K in A Layered Transition Metal Disulphide MoS_2," *Applied Physics Letters*, 2012, **101**(4): 042603.
28. Zhu, H., Y. Wang, J. Xiao, et al., "Observation of Piezoelectricity in Free-standing Monolayer MoS_2," *Nature Nanotechnology*, 2015, **10**(2):151–155.
29. Mak, K. F., K. L. McGill, J. Park, et al., "The Valley Hall Effect in MoS_2 Transistors," *Science*, 2014, **344**(6191):1489–1492.
30. Wu, J., H. Schmidt, K. K. Amara, et al., "Large Thermoelectricity Via Variable Range Hopping in Chemical Vapor Deposition Grown Single-layer MoS_2," *Nano Letters*, 2014, **14**(5):2730–2734.
31. Xia, F., H. Wang, and Y. Jia, "Rediscovering Black Phosphorus as An Anisotropic Layered Material for Optoelectronics and Electronics," *Nature Communications*, 2014, **5**: 4458.
32. Novoselov, K. S., A. K. Geim, S. V. Morozov, et al., "Electric Field Effect in Atomically Thin Carbon Films," [J], *Science*, 2004, **306**(5696): 666–669.
33. Li, J., L. Niu, Z. Zheng, and F. Yan, "Photosensitive Graphene Transistors," *Advanced Materials*, 2014, **26**(31): 5239–5273.
34. Xia, F., T. Mueller, Y. M. Lin, A. Valdes-Garcia, et al., "Ultrafast Graphene Photodetector," *Nature Nanotechnology*, 2009, **4**(12): 839.
35. Han, S. W., H. Kwon, S. K. Kim, et al., Band-gap Transition Induced by Interlayer van der Waals Interaction in MoS_2," *Physical Review B*, 2011, **84**(4): 045409.
36. Wang, Q. H., K. Kalantar-Zadeh, A. Kis, J. N. Coleman, et al., "Electronics and Optoelectronics of Two-Dimensional Transition Metal Dichalcogenides," *Nature Nanotechnology*, 2012, **7**(11), 699.
37. Radisavljevic, B., A. Radenovic, J. Brivio, et al., "Single-layer MoS_2 Transistors," *Nature Nanotechnology*, 2011, **6**(3): 147–150.
38. Mak, K. F., C. Lee, J. Hone, et al., "Atomically Thin MoS_2: A New Direct-gap Semiconductor," *Physical Review Letters*, 2010, **105**(13): 136805.
39. Splendiani, A., L. Sun, Y. Zhang, et al., "Emerging Photoluminescence in Monolayer MoS_2," *Nano Letters*, 2010, **10**(4): 1271–1275.
40. Castellanos-Gomez, A., M. Poot, G. A. Steele, et al., "Elastic Properties of Freely Suspended MoS_2 Nanosheets," *Advanced Materials*, 2012, **24**(6): 772–775.
41. Lembke, D., and A. Kis, "Breakdown of High-Performance Monolayer MoS_2 Transistors," *ACS Nano*, 2012, **6**(11): 10070–10075.
42. Li, L., Y. Yu, G. J. Ye, et al., "Black Phosphorus Field-Effect Transistors," *Nature Nanotechnology*, 2014, **9**(5): 372–377.
43. Tran, V., R. Soklaski, Y. Liang, et al., "Layer-Controlled Band Gap and Anisotropic Excitons in Few-Layer Black Phosphorus," *Physical Review B*, 2014, **89**(23): 235319.
44. Fei, R., A. Faghaninia, R. Soklaski, et al., "Enhanced Thermoelectric Efficiency Via Orthogonal Electrical and Thermal Conductances in Phosphorene," *Nano Letters*, 2014, **14**(11): 6393–6399.
45. Geim, A. K., and I. V. Grigorieva,. "Van der Waals Heterostructures," *Nature*, 2013, **499**(7459): 419–425.
46. Liu Y., N. O. Weiss, X. Duan, et al., "Van der Waals Heterostructures and Devices," *Nature Reviews Materials*, 2016, **1**(9): 16042.
47. Novoselov, K. S., A. Mishchenko, A. Carvalho, et al., "2D Materials and Van Der Waals Heterostructures," *Science*, 2016, **353**(6298): aac9439.
48. Xia F., H. Wang, D. Xiao, et al., "Two-dimensional Material Nanophotonics," *Nature Photonics*, 2014, **8**(12): 899–907.
49. Bube, R. H., 1992, *Photoelectronic Properties of Semiconductors*, Cambridge: Cambridge University Press.
50. Xie C., C. Mak, X. Tao, et al., "Photodetectors Based on Two-Dimensional Layered Materials Beyond Graphene," *Advanced Functional Materials*, 2017, **27**(19): 1603886.
51. Furchi, M. M., D. K. Polyushkin, A. Pospischil, et al., "Mechanisms of Photoconductivity in Atomically Thin MoS_2," *Nano Letters*, 2014, **14**(11): 6165–6170.
52. S. M. Sze, and Ng, K. K., 1981, *Physics of Semiconductor Devices*, 2nd Edition, New York: Wiley.
53. Song, J. C., M. S. Rudner, C. M. Marcus, et al., "Hot Carrier Transport and Photocurrent Response in Graphene," *Nano Letters*, 2011, **11**(11): 4688–4692.

54. Low, T., A. S. Rodin, A. Carvalho, Y. Jiang, H. Wang, F. Xia, et al., "Tunable Optical Properties of Multilayer Black Phosphorus Thin Films," *Physical Review B*, 2014, **90**(7): 075434.
55. Bernardi, M., M. Palummo, and J. C. Grossman, "Extraordinary Sunlight Absorption and One Nanometer Thick Photovoltaics Using Two-dimensional Monolayer Materials," *Nano Letters,* 2013, **13**(8): 3664–3670.
56. Yin, Z., H. Li, et al., "Single-Layer MoS_2 Phototransistors," *ACS Nano*, 2011, **6**(1): 74–80.
57. Lee H. S., S. W. Min, Y. G. Chang, et al., "MoS_2 Nanosheet Phototransistors with Thickness-Modulated Optical Energy Gap," *Nano Letters*, 2012, **12**(7): 3695–3700.
58. Choi, W., M. Y. Cho, A. Konar, et al., "High-Detectivity Multilayer MoS_2 Phototransistors with Spectral Response from Ultraviolet to Infrared," *Advanced Materials*, 2012, **24**(43): 5832–5836.
59. Lopez-Sanchez, O., D. Lembke, M. Kayci, et al. "Ultrasensitive Photodetectors Based on Monolayer MoS_2," *Nature Nanotechnology*, 2013, **8**(7): 497–501.
60. Zhang, W., J. K. Huang, C. H. Chen, et al., "High-Gain Phototransistors Based on a CVD MoS_2 Monolayer," *Advanced Materials*, 2013, **25**(25): 3456–3461.
61. Lu, J., J. H. Lu, H. Liu, et al., "Improved Photoelectrical Properties of MoS_2 Films After Laser Micromachining," *ACS Nano*, 2014, **8**(6): 6334–6343.
62. Kwon J., Y. K. Hong, G. Han, et al., "Giant Photoamplification in Indirect-Bandgap Multilayer MoS_2 Phototransistors with Local Bottom-Gate Structures," *Advanced Materials*, 2015, **27**(13): 2224–2230.
63. Kufer D., Konstantatos G.. "Highly Sensitive, Encapsulated MoS_2 Photodetector with Gate Controllable Gain and Speed," *Nano Letters*, 2015, 15(11): 7307–7313.
64. Wang X, P. Wang, J. Wang, et al., "Ultrasensitive and Broadband MoS_2 Photodetector Driven by Ferroelectrics," *Advanced Materials*, 2015, **27**(42): 6575–6581.
65. Kang, D. H., M. S. Kim, J. Shim, et al., "High-Performance Transition Metal Dichalcogenide Photodetectors Enhanced by Self-Assembled Monolayer Doping," *Advanced Functional Materials*, 2015, **25**(27): 4219–4227.
66. Li, H. M., D. Lee, D. Qu, et al., "Ultimate Thin Vertical p-n Junction Composed of Two-Dimensional Layered Molybdenum Disulfide," *Nature Communications*, 2015, **6**: 6564.
67. Sobhani, A., A. Lauchner, S. Najmaei, et al., "Enhancing the Photocurrent and Photoluminescence of Single Crystal Monolayer MoS_2 with Resonant Plasmonic Nanoshells," *Applied Physics Letters*, 2014, **104**(3): 031112.
68. Miao, J., W. Hu, Y. Jing, et al., "Surface Plasmon-Enhanced Photodetection in Few Layers MoS_2 Phototransistors with Au Nanostructure Arrays," *Small*, 2015, **11**(20): 2392–2398.
69. Wang, W., A. Klots, Prasai D, et al., "Hot Electron-based Near-Infrared Photodetection Using Bilayer MoS_2," *Nano Letters*, 2015, **15**(11): 7440–7444.
70. Hou, C., Y. Wang, L. Yang, et al., "Position Sensitivity of Optical Nano-Antenna Arrays on Optoelectronic Devices," *Nano Energy*, 2018, **53**: 734–744.
71. Buscema, M., D. J. Groenendijk, S. I. Blanter, et al., "Fast and Broadband Photoresponse of Few-Layer Black Phosphorus Field-Effect Transistors," *Nano Letters*, 2014, **14**(6): 3347–3352.
72. Wu, J., G. K. W. Koon, D. Xiang, et al., "Colossal Ultraviolet Photoresponsivity of Few-Layer Black Phosphorus," *ACS Nano*, 2015, **9**(8): 8070–8077.
73. Huang, M., M. Wang, C. Chen, et al., "Broadband Black-Phosphorus Photodetectors with High Responsivity," *Advanced Materials*, 2016, **28**(18): 3481–3485.
74. Guo, Q., A. Pospischil, M. Bhuiyan, et al., "Black Phosphorus Mid-Infrared Photodetectors with High Gain," *Nano Letters*, 2016, **16**(7): 4648–4655.
75. Hou, C., L. Yang, B. Li , et al., "Multilayer Black Phosphorus Near-Infrared Photodetectors," *Sensors*, 2018, **18**(6): 1668.
76. Youngblood, N., C. Chen, S. J. Koester, et al., "Waveguide-Integrated Black Phosphorus Photodetector with High Responsivity and Low Dark Current," *Nature Photonics*, 2015, **9**(4): 247–252.
77. Chen, C., N. Youngblood, R. Peng, et al., "Three-Dimensional Integration of Black Phosphorus Photodetector with Silicon Photonics and Nanoplasmonics," *Nano Letters*, 2017, **17**(2): 985–991.
78. Liu, Y., Y. Cai, G. Zhang, et al., "Al-Doped Black Phosphorus p-n Homojunction Diode for High Performance Photovoltaic," *Advanced Functional Materials*, 2017, **27**(7): 1604638.
79. Buscema, M., D. J. Groenendijk, G. A. Steele, et al., "Photovoltaic Effect in Few-Layer Black Phosphorus PN Junctions Defined by Local Electrostatic Gating," *Nature Communications*, 2014, **5**: 4651.

80. Britnell, L., R. M. Ribeiro, A. Eckmann, et al., "Strong Light-Matter Interactions in Heterostructures of Atomically Thin Films," *Science*, 2013, **340**(6138): 1311–1314.
81. Roy, K., M. Padmanabhan, S. Goswami, et al., "Graphene-MoS_2 Hybrid Structures for Multifunctional Photoresponsive Memory Devices," *Nature Nanotechnology*, 2013, **8**(11): 826–830.
82. Yu, W. J., Y. Liu, H. Zhou, et al., "Highly Efficient Gate-Tunable Photocurrent Generation in Vertical Heterostructures of Layered Materials," *Nature Nanotechnology*, 2013, **8**(12): 952–958.
83. Zhang, W., C. P. Chu, J. K. Huang, et al. "Ultrahigh-gain Photodetectors Based on Atomically Thin Graphene-MoS_2 Heterostructures," *Scientific Reports*, 2014, **4**: 3826.
84. Massicotte, M., P. Schmidt, F. Vialla, et al., "Picosecond Photoresponse in Van Der Waals Heterostructures," *Nature Nanotechnology*, 2016, **11**(1): 42–46.
85. Lee, C. H., G. H. Lee, A. M. Van Der Zande, et al., "Atomically Thin p-n Junctions with Van Der Waals Heterointerfaces," *Nature Nanotechnology*, 2014, **9**(9): 676–681.
86. Xue, Y., Y. Zhang, Y. Liu, et al., "Scalable Production of a Few-layer MoS_2/WS_2 Vertical Heterojunction Array and Its Application for Photodetectors," *ACS Nano*, 2015, **10**(1): 573–580.
87. Huo, N., J. Yang, L. Huang, et al., "Tunable Polarity Behavior and Self-Driven Photoswitching in p-WSe_2/n-WS_2 Heterojunctions," *Small*, 2015, **11**(40): 5430–5438.
88. Flöry, N., A. Jain, P. Bharadwaj, et al., "A $WSe_2/MoSe_2$ Heterostructure Photovoltaic Device," *Applied Physics Letters*, 2015, **107**(12): 123106.
89. Pezeshki, A., S. H. H. Shokouh, T. Nazari, et al., "Electric and Photovoltaic Behavior of a Few-Layer α-$MoTe_2/MoS_2$ dichalcogenide heterojunction," *Advanced Materials*, 2016, **28**(16): 3216–3222.
90. Ye, L., H. Li, Z. Chen, et al., "Near-Infrared Photodetector Based on MoS_2/black phosphorus heterojunction," *ACS Photonics*, 2016, **3**(4): 692–699.
91. Kwak, D. H., H. S. Ra, M. H. Jeong, et al., "High-Performance Photovoltaic Effect with Electrically Balanced Charge Carriers in Black Phosphorus and WS_2 Heterojunction," *Advanced Materials Interfaces*, 2018, **5**(18): 1800671.
92. Long, M., E. Liu, P. Wang, et al., "Broadband Photovoltaic Detectors Based on An Atomically Thin Heterostructure," *Nano Letters*, 2016, **16**(4): 2254–2259.
93. Li, H., L. Ye, and J. Xu, "High-Performance Broadband Floating-base Bipolar Phototransistor Based on WSe_2/BP/MoS_2 heterostructure," *ACS Photonics*, 2017, **4**(4): 823-829.
94. Li, M. Y., Y. Shi, C. C. Cheng, et al., "Epitaxial Growth of a Monolayer WSe_2-MoS_2 Lateral pn Junction with An Atomically Sharp Interface," *Science*, 2015, **349**(6247): 524–528.
95. Gong, Y., S. Lei, G. Ye, et al., "Two-Step Growth of Two-Dimensional $WSe_2/MoSe_2$ Heterostructures," *Nano Letters*, 2015, **15**(9): 6135–6141.
96. Huang, C., S. Wu, A. M. Sanchez, et al., "Lateral Heterojunctions within Monolayer $MoSe_2$-WSe_2 Semiconductors," *Nature Materials*, 2014, **13**(12): 1096–1101.
97. Duan, X., C. Wang, J. C. Shaw, et al., "Lateral Epitaxial Growth of Two-Dimensional Layered Semiconductor Heterojunctions," *Nature Nanotechnology*, 2014, **9**(12): 1024–1030.
98. Gong, Y., J. Lin, X. Wang, et al., Vertical and In-plane Heterostructures from WS_2/MoS_2 Monolayers," *Nature Materials*, 2014, **13**(12): 1135–1142.

3 Room-Temperature Self-Powered Nanostructure Organo-Metallic Superconductive Devices

Ellen. T. Chen

President/Chief Technology Officer, Advanced Biomimetic Sensors, Inc. Rockville, Maryland, USA

CONTENTS

3.1 THE ENGINEERING DESIGN

3.1.1 Single-Layer Toroidal Membrane Device with Josephson Junction

The quest for room temperature superconductive materials or devices has gripped researchers all over the world for decades, since they first saw the possibility that electronic devices could operate more efficiently without energy dissipation [1-4]. The *room-temperature superconductor* may revolutionize the electronic industries [1-4]. Superconductor quantum bits (qubits) are vulnerable to low-frequency noise with sources that come from 1/f noise, wave dephasing noise, flux noise, critical current noise, quasiparticle tunneling noise, and capacitance noise [5-6]. The nature of the qubits operating multiple states at the same time is more sensitive to decoherence caused by the control and readout circuitry, and the environmental noise as well as to the qubits's intrinsic reaction to the low-frequency noise of conventional computing [5-6]. Building error-tolerant qubits are one of the leading directions to achieve quantum computing [7]. It is well recognized that controlling the fractional quantization has been greatly anticipated for enhancing quantum computing and for advancing quantum science and technology [8-13]. In general, *the fractional Josephson effect* possesses a well-known 4π periodicity vortex rather than a 2π periodicity vortex as in the *long Josephson junction (LJJ)* vortex with or without *Majorana zero mode (MZM)* [7]. In recent decades, more fractional Josephson effect systems were theoretically or experimentally reported, which are comprised of one or two vortices [8-9, 14]. Driving by the applications of fractional Josephson junction, such as a two-vortex model system, was proposed for a macroscopic qubit for quantum computing [9]. Amongst the discoveries, p-wave and d-wave molecules were chosen to build the fractional Josephson π-junction [9, 12-13, 15].

Collagen is the primary structural component of the *extracellular matrix (ECM),* which is responsible for the physical maintenance of all cells [16]. The triple-helical structure of collagen assembles into insoluble collagen fibrils to strengthen the bones and tissues and prevent proteinase from engaging [16-17]. Collagen is a double-edged sword. Not only does it actively pave the road for physiologically normal cell and pathologically abnormal cell adhesion, and migration and intracellular communication, but it also activates some receptors for either the overproduction of or the failure of the matrix. This degradation is caused by either bacterial collagenase or abnormal fibrosis from fibroblast cells, endothelial cells or epithelial cells, and hence many diseases are associated with the malfunction of collagen [16-20]. The long history of a traditional approach utilized and used to be called the *protein denaturing step*; hence collagen was the candidate to be used as a substrate to probe collagen degradation or to study *matrix metalloproteinase (MMP)* activity [21-22]. However, a clinically useful detection range at the low end for collagen-1 is difficult to accomplish due to the denaturing process.

The Josephson coupled-superconductor effect is inherent in any *superconductor-insulator-superconductor (SIS)* tunnel junction if the two sides of the barriers are sufficiently thin enough to allow the coupling energy from the Cooper pair tunneling at the coherent wave state between the two superconductors in order to exceed thermal fluctuations [23-25]. Inspired by several reports and predictions [26-28], we thought the unique coherent wave state produced in the S-I-S module at a toroidal Josephson junction (JJ) may be a key feature to help accomplish the goal of our project, where we attempt to utilize a circular JJ to solve our biosensing problem. Inspired by another theoretical prediction, a π-phase difference on a *topological-superconductor (TSC)/normal metal (NM)* can arise, induced by Majorana spin-triplet pairing, which exhibit a Josephson phase of 0 and π-junction in its ground state without any applied magnetic flux [27].

Our prior published works in developing a nanostructure biomimetic *Matrix Metalloproteinase-2 (MMP-2)* superconductive, memristive memcapacitive and meminductive device for sensing fg/mL collagen-1(7.3×10^{-18}M, i.e., 7.3 attoM) under antibody-free and reagent-free conditions utilized a unique approach that used collagen as an analyte and also used Josephson toroidal junction barriers in order to promote *Direct Electron-Relay (DER)* in a superlattice organo-metallic membrane [29-30] at an innate state without denaturing the protein. Further exploration of the utility of the fractional Josephson vortices arrays in the biomimetic MMP-2 Flexible Fractional Toroidal Josephson Junction (FFTJJ) device for multiple types of sensing, energy storage, and quantum computing at room temperature without an external microwave power supply will be discussed in the following chapters; the circular self-oscillating innate protein comprised superconductive/memristive/meminductive slqubit device will also be discussed.

3.1.2 Single-Layer Membrane Superlattice (Sl) Qubits with Josephson Junction (JJ)

3.1.2.1 The Flexible Fractional Toroidal Josephson Junction (FFTJJ) Device

The Flexible Fractional Toroidal Josephson Junction (FFTJJ) was used for the following purposes:

1. As a *magnetic flux quantum system*, which has a non $2\pi\, x$ integer phase winding, i.e., fractional phase winding in heterogeneous interfaces between two superconductors with a barrier through a weak Josephson junction;
2. As a *uniform multicomponent superconductor*, which allows integer phase winding $2\pi N$, where $N = \pm 1, \pm 2 \ldots$, which carry arbitrarily fractional quantized magnetic flux [31-32].

Figure 3.1 shows the symbol of our proposed FFTJJ device. It is a superconductive system, which has one of these configurations—superconductive-insulator-superconductive (SIS), superconductive-normal metal-superconductive (SNS), or superconductor-insulator-mem-element (SIM)—with barriers such as dielectric insulator, air, transitional metal, and conductive polymer with at least one or more inserting spatial locations. This section explains the concept of the FFTJJ throughout the contents and illustrates its applications and uniqueness.

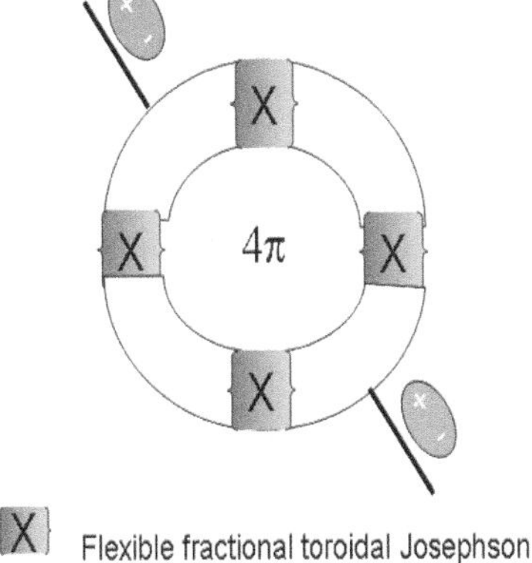

FIGURE 3.1 This figure depicts the FFTJJ symbol with reversible connection leads. (Chen, 2018, with permission.)

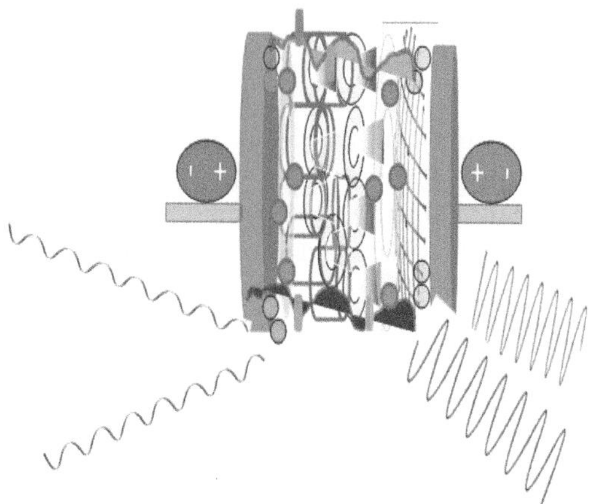

FIGURE 3.2 This figure depicts the art module of the FFTJJ device. (Chen, 2018, with permission.)

3.1.2.2 The Embedded Transition Metal Biomimetic Innate SIqubit Device

Figure 3.2 shows an art model of the FFTJJ device. The *Membrane Electrode Assembly (MEA)* has a self-assembled organo-metallic polymer comprised of triacetyl-ß-cyclodextrin (TCD), polyethylene glycol diglycidyl ether (PEG), poly(4-vinylpyridine) (PVP), bis-substituted imidazole dimethyl-β-cyclodextrin (bM-β-DMCD), cysteine, and embedded zinc chloride on a gold chip. The membrane mimics the innate matrix metalloproteinase-2's (MMP-2's) function and structure. The nanostructure superlattice image is shown in Figure 3.4. Zinc atom clusters are seen in the atomic force microscopy (AFM) images. Zinc atoms serve as the superconducting barrier; and another dielectric collagen barrier, a collagen solution was injected into the assembly before conducting quantitative measurement of collagen, wherein the components of the MEA can be reversibly connected including the Josephson junction (JJ) barriers.

Friedel *oscillations* is a phenomenon that occurs during long-range indirect interactions between electrons on the surface based on AFM images [33]. It is well known that superlattice membranes having Friedel oscillations can be used as candidates for applications in superconductivity [33]. The observation of Friedel oscillations in the membrane superlattice was a confirmation of the Cooper-pairs' hopping-flip existence. Details are referenced in the literature [29-30].

3.1.2.3 The Nanostructured Protein MMP-2 Toroidal Oscillating Array SLQubit Device

Figure 3.3 shows the art model of the protein MMP-2 device, and it was compared with the single-layer biomimetic superlattice model in Figure 3.2. The MMP-2 protein is a family of zinc-dependent endopeptidases. The enzymes play a key role in human health for promoting newborn growth and nervous system growth, as well as for promoting various human diseases, such as cancer invasion, osteoarthritis, tissue destruction, diabetes, coronary malfunction, epilepsy, and Alzheimer's [34-37]. The major role of MMP-2 is in the degradation of the extracellular matrix as a double-edged sword. MMP-2 has been identified as a critical biomarker for diagnosing, monitoring, and predicting multiple types of human diseases [37-42]. However, direct detection of MMP-2 is impossible if the conserved cysteine residue bonding to the zinc in the active site is not turned "off"; in order to activate MMP-2, we must turn off the "cysteine switch" according to conventional teaching [43-44]. Inspired by our prior reported work using an organic biomimetic polarizable microtubule memristive/memcapacitive device to enable direct detection of MMP-2 with an Ag/mL level sensitivity under antibody-free, tracer-free, and reagent-free conditions [45-46], we were able to develop superconductive quantum devices with a superlattice structure based on the FFTJJ junction [29-30].

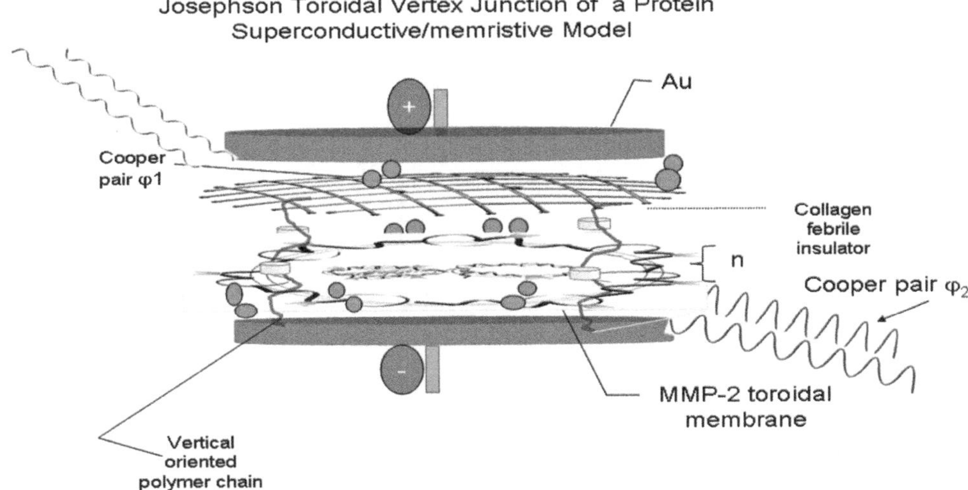

FIGURE 3.3 This figure depicts the engineering design of the protein device with the toroidal nanostructure membrane. (Chen, 2018, with permission.)

Activation of MMP-2 without turning off the "cysteine-switch" avoided interferences that enabled us to directly detect collagen-1 at extreme sensitivity [29-30].

Later on we discovered that the model device we designed, as shown in Figure 3.3, can be a good superconductor even without the presence of collagen due to its circular oscillation behavior, which is different from the design of model 1 in Figure 3.2. The protein model comprised of the innate native MMP-2 with TCD, PEG, and PVP crossed-linked organic polymers formed the toroidal protein slqubit device 2 with a well-defined toroidal array structure that facilitates Friedel oscillations and induced superconductivity reported at the first time in literature [29-30] with images shown in Figures 3.5 and 3.6.

3.1.3 The Multiple-Layer Superconductive/Memristive/Meminductive Device with Josephson Junction

3.1.3.1 An Embedded Transition Metallic Biomimetic Activated MMP-2 Chelated Superlattice Membrane Device

Figure 3.7 depicts the AFM image of the embedded transition metallic biomimetic activated MMP-2 chelated superlattice device, and the membrane layer was built upon the first layer of the nano-islands membrane, as shown in Figure 3.7. For detailed information, refer to literature 29, It will also be explained in the following section 3.2.

3.2 EVALUATION OF FRIEDEL OSCILLATIONS IN THE SUPERLATTICE MEMBRANES

The superlattice membrane with Friedel oscillations as a candidate for applications in superconductivity is very encouraging news [33]. The AFM image revealed in Figure 3.4 shows that Friedel oscillations promoted the Cooper pairs electrons' mobility. Many mobile electron clouds, shown as "tails" surrounding zinc atoms that headed toward the same direction, were observed in the image for device #1. Figure 3.5 depicts the native MMP-2 sensor's single-wall nanotubule toroidal structures with zinc atoms like diamonds on a ring. All rings have the same thickness of 90 nm in the circular nanotubes, the diameters of the toroidals are in the range of 2.3 to 5.5 μm, and the height of

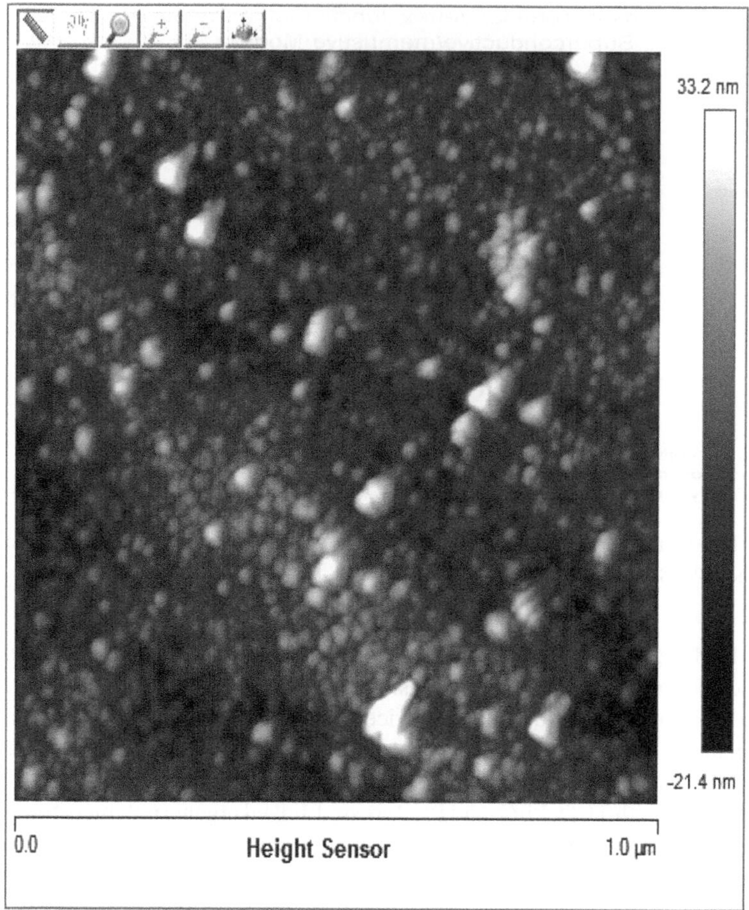

FIGURE 3.4 This figure depicts the AFM image of the single-layer biomimetic MMP-2 FFTJJ device # 1 at the innate state.

the toroidal is in the range of 0.5 to 0.9 μm. In Figure 3.5, Friedel oscillations were also observed for protein device #2 in 81 μm^2 and in Figure 3.6 in 1 μm^2. The zinc atoms were oscillating on the edge or in the center of the rings. The migration of zinc atoms from the original cluster to other rings was observed. Figure 3.7 depicts the superlattice morphology with the Friedel oscillations of the biomimetic double-layer membrane in device #3 in an area of 25 μm^2. Figure 3.8 depicts the AFM of the first layer of the nanoisland structure that Figure 3.7's AFM image was built upon.

3.3 EVALUATION OF THE SUPERCONDUCTIVITY/MEM-ELEMENT CHARACTERISTICS

3.3.1 The Device with the Flexible Fractional Toroidal Josephson Junction (FFTJJ) Configuration in a Single-Layer Membrane

3.3.1.1 The Single-Layer Membrane Mimics of MMP-2

Before the evaluation of the device's superconductivity and the Mem-element characteristics, Figure 3.9 shows the FFTJJ configuration with the waves at 300 Hz having three cross-points at phase 1.45π, 2.22π, and 2.55π, respectively. Notice, that none of the cross-points of the phase discontinuity happened at the 0 and π spatial position according to the conventional fractional 4π

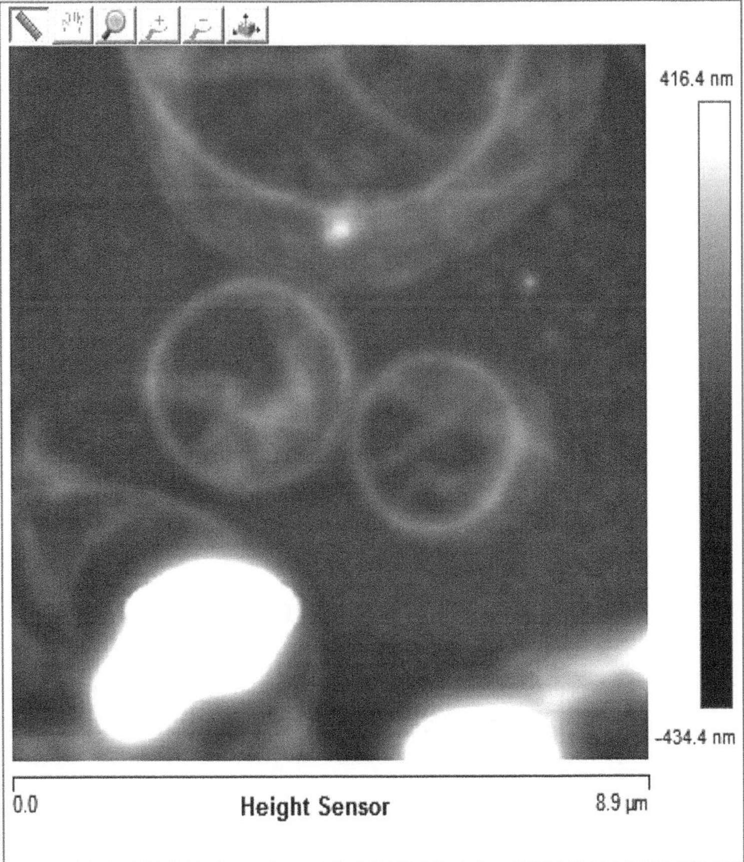

FIGURE 3.5 This figure depicts the single-layer superlattice with the mobile zinc atoms in the innate protein MMP-2 device # 2 in the tapping mode.

Josephson effect's request. The intrinsic bidirectional polarity of the electron-relay (evidence given in the Section 3.3.1.2) around the discontinuity point was asymmetric; hence, the overall vortices array displayed irregular phase change.

A conventional approach that has been used for a long time is to denature collagen by means of heating it in order to eliminate the cysteine group and use collagen as a substrate to probe collagen degradation or to study matrix metalloproteinase (MMP) activity [47-49]. Using a non-denatured protein MMP-2 device and a biomimetic MMP-2 device, our prior work was able to directly detect extremely low collagen -1 concentration based on the Josephson toroidal junction approach and the organo-metallic superlattice membranes [29-30]. The innate biomimetic single-layer slqubit device # 1 increased superconductivity when the collagen concentration increased in biological specimens without denaturing the membrane that was solely based upon the FFTJJ's characteristics. Its facilitating of the sensing function will be discussed in section 3.6. This section further emphasized and confirmed that a denatured biomimetic MMP-2 device will have a higher superconducting power at room temperature with a higher energy band occurring than that of innate device # 1. We will also show how the Cooper pairs penetrate the FFTJJ barrier.

Figure 3.10 demonstrates different media effects on the supercurrent at the activated biomimetic MMP-2 device in the presence of 50 ng/mL collagen in the 50 ng/mL MCD media and compared the results without the MCD media, against controls, respectively. It demonstrated that in the presence of collagen in the MCD media, the current intensity increased 3571-fold compared with

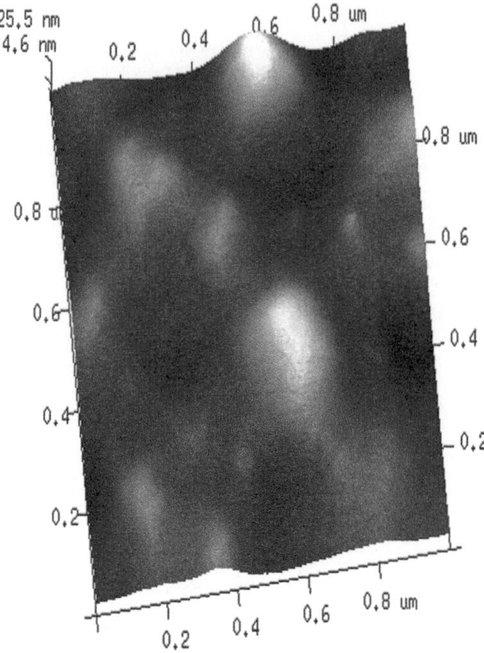

FIGURE 3.6 This figure depicts the AFM image of the zinc atoms in the innate MMP-2 protein membrane at 100 μm^2 view size. (Chen, 2018, with permission.)

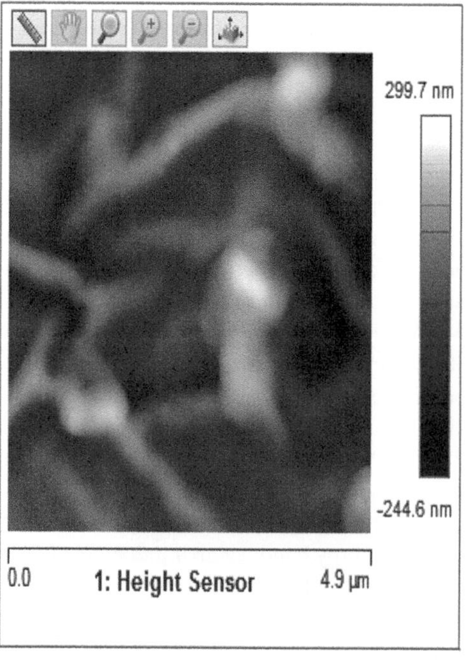

FIGURE 3.7 This figure depicts the superlattice multiple-layer biomimetic membrane for device # 3. (Chen, 2018, with permission.)

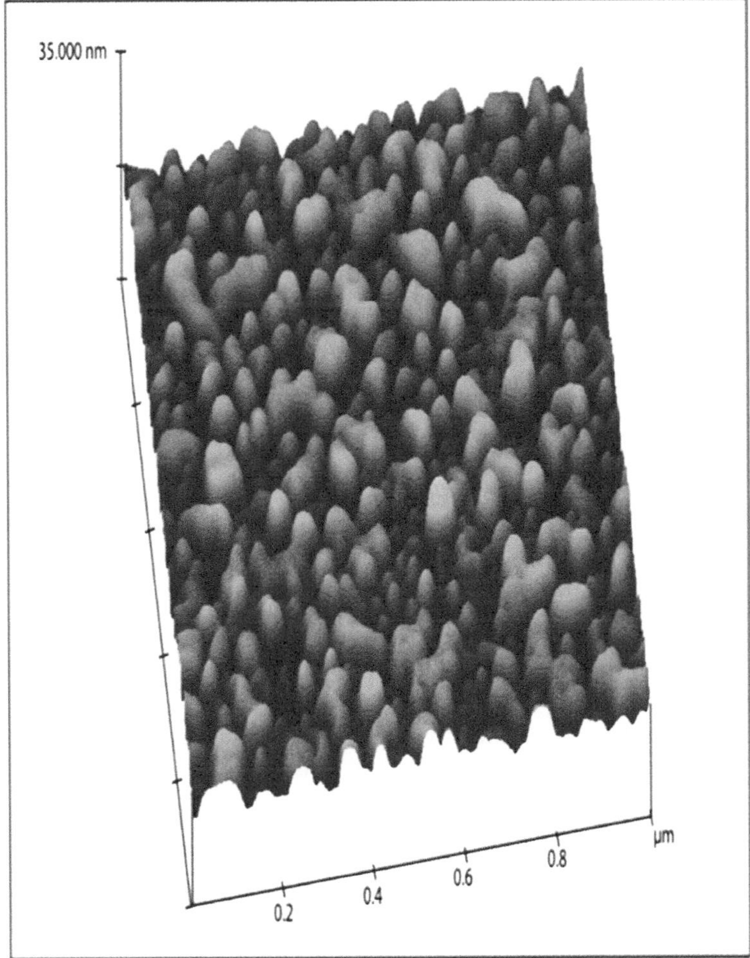

FIGURE 3.8 This figure depicts the first layer nanoisland membrane. (Chen, 2018, with permission.)

collagen alone; this confirmed our proposed Direct-Electron Relay (DER) mechanism. In the presence of collagen at the activated state, two supercurrent waves were observed and were labeled as point $P_1(0V, -0.049A)$ and point $P_2(0V, 0.10A)$ at zero-bias, as shown in Figure 3.10. This indicates that the activated biomimetic MMP-2 device has a quantized conductance intensity of 7344.6-fold higher in the presence of collagen with MCD than with collagen alone; hence MCD promoted the *Direct Electron-Transfer (DET)* and formed long-range delocalized DER. The 50 ng/mL MCD control alone showed a small 4.8 μA background current in Phosphate-buffered solution (PBS). It was shown in our prior work that the well-aligned cyclodextrin truncated donut-like cavities formed large nanopore toroidal wells with dipole polarized circular current flow in opposite directions and induced a non-ferromagnetic field [50-51].

Figure 3.11 depicts 3D Cooper Pair's dynamic tunneling and showed crossing waves through the FFTJJ barrier at zero-bias with more than $\pm$2000 S/cm^2 differential conductance for the activated device # 1. Figure 3.12 depicts an enlarged view of the zero-bias superconducting peaks in the forward and backward scan at zero-bias in the presence of 50 ng/mL collagen. Figure 3.13 depicts the image of the superconducting illuminating bands at zero bias $\pm 1\Delta$; light-dark alternating subbands were also observed.

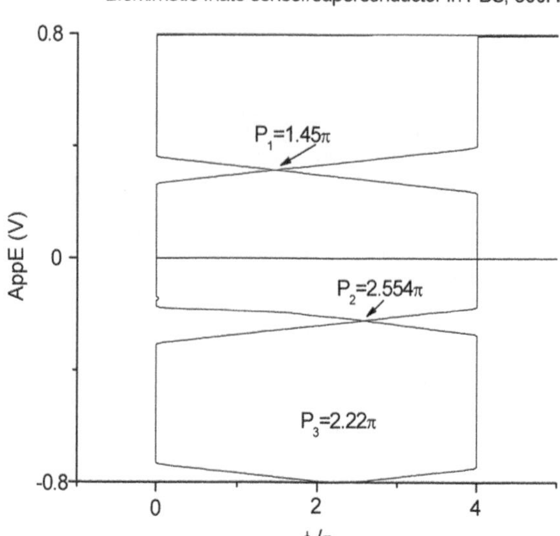

FIGURE 3.9 This figure depicts the phase change of device # 1 in PBS solution with scan rate 300Hz. (Chen, 2018, with permission.)

3.3.1.2 The Single-Layer Protein MMP-2 Membrane

The scan rate affecting the current intensity of the innate protein device demonstrated abnormal memristive characteristics, as shown in Figure 3.14. The Direct Electron Transfer $(DET)_{red}$ peak

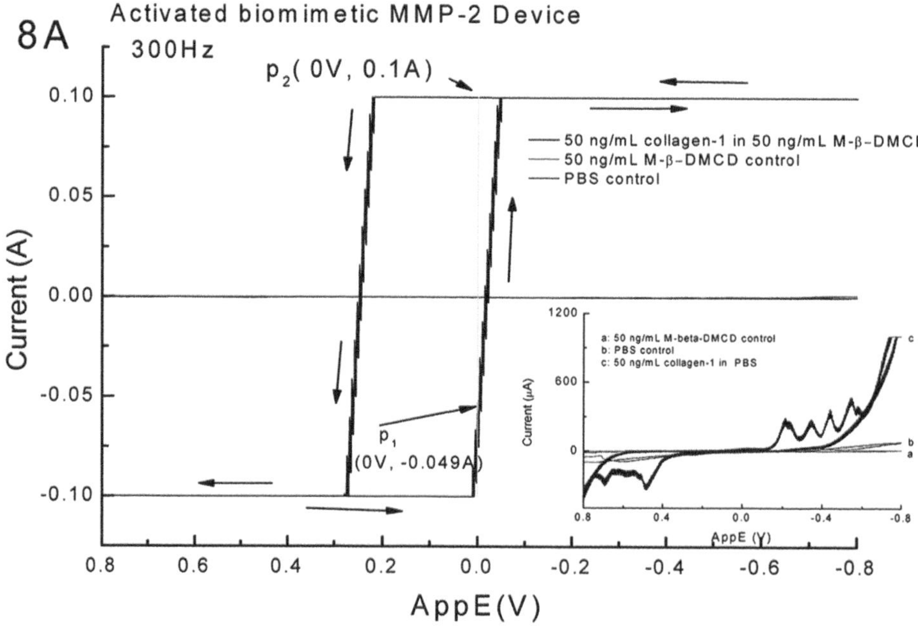

FIGURE 3.10 This figure depicts denatured device # 1's I-V curve in the presence of 50 ng/mL collagen-1 with 50 ng/mL MCD in the PBS solution compared with the MCD control and the pure PBS control, respectively. Insert is the curve with 50 ng/mL collagen-1 alone compared with the curve of MCD control and PBS controls after activated the device. (Chen, 2018, with permission.)

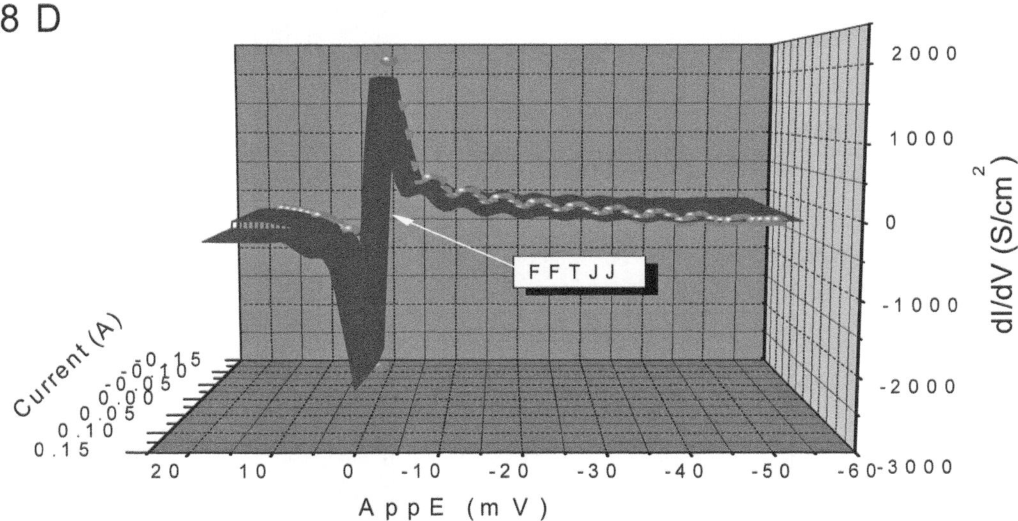

FIGURE 3.11 This figure depicts a 3D dynamic plot of the Cooper pair's dynamic tunneling and its crossing of the FFTJJ barrier based on the 2D curves. (Chen, 2018, with permission.)

current intensity increased exponentially as the scan rate increased from 1 Hz to 10 kHz. The first-order exponential decay model results, as shown in Figure 3.15, indicate that neither the DET_{red} nor the DET_{ox} follows the typical memristive curve behavior, which is that the peak intensity non-linearly increases at the highest scan rate. Instead the signal went against normal expectations and intensity reduced [52]. In contrast, both curves are increased or decreased exponentially and reached

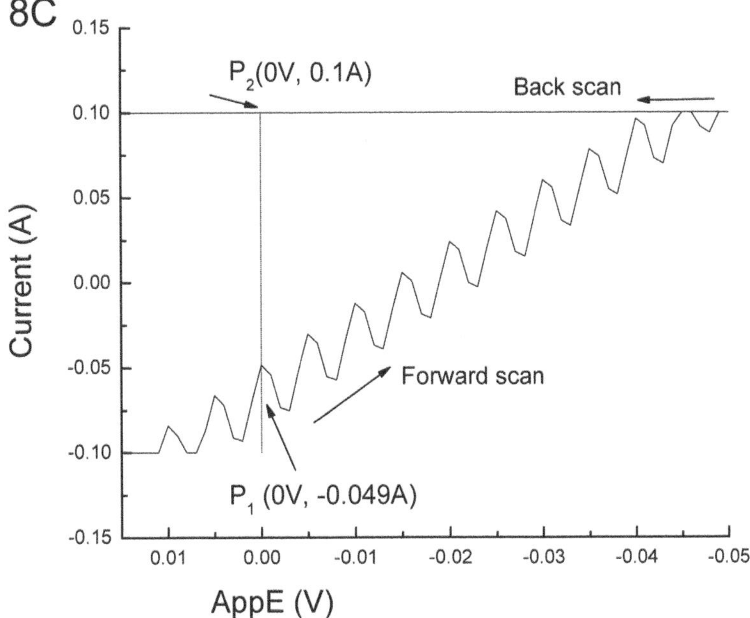

FIGURE 3.12 This figure depicts the enlarged view of zero-bias superconducting peak in the presence of 50 ng/mL collagen and 50 ng/mL MCD in the PBS solution, as shown in Figure 3.10. (Chen, 2018, with permission.)

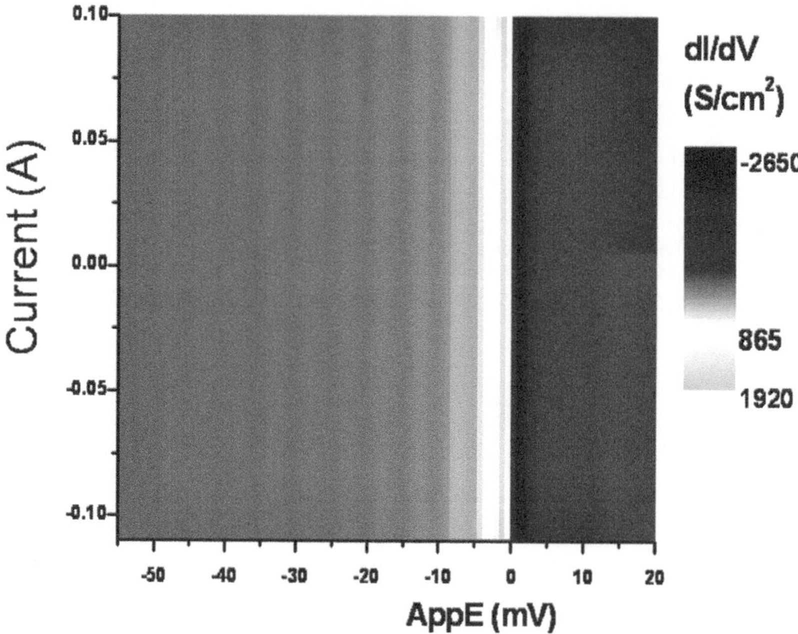

FIGURE 3.13 This figure depicts the image of the superconducting bands at zero bias ± 1Δ and sub-bands. (Chen, 2018, with permission.)

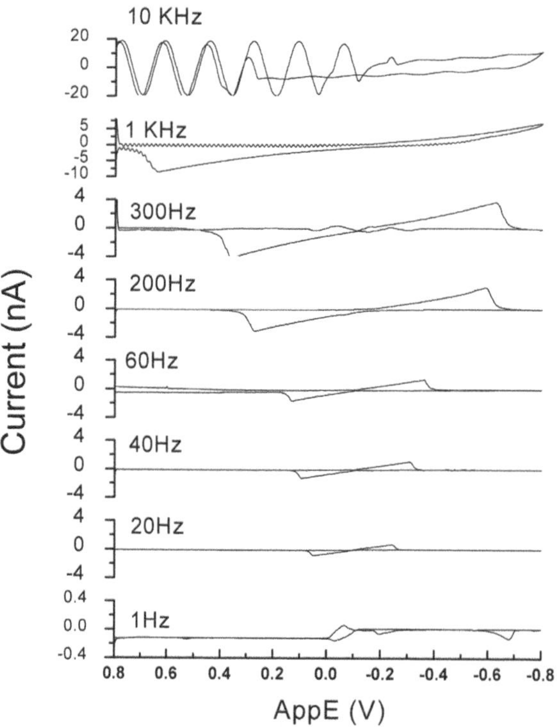

FIGURE 3.14 This figure depicts the scan rate impacts on CV curves of the innate protein device # 3 in PBS solution. (Chen, 2018, with permission.)

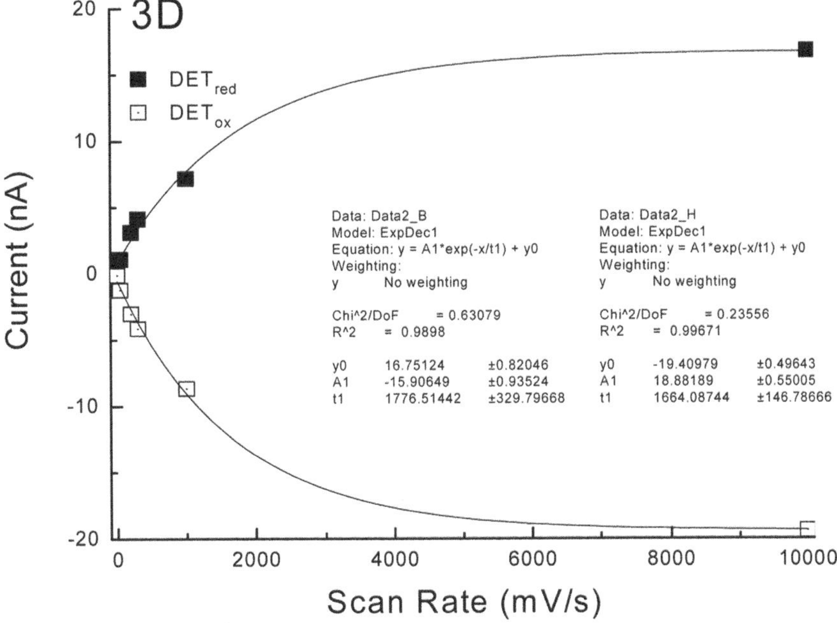

FIGURE 3.15 This figure depicts the plots of current versus scan rate of peak DET_{red} compared with peak DET_{ox} and its fitting curves. (Chen, 2018, with permission.)

an s-s state at 10 KHz of the highest scan rate. The rate of the DET_{ox} peak intensity versus scan rate over 1 Hz to 10 kHz is only higher than that of the DET_{red} peak by 7 percent. This is strong evidence that the existence of bidirectional circular accelerating vortex in the innate protein device in the control solution became a powerful force to turn the memristivity to sine wave oscillating superconductivity at 10 kHz. However, the quantum conductance from the activated protein device at 1 Hz, 1 kHz, and 10 kHz at zero-bias increased significantly, as shown in Figure 3.16, compared

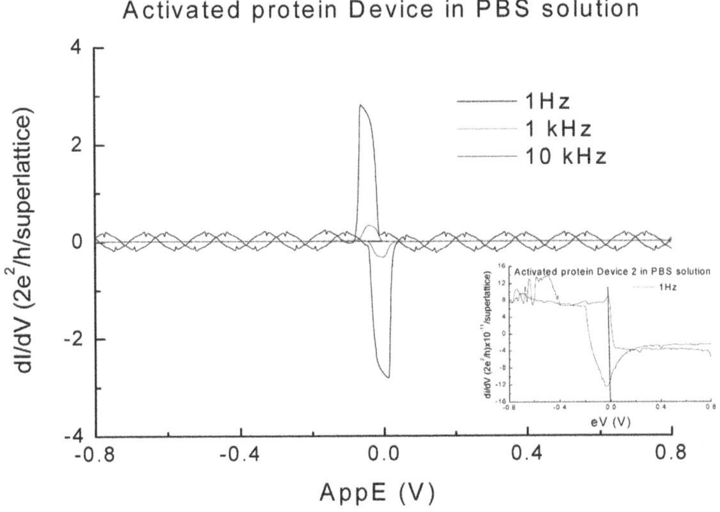

FIGURE 3.16 This figure depicts the quantum conductance per superlattice of the activated protein device versus applied potential compared with the innate protein device shown in Figure 3.17. (Chen, 2018, with permission.)

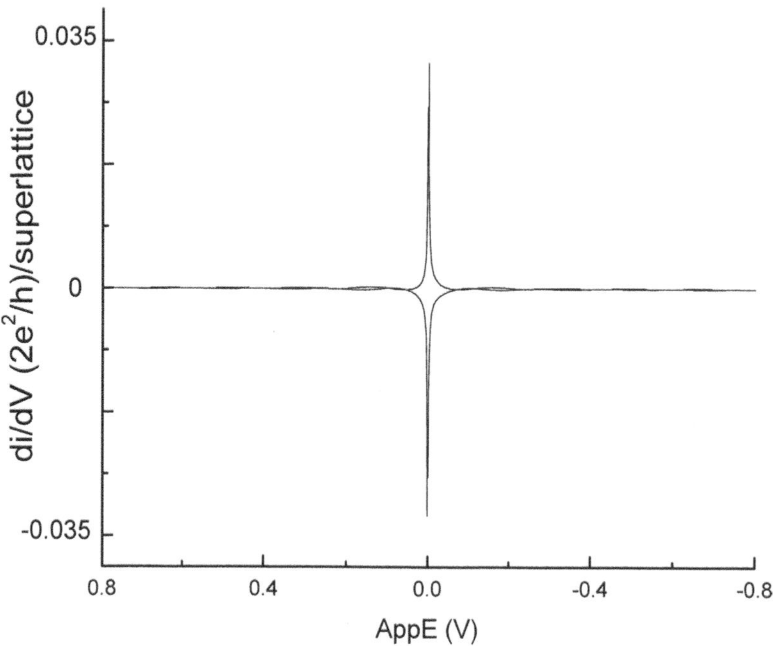

FIGURE 3.17 This figure depicts the innate protein device. (Chen, 2018, with permission.)

with the innate device at 10 kHz in the control solution, as shown in Figure 3.17. The quantum conductance at 10 kHz at the activated state is 2 orders of magnitude higher than the innate state. Both states comprise an absence of collagen, but have the superconductivity at high scan frequency. This indicates that there is a tunneling memristivity to a self-oscillating superconductivity path, which asserts its uniqueness over the single-layer biomimetic device. The toroidal array structure may shorten the path for biocommunication between the functional groups in the MMP-2 protein, the functional groups in the polymers, and the transitional metal ions in the protein. Therefore, the "flexibility mobility" of the zinc ions is an advantage compared to the zinc ions alone in the biomimetic device, as shown in the AFM image Figure 3.5 compared with Figure 3.7. The mobility of the zinc atoms observed in the protein lattice confirms its flexibility and "visionary capability" over the zinc atoms alone in the superlattice of the biomimetic membrane.

3.3.2 The Device with the FFJV Configuration in the Double Layer

Device #3 is a double-layer SIS/SIN quantum slqubit device, which utilized the first bottom layer of biomimetic *Choline acetyltransferase* (ChAT) as the pseudo TSC layer under the assumption that the direct super vortex current flow might be promoted by the presence of collagen-1. The top layer of [(bis-substituted dimethyl-β-cyclodextrin)/(triacetyl-β-cyclodextrin)/(polyethylene glycol diglycidyl ether) /(poly4-vinylpyridine) (zinc chloride)]—[(bM-β-DMCD)/(TCD)/(PEG)/(PVP)/($ZnCl_2$)]—an activated biomimetic 3D collagen-1...MMP-2...ChAT relay tunnel may form an SIS/SIN device with the TSC/QMR function peak at zero-bias. Because the ChAT regulates the MMP-2 function and its activity, the MMP-2...ChAT...collagen controllable DER could slow down the Cooper-pair transmission rate when the pair passes through the barriers [53]. Hence this approach might revolutionarily enhance slqubit's memory storage and energy storage capability, including a capability of sensitive sensing. Similar superconducting current results from device #3 compared with that of device #1 at its activated state confirmed our approach [30]. It confirmed the fact that alignment of the large and small "donuts" or toroidal rings in the superlattice promote the Cooper pairs' hopping cross the holes at the junction tunnel.

3.4 EVALUATION OF THE COOPER PAIRS' CHARACTERISTICS

3.4.1 The Flexible Fractional Toroidal Josephson Junction (FFTJJ) Toroidal Vortex Characteristics

The hallmarks of the Josephson junction (JJ) characteristics are

(1) at a DC voltage = 0:

$$I_s = I_c \sin(\Delta\varphi) \quad (3.1)$$

where, I_s is the supercurrent, I_c is critical current, and $\Delta\varphi$ is the phase difference between the waves of two superconductors appears at the DC Josephson junction;

(2) at a finite DC voltage, the phase change of the superconducting wave versus time caused oscillation at the AC Josephson junction is proportional to $2eV_{DC}$, i.e.,

$$\partial\varphi/\partial t = (2e/h)\, V_{DC} \quad (3.2)$$ [23, 30].

Here in Equation 3.2, φ is the phase difference between the waves of two superconductors that appears at two sides of the Josephson barrier, V_{DC} is the potential difference across the barrier, h is the Plank constant (1.055×10^{-34} Js), and e is the electron charge.

Figure 3.18 depicts the amplitude comparison of the super AC current among the three devices The magnitude of the activated biomimetic single-layer device # 1 in the PBS control solution is 29-fold higher than the double-layer biomimetic device # 3 and 260-fold higher than the innate protein device # 2 by the Chronoamperometry (CA) method under zero potential for each of the two steps with fixed 10 kHz data rate. Device # 3's supercurrent is 9.5-fold higher than that of the innate protein device # 2. A method was developed to quantify the Friedel oscillations observed in the AFM images, described in Section 3.3. The results obtained from the CA method under the

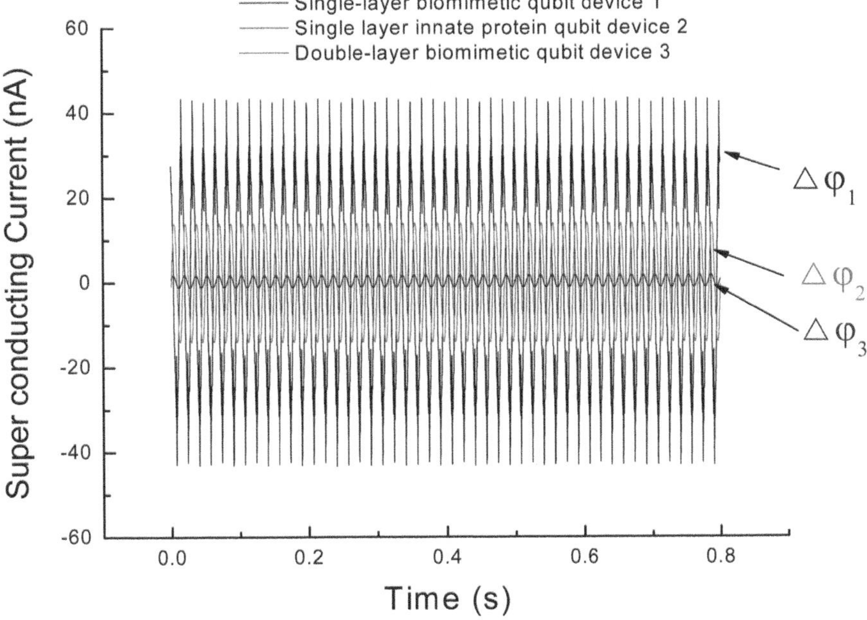

FIGURE 3.18 This figure depicts the AC current oscillating with a time span in 0.4 s for 4000 data points per step measurement compared among the activated single-layer biomimetic device # 1, the double-layer biomimetic device # 3, and the innate protein device # 2 with 13 ms, 18 ms, and 12 ms per peak spent for oscillation at zero potential in each of the two steps in the PBS solution using the CA method.

double-step zero-potential approach was linked to the AFM images; herein the order of the Friedel oscillations frequency among the three devices was in the order of protein device # 2 > activated biomimetic single-layer device # 1 > biomimetic double-layer device # 3, i.e., 83.33 Hz/peak > 76.9 Hz/peak > 55.56 Hz/peak, respectively. Under the same data rate, within the 0.8s period, device # 2 has 66.7 peaks, the activated device # 1 has 61.5 peaks, and device # 3 has 44 peaks. It was verified that our CHAT...biomimetic MMP-2...collagen-1 approach has paved a road to reduce the JJ tunnel strong oscillation in order to enable detecting extremely low concentrations of collagen-1 by the CA method, as demonstrated by device # 1 and device # 2, which failed using the CA method to detect collagen-1 due to the very strong oscillation in both with the control solution and in the presence of collagen.

3.4.2 Hearing the Cooper-pairs Second "Sound"

Several decades ago, the Nobel Laureate Brian D. Josephson theoretically predicted the presence of a second Cooper-pair wave that might coexist with the first Cooper-pair wave penetrating the Josephson junction (JJ), i.e., scientists should be able to observe and hear the second "sound" [23]. So far there has been no report showing such an event. Problems are associated with a lack of appropriate testing methods, the limitation of the instrument, and the random nature of the second wave. Multiple researchers in our laboratory had observed the strange second waves phenomena starting eight years ago, but we were unable to immediately record the second Cooper-pair wave because of its appearances in a random manner, and we did not know the exact time and manner in which the wave should show up. Plus, the event, which is beyond conventional teaching, was totally new to us. Since then, it has become a long-time pursuit to catch the second wave of the Cooper-pair and hear its second sound.

The key elements to building the slqubit device comprise a superlattice structure membrane, a flexible window barrier, the membrane electrode assembly (MEA), and the JJ cliff. We understood the Cooper pairs are high superconducting paired electrons that are strongly inclined to cross barriers at the JJ cliff under zero-energy state, and their waves come with changing phases and their voices are a symphony. The curious and elusive second wave occurs in a random manner. As our knowledge deepened, finally our approach was to use the electrochemical detector with newly designed membranes to detect the open circuit potential of the device. If there are two waves penetrating the junction, then we would be able to detect the two waves with flip phases under a zero-current condition. Indeed, finally, we were able to see Cooper-pair's second wave with its phase flipped compared with the phase from the first wave. We recorded the wave's voltage as time elapsed; the second wave of the Cooper pair showed up spontaneously and randomly. We still cannot control the time and manner of the second Cooper-pair wave sensed by our detector. There are two curves from the two Cooper-pair waves with the phase flip. Wave a's raw data curve from the first Cooper-pair wave was fitted with an exponential decay model in the equation:

$$y = \mathrm{A1} * \exp(-x/t1) + y0 \text{ with } y0 = 0.094,\ \mathrm{A1} = -0.076,\ t1 = 101.209$$

The $\mathrm{Chi}^2/\mathrm{DOF} = 0.00001$, $r = 0.9839$ with a rate constant = 0.01 $\mathrm{V.s}^{-1}$ with triplicate runs compared with the "flip" second Cooper-pair wave "b"'s raw experimental curve fitting with an exponential decay second-order model:

$$y = \mathrm{A1} * \exp(-x/t1) + \mathrm{A2} * \exp(-x/t2) + y0$$

with $y0 = 0.2393$, $A1 = -0.03944$, $t1 = 6.09$, $A2 = 0.0129$, $t2 = 0.826$; $\mathrm{Chi}^2/\mathrm{DOF} = 4.53e^{-6}$, $r = 0.8133$.

The first-rate constant = 0.164 s^{-1}, and second-rate constant = 1.22 s^{-1}, as shown in Figure 3.19 for the innate protein device. There is a 122-fold rate difference between the two Cooper-pair waves. The signal strength at the s-s state between the two Cooper-pair waves has an 8-fold difference,

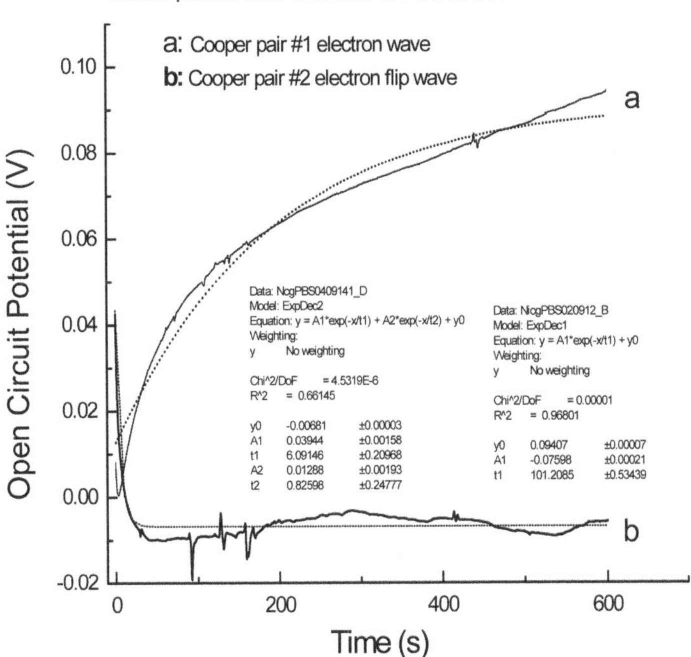

FIGURE 3.19 This figure depicts the curves in open circuit potential for the innate protein device in the PBS solution and the water media, respectively. (Chen, 2018, with permission).

which is direct evidence of "hearing" the two-wave sounds from the Cooper-pair magnetic flux. The intrinsic electro-relay system within the toroidal cavities is activated under the zero-current open circuit condition, which is the driving force of the electromagnetic flux strength.

Figure 3.20 depicts the two Cooper-pair's non-linear amplification waves of increase potential in the tunnel of Josephson junction (JJ) using the same OPO method for the double-layer biomimetic superconductor device in pure water and in phosphate-buffered solution (PBS), respectively. Curve "a" exhibits a first-order exponential decrease by the second wave Cooper-pair with the model equation:

$$y = \mathrm{A1} * \exp(-x/t1) + y0$$

with $y0 = -0.0237$, $\mathrm{A1} = 3.6137$; $t1 = 95.9268$, with the rate constant 0.01 $\mathrm{V.s^{-1}}$ in pure water. Curve "b" and "c" in pure water and in PBS solution, respectively, has the first-order exponential rate of 0.02 s^{-1} and 0.014 s^{-1}, respectively. The peak voltage between curve a, b, and c has no significant difference at the s-s state; but at the beginning of the discharge, there is a 37-fold large difference of the peak voltage strength between curve a and b or between a and c. The rate between the increasing curve and the decreasing curve is 30 percent higher in pure water than in PBS solution. The Cooper pairs are coherent when tunneling within the barrier and then suddenly the coherence was broken, which caused a high open circuit potential to occur. That means the intrinsic waves are able to penetrate the Josephson junction (JJ) barriers, such as zinc atoms. The double-layer biomimetic device offered advantage for quantum computing, because its 122-fold slower decrease rate of the second Cooper-pair wave compared with that of the innate protein device's second wave Cooper-pair, i.e., the Cooper-pair second wave has a longer half-life of 66.63s when it stayed in the Josephine junction than the Cooper-pair in the protein device having 0.57s for the half-life in the JJ location, which enhanced the capability for nonvolatile memory storage in the double-layered biomimetic device # 3. The design of the biomimetic device with the emphasis of the DER effect and

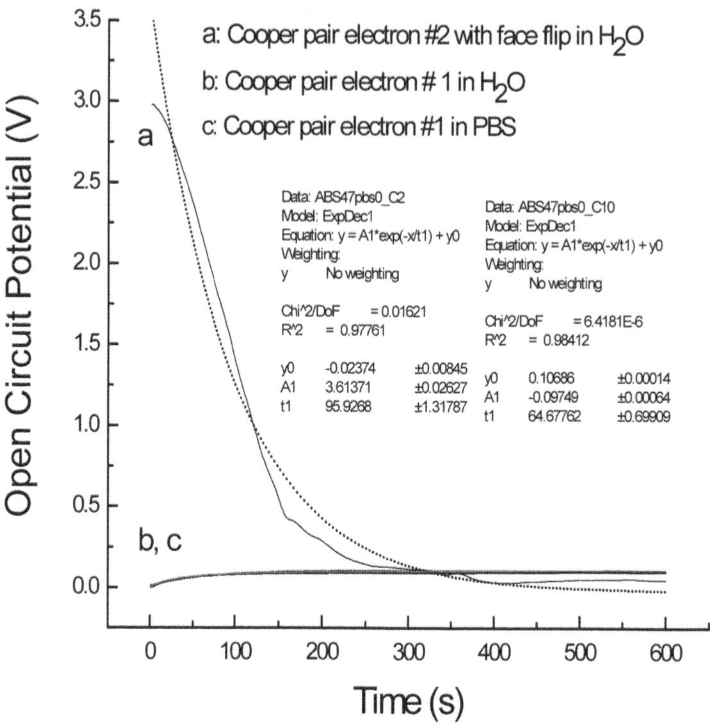

FIGURE 3.20 This figure depicts the two Cooper-pair waves open circuit potential curves for the double-layer biomimetic device # 3. (Chen, 2018, with permission).

the superlattice structure enabled the device to gain the power for sensing, and it will be presented in Sections 3.4.4 and 3.6.

3.4.3 Calibration of the Shapiro Step

Inspired by the theoretical prediction of a π-phase difference on a topological-superconductor, (TSC)/normal metal (NM) induced by Majorana spin-triplet pairing, exhibits a Josephson phase of 0 and a π-junction in its ground state without any applied magnetic flux [27]. Use of superconducting quantum interference devices (SQUIDs) to sense the magnetic flux, for example, magnetocardiogram signals, was reported everywhere; however, under the cryogenic condition, it is not so popular for widespread clinical use [54]. A DC SQUID utilized d-wave $d_{x^2-y^2}$ symmetry and two-π junction interferometer was experimentally realized [55]. The self-induced oscillation between the zero bias potential supercurrent by d-wave and the Josephson frequency at the Fractional Josephson Vortices (FJV) observed in the I-V curve reported in the above-cited literature was 4.2-77K. Our experiment results show that the d-wave hysteretic cross points occurred with the scan rate 1 Hz, 60 Hz, and 300 Hz, respectively, and that the Shapiro steps in PBS solution under zero external applied magnetic field at room temperature were observed in Figure 3.21.

The *Shapiro step* is a phenomenon that the Cooper pairs tunnel in the JJ tunnel with steps that look like a sawtooth. It arises from the response of the supercurrent's oscillating that occurred at a voltage equal to $nhf/(2e)$, where n is an integer, h is Plank's constant, e is the electron when a photon of frequency f or a high-frequency current is applied on the JJ tunnel. A DC I-V curve or an AC Josephson current oscillating with the applied frequency gave rise to constant voltage in the I-V curve. The Shapiro step occurs at voltages = nf/K_J, where K_J is the Josephson constant, f is the

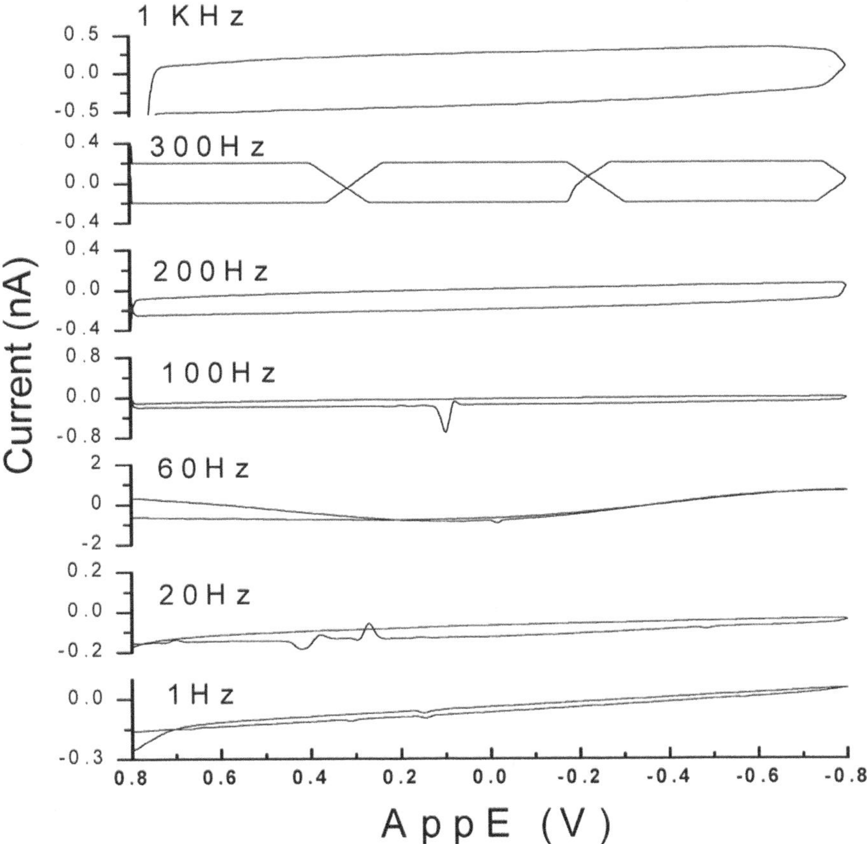

FIGURE 3.21 This figure depicts the I-V curve profiles in the PBS control solution with the scan-rate change from 1 Hz to 1 KHz for the innate single-layer biomimetic slqubit device. Shapiro steps can be seen. (Chen, 2018, with permission.)

Josephson frequency. $K_J = 483.5979$ THz/V is an internationally defined constant, equal to 2e/h, which is the inverse of the single magnetic flux quanta, i.e., $1/\Phi_0$. This effect is used as the Josephson voltage standard [56-57].

Under an FFTJJ situation, Figure 3.22 depicts the calibration curve of the Josephson frequency in the range between 6.5 MHz to 0.77 THz versus Shapiro step voltage between 23 nV to 3 mV over scan rate 1 to 300 Hz. The slope gave the K_J value indicates that a 4π FFTJJ vortex system existed [58-69].

3.4.4 The Flexible Fractional Toroidal Josephson Junction (FFTJJ) Dynamic Multiple-Variable Study

Stern's group reported the observation of Majorana bound states of Josephson vortices in topological superconductors, and the equations of three types of energy contributions to the Josephson vortices in a long circular junction in a Sine-Gordon system was published [60]. The Josephson junction (JJ) energy was from the Cooper pair, the magnetic energy was from the inductivity of the circular vortex, and the charge energy was from the SIS quantum capacitor-like device [60]. The vortex suppression of the supercurrent effect also was considered in the equation. However, there was no further analysis of how each component energy contributes to the system superconductivity from the experimental data.

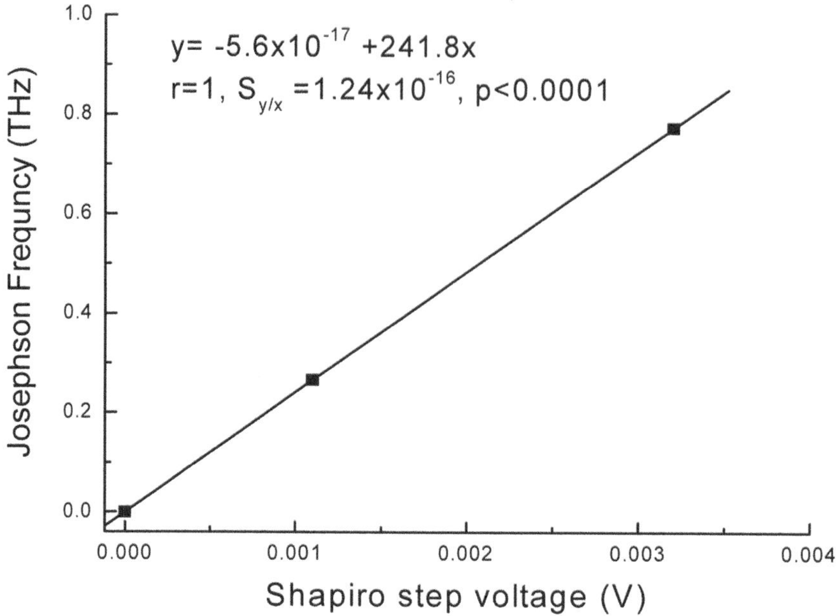

FIGURE 3.22 This figure depicts the plot of Josephson frequency versus Shapiro step voltage over scan rate 1, 60 to 300 Hz that has cross-points of FFTJJ. (Chen, 2018, with permission.)

Cosmic's group reported seeing the vortex in a Josephson array based on a fractional Josephson effect in the vortex lattice [61]. The Hamiltonian of the Josephson Junction Array (JJA) was given in the combinations of the first part of charging energy obtained from all arrays and the second part of the Josephson effect energy [61]. Still, no reports were given on how the energies impacted on one another in their experiment. Inspired by their experimental works, our attempt was, by using the 3D dynamic map method, to further seek a method to elucidate the reactions between the component energies to the superconductivity of the vortex array system at room temperature without external magnetic field applied. Our experimental data were shown on the I-V curves and the AFM structure of the superlattice array. The modified Sine-Gordon system energy for our d-wave vortex array is:

$$E^{n}_{JJA} = \left({}^{1}/_{2}\right) C_{i}^{-1} \left(Q - en_{1\ldots i}\right)^{2} \tag{3.3}$$

$$E^{n}_{L} = \left({}^{1}/_{2}\right) \mu_{0} N^{2}_{n=1..i}.A.L^{-1}_{n=1..i}.I^{2}_{n=1..i} \tag{3.4}$$

where,

- $E^{n}{}_{jjA}$ is the charge energy of Josephson Junction arrays at $n = 1..i$.
- Q is the charge.
- C is the total capacitance at $n = 1..i$.
- en is the n quantum particles at $1..i$ data point with an energy periodic in h/e for Josephson effect for d-wave [62].
- E^{n}_{L} is the Inductive energy induced by the circular toroidal array.
- N is the turning number around the toroidal porous at $n = 1..i$.
- A is the cross-sectional area of the porous.
- L is the length of the wending, μ_0 is the magnetic permeability constant in free space.
- I is the current.

The toroidal arrays are in series connected. Recent publications regarding our FFTJJ multiple-variable study results in 3D dynamic maps was presented in the literature [63].

3.5 QUANTUM COMPUTATION

3.5.1 Superpositioning Multiple-state Characteristics of Quantum Computing

Quantum computing is computing using quantum-mechanical phenomena, such as superposition and entanglement [23]. *Superconducting flux qubit* has two states that can be effectively separated from the other states is the basic building block of quantum computers. Current DC or RF superconducting SQUID were made in advance for a faster switch time; however, hundreds of MHz electromagnetic field applied onto a tank circuit coupled to the SQUID are needed for the system to work under cryogenic conditions [54, 64-65]. The *RF-SQUID* consists of a superconducting ring of inductance L interrupted by a JJ, the potential energy of the SQUID and the Hamiltonian equations are given by:

$$U(\Phi) = (\Phi - \Phi_e)^2/2L - E_J\cos(2\pi\Phi/\Phi_0) \qquad (3.5)\ [23]$$

$$H = Q^2/2C + (\Phi - \Phi_e)^2/2L - E_J\cos(2\pi\Phi/\Phi_0) \qquad (3.6)\ [23]$$

where,

- Φ_e is the applied magnetic flux penetrating the SQUID ring.
  Φ is the total magnetic flux threading the SQUID ring.
- L is the inductance.
- E_J represents the Josephson coupling energy.
- Φ_0 is the superconducting magnetic flux quantum.
- Q is the charge on junction's shunt capacitance satisfying $[\Phi, Q] = ih/2\pi$.
- h is the Planck constant.

As a consequence, superconductor qubits are vulnerable to low-frequency noise with sources coming from 1/f noise, wave dephasing noise, flux noise, critical current noise, quasiparticle tunneling noise, and capacitance noise [67-68]. The Josephson junction (JJ) is a key element in the broader area of superconductivity devices, including SQUIDs [54, 64-66]. That qubit devices are vulnerable to decoherence caused by the control and readout circuitry and by the environmental noise, as well as by the qubits' own intrinsic low-frequency noise compared to the conventional computing has been well known because of qubits operating simultaneously at multiple states [66-67]. Nevertheless, so far, the DC-SQUID or RF-SQUID devices have not demonstrated their function as a memory cell, i.e., devices that remember past events; hence, the SQUID devices have to be connected with memory devices and other auxiliary devices in order to function as a large-scale quantum processor for computational function.

Figures 3.23 and 3.24 show the multiple states of superposition at the zero-bias in the -V curves in the presence of collagen. In the presence of 200 ng/mL collagen, it offers quantum computing: at "1" state with the spatial location at (0 mV, 0.59 μA) and "-1" state with location at (0 mV, -0.69 μA), and the "0" state window is located from (0 mV, 0 μA) to (-1 mV, 0 μA), respectively, as shown in Figure 3.23. This origin point contained in the rectangle loop of the I-V curve is the signature feature of an "O-type" loop memcapacitor whose hysteresis pinch point does not directly pass origin [68]. The loop contains origin because the permeability of the materials cannot instantaneously follow the current (flux) across the inductor due to a negative charge or negative capacitance [68]. These "O-type" micro-electro-mechanical (MEM) devices "remembered" the past event offer advantages. In that observation, the "0," "1," and "-1" states simultaneously occurred with its Shapiro step of 3 mV, as shown in Figure 3.24 having a switch time of 1.4 ns. There was no energy loss for operations at these states at the zero-bias condition.

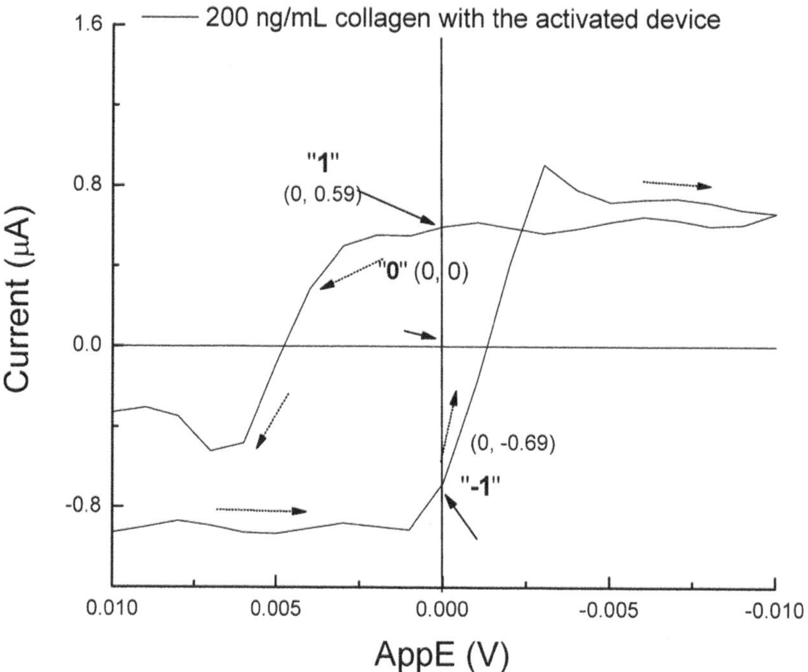

FIGURE 3.23 This figure depicts the detail assignments for quantum computing "0," "1," and "−1," three states at zero-bias in the presence of 200 ng/mL collagen over 0.010 to -0.01V. (Chen, 2018, with permission.)

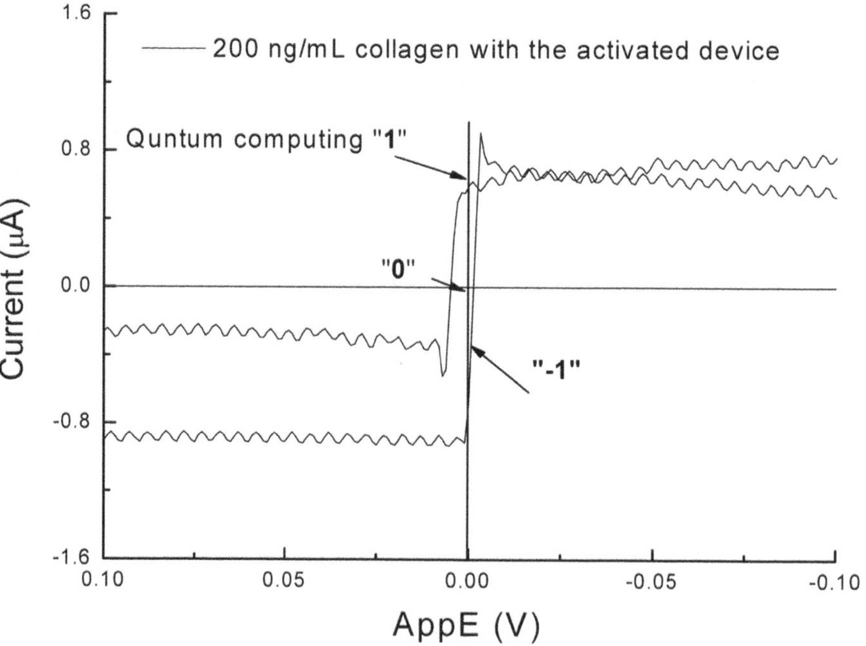

FIGURE 3.24 This figure depicts the Shapiro steps that occur in the I-V curve covers from 0.10 to −0.10V. (Chen, 2018, with permission).

3.5.2 Characteristics of Quantum Computing Without Switch Time Delay

In section 3.5.1, we used the *activated single-layer biomimetic slqubit device* to study quantum computation with memristive characteristics by a heating method to eliminate the cysteine group. In this section, we use the *innate single-layer biomimetic slqubit device* to study quantum computation without the switch time delay. We used the CV method to define the quantum computing in "0" and "1" states at the zero-bias condition. We also defined the "0" and "1" states at the zero current condition by a voltage method— i.e., the Double Step Chronopotentiometry (DSCPO) method—by setting up each of the two steps at current = 0 for the innate single-layer biomimetic qubit device. Each step time period was set in one of the constant values, such as 1ms, 4 ms, 40 ms to 800 ms under conditions (1) with 150 ng/mL collagen and (2) with 150 ng/mL collagen in 50 ng/mL Mono Imidazole Derivative Dimethyl ß-cyclodextrin (mM-ß-DMCD), in short, MCD in the PBS solution compared with the control, respectively. Figures 3.25(a), (b), (c), and (d)show the step time impacting on the intensity of voltage curves at current = 0 in different media compared with or without collagen using the DSCPO method for the innate single-layer biomimetic slqubit device # 1. Curves with collagen in the MCD/PBS media have the highest voltage amplitude compared with collagen alone in the PBS solution. Curves in the control have the lowest voltage intensity. We observed the phase changes between the curves in Figures 3.25(b), (c), and (d) compared with the phase of the control sample. The "0" value assigned at the curves with voltage = 0 and current = 0, the "1" assigned at

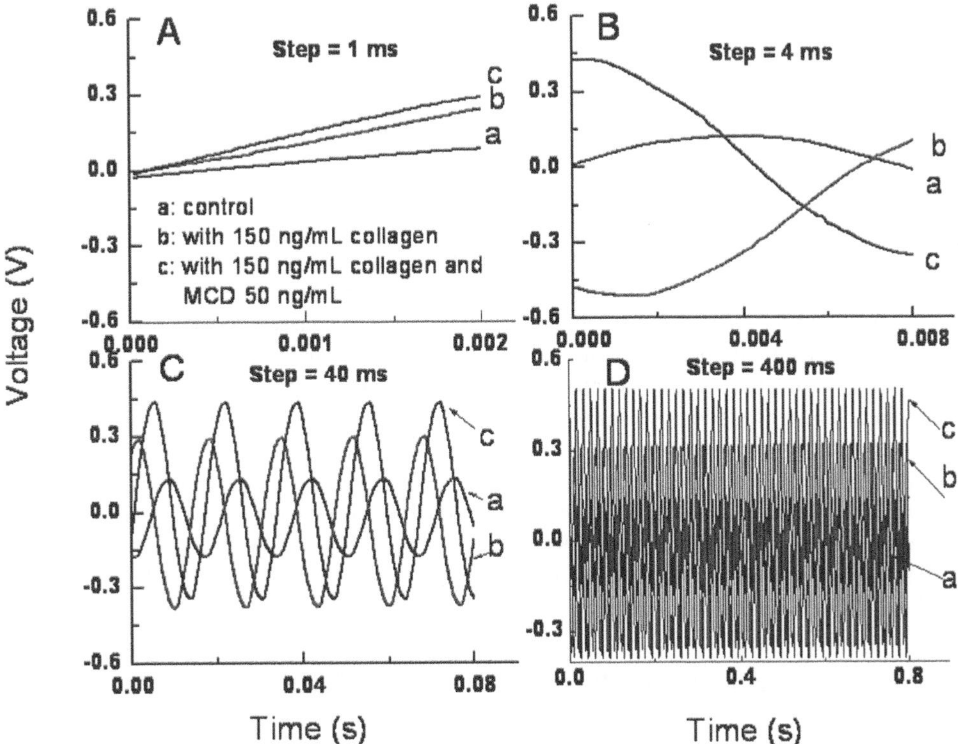

FIGURE 3.25 These figures depict the step time impacts on the intensity of voltage at current = 0 versus time curves using the DSCPO method for the innate single-layer biomimetic slqubit device # 1 for the PBS control solution (a); for with 150 ng/mL collagen (b), except the same condition in (b), it also used 50 ng/mL MCD (c) compared with the PBS control. 1 ms step time was in Figure 3.25(a); 4 ms was in Figure 3.25(b); **g** 40 ms time was in Figure 3.25(c); and 800 ms time was in Figure 3.25(d) at current = 0. (Except for Figure 3.25(d), Figures 3.25(a) to (c) are from Chen, 2018, with permission.)

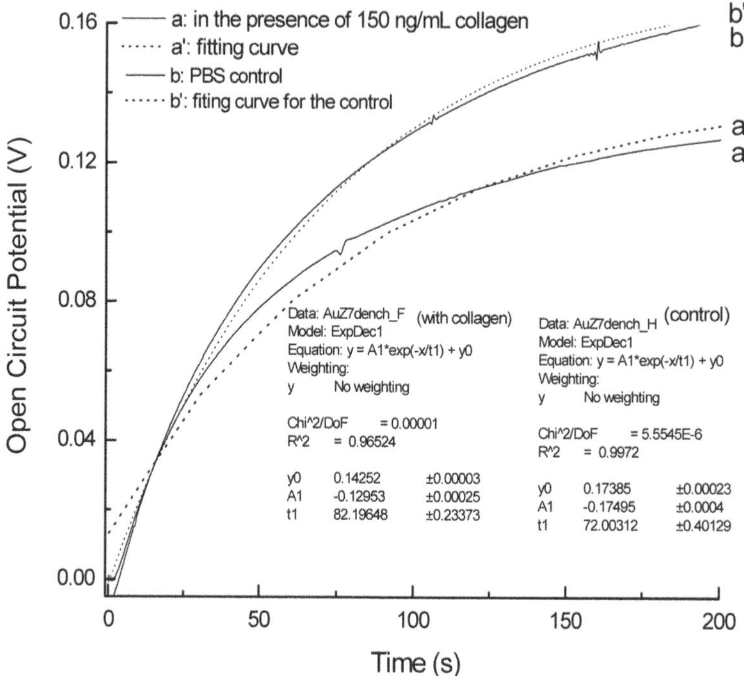

FIGURE 3.26 This figure depicts the open circuit potential curve versus time at current = 0 in the presence of 150 ng/mL collagen in the PBS solution for the activated single-layer biomimetic slqubit device compared with the control using the OPO method. (Chen, 2018, with permission.)

voltage larger than 0 with current = 0; and "−1" can be seen at voltage less than 0 with current = 0. Figure 3.25(d) shows that with the superposition of the switch time over "0" and "1" states or "−1" state, there was no time delay compared with the other Figures. It revealed that in the presence of collagen and MCD, the DER and superconducting effects were promoted under 400 ms stepping time at the zero-current condition. Herein, the switch time between states was reduced, which offered an advantage for solving the lagging switch time problems. The spontaneous discharged voltage pulses indicate there was embedded power in the device, which could have a self-powering capability for quantum computing in the presence of collagen. Therefore, this leads to a suggestion that there is no need for a microwave power supply because the device does not consume any energy, as shown in Figure 3.26. Figure 3.26 shows the open circuit potential curve versus time at current = 0 in the presence of 150 ng/mL collagen in the PBS solution for the activated state compared with the control. The spontaneous first-order rate constant of the open circuit potential results of 0.0139 V/s from the control and 0.012 $V/s^{-1}$for with collagen were reported.

3.6 APPLICATIONS IN SELF-POWERING, SENSING, AND ENERGY STORAGE

3.6.1 Self-Powering, Energy Storage, and Sensing of Microwave Frequency

In Figure 3.22 demonstrates that the Josephson frequency can be sensitively detected by the single-layer biomimetic device # 1 at its innate state with the FFTJJ qubits based on the Shapiro step voltage. The embedded FFTJJ in the device not only enables sensing the presence of Josephson frequency change but it also enables self-powering without a need for a microwave power supply. The self-powering and energy storage features from the device # 1 at its innate state demonstrates by the voltage method under zero-current, as shown in Figures 3.25(a) to (d). Figure 3.26 shows that the

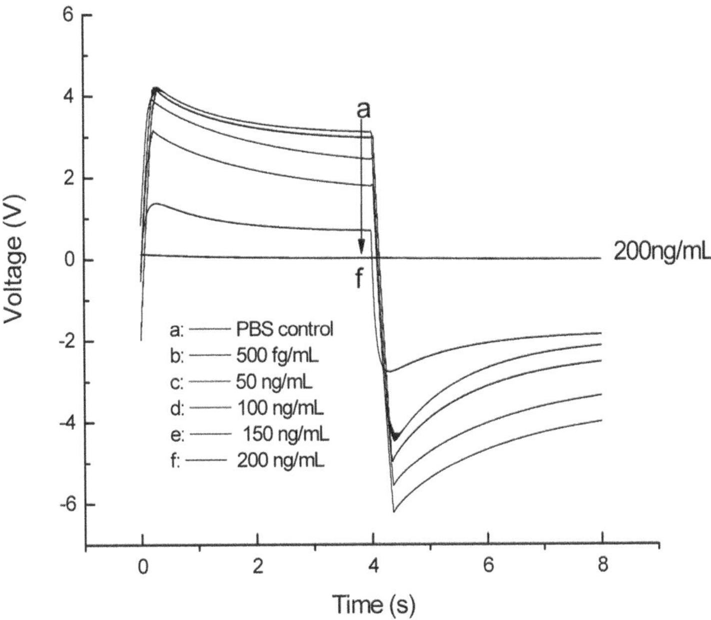

FIGURE 3.27 This figure depicts the innate single-layer biomimetic device # 1's voltage curves using the DSCPO method over collagen level from 0.5 pg/mL to 200 ng/mL in PBS solution at 0.25Hz. (Chen, 2018, with permission).

spontaneous energy pulse rate and the intensity using the OPO method in the presence of 150 ng/mL collagen was reduced by 13.7 percent and 21 percent, respectively, compared with the control.

3.6.2 Sensing of Biological Substances

Three types of slqubit devices have unique features in the sensing of biological substances, such as for direct detection of collagen-1. The innate single-layer biomimetic device # 1 can be viewed as a perfect superconductor/memcapacitor device that directly detects collagen-1 with extreme sensitivity, as shown in Figure 3.27. Figure 3.27 depicts the curves of voltage versus time at 0.25 Hz at $\pm$10 nA over 0.5 pg/mL (3.4 pM) to 200 ng/mL (1.4 nM) collagen-1 concentrations against the control samples in pH 7.4 PBS solution. The calibration curve produced a linear regression equation $Y = 3.1 - 0.015x$, $r = 0.995$ ($n = 18$, 6 levels), $P < 0.0001$, $Sy/x = 0.12$, with a pooled relative sum of squares pure error (PRSSPE) of 2.0%. The recovery results using spiked NIST human standard serum specimens have an agreement of 92 $\pm$ 0.03% over the studied range. The imprecision of the PRSSPE error was 0.3% ($n = 15$) [30].

The other two types of devices did not use the DSCPO method. The direct sensing of collagen in human blood serum samples using the innate device # 1 also was evaluated using another analytical method, the Cyclic Voltammetry (CV) method. Using the innate device # 1, the superconducting current exponentially increased when the collagen concentration increased in the spiked NIST standard human serum samples with a first-order rate constant of 22 pA $(ng/mL)^{-1}$ over the collagen concentration 1 ng/mL to 200 ng/mL in the I-V curves at the zero bias, as shown in Figure 3.28 [30]. For comparison, the activated protein device # 2 has the CV profiles reported in literature [30], which only has the peak intensity of 13% of the innate device # 1, but the rate constant is 22 nA $(ng/mL)^{-1}$, which is 1000-fold faster decrease than the innate device # 1's increase rate over 0.5 pg/mL to 100 ng/mL in human serum. The reason may be due to the protein device having a very fast rate for the second Cooper-pair penetrating the zinc barrier, as described in Figure 3.19, **or** due to the

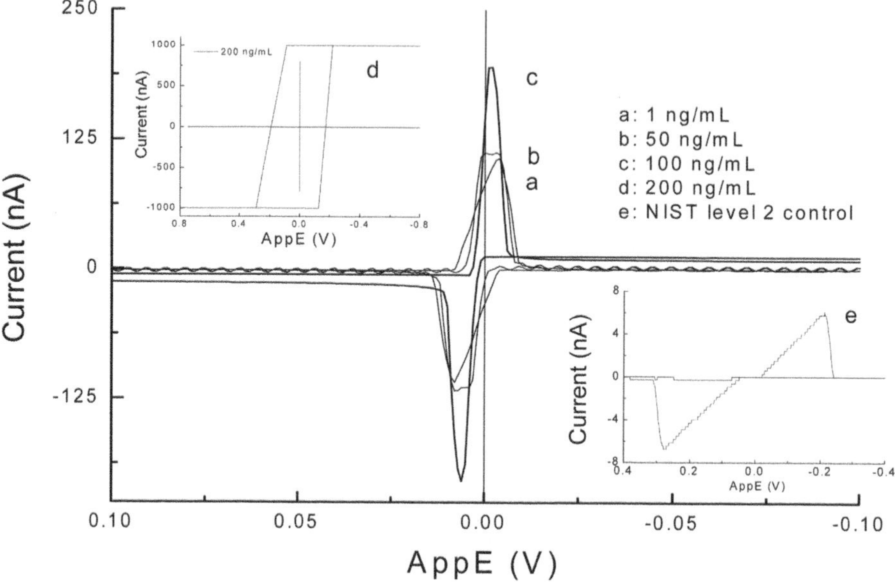

FIGURE 3.28 This figure depicts the plots of the I-V curves of the innate single-layer biomimetic device 1 at zero-bias at different collegen-1 concentrations from 1 to 100 ng/mL with the top insert curve at 200 ng/mL and the bottom insert is the control using the CV method. (Chen, 2018, with permission.)

multiple-ring structured toroidal superlattice in the membrane, which stimulates the localized biological zinc atoms to become mobile and causes Friedel oscillations, as shown in Figure 3.5. Figure 3.29 depicts the curves of current versus time under −0.3 V fixed potential over collagen-1 concentrations 5.0 fg/mL (34 attoM) to 200 ng/mL (1.4 nM), compared with the control in PBS solution for the double-layer biomimetic device # 3 using the CA method. Inserts are the enlarged

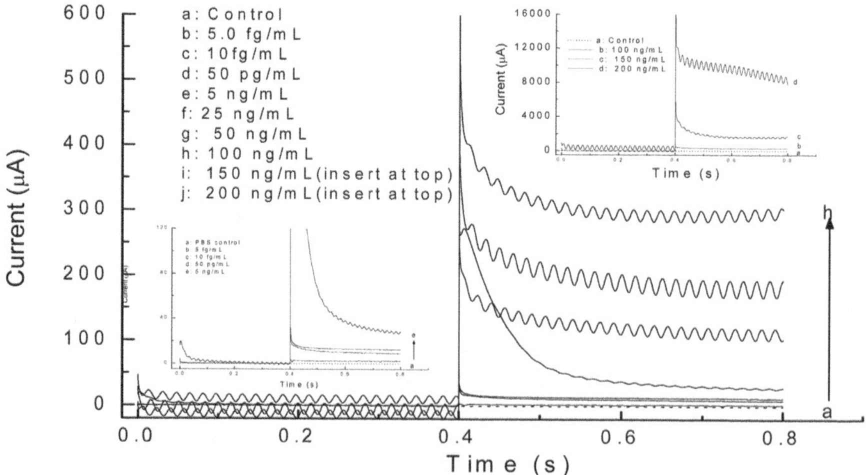

FIGURE 3.29 This figure depicts the double-layer biomimetic device # 3's current profiles over collagen level 5.0 fg/mL to 200 ng/mL (9 levels) versus controls in 10 mM PBS solutions. Samples run triplicates. Inserts are for the enlarged view for the results at high- and low-end concentration levels compared with controls. (Chen, 2018, with permission).

view of the profiles at low and high levels, respectively. All curves oscillating at the AC JJ were observed. The biomimetic device # 3's exponential current increase versus concentration curve over 5.0 fg/mL to 200 ng/mL (9 levels, $n = 27$), with a Detection Limits (DOL) of 0.43 pg/mL/cm^2 (14 fg/mL, i.e. 0.1 fM for this sensor) with a relative percent of sum of squares pure error (SSPE) of 0.05% at the high end and 0.5% at the low end, respectively. The recovery results between the standards and the human fresh finger capillary whole blood (CPWB) serum specimens were 96.4 ± 3.4% and 97.9 ± 0.73%, with the imprecision of 4.9% and 0.8% at 2.5 pg/mL and 166 ng/mL level, respectively. Other features of sensing of collagen under various conditions and media were reported in the literature [29-30].

3.6.3 Applications in Toroidal Array Self-Oscillation (TASO)

This section emphasizes the application of the Flexible Fractional Toroidal Josephson Junction (FFTJJ) in Toroidal Array Self-Oscillation (TASO) by the innate protein device. The results were presented in the kinetic study using the CV method by 10 cycles of consecutive scans at the scan rate 300Hz, 1 kHz and 10 kHz, respectively in the Josephson toroidal array junctions. At zero-bias potential, there is a negative supercurrent observed at 300 Hz scan rate for the innate protein device, as shown in Figure 3.30(a). The hysteresis point far from origin was observed. There are many oscillation sine waves at 1 kHz observed in Figure 3.30(b) that included the origin located in the curve envelop. There was no peak intensity change observed in the 10 scan cycles. Figure 3.30(b) shows a significant event that happened in the first forward scan. The sine wave has an absolute peak height of 1.592 nA and the λ length within the adjacent peak is 0.0114 V; the backward scan has a cosine wave with an absolute peak height of 1.56 nA and the λ value is 0.0166 V. As the scan cycle increases, the peak location has a small movement toward to the lower field by 0.9 mV for each consecutive cycle and the overall peak intensity did not change over the entire scan range. For a clear view, Figure 3.30(b) only shows the range between 0.06V to −0.06V. Figure 3.30(c) depicts the forward

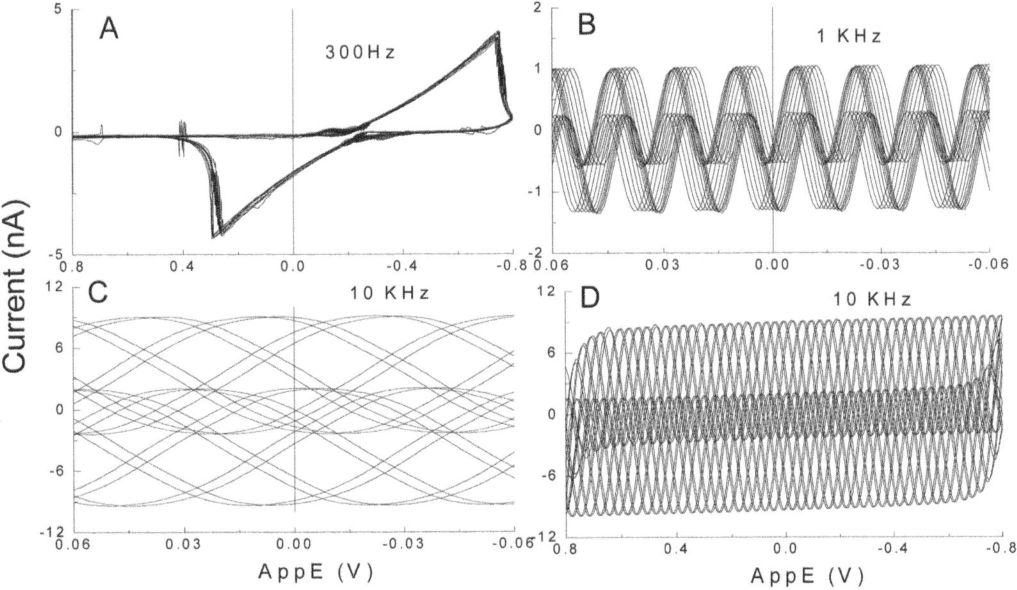

FIGURE 3.30 This figure depicts the I-V curves for 10 consecutive scan cycles at 300 Hz for the innate protein device in 10 mM pH 7.4 PBS solutions. (b) This figure depicts the I-V curves of 10 cycles at 1 kHz. (c) This figure shows the I-V curves of 10 consecutive scan cycles at 10 kHz over 0.06 to −0.06V. (d) This figure shows the I-V curves for 10 scan cycles over 0.8V to −0.8V at 10 kHz. (Chen, 2018, with permission.)

sine wave and the backward cosine wave, which has a similar peak height of 11.42 nA and the between peak λ value is 0.1643V over 0.06V to −0.06V at 10 kHz for 10 cycles. Figure 3.30(d) shows the 10 kHz wave over −0.8 to 0.8 V. The peak height remained unchanged, and the only change was the phase change that indicated the innate protein device at 10 kHz high frequency was becoming a strong oscillating circular accelerator, which kept periodic phase changes without energy dissipation at the zero-bias potential. This led to a unique opportunity for superpositioning of multiple states in quantum computing at states "+1," "0," and "−1," yet to be able to distinguish between these states at zero-bias simultaneously, as shown in the above figures. The superconductivity at zero-bias is also observed in these figures. Without the presence of collagen, only the presence of the self-circular oscillation phenomena could provide enough attractive forces that pave the road for direct sensing of collagen without using denaturing step, which also is a significant benefit in medical diagnosing of diseases.

3.7 CONCLUSION AND DISCUSSIONS

We demonstrated the utilities of the slqubit devices for sensing, memory storage, and quantum computing under the conditions of operation at room temperature and without applying an external magnetic field. The Flexible Fractional Toroidal Josephson Junction (FFTJJ) membrane-initiated the intrinsic magnetic flux accompanied by a phase change that promoted the quantum computing function in the presence of collagen, and the collagen acted as an analyte and as an insulator. Cooper pair's second wave sound was "heard" and observed in the FFTJJ devices. The double-layer biomimetic slqubit device offered higher superconductivity and sensitivity to detect collagen due to the longer half-life of the second Cooper-pair wave that stayed in the FFTJJ spatial barrier compared with that of the protein slqubits. The protein made Matrix Metalloproteinase-2 (MMP-2) toroidal array qubits with self-oscillating characteristics based on the flexible mobility of the MMP-2's transitional metal ions in the superlattice membrane, as the flexible barriers, were demonstrated. Its superconductivity—produced by the high-frequency self-oscillation without the presence of collagen—has advanced utilities, which could open broad applications in the future. However, currently due to the very short half-life during the Cooper-pair electrons tunneling through the FFTJJ barrier, the realistic application is limited. Therefore, further innovation is needed in order to promote the applications of protein qubits in the quantum computing feild.

3.8 ACKNOWLEDGMENTS

The author would like to thank all the contributors who made this chapter possible: Show-hong Duh, John Thornton, and Peter T. Kissinger. E. Chen would like to thank the Nanofab of National Institute of Science and Standards (NIST) for their professional guidance and assistance in the cleanroom facility and operating instruments. Dr. R. Kasica's discussions and help are greatly appreciated. E. Chen also thanks Mrs. V. Kitts for the editing work of the manuscript.

REFERENCES

1. University of Cambridge, "Room Temperature Superconductivity: One Step Closer to the Holy Grail of Physics, July 9, 2008, www.phys.org.
2. Tantillo, A., Room-Temp Superconductors Could Be Possible, Brookhaven National Laboratory, September 29, 2016, www.phys.org.
3. Stone, M., "'Holy Grail' of Superconductors Could Revolutionize Electronics, August 18, 2015, www.gizmodo.com.
4. Nicodemo, A., "What if Superconductors Could Work at Room Temperature?" April 4, 2018, news @Northeastern University.
5. Likharev, K. K., *Dynamics of Josephson Junctions and Circuits*, Third printing, the Netherlands: Gordon and Breach Publishers, 1996.

6. Zagoskin, A. M., A. Chipouline, E. Illichev, et al., "Toroidal Qubits: Naturally-Decoupled Quiet Artificial Atoms, Arxiv: 1406.7678v2, 2014.
7. Chiu, C. K., and S. Das Sarma, "Fractional Josephson Effect with and without Majorana Zero Modes, *arXiv*: 1806.02224v1, 2018.
8. Tanaka, Y., H. Yamamori, T. Yanagisawa, et al., "Experimental Formation of a Fractional Vortex in a Superconducting Bi-layer, *Physica C: Superconductivity and its Applications*, **548**, 44–49, 2018.
9. Kienzle, U., J. M. Meckbach, K. Buckenmaier, et al., "Spectroscopy of a Fractional Josephson Vortex Molecule, *arXiv*: 1110.4229v1, 2011.
10. Liu, X., X. Li, D-L. Deng, et al., Majorana Spintronics, *arXiv*: 1602.08093v2, 2016.
11. Li, Y-H., J. Song, J. Liu, et al., Doubled Shapiro Steps in a Topological Josephson Junction," *arXiv:* 1709.08355v2, 2018.
12. Wiedenmann, V. J., "Induced Topological Superconductivity in HgTe Based Nanostructures, (Dissertation), Wurzburg, Eingereicht am, 2017.
13. K. Le Calvez, "Signatures of a 4 pi Periodic Abdreev Bound State in Topological Josephson Junctions, Universite-Grenoble Alpes, (Dissertation), HAL archiyes-ouvertes, fr., 2017.
14. Goldobin, E., D. Koelle, R. Kleiner, "Ground States of One and Two Fractional Vortices in Long Josephson 0–κ Junctions," *Physical Review B*, **70**, 1745192004.
15. Kim, H-T., "High-T_c Mechanism through Analysis of Diverging Effective Mass for $YaBa_2Cu_3O_{6+x}$ and Pairing Symmetry in Cuprate Superconductors," *arXiv*: 1710.07754, 2017.
16. Kular, J. K., S. Basu, and R. I. Sharma, "The Extracellular Matrix: Structure, Composition, Age-Related Differences, Tools for Analysis and Applications for Tissue Engineering, *Journal of Tissue Engineering*, **5**, 1–17, 2014.
17. Watanabe-Nakayama, T., M. Itami, N. Kodera, et al., High-Speed Atomic Force Microscopy Reveals Strongly Polarized Movement of Clostridial Collagenase along Collagen Fibrils, *Scientific Reports*, **6**, 28975, 2016.
18. Najafi, M. F., S. Zahri, F. Vahedi, et al., "Which Form of Collagen is Suitable for Nerve Cell Culture?" *Neural Regeneration Research*, **8**(23), 2165–2170, 2013.
19. McKleroy, W., T-H Lee, and K. Atahai, "Always Cleave Up Mess: Targeting Collagen Degradation to Treat Tissue Fibrosis," *American Journal of* Physiology-*Lung Cellular Molecular Physiology*, **304**, L709–L721, 2013.
20. Takai, E., K. D. Costa, A. Shaheen, et al., "Osteoblast Elastic Modulus Measure by Atomic Force Microscopy is Substrate Dependent," *Annals of Biomedical Engineering*, **33**(7), 963–971, 2005.
21. San, B. H., Y. Li, E. B. Tarbet, and S. M. Yu, "Nanoparticle Assembly and Gelatin Binding Mediated by Triple Helical Collagen Mimetic Peptide, *ACS Applied Materials and Interfaces*, **8**, 19907–19915, 2016.
22. Seo, E., K. W. Seo, J-E Gil, Y-R Ha, et al., "Biophysiochemical Properties of Endothelial Cells Cultured on Bio-Inspired Collagen Films," *BioMed Central BMC Biotechnology*, **14**, 61, 2014.
23. Wolf, E., G. Arnold, M. Gurvitch, and J. Zasadzinski, Editors, "Preface," *Josephson Junctions, History, Devices, and Applications*, Pan Stanford Publishing Pte, Singapore, Ltd, 2017.
24. Frolov, S., "Quantum Transport," www.sergeyfrolov.wordpress.com/teaching.
25. Kivelson, S., Superconductivity and Quantum Mechanics at Micro-Scale, Stanford University, www.youtube.com/watch?v=yx666k2xH8E.
26. Grosfeld, E. and A. Stern, "Observing Majorana Bound States of Josephson Vortices in Topological Superconductors," Proceedings of the National Academy of Sciences of the USA (PNAS), **108**(29), 11810–11814, 2011.
27. Liu, X., X. Li, D-L Deng, X-J Liu, and S. Das Sarma, "Majorana Spintronics," *arXiv*, 1602.08093v2, 2016.
28. Li, J., T. Neupert, B. A. Bernevig, and A. Yazdani, "Manipulating Majorana Zero Modes on Atomic Rings with an External Magnetic Field," *Nature Communications*, **7**, 10395, 2016.
29. Chen,E. T., J. T. Thornton, P.T. Kissinger, and S-H. Duh, "Nanobiomimetic Structured Superconductive/Memristive Organo-Metal Devices at Room Temperature Serve as Amperometric Sensors for sub fg/mL Collagen-1," *Biotech, Biomaterials and Biomedical: TechConnect Briefs*, 107–110, 2018.
30. (a) Chen, E. T.,J. T. Thornton, P. T. Kissinger, and S-H. Duh, "Discovering of Collagen-1's Role in Producing Superconducting Current in Nanobiomimetic Superlattice Structured Organometallic Devices at Room Temperature Enabled Direct Quantitation of Sub pg/mL Collagen-1," *Informatics, Electronics and Microsystems: TechConnect Briefs*, 43–46, 2018.

(b) E. T. Chen, J. T. Thornton, P. T. Kissinger,S-H. Duh, "Nanostructure Protein Toroidal Array Devices Promoted Room Temperature Superconductivity and Direct Sensing of Collagen-1," *Sensors and Transducers,* **228**(12), 48–62, 2018.

31. http://en.wikipedia.org/fractional_vortices.
32. Le Calvez, K., *Signature of a 4 pi Periodic Andreev Bound State in Topological Josephson Junction,* Dissertation, Universite Grenoble Press, 2017.
33. Ternes, M., M. Pivetta, F. Patthey, and W-D Schneider, "Creation, Electronic Properties, Disorder, and Melting of Two-Dimensional Surface-State-Mediated Adatom, *Progress in Surface Science,* **85**, 1–27, 2010.
34. Parks, W. C., and R. P. Mecham, *Matrix Metalloproteinases*, Academic Press, San Diego, 1998.
35. Siemianowicz, K., W. Likus, and J. Markowski, Metalloproteinases in Brain Tumors, 2015. http://dx.doi.org/10.5772/58964.
36. Schnaeker, E-M, R. Ossig, T. Ludwig, et al., "Microtubule-Dependent Matrix Metalloproteinases-2/ matrix metalloproteinases-9 exocytosis: Prerequisite in Human Melanoma Cell Invasion, *Cancer Research,* **64**, 8924–8931, 2004.
37. Rydlova, M., L. Holubec Jr., M. Ludvikova Jr, and D. Kalfert, "Biological Activity and Clinical Implications of the Matrix Metalloproteinases," *Anticancer Research* **28**, 1389–1398, 2008.
38. Riedel, F., K. Gotte, J. Schuwalb, and K. Hormann, "Serum Levels of Matrix Metalloproteinase-2 and -9 in Patients with Head and Neck Squamous Cell Carcinoma, *Anticancer Research,* **20**(5A), 3045–3049, 2000.
39. Dahiya, K., and R. Dhankhar, "Updated Overview of Current Biomarkers in Head and Neck Carcinoma, *World Journal of Methodology*, **6**(1), 77–86, 2016.
40. Ghargozlian, S., K. Svennevig, H-J. Bangstad, et al., "Matrix Metalloproteinases in Subjects with Type 1 Diabetes, *BMC Clinical Pathology*, **9**(7), 2009. doi: 10.1186/1472-6890-9-7.
41. Hermandez-Guillamon, M., S. Mawhirt, S. Blais, et al., Sequential Beta-amyloidal Degradation by the Matrix Metalloproteinases MMP-2 and MMP-9, *Journal of Biological Chemistry*, **290**(24), 15078–15091, 2015.
42. Wang, R., G. Q., Zeng, R. Z. Tong, D. Zhou, and Z. Hong, "Serum Matrix Metalloproteinase-2: A Potential Biomarker for Diagnosis of Epilepsy, *Epilepsy Research,* **122**, 114–119, 2016.
43. Yang, G., L. Li, R. K. Rana, and J-J. Zhu, "Assembled Gold Nanoparticles on Nitrogen-doped Graphene for Ultrasensitive Electrochemical Detection of Matrix Metalloproteinase-2, *Carbon 61*, 367–366, 2013.
44. Conant, K., and P. E. Gottschall, *Matrix Metalloproteinases in the Central Nerve System,* London: Imperial College Press, 2005.
45. Chen, E. T., J. T. Thornton, S-H. Duh, and P. T. Kissinger, "Observation of Fermi Arc Surface States Induced by Organic Memristive/Memcapacitive Devices with a Double-Helical Polarized Single-Wall Nanotube Membrane for Direct Chelating with Matrix Matelloproteinase-2," *Sensors and Transducers,* **214**(7), 69–84, 2017.
46. E. T. Chen, J. T. Thornton, S-H. Duh, and P. T. Kissinger, "Organic Nanobiomimetic Memristive/ Memcapacitive Devices Ultrasensitive Direct Detect Matrix Matelloproteinase-2 in Human Serum," *Biotech, Biomaterials and Biomedical*: *TechConnect Briefs*, 271–274, 2017.
47. Takai, E., K. D. Costa, A. Shaheen, et al., "Osteoblast Elastic Modulus Measured by Atomic Force Microscopy Is Substrate Dependent, *Annals of Biomedical Engineering,* **33**(7), 963–971, 2005.
48. San, B. H., Y. Li, E. B. Tarbet, and S. M. Yu, "Nanoparticle Assembly and Gelatin Binding Mediated by Triple Helical Collagen Mimetic Peptide," *ACS Applied Materials and Interfaces*, **8**, 19907–19915, 2016.
49. Seo, E., K. W. Seo, J-E. Gil, Y-R. Ha, et al., "Biophysiochemical Properties of Endothelial Cells Cultured on Bio-Inspired Collagen Films," *BioMed Central Biotechnology,* 2014, doi: 10.1186/1472-6750-14-61.
50. Chen, E. T., and H. L. Pardue, "Analytical Applications of Catalytic Properties of Modified-Cydodextrins," *Analytical Chemistry*, 65(19), 2583–2587, 1993.
51. Cosmic, R., H. Ikegami, A. Lin, et al., "Circuit QED-based Measurement of Vortex Lattice Order in a Josephson Junction Array," arXiv: 1803.04113v1 March 12, 2018.
52. Pickett, M. D., G. Medeiros-Ribeiro, and R. S. Williams, "A Scalable Neuristor Built with Mott Memristors, *Nature Materials*, **12**, 114–117, 2013.
53. Davari, S., S. A. Talaei, H. Alaei, and M. Salami, "Probiotics Treatment Improves Diabetes Induced Impairment of Synaptic Activity and Cognitive Function: Behavioral and Electrophysiological Proofs for Microbiom-gut-brain axis, *Neuroscience*, **240**, 287–296, 2013.

54. Drung, D., and J. Beyer, "Preface," and chap. 8, "Applications in Superconducting Quantum Interference Devices SQUIDs," in E. Wolf, G. Arnold, M. Gurvitch, and J. Zasadzinski, Editors, *Josephson Junctions: History, Devices, and Applications,* Pan Stanford Publishing Pte, Singapore, Ltd, 2017.
55. Askerzade, I., A. Bonzby, and M. Canturk, chap. 2.3, "Superconducting Samples," in *Modern Aspects of Josephson Dynamics and Superconductivity Electronics*, Springer, 2017.
56. https://Physics.nist.gov/cgi-bin/cuu/value/kjos. The NIST reference on Constants, Units and Uncertainty.
57. C. A. Hamilton, C. Burroughs, and K. Chieh, "Operation of NIST Josephson Array Voltage Standards," *Journal of Research of the NIST*, 95(3), 219–235, 1990.
58. Loder, F., A. P. Kampf, T. Kopp, et. al., "Magnetic Flux Periodicity of h/e in Superconducting Loops, *Nature Physics,* **4**, 1.12–115, 2008.
59. http://en.wikipedia.org/fractional_vortices.
60. Grosfeld, E., and A. Stern, "Observing Majorana Bound States of Josephson Vortices in Topological Superconductors," *PNAS*, **108**(29), 11810–11814, 2011.
61. Cosmic, R., H. Ikegami, A. Lin, and K. Inomata, "Circuit QED-based Measurement of Vortex Lattice Order in a Josephson Junction Array," *Physical Review B*, 98(6), 060501–060505, 2018.
62. Loder, F., A. P. Kampf, T. Kopp, et al., "Magnetic Flux Periodicity of h/e in Superconducting Loops, *Nature Physics,***4**, 112–115, 2008.
63. Chen, E. T., J. T. Thornton, P. T. Kissinger, and S-H. Duh, "Utilization of the Flexible Fractional Josephson Toroidal Arrays for Sensing, Memory Storage and Quantum Computing," *Sensors and Transducers*, **228**(12), 30–47, 2018.
64. Clarke, J., and A. I. Braginski, *The SQUID Handbook, Volume 1 and 2,* Weinheim, Germany, Wiley-VCH, 2004.
65. Back, B., W. H., Rippard, M. R. Pufall, et al., Spin-Transfer Torque Switch in Nanopillar Superconducting-Magnetic Hybrid Josephson Junction, *Physical Review Applied,* **3**, 011001, 2015.
66. Likharev, K. K., *Dynamics of Josephson Junctions and Circuits*, the Netherlands, Gordon and Breach Publishers, **1,** 30–47, 1996.
67. Zagoskin, A. M., A.Chipouline, E., Illichev, et al., "Toroidal Qubits: Naturally-decoupled Quiet Artificial atoms, *Scientific Reports*, **5**, 16934; Doi: 101038/srep16934, 2015.
68. Di Ventra, M. and Y. V. Pershin, "On the Physical Properties of Memristive, Memcapacitive, and Meminductive Systems," *Nanotechnology* 24, 255201, 2013.

4 Flexible Supercapacitors with Vertically Aligned Carbon Nanotubes

Runzhi Zhang, Anthony Palumbo, Jian Xu, and Eui-Hyeok Yang

Department of Mechanical Engineering, Stevens Institute of Technology, Hoboken, New Jersey, USA

CONTENTS

4.1 INTRODUCTION

Supercapacitors attract considerable attention in the research and development toward applications requiring many rapid charge–discharge cycles. Distinct advantages of supercapacitors compared with alternative electrochemical energy storage–conversion devices such as batteries and fuel cells include its high-power density, rapid charge–discharge, and long cycle life (1,2). Ongoing efforts to enhance the performance parameters of supercapacitors and adapt them to varying applications and environments expand their usage in portable electronics, hybrid electric vehicles, and industrial energy systems applications (3,4).

Various types of supercapacitors are realized, including electric double-layer capacitors (EDLCs), pseudocapacitors, or a combination of the two, known as hybrid supercapacitors (5). The working mechanism of EDLCs is governed by the collection of electrostatic charge at the electrode–electrolyte interface, enabled by high surface area carbon-based materials. EDLCs are characterized by a non-Faradaic current, which is reversible for millions of cycles due to the lack of chemical reactions involved. Unlike EDLCs, pseudocapacitors are characterized by a chemical reaction at the electrode; storage of electrical charge in pseudocapacitors operates faradaically via electron transfer between electrode and electrolyte, by mechanisms of electrosorption, reduction–oxidation reactions (redox reaction), or intercalation processes (6–8). Conjugated polymers, such as polypyrrole (PPy), polyaniline (PANI), polythiophene (PTh), and poly(3,4-ethylenedioxythiphene), are commonly employed electrode materials in pseudocapacitors due to their high cycling stability and tunable electronic conductivity. Initiation of the redox reaction at the interface thus permits the capacitance to be stored as accumulated charge in the bulk material (9). These pseudocapacitors exhibit a large capacitance (on the order of 10–100 times higher than EDLCs). Hybrid supercapacitors combine an electrode with a high level of pseudocapacitance with an electrode with high double-layer capacitance to provide high specific energy and high specific power, respectively. The combination of the two types of working mechanisms in one device allows a higher specific capacitance and higher specific energy, when compared with symmetrical EDLCs (10). Since their relatively recent commercial emergence, these supercapacitors now replace or complement pre-existing batteries in widespread applications, such as uninterruptible power supplies and load-leveling applications (11–13). Many more applications remain to be realized on a commercial level, and research focus persists on further developing supercapacitors toward diverse environments and greater performances, including wearable and portable electronic devices, such as flexible displays, smart watches and clothing, electronic paper, and wearable health monitoring devices (14–17).

Combining supercapacitor technology with mechanical flexibility enables multi-axial stretching, bending, or folding toward new applications and cost-effective large-scale manufacturing methods, such as roll-to-roll manufacturing. To achieve flexible supercapacitors, the electrodes require high flexibility in addition to high conductivity properties, and various fabrication strategies are pursued, including free-standing films of active materials or supporting active materials on flexible substrates (18). Selection of materials and their configurations are crucial in the development of flexible supercapacitors, and various strategies and combinations of materials are reported in the literature including microstructures and macroscopic patterns (19), carbon nanomaterials (5), two-dimensional (2D) nanomaterials (20), graphene-based materials (21,22), conjugated polymers (23), nanohybrid materials of conjugated polymers and carbon (24), electrode materials (25), paper substrates (26), and flexible substrates (18).

Carbon nanotubes (CNTs) in various configurations, most notably VACNTs, are promising electrode material candidates in flexible supercapacitor designs due to their advantages of high mechanical flexibility, electrical conductivity, electrochemical stability, and desirable porosity (5). CNT porosity is characterized by readily accessible mesopores, which increase the effective surface area at the interface of the electrolyte and electrode, allowing greater access for ions to form an electrical double layer (27). In addition, individual VACNTs are in close proximity to neighboring CNTs, which altogether contribute to a combined charge capacity and increase the energy density of supercapacitors (28–32). To further increase the capacitance and energy density, CNTs are combined with other materials such as conjugated polymers to form hybrid electrodes (33–35). While CNTs provide superior electron–conduction pathways, conjugated polymers enable redox reactions, which store electric charge in the bulk material for greater performance. The combination of conjugated polymers with CNTs is a promising emergence as a hybrid electrode in flexible supercapacitors, which requires further design and fabrication considerations toward widespread flexible supercapacitor commercial applications.

In this chapter, we present a facile fabrication of flexible supercapacitors with an electrode composed of either VACNT or PPy(DBS)-VACNT conjugated polymer on a PDMS substrate. In the first section of this chapter, flexible electrodes composed of PPy and/or VACNTs and their properties toward usage as pseudocapacitors are briefly presented. In the second section of this chapter, the electrochemical properties of VACNT-based supercapacitor and PPy(DBS)-VACNT-based supercapacitor are presented. Then, in the third section of this chapter, the mechanical flexibility of those supercapacitors is demonstrated. The developed fabrication process enables a large range of strains (i.e., stretching and bending) with stable performance for the flexible supercapacitor.

4.2 FLEXIBLE ELECTRODES FOR SUPERCAPACITORS

Flexible supercapacitors need to be comparable or better than conventional capacitors in performance with the expectation to provide high flexibility. Development of flexible electronics is sparked by a combination of capabilities (i.e., recent advances in material science and technology) and the societal high demand for portable, thin, and flexible devices. Developing applications poised to revolutionize the consumer market include wearable sensors (36), smart clothing (16), flexible displays (37), and electronic paper (38). Despite these advances, research efforts are warranted to provide cost-effective supercapacitors, which are durable, flexible, and result in high-capacitance performance indicators. Electrode materials and their arrangements are one of the most critical considerations for designing flexible supercapacitors. Selection of electrode materials is likewise directed by achieving stable electrochemical properties under various strains (e.g., stretching and bending). The most extensively studied classes of electrode materials include transition metal oxides (39–42), carbon-based high surface area materials (e.g., CNTs, carbon nanofibers, graphene, and graphene oxide) (43–47), and conjugated polymers (e.g., PPy, PANI, and polythiophiene) (48–52). Conjugated polymer, PPy(DBS), and CNTs are used as composite electrode materials in the pseudocapacitor described herein, and each material is discussed in this section concerning its state-of-the-art usage toward supercapacitors.

4.2.1 CNT-Based Flexible Electrodes

CNTs are promising candidates for electrode materials in flexible supercapacitors owing to their high mechanical flexibility, electrical conductivity, electrochemical stability, and porosity with high effective surface area (5). In addition, the geometric arrangement of CNTs affects the electrical properties; vertical alignment of CNTs results in increased electrical conductivity (53). Relating to CNT geometry and size scale, CNTs can be modeled as mesoporous structures that increase the effective surface area at the electrolyte and electrode interface where ion exchange occurs (27). Increased effective surface area is important in the supercapacitor design, because it relates to increased electroactive sites, resulting in higher capacitance. In combination, individual VACNTs are in close proximity to one another to exchange charge laterally, contributing to a combined charge capacity and increased energy density (28–31).

4.2.2 PPy-Based Flexible Electrodes

Conjugated polymers, particularly PPy, are commonly used materials in electrochemical supercapacitors due to their low cost, high redox active capacitance, high conductivity, and high intrinsic flexibility (54–56). Organic conjugated polymers conduct charge through conjugated bonds along the polymer chain and result in reversible Faradaic redox reactions (2,57,58). In comparison to PANI and PTh, PPy has superior flexibility and higher conductivity. The pseudocapacitor performance of doped PPy results in relatively high specific capacitance due to its intrinsic redox reactions (i.e., Faradaic charge storage mechanism). PPy films are used as materials in pseudocapacitor electrodes

(59–62), as well as in composites (i.e., in combination with other materials) (63–65). The morphology of PPy is a contributing factor to the electrochemical performance as it relates to high surface area and increased electroactive sites.

4.2.3 PPy Combined with CNT as Electrode

Conjugated polymers used in pseudocapacitors, such as PPy(DBS), have intrinsic redox reactions that result in a Faradaic charge storage mechanism. Specific capacitances of EDLCs, as in the case of pure CNTs, depend on a combination of specific surface area and charge storage capacity of the electroactive electrode materials. Due to rapid ion absorption–desorption at the electrode–electrolyte interface, EDLCs demonstrate high-power capability in addition to high stability. In contrast, pseudocapacitors are demonstrated with the redox reaction (i.e., Faradaic charge storage), exhibiting up to 100 times higher specific capacitance than EDLCs. Disadvantages of conjugated polymers such as PPy include their low mechanical stability, resulting in poor cycle lifetimes. In order to combine the advantages of both electrode materials, composite electrodes consisting of PPy and CNT are common in current state of the art. To overcome the relatively poor cycle lifetime of PPy-based pseudocapacitors and relatively limited specific capacitance of CNT-based supercapacitors, combinations of PPy with CNTs in various composites are utilized, taking advantage of CNTs' intrinsic mechanical support to the conjugated polymer (resulting in higher cycle lifetimes) and PPy's Faradaic charge storage mechanism with increased specific capacitance (66,67). Pseudocapacitor electrodes combining conjugated polymers with CNTs commonly incorporate CNTs via dip-coating (51,64), vacuum filtration (68), precipitation from aqueous solutions (69), and doping via polymerization (66,67). In contrast, VACNTs presented herein are directly transferred onto partially cured PDMS, after which PPy is electropolymerized with DBS^- molecules (i.e., surfactant that reduces interfacial tension), facilitating the coating of the polymer conformably to individual CNTs and resulting in a large effective surface area and higher areal capacitance.

4.2.4 PPy(DBS)-VACNT/PDMS Electrode

Here, we show one example of the fabrication of flexible electrodes comprised of PPy(DBS)-VACNT hybrid embedded into the PDMS structure.

4.2.5 Carbon Nanotube Synthesis

VACNTs are grown via atmospheric-pressure chemical vapor deposition from the catalyst chip (i.e., Al [5 nm] and Fe [3 nm] films deposited via physical vapor deposition on a Si chip). The catalyst chip is heated to 750°C under Ar gas flow of 500 sccm, at which point H_2 (60 sccm) and C_2H_4 (100 sccm) are flowed for 15 min at 750°C for CNT growth (Figures 4.1a and b). Cooling proceeds to room temperature while maintaining Ar flow.

4.2.6 PDMS Synthesis and CNT Transfer

PDMS base and curing agent (Sylgard 184 Silicone Elastomer, Dow Corning) are mixed with a ratio of 10:1 and degassed under vacuum in a desiccator for bubble removal. To induce partial curing, the liquid PDMS is heated at 65°C for 30 min on a hotplate, until the PDMS hardens to a sufficient mechanical stiffness that a tweezer can lift a "strand" from bulk PDMS up to several millimeters without separation or wetting. At this point, the PDMS is considered partially cured, and the VACNT on Si is transferred directly onto the PDMS, which penetrates between individual carbon nanotubes and wets individual VACNT tips; the immersed VACNT tips then become mechanically anchored into the PDMS, as it is fully cured for an additional 15 min at 65°C. This method with partially cured PDMS results in a strong adhesion between VACNT and PDMS. Si is removed following the

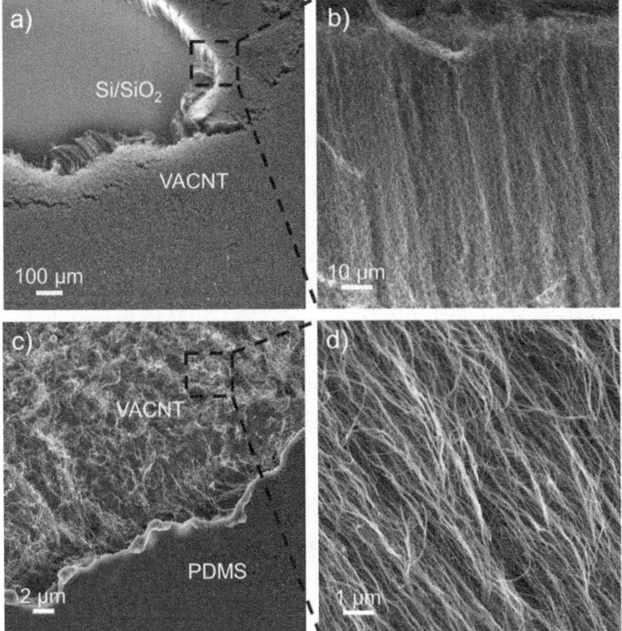

FIGURE 4.1 SEM images of (a, b) VACNTs grown on Si/SiO$_2$ substrate, and (c, d) VACNTs transferred to PDMS surface as electrode of flexible supercapacitor.

full curing of PDMS, and PDMS/VACNT (Figures 4.1c and d) remains due to its stronger formed adhesion than that of Si/VACNT.

4.2.7 Polymerization of PPy(DBS) on VACNT/PDMS

A thin film of PPy(DBS) is electropolymerized on the surfaces of exposed VACNTs of the VACNT/PDMS (Figure 4.2a–c). VACNT/PDMS serves as the working electrode in a solution consisting of 1 mL of pyrrole (reagent grade, 98%, Sigma-Aldrich, St. Louis, MO) and 150 mL of 0.1 M NaDBS (technical grade, Sigma-Aldrich). The reference electrode used is a saturated calomel electrode (SCE), and a 5 × 5 cm Si substrate coated with Cr (5 nm) and Au (3 nm) serves as the counter electrode. A constant potential (0.8 V vs. SCE) is applied until reaching a surface charge density of 1000 mC/cm^2. As evidenced by Figure 4.2c, the PPy(DBS) films are uniformly coated on individual

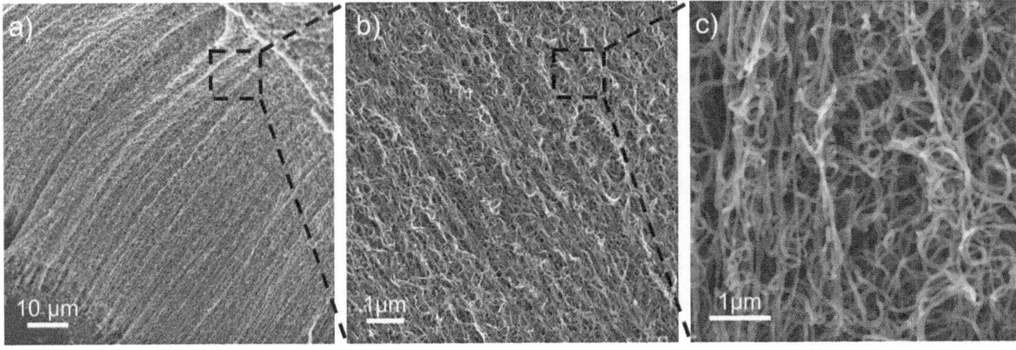

FIGURE 4.2 SEM images of PPy(DBS)-VACNT/PDMS as electrode of hybrid flexible supercapacitor.

VACNTs, thanks to the introduction of DBS^- into the pyrrole monolayer solution, which lowers the interfacial tension, allowing conformal coating of the PPy(DBS) films on individual VACNTs, facilitating a large effective surface area, and enabling higher capacitance (discussed later).

4.2.8 Fabrication of Flexible Supercapacitor

To fabricate all solid-state flexible supercapacitors using PPy(DBS)-VACNT/PDMS electrodes, two electrodes are sandwiched between a solid electrolyte gel of poly(vinyl alcohol) (PVA)–KOH. PVA-KOH is synthesized by adding 1 g of PVA powder and 1 g of KOH powder to 10 mL of deionized water as the solvent. Mixing is performed for 1 h at 60°C until the mixture is formed. PVA-KOH gel is applied onto the surface of each PPy(DBS)-VACNT/PDMS electrode to function as both the electrolyte source and separator in the supercapacitor. VACNT/PDMS electrodes that do not undergo electropolymerization of PPy(DBS) are fabricated by the same method for comparison of performance with and without PPy(DBS). Each electrode of the supercapacitor is connected by a thin Cu wire to a potentiostat (263A, Princeton Applied Research, Oak Ridge, TN) for electrical characterization.

4.3 COMPARISON BETWEEN CNT-BASED FLEXIBLE SUPERCAPACITORS AND PPy(DBS)-BASED FLEXIBLE SUPERCAPACITORS

The morphology of the electrodes and their constituents are characterized via scanning electron microscopy (SEM). Electrical characterizations are carried out with a potentiostat (263A, Princeton Applied Research), including cyclic voltammetry (CV) with a potential range of 0–0.5 V at scan rates ranging from 50 to 1000 mV/s. The electrode's areal capacitance is calculated as capacitance per area in unit of F/cm^2. Average capacitance is normalized per area and estimated as calculated elsewhere (70,71) by the following equation: $C = \frac{\int_0^V IdV}{V \times \Delta V \times A}$, where A is the electrode area, I is the current, V is the applied voltage, and ΔV is the scanning rate. The energy density, E, and power density, P, are calculated as reported elsewhere (70,71) with the following equations: $E = \frac{1}{3600 \times A \times \Delta V} \int_V^0 I \times VdV$ and $P = \frac{\int_V^0 I \times VdV}{V \times A}$, respectively, where V is the initial discharge voltage, and I is the discharge current.

4.3.1 All-Solid-State Flexible EDLCs of VACNT/PDMS Based on PVA-KOH Electrolyte Gel

The all-solid-state flexible supercapacitor of VACNT/PDMS with PVA-KOH gel electrolyte between two face-to-face electrodes is characterized via CV measurements (Figure 4.3a) at scan rates from 300 to 1000 mV/s. At 300 mV/s, the capacitance attained is 1.6 mF/cm^2 (Figure 4.3b), indicating

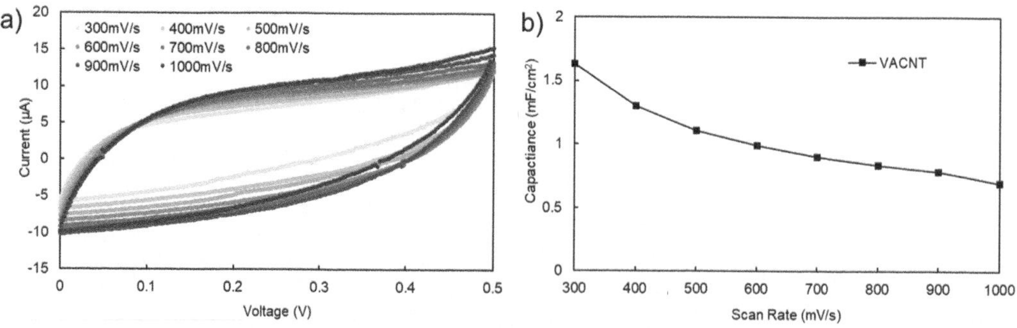

FIGURE 4.3 (a) CV curves and (b) calculated capacitance of VACNT/PDMS supercapacitor at scan rates ranging from 300 to 1000 mV/s.

good electrochemical properties of VACNT/PDMS. The measured maximum voltage and current are 1 V and ~30 μA, respectively, with PVA-KOH gel electrolyte. The measured power density is between 5.5 and 6.0×10^{-5} W/cm^2, and the measured energy density is between 4.3 and 5.1×10^{-8} Wh/cm^2.

4.3.2 All-Solid-State Flexible Supercapacitors Based on PPy(DBS)-VACNT/PDMS

The all-solid-state flexible supercapacitors based on PPy(DBS)-VACNT hybrids are realized using CNTs embedded into PDMS structure and coated with PPy(DBS) as a composite electrode toward enhanced pseudocapacitor performance. The electrochemical behavior of PPy(DBS)-VACNT/PDMS is characterized via CV (Figure 4.4a) with a two-electrode setup at scan rates ranging from 300 to 1000 mV/s. At 300 mV/s, the areal capacitance attained is 3.5 mF/cm^2 (Figure 4.4b). This capacitance is 113% times greater than the same structure without PPy(DBS), which results in 1.6 mF/cm^2. The measured maximum voltage and current are 1 V and ~30 μA, respectively, with PVA-KOH gel electrolyte. The measured power density is between 7.9 and 9.3×10^{-4} W/cm^2, and the energy density is between 2.6 and 3.1×10^{-7} Wh/cm^2, which are both orders of magnitude greater than the EDLC without PPy(DBS). These energy densities and power densities are significantly higher than flexible supercapacitors with electrodes composed of graphene microfibers (72) and CNT fiber (73), which have power densities of $6.0–100 \times 10^{-6}$ W/cm^2 and 4.9×10^{-4} W/cm^2, respectively, and energy densities of $0.4–1.7 \times 10^{-7}$ Wh/cm^2 and $0.8–2.3 \times 10^{-7}$ Wh/cm^2, respectively. The PPy(DBS)-VACNT/PDMS supercapacitor resulted in energy density and power density values that are comparable to supercapacitors based on electrodes of composites with similar materials, including PPy/graphene oxide nanocomposite electrodes, which exhibited power density of 9.5×10^{-4} W/cm^2 and energy density of 1.29×10^{-5} Wh/cm^2 (74). The supercapacitor based on PPy-MnO_2-CNT electrode yields a power density of 6.7×10^{-4} W/cm^2 and energy density of 3.3×10^{-5} Wh/cm^2 (75). Of similar magnitude, Mn_3O_4/RGO/CNT hybrid film exhibited a power density of 3.9×10^{-4} mW/cm^2 and energy density of 3.2×10^{-5} Wh/cm^2 (76).

Compared with commercially available energy storage devices, such as from Sony Corporation and NEC Corporation, with typical power densities in the range of 10–1500 W/kg, the supercapacitors result in significantly higher power densities (e.g., 3000–40,000 W/kg) than electrolytic capacitors and batteries; furthermore, significantly larger cycles are achieved by supercapacitors (i.e., 500,000–20,000,000) than a typical Li-ion battery (i.e., 500–1000), as well as greater lifetime (10–15 years vs. 5–10 years) and higher efficiency (98% vs. 75–90%) (13,77). The vast majority of commercially available capacitors are rigid and lack flexibility (78), whereas the PPy(DBS)-VACNT/PDMS supercapacitor is highly flexible, enabling flexible electronic applications. Although flexible supercapacitors are not yet commercially available, flexible energy storage devices such as

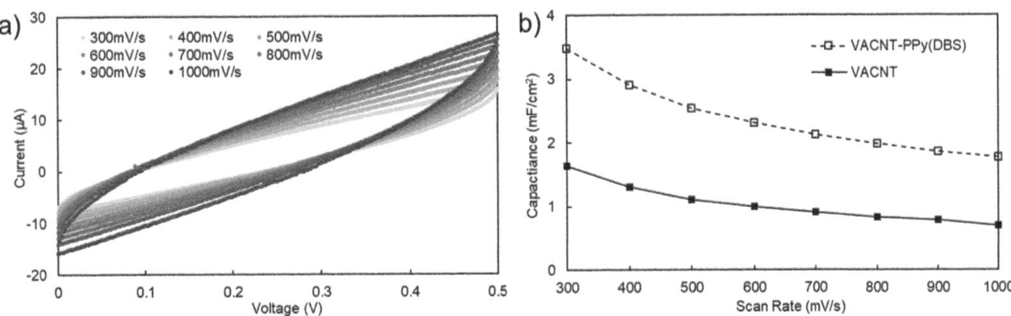

FIGURE 4.4 (a) CV curves and (b) calculated capacitance of PPy(DBS)-VACNT/PDMS hybrid supercapacitor at scan rates ranging from 300 to 1000 mV/s. Results show the coating of PPy(DBS) increased the capacitance of flexible supercapacitor.

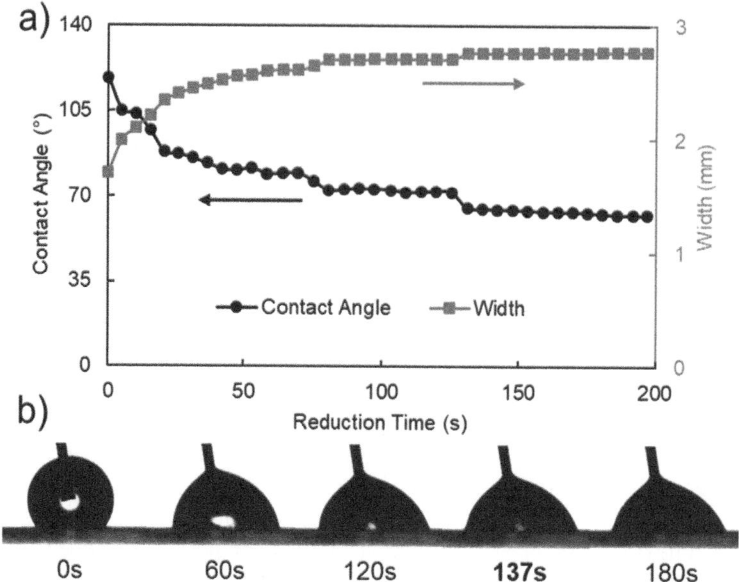

FIGURE 4.5 (a) The change of contact angle and width of water droplet (0.1 M $NaNO_3$) on PPy(DBS)-VACNT/PDMS hybrid supercapacitor electrode during surface reduction. (b) The droplet experienced the spread-out process (characterized by the decrease in droplet contact and increase in droplet width) due to the wettability switch of PPy(DBS) surfaces. After 137 s of surface reduction, the shape of the droplet reached a stable state (also characterized by the droplet contact and width reaching a plateau), and this time is characterized as switch time. The PPy(DBS) surface is coated using 750 mC/cm^2 and under 30% tensile strain.

batteries are emerging. Panasonic Corporation is developing flexible Li-ion batteries with an initial capacity retention of ~80% after 1000 times of charge and discharge cycles, following either 1000 times of bending or twisting. NEC Corporation is developing flexible organic radical batteries that result in capacitance retention of only 75% after 500 charge–discharge cycles. In comparison, the PPy(DBS)-VACNT/PDMS supercapacitor demonstrates a capacity retention of 109% after 10,000 charge–discharge cycles. LG Chem Corporation developed the first flexible cable-type Li-ion battery, which is currently under further development for commercialization.

The wettability switch of PPy(DBS)-VACNT/PDMS surfaces during surface reduction is shown in Figure 4.5. PPy(DBS) undergoes a switch from hydrophobic (oleophilic) state to hydrophilic (oleophobic) state (and vice versa), which is attributed to the re-orientation and release of DBS$^-$ molecules (79–83). For PPy(DBS)-VACNT/PDMS surfaces, the contact angle of a water droplet is decreased by 39% (from 119° to 73°) and the width of the water droplet is increased by 58% (from 1.7 to 2.7 mm) after 100 s of surface reduction, indicating that the surface becomes more hydrophilic. This wettability switch of PPy(DBS) enhances the access of ions (i.e., K^+ from PVA-KOH gel) to the PPy(DBS)-VACNTs/PDMS surface during reduction, further improving the device performance (Figure 4.4b). After 137 s of surface reduction, the droplet contact angle and width reached a plateau (Figure 4.5a), and the shape of water droplet remained unchanged (Figure 4.5b), indicating the completion of surface reduction. The time (137 s) is characterized as the switch time of PPy(DBS)-VACNT/PDMS and is affected by PPy(DBS) coating thicknesses and tensile strain.

The speed of PPy(DBS) wettability switch and capacitance enhancement from PPy(DBS) coating is affected by the thickness of PPy(DBS) and the mechanical loading on the flexible supercapacitor (e.g., stretching and bending). The wettability switch and enhanced capacitance are both proportional to the ion diffusion during PPy(DBS) reduction, and switch time of PPy(DBS)-VACNT/PDMS plays an important role in PPy(DBS)-based supercapacitor performance. As shown in Figure 4.6,

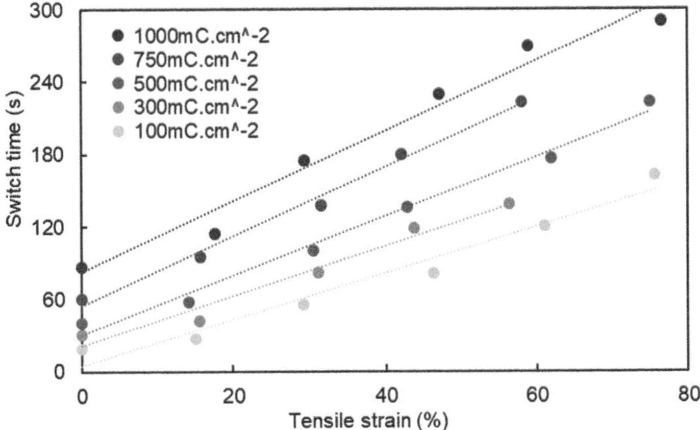

FIGURE 4.6 Switch times of PPy(DBS) surfaces coated on VACNT/PDMS with different thickness and under various tensile strains.

the switch time increases with thicker PPy(DBS) coating (from 100 to 1000 mC/cm^2) under higher tensile strain loading. The detailed mechanism behind the increase in switch time increase and capacitance requires further study. However, thicker PPy(DBS) coatings on VACNT/PDMS result in higher capacitance (79,84), and higher tensile strain results in slightly enhanced capacitance, indicating the positive correlation between switch time and capacitance (longer switch time and higher capacitance).

4.4 STRAIN EFFECT ON THE PERFORMANCE OF SUPERCAPACITORS

To evaluate the mechanical properties such as flexibility and durability, tensile and bending strain measurements are performed in ranges of 0–150% stretching and 0°–180° bending for all-solid-state flexible EDLC (VACNT/PDMS) with solid PVA-KOH gel electrolyte and all-solid-state-state flexible supercapacitor (PPy(DBS)-VACNT/PDMS) with solid PVA-KOH gel electrolyte.

4.4.1 All-Solid-State Flexible EDLCs of VACNT/PDMS Based on PVA-KOH Electrolyte Gel

The all-solid-state flexible EDLCs can be stretched up to 150% to prevent stretching to the VACNT/PDMS electrode's tensile strength limit. At 150% stretching, the areal capacitance is increased by 10% of the capacitance observed before stretching (Figure 4.7a), using a scan rate of

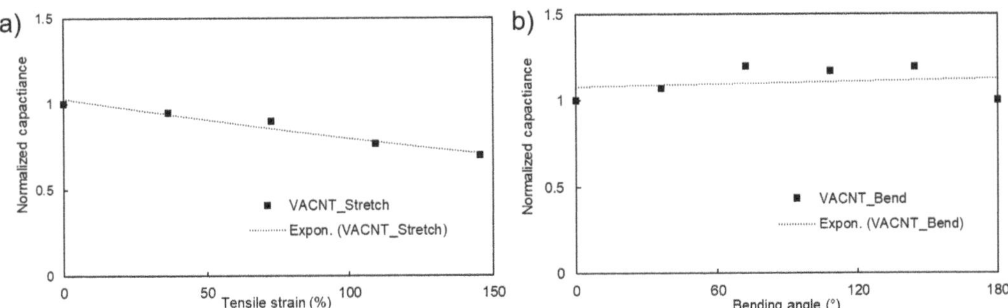

FIGURE 4.7 Normalized capacitance of VACNT/PDMS supercapacitor (a) under tensile strains from 0% to 150% or (b) at bending angles from 0° to 180°.

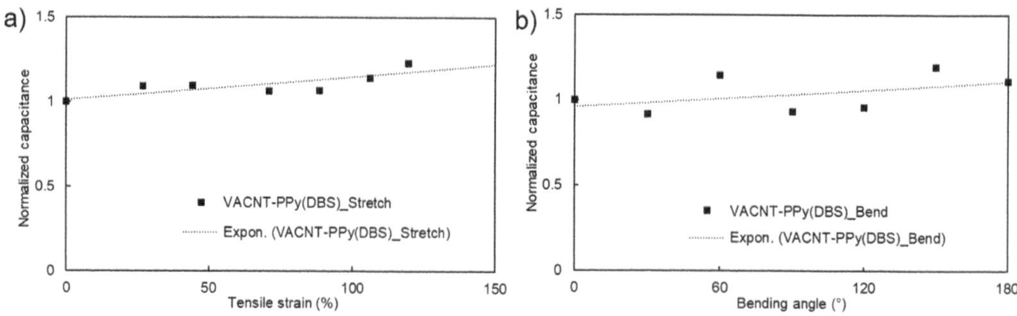

FIGURE 4.8 Normalized capacitance of PPy(DBS)-VACNT/PDMS supercapacitor (a) under various tensile strains from 0% to 150% or (b) at various bending angles from 0° to 180°.

1000 mV/s. In addition, the areal capacitance of the all-solid-state flexible EDLC results in a stable performance without any significant changes when undergoing bending angles between 0° and 180° (Figure 4.7b).

4.4.2 All-Solid-State Flexible Supercapacitors Based on PPy(DBS)-VACNT/PDMS

The supercapacitor experienced stretching from 0% to 150% and bending from 0° to 180°, while maintaining the areal capacitance before stretching (Figure 4.8a) or bending (Figure 4.8b).

4.5 CONCLUSIONS

A facile fabrication of flexible supercapacitors using VACNTs or PPy(DBS)-VACNT on PDMS substrates is been presented. The areal capacitance is characterized under varying types and quantities of strains. Unlike other PPy and CNT composites, the unique fabrication of the pseudocapacitor presented here enables a high level of integrity at various scan rates, while maintaining stable electrochemical properties under various types and amounts of strains. The PPy(DBS)-VACNT/PDMS structure takes advantage of a direct transfer method with partially cured PDMS to form a strong adhesion between CNTs and PDMS, enabling the VACNT/PDMS structure to sustain a vast range of various strains. The PPy(DBS) is electropolymerized directly on the VACNTs to synthesize a thin film polymer layer exposed at the electrolyte–electrode interface for Faradaic charge storage. The PPy is doped with DBS molecules (i.e., surfactants) to lower the interfacial tension, allowing the PPy(DBS) films to be conformally deposited on the surfaces of individual CNTs and resulting in a high specific area and high areal capacitance. At scan rates from 300 to 1000 mV/s, VACNT/PDMS shows electrochemical stability and repeatability with PVA-KOH gel electrolyte. In comparison to the PPy(DBS)-VACNT supercapacitor, the VACNT-based flexible supercapacitor shows 113% higher capacitance and high flexibility with similar capacitance retention under similar strain values. The measured maximum voltage and current are 1 V and ~30 μA, respectively, with PVA-KOH gel electrolyte. The power density is measured between 7.9 and 9.3×10^{-4} W/cm^2, and the energy density is between 2.6 and 3.1×10^{-7} Wh/cm^2, which are both orders of magnitude greater than the EDLC without PPy(DBS). In addition, PPy(DBS)-CNT on PDMS results in a platform that allows an immense range of applied tensile strain (0–150%) and bending angles (0°–180°) with high stability. This combination of VACNT embedded into partially cured PDMS can be further applied to other materials in a similar configuration for promising applications in emerging flexible electronics.

REFERENCES

1. Conway BE. Electrochemical supercapacitors: scientific fundamentals and technological applications. Springer Science & Business Media; NewYork;2013.
2. Burke A. R&D considerations for the performance and application of electrochemical capacitors. Electrochim Acta. 2007; 53(3 SPEC. ISS.): 1083–91.
3. Li X, Wei B. Supercapacitors based on nanostructured carbon. Nano Energy. 2013; 2(2):159–73.
4. Beidaghi M, Gogotsi Y. Capacitive energy storage in micro-scale devices: recent advances in design and fabrication of micro-supercapacitors. Energy Environ Sci. 2014; 7(3):867–84.
5. Chen T, Dai L. Flexible supercapacitors based on carbon nanomaterials. J Mater Chem A. 2014; 2(28):10756–75.
6. Mefford JT, Hardin WG, Dai S, Johnston KP, Stevenson KJ. Anion charge storage through oxygen intercalation in $LaMnO_3$ perovskite pseudocapacitor electrodes. Nat Mater. 2014; 13(7):726–32.
7. Conway BE, Birss V, Wojtowicz J. The role and utilization of pseudocapacitance for energy storage by supercapacitors. J Power Sources. 1997; 66(1–2):1–14.
8. Augustyn V, Simon P, Dunn B. Pseudocapacitive oxide materials for high-rate electrochemical energy storage. Energy Environ Sci. 2014; 7(5):1597–614.
9. Abdelhamid ME, O'Mullane AP, Snook GA. Storing energy in plastics: a review on conducting polymers & their role in electrochemical energy storage. RSC Adv. 2015; 5(15):11611–26.
10. Volfkovich YM, Mikhailin AA, Bograchev DA, Sosenkin VE, Bagotsky VS. Studies of supercapacitor carbon electrodes with high pseudocapacitance. Recent Trend Electrochem Sci Technol. 2012; 159–82.
11. Lahyani A, Venet P, Guermazi A, Troudi A. Battery/supercapacitors combination in uninterruptible power supply (UPS). IEEE Trans Power Electron. 2013; 28(4):1509–22.
12. Casadel D, Grandi G, Rossi C. A supercapacitor based power conditioning system for PQ improvement and UPS. IEEE Trans Power Electron. 2002.
13. Miller JR, Simon P. Electrochemical capacitors for energy management. Science. 2008;321(5889):651–2.
14. Niu S, Wang X, Yi F, Zhou YS, Wang ZL. A universal self-charging system driven by random biomechanical energy for sustainable operation of mobile electronics. Nat Commun. 2015;6:1–8.
15. Park S, Xiong Y, Kim R, Elvikis P, Meitl M, et al. Printed assemblies of inorganic light-emitting diodes for deformable and semitransparent displays. Nat Mater. 2009;325(6936):977–81.
16. Lukowicz P, Kirstein T, Tröster G. Wearable systems for health care applications. Methods Inf Med. 2004;43(3):232–8.
17. Park S, Jayaraman S. Enhancing the quality of life through wearable technology. IEEE Eng Med Biol Mag. 2003;22(3):41–8.
18. Dubal DP, Kim JG, Kim Y, Holze R, Lokhande CD, Kim WB. Supercapacitors based on flexible substrates: an overview. Energy Technol. 2014;2(4):325–41.
19. Dong L, Xu C, Li Y, Huang Z-H, Kang F, Yang Q-H, et al. Flexible electrodes and supercapacitors for wearable energy storage: a review by category. J Mater ChemA. 2016;4(13):4659–85.
20. Peng X, Peng L, Wu C, Xie Y. Two dimensional nanomaterials for flexible supercapacitors. Chem Soc Rev. 2014;43(10):3303–23.
21. Shao Y, El-Kady MF, Wang LJ, Zhang Q, Li Y, Wang H, et al. Graphene-based materials for flexible supercapacitors. Chem Soc Rev. 2015;44(11):3639–65.
22. Ke Q, Wang J. Graphene-based materials for supercapacitor electrodes – a review. J Mater. 2016;2(1): 37–54.
23. Shown I, Ganguly A, Chen L-C, Chen K-H. Conducting polymer-based flexible supercapacitor. Energy Sci Eng. 2015;3(1):2–26.
24. Basnayaka PA, Ram MK. A review of supercapacitor energy storage using nanohybrid conducting polymers and carbon electrode materials. In: Kumar V., Kalia S., Swart H. (eds) Conducting Polymer Hybrids. Springer; 2017. p. 165–92.
25. Yang P, Mai W. Flexible solid-state electrochemical supercapacitors. Nano Energy. 2014;8:274–90.
26. Zhang Y-Z, Wang Y, Cheng T, Lai W-Y, Pang H, Huang W. Flexible supercapacitors based on paper substrates: a new paradigm for low-cost energy storage. Chem Soc Rev. 2015;44(15):5181–99.
27. Lu W, Dai L. Carbon nanotube supercapacitors. INTECH Open Access Publisher; 2010.

28. Hiraoka T, Izadi-Najafabadi A, Yamada T, Futaba DN, Yasuda S, Tanaike O, et al. Compact and light supercapacitor electrodes from a surface-only solid by opened carbon nanotubes with 2200 m^2 g^{-1} surface area. Adv Funct Mater. 2010; 20(3):422–8.
29. Jiang Y, Wang P, Zang X, Yang Y, Kozinda A, Lin L. Uniformly embedded metal oxide nanoparticles in VACNT forests as pseudocapacitor electrodes for enhanced energy storage. Nano Lett. 2013; 13(8): 3524–30.
30. Kim B, Chung H, Kim W. High-performance supercapacitors based on VACNTs and nonaqueous electrolytes. Nanotechnology. 2012; 23(15):155401.
31. Lin Z, Li Z, Moon K, Fang Y, Yao Y, Li L, et al. Robust VACNT – carbon fiber paper hybrid as versatile electrodes for supercapacitors and capacitive deionization. Carbon NY. 2013; 63:547–53.
32. Zhang R, Ding J, Liu C, Yang E-H. Highly stretchable supercapacitors enabled by interwoven CNTs partially embedded in PDMS. ACS Appl Energy Mater. 2018; 1(5):2048–55.
33. Hu L, Chen W, Xie X, Liu N, Yang Y, Wu H, et al. Symmetrical MnO_2 – carbon nanotube – textile nanostructures for wearable pseudocapacitors with high mass loading. ACS Nano. 2011; 5(11):8904–13.
34. Li P, Yang Y, Shi E, Shen Q, Shang Y, Wu S, et al. Core-double-shell, carbon nanotube@ polypyrrole@ MnO_2 sponge as freestanding, compressible supercapacitor electrode. ACS Appl Mater Interfaces. 2014; 6(7):5228–34.
35. Yesi Y, Shown I, Ganguly A, Ngo TT, Chen L-C, Chen K-H. Directly-grown hierarchical carbon nanotube@ polypyrrole core – shell hybrid for high-performance flexible supercapacitors. ChemSusChem. 2016; 9(4):370–8.
36. Patel S, Park H, Bonato P, Chan L, Rodgers M. A review of wearable sensors and systems with application in rehabilitation. J Neuroeng Rehab. 2012; 9(1):21.
37. Khan S, Lorenzelli L, Dahiya RS. Technologies for printing sensors and electronics over large flexible substrates: a review. IEEE Sensors J. 2015; 15(6):3164–85.
38. Heikenfeld J, Drzaic P, Yeo J-S, Koch T. Review paper: a critical review of the present and future prospects for electronic paper. J Soc Inf Disp. 2011; 19(2):129.
39. Li SM, Wang YS, Yang SY, Liu CH, Chang KH, Tien HW, et al. Electrochemical deposition of nanostructured manganese oxide on hierarchically porous graphene–carbon nanotube structure for ultrahigh-performance electrochemical capacitors. J Power Sources. 2013; 225:347–55.
40. Chang KH, Hu CC, Huang CM, Liu YL, Chang CI. Microwave-assisted hydrothermal synthesis of crystalline WO3-WO3·0.5H_2O mixtures for pseudocapacitors of the asymmetric type. J Power Sources. 2011; 196(4):2387–92.
41. Chen YC, Hsu YK, Lin YG, Lin YK, Horng YY, Chen LC, et al. Highly flexible supercapacitors with manganese oxide nanosheet/carbon cloth electrode. Electrochim Acta. 2011; 56(20):7124–30.
42. Srinivasan V, Weidner JW. Studies on the capacitance of nickel oxide films: effect of heating temperature and electrolyte concentration. J Electrochem Soc. 2000; 147(3):880.
43. Chandra V, Kim KS. Highly selective adsorption of Hg^{2+} by a polypyrrole-reduced graphene oxide composite. Chem Commu. 2011; 47(i):3942–4.
44. Xu Y, Hennig I, Freyberg D, James Strudwick A, Georg Schwab M, Weitz T, et al. Inkjet-printed energy storage device using graphene/polyaniline inks. J Power Sources. 2014; 248:483–8.
45. Zhang K, Mao L, Zhang LL, On Chan HS, Zhao XS, Wu J. Surfactant-intercalated, chemically reduced graphene oxide for high performance supercapacitor electrodes. J Mater Chem. 2011; 21(20):7302–7.
46. Lee JY, Liang K, An KH, Lee YH. Nickel oxide/carbon nanotubes nanocomposite for electrochemical capacitance. Synth Met. 2005; 150(2):153–7.
47. Peng C, Jin J, Chen GZ. A comparative study on electrochemical co-deposition and capacitance of composite films of conducting polymers and carbon nanotubes. Electrochim Acta. 2007; 53(2):525–37.
48. Meng C, Liu C, Chen L, Hu C, Fan S. Highly flexible and all-solid-state paperlike polymer supercapacitors. Nano Lett. 2010; 10(10):4025–31.
49. Kim BC, Ko JM, Wallace GG. A novel capacitor material based on Nafion-doped polypyrrole. J Power Sources. 2008; 177(2):665–8.
50. Gao H, Xiao F, Ching CB, Duan H. Flexible all-solid-state asymmetric supercapacitors based on free-standing carbon nanotube/graphene and Mn_3O_4 nanoparticle/graphene paper electrodes. ACS Appl Mater Interfaces. 2012; 4(12):7020–6.

51. Yun TG, Hwang B IL, Kim D, Hyun S, Han SM. Polypyrrole – MnO_2-coated textile-based flexible-stretchable supercapacitor with high electrochemical and mechanical reliability. ACS Appl Mater Interfaces. 2015; 7(17):9228–34.
52. Yuan W, Han G, Xiao Y, Chang Y, Liu C, Li M, et al. Flexible electrochemical capacitors based on polypyrrole/carbon fibers via chemical polymerization of pyrrole vapor. Appl Surf Sci. 2016; 377: 274–82.
53. Jakubinek MB, White MA, Li G, Jayasinghe C, Cho W, Schulz MJ, et al. Thermal and electrical conductivity of tall, VACNT arrays. Carbon NY. 2010; 48(13):3947–52.
54. Snook GA, Kao P, Best AS. Conducting-polymer-based supercapacitor devices and electrodes. J Power Sources. 2011; 196(1):1–12.
55. Mike JF, Lutkenhaus JL. Recent advances in conjugated polymer energy storage. J Polym Sci B Polym Phys. 2013; 51(7):468–80.
56. Ramya R, Sivasubramanian R, Sangaranarayanan MV. Conducting polymers-based electrochemical supercapacitors. Electrochim Acta. 2013; 101:109–29.
57. Ryu KS, Kim MK, Park N-G, Park YJ, Chang SH. Symmetric redox supercapacitor with conducting polyanili. J Power Sources. 2002; 103(2):2002.
58. Rudge A, Raistrick I, Gottesfeld S, Ferraris JP. A study of the electrochemical properties of conducting polymers for application in electrochemical. Electrochim Acta. 1994; 39(2):273–87.
59. Amarnath CA, Chang JH, Kim D, Mane RS, Han S-H, Sohn D. Electrochemical supercapacitor application of electroless surface polymerization of polyaniline nanostructures. Mater Chem Phys. 2009; 113(1):14–7.
60. Zhang H, Zhao Q, Zhou S, Liu N, Wang X, Li J, et al. Aqueous dispersed conducting polyaniline nanofibers: promising high specific capacity electrode materials for supercapacitor. J Power Sources. 2011; 196(23):10484–9.
61. Park JH, Park OO. Hybrid electrochemical capacitors based on polyaniline and activated carbon electrodes. J Power Sources. 2002; 111(1):185–90.
62. Prasad KR, Munichandraiah N. Fabrication and evaluation of 450 F electrochemical redox supercapacitors using inexpensive and high-performance, polyaniline coated, stainless-steel electrodes. J Power Sources. 2002; 112(2):443–51.
63. Hughes M, Shaffer MSP, Renouf AC, Singh C, Chen GZ, Fray DJ, et al. Electrochemical capacitance of nanocomposite films formed by coating aligned arrays of carbon nanotubes with polypyrrole. Adv Mater. 2002; 14(5):382–5.
64. Sun J, Huang Y, Fu C, Wang Z, Huang Y, Zhu M, et al. High-performance stretchable yarn supercapacitor based on PPy@CNTs@urethane elastic fiber core spun yarn. Nano Energy. 2016; 27:230–7.
65. Zhang S, Shao Y, Liu J, Aksay I, Lin Y. Graphene-polypyrrole nanocomposite as a highly efficient and low cost electrically switched ion exchanger for removing ClO_4^- from wastewater. ACS Appl Mater Interfaces. 2011; 3:3633–7.
66. Hughes M, Chen GZ, Shaffer MSP, Fray DJ, Windle AH. Electrochemical capacitance of a nanoporous composite of carbon nanotubes and polypyrrole. Chem Mater. 2002; 14(4):1610–3.
67. An KH, Jeon KK, Heo JK, Lim SC, Bae DJ, Lee YH. High-capacitance supercapacitor using a nanocomposite electrode of single-walled carbon nanotube and polypyrrole. J Electrochem Soc. 2002; 149(8):A1058.
68. Chen Y, Du L, Yang P, Sun P, Yu X, Mai W. Significantly enhanced robustness and electrochemical performance of flexible carbon nanotube-based supercapacitors by electrodepositing polypyrrole. J Power Sources. 2015; 287:68–74.
69. Sivakkumar SR, Ko JM, Kim DY, Kim BC, Wallace GG. Performance evaluation of CNT/polypyrrole/MnO_2 composite electrodes for electrochemical capacitors. Electrochim Acta. 2007; 52(25):7377–85.
70. Chen T, Peng H, Durstock M, Dai L. High-performance transparent and stretchable all-solid supercapacitors based on highly aligned carbon nanotube sheets. Sci Rep. 2014; 4:3612.
71. Lee JA, Shin MK, Kim SH, Cho HU, Spinks GM, Wallace GG, et al. Ultrafast charge and discharge biscrolled yarn supercapacitors for textiles and microdevices. Nat Commun. 2013; 4.
72. Meng Y, Zhao Y, Hu C, Cheng H, Hu Y, Zhang Z, et al. All-graphene core-sheath microfibers for all-solid-state, stretchable fibriform supercapacitors and wearable electronic textiles. Adv Mater. 2013; 25(16):2326–31.

73. Xu P, Gu T, Cao Z, Wei B, Yu J, Li F, et al. Carbon nanotube fiber based stretchable wire-shaped supercapacitors. Adv Energy Mater. 2014; 4(3).
74. Zhou H, Han G, Xiao Y, Chang Y, Zhai HJ. Facile preparation of polypyrrole/graphene oxide nanocomposites with large areal capacitance using electrochemical codeposition for supercapacitors. J Power Sources. 2014; 263:259–67.
75. Liu N, Ma W, Tao J, Zhang X, Su J, Li L, et al. Cable-type supercapacitors of three-dimensional cotton thread based multi-grade nanostructures for wearable energy storage. Adv Mater. 2013; 25(35):4925–31.
76. He J, Yang D, Li H, Cao X, Kang L, He X, et al. Mn3O4/RGO/SWCNT hybrid film for all-solid-state flexible supercapacitor with high energy density. Electrochim Acta. 2018; 283:174–82.
77. Johnson BA, White RE. Characterization of commercially available lithium-ion batteries. J Power Sources. 1998; 70(1):48–54.
78. Yassine M, Fabris D. Performance of commercially available supercapacitors. Energies. 2017;10(9); 1340.
79. Xu J, Palumbo A, Xu W, Yang EH. Effects of electropolymerization parameters of PPy(DBS) surfaces on the droplet flattening behaviors during redox. J Phys Chem B. 2016; 120(39):10381–10386.
80. Xu W, Xu J, Choi C-H, Yang EH. In situ control of underwater-pinning of organic droplets on a surfactant-doped conjugated polymer surface. ACS Appl Mater Interfaces. 2015; 7(46):25608–17.
81. Xu W, Xu J, Li X, Tian Y, Choi C-H, Yang EH. Lateral actuation of an organic droplet on conjugated polymer electrodes via imbalanced interfacial tensions. Soft Matter. 2016;12(33):6902–9.
82. Xu W, Palumbo A, Xu J, Jiang Y, Choi C-H, Yang EH. On-demand capture and release of organic droplets using surfactant-doped polypyrrole surfaces. ACS Appl Mater Interfaces. 2017; 9(27):23119–27.
83. Xu J, Fu S, Xu W, Yang EH. A carbon nanotube-embedded conjugated polymer mesh with controlled oil absorption and surface regeneration via in situ wettability switch. J Colloid Interface Sci. 2018; 532:790–7.
84. Jiang Y, Xu J, Lee J, Du K, Yang EH, Moon MW, et al. Nanotexturing of conjugated polymers via one-step maskless oxygen plasma etching for enhanced tunable wettability. Langmuir. 2017;33(27):6885–94.

5 Eumelanin: Semiconductor, Protonic Conductor, or Mixed Electronic-Ionic Conductor?

Manuel Reali and Clara Santato

Department of Engineering Physics, Polytechnique Montréal, Montréal Québec, Canada

CONTENTS

5.1 INTRODUCTION

We live in an era where electronic devices are essential tools of everyday life. The tremendous success of electronics is the result of decades of research primarily focused on inorganic crystalline materials such as silicon, gallium, germanium, and II-VI, IV-VI, and III-V semiconductors (e.g., CdTe, PbSe, and GaAs). Silicon remains predominant in modern technology due to its excellent physical and electrical properties, abundance in nature, outstanding performance, and compatibility.[1]

Despite the obvious benefits of versatile and multifunctional devices, the rapid growth of electronics' demand is giving rise to manufacturing processes that are unsustainable in terms of energy consumption, electronic waste, and depletion of natural sources. Furthermore, the silicon industry often uses non–environmentally friendly materials (e.g., etchants, toxic solvents, and resists), which increase the anthropogenic effects on climate change with an expected impact on future generations. *Green organic electronics* is emerging as a new scientific research field within organic electronics. Green electronics aims to implement *sustainable electronics* making use of carbon-based, natural and bioinspired, non-toxic, low-energy-consumption materials.[2,3] Although green electronic devices cannot currently compete with silicon technology, they represent its most promising complement to achieving long-term, zero environmental impact. In fact, organic natural and synthetic semiconductors (e.g., semiconducting conjugated polymers) display simple processability, biocompatibility, biodegradability, high conformability, and flexibility.[4] Organic light-emitting diodes (OLEDs), based on stacked donor-acceptor layers of organic semiconductors (e.g., Poly(9,9-dioctylfluorene-alt-N-(4-sec-butylphenyl)-diphenylamine (TFB) and

Poly(9,9-dioctlylfluorene-alt-benzothiadiazole) (F8BT)), have successfully accessed the market of TV displays and laptop screens. Today, OLEDs are the top application delivered by organic electronics with production on a global scale. Air instability, poor performance and charge carrier mobility are the major drawbacks preventing organic photovoltaics (OPVs) and field effect transistors from emerging as a real alternative to silicon technology anytime soon. For example, OPVs, suffer from photo-oxidation of the active layer under illumination (the active layer is area of the junction where excitons are separated into electrons and holes), and water and oxygen generate superoxides that, by chemically attacking the active layer, decrease the output performance.[5]

Low-charge carrier mobility in organic semiconductors is a consequence of the weak nature of the van der Waals bonds compared with covalent ones. Weak intermolecular interactions usually generate short-range-order systems where, in principle, charge carrier transport is few. However, there exist organic molecules that display inter- and intramolecular hydrogen bonds and usually organize into more ordered structures. Organic materials featuring hydrogen-bonded structures can also feature electronic and protonic currents. This aspect is underpinning to connect biology, which speaks the language of protons, with conventional electronics based on electron and hole transport.

In this regard, eumelanin is a natural, nontoxic, biocompatible, and solution-processable conjugated biomacromolecule that exhibits mixed ionic-electronic transport when hydrated. Eumelanin and pheomelanin are the two main subgroups of melanins, natural pigments responsible for manifold functionalities in living organisms.[6]

Besides human skin, birds' feathers, and mammalian hair, melanins are also present in two areas of the human brain, the *substantia nigra* and the *locus coeruleus*. This pigmental constituent, known as neuromelanin, is composed of a pheomelanin-core/eumelanin-shell structure.[7] A rapid interest is growing around the biorole of neuromelanin. It is indeed posited that the electrical activity and the metal-ion binding properties of the eumelanin shell are related to Parkinson's disease and brain aging.[8] Therefore, elucidating eumelanin's transport physics is of utmost importance to revealing certain poorly understood aspects of Parkinson's syndrome. Apart from its biological importance,[*9] eumelanin has shown a wide range of applications (e.g., in memory devices, porous Si/eumelanin heterojunctions, batteries, energy storage devices, and electrochemical transistors,[10–13] establishing its tremendous potential as an environmentally benign candidate for green electronics and its powering elements.

The conjugated sp^2 backbone of eumelanin points to the question of whether it is a naturally occurring organic semiconductor. The semiconductor hypothesis was formulated for the first time in the 1960s, and it has been recently questioned by mixed *protonic–electronic* as well as *proton membrane* models.[14–16]

In this chapter, we review eumelanin's physical and optical properties and present the basic concepts of transport physics in disordered amorphous semiconductors (AScs). With this set of fundamental concepts, we aim to provide, from previous work and recent insights of our group, the basis for a discussion of the electrical response of the biopigment. A conclusion and perspectives will follow.

5.2 EUMELANIN'S STRUCTURAL PROPERTIES

Eumelanin originates from the oxidative polymerization of L-3-(3,4-dihydroxyphenyl)-alanine via 5,6-dihydroxyindole (DHI) and 5,6-dyhydroxyindole-2 carboxyl acid (DHICA) building blocks (Figure 5.1); pheomelanin generates from the oxidative polymerization of cysteinyldopa via benzothiazine and benzothiazole intermediates.[6,17] As stated in the introduction, eumelanin is the form of melanin most studied by materials scientists thanks to (i) the extensive occurrence in the biosphere and (ii) the peculiar physicochemical properties, i.e., broadband optical absorption,

* For details, refer to the review paragraphs "Eumelanin's Structural Properties" and "Eumelanin's Optical Properties."

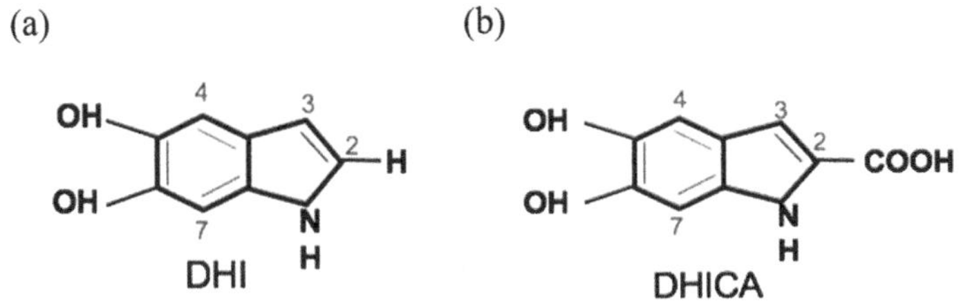

(a) (b)

FIGURE 5.1 Structure and binding sites of (a) DHI and (b) DHICA building blocks of eumelanin.

radical scavenging, redox activities, metal-ion binding , hydration-dependent electrical response, and photoresponse.[9]

Significant efforts have been devoted so far to unraveling the macro- and supramolecular structures of eumelanin, mainly to explain the broadband optical absorption. Two models have been proposed to depict the supramolecular (secondary) structure of eumelanin: the chemical disorder model[9,18] and the geometrical disorder model.[19] The former model posits that the oxidative polymerization forms groups of oligomers exhibiting different bonding configurations, the wide variety of which generates ensembles of oligomers with different secondary structures. In this picture, the broadband absorption is a consequence of superimposed, non-homogeneously broadened Gaussian transitions associated to each secondary structure composing the ensemble. According to the latter model, the secondary structure is composed of covalently bonded planar sheets of a few oligomers, featuring lateral extension of about 15 Å and stacking noncovalently (edge-to-edge π–π interaction) along the out-of-plane direction, with typical spacing of 3.8 Å.[20,21] Therefore, in the geometrical model, the excitonic couplings between eumelanin oligomers and the interplay with ordered and disordered parts of the structure explain the UV-Visible absorption properties of the biopigment.

The polymerization and solid-state aggregation in eumelanin are still under investigation. Several key features can be emphasized: (i) DHI polymerizes preferentially forming dimers and tetramers cross linking at multiple binding sites (2, 4, and 7 positions in Figure 5.1); (ii) oligomers of DHICA favorably cross link at the 4 and 7 positions, and DHICA also features steric hindrance due to the presence of the COOH carboxyl group; and (iii) co-existing redox active species [fully reduced hydroquinone (H2Q), semi-oxidized/semi-reduced semiquinone (SQ), and fully oxidized quinone (Q)] contribute to the geometrical and chemical disorder of the structure[20–22] (Scheme 5.1).

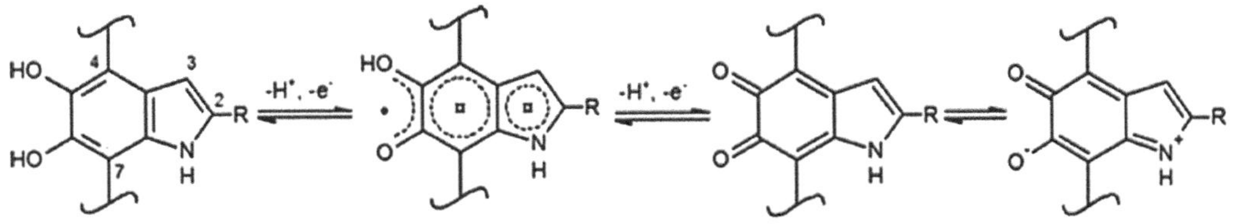

Hydroquinone (H2Q) Semiquinone (SQ) Quinone (Q) Quinone imine (QI)

SCHEME 5.1 The redox states of DHI and DHICA: H2Q, SQ, and Q. The quinone imine (QI) form is the tautomer of Q.

5.3 EUMELANIN'S OPTICAL PROPERTIES

5.3.1 Broadband Optical Absorption

Eumelanin displays broadband optical absorption in both solution and solid-state form.[9,23] The lack of any absorption peak in the excitation and absorption spectra, usually associated with individual and/or combined vibro-electronic transitions of molecules, makes eumelanin an atypical and rather exotic pigment.

Several explanations have been proposed to assess the origin of such a monotonic feature: (i) the absorbance being largely dominated by scattering rather than absorption, (ii) the chemical disorder model, (iii) the geometrical disorder model, and (iv) eumelanin being an organic ASc, due to the similarity between the optical absorption spectra of eumelanin and those of AScs.

Regarding explanations (ii) and (iii), the geometrical disorder provides a reasonable cause for the monotonic increase of the absorption in the ultraviolet region, *de facto* rendering unnecessary the chemical disorder model.[19] Yet, it is not presently clear how the visible light interacts with the pigment as a function of its structure to eventually generate such a featureless spectrum. According to explanation (i), due to the poor solubility and the presence of aggregates, Rayleigh and Mie scattering has been proposed. However, it was found that scattering contributes <6% to the total light attenuation.[24] Explanation (iv) will be widely discussed in the section 5.4.2 *Electrical resistive switching behavior of eumelanin: the ASc model Electrical.*

5.3.2 Optical Excitation and Emission

Countless steady-state and time-resolved investigations have been carried out to clarify the mechanisms of relaxation after the optical excitation of both natural and synthetic eumelanin.[25–28] Eumelanin does show radiative emission (fluorescence), with a typical quantum yield of 10^{-4}.[29–32] These studies essentially proved that eumelanin's optical properties do not respect the well-known *Kasha's rule*, which posits that absorption lines are mirror images of emission lines. Over the last 20 years, picoseconds and femtoseconds time-resolved spectroscopy measurements enabled researchers to explain the dynamics of excitation and relaxation in the biopigment.[33–35] In summary, the conclusions of these studies are as follows:

i) the radiative emission is composed of four main components, each one decaying on the order of nanoseconds;
ii) the radiative emission energy constitutes a small portion of the total energy of the absorbed photons; and
iii) all the remaining excited states decay non-radiatively, and a repopulation of the ground state takes place via vibrational relaxation and/or internal conversion.

5.4 EUMELANIN'S ELECTRICAL PROPERTIES AND FUNDAMENTALS OF TRANSPORT PHYSICS IN AScs

During the 1960s and the early 1970s, the resemblance between the optical properties of eumelanin and those of amorphous inorganic semiconductors (e.g., chalcogenide glasses), and the observation of *an electrical resistive switching* in eumelanin pressed pellets[36] set the basis toward the interpretation of eumelanin mesoscopic properties by the ASc model.

This theoretical approach can be comprehensively understood by recalling the *Davis–Mott theory* of AScs.[37,38] Over time, different electrical resistive switching behaviors have been reported for a large class of solid and liquid AScs by several research groups.[39] Among them, one is of particular interest in this chapter, as correlated to the electrical behavior of eumelanin pressed pellets: the *threshold switch* device.[40,41] With the eumelanin optical properties already reviewed, a short survey of the fundamental optical and electrical properties of AScs will be provided. An explanation proposed for the resistive switching behavior in these systems will be presented as well. All in all, what follows aims to provide the reader with the necessary elements to understand the current challenges faced when describing charge transport in eumelanin.

5.4.1 Broadband Absorption and Transport Physics in AScs

In inorganic crystalline semiconductors, the density of states, $N(E)$, defined as the number of electronic states in the volume of the reciprocal space, is a parabolic function of the energy. A well-defined gap separates the top of the valence band (VB) from the bottom of the conduction band (CB). In these systems, the periodic arrangement of atoms allows us to predict the VB and the CB energies by solving the Bloch's equation. The gap is defined as the minimum energy ($h\nu$) provided by a photon to an electron to let it jump to the CB from the VB during an excitation process. In undoped crystalline semiconductors, the gap is an empty forbidden energy region. If the VB and CB edges lay at the same point of the k-space, the transition is *direct*; otherwise, it is indirect (*phonon assisted*). Both transitions obey the law of the conservation of quasi-momentum. The excitations turn into specific peaks in absorbance and into a well-defined absorption edge in the *Tauc*'s plot.[42] This is a plot of $(\alpha h\nu)^n$ as a function of $h\nu$, where α is the absorption coefficient in cm^{-1}, $h\nu$ is the photon energy in eV, and n is an exponent dependent on the type of transition (direct, indirect, allowed and forbidden). Over the years, a series of studies conducted by Tauc contributed to the computation of the optical band gap of most of the crystalline semiconductors. Today, Tauc plots are powerful tools to evaluate the band gap of semiconducting quantum dots and novel nanostructures and nanohybrids.[43] Indeed, in ordered crystalline semiconductors, the distinct absorption edge allows for the evaluation of the band gap by extrapolating from the linear portion of the Tauc plot. Conversely, in AScs, the long-range disorder breaks the periodic order of atoms; this ultimately makes the mathematical treatment of the electronic states challenging. However, it has been shown that the Hartree–Fock approach[†] predicts the electronic structure of a wide range of AScs with good approximation.[44] It is important to stress that an intrinsic feature of AScs is the absence of a clear absorption edge because of their electronic structure (Scheme 5.2). Typically, in AScs like chalcogenide glasses, a linear combination of nonbonding states generates the VB, while the CB arises from a linear combination of antibonding states. A high degree of structural disorder, given by spatial fluctuations of bond length and potential fluctuations related to the nonperiodic arrangement of the constituent atoms, characterizes the AScs. This structural disorder localizes the wave functions of the charge carriers, resulting in extended tails of valence and CB states, having a Maxwell–Boltzmann distribution whose maximum is centered around the Fermi level (E_F). As can be seen from Scheme 5.2, this continuum of states is responsible for the presence of a set of activation energies [e.g., electronic transitions between extended states in VB (below E_v) and localized states in CB between (E_c and $E_F + E_0$)] and is the primary cause of the observed broadband absorption in AScs. The Mott–Davis description[37,38] provides a method to compute the density of extended and localized states $N(E)$ and the mobility $\mu(E)$ as a function of the energy. If $f(E)$ is the Fermi–Dirac distribution, the conductivity σ can be written as

$$\sigma(E) = \int N(E')\mu(E')f(E')(1 - f(E'))dE' \tag{5.1}$$

[†] For further information: Ramachandran, K.I., Gopakumar, D., & Namboori, K. *Computational Chemistry and Molecular Modeling. Principles and Applications* (Springer, Berlin, Helder, 2008), Chapter 5, doi: https://doi.org/10.1007/978-3-540-77304-7

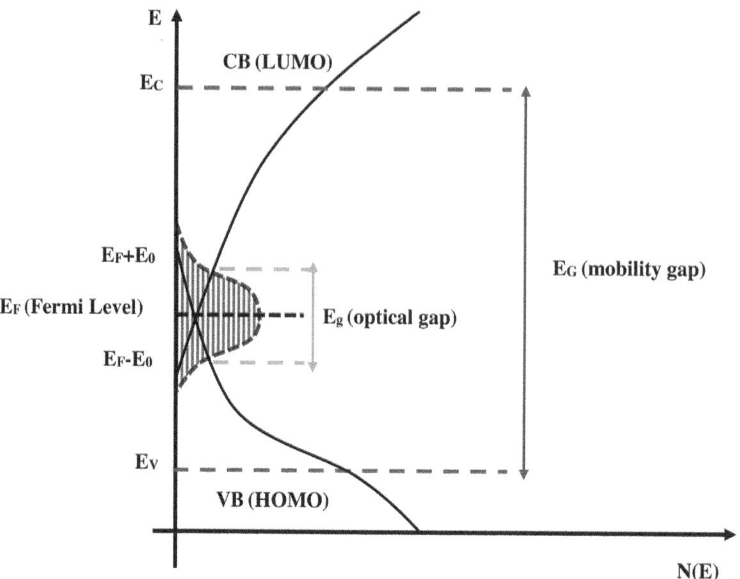

SCHEME 5.2 Typical electronic structure of ASc shows the energy as a function of the energy density: tails of states (shaded Maxwellian distribution) form a continuum spectrum with an energy width of $2E_0$ (optical gap E_g) around the Fermi level E_F. E_c and E_v, which are the conduction and VB energy, respectively, define the mobility gap E_G.

This model works well for AScs even though correlation effects between localized states are neglected. For electrons, the conductivity can thus be described as follows:

1) *Band-like conduction* of high-mobility electrons excited above the mobility edge E_C. If the extended states are not extensively affected by the disorder, the mobility describes a motion of electrons with occasional scattering. The conductivity is given as

$$\sigma = \sigma_0 e^{-(E_C - E_F)/K_B T} \tag{5.2}$$

where σ_0 is a pre-exponential factor, K_B is the Boltzmann constant, and T is the temperature. In this regime of conduction, the separation of the Fermi level from the mobility edge is a linear function of the temperature. If $N(E)$ slightly changes with the temperature above E_C, the relation (Eq. 5.2) becomes:

$$\sigma = q\mu_C N(E_C) K_B T e^{-(\Delta E)/K_B T} e^{\eta/K_B} \tag{5.3}$$

where q is the electron charge, μ_C is the average electron mobility above E_C, and η is a characteristic constant. In the case that μ_C is not significantly affected by the disorder, it has been proposed that the charge transport takes places via Brownian motion. In this case, the extended state wave functions can be taken as a linear combination of atomic wave functions. If the phase coefficients of this linear combination show no phase relation from one site to another, the mobility near E_C can be written as

$$\mu = k \frac{(J)^2 q a^5}{K_B T} s N(E_C) \tag{5.4}$$

where k is a constant, $J = h\nu$ is the interatomic transfer integral, $\nu \approx 10^{15} s^{-1}$ is the typical jump frequency, s is the coordination number, and a is the interatomic spacing.

2) *Thermally assisted tunneling* in localized states close to the mobility edge. As in Scheme 5.2, $2E_0$ is the energy width of the localized states around the Fermi level, and

the main contribution to tunneling is given by hopping from filled to empty localized states within the $(E_F + E_0, E_C)$ energy interval. Thus, the hopping mechanism brings another contribution ΔE_1^h to the thermal activation energy to promote the charge carrier to a given localized state. If the activation energy is written as $E-E_F$, with $E = E_C-E_0$, the conductivity becomes:

$$\sigma = \sigma_1 e^{-(E-E_F+\Delta E_1^h)/K_BT} \tag{5.5}$$

with $\sigma_1 \sim 10^{-3}\sigma_0$ and ΔE_1^h is the hopping energy due to thermally assisted tunneling. Here, σ_0 is the same pre-exponential factor defined in Equation 5.2.

3) *Tunneling conduction near the Fermi level* given by the hopping from strongly localized filled states within $(E_F - E_0, E_F)$ and empty states within $(E_F, E_F + E_0)$. As in thermally assisted tunneling, another hopping barrier, ΔE_2^h, contributes to the thermally activated conduction. Furthermore, the thermal activation energy barrier depends on: (i) the average one electron transfer integral $\langle E_{ij}\rangle$, (ii) the hopping distance between the generic states i and j $(r_{i,j})$, and (iii) the localization radius r_0. The complexity of the charge transport mechanism is also given by the fact that r_0 depends on the Columbic screen, i.e., on the local dielectric constant of the material. Including all these contributions, the tunneling conduction can be expressed as follows:

$$\sigma = \sigma_2 e^{-(\Delta E_{ij}+\Delta E_2^h)/K_BT} e^{-r_{ij}/r_0} \tag{5.6}$$

with $\sigma_2 \ll \sigma_1 < \sigma_0$. With the transfer integral proportional to $e^{-r_{ij}/2r_0}$, the hopping probability P is given as

$$P(r_0, r_{ij}, T) \approx Ce^{-\left(r_{ij}/2r_0+\Delta E_{ij}/K_BT\right)} \tag{5.7}$$

where C is a constant.

If $\Delta E_{ij} = \langle \Delta E_{ij}\rangle \approx E_0$, the number of electronic states around the Fermi level is

$$\tilde{N}(E_F) = N(E_F) * 2E_0 \approx D\langle r_{ij}\rangle^{-3} \tag{5.8}$$

where D is a constant. By plugging $\langle r_{ij}\rangle$ into Equation 5.7, the hopping probability is found as a function of E_0:

$$P(E_0, T) \approx Fe^{-\left[\left(N(E_F * 2E_0)^{-1/3}\right)/r_0+E_0/K_BT\right]} \tag{5.9}$$

where F is a constant. From Equations 5.8 and 5.9, an optimal band width $2E_0$ can be deduced. Indeed, a large value of E_0 implies more available sites for hopping and consequently an increased hopping probability.

On the other hand, since $\Delta E_{ij} = \langle \Delta E_{ij}\rangle \approx E_0$, for larger E_0, the activation energy barrier for hopping increases. This is obviously accompanied by a reduction of the hopping probability. A trade-off between these two effects can be established in such a way as to define an optimal bandwidth $E_0^{\text{opt}}(T) \approx \left[\frac{(K_BT)^{-3}}{N(E_F)r_0^3}\right]^{1/4}$. Since the hopping rate is proportional to the conductivity and to the hopping probability, these are proportional to each other. From the evaluation of an optimal bandwidth, it is possible to show that this hopping regime features the following temperature variation:

$$\sigma = Ge^{-(B/T)^{1/4}} \tag{5.10}$$

where G and B are two constants. According to Equation 5.10, by raising the temperature, the conductivity of the system improves because the hopping distance decreases. Finally, since the hopping distance changes with the temperature, this regime is also known as *variable range hopping* conduction.

It should now be clear that AScs exhibit very different charge transport properties depending on the population of charge carriers that can be eventually promoted to the band states. Ultimately, this consideration has been ascribed as the underlying cause of a unique feature of these materials, the *electrical resistive switching*. In most of materials biased with electric fields exceeding 10^5–10^6 V/cm, a disruptive breakdown is observed. Conversely, some crystalline and amorphous materials, if such high electric fields are applied, do not experience a disruptive breakdown, but they instead switch from high- to low-resistance states, respectively, called OFF and ON states. This is, for example, the case of AScs.

In these devices, the threshold voltage (V_t) is defined as the voltage beyond which the switching takes place; V_t depends on the electrode distance (hence to the applied electric field), but a specific functional dependence has never been found. After the switch to the high conducting state, a voltage drop is usually observed because of the decrease in resistance by several orders of magnitude. The low resistance state can be kept if a minimum current I_H, called a holding current, is provided. Therefore, these materials can reversibly switch from the OFF to ON state depending on the minimum current we can provide from an external circuit. The practical interest of these devices is that they set up a self-sustained high conducting state that prevents breakdown, making them suitable for applications in memory and switching devices. This fascinating feature was clarified mainly by Mott and co-workers during the 1970s. Briefly, the breakdown (or electrical failure) is a spatial instability limiting the current flow to a thin high-current density filament in the material. Near this filament, an intense heating process takes place. If the breakdown occurs, the material surrounding the filament burns out and a short circuit is observed. With respect to dielectrics, in AScs, the region in proximity to the filament remains intact thanks to the sudden increase in conductance, preventing the generation of excessive heat. This self-sustained conductance has been explained as due to thermally modulated electronic processes. Accordingly, intense applied electric fields promote the formation of high tunneling fields and can generate electron avalanches before the thermal breakdown effects occur. Indeed, the distortion introduced by the electric field causes a nonequilibrium condition in which a high space charge density accumulates near the electrodes. Intense electric fields build up at the electrodes, promoting tunnel injection and providing the material with an "extra charge sink" able to sustain the high conducting state. Avalanches of electrons can be generated as well if the rate of energy gain of the charge carriers (from the electric field) is faster than the rate of energy loss by phonon emission.

5.4.2 Electrical Resistive Switching Behavior of Eumelanin: The ASc Model

The intriguing idea of describing eumelanin as an ASc triggered a sequence of quantum chemical studies, combined with electrical and photoelectrical characterizations, looking for evidence in support of the ASc hypothesis.[45] Longuet-Higgins[46] and Pullman and Pullman[47] for the first time applied the *Huckel* theory to the indolic structures of eumelanin and found that conjugation confers one-dimensional semiconducting properties to the structure. Following these results, McGinness and co-workers proposed a theoretical model by which eumelanin would display mobility gaps.[45] During the late 1980s and the early 1990s, Galvão and Caldas showed how five to six oligomeric units are enough to set up stable semiconducting properties. They also suggested that, in eumelanin, a short-range-ordered structure made of a few subunits may predominate on an overall disordered structure conveying it semiconducting properties.[48,49] Unexpectedly, despite the simplicity of Galvão and Caldas's theoretical model (e.g., interchain interactions and boundary condition effects considered as defects of the structure), their results agree with the latest accepted model of geometrical (packing) disorder.[19] Over the past two decades, more sophisticated steady-state and time-dependent density functional theories have also been explored. Theoretical vibrational spectra, confirmed by the experiments, and the Highest Occupied Molecular Orbital-Lowest Unoccupied Molecular Orbital (HOMO–LUMO) band gap of H2Q-, SQ-, and Q-based dimers, tetramers, and ultimately hexamers, have been calculated.[50,51]

Eumelanin features electrical conductivity and photoconductivity as well. The first photoconductivity response of the biopigment was reported by Rosenberg and Postow.[52] A few years later, Powell and Rosenberg discovered that eumelanin's conductivity is strongly affected by the content of adsorbed water.[53] Room temperature conductivity of hydrated samples showed lower resistance compared with dry ones. In the hypothesis of ASc, it was suggested that the water locally modifies the dielectric constant via ionization of acid groups. The effect of the water has been considered with the following modified version of Equations 5.2–5.5:

$$\sigma = \sigma_0 e^{-E_D/K_BT} e^{-q^2/2K_BTR\left\{\frac{1}{\varepsilon_d}-\frac{1}{\varepsilon_w}\right\}} \tag{5.11}$$

where E_D is the activation energy in dry conditions; ε_d and ε_w are the dry and hydrated state dielectric constants, respectively; and R is the effective screening radius of absorbed water molecules (about 1 Å). In 1974, McGinness et al.[36] fabricated natural and synthetic eumelanin pressed pellets of different thicknesses in sandwich architecture, between Cu, Al, and carbon electrodes. High-voltage tests demonstrated that hydrated samples behaved as threshold switch devices. The electrical resistive switching was not observed when the samples were dried, and this was interpreted invoking the Rosenberg theories. The authors stated that the overall increase of the dielectric constant, due to the presence of absorbed water, provided the samples with extra charge carriers to sustain the ON state. These results were considered as the final evidence pointing toward the ASc model, triggering a sort of *paradigm* in eumelanin literature.

5.4.2.1 Hydration-Dependent Electrical Response: New Transport Models

Assuming the ASc model, the role of the absorbed water as correlated to electrical and photoelectrical properties was consequently investigated to look for any change in the temperature activation energy of samples hydrated at different percentages of relative humidity (%RH). The main results are: (i) in eumelanin, two types of water exist, i.e., weakly bound (surface) and strongly bound water (molecular structure level)[54]; (ii) eight orders of magnitude constitute the typical difference in conductivity in eumelanin samples characterized at 0% RH and 100% RH[55]; (iii) studies performed as a function of limited ranges of temperature (ca. between 273 and 315 K) showed that the conductivity of eumelanin samples increases if the temperature is raised, as for AScs[56,57]; (iv) from temperature-activated conductivity measurements, optical gap and band gap have been obtained (in this regard, de Albuquerque et al. computed the band gap of synthetic and natural eumelanin via photopyroelectric spectroscopy, finding a value of 1.7 eV[58,59]); and (v) eumelanin dielectric response is modulated by clusters of percolating water, which contribute to decrease the charge carrier hopping distance by increasing the local Coulomb screen.[60]

Despite the above discussed evidence, a few recent observations are casting doubt on the ASc model.[14–16,61,62] In regard to the optical properties, it has been posited that the ASc model is not necessary to explain the broadband absorption: chemical and geometrical disorder models can also explain it. Regarding the electrical and photoelectrical results, there is general agreement that no conclusion can be drawn without detailed knowledge on the eumelanin water absorption isotherms, namely the weight percent gained versus relative humidity percentage [obtained from Brunauer–Emmett–Teller (BET) plots]. The lack of a detailed water intake process would make it impossible to control the absolute water content in the samples under electrical tests. Mostert et al.[14,62] assembled 2-mm-thick synthetic eumelanin pellets between gold electrodes in sandwich and van der Pauw (vdP) geometries and followed the absorption isotherms in the two configurations. The samples were hydrated at different %RH for 1 h and the samples biased at 15 V, both in sandwich and vdP geometries. The results showed that the sandwich geometry fits the Mott–Davis ASc–BET plot, whereas the vdP does not. This was interpreted as evidence that, when dealing with the electrical properties of the hydrated eumelanin, the geometry of the electrical contacts can lead to misleading results. The sandwich geometry would take a long time to reach the hydration equilibrium because

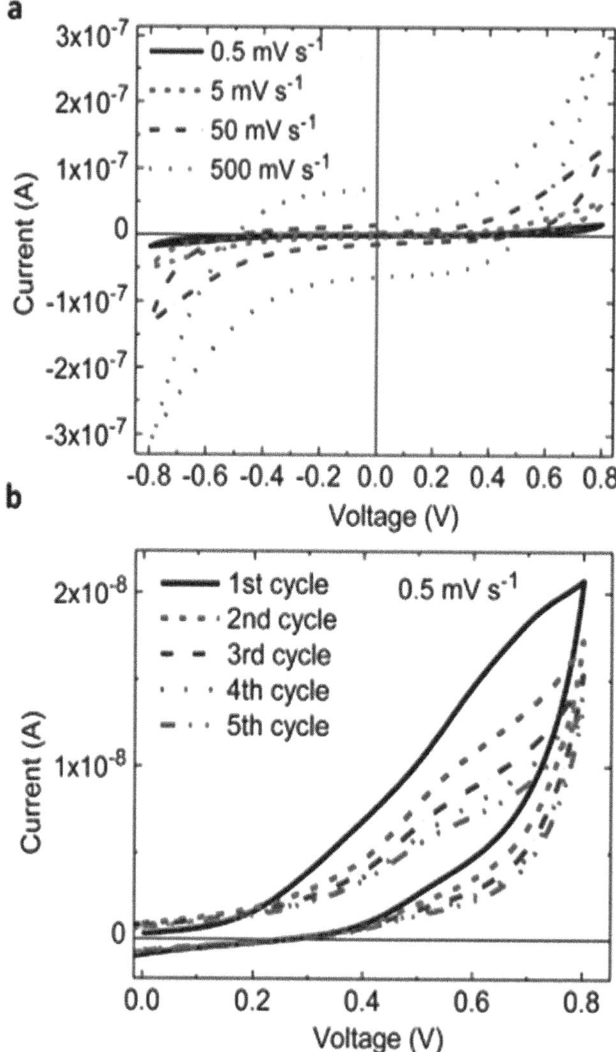

FIGURE 5.2 *I–V* characteristics of 90% RH eumelanin films (t = 50 nm) casted on coplanar electrodes (W = 24.5 mm, L = 10 μm). (a) Voltage sweep rate dependence of the first cycle and (b) first five cycles (0.5 mV s^{-1}) for positive voltages. Reprinted with permission from Wünsche et al.[64] Copyright (2015) American Chemical Society.

only parts of the sample are exposed to water vapor. Furthermore, from muon-spin relaxation spectroscopy measurements (μSR), the same authors found that the number of protons and free spin electrons increases if the %RH is increased while the proton hopping rate is constant. Accordingly, the increase in conductivity with the %RH would be due to an increased population of the charge carriers in form of semiquinone (SQ) free radicals and not to a decreased activation energy barrier for hopping (Scheme 5.1). This could be an indication that the conductivity of hydrated eumelanin samples is higher than that of dry ones simply because the carrier density increases (and not due to the decrease of the activation energy transport). In conclusion, Mostert et al. offered a diverse interpretation of eumelanin's electrical properties by which electrons and/or protons are released during a comproportionation equilibrium reaction:

$$H_2Q + Q + 2H_2O \rightleftharpoons 2H_3O^+ + 2SQ^-$$

The reaction takes place between H2Q and Q species with water molecules. Absorbed water would provide a percolation path for ions and protons that would be transported via Grotthuss mechanism.[63]

In our research group, we investigated[64] the electrical response of synthetic eumelanin thin films drop-casted on Pt electrode-patterned substrates (planar configuration, Figures 5.2a and b). The *I–V* characteristics of hydrated films as a function of the voltage sweeping rate show a capacitive behavior for $|V| < 0.2$ V. From a linear fit of the voltage rate dependence, the authors extracted a capacitance value way higher than that expected in dielectric polarization processes, an indication of bulk ionic conductivity. For $|V| > 0.2$ V, the increase of the current with the voltage was exponential, suggesting a noncapacitive behavior with redox processes taking place at Pt/eumelanin interfaces. The contribution of protonic transport with respect to the electronic one was assessed in planar configuration by studying transient current measurements on synthetic eumelanin films, making use of Pd and PdH_x contacts. Pd, if exposed to H_2, forms PdH_x, which can inject both electrons and protons. The formation of PdH_x from Pd and H_2 is given by the following reaction:

$$Pd + H^+ + e^- \rightarrow PdH$$

The results in Figure 5.3 show that the steady-state current with PdH_x contacts is higher than that of Pd electrodes, at the same %RH conditions. Furthermore, this difference increases with the relative humidity, suggesting that the protonic contribution to the total conductivity becomes more and more prominent with the content of absorbed water.

In this context, depending on the hydration level of the biopigment, *standard* (ON/OFF ratio of about 10^4) and *hybrid* (ON/OFF ratio of 10^2) electrical resistive switchings in planar Au/eumelanin/Au configuration have been reported by Di Mauro et al.[65] However, this resistive switching is different from that discussed in Sections 5.3.1 and 5.3.2 because it is not reversible. In this case, the dissolution of metal electrodes caused the formation of dendrites bridging the electrodes during the electrical tests, pointing to the extremely important role of metal–eumelanin interfaces and electrochemical processes during studies pertaining to the charge transport mechanism of the pigment. The content of chloride anions, known to form gold complexes, and the presence of chelating groups in the molecular structure of eumelanin were pinpointed as fundamental factors for the formation of the dendritic structures reported in Figure 5.4.

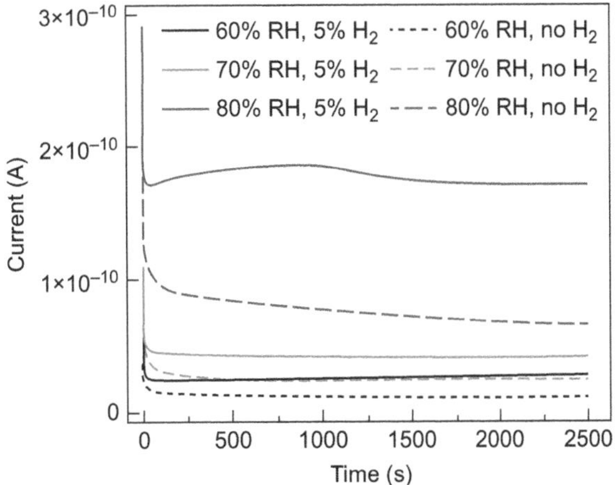

FIGURE 5.3 Transient current measurements (0.5 V) of a eumelanin thin film (t = 50 nm) with Pd electrodes and PdH_x contacts, respectively, electrons and electron and proton injectors, at 60%, 70%, and 80% RH. Reprinted with permission from Wünsche et al.[64] Copyright (2015) American Chemical Society.

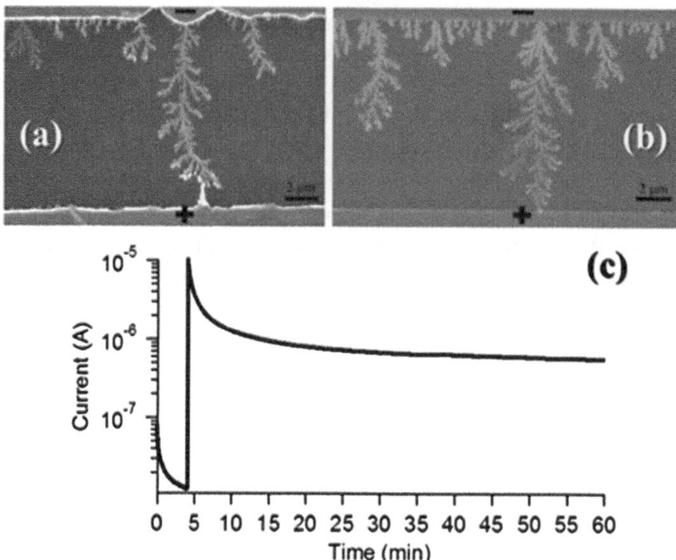

FIGURE 5.4 SEM images of dendrites bridging two electrodes for samples hydrated 1 h at 80% RH. The resistive switching took place after (a) 4 min and (b) 12 min of biasing at 1 V. (c) Transient current measurements of sample (a) during and after the dendrites' formation. Reprinted from Di Mauro et al.[22] Copyright (2016) Royal Society of Chemistry.

5.4.2.2 Mixed Ionic—Electronic Conductor or ASc: A Debate Still Open

Several works recently interpreted the charge transport properties of eumelanin. An interplay of contributions, such as interfacial electrochemical processes, mixed ionic—electronic conduction, proton migration, and redox processes, was described. These investigations certainly opened the debate regarding whether for a long time a predominant and apparently solid literature could have been framed within the paradigm of the ASc model. At this stage, it should be clear that the physical description of the charge transport properties of eumelanin is still partly undiscovered and deserves more attention, as several reliable works point to different interpretations.

It is also important to stress that both mixed ionic—electronic and Mott–Davis ASc models show limitations. The former has not been tested in regard to the electrical resistive switching (only low electric fields have been considered) and cannot currently explain it. It is worth noticing that crucial details were not reported in McGinness's experiment, i.e., the %RH, hydration time, and the *I–V* response of dry eumelanin samples. Besides that, in most of the works published between the 1960s and the 1970s, the quantitative water intake before and after tests supporting the ASc model is commonly missing.

The debate is still open, and despite the huge effort expended thus far, there is still an urgent need to fill this knowledge gap. Providing new insights into the resistive switching behavior of hydrated and dry eumelanin samples, with the current awareness of its sensitivity to the water content, is an important path to follow.

5.5 CONCLUSION AND PERSPECTIVES

Eumelanin is a conjugated biomacromolecule whose macromolecular and supramolecular structures are capable of electrical conduction. The chemical and structural features of eumelanin match the requirements for applications in bioelectronics, a flourishing branch of organic electronics. Bioelectronics is a research field at the interface between biology and electronics. Bioelectronic devices can be integrated with living tissues and are commonly made of quite environmentally benign biocompatible materials, sustaining both ionic and electronic signals. Being solution-processable and biocompatible, eumelanin is an excellent candidate for bioelectronic applications, beyond well-established conjugated materials such as PEDOT:PSS.

In this context, describing the charge transport properties of the biopigment with the correct model is of utmost importance to demonstrate, for example, ion-to-electron transducers (in the case of mixed ionic—electronic conductors, proton membranes) or write-once-read-many organic semiconductor-based memory devices.

In conclusion, this chapter aimed to provide an insight into the complexity of the appropriate physical description of eumelanin electrical properties. A short discussion of the optical and electrical properties of chalcogenide AScs as compared with those of eumelanin has been presented. Attention has been given to the long-standing debate about eumelanin being a mixed ionic—electronic conductor or an ASc. The mixed- ionic–electronic conductor model cannot alone explain the electrical resistive switching behavior of eumelanin. To assess the potential of eumelanin-based technologies, fundamental aspects of the charge transport in eumelanin pellets and thin films need to be reconsidered. In doing so, temperature-activated conductivity measurements of dry eumelanin pellets or melanins resulting from only one of the building blocks DHI and DHICA would help to disentangle electronic and protonic currents in correlation with their supramolecular aggregation.

REFERENCES

1. Willner, I. & Katz, E. *Bioelectronics: From Theory to Applications* (John Wiley & Sons, 2006).
2. Irimia-Vladu, M., Głowacki, E. D., Voss, G., Bauer, S. & Sariciftci, N. S. Green and biodegradable electronics. *Mater. Today* **15**, 340–346 (2012).
3. Irimia-Vladu, M. 'Green' electronics: biodegradable and biocompatible materials and devices for sustainable future. *Chem. Soc. Rev.* **43**, 588–610 (2014).
4. Kenry, K. & Liu, B. Recent advances in biodegradable conducting polymers and their biomedical applications. *Biomacromolecules* **19**, 1783–1803 (2018).
5. Jørgensen, M., Norrman, K. & Krebs, F. C. Stability/degradation of polymer solar cells. *Sol. Energy Mater. Sol. Cells* **92**, 686–714 (2008).
6. d'Ischia, M. *et al.* Melanins and melanogenesis: from pigment cells to human health and technological applications. *Pigment Cell Melanoma Res.* (2015). doi:10.1111/pcmr.12393
7. Bush, W. D. *et al.* The surface oxidation potential of human neuromelanin reveals a spherical architecture with a pheomelanin core and a eumelanin surface. *Proc. Natl. Acad. Sci. USA* **103**, 14785–14789 (2006).
8. Segura-Aguilar, J. *et al.* Protective and toxic roles of dopamine in Parkinson's disease. *J. Neurochem.* **129**, 898–915 (2014).
9. Meredith, P. & Sarna, T. The physical and chemical properties of eumelanin. *Pigment Cell Res.* (2006). doi:10.1111/j.1600-0749.2006.00345.x
10. Ambrico, M. *et al.* Memory-like behavior as a feature of electrical signal transmission in melanin-like bio-polymers. *Appl. Phys. Lett.* (2012). doi:10.1063/1.4729754
11. Mula, G., Manca, L., Setzu, S. & Pezzella, A. Photovoltaic properties of PSi impregnated with eumelanin. *Nanoscale Res. Lett.* **7**, 377 (2012).
12. Kim, Y. J., Wu, W., Chun, S.-E., Whitacre, J. F. & Bettinger, C. J. Biologically derived melanin electrodes in aqueous sodium-ion energy storage devices. *Proc. Natl. Acad. Sci. USA* **110**, 20912–20917 (2013).
13. Sheliakina, M., Mostert, A. B. & Meredith, P. An all-solid-state biocompatible ion-to-electron transducer for bioelectronics. *Mater. Horizons* **5**, 256–263 (2018).
14. Mostert, A. B., Powell, B. J., Gentle, I. R. & Meredith, P. On the origin of electrical conductivity in the bio-electronic material melanin.*Appl. Phys. Lett.* **100**, 093701 (2012).
15. Mostert, A. B., Rienecker, S. B., Noble, C., Hanson, G. R. & Meredith, P. The photoreactive free radical in eumelanin. *Sci. Adv.* **28**, eaaq1293 (2018).
16. Sheliakina, M., Mostert, A. B. & Meredith, P. Decoupling ionic and electronic currents in melanin. *Adv. Funct. Mater.* (2018). doi:10.1002/adfm.201805514

17. Prota, G. Melanins, melanogenesis and melanocytes: looking at their functional significance from the chemist's viewpoint. *Pigment Cell Res.* **13**, 283–293 (2000).
18. Tran, M. L., Powell, B. J. & Meredith, P. Chemical and structural disorder in eumelanins: a possible explanation for broadband absorbance. *Biophys. J.* **90**, 743–752 (2006).
19. Chen, C. T. *et al.* Excitonic effects from geometric order and disorder explain broadband optical absorption in eumelanin. *Nat. Commun.* **5**, 1–10 (2014).
20. Pezzella, A. *et al.* 5,6-Dihydroxyindole tetramers with 'anomalous' interunit bonding patterns by oxidative coupling of 5,5′,6,6′-tetrahydroxy-2,7′-biindolyl: emerging complexities on the way toward an improved model of eumelanin buildup. *J. Org. Chem.* **72**, 9225–9230 (2007).
21. Panzella, L. *et al.* Atypical structural and π-electron features of a melanin polymer that lead to superior free-radical-scavenging properties. *Angew. Chem. Int. Ed.* **52**, 12684–12687 (2013).
22. Di Mauro, E., Xu, R., Soliveri, G. & Santato, C. Natural melanin pigments and their interfaces with metal ions and oxides: emerging concepts and technologies. *MRS Commun.* (2017). doi:10.1557/mrc.2017.33
23. Micillo, R. *et al.* Eumelanin broadband absorption develops from aggregation-modulated chromophore interactions under structural and redox control. *Nat. Publ. Gr.* (2017). doi:10.1038/srep41532
24. Riesz, J. *The Spectroscopic Properties of Melanin* (Queensland University, 2007).
25. Gallas, J. M. & Eisner, M. Fluorescence of melanin? Dependence upon excitation wavelength and concentration. *Photochem. Photobiol.* **45**, 595–600 (1987).
26. Ito, A. S., Azzellini, G. C., Silva, S. C., Serra, O. & Szabo, A. G. Optical absorption and fluorescence spectroscopy studies of ground state melanin-cationic porphyrins complexes. *Biophys. Chem.* **45**, 79–89 (1992).
27. Mosca, L., De Marco, C., Fontana, M. & Rosei, M. A. Fluorescence properties of melanins from opioid peptides. *Arch. Biochem. Biophys.* **371**, 63–69 (1999).
28. Brian Nofsinger, J. & Simon, J. D. Radiative relaxation of sepia eumelanin is affected by aggregation. *Photochem. Photobiol.* **74**, 31–37 (2007).
29. Riesz, J. The spectroscopic properties of melanin. *Tese - Univ. Queensl.* 1–239 (2007).
30. Nighswander-Rempel, S. P., Riesz, J., Gilmore, J., Bothma, J. P. & Meredith, P. Quantitative fluorescence excitation spectra of synthetic eumelanin. *J. Phys. Chem. B* **109**, 20629–20635 (2005).
31. Nighswander-Rempel, S. P., Riesz, J., Gilmore, J. & Meredith, P. A quantum yield map for synthetic eumelanin. *J. Chem. Phys.* **123** (2005).
32. Sutter, J. U. & Birch, D. J. S. Metal ion influence on eumelanin fluorescence and structure. *Methods Appl. Fluoresc.* **2** (2014).
33. Meng, S. & Kaxiras, E. Mechanisms for ultrafast nonradiative relaxation in electronically excited eumelanin constituents. *Biophys. J.* **95**, 4396–4402 (2008).
34. Gauden, M. *et al.* Ultrafast excited state dynamics of 5,6-dihydroxyindole, a key eumelanin building block: nonradiative decay mechanism. *J. Phys. Chem. B* **113**, 12575–12580 (2009).
35. Forest, S. E., Lam, W. C., Millar, D. P., Nofsinger, J. B. & Simon, J. D. A model for the activated energy transfer within eumelanin aggregates. *J. Phys. Chem. B* **104**, 811–814 (2000).
36. McGinness, J., Corry, P. & Proctor, P. Amorphous semiconductor switching in melanins. *Science.* (1974). doi:10.1126/science.183.4127.853
37. Davis, E. A. & Mott, N. F. Conduction in non-crystalline systems V. Conductivity, optical absorption and photoconductivity in amorphous semiconductors. *Philos. Mag.* **22**, 0903–0922 (1970).
38. N.F. Mott and E.A. Davis. *Electronic Processes in Non-Crystalline Materials,* The international monographs on physics, N.F. Mott and E. A. Davis Editors, W. Marshall and D. H. Wilkinson General Editors (Oxford University Press, Great Clarendon Street, Oxford, OX2 6DP, ISBN 978-o-I9-964533-6 (pbk), 1979).
39. Pearson, A. D. Memory and switching in semiconducting glasses. A review. *J. Non Cryst. Solids* **2**, 1–15 (1970).
40. Ovshinsky, S. Reversible electrical switching phenomena in disordered structures. *Phys. Rev. Lett.* **21**, 1450–1453 (1968).
41. Ovshinsky, S. R., Nelson, D. L., Fritzsche, H. & Evans, E. J. Radiation hardness of ovonic devices. *IEEE Trans. Nucl. Sci.* **15**, 311–321 (1968).
42. Greenwood, D. A. Amorphous and liquid semiconductors. *Opt. Acta Int. J. Opt.* (1970). doi:10.1080/715120893

43. Papikyan, A. K., Gevorgyan, V. A., Mangasaryan, N. R. & Gladyshev, P. P. Characterization of vacuum flash evaporated CdTe thin films for solar cell application. *J. Phys. Conf. Ser.* **945**, 153–163 (2018).
44. Morigaki, K. & Ogihara, C. *Springer Handbook of Electronic and Photonic Materials* (Springer, Cham, 2017). doi:https://doi.org/10.1007/978-3-319-48933-9
45. McGinness, J. E. Mobility gaps: a mechanism for band gaps in melanins. *Science.* **177**, 896–897 (1972).
46. Longuet-Higgins, H. C. On the origin of the free radical property of melanins. *Arch. Biochem. Biophys.* **86**, 231–232 (1960).
47. Pullman, A. & Pullman, B. The band structure of melanins. *Biochim. Biophys. Acta* **54**, 384–385 (1961).
48. Galvão, D. S. & Caldas, M. J. Theoretical investigation of model polymers for eumelanins. II. Isolated defects. *J. Chem. Phys.* **93**, 2848–2853 (1990).
49. Galvão, D. S. & Caldas, M. J. Polymerization of 5,6-indolequinone: a view into the band structure of melanins. *J. Chem. Phys.* **88**, 4088–4091 (1988).
50. Stark, K. B., Gallas, J. M., Zajac, G. W., Eisner, M. & Golab, J. T. Spectroscopic study and simulation from recent structural models for eumelanin: I. Monomer, dimers. *J. Phys. Chem. B* **107**, 3061–3067 (2003).
51. Stark, K. B. *et al.* Effect of stacking and redox state on optical absorption spectra of melanins-comparison of theoretical and experimental results. *J. Phys. Chem. B* **109**, 1970–1977 (2005).
52. Rosenberg, B. & Postow, E. Semiconduction in proteins and lipids: its possible biological import. *Annals of the New York Academy of Science,* 161-190 (1969).
53. Powell, M. R. & Rosenberg, B. The nature of the charge carriers in solvated biomacromolecules. *J. Bioenerg.* **1**, 493–509 (1970).
54. Albanese, G., Bridelli, M.G. & Deriu, A. Structural dynamics of melanins investigated by rayleigh scattering of mossbauer radiation. *Biopolymers* **23**, 1481–1498 (1984).
55. Jastrzebska, M. M., Isotalo, H., Paloheimo, J. & Stubb, H. Electrical conductivity of synthetic DOPA-melanin polymer for different hydration states and temperatures. *J. Biomater. Sci. Polym. Ed.* (1995). doi:10.1163/156856295X00490
56. Jastrzebska, M., Kocot, A. & Tajber, L. Photoconductivity of synthetic DOPA-melanin polymer. *J. Photochem. Photobiol. B Biol.* (2002). doi:10.1016/S1011-1344(02)00268-3
57. Crippa, P. R., Cristofoletti, V. & Romeo, N. A band model for melanin deduced from optical absorption and photoconductivity experiments. *Biochim. Biophys. Acta* **538**, 164–170 (1978).
58. De Albuquerque, J. E., Giacomantonio, C., White, A. G. & Meredith, P. Study of optical properties of electropolymerized melanin films by photopyroelectric spectroscopy. *Eur. Biophys. J.* **35**, 190–195 (2006).
59. De Albuquerque, J. E., Giacomantonio, C., White, A. G. & Meredith, P. Determination of thermal and optical parameters of melanins by photopyroelectric spectroscopy. *Appl. Phys. Lett.* **87**, 061920 (2005).
60. Jastrzebska, M. M., Jussila, S. & Isotalo, H. Dielectric response and a.c. conductivity of synthetic DOPA-melanin polymer. *J. Mater. Sci.* **33**, 4023–4028 (1998).
61. Bernardus Mostert, A. *et al.* Gaseous adsorption in melanins: hydrophilic biomacromolecules with high electrical conductivities. *Langmuir* **26**, 412–416 (2010).
62. Mostert, A. B. *et al.* Role of semiconductivity and ion transport in the electrical conduction of melanin. *Proc. Natl. Acad. Sci. USA* (2012). doi:10.1073/pnas.1119948109
63. Gomez, E. F. & Steckl, A. J. *Green Materials for Electronics. Green Materials for Electronics* (Wiley-VCH, Verlag GmbH&Co.KGaA, Boschstr. 12, 69469, Weinheim, Germany, 2018).
64. Wünsche, J. *et al.* Protonic and electronic transport in hydrated thin films of the pigment eumelanin. *Chem. Mater.* **27**, 436–442 (2015).
65. Di Mauro, E. *et al.* Resistive switching controlled by the hydration level in thin films of the biopigment eumelanin. *J. Mater. Chem. C* (2016). doi:10.1039/C6TC02793H

6 Aqueous Stable Colloidal Gold Nanoparticles from Synthesis and Conjugation to Biomedical Application

Kamil Rahme[a,b] *and Justin D. Holmes*[b,c]

[a]Department of Sciences, Faculty of Natural and Applied Science, Notre Dame University-Louaize Zouk Mosbeh, Lebanon
[b]School of Chemistry, University College Cork, Ireland
[c]Advanced Materials and Bioengineering Research (AMBER) Centre, ERI, University College Cork, Ireland

CONTENTS

6.1 INTRODUCTION

Gold nanoparticles (AuNPs) have a strong absorption band and high luminescent properties, due to the surface plasmon resonance (SPR), originated from the coherent oscillations of conduction-band electrons on nanoparticle surface upon interaction with an electromagnetic radiation of appropriate wavelength.[1,2] The SPR optical absorption and scattering properties of the AuNPs are also known to depend on the final size and shape of the nanoparticles and to be very sensitive to the surrounding media and the aggregation state of the nanoparticles.[3–7] Therefore, AuNPs have attracted researchers' attention for their use as a very useful tool for a wide range of applications, including chemical sensing, optoelectronics, diagnostics, thermal therapy, drugs, and gene delivery.[3,4,6,8–20] Furthermore, the low cytotoxicity of AuNPs and their ease of bioconjuguation have contributed to the boom of

AuNPs in bionanotechnology,[21,22] and many publications on the studies of the effect of nanoparticle size and shape on cytotoxicity, biodistribution, and interaction with the biosystem have appeared recently.[23–28] Therefore, the synthesis of AuNPs of different sizes and shapes has attracted much interest over the past few decades. Different ways to synthesize AuNPs are now available in the literature, among which AuNPs are mainly produced through chemical reduction of gold precursor (typically $HAuCl_4$). At first, Turkevish in 1951 reported on the formation of colloidal AuNPs using trisodium citrate to reduce tetrachloroauric acid in water; later, in 1971, Frens published an improved and slightly modified method still popular today. Unfortunately, AuNPs obtained by the above methods are between 10 and 40 nm, with a larger size making the NPs less monodisperse and irregular in shape.[29,30]

Methods to improve monodispersity and shape of AuNPs when their diameter exceeds 50 nm use a seeding growth strategy.[31–33] Large AuNPs were formed by reducing gold $HAuCl_4$ or A(III)-surfactants (i.e., CTAB or Triton X-100) complexes in the presence of AuNPs seeds with hydroxylamine hydrochloride or ascorbic, respectively.[34,35] When CTAB is present, various shapes, such as nanorods, nanoprisms, and so on, can be obtained by varying the synthesis conditions and mainly the CTAB concentration.[35] However, AuNPs-CTAB are known to be toxic, and excess of CTAB in the media has to be removed prior to biological assays.[36] Quasispherical AuNPs of diameter ranging from 15 to 300 nm were synthesized by different authors, including us, through seeded growth methods using 2-mercaptosuccinic acid, hydroquinone, hydroxylamine hydrochloride, and H_2O_2 · [33,37–39] Moreover, seeding growth methods may not always give a homogeneous growth of all the seeds, and some smaller nanoparticles can sometimes be found in solution (usually <5%).

Nanotechnology and nanomedicine require improved synthetic methods for nearly monodisperse conjugated and stable colloidal solutions, avoiding the use of cytotoxic chemicals. Amine-containing molecules are well known to be widely used as reducing agents for nanoparticle synthesis and stabilization.[40] For example, when polyelectrolytes such as chitosan, polyallylamine, and polyethylenimine (PEI) are present during the synthesis of nanoparticles, positively charged nanoparticles can be obtained.[41] The presence of this positive charge on the surface of the nanoparticles may be of benefit for electrostatic complexation for biomolecules such as the phosphate ester backbones of the nucleic acid,[42–44] whereas the more general process for the bioconjugation of nanoparticles is known to occur through formation of a biologically stable amide linkage between the amino groups (NH_2) of the biomolecules and carboxylic acid at the surface of the nanoparticles or vice versa.[45–47] The attachment of biomolecules such as peptides, proteins, and other targeted ligands containing specific molecules that recognize receptors on various cell lines on the surface of nanoparticles plays a key role not only for their interactions with biological tissues, but also in their biological functions and cellular uptake, which can facilitate their use in drug delivery, diagnosis, and therapeutics.[48–53]

Here we report a fast and simple protocol for the synthesis of seedless AuNPs colloidal solutions with size range between 4 and 150 nm in water.[54,55] We also show in this study the failure of the attempt to obtain AuNPs with diameter larger than 30–40 nm via the citrate reduction, due to an increase in the polydispersity and the shape diversity of the final AuNPs.[31] This result indicates the importance of new reducing agents in the production of AuNPs, especially for diameters above 40 nm. Therefore, we demonstrate for the first time that we can obtain AuNPs with diameters of ~60, ~80, and ~150 nm by using hydroxylamine-*o*-sulfonic acid (HOSA) as a new reducing agent for $HAuCl_4{\cdot}3H_2O$ at RT. Our results from dynamic light scattering (DLS) based on the Rayleigh scattering, known for spherical nanoparticles, show that the obtained nanoparticles were not polydisperse (PdI < 0.2), as further justified by electron microscopy and UV–visible (UV–vis) spectroscopy. The simple one-step method can serve as a complement for the commonly used Turkevich and Frens method for direct production of AuNPs. On the other hand, we also show that similarly nearly monodisperse, positively charged AuNPs-PEI with different diameters could be also obtained via one pot synthesis by simply reducing $HAuCl_4{\cdot}3H_2O$ with ascorbic acid or HOSA in the presence of PEI as a capping ligand. Different PEI molecular weights (1.3, 2, and 25 kDa) could be used; however, in this chapter, only AuNPs obtained with PEI 2 kDa are presented. Further bioconjugation on 15, 30, and ~90 nm PEGylated AuNPs was performed by grafting apolipoprotein E (ApoE) and bovine serum albumin (BSA), via an amide linkage using an amine reactive functional group *N*-hydroxysuccinimide, that easily reacts with amino groups from the protein in aqueous media with pH between 8 and 10. Using a similar protocol, we were also able to attach transferrin onto AuNPs to obtain AuNP-PEG-Tf. The resulting AuNPs-PEG-Tf were highly biocompatible and not cytotoxic. Furthermore, AuNPs-PEG-Tf have also demonstrated enhanced cell uptake in prostate cancer cells via receptor-mediated pathway.[56] Moreover, positively charged AuNPs-PEI 2 kDa were successfully conjugated by covalent attachment of anisic acid (AA) and folic acid (FA), both known as targeting ligands for prostate cancer, using *N*-hydroxy succinimide and 1-ethyl-3-(3-dimethylaminopropyl)carbodiimide hydrochloride) (NHS)/EDC coupling chemistry. Our AuNPs-PEI-targeting ligand were shown to efficiently complex with small interfering RNA (siRNA) and were not cytotoxic; they have also demonstrated a high in vitro–specific uptake, successful endosomal escape, and gene knockdown in prostate cancer cell, making them very useful for siRNA delivery and further biomedical applications.[56–58] More work still needs to be performed in order to understand the effect of temperature and pH on the synthesis of AuNPs via hydroxyl-*o*-sulfonic acid reduction, and to graft other PEG-biomolecules (i.e., PEG-FA, PEG-AA, PEG-mannose, PEG-fluorescent dye, and PEG-doxorubicin) onto AuNPs, as well as on testing these AuNPs-biomolecule conjugates in biology, especially in vivo.

6.2 MATERIALS AND METHODS

6.2.1 Chemicals and Materials

Tetrachloroauric(III) acid trihydrate ($HAuCl_4 \cdot 3H_2O$), sodium citrate ($C_6H_5Na_3O_7 \cdot 2H_2O$), sodium borohydride ($NaBH_4$), HOSA ($NH_2SO_4H$), sodium hydroxide (NaOH), 4,7,10,13,16,19,22,25,32–35,38,41,44,47,50,53-hexadecaoxa-28,29-dithiahexapentacontanedioic acid di-*N*-succinimidyl ester (NHS-PEG-S-S-PEG-NHS, PEG NHS ester disulfide ($n = 7$), BSA, L-ascorbic acid, branched PEI with molecular weight (2 kDa) solution 50% (w/v), AA, FA, dry dimethyl sulfoxide (DMSO), triethylamine, dicyclohexylcarbodiimide, and *N*-hydroxysuccinimide were purchased from Sigma-Aldrich. Thiol terminated poly(ethylene glycol) methyl ether, $M_w = 2100$, 5400, and 10,800 g mol^{-1}, was purchased from Polymer Source. ApoE human plasma very-low-density lipoprotein was purchased from peptide Startech Scientific; 50 mM NH_4HCO_3 medium was exchanged with citric buffer (pH ~7) by dialysis at 4°C for 24 h prior to use.

6.2.2 Preparation and Bioconjugation of AuNPs

Diameter of 4 ± 1 nm AuNPs-citrate: To an aqueous solution (150 mL) of $HAuCl_4 \cdot 3H_2O$ (0.25 mmol L^{-1}) was added 0.22 mL of a 340 mmol L^{-1} sodium citrate aqueous solution, and the mixture was stirred vigorously in an ice bath, followed by the addition of 0.375 mL of an ice-cold solution of $NaBH_4$ (100 mmol L^{-1}). An instantaneous color change from pale yellow to deep red-orange was noted after addition of $NaBH_4$. The AuNPs obtained with this procedure were ~4 ± 1 nm.

Diameter of ~15 nm AuNPs-citrate: 50 mL of an aqueous solution of $HAuCl_4 \cdot 3H_2O$ (0.25 mmol L^{-1}) was heated to 95°C while stirring; 0.17 mL of 340 mmol L^{-1} sodium citrate aqueous solution was rapidly added. The color of the solution changed from pale yellow to dark blue, and then to deep red-burgundy within about 8 min. Stirring and heating were maintained during 1 h after addition of sodium citrate. The heat was then removed and the solution was kept under stirring, until cooled to RT. The AuNPs obtained with this procedure were ~15 ± 2.5 nm.[33]

Diameter of ~30 nm AuNPs-citrate: A similar method was used for 15 nm AuNPs with a slight change where the pH of the solution was adjusted to ~7 with NaOH before heating, and the heat was maintained for 4 h after addition of sodium citrate. The AuNPs obtained with this procedure were ~30 ± 4 nm.[55]

Diameters of ~60 nm, ~90 nm, and ~150 nm AuNPs-citrate: For AuNPs larger than 30 nm, a very weak reducing agent, HOSA, was used in this study for the first time to reduce $HAuCl_4{\cdot}3H_2O$ in the presence of sodium citrate at RT. For ~60 nm AuNPs, 150 mL of an aqueous solution of 0.1 mmol L^{-1} $HAuCl_4{\cdot}3H_2O$ and 0.48 mmol L^{-1} sodium citrate was added to HOSA (final concentration 0.15 mmol L^{-1}). After addition of NH_2SO_4H, the color of the solution changed from pale yellow to colorless in 5 min, then to gray, blue, and then slight red-pink after several hours; the solution was kept under stirring for about 18 h. The AuNPs obtained were 61 ± 6.5 nm. By increasing the concentration of $HAuCl_4$ from 0.1 to 0.2 and 0.5 mmol L^{-1}, while keeping all the other conditions fixed, we found that the size of AuNPs increased from 61 ± 6.5 to ~92 ± 12 nm and ~148 ± 22 nm, respectively.

Diameter of ~25 nm AuNPs-PEI 2 kDa: For 128 mL of an aqueous solution of $HAuCl_4{\cdot}3H_2O$ (0.25 mmol L^{-1}) at RT, 2.23 mL of a 2.3 mmol L^{-1} PEI 2 kDa aqueous solution was added. The color of the solution changed from pale yellow to deep yellow upon addition of PEI; afterward, 0.29 mL of 109 mmol L^{-1} of ascorbic acid was quickly added, the solution turned deep red within about 1 min, and the solution was left under stirring for 16 h. The AuNPs obtained with this procedure were ~26 ± 5 nm.

Preparation of larger AuNPs-PEI 2 kDa: For the synthesis of larger AuNPs-PEI, a similar protocol was used while decreasing the PEI concentration.

PEGylation of AuNPs-citrate: PEGylation was performed by adding PEG_{280}-S-S-PEG_{280}-NHS to the AuNP solution, and stirring was maintained for 2 h.

Protein bioconjugation: The pH of the AuNPs-PEG-NHS solution was adjusted to between 7.5 and 9 by adding few drops of NaOH (0.1 mol L^{-1}) solution, followed by the addition of the protein solution (~50 protein/particle) in citric buffer to the AuNP colloidal solution. The solution was left to react under shaking for 4 h prior to DLS measurement.

6.2.2.1 Conjugation of AuNPs-PEI with AA and FA

The carboxylic acid group in FA and AA was first activated to form *N*-hydroxysuccinimide ester. The protocols and characterization of the resulting anisic-NHS and FA-NHS are described in Refs. 56 and 57, respectively.

6.2.2.2 Synthesis of AuNPs-PEI-FA

For 3 mL of ~110 nm AuNP-PEI 30 μL of AA-NHS (25 mg mL^{-1}) were added and incubated with stirring at RT for 10 min. Following incubation, 50 μL of 0.1 mol L^{-1} NaOH were added to the solution and left under stirring at RT for 48 h in dark, achieving AuNP-PEI-FA (thereafter referred to as AuNP-PEI-FA). The resultant AuNP-PEI-FA solution was purified using centrifugation at 12,000 rpm for 5 min, and the free FA-NHS in the supernatant was quantified using a CARY UV–vis spectrophotometer with a Xenon lamp (300–900 nm range, 0.5 nm resolution).[56]

6.2.2.3 Synthesis of AuNPs-PEI-AA

Two hundred and fifty microliters of NaOH 0.1 M were added to 25 mL of presynthesized ~60 nm Au-PEI NPs (~0.25 mmol L^{-1}–11.2 μmol L^{-1}), followed by the addition of 0.5 mL of anisic-NHS (25 mg mL^{-1}) in dry DMSO, and the solution was left under stirring for 21 h at RT to achieve Au-PEI-AA NPs. The resulting Au-PEI-AA solution was purified using centrifugation at 12,000 rpm for 15 min, and the AA -NHS in the supernatant was quantified using a CARY UV–vis spectrophotometer with a Xenon lamp (190–900 nm range, 0.5 nm resolution).[57]

PEGylation of AuNPs-protein conjugate and AuNPs-PEI: A solution of mPEG-SH of the desired molecular weight was added to a solution of citrate-capped AuNPs or PEI-capped AuNPs with stirring. The solution was stirred for ~1 h, allowing mPEG-SH to be grafted on the AuNPs surface prior to DLS measurement.

6.3 INSTRUMENTATION

Optical spectra were obtained on a CARY UV–vis spectrophotometer with a Xenon lamp (300–900 nm range, 0.5 nm resolution).

DLS and zeta potential: Measurements were carried out with the Malvern instrument (Zeta sizer Nano series) at 25°C; measurements on each sample were performed in triplicate.

Transmission electron microscopy (TEM): A drop of nanoparticles dispersion was first placed on a carbon-coated TEM copper grid (Quantifoil, Germany) and left to air-dry, before being introduced into the sample chamber of the TEM. Samples were analyzed using a JEOL JEM-2100 TEM operating at 200 kV. All images were recorded on a Gatan 1.35 K × 1.04 K × 12 bit ES500W CCD camera. TEM images were analyzed using *Image J* software.

6.4 RESULTS AND DISCUSSION

6.4.1 Chemical Synthesis and Characterization of AuNPs

AuNPs with size-dependent optical properties are now very useful in different fields of nanotechnology.[51,59,60] Different ways to produce AuNPs are now available in the literature.[34,41,61–65] The most frequently used techniques are based on the reduction of gold chloride precursor, mainly of $HAuCl_4$, by different reducing agents.[29,30,33,34,41,61–66] However, no universal reducing agent is known yet in order to produce monodisperse AuNPs of different sizes; instead, different ways such as seeding growth methods and different reducing agents with variable strength are usually used.[31–33,37,54] Chemical reduction of $HAuCl_4$ with a strong reducing agent such as sodium borohydride or hydrazine occurs very fast and in a non-controllable manner to produce AuNPs usually with diameter of less than 5 nm.[54] The reduction with sodium citrate produces AuNPs with diameters larger than 8–9 nm and is more widely used to date for the synthesis of AuNPs and more widely studied.[55,67,68] The mechanism of the formation of AuNPs obtained by citrate reduction of $HAuCl_4$ was found to contain four steps as concluded by Polte *et al.* from their study using in situ nanoparticle growth monitoring via XANES and SAXS.[67] Moreover, Ji *et al.* have described two pH-dependent mechanisms for the formation of AuNPs with citrate, whereby AuNPs were found to form by fast, random particle attachment and ripening for solutions with pH <6.5, whereas slower nucleation and ripening were observed for synthesis solutions with pH 6.5–7.7.[55] In our study, we have synthesized AuNPs citrate with diameters ranging between 4 and 150 nm using three different reducing agents. The produced AuNPs have size-dependent optical absorption as shown from the characterization using UV–vis absorption spectroscopy (Figure 6.1), TEM (Figures 6.2 and 6.3), and DLS (Figure 6.4). Figure 6.1 shows the UV–vis spectra of the resulting nearly monodisperse different diameter AuNPs (~4, ~15, ~30, ~60, ~90, and ~150 nm), and a clear red shift of the plasmon absorption maximum (λ_{max}) from 513 to 600 nm accompanied with a further broadening, of the absorbance band from 50 to 380 nm with increasing AuNP diameter from ~4 ± 1, to ~148 ± 22 nm, respectively, is observed. The red shift and the SPR broadening are due to higher oscillation modes (quadrupole, octopole absorption, and scattering) that also affect the extinction cross-section with increasing size.[69] The possible chemical reduction reactions that take place in each case are also shown in Figure 6.1. The use of the strong reducing agent sodium borohydride in the presence of sodium citrate as stabilizing agent at RT allowed us to produce AuNPs with diameter of 4 ± 1 nm, as resulted from TEM analysis (Figure 6.2a), while the 15 nm AuNPs were obtained by

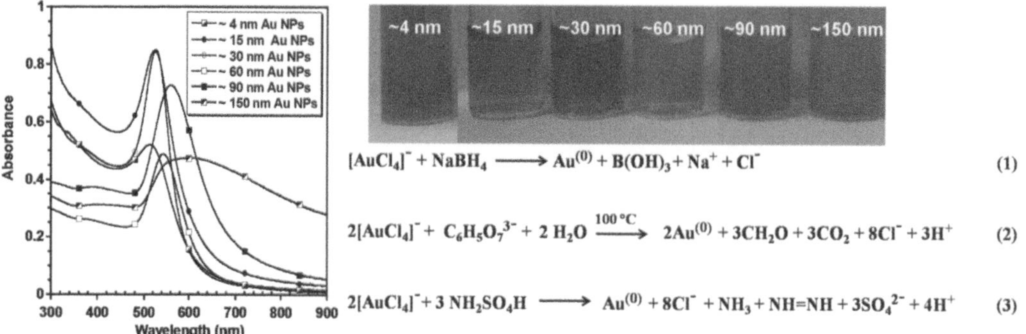

FIGURE 6.1 UV–visible spectra (left) of citrate-capped AuNPs-citrate synthesized in this work; the corresponding colloidal solution (right) and the possible chemical reactions for reduction of $HAuCl_4$ with the different reducing agents (sodium borohydride [1], sodium citrate [2], and HOSA) used in this study are also shown.

chemical reduction of $HAuCl_4$ with sodium citrate at pH <6.5 and at a temperature of about 95°C, suggesting that AuNPs form by fast, random particle attachment and ripening, as outlined by Ji *et al.* (Figure 6.2b).[55] Histograms of the nanoparticle diameters presented on the right of each TEM image are obtained from TEM image analysis using *Image J* software. An attempt to produce larger AuNPs at pH <6.5 was also performed through the decrease of both the citrate/$HAuCl_4$ ratio and

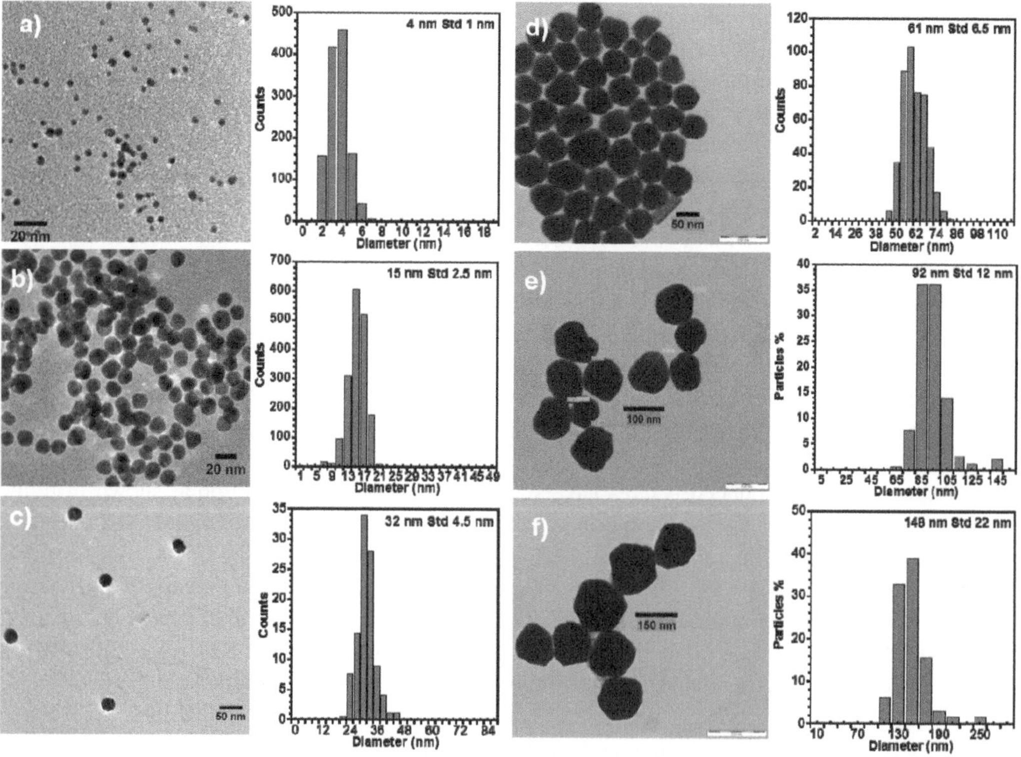

FIGURE 6.2 TEM micrographs of the AuNPs with diameter ~4 nm (a) obtained from $NaBH_4$, ~15 nm (b); ~30 nm (c) obtained from citrate; and ~60 nm (d); ~90 nm (e), and ~150 (f) nm obtained from NH_2OSO_2OH reduction. To the right of each picture is a histogram showing the AuNP diameter distribution obtained from TEM image analysis using *Image J* software.

the temperature. Our results here show clearly that when the nanoparticles increase above 35 nm for a citrate/$HAuCl_4$ ratio of 0.76, different shapes with mostly elongated particles were obtained in the final solution, as shown in Figure 6.3a. However, 30 nm AuNPs with better circularity and polydispersity used in this study were obtained by a previously reported method by Ji *et al.* through the chemical reduction of $HAuCl_4$ with sodium citrate at pH >6.5 (pH ~7 adjusted by addition of NaOH 1 M to the solution mixture) and at a temperature of about 95°C; the increase in pH to about 7 results in a more controlled synthesis based on slower nucleation and ripening, and the resulting nearly spherical AuNPs with diameter of 32 ± 4.5 nm are shown in Figure 6.2c.[55] Further decreasing in the ratio of citrate/$HAuCl_4$ to 0.35 and 0.26 as well as the temperature from 100 to about 70°C led only to a slight increase in the size of the nanoparticles with a large increase in polydispersity and shape diversity, as shown in the TEM images presented in Figure 6.3c and d. In fact, we show here that it is not possible to obtain large AuNPs by simply using the Turkevich and Frens method via a simultaneous decrease of the citrate/$HAuCl_4$ ratio and the temperature.

Large AuNPs are usually obtained through a seeding growth method.[31,37,39] We have reported recently on the synthesis of AuNPs with diameters ranging between 15 and 170 nm in aqueous solution using a seed-mediated growth method, employing hydroxylamine hydrochloride as a reducing agent.[33] Here we report on a new method for the synthesis of AuNPs of diameters above 50 nm using a simple one-step synthesis based on the use of HOSA as a new reducing agent of $HAuCl_4$ at RT and in the presence of sodium citrate as stabilizing agent. In fact, Zou *et al.* reported a seed-mediated synthesis of branched AuNPs with the assistance of citrate using hydroxylamine sulfate in the presence of AuNPs-citrate seeds.[70] Rozenkranz found in 1973 that exposure of DNA solutions to low levels (2 mmol L^{-1}) of HOSA resulted in limited degradation without significant change to the thermal helix-coil profile of the DNA[71]; these results make HOSA a biologically

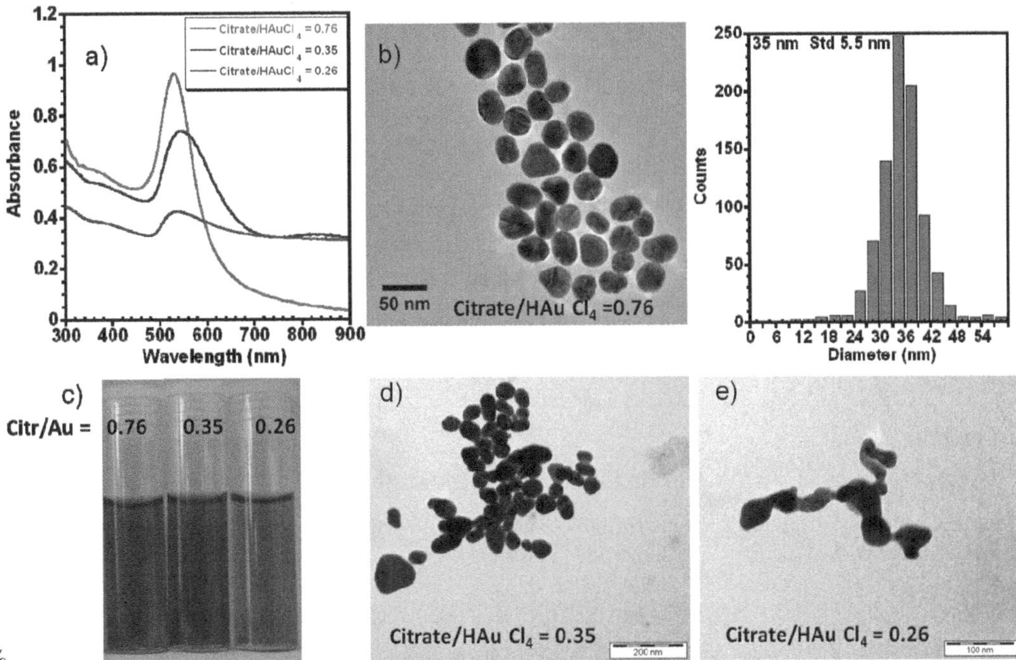

FIGURE 6.3 UV–visible spectra (a) of AuNPs obtained from the reduction with citrate at different citrate/$HAuCl_4$ at ~80°C. The corresponding TEM micrographs [(b), (d), and (e)] and colloidal solution of each ratio are also presented (c). The histogram shows the nanoparticle diameter distribution obtained from TEM image analysis using *Image J* software for the citrate/$HAuCl_4$ ratio of 0.76.

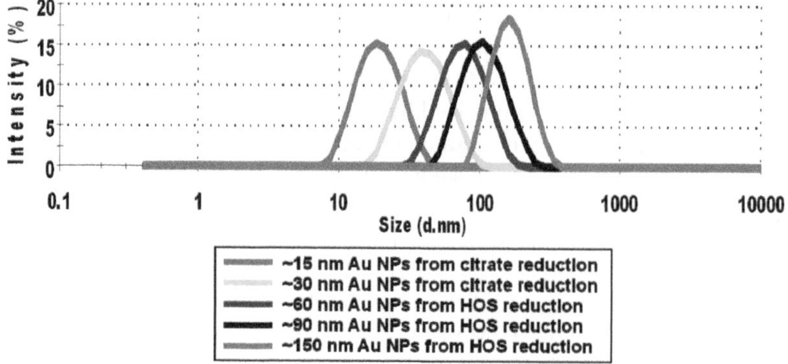

FIGURE 6.4 Size distribution by intensity obtained from DLS on the nearly monodisperse AuNPs obtained in this work, AuNPs with diameter ~60; and AuNPs of ~90, and ~150 nm obtained by reduction of $HAuCl_4$ (0.1, 0.2, and 0.5 mmol L^{-1}, respectively) with NH_2OSO_2OH (HOSA) at room temperature (PdI < 0.2).

friendly reducing agent and a candidate of choice. Nearly monodisperse AuNPs with diameters of about ~60, ~90, and ~150 nm with size-dependent optical properties (Figure 6.1) were obtained in this study, as shown in Figure 6.2d–f, from the electron micrographs of the AuNPs obtained via the reduction of 0.1, 0.2, and 0.5 mmol L^{-1} $HAuCl_4{\cdot}3H_2O$ with NH_2SO_4H at 0.155, 0.31, and 0.775 mmol L^{-1}, respectively. Our results from TEM image analysis using *Image J* software show that the size was found to increase from 61 ± 6.5 to 92 ± 12 nm and 148 ± 22 nm when the concentration of both $HAuCl_4$ and NH_2SO_4H increases, while keeping the ratio of $NH_2SO_4H/HAuCl_4$ constant at 1.55. The circularity of the resultant nanoparticles was also not very affected, and nearly spherical shapes were obtained. In fact, NH_2SO_4H seems to be a very weak reducing agent, thus allowing a slow reduction of $HAuCl_4$ (more than 18 h) that leads to a nearly monodisperse colloidal solution. Figure 6.4 shows the diameter distribution by intensity of synthesized citrate-capped AuNPs with mean diameters of ~15, ~30, ~60, ~90, and ~150 nm before conjugation. The size distribution peak by intensity was found to shift from ~20 nm (Zav 17.5 nm) to ~170 nm (Zav = 161 nm) when the size increases from ~15 ± 2.5 to 148 ± 22 nm (results from TEM). The zeta potential of all the AuNPs was ranged from −31 mV for 15 nm AuNPs to −40 mV for ~30 nm, as shown in Table 6.1. As discussed previously, we note here that the final size of the obtained nanoparticles from HOSA was found to depend on both the reducing agent and $HAuCl_4$ concentrations. This new method presented in this study using NH_2SO_4H can be applied as an extension of the approaches described to produce AuNPs-citrate with diameter above 40 nm by direct reduction. However, since

TABLE 6.1
Characteristics of AuNPs-Citrate Obtained in This Study

Reducing Agent and Conditions	UV–vis λ_{max} (nm)	Plasmon Width $\Delta\lambda$ (nm)	Diameter (nm) from EM	Diameter, Zav (nm) from DLS	PDI from DLS	Zeta Potential (mV)
Sodium borohydride/(RT)	513	84	4 ± 1	–	–	–
Citrate (95°C)	521	104	15 ± 2.5	17.5 ± 0.3	0.12	−30 ± 0.8
Citrate (95°C)/ pH ~7	527	112	30 ± 4	34 ± 0.1	0.21	−46 ± 0.1
HOSA	543	120	61 ± 6.5	65 ± 0.4	0.2	−25 ± 1
HOSA	560	158	92 ± 12	92 ± 0.5	0.2	−30 ± 0.2
HOSA	600	350	148 ± 22	161 ± 1	0.13	−33 ± 1

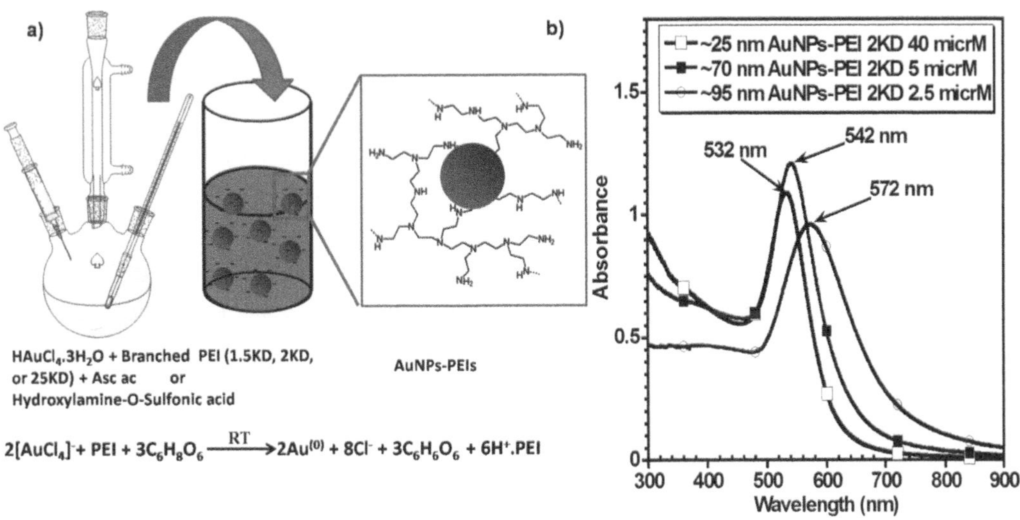

FIGURE 6.5 (a) Scheme of synthesis procedure of AuNPs-PEIs and possible chemical reduction. (b) UV–visible spectra of PEI-capped AuNPs used in this work.

both the gold precursor and the reducing agents are acidic, the pH is expected to decrease when their concentration increases.[15,55,72,73] More studies are now in progress to study the effect of pH and temperature on the final nanoparticles size distribution as well as the yield of this reduction. Similarly, AuNP-PEI with diameters ranging between 25 and 150 nm could be obtained in aqueous solutions using PEI (1.3, 25, and 2 kDa) as capping ligands and HOSA or ascorbic acid as reducing agents. Our results demonstrated that different parameters such as gold salt concentrations, PEI molecular weight/concentrations, and temperature can affect the final size of AuNPs. Figure 6.5a shows the schematic of the synthesis procedure of AuNPs-PEIs, while Figure 6.5b shows the UV–vis spectra of some selected different sizes of AuNPs-PEI 2 kDa obtained through variation of the PEI 2 kDa concentration, while fixing the concentrations of AA (0.388 mmol L^{-1}) and $HAuCl_4 \cdot 3H_2O$ (0.25 mmol L^{-1}). It is clearly seen from UV–vis spectra that modifying the concentration of PEI 2 kDa affect the position of the plasmon resonance band of the resulting AuNPs. The plasmon band was shown to shift by about 40 nm from 532 nm when 40 μM PEI 2 kDa was used to 542 and 572 nm when the concentration of PEI 2 kDa was decreased to 5 and 2.5 μmol L^{-1}, respectively. This plasmon shift also translated into a change in the color of the colloidal solutions of AuNPs-PEI 2 kDa. Therefore, the AuNPs-PEI offer size-dependent optical properties with a plasmon band shift to longer wavelengths related to an increase in the size of the AuNPs with decreasing PEI concentrations as demonstrated from scanning electron microscopy (SEM) micrograph analysis (Figure 6.6) and DLS measurements (Figure 6.7a). Full characteristics of AuNPs-PEI 2kDa obtained in this study are represented in Table 6.2. Figure 6.6 (upper left) shows that the AuNPs-PEI obtained with a high concentration of PEI 2 kDa (40 μmol L^{-1}) are well dispersed, with a diameter of $\sim 26 \pm 5$ nm as estimated from *Image J* software analysis (histogram lower left), whereas the size was found to increase to 70 ± 18 nm (upper and lower middle) and to $\sim 96 \pm 20$ (upper and lower right) when the PEI concentration was decreased to 5 and 2.5 μmol L^{-1}, respectively. In addition, DLS analysis was used to determine the hydrodynamic diameter (D_h) of AuNPs-PEI hybrid particles in colloidal solution. DLS measurements shown in Figure 6.7 represent the size distribution by intensity of the samples in Figure 6.6. DLS indicated that all samples were nearly monodisperse with one size distribution. It is clearly seen that the hydrodynamic diameter increases from around 55 nm for 40 μmol L^{-1} PEI to 97 nm with 5 μmol L^{-1} PEI and 123 nm with 2.5 μmol L^{-1} PEI, further confirming that the size of the AuNPs-PEI increases when the concentration of the PEI decreases. It was also noticed here that the

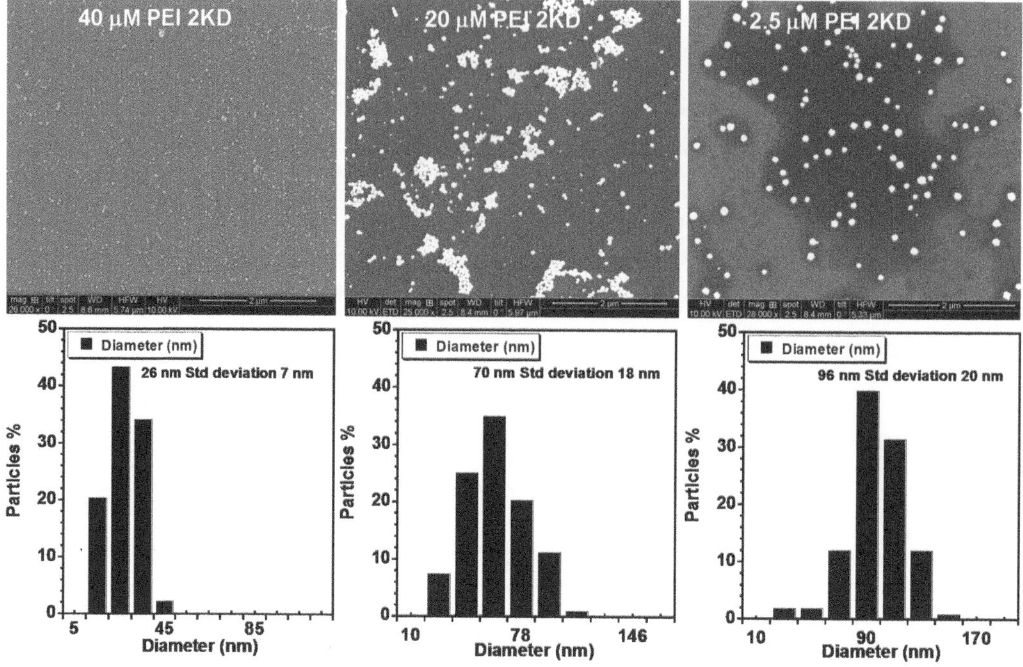

FIGURE 6.6 SEM micrographs of the AuNPs-PEI 2 kDa show that the AuNPs core size increases with decreasing PEI 2 kDa concentration; the AuNPs are well dispersed and nearly spherical in shape with diameters ~25 nm (upper left), ~70 nm (upper middle), and ~96 nm (upper right) obtained from AA reduction. Below each picture is a histogram showing the AuNPs diameter distribution obtained from SEM images analysis using *Image J* software.

polydispersity index (PDI) decreased with the concentration of PEI (i.e., from 0.266 to 0.109 when the concentration of PEI decreased from 40 to 2.5 μmol L^{-1}). Furthermore, the size distribution by intensity from DLS was also found to be much larger than the size measured by SEM, indicating that the AuNPs were successfully coated with a layer of branched PEI. To further confirm that the nanoparticles are capped with branched PEI, the zeta potential (ζ) was measured, and the results are shown in Figure 6.7b. Zeta potential measurements showed that all AuNPs-PEI samples were positively charged with a ζ potential in the range of 37 ± 3 mV, leading to a very high stability of the colloidal solution for several months when stored at 4°C. In fact, AuNPs produced in this study are electrostatically stabilized, and as such, they are very sensitive to any change in the ionic strength and/or pH of the medium in which they are dispersed, which can induce nanoparticle aggregation. We have demonstrated previously that conjugation on the surface of AuNPs-citrate, either through Au-SH chemical bonding with PEG or by auto-assembly of triblock and diblock surfactant polymers

TABLE 6.2
Characteristics of AuNPs-PEI 2 kDa Obtained in This Study

Reducing Agent and Conditions	UV–vis λ_{max}(nm)	Plasmon Width $\Delta\lambda$ (nm)	Diameter (nm) from EM	Diameter, Zav (nm) from DLS	PDI from DLS	Zeta Potential (mV)
Ascorbic acid/(RT)	532	106	25 ± 7	46 ± 0.1	0.266	+40 ± 0.5
Ascorbic acid/(RT)	542	124	70 ± 18	85 ± 0.3	0.125	+38 ± 0.1
Ascorbic acid/(RT)	572	174	96 ± 20	109 ± 0.1	0.109	+35 ± 0.3

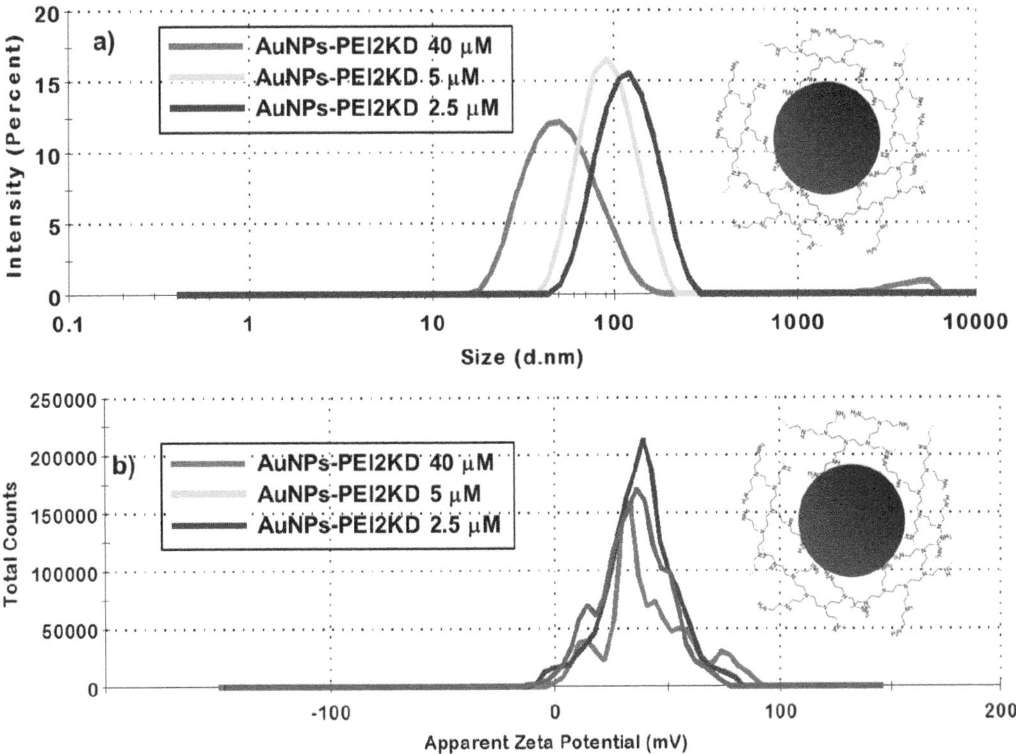

FIGURE 6.7 (a) Size distribution by intensity of AuNPs-PEI 2 kDa from DLS measurements, showing that the size increases from 55 to 123 nm when the concentration of PEI 2 kDa decreases from 40 to 2.5 μM. (b) Zeta potential for the corresponding AuNPs-PEI 2 kDa showing that all AuNPs-PEI samples were positively charged with a ζ-potential in the range of 37 $\pm$ 3 mV.

based on polyethylene oxide and polypropylene oxide, can improve the stability and biocompatibility of the nanoparticles.[7,33,68,74,75] Moreover, proteins such as BSA and other biomolecules are also known to increase the stability of nanoparticles under physiological condition.[50,76–79]

6.4.2 Bioconjugation of AuNPs

The presynthesized AuNPs-citrate in water of mainly 15 and 30 nm (obtained through the Turkevich/Frens method) and ~90 nm obtained by NH_2OSO_2OH reduction were used for further bioconjugation with different proteins attached through a biologically stable amide linkage in water. In this study, ApoE protein, which is known to enhance the permeability of nanoparticles across the blood–brain barrier, was attached on 30 nm AuNPs (Scheme 6.1)[80,81]; a similar method was also used to conjugate 15 nm AuNPs BSA.[47,82] As we previously reported, the stability of the AuNPs under physiological conditions (0.157 mol L^{-1} NaCl) can be well improved by grafting a PEG-based thiol polymer shell on their surface.[33,75] We have also determined the number of PEG-SH ligands needed to coat the surface of an AuNP of a particular diameter and estimated the grafting density using thermal gravimetric analysis and TEM.[33] In this study, we have chosen to graft first a functional oligomer NHS-PEG_{280}-S-S-PEG_{280}-NHS, based on 14 ethylene glycol units, thiol disulfide, and NHS reactive ester group as an active group on the AuNPs surface (Scheme 6.1) used as both a stabilizing agent and chemical linker that is able to react with the amino group of the protein in water.[21,47,81,83,84] Therefore, the addition of NHS-PEG_{280}-S-S-PEG_{280}-NHS to the presynthesized AuNPs colloidal solution resulted in the formation of a NHS functional PEG layer grafted on the

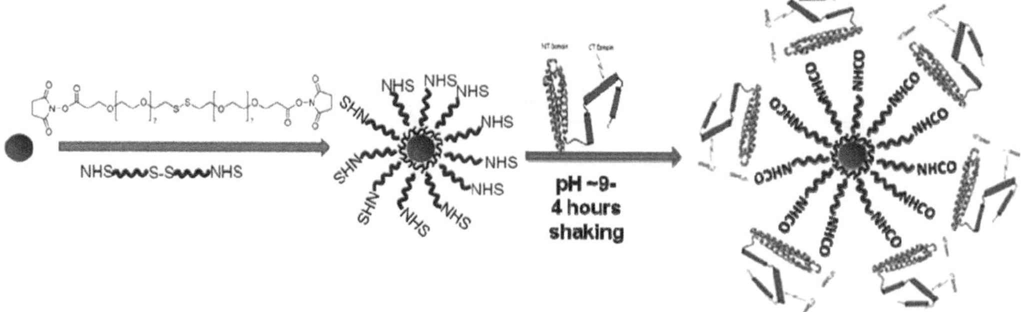

SCHEME 6.1 Formation of AuNPs-PEG-NHS and grafting of ApoE protein on AuNPs-PEG-NHS in water; a similar synthesis was used to graft BSA on 15 nm AuNPs.

AuNPs surface, where NHS-PEG$_{280}$-S is attached through a gold thiol (Au–S) chemical bond, as shown in Scheme 6.1. The successful PEGylation of the AuNPs was confirmed by DLS/zeta potential (ξ) measurements, where a slight increase in the size of nanoparticles of about 2–3 nm was detected (Figure 6.8), and ξ was found to decrease slightly from −40 to about −35 mV. After PEGylation, ApoE protein dispersed in a citric buffer solution (pH ~7) was added to the AuNPs-PEG-NHS solution (Scheme 6.1), the pH was adjusted to ~9, and the solution was left to react for about 4 h under shaking. The protein was able to be grafted on the PEG layer of the AuNPs through formation of an amide linkage (NHCO) between the amino groups of the protein and the NHS at the surface of the AuNPs-PEG-NHS, as confirmed by DLS (Figure 6.8). Figure 6.8 shows a size increase of about 5 nm is after protein conjugation. Similarly, the BSA was also grafted on the 15 nm AuNPs (Figure 6.9), and the size was found to increase by about 4 nm after PEG-NHS attachment and by 7 nm after protein attachment. However, we note here that the conjugation of 15 nm AuNPs-PEG-NHS seems to cause a slight flocculation, as can be detected from DLS measurements; indeed, no change in the initial red color of the colloidal solution was observed. The filtration of the AuNPs-PEG-NHCO-BSA colloidal solution through a 0.2-μm nylon filter was found to remove the larger size flocks observed in DLS (Figure 6.9). Finally, we note that the number of proteins used in this study was estimated to be 55 BSA proteins per 15 nm AuNPs and about 158 ApoE per 30 nm AuNPs.

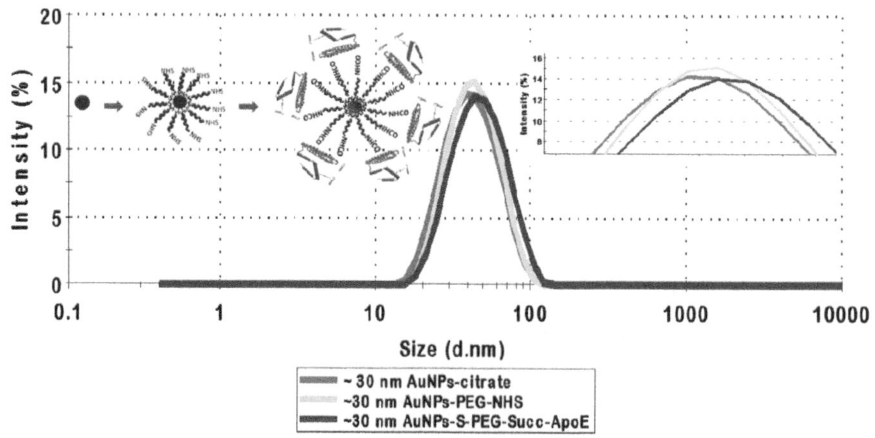

FIGURE 6.8 Size distribution by intensity obtained from DLS on 30 nm AuNPs-citrate before and after PEGylation with NHS-PEG-S-S-PEG-NHS and further attachment of ApoE protein through a biologically stable amide linkage; inset left (overall bioconjugation process Schemes 6.1 and 6.2) and inset right show the AuNPs size increase after each step.

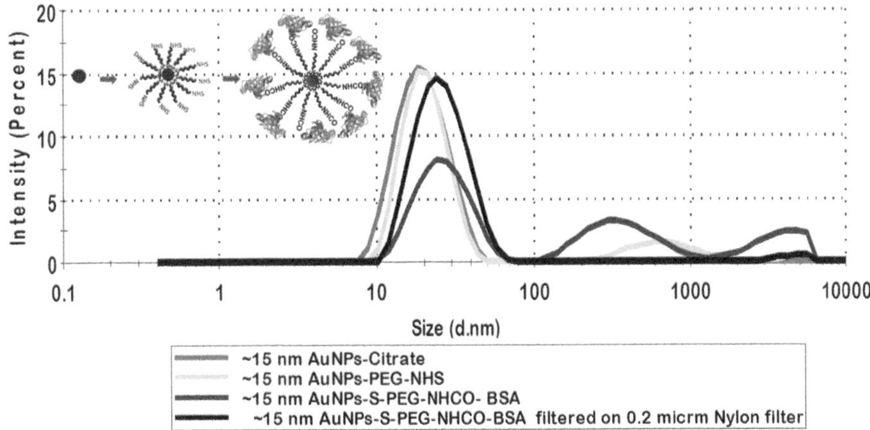

FIGURE 6.9 Size distribution by intensity obtained from DLS on 15 nm AuNPs-citrate before and after PEGylation with NHS-PEG-S-S-PEG-NHS and further attachment of BSA protein through a biologically stable amide linkage; inset left (overall bioconjugation process Scheme 6.1), conjugation on 15 nm AuNPs caused slight aggregation, and the aggregated AuNPs were removed via filtration of the colloidal solution on a 0.2-μm nylon filter.

Further stabilization of the AuNPs-protein was also performed using a tiny amount of mPEG-SH 2000, 5000, and 10,000 (Scheme 6.2). PEGylation of nanoparticles is well known to increase the stability of nanoparticles in biological fluids and to increase the circulation in the bloodstream. However, the addition of the SH-PEG may also remove some of the protein coating from the surface of the nanoparticles by ligand exchange. Therefore, the amount of SH-PEG added here was estimated in order to have a partial coverage of the AuNPs surface. Figure 6.10 shows the results from DLS on the PEGylation of the 30 nm AuNPs-PEG-ApoE, which demonstrated an increase in the size of the AuNPs-PEG-ApoE by ~2, 7, and 12 nm accompanied by an increase in the zeta potential of −34, −28, and −20 mV (results not shown) with an increase in the PEG length from 47 to 122 and 245 EO units, respectively. This increase in the size and zeta potential indicates the successful attachment of the PEG on the AuNPs surface, which might increase their stability.[75] We note that from our results obtained in our previous study on PEGylation with a maximum coverage of the AuNPs surface with mPEG-SH, a larger increase in the nanoparticles size by about, ~8, ~15, and ~23 nm with an increase of the PEG length from 47 to 122 and 245 EO units, respectively, was observed, and the zeta potential was also found to increase to $\sim -8 \pm 2$ mV for the PEG with 245 EO units.[28] Therefore,

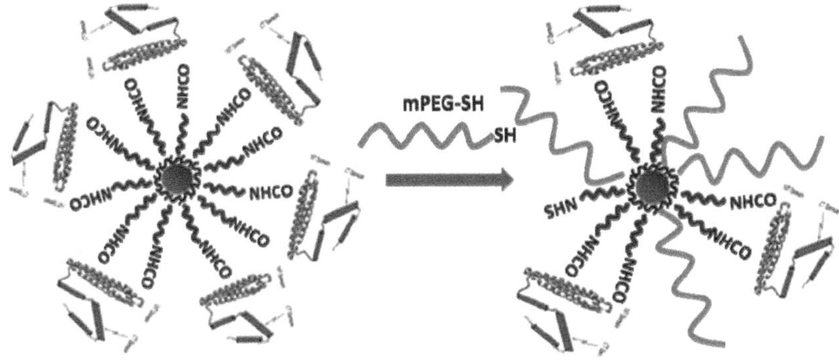

SCHEME 6.2 Further PEGylation of ~30 nm AuNPs-PEG-NHCO-APoE in water (mPEG-SH with M_w of 2100, 5400, and 10,800 g mol^{-1}) was used.

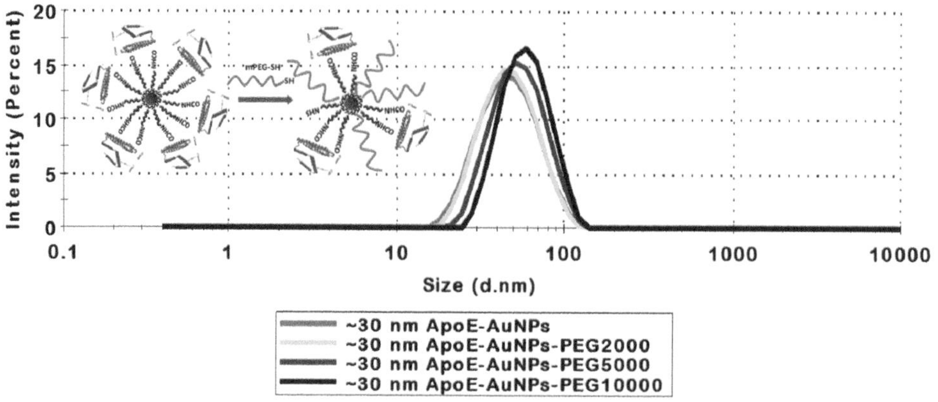

FIGURE 6.10 Size distribution by intensity obtained from DLS on 30 nm AuNPs-PEG-NHCO-ApoE (ApoE-AuNPs) before and after PEGylation with mPEG-SH (M_w ~2100, 5400, and 10,000 g mol^{-1}); inset shows the overall process (Scheme 6.2).

only a partial coverage with mPEG-SH is supposed here, leaving the proteins grafted on the AuNPs surface. The effect of the PEGylation on biological applications (uptake/delivery protein stability and conformation) will be evaluated in a future study.

DLS was found here to be a very sensitive technique for the characterization of dispersions of nanoparticles and nanoparticle–protein/polymer hybrids. In fact, DLS is also known as a method of choice for protein (and other biomolecule) molecular weight distributions in solution.[85] Consequently, the conjugation of protein and PEG can dramatically affect the Brownian motion of particles by introducing additional frictional drag and thus reducing the nanoparticle diffusivity that allowed us to follow each step of conjugation (PEGylation/protein attachment). As shown in Figures 6.5, 6.6, and 6.7, a shift in the size distribution of AuNPs to bigger sizes after each step can confirm the successful grafting of PEG-NHS and proteins on the AuNPs-PEG surface. The method presented herein is successful for bioconjugation of AuNPs in water and can also be applied to other biologically active molecules, including but not limited to biotin, small molecular ligand of prostate-specific membrane antigen (urea derivatives, etc.), fluorescent dyes such as fluorescein and 5-carboxytetramethylrhodamine, and near-infrared–emitting fluorescent dye molecules (i.e., Cy7), making them very useful candidates for cell targeting, sensing, and bioimaging.

6.4.2.1 Synthesis of AuNPs-PEI 2 kDa-FA and AuNPs-PEI-AA

The synthesis of AuNP-PEI-FA and AuNPs-AA adopted in this study (Scheme 6.3) was similar to a method previously reported for the attachment of fluorescein onto gadolinium oxide NPs capped

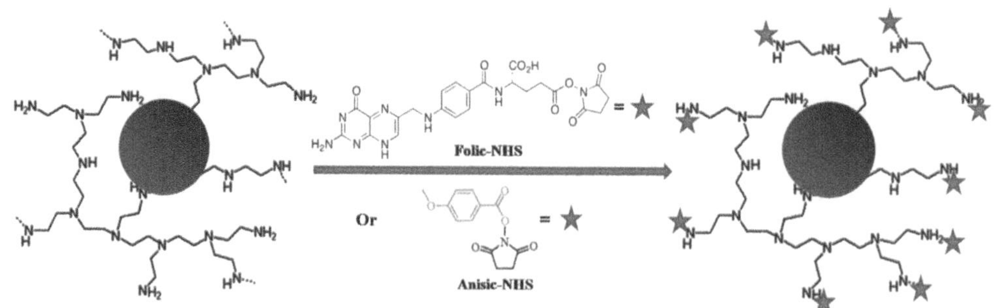

SCHEME 6.3 Formation of AuNPs-PEI-AA and AuNPs-PEI-FA by covalent attachment of anisic-NHS and folic-NHS onto AuNPs-PEI in water.

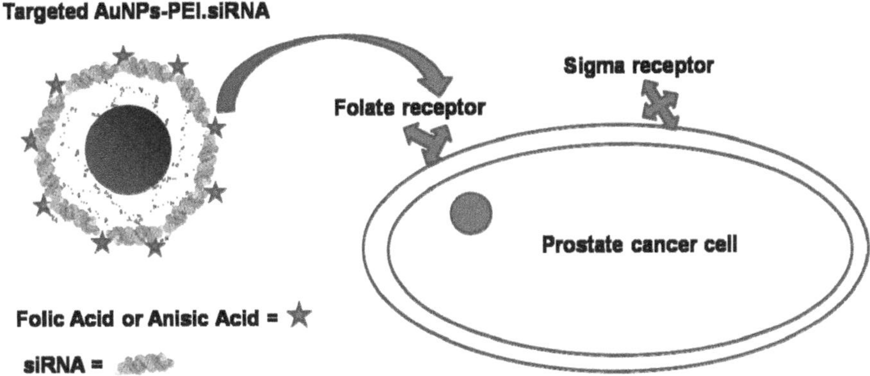

FIGURE 6.11 A schematic representation of the siRNA complexed to AA- or FA-conjugated AuNPs-PEI (targeted AuNPs-PEI siRNA) achieved through electrostatic interaction, and the specific delivery of siRNA to prostate cancer cells via receptor-mediated pathway.

with PEI.[86] The amines of PEI were reacted with activated FA-NHS and anisic-NHS to produce an amide linkage. Data from DLS and zeta potential confirmed the successful attachment of FA and AA onto AuNPs-PEI (results not shown here). In addition, the attachment of FA and AA onto AuNP-PEI was also confirmed using UV–visible spectroscopy, indicating that ~50% of the initially added AA of FA was grafted to the AuNP-PEI surface.[56,57] Finally, we have also demonstrated that the positively charged AuNPs-PEI and the bioconjugated AuNPs-PEI-AA and AuNPs-PEI-FA could complex and deliver negatively charged siRNA specifically to PC3 prostate cancer cells via a receptor-mediated pathway, inducing efficient endosomal escape of siRNA and effective downregulation of the *RelA* gene (Figure 6.11).[56,57]

To enhance stability in physiological environments and prevent extensive aggregation, further stabilization of the AuNPs-PEI 2 kDa was also performed using a tiny amount of mPEG-SH 2000, 5000, and 10,000 (Scheme 6.4). Similarly to the PEGylation of AuNPs-PEG-ApoE, DLS also shows that the size increases when the PEG length increases (Figure 6.12), and the zeta potential was also found to decrease (data not shown), both indicating successful PEGylation of AuNPs-PEI that might cause possible reduction of nonspecific binding to serum proteins and enhance their blood circulation during in vivo applications.

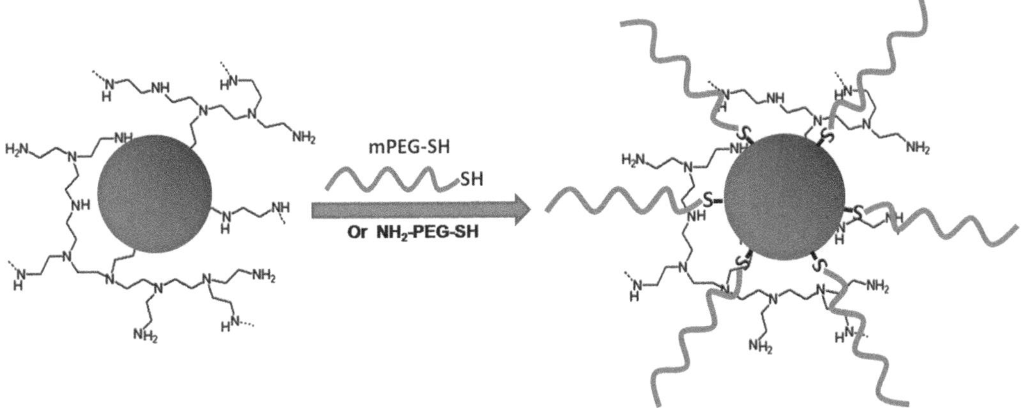

SCHEME 6.4 PEGylation of AuNPs-PEI in water, mPEG-SH, or NH_2-PEG-SH with M_w of 3500, 5000, and 7500 g mol^{-1} was used.

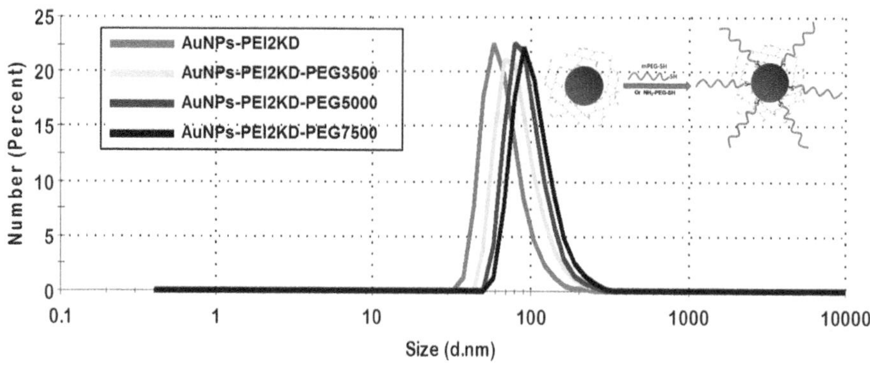

FIGURE 6.12 Size distribution by number obtained from DLS measurements on ~70 nm AuNPs-PEI before and after PEGylation with PEG-SH (M_w ~3500, 5000, and 7500 g mol^{-1}) inset shows the overall process (Scheme 6.2).

6.5 CONCLUSION

The successful applications of nanoparticles in biomedical science require a stable colloidal solution in complex media with a high ionic strength, a very low or no cytotoxic effect, as well as coating with biomolecules of interest. In this study, a new method for synthesizing large AuNPs between 25 and 150 nm through direct reduction of gold precursor using an HOSA as weak reducing agent at RT and citrate or PEI as stabilizers was performed. The obtained nanoparticles were not polydisperse (PdI <0.2) and had size-dependent optical properties. In contrast, we show here that the citrate reduction method is not successful in producing AuNPs with sizes more than 40–50 nm. Therefore, the above method represents an extension of the approaches such as described by Turkevich, Frens, and others to produce AuNPs with a diameter above 50 nm without increasing polydispersity of the colloidal solution. Finally, bioconjugation of the obtained AuNPs-citrate with proteins and AuNPs-PEI with target ligands (AA and FA) makes them very attractive for applications in targeted drug and siRNA delivery. PEGylation of the AuNPs was also shown to be possible, and the latter is required to avoid interaction with serum proteins and thus to enhance the potential for in vivo application, which is a concept to be investigated in future work.

Acknowledgment

Microscopy analysis was undertaken at the Electron Microscopy and Analysis Facility (EMAF) at the Tyndall National Institute, Cork, Ireland.

REFERENCES

1. Zheng, J., Zhou, C., Yu, M. X., Liu, J. B.: Different sized luminescent gold nanoparticles. *Nanoscale* **2012**, *4*, 4073-4083.
2. Amendola, V., Pilot, R., Frasconi, M., Marago, O. M., Iati, M. A.: Surface plasmon resonance in gold nanoparticles: a review. *Journal of Physics. Condensed Matter: An Institute of Physics Journal* **2017**, *29*, 203002.
3. Liu, C. H., Mi, C. C., Li, B. Q.: The plasmon resonance of a multilayered gold nanoshell and its potential bioapplications. *IEEE Transactions on Nanotechnology* **2011**, *10*, 797-805.
4. Li, Y., Wu, P., Xu, H., Zhang, H., Zhong, X. H.: Anti-aggregation of gold nanoparticle-based colorimetric sensor for glutathione with excellent selectivity and sensitivity. *Analyst* **2011**, *136*, 196-200.
5. Zhang, R. Y., Hummelgard, M., Olin, H.: Simple synthesis of clay-gold nanocomposites with tunable color. *Langmuir* **2010**, *26*, 5823-5828.

6. Liu, S. Q., Tang, Z. Y.: Nanoparticle assemblies for biological and chemical sensing. *Journal of Materials Chemistry* **2010**, *20*, 24-35.
7. Rahme, K., Gauffre, F., Marty, J. D., Payre, B., Mingotaud, C.: A systematic study of the stabilization in water of gold nanoparticles by poly(ethylene oxide)-poly(propylene oxide)-poly(ethylene oxide) triblock copolymers. *Journal of Physical Chemistry C* **2007**, *111*, 7273-7279.
8. Li, W. Y., Brown, P. K., Wang, L. H. V., Xia, Y. N.: Gold nanocages as contrast agents for photoacoustic imaging. *Contrast Media & Molecular Imaging* **2011**, *6*, 370-377.
9. Khlebtsov, N. G., Dykman, L. A.: Optical properties and biomedical applications of plasmonic nanoparticles. *Journal of Quantitative Spectroscopy & Radiative Transfer* **2010**, *111*, 1-35.
10. Giljohann, D. A., Seferos, D. S., Daniel, W. L., Massich, M. D., Patel, P. C., Mirkin, C. A.: Gold nanoparticles for biology and medicine. *Angewandte Chemie – International Edition* **2010**, *49*, 3280-3294.
11. Yuan, J., Deng, D. W., Lauren, D. R., Aguilar, M. I., Wu, Y. Q.: Surface plasmon resonance biosensor for the detection of ochratoxin A in cereals and beverages. *Analytica Chimica Acta* **2009**, *656*, 63-71.
12. Wang, Z. X., Ma, L. N.: Gold nanoparticle probes. *Coordination Chemistry Reviews* **2009**, *253*, 1607-1618.
13. Tong, L., Wei, Q. S., Wei, A., Cheng, J. X.: Gold nanorods as contrast agents for biological imaging: optical properties, surface conjugation and photothermal effects. *Photochemistry and Photobiology* **2009**, *85*, 21-32.
14. Ma, Z. F., Tian, L., Di, J., Ding, T.: Bio-detection, cellular imaging and cancer photothermal therapy based on gold nanorods. *Progress in Chemistry* **2009**, *21*, 134-142.
15. Zhu, L., Xue, D. H., Wang, Z. X.: Gold nanoparticle-based colorimetric sensor for pH sensing. *Chemical Research in Chinese Universities* **2008**, *24*, 537-540.
16. Zhao, W., Brook, M. A., Li, Y. F.: Design of gold nanoparticle-based colorimetric biosensing assays. *ChemBioChem* **2008**, *9*, 2363-2371.
17. Ghosh, P., Han, G., De, M., Kim, C. K., Rotello, V. M.: Gold nanoparticles in delivery applications. *Advanced Drug Delivery Reviews* **2008**, *60*, 1307-1315.
18. Guo, J., Rahme, K., He, Y., Li, L. L., Holmes, J. D., O'Driscoll, C. M.: Gold nanoparticles enlighten the future of cancer theranostics. *International Journal of Nanomedicine* **2017**, *12*, 6131-6152.
19. Carabineiro, S. A. C.: Applications of gold nanoparticles in nanomedicine: recent advances in vaccines. *Molecules* **2017**, *22*.
20. Singh, P., Pandit, S., Mokkapati, V., Garg, A., Ravikumar, V., Mijakovic, I.: Gold nanoparticles in diagnostics and therapeutics for human cancer. *International Journal of Molecular Sciences* **2018**, *19*.
21. Sperling, R. A., Parak, W. J.: Surface modification, functionalization and bioconjugation of colloidal inorganic nanoparticles. *Philosophical Transactions of the Royal Society A - Mathematical Physical and Engineering Sciences* **2010**, *368*, 1333-1383.
22. Alkilany, A. M., Murphy, C. J.: Toxicity and cellular uptake of gold nanoparticles: what we have learned so far? *Journal of Nanoparticle Research* **2010**, *12*, 2313-2333.
23. Nikolai Khlebtsov, Dykmana, L.: Biodistribution and toxicity of engineered gold nanoparticles: a review of in vitro and in vivo studies. *Chemical Society Reviews* **2011**, *40*, 1647–1671.
24. Xiao-Dong Zhang, Di Wu, X. S., Pei-Xun Liu, N. Y., Bin Zhao, H. Z., Yuan-Ming Sun, Liang-An Zhang, Fan, F.-Y.: Size-dependent in vivo toxicity of PEG-coated gold nanoparticles. *International Journal of Nanomedicine* **2011**, *6*, 2071—2081.
25. Vijayakumar, S., Ganesan, S.: Size-dependent in vitro cytotoxicity assay of gold nanoparticles. *Toxicological & Environmental Chemistry* **2013**, *95*, 277-287.
26. Carrie, A. S., Kenneth, J. S., David, E. C., Daniel, L. F.: In vivo toxicity, biodistribution, and clearance of glutathione-coated gold nanoparticles. *Nanomedicine: Nanotechnology, Biology and Medicine* **2013**, *9*, 257–263.
27. He, C. B., Hu, Y. P., Yin, L. C., Tang, C., Yin, C. H.: Effects of particle size and surface charge on cellular uptake and biodistribution of polymeric nanoparticles. *Biomaterials* **2010**, *31*, 3657-3666.
28. Takeuchi, I., Nobata, S., Oiri, N., Tomoda, K., Makino, K.: Biodistribution and excretion of colloidal gold nanoparticles after intravenous injection: effects of particle size. *Biomedical Materials and Engineering* **2017**, *28*, 315-323.
29. Turkevich, J., Stevenson, P. C., Hillier, J.: A study of the nucleation and growth process in the synthesis of colloidal gold. *Discussions of the Faraday Society* **1951**, *11*, 55-75.

30. Frens, G.: Controlled nucleation for regulation of particle-size in monodisperse gold suspensions. *Nature-Physical Science* **1973**, *241*, 20-22.
31. Brown, K. R., Natan, M. J.: Hydroxylamine seeding of colloidal Au nanoparticles in solution and on surfaces. *Langmuir* **1998**, *14*, 726-728.
32. Brown, K. R., Walter, D. G., Natan, M. J.: Seeding of colloidal Au nanoparticle solutions. 2. Improved control of particle size and shape. *Chemistry of Materials* **2000**, *12*, 306-313.
33. Rahme, K., Chen, L., Hobbs, R. G., Morris, M. A., O'Driscoll, C., Holmes, A. J. D.: PEGylated gold nanoparticles: polymer quantification as a function of PEG lengths and nanoparticle dimensions. *RSC Advances* **2013**, *3*, 6085-6094.
34. Sau, T. K., Pal, A., Jana, N. R., Wang, Z. L., Pal, T.: Size controlled synthesis of gold nanoparticles using photochemically prepared seed particles. *Journal of Nanoparticle Research* **2001**, *3*, 257-261.
35. Sau, T. K., Murphy, C. J.: Room temperature, high-yield synthesis of multiple shapes of gold nanoparticles in aqueous solution. *Journal of the American Chemical Society* **2004**, *126*, 8648–8649.
36. Choi, B. S., Iqbal, M., Lee, T., Kim, Y. H., Tae, G.: Removal of cetyltrimethylammonium bromide to enhance the biocompatibility of Au nanorods synthesized by a modified seed mediated growth process. *Journal of Nanoscience and Nanotechnology* **2008**, *8*, 4670-4674.
37. Niu, J. L., Zhu, T., Liu, Z. F.: One-step seed-mediated growth of 30-150 nm quasispherical gold nanoparticles with 2-mercaptosuccinic acid as a new reducing agent. *Nanotechnology* **2007**, *18*, -.
38. Perrault, S. D., Chan, W. C. W.: Synthesis and surface modification of highly monodispersed, spherical gold nanoparticles of 50-200 nm. *Journal of the American Chemical Society* **2009**, *131*, 17042-+.
39. Xiaokong, L., Haolan, X., Haibing, X., Wang, D.: Rapid seeded growth of monodisperse, quasi-spherical, citrate-stabilized gold nanoparticles via H_2O_2 reduction. *Langmuir* **2012**, *28*, 13720-13726.
40. Newman, J. D. S., Blanchard, G. J.: Formation of gold nanoparticles using amine reducing agents. *Langmuir* **2006**, *22*, 5882-5887.
41. Bhumkar, D. R., Joshi, H. M., Sastry, M., Pokharkar, V. B.: Chitosan reduced gold nanoparticles as novel carriers for transmucosal delivery of insulin. *Pharmaceutical Research* **2007**, *24*, 1415-1426.
42. Germershaus, O., Mao, S. R., Sitterberg, J., Bakowsky, U., Kissel, T.: Gene delivery using chitosan, trimethyl chitosan or polyethylenglycol-graft-trimethyl chitosan block copolymers: establishment of structure-activity relationships in vitro. *Journal of Controlled Release* **2008**, *125*, 145-154.
43. Guo, J. F., O'Mahony, A. M., Cheng, W. P., O'Driscoll, C. M.: Amphiphilic polyallylamine based polymeric micelles for siRNA delivery to the gastrointestinal tract: In vitro investigations. *International Journal of Pharmaceutics* **2013**, *447*, 150-157.
44. Lou, Y. L., Peng, Y. S., Chen, B. H., Wang, L. F., Leong, K. W.: Poly(ethylene imine)-g-chitosan using EX-810 as a spacer for nonviral gene delivery vectors. *Journal of Biomedical Materials Research Part A* **2009**, *88A*, 1058-1068.
45. Retnakumari, A., Setua, S., Menon, D., Ravindran, P., Muhammed, H., Pradeep, T., Nair, S., Koyakutty, M.: Molecular-receptor-specific, non-toxic, near-infrared-emitting Au cluster-protein nanoconjugates for targeted cancer imaging. *Nanotechnology* **2010**, *21*.
46. De la Fuente, J. M., Penades, S.: Glyconanoparticles: Types, synthesis and applications in glycoscience, biomedicine and material science. *Biochimica Et Biophysica Acta-General Subjects* **2006**, *1760*, 636-651.
47. Vasudevanpillai, B.: Chemical modifications and bioconjugate reactions of nanomaterials for sensing, imaging, drug delivery and therapy. *Chemical Society Reviews* **2014**, *43*, 737-962.
48. Kogan, M. J., Olmedo, I., Hosta, L., Guerrero, A. R., Cruz, L. J., Albericio, F.: Peptides and metallic nanoparticles for biomedical applications. *Nanomedicine* **2007**, *2*, 287-306.
49. Liu, Y. L., Shipton, M. K., Ryan, J., Kaufman, E. D., Franzen, S., Feldheim, D. L.: Synthesis, stability, and cellular internalization of gold nanoparticles containing mixed peptide-poly(ethylene glycol) monolayers. *Analytical Chemistry* **2007**, *79*, 2221-2229.
50. de la Rica, R., Matsui, H.: Applications of peptide and protein-based materials in bionanotechnology. *Chemical Society Reviews* **2010**, *39*, 3499-3509.
51. Mallick, K., Witcomb, M. J.: Gold nanoparticles as a delivery vehicle in biomedical applications. *Gold Nanoparticles: Properties, Characterization and Fabrication* **2010**, 225-243.
52. Pearce, M. E., Melanko, J. B., Salem, A. K.: Multifunctional nanorods for biomedical applications. *Pharmaceutical Research* **2007**, *24*, 2335-2352.

53. Pellegrino, T., Kudera, S., Liedl, T., Javier, A. M., Manna, L., Parak, W. J.: On the development of colloidal nanoparticles towards multifunctional structures and their possible use for biological applications. *Small* **2005**, *1*, 48-63.
54. Jana, N. R., Gearheart, L., Murphy, C. J.: Seeding growth for size control of 5-40 nm diameter gold nanoparticles. *Langmuir* **2001**, *17*, 6782-6786.
55. Ji, X. H., Song, X. N., Li, J., Bai, Y. B., Yang, W. S., Peng, X. G.: Size control of gold nanocrystals in citrate reduction: The third role of citrate. *Journal of the American Chemical Society* **2007**, *129*, 13939-13948.
56. Guo, J., O'Driscoll, C. M., Holmes, J. D., Rahme, K.: Bioconjugated gold nanoparticles enhance cellular uptake: a proof of concept study for siRNA delivery in prostate cancer cells. *International Journal of Pharmaceutics* **2016**, *509*, 16-27.
57. Fitzgerald, K. A, Rahme, K., Guo, J., Holmes, J. D, O'Driscoll, C. M.: Anisamide-targeted gold nanoparticles for siRNA delivery in prostate cancer – synthesis, physicochemical characterisation and in vitro evaluation. *Journal of Materials Chemistry B* **2016**, *4* 2242–2252.
58. Guo, J., Rahme, K., Fitzgerald, K. A., Holmes, J. D., O'Driscoll, C. M.: Biomimetic gold nanocomplexes for gene knockdown – will gold deliver dividends for siRNA nanomedicines? *Nano Research* **2015**, *8*, 3111-3140.
59. Boisselier, E., Astruc, D.: Gold nanoparticles in nanomedicine: preparations, imaging, diagnostics, therapies and toxicity. *Chemical Society Reviews* **2009**, *38*, 1759-1782.
60. Zeng, S. W., Yong, K. T., Roy, I., Dinh, X. Q., Yu, X., Luan, F.: A review on functionalized gold nanoparticles for biosensing applications. *Plasmonics* **2011**, *6*, 491-506.
61. Fleming, D. A., Williams, M. E.: Size-controlled synthesis of gold nanoparticles via high-temperature reduction. *Langmuir* **2004**, *20*, 3021-3023.
62. Engelbrekt, C., Sørensen, K. H., Zhang, J., Welinder, A. C., Jensen, P. S., Ulstrup, J.: Green synthesis of gold nanoparticles with starch–glucose and application in bioelectrochemistry. *Journal of Materials Chemistry* **2009**, *19*, 7839-7847.
63. Yin, X. J., Chen, S. G., Wu, A. G.: Green chemistry synthesis of gold nanoparticles using lactic acid as a reducing agent. *Micro & Nano Letters* **2010**, *5*, 270-273.
64. Lee, Y., Park, T. G.: Facile fabrication of branched gold nanoparticles by reductive hydroxyphenol derivatives. *Langmuir* **2011**, *27*, 2965-2971.
65. Panda, T., Deepa, K.: Biosynthesis of gold nanoparticles. *Journal of Nanoscience and Nanotechnology* **2011**, *11*, 10279-10294.
66. Patel, J., Němcová, L., Maguire, P., Graham, W. G., Mariotti, D.: Synthesis of surfactant-free electrostatically stabilized gold nanoparticles by plasma-induced liquid chemistry. *Nanotechnology* **2013**, *24* 245604. doi:10.1088/0957-4484/24/24/245604.
67. Polte, J., Ahner, T. T., Delissen, F., Sokolov, S., Emmerling, F., Thunemann, A. F., Kraehnert, R.: Mechanism of gold nanoparticle formation in the classical citrate synthesis method derived from coupled in situ XANES and SAXS evaluation. *Journal of the American Chemical Society* **2010**, *132*, 1296-1301.
68. Rahme, K., Vicendo, P., Ayela, C., Gaillard, C., Payre, B., Mingotaud, C., Gauffre, F.: A simple protocol to stabilize gold nanoparticles using amphiphilic block copolymers: stability studies and viable cellular uptake. *Chemistry-A European Journal* **2009**, *15*, 11151-11159.
69. Link, S., El-Sayed, M. A.: Size and temperature dependence of the plasmon absorption of colloidal gold nanoparticles. *Journal of Physical Chemistry B* **1999**, *103*, 4212-4217.
70. Zou, X. Q., Ying, E. B., Dong, S. J.: Seed-mediated synthesis of branched gold nanoparticles with the assistance of citrate and their surface-enhanced Raman scattering properties. *Nanotechnology* **2006**, *17*, 4758-4764.
71. Rozenkranz, H. S.: Hydroxylamine-O-sulfonic acid: in vitro and possible in vivo reaction with DNA. *Chemico-Biological Interactions* **1973**, *7*, 195-204.
72. Chah, S., Hammond, M. R., Zare, R. N.: Gold nanoparticles as a colorimetric sensor for protein conformational changes. *Chemistry & Biology* **2005**, *12*, 323-328.
73. Sistach, S., Rahme, K., Perignon, N., Marty, J. D., Viguerie, N. L. D., Gauffre, F., Mingotaud, C.: Bolaamphiphile surfactants as nanoparticle stabilizers: application to reversible aggregation of gold nanoparticles. *Chemistry of Materials* **2008**, *20*, 1221-1223.

74. Rahme, K., Oberdisse, J., Schweins, R., Gaillard, C., Marty, J. D., Mingotaud, C., Gauffre, F.: Pluronics-stabilized gold nanoparticles: investigation of the structure of the polymer-particle hybrid. *ChemPhysChem: A European Journal of Chemical Physics and Physical Chemistry* **2008**, *9*, 2230-2236.
75. Rahme, K., Nolan, M. T., Doody, T., McGlacken, G. P., Morris, M. A., O'Driscoll, C., Holmes, A. J. D.: Highly stable PEGylated gold nanoparticles in water: applications in biology and catalysis. *RSC Advances* **2013**, *3*, 21016-21024.
76. Wangoo, N., Bhasin, K. K., Boro, R., Suri, C. R.: Facile synthesis and functionalization of water-soluble gold nanoparticles for a bioprobe. *Analytica Chimica Acta* **2008**, *610*, 142-148.
77. Huang, P., Bao, L., Yang, D. P., Gao, G., Lin, J., Li, Z. M., Zhang, C. L., Cui, D. X.: Protein-directed solution-phase green synthesis of BSA-conjugated MxSey (M = Ag, Cd, Pb, Cu) nanomaterials. *Chemistry - An Asian Journal* **2011**, *6*, 1156-1162.
78. Focsan, M., Gabudean, A. M., Canpean, V., Maniu, D., Astilean, S.: Formation of size and shape tunable gold nanoparticles in solution by bio-assisted synthesis with bovine serum albumin in native and denaturated state. *Materials Chemistry and Physics* **2011**, *129*, 939-942.
79. Jie, Z., Xiaoming, M., Yuming, G., Lin, Y., Qingming, S., Huajie, W., Ma, Z.: Size-controllable preparation of bovine serum albumin-conjugated PbS nanoparticles. *Materials Chemistry and Physics* **2010**, *119*, 112-117.
80. Sylvia, W., Anja, Z., Sascha, L. W., Sabrina, E. T., Wladislaw, M., Tikva, V., Franz, W., Claus, U. P., Jörg, K., Briesen, H. V.: Uptake mechanism of ApoE-modified nanoparticles on brain capillary endothelial cells as a blood-brain barrier model. *PLoS ONE* **2012**, *7*, e32568.
81. Björn, J., Theato, P.: Chemical strategies for the synthesis of protein–polymer conjugates. *Advances in Polymer Science* **2013**, *253*, 37-70.
82. Poonam, K., Vijender, S., Aabroo, M., Pragnesh, N. D., Sourbh, T., Gurinder, K., Jatinder, S., Sukhdev S. K., Baksh, M. S.: Bovine serum albumin bioconjugated gold nanoparticles: synthesis, hemolysis, and cytotoxicity toward cancer cell lines. *Journal of Physical Chemistry C* **2012**, *116*, 8834-8843.
83. Martin, S., Fernanda, S., Alexander, W., Leopoldo, S., Stephanie, H., Carsten, S., Nadine, H., Martina, V., Mara, C., Patrizia, A., Mario, S., Paolo, B., Wolfgang, G. K., Silke, K. Blood protein coating of gold nanoparticles as potential tool for organ targeting. *Biomaterials* **2014**, *35*, 3455-3466.
84. Pawan, K., Parveen, K., Akash, D., Bharadwaj, L. M.: Synthesis and conjugation of ZnO nanoparticles with bovine serum albumin for biological applications. *Applied Nanoscience* **2013**, *3*, 141-144.
85. Gun'ko, V. M., Klyueva, A. V., Levchuk, Y. N., Leboda, R.: Photon correlation spectroscopy investigations of proteins. *Advances in Colloid and Interface Science* **2003**, *105*, 201-328.
86. Xu, W., Park, J. Y., Kattel, K., Ahmad, M. W., Bony, B. A., Heo, W. C., Jin, S. Park, J. W. Chang, Y., Kim, T. J., Park, J. A., Do, J. Y., Chae, K. S., Lee G. H.: Fluorescein-polyethyleneimine coated gadolinium oxide nanoparticles as T-1 magnetic resonance imaging (MRI)-cell labeling (CL) dual agents. *RSC Advances* **2012**, *2*, 10907-10915.

7 Oxygen Functionalization of Carbon Black as a Surrogate Carbon for Composites, Health, and Environmental Studies

Madhu Singh[a], Tawanda J. Zimudzi[b], Jennifer L. Gray[b], Jeffrey Shallenberger[b], Randy L. Vander Wal[a]

[a]John and Willie Leone Family Department of Energy and Mineral Engineering and the EMS Energy Institute, Penn State University, University Park, Pennsylvania, USA

[b]Materials Research Institute, Penn State University, University Park, Pennsylvania, USA

CONTENTS

7.1 OUTLINE

Carbon black is a commercially available product with broad and viable uses, especially in the automotive and rubber industry. It is manufactured with tailored consistency in particle size and aggregate morphology, without ash or hetero-element content, and presents a balanced mix of edge and basal plane carbon sites toward a cheap and easily available representative carbon allotrope. Carbon black has been functionalized using a wet (nitric acid) and dry (ozone) chemical treatment to introduce oxygen functionality and subsequently analyzed for the amount introduced as its degree of functionalization. Wet and dry chemical treatments have different intensity of functionalization, with nitric acid being a harsh oxidant when compared with mild ozone treatment. Consequently, the amount and type of oxygen functional groups introduced onto the carbon black differ, giving rise to different signature peaks when analyzed spectroscopically. Electron microscopy, X-ray diffraction, and Raman spectroscopy have been used for material characterization. Functionalization techniques have been observed to differ in their degree of functionalization, i.e., volumetric vs. surface functionalization, and number of functional groups introduced. Results from X-ray photoelectron and infrared spectroscopy along with thermogravimetric analysis (TGA) have been compared for consistency in functional group detection, with the surface and bulk analysis complimenting one another toward a holistic understanding of the functionalized material. Material characterization and functional group quantification techniques show a consistent and complimentary comparison of the materials across analytical techniques. Current and potential applications of the material have also been discussed. The flowchart in Figure 7.1 provides an overview and a layout of this work.

Testing the varied quantification techniques for oxygen functional groups in this study and documenting a consistent and complimentary comparison across analytical techniques will be a reference guide for carbon nanomaterials in present and future applications.

7.2 INTRODUCTION

Functional groups impart surface reactivity to an otherwise relatively unreactive material like carbon. Typically, carbon functionalization is done in a controlled manner toward a specific application of the functionalized material. However, we come across functional forms of carbon in our everyday life as well in the form of particulate matter, which is largely invisible but present in developed countries and wreaks havoc on the health of those in developing countries [1–3]. The carbon, along with its surface functionalities, is regarded as the leading cause of lung diseases and reduced life expectancy in populations that are routinely exposed to this material for prolonged durations of time [4,5]. In a positive light, functionalized carbon materials, given their organic nature and compatibility with the human body, are being explored for drug delivery and other medical applications [6]. Given this diversity in functionalized forms of carbon and their applications and implications in the real world, this chapter will provide quantification of these functional groups by controlled treatment and subsequent microscopic and spectroscopic analysis preceded by a brief overview of the landscape of functionalized carbon as encountered in the scientific and everyday world.

Carbon is one of the earliest known materials, and its surface has been modified for decades to enhance its properties. Its allotropic forms, such as carbon black, carbon nanotubes (CNTs), and

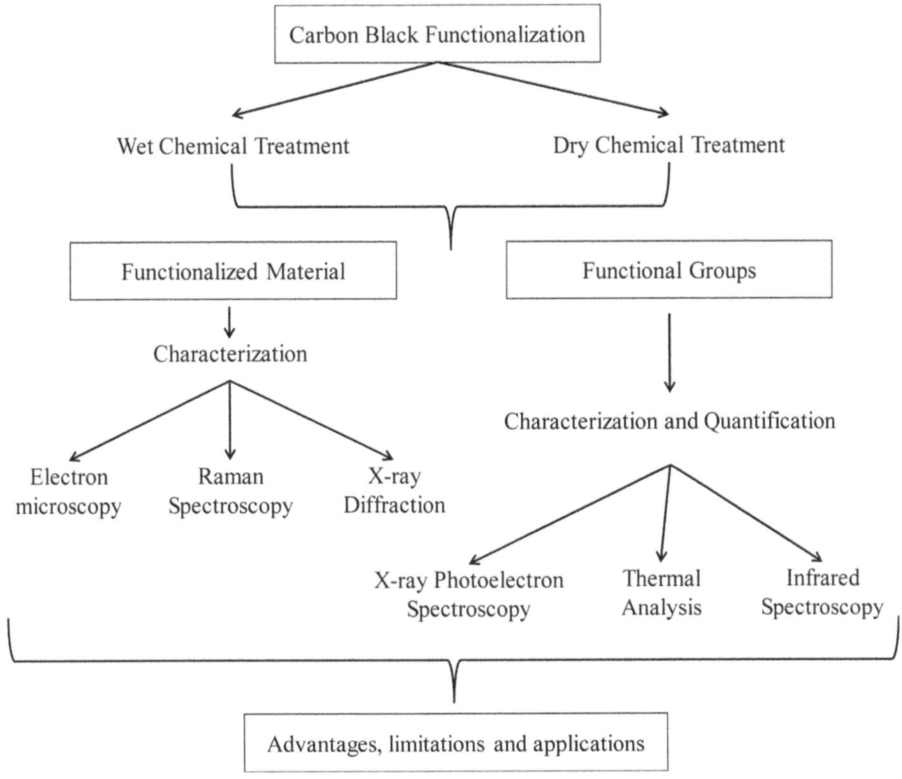

FIGURE 7.1 The flowchart outlines the chemical treatment and subsequent analytical characterization.

graphene, have found a wide variety of applications in recent years [7–11]. For instance, carbon black, given its age-old established bulk manufacturing process, has extensive use in the tire and rubber industry. It is manufactured free of ash, hetero-elements (S, O, and N), mineral matter, or metal catalysts. However, carbon black tends to aggregate or agglomerate when dispersed in solvents or polymers owing to strong van der Waals interactive forces, limiting their applicability when used in its nascent state [12]. Its uses, which range from use as a binding agent to a conductive filler, often require functionalization [10]. Though not formally considered a nanoscale material, given its aggregate structure, the substructures, primary particles, and their nanostructure are directly relatable to that of the prized carbon allotropes. The mixed edge and basal sites make it sufficiently representative of carbon sp^2 forms like nanotubes and graphene [13]. Nanotubes and graphene also need to be functionalized for their various applications. Thus, oxygen functionalization is widely used as a gateway to enhance surface reactivity for carbon to further tailor its chemistry with other atypical functional groups [14–18].

Based on the type of functionalization used, carbon black can be modified volumetrically, or functionality can be surface limited. For instance, a strong oxidant like nitric acid functionalizes the bulk of the particle and thereby introduces a lot more oxygen groups as compared to ozone, which reacts with the available edge sites on the particle's surface and is not aggressive enough to attack the more stable carbon basal planes. Functionalized carbon black makes the material hydrophilic, while originally, carbon is hydrophobic. This property can then be used toward applications that require better dispersion of the material, such as in a polymer matrix for carbon–polymer composites. Quantifying functional groups and potentially associating enhanced properties of the material to quantities and specific types of functional groups can help optimize current carbon functionalization methodology.

7.3 FUNCTIONALIZATION METHODS

Oxygen functional groups can be introduced on a variety of carbon allotropes by dry or wet chemical treatments [19–24]. The basic nature of π–electrons allows mild acids to functionalize carbon blacks, for instance, and the presence of surface oxides makes the otherwise hydrophobic material hydrophilic, with enhanced water solubility after functionalization [25].

7.3.1 Wet Chemical Treatment

The surface of carbon has been modified by treatment with nitric acid, resulting in carbonyl functionalization [24], with hydrogen peroxide and ammonium peroxidisulfate functionalizing introducing ketone and ether groups [23]. Quinone functionalization of carbon blacks has been done by immersion in a solution of acetonitrile [20]. Wet chemical treatments have also been used for purification of nanotubes [26], for instance, with acidic and nonacidic oxidative reagents employed owing to differences in their oxidation capability, thereby changing the quantity and composition of functional groups introduced [27].

7.3.2 Dry Chemical Treatment

Dry surface modification includes treatment with microwave radiation [28], microwave oxygen plasma introducing keto-enol groups [22], or nitric acid vapor for carboxylic groups [29]. Radiofrequency plasma treatment with oxygen for hydroxyl, carbonyl, and carboxyl groups; ammonia for amine groups; and carbon tetrafluoride for fluorine groups has been studied for the effect of treatment time, gas pressure, and plasma power [30]. Ozonolysis, gaseous treatment with ozone, has been observed to influence carbon black pH and tensile strength in rubber [19], although treatment with ozone is a relatively milder functionalization as compared to nitric acid, as observed by the atomic percentages of functional groups reported in the literature [25], also shown in this work.

7.4 CONTROL OF FUNCTIONAL GROUPS

Controlling the type of functional groups introduced onto the carbon material is a challenge, given the inert chemical nature of sp^2 forms of carbon. Acid functionalization primarily introduces carboxylic acid groups, while milder oxidative treatments show an abundance of phenolic groups. The reagent, treatment time, and temperature also influence the functional groups.

The degree of functionalization can be determined by quantifying the amount of acidic or basic functional groups by titration and neutralization with a base or an acid, respectively [31]. However, titration may require a large quantity of the functionalized material, and therefore, other spectroscopic techniques have been employed for quantitative and qualitative assessment of functionalization. For instance, functionalized carbon materials have largely been quantitatively assessed by X-ray photoelectron spectroscopy (XPS) [19,31,32], Fourier transform infrared (FT-IR) spectroscopy [19,33,34], thermal analysis such as TGA [27,28,35], and temperature programmed desorption (TPD) [36], along with electron microscopy such as scanning electron microscopy (SEM) [27,28,37,38], transmission electron microscopy (TEM) [26,35,37,39], and/or atomic force microscopy [15]. The area under the curve of the peaks after XPS spectra deconvolution has been used to quantitatively estimate relative atomic percentages of the functional groups on the material's surface [32]. FT-IR has been used qualitatively to identify functional groups [33]. FT-IR peak assignments vary with the interpretation of what the functional groups may be and the potential shift in the FT-IR spectra due to the carbon structure/morphology [31]; consequently, it is largely used to identify groups qualitatively. FT-IR is somewhat limited in its detection capability due to the high absorbance of carbon [31]. It is useful to identify groups on highly functionalized carbon because of sufficient absorption intensity, and when present, the intensity of an absorption band can have contributions from multiple groups, making it challenging to delineate quantitatively [36]. Carbonaceous material subject to thermal treatment has been well documented to result in the evolution of carbon dioxide (CO_2) and carbon monoxide (CO) as off gases, with CO_2 typically resulting from carboxylic and lactonic groups, while CO is from phenolic, quinone, pyrone, and anhydride groups. CO_2 is observed to evolve at relatively lower temperatures (300°C) [31,36] as compared to CO, which evolves from these acidic groups at higher temperatures (600°C) [23,24,31,33,36]. In intermediate temperature regions where both off-gases may be evolved, deconvolution of the TPD spectrum becomes important to analyze surface chemistry [36]. SEM and TEM are routinely used to study morphology and nanostructure of carbons and have extensively been used to visualize carbon black or CNTs for their structural integrity after functionalization [27–29,38].

7.5 EXPERIMENTAL METHODS FOR CARBON BLACK FUNCTIONALIZATION

Commercially produced carbon black (Regal 250 of Cabot Corp.) has been used here as the model carbon black for its chemical purity and the absence of organic content. To characterize and quantify functional groups on carbon, R250 was subjected to controlled oxidation by the following two methods.

7.5.1 Wet Chemical Treatment

A gram of carbon black was treated with 100 mL of laboratory-grade concentrated nitric acid (HNO_3, >90%) at 80°C under reflux for 24 h at 80°C, just below the acid's boiling point of 83°C. The carbon–acid mixture was continuously stirred using a magnetic stirrer for uniform oxidation and functionalization. The mixture was maintained at a consistent simmer and was washed with distilled water, filtered, and dried to obtain functionalized carbon black as synthetic soot. The effect of treatment time on the degree of functionalization of the material was assessed by exposing the raw material to wet chemical treatment for 6, 12, 18, 24, 48, and 72 h.

7.5.2 Dry Gaseous Treatment

Carbon black was exposed to ozone (O_3) generated by the action of ultraviolet (UV) light on oxygen (O_2). O_3, being a reactive gas, interacts with the carbon at room temperature and mildly oxidizes it, thereby functionalizing the carbon in the process. This method is a comparatively mild oxidative treatment compared with wet acid reflux. Only one exposure time of ~45 min was used.

Carbon black generated by the wet and dry treatments has been characterized for its atomic oxygen content and functional groups introduced onto the carbon surface. Material characterization was conducted by TEM, EDS, Raman, and XRD. Nascent (untreated) and graphitized R250 are used as reference materials. The nascent material serves as a base case raw material consisting of edge and basal-plane sites, making it prone to functionalization, while the heat-treated graphitized carbon is an extreme case with no available edge sites. Comparative characterization results for nascent, heat-treated, and functionalized carbon black are shown and discussed. Functional groups on the oxygenated material are quantified by XPS, TGA-MS, and FT-IR. The characterization and functional group quantification methods and their results are as follows.

7.6 MATERIAL CHARACTERIZATION

7.6.1 Transmission Electron Microscopy

The instrument: TEM was done using the 200 keV FEG source of an FEI Talos F200X with a resolution of 0.12 Å. Samples were dispersed and sonicated in methanol before being dropped onto 300 mesh C/Cu lacey TEM grids.

A TEM of a carbon black aggregate and primary particle is shown in Figure 7.2. (The web in the background of Figure 7.2A is part of the lacy mesh of the TEM grid support.) The carbon black,

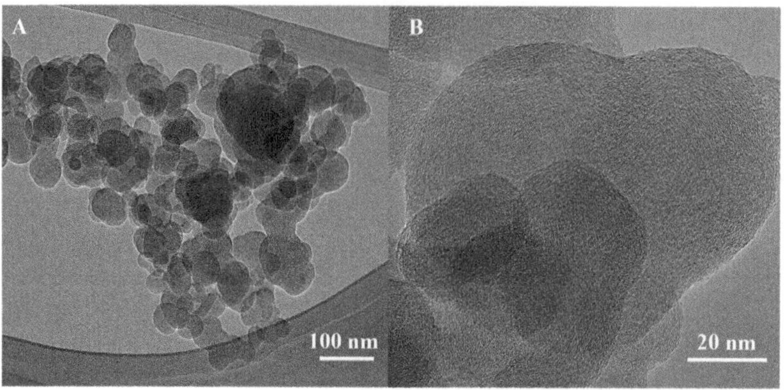

FIGURE 7.2 TEM image of a carbon black (A) aggregate and (B) primary particle.

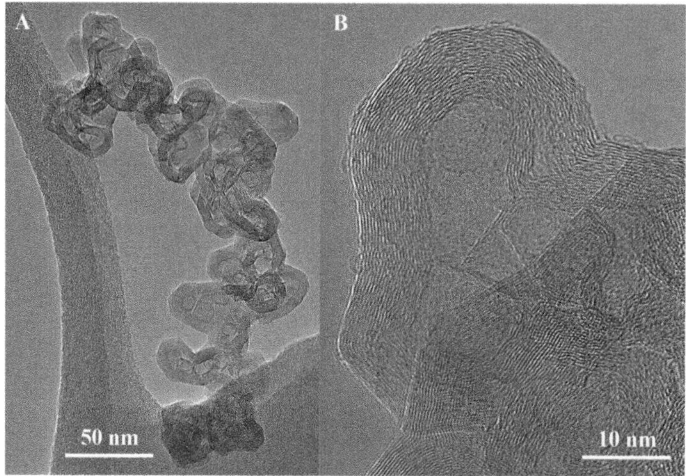

FIGURE 7.3 TEM images of an (A) aggregate and (B) primary particle of graphitized carbon black.

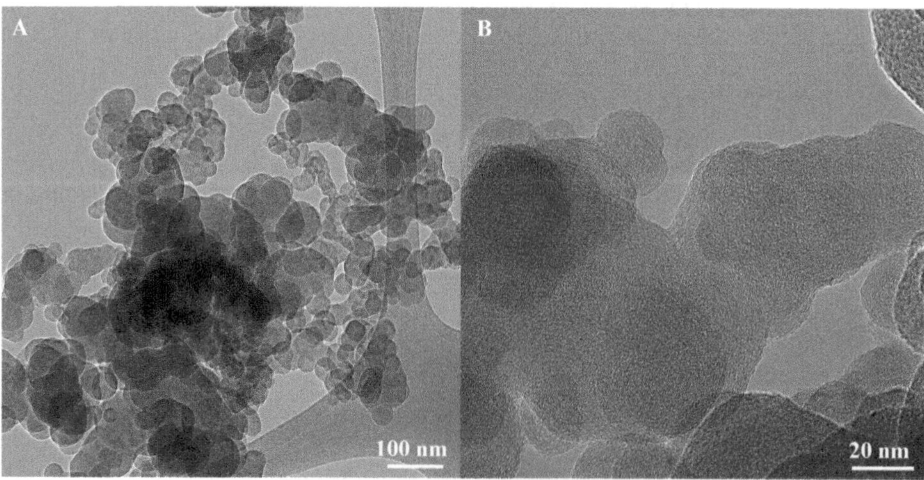

FIGURE 7.4 TEM micrographs showing (A) aggregate and (B) primary particles of R250 after exposure to ozone.

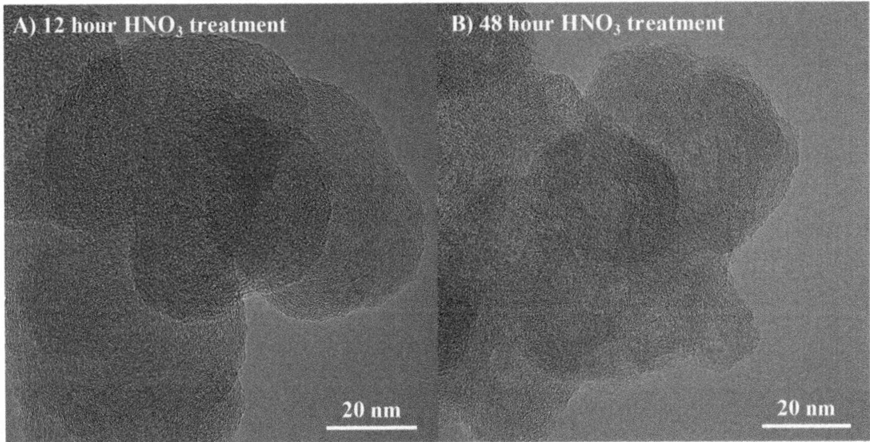

FIGURE 7.5 TEM micrographs showing R250 after treatment with nitric acid for (A) 12 h and (B) 48 h.

R250, was subjected to high-temperature heat treatment in a graphitization furnace at 3000°C. TEM images of an aggregate and a primary particle of graphitized R250 are shown in Figure 7.3, and those of nitric acid and ozone-treated and graphitized carbon black follow in Figures 7.4 and 7.5, respectively.

The images show the nanostructure evolution from well-defined commercially manufactured and untreated carbon black particles in Figure 7.2 to its gradual evolution into a more fused, real soot-like morphology as the nitric acid functionalization proceeds, with the images taken after the first 12 h and then 48 h of treatment. Figure 7.5B shows an eroded particle surface and highly merged R250 particles after functionalization. This is based on surveying multiple (>20) areas on the grid and >100 aggregates.

7.6.2 Energy Dispersive X-ray Spectroscopy

The instrument: EDS for elemental analysis and mapping has been performed in the TEM (FEI Talos) in scanning transmission electron microscopy (STEM) mode with a sample holder designed to provide low background signal for EDS. STEM mode offers a high spatial resolution on the order of the minimum probe size, which is 1.6 Å. The Talos uses the Super X-EDS system, which is composed of four silicon drift detectors that produce very large X-ray counts to allow for a better signal-to-noise ratio. The high counts from the large area of the detectors also provide for very low detection limits of typically <1 atomic percent (at.%) depending on collection parameters. In these experiments, ~7–10 regions of each material were sampled, and the data averaged for representative elemental quantification.

EDS has been used here to identify and preliminarily quantify elemental carbon and oxygen content. Figure 7.6 shows high-angle annular dark field (HAADF) images and EDS maps for untreated and graphitized carbon black. Figures 7.7 and 7.8 show these for ozone and nitric acid functionalized carbon black, respectively. Quantification of relative atomic percentages of elemental carbon and oxygen are tabulated in Table 7.1. One-dimensional line scans extracted from EDS maps, comparing relative intensities of carbon and oxygen along the diameter of a particle in each of the four materials, are shown in Figure 7.9.

Untreated R250 shows a layer of oxygen accumulated at the particle periphery, while graphitized R250 shows no such oxygen buildup. R250, without any treatment, contains edge sites and, therefore, tends to attract atmospheric oxygen, resulting in the formation of an oxygen shell around the carbon particle. Oxygen detected amounts to ~2 at.% relative to graphitized carbon with no available edge

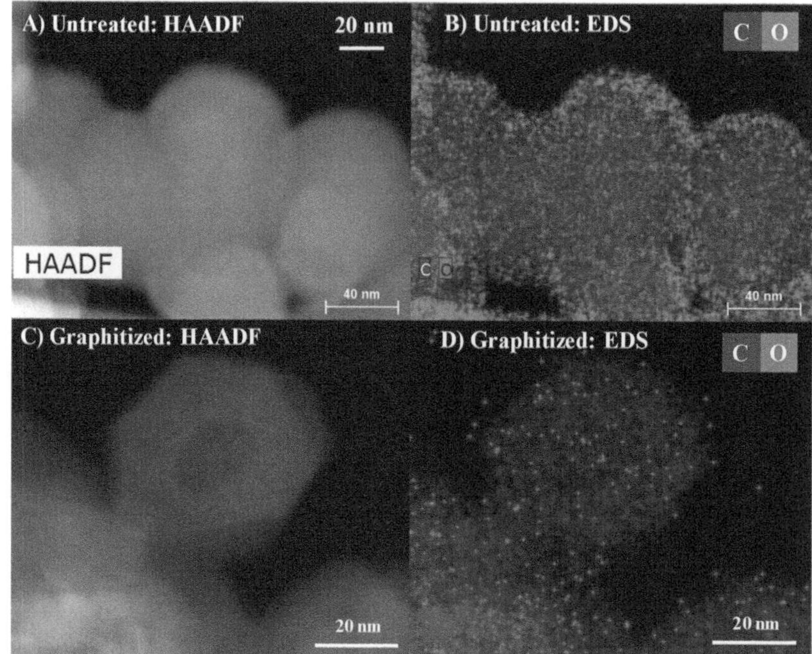

FIGURE 7.6 Image panel shows HAADF images and EDS maps, respectively, of (A, B) untreated and (C, D) graphitized carbon black. Oxygen is represented by the bright pixels along the edge of the carbon particles in image B, and scattered across the particle as noise in image D. For similar colored images, see M. Singh, 2019. *Measurement, Characterization, Identification and Control of Combustion Produced Soot*. Ph.D. Thesis. The Pennsylvania State University.

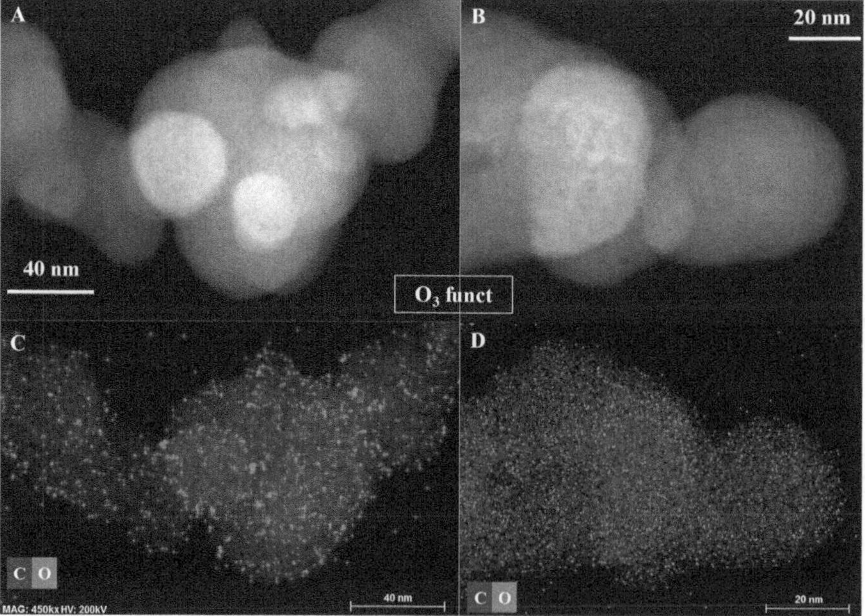

FIGURE 7.7 Image panel shows (A, B) HAADF images and (C, D) respective EDS maps. Oxygen is represented by the bright pixels in images C and D, scattered across the carbon particles.

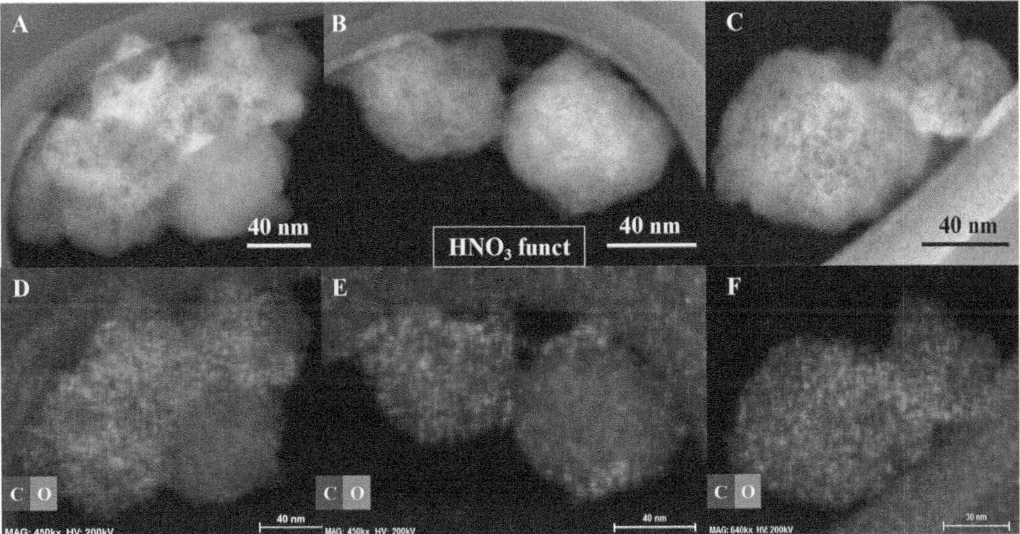

FIGURE 7.8 Image panel shows (A–C) HAADF images and (D–F) respective EDS maps showing carbon (grey) and oxygen (bright grey) in nitric acid functionalized carbon black. Oxygen is represented by the bright pixels along the perimeter and the bulk of the carbon particles in dull grey pixels.

sites and no detectable oxygen content, having being subjected to high-temperature heat treatment. Oxygen pixels observed for graphitized carbon in Figure 7.6D are noise, given the detection limit of the instrument. No corresponding peak for oxygen is observed in the EDS spectrum for graphitized carbon black.

Ozone-treated carbon black shows a scattered presence of oxygen around the particle's perimeter. Ozone has likely reacted with available edge sites and reduced the amount of atmospheric oxygen adsorbed by "cleaning" the surface of the carbon particle, thereby showing a reduced concentration along the particle perimeter, resulting in an overall scattered presence. Oxygen is quantified at ~1 at.% for this material. In contrast, nitric acid functionalized carbon (Figure 7.8) with an ~7 at.% oxygen content shows oxygen distributed throughout many of the particles instead of just being concentrated only on the edges. While EDS cannot point to a definitive volumetric vs. surface oxygen presence given its 2D nature, the presence of volumetric oxygen for this material can be well correlated with its respective HAADF image. The variation in contrast within each particle corresponds to the presence of holes or pores created in the particle, now occupied by oxygen. The surface of the particle was etched, owing to the vigorous acid treatment, and resulted in oxygen penetrating throughout for volumetric functionalization of the material.

Nitric acid–treated carbon black shows an increase in the relative amount of elemental oxygen compared with untreated carbon, while ozone-treated carbon shows a reduced amount of oxygen as it etches edge sites off of the particle's surface. Thus, oxygen content changes with the type of functionalization used (dry or acid chemical treatment) and with treatment time within a particular functionalization method, such as acid functionalization in this case. High-temperature graphitization removes all oxygen content, making it almost pure carbon. The graphitized carbon does not show a significant affinity to atmospheric oxygen either and therefore registers as a near 100% carbon content. A snapshot of comparative oxygen atomic percent is shown in Table 7.1.

The four materials show a significant visual and quantitative difference in the oxygen detected by EDS. Given the electron transparent nature of a TEM sample owing to its thickness of ~100 nm, the beam interacts with the entire thickness and, therefore, volume of the sample while collecting EDS data in a TEM. A relatively thin sample, such as that used for TEM analysis, does not cause the beam to spread, allows low-energy X-rays to pass through to the detector, and therefore, gives a

TABLE 7.1
Relative Carbon and Oxygen Weight Percent from EDS

Treatment	Carbon (Rel. at. %) (±3%)	Oxygen (Rel. at. %)
Untreated	98	2 (±0.10)
Graphitized	100	0 (±0.10)
O_3	99	1 (±0.15)
HNO_3 (12 h)	97	3 (±0.15)
HNO_3 (48 h)	93	7 (±0.20)

[1]Value is at the instrument's detection limit. No actual oxygen peak is observed in the material's EDS spectrum.

high-resolution elemental mapping from the region illuminated by the beam. Figure 7.9 shows line scans of primary particles from each of the four materials, with relative carbon and oxygen intensity plotted across the particle's diameter.

Figure 7.9A for untreated carbon black shows a higher oxygen content along the particle edges relative to carbon, shown in Figure 7.6B. This likely corresponds to atmospheric oxygen being adsorbed on the surface of the untreated material, helping satisfy any dangling bonds or edge sites.

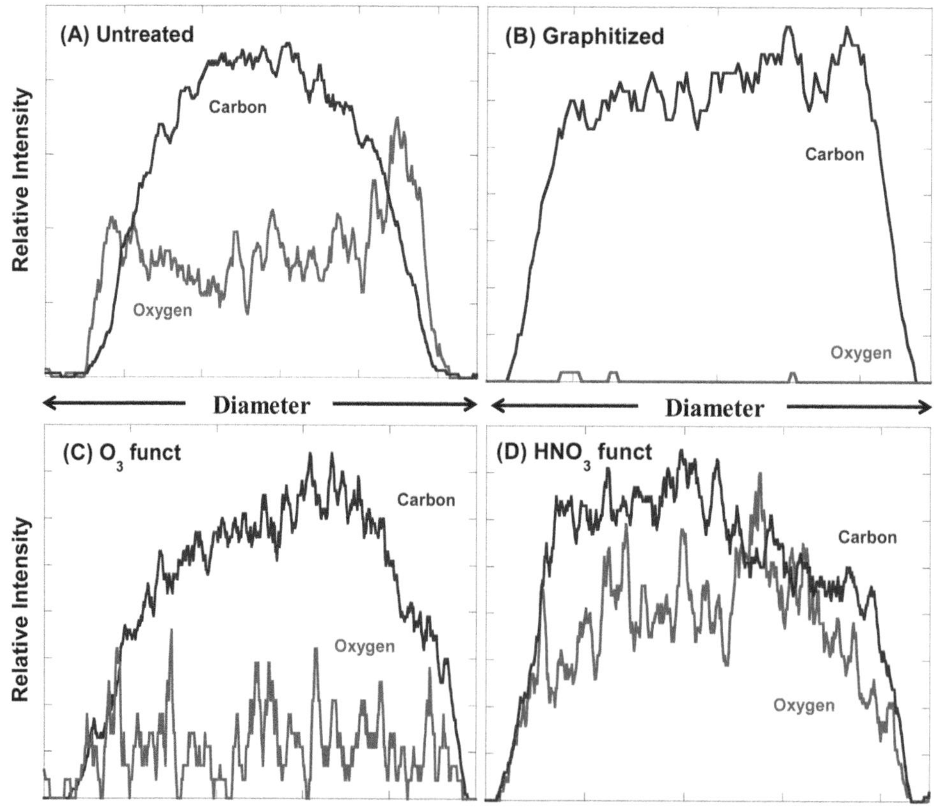

FIGURE 7.9 Image panel shows line scans of (A) untreated, (B) graphitized, (C) ozone, and (D) nitric acid functionalized carbon black showing relative intensities of carbon and oxygen along the diameter of the primary particle.

Oxygen intensity for graphitized carbon black in Figure 7.9B is negligible. Graphitization results in the elimination of edge sites present in the untreated material, thereby resulting in even less atmospheric oxygen clinging to the particle (Figure 7.6D). Ozone functionalization results in increased oxygen content relative to graphitized but decreased oxygen relative to nascent carbon, especially at the particle perimeter (Figure 7.9C). Figure 7.9D shows line scans from nitric acid–treated carbon, showing a significantly high oxygen intensity, nearly equal to carbon across the particle's diameter, indicating volumetric functionalization of the material, supported by its HAADF and EDS maps in Figure 7.8. The high intensity of the EDS line scans for acid functionalized carbon necessarily implies that the oxygen content is volumetrically distributed throughout the particle. It is interesting to note the volumetric elemental oxygen detected throughout the functionalized material when compared with the surface-bound oxygen for the untreated carbon and virtually no oxygen for its graphitized form, evidenced by a scattered presence of elemental oxygen. Thus, acid functionalization introduces oxygen through the bulk of the material and not just the surface, as seen by its respective EDS map. However, it must be noted that EDS shows relative amounts of elemental carbon and oxygen and does not give information on what functional groups are present or, in other words, how carbon and oxygen are bonded to one another.

7.6.3 Raman Spectroscopy

The instrument: A Horiba LabRam Raman microscope was used to obtain Raman spectra for the samples when exposed to a 488-nm, 100-mW laser with a 300 grooves/mm grating providing a spectral resolution of 4 cm^{-1}.

Raman spectroscopy has been used to infer changes in the carbon's structure after functionalization. Figure 7.10 shows overlaid and offset Raman spectra for nascent (untreated), O_3-exposed, and HNO_3-treated R250. The spectrum for O_3- and HNO_3-treated carbon black has been offset to help visualize these spectra when overlaid. As a comparison, Raman spectrum for graphitized R250 is shown on a different graph due to the order of magnitude change in intensity. Its Raman spectrum is compared to untreated R250 as a baseline reference and to give the reader a sense of the high-intensity counts for graphitized R250.

As observed from the comparative Raman spectra, O_3 treatment mildly modifies the structure of the carbon black and has a spectrum similar to nascent R250. Treatment with a strong oxidizing agent like HNO_3 results in a significant change in carbon structure, seen by the appearance of the second-order peaks in the 2400–3200 cm^{-1} wavenumber regions. Peak intensities have been tabulated in Table 7.2. Peak positions and the ratio of the D/G and 2D/G peaks compare structural changes in R250 when oxidized with the two reagents and, as a comparison, when graphitized at 3000°C.

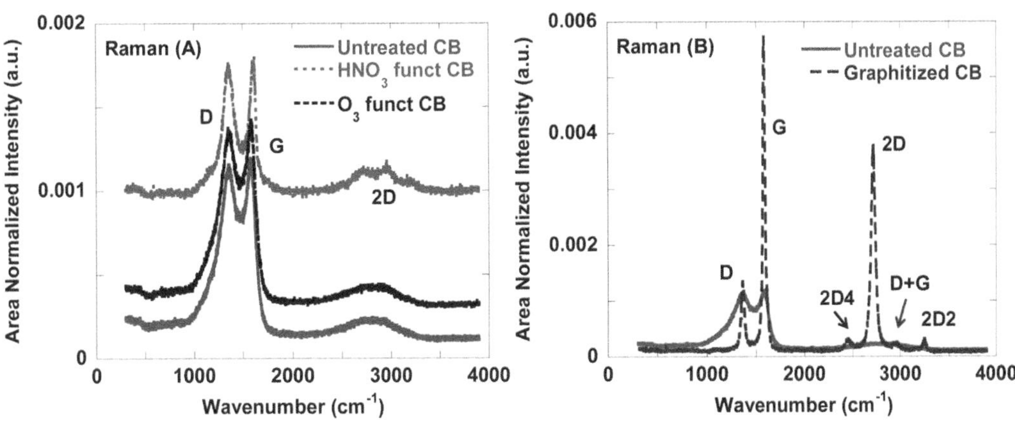

FIGURE 7.10 Raman spectra for nascent, ozone-exposed, and nitric acid–treated carbon black.

TABLE 7.2
Peak Ratios from Raman Spectroscopy

	Peak Position (cm^{-1})				Peak Intensity Ratio		
Sample	D	G	D'	$2D$	I_D/I_G	$I_{D'}/I_G$	I_{2D}/I_G
Untreated	1359	1589	–	2682	1.23	0	0.10
HNO_3	1358	1595	1620	2711	1.38	0.75	0.18
O_3	1360	1583	1619	2700	1.52	0.52	0.15
Graphitized	1359	1581	1620	2717	0.21	0.05	0.66

I_D/I_G ratio: The successive increase in I_D/I_G with functionalization is attributed to the creation of edge sites due to chemical treatment. Edge sites in graphene layers register as "defects" and activate D (~1360 cm^{-1}) and the D' (~1620 cm^{-1}) peaks [40]. This is supported by the intensity of the D' peaks for functionalized carbon showing the presence of defects or edge sites when compared with graphitized or untreated R250 for which there is no significant D' peak, pointing to the absence of detectable edge sites. Graphitized R250 has the lowest I_D/I_G, supported by its TEM image in Figure 7.3, and exhibits the least number of lamellae defects of the four materials.

I_{2D}/I_G ratio: Graphitized R250 shows an intense 2D peak, typically considered a signature for pristine single-layer graphene, which is due to the change in the electronic band structure of the material and relates to the number of stacked layers, their relative orientation, and spacing. Graphitized carbon black shows a d_{002} ~3.44 Å (see Section 7.6.4, Table 7.3), typical of turbostratic graphite. Turbostratic graphite is known to have a sharp 2D peak despite it being graphite and not few-layer graphene [40]. The twist between the layer planes changes the material band structure, giving rise to a prominent 2D at ~2700 cm^{-1}. For the functionalized material, the relative intensity of the 2D peak increases as nascent R250 is mildly oxidized with O_3. I_{2D}/I_G is larger with strong HNO_3 treatment, with the broad peak at 2700 cm^{-1} splitting into four peaks. This is due to oxygen intercalation in the carbon black lattice structure and alteration of the material's electronic band structure in a similar manner to that seen in turbostratic (d_{002} ~3.44 Å) vs. regular (d_{002} ~3.35 Å) graphite. Considering that graphitization and functionalization are two different treatments, a comparison is presented here only as an example of interpretation.

The intensity of the 2D peak is not a result of the functionalization per se but rather is due to the structural modification of carbon blacks as a consequence of chemical or thermal treatment. A turbostratic layer plane happens to have a favorable electronic band structure for the graphitized carbon to exhibit a 2D peak. Again, the 2D peak is not because of turbostratic layer planes either. It exists because this particular orientation is favorable. The presence of a sharp 2D peak is necessarily observed in pristine single-layer graphene, but a favorable combination of some layer planes, their spacing, and relative orientation can also result in a similar sharp peak because of the altered electron band structure. Thus, exercising caution is important while interpreting Raman spectra. This has been well summarized in the review by Ferrari and Basko [40].

7.6.4 X-RAY DIFFRACTION

The instrument: XRD measurements of the powdered samples were carried out in a PANalytical Empyrean X-ray Diffractometer using a Cu source, para-focusing optics, and a PIXcel 3D detector. Background subtraction and peak deconvolution of the XRD pattern were performed using JADE software, which also calculated the d_{002}-layered plane spacing as per the Bragg equation. The carbon layer plane stacking height (L_c) was calculated using the (002) peak at 26° by applying the Scherrer equation [41].

XRD has been used to characterize functionalized carbon black for its structure and the degree of change in its crystallinity due to the functionalization process relative to the untreated carbon

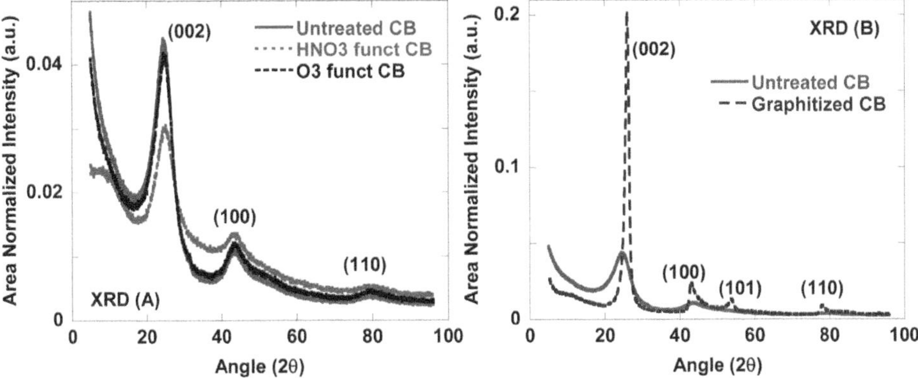

FIGURE 7.11 Comparison of XRD spectra of untreated, nitric acid–treated, and ozone-treated R250.

black. Figure 7.11 shows comparative XRD line shapes of untreated, O_3-treated, HNO_3-treated, and graphitized R250. Lattice parameters after peak deconvolution are tabulated in Table 7.3.

While nascent and O_3-treated R250 show a significant similarity in their line shape owing to the mild oxidation process, the XRD spectrum of HNO_3-treated R250 deviates in that there is development of a peak at ~10° (2θ) and a broadening of the peak ~26° (2θ), also marked by a relatively lower peak intensity. This is likely due to the incorporation of oxygen into its crystal structure given the strong and corrosive nature of nitric acid causing a change in the overall material crystallinity. This is consistent with *D*, *G*, and 2*D* peak intensity ratios observed by Raman spectroscopy, where incorporation of the oxygen increases the lamella spacing (d_{002}) seen by XRD, and consequently, the resultant vibrational modes that manifest themselves as changed *D*/*G* and 2*D*/*G* peak ratios by Raman.

Graphitized R250 is significantly different from untreated and oxidized carbon black, with a reduced *d*-spacing, close to that of highly ordered turbostratic graphite (3.44 Å) [42], and long lamellae, also seen by its TEM. Lattice parameters calculated from XRD show a somewhat reduced *d*-spacing of functionalized R250 relative to the nascent material and similar L_c and a reduced L_a, in line with the breaking up of carbon lamellae, creating more fragments and edge sites that are then functionalized.

It is interesting to note the absence of the (101) peak at $2\theta = 44°$ in Figure 7.11A, and its presence in Figure 7.11B for graphitized R250. This peak is a result of *c*-axis stacking and is usually challenging to deconvolve from its neighboring (100) peak at $2\theta = 42°$, sometimes even for heat-treated carbons with an ordered structure. Its prominence in Figure 7.11B shows the extensive *c*-axis stacks present after graphitization relative to nascent and functionalized R250. Its seeming "absence" from Figure 7.11A does not indicate complete lack of stacks along the *c*-axis. Short and stacked carbon lamellae exist for carbon black as seen by TEM but lack the intensity for detection by XRD, reflecting misaligned lamellae, and thus, they do not fulfill the criteria required for Bragg diffraction from the (101) planes. Although present in Figure 7.11A, the sharp (110) peak in Figure 7.11B relative to untreated R250 is an additional indicator of long and expansive lamellae after graphitization as compared to the other carbon materials.

TABLE 7.3
Lattice Parameter Calculations from X-ray Diffraction

Treatment/Lattice Parameter	Untreated	HNO_3	O_3	Graphitized
d_{002}, Å	3.60	3.54	3.58	3.39
L_c (using 002), nm	1.7	2.2	1.8	6.7
L_a (using 100), nm	5.6	4.0	3.2	15.7

7.7 FUNCTIONAL GROUP CHARACTERIZATION AND QUANTIFICATION

XPS, TGA, and FT-IR have been used to characterize and quantify functional groups on carbon black. These methods, their results, and inferences from the data collected follow.

7.7.1 X-RAY PHOTOELECTRON SPECTROSCOPY

Operating principle: XPS is a surface analysis technique based on the photoelectric effect where incoming X-rays are used to eject electrons. With the known total energy of a photon, the kinetic energy (KE) of the ejected electron is measured, and binding energy (BE) is calculated. BE of a core-shell electron is characteristic of the element from which it is ejected and is used to identify the elements present. Thus, XPS is used here to identify and quantify possible different functional groups on the carbon surface, typically at a depth of 1–5 nm, and their contribution to the total surface atomic oxygen.

The instrument: XPS experiments were performed using a Physical Electronics VersaProbe II instrument equipped with a monochromatic Al K_α X-ray source ($h\nu$ = 1.4867 keV) and a concentric hemispherical analyzer. Charge neutralization was performed using both low-energy electrons (<5 eV) and argon ions. The BE axis was calibrated using sputter cleaned Cu foil (Cu $2p_{3/2}$ = 932.7 eV, Cu $2p_{3/2}$ = 75.1 eV). Peaks were charge referenced to the C–C band in the carbon 1s (C1s) spectra at 284.5 eV. Measurements were made at a takeoff angle of 45° with respect to the sample surface plane. This resulted in a typical sampling depth of 3–6 nm (95% of the signal originated from this depth or shallower). Quantification was done using instrumental relative sensitivity factors (RSFs) that account for the X-ray cross-section and inelastic mean free path of the electrons. The accuracy of the O1s and C1s RSFs was checked by analyzing a polyethylene terephthalate (PET) film in triplicate. The measured PET composition was within 1 at.% for carbon and 1.5 at.% for oxygen. The repeatability (standard deviation) was <1%. Similar accuracy is expected on these functionalized carbon samples. However, the repeatability of oxygen may be slightly poorer due to lower signal to noise.

Untreated, HNO_3 and O_3 functionalized, and graphitized carbon blacks were characterized with XPS using survey and high-resolution scans. These are shown in Figure 7.12.

Survey scans are used as a quick analysis tool to identify the elements present on the surface, after which high-resolution scans are done focusing only on the relevant range of BE for the elements of interest. Quantification of the elements is done by appropriate curve-fitting of the high-resolution

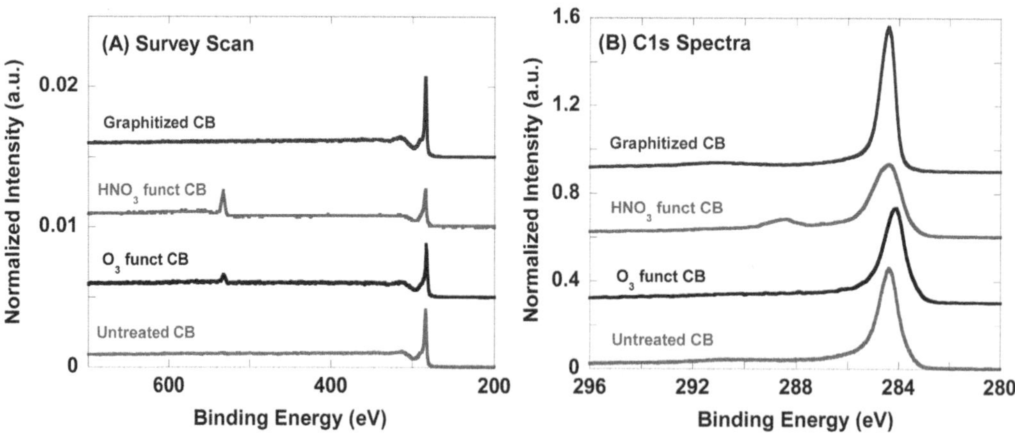

FIGURE 7.12 XPS (A) survey scans and (B) C1s spectra for untreated and HNO_3 and O_3 functionalized and graphitized carbon black.

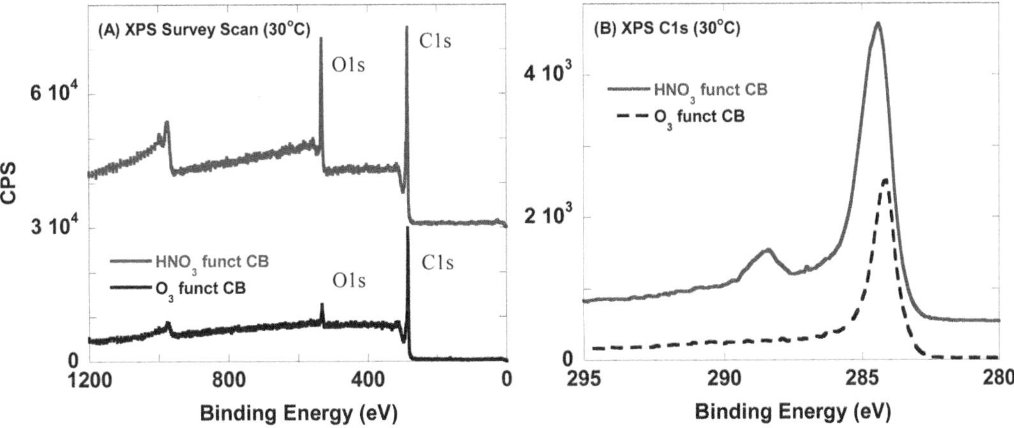

FIGURE 7.13 XPS (A) survey scans and (B) C1s spectra for HNO_3- and O_3-treated carbon black. (Top spectra is for HNO_3 funct CB and bottom spectrum is for O_3 funct CB in both graphs.)

scan using Casa XPS version 2.3.19. As a baseline, nascent (untreated) carbon black is used as a "blank" sample with its line shape being a reference for curve fitting and subsequent peak assignment for the functionalized material. No surface functionalities were observed after high-temperature graphitization treatment of the material, and therefore, graphitized carbon black is ~100% carbon, as seen by the intense C1s peak and lack of any oxygen peak and also verified by EDS analysis. Survey and C1s spectra for functionalized carbon blacks are shown in Figure 7.13 followed by curve-fitted high-resolution scans in Figure 7.14.

Survey scans showing C1s and O1s peaks for both functionalized materials show the significant extent of oxygen introduced onto the HNO_3 carbon (Figure 7.13A). The change in the line shape of the C1s envelope is compared in Figure 7.13B, with a prominent O–C=O peak for acid-treated carbon. Peak deconvolution of the C1s spectrum for both materials is shown in Figure 7.14.

Exposure to a strong acid such as HNO_3 results in the presence of a significant number of carboxylic acid (–COOH) groups, seen from the peak at ~288 eV. The change in line shape of the O_3-treated carbon spectrum is less significant as compared to acid functionalized carbon, with most functional groups being phenolic.

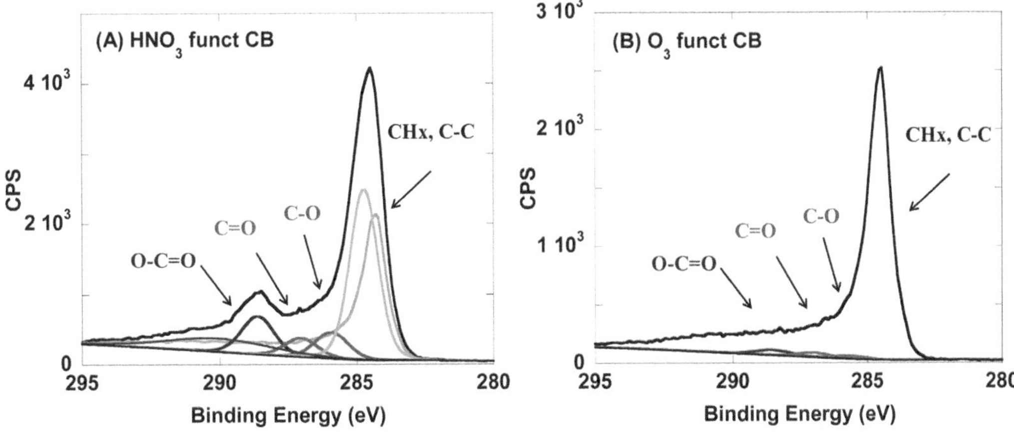

FIGURE 7.14 High-resolution scans showing C1s spectrum and its respective peak deconvolution for (A) HNO_3-treated and (B) O_3-treated carbon black.

TABLE 7.4
Atomic % of Elements Using XPS

Treatment	*Measured Atomic %*		*Calculated Atomic % After Curve-Fitting Spectra*					
	C	O	C–C	C–O	C=O	O–C=O	Aromatic	Total O
Untreated	97.2	1.3	—	—	—	—	—	1.3
HNO_3	67.8	31.5	41.1	10.2	4.9	9.4	1.6	34.0
O_3	90	9.4	84	2.2	1.4	2.4	—	8.3

Curve fitting of the C1s spectra for both materials has been done by using the minimum number of peaks that are necessary and sufficient to describe the functional groups present. For instance, COOH may also be present as an interchangeable lactol or other functional group forms [25], and the presence of other lactone and quinone groups has been reported [43]. However, in keeping with what is historically known [31,44], minimizing the number of peaks involves the least assumptions and avoids overfitting the spectrum with peaks that may not correspond to real groups.

Atomic percentages of the elements present in the materials analyzed are summarized in Table 7.4. The table is divided into the measured and calculated atomic percentages. Measured numbers are obtained directly from counts observed by the detector. Curve fitting is then used to help delineate the contribution of oxygen from the various functional groups present. These numbers are shown in the second half of the table. A good agreement (± 10%) in the measured and calculated value of atomic oxygen indicates an appropriate curve-fitting procedure for functional group identification.

Thus, wet acid reflux treatment of carbon black results in ~32 at.% oxygen compared with ~10 at.% from dry gaseous treatment; the latter is a relatively mild oxidant with oxidation performed at room temperature, thus explaining this difference in the type and degree of functionalization. Higher oxygen content changes and distorts the layer plane arrangement of individual graphene sheets. Oxygen atoms attach to the carbons and keep consecutive layers parted from one another, and the resultant intermolecular interactions cause the layer planes to twist with respect to one another with higher *d*-spacing as a consequence, confirmed by XRD and Raman.

It should be noted that the total O-content percentages as listed in Table 7.4 are significantly higher than those measured by EDS. XPS is a surface analysis technique, and for a disordered carbon, such as the functionalized carbon blacks considered here, the effective depth probed is ~2 nm. Intuitively, the highest oxygen content would be expected near the particle surface. By contrast, EDS in the TEM instrument provides a volumetric measure, with the lower value reflecting total O-atom content. Therein lies the information value for these complementary techniques, as only EDS shows the volumetric O-atom content. Although this carbon has very low porosity, the aggressive oxidation apparently can open pores, accessing particle interior.

In addition, a time-series HNO_3 functionalization of R250 was done to analyze the effect of treatment time on the atomic percentages of functional groups as determined by XPS. R250 was treated in reflux of HNO_3 for 6, 12, 18, 24, 48, and 72 h, and XPS spectra were recorded for these. These spectra have been analyzed for the concentration of oxygen and the contribution of different functional groups. These data are presented graphically in Figure 7.15 as well as tabulated in Table 7.5.

The concentration of oxygen as measured by XPS is presented in Figure 7.15. Values calculated by a mole balance on oxygen using the bonding information about the other detected elements are in good agreement with measured values and almost always within a 5% error margin shown by the error bars. With increasing treatment time, the concentration of oxygen functionalization increases rapidly at first up to 20 h of treatment time, but eventually, it reaches a plateau, and increasing treatment time beyond this threshold does not add more functional groups or oxygen content to the carbon black. Contribution from the different functional groups is shown in Figure 7.16. Values for the analysis presented are tabulated in Table 7.5 for reference.

TABLE 7.5
XPS Data Showing the Effect of Treatment Time for Nitric Acid Functionalization of R250

Treatment Time in HNO_3 (h)	*C*	*O*	*O–C Measured*	*C As C(C,H)*	*C As C–O*	*C As C=O*	*C As COO*	*O Calc'd*
0	97.8	1.3						
6	81.5	16.9	14.1	72.5	2.3	1.6	5.0	13.9
12	84.9	14.3	12.3	76.1	3.2	1.2	4.4	13.3
18	74.0	23.8	20.0	61.1	3.3	2.3	7.3	20.2
24	71.8	25.9	20.9	59.6	2.9	2.3	7.0	19.3
48	72.4	27.1	25.9	60.2	3.1	0.1	9.0	21.3
72	65.6	30.5	22.9	51.4	4.1	0.6	9.6	23.8

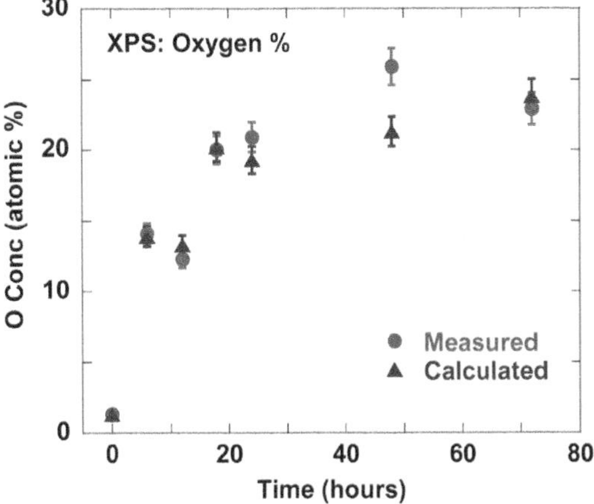

FIGURE 7.15 Percentage of atomic oxygen measured by XPS at different treatment times with nitric acid.

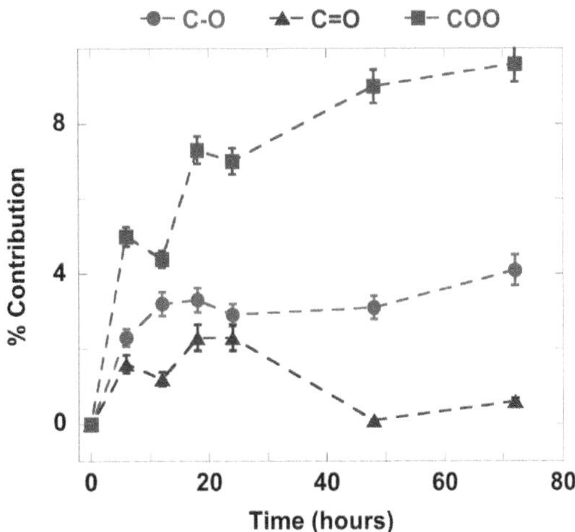

FIGURE 7.16 Relative percentage contributions from the likely carbon–oxygen groups to the total atomic oxygen.

7.7.1.1 In Situ Heating Stage XPS

HNO_3- and O_3-treated carbon were analyzed with in situ vacuum heating using a hot stage attachment in the X-ray spectrometer. Spectra were collected at room temperature, ~30°C, and subsequently heated to 300 and 500°C as measured by the thermocouple accompanying the hot stage apparatus. Changes in the XPS C1s line shape of the samples with increasing temperature are shown in Figure 7.17, and measured atomic percentages of carbon and oxygen are compared with calculated amounts in Table 7.6.

Figure 7.17 shows the change in line shape as the acid functionalized carbon black is heat treated in situ with a heating stage apparatus. The decrease in the peak corresponding to the acidic COOH group is evident. Given the short heating duration, a small quantity may be left over, seen by the small peak at ~288 eV after in situ heating at 300°C, but this peak virtually disappears at 500°C. Atomic percentages of carbon and oxygen after peak deconvolution of the in situ heat-treated carbon materials are summarized in Table 7.6.

Oxygen atomic percentage shows a consistent decline with increasing temperature for both materials in situ. However, being a surface technique, most of the atomic percentage value is the amount of oxygen present in the top 5 nm surface of the functionalized material. The amount of oxygen present is calculated based on the numbers for carbon that are a result of the C1s peak deconvolution. The nascent (untreated) R250 XPS line shape was used as reference, and deviations from this line shape are assigned to the presence of possible functional groups based on the BE signatures of C–O, C–H, or C=O bonds.

7.7.2 Thermogravimetric Analysis

Operating principle: Thermal analysis is used to study the response of a material to controlled heat treatment and draw inferences regarding its composition or thermal properties based on its weight loss as a function of temperature. It is a bulk material analysis technique and gives a qualitative volumetric assessment of the material. It can often be coupled with an auxiliary instrument, such as a mass spectrometer (MS), to identify evolved gases toward a component analysis as the material changes with heat treatment.

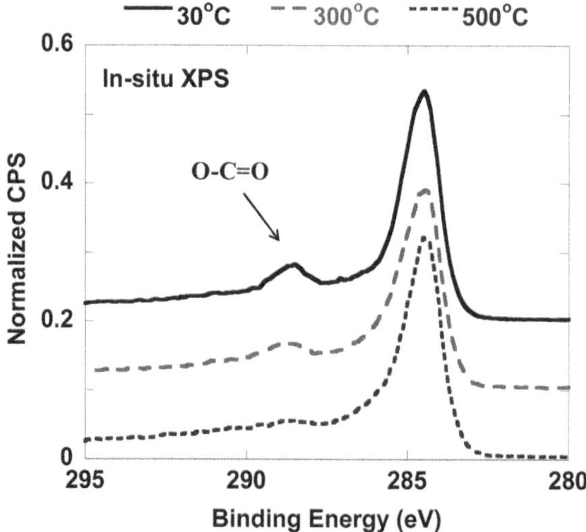

FIGURE 7.17 XPS spectra comparison for acid functionalized carbon heated in situ showing change in line shape of C1s spectra with temperature.

TABLE 7.6
Variation in C and O Atomic % with in Situ Heat Treatment

Treatment Reagent	*Temperature* (°C)	*Measured* (at.%)		*Calculated* (at.%)				
		C	O	C As C(C,H)	C As C–O	C As C=O	C As COO	Total O
O_3	30	93.6	6.1	89.7	0.7	1.1	1.5	4.7
	300	93.7	5.7	91.0	0.1	1.0	0.5	2.0
	500	96.1	3.6	95.8	0.0	0.1	0.0	0.0
HNO_3	30	78.7	21.1	62.3	4.5	3.1	6.5	20.5
	300	83.5	16.2	58.1	7.6	2.7	4.6	19.5
	500	88.1	11.6	68.6	7.5	2.2	2.6	14.9

The instrument: A TA Instruments Thermogravimetric Analyzer TA 5500 coupled with their Discovery Mass Spectrometer (MS) was used to analyze weight loss and the composition of the evolved gases. The temperature was ramped up at 5°C/min. The measurements were performed under conditions of air oxidation diluted with an inert, so results at temperatures >600°C compare the different reactivities of the carbon forms as compared to lower temperatures (<600°C) that show weight loss by the evolution of pre-existing oxygen functional groups. In this manner, hybrid data were effectively taken in one TGA scan.

TGA is used as a bulk material characterization tool to complement results observed by XPS. When subject to a steady temperature ramp, functional groups on the carbon oxidize (leave) at different temperatures. Inflections on the subsequent weight loss curve can be used to identify the functional groups present on the carbon black qualitatively. For the significant functionalization observed by HNO_3, its TGA curve shows distinct regions of weight loss owing to functional groups leaving as temperature increases. The TGA and MS curve for HNO_3 functionalized R250 is shown in Figure 7.18.

The weight loss curve in Figure 7.18A and its respective derivative show a significant loss at 300°C, a temperature that corresponds to the loss of COOH groups [31,36] observed to leave as CO_2, as corroborated by the time-resolved mass spectrum at 44 AMU (Figure 7.18B). Other time-resolved mass channels such as CO and its overlap with that of N_2 make it challenging to delineate

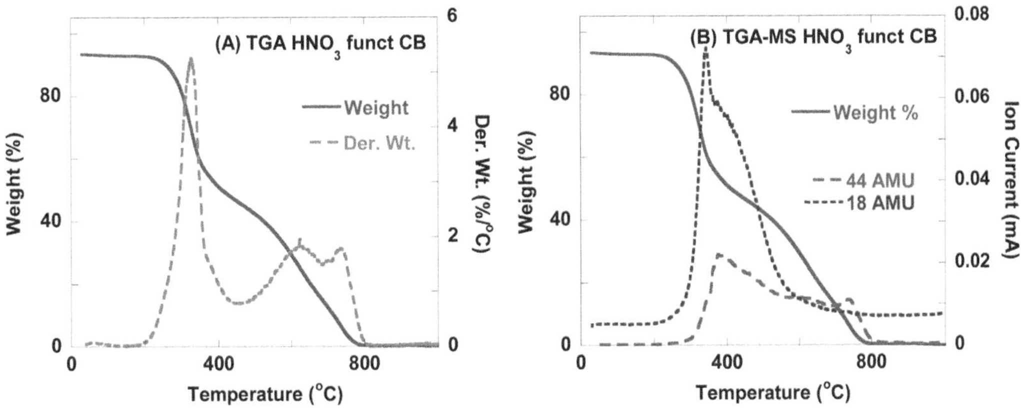

FIGURE 7.18 TGA–MS analysis for HNO_3-treated carbon black.

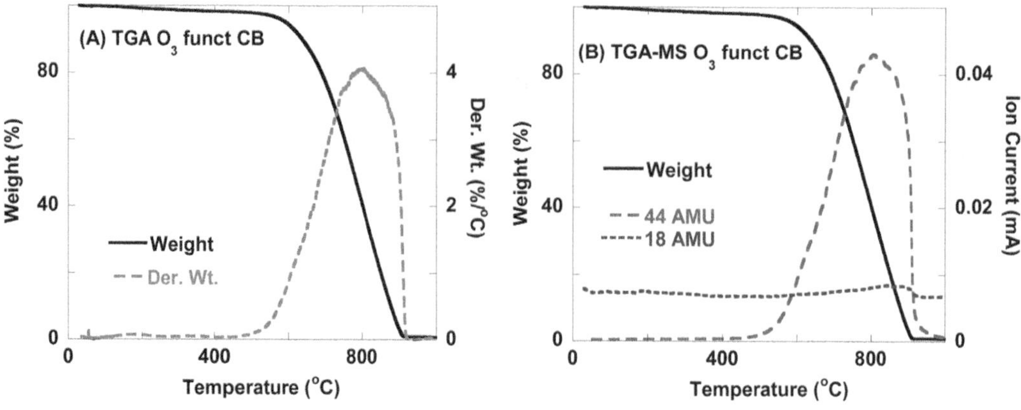

FIGURE 7.19 TGA–MS curve for O_3-treated carbon black.

the two gases. The mass channel for water, 18 AMU, follows the trend of evolved CO_2 as detected by the MS. Thus, CO_2 and H_2O evolve preferentially with significant detectable quantities. It should be noted that the elevated H_2O baseline as observed by the MS is likely due to the moisture present in atmospheric gases used for the experiment.

Past 600°C, there is a loss of the remaining phenolic functional groups that overlap with pure carbon black oxidation, and thus, functional groups have not been further demarcated owing to this experimental uncertainty. O_3 oxidized carbon shows no peak at 300°C, confirming the absence of significant COOH group content in this mildly oxidized material, consistent with XPS analyses. However, it does contain phenolic groups, and their overlapping oxidation with pure carbon after 600°C again makes it challenging to point to the specificity of the phenolic groups in the material. TGA weight loss and derivative weight curves are shown in Figure 7.19A, with corresponding mass spectral data showing 44 AMU evolution, in correspondence with the weight loss in Figure 7.19B. The 18 AMU decomposition product channel is also shown, but a significant change in the ion current from its detected baseline value is not observed.

TGA and its weight derivative with respect to temperature for the four materials discussed are shown in Figure 7.20, with HNO_3 functionalized R250 showing a stark difference in its response to thermal treatment with multiple temporal regions of varying weight loss relative to the other carbon blacks, with one narrow range of temperature over which all mass is lost. HNO_3–carbon shows the greatest weight loss at 300°C in the form of CO_2 and H_2O, consistent with XPS measurements of the material heated in situ, showing a decrease in atomic percent of oxygen at the 300°C mark. It should be noted that the two thermal treatments differ; i.e., XPS in situ heating is in a vacuum, while TGA is at atmospheric pressure.

The comparative derivative weight plot shown in Figure 7.20 is telling of the differences between the nascent material and oxidized (functionalized) forms, with reference to the graphitized form. HNO_3-treated R250 has a highly oxygenated and thereby unstable surface with multiple functional groups (likely –COOH) attached to the particle corresponding to the first weight loss peak at ~300°C followed by a peak at 600°C, coinciding with the weight loss for the untreated R250 (due to residual oxidation). A third peak is observed at ~700°C, likely corresponding to the graphitic structures/periphery of the particles, analogous to the derivative of graphitized R250 at ~800°C, given its resilient particle nanostructure. The derivative peak for O_3 functionalized R250 is broad and spans a 600–900°C range, corresponding to oxidation of primary particles resembling those of R250, more graphitic particles owing to the oxidative treatment, and finally, weight loss of all the remaining carbon from that sample by ~950°C. Thus, three distinct regions of weight loss have been identified based on the derivative curves, and percentage weight loss has been calculated from the

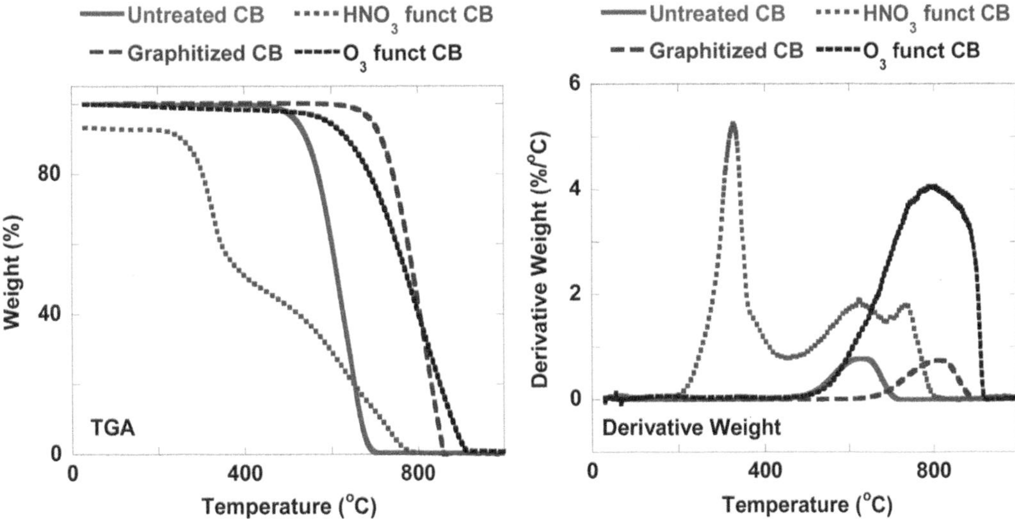

FIGURE 7.20 Comparative weight loss and derivative weight graphs for untreated, functionalized, and graphitized carbon black.

area under the curve for each region with respect to the total area under the derivative curve. These self-normalized values are tabulated in Table 7.7 below.

Untreated and graphitized forms of carbon black show weight loss in only one of the three identified regions given the material's structural uniformity and chemical purity. Acid and ozone functionalized carbon black, on the other hand, show weight loss over two or all three regions, supporting the presence of functional groups that are lost first at temperatures <600°C while the material yet follows a pyrolysis treatment given the reduced reactivity of (external) oxygen at low temperatures, followed by loss of the carbon matrix itself. Finally, the relative weight loss between 700 and 900°C corresponds to oxidation, and as such, the differences between the carbons reflect the relative reactivities, with the nascent and graphitized forms serving as references by which to gauge the increased reactivity of the functionalized forms after thermal removal of oxygen groups.

7.7.3 FT-IR Spectroscopy

Operating principle: FT-IR works on the principle of absorption of IR radiation by a material. For a material to be IR active, a change in the dipole moment of the molecule is required for it to generate a

TABLE 7.7
Percent Weight Loss as a Function of the Range of Treatment Temperature

Temperature	Contribution to Weight Loss (%)			
Range, °C	Untreated	HNO_3	O_3	Graphitized
200–400	—	35	—	—
400–700	100	41	24	—
700–900	—	24	76	100

signature absorption peak that corresponds to the wavenumbers causing that dipole moment change. Thus, an asymmetric molecule, such as functional groups, will be IR active, while the backbone, carbon in this case, is IR inactive and tends to absorb all wavelengths of incident radiation equally.

The instrument: All FT-IR spectra were collected on a Bruker Vertex 70v FT-IR Spectrometer with a liquid nitrogen–cooled narrow band mercury cadmium telluride detector. These measurements were taken using a Harrick Praying Mantis Diffuse Reflectance Infrared Fourier Transform Spectroscopy accessory equipped with an in situ heating cell. The instrument was constantly purged using nitrogen, while the in situ heating cell was purged using argon. Each spectrum was an average of a total of 800 scans collected at a resolution of 4 cm^{-1} and absorbance calculated using KBr as the reference material.

FT-IR spectroscopy was used to compare the degree of functionalization of the HNO_3 and O_3-treated carbons. Given the absence of functional groups on untreated and graphitized carbon forms, these materials were not analyzed with FT-IR. Carbon sp^2 forms absorb IR radiation, and therefore, to make the measurement, the sample was prepared by diluting the carbon with KBr to increase the signal-to-noise ratio. Spectra were first collected at room temperature of ~30°C and subsequently with in situ heating. Each measurement was performed after collecting a KBr reference spectrum to which the measured sample spectra were normalized after atmospheric compensation, to correct for water and CO_2, and baseline correction. Peak assignments for all spectra have been done based on functional carbon analyzed in the literature [28,33,36]. Room temperature (30°C) FT-IR for the wet and dry functionalized carbons is shown in Figure 7.21. In situ heated spectra for both materials and their evolution with temperature were as follows.

The 1000–1800 cm^{-1} regions are indicative of different functional groups, whose contributions to absorbance intensity often overlap [36]. Peaks in the ~1000–1200 cm^{-1} region indicate the presence of carbon–oxygen single bonds, such as ethers, alcohols, and phenols, i.e., a C–OH or a C–O–C group, along with the O–H from water whose bending mode occurs at ~1500 cm^{-1}, which is unclear here given overlapping peaks. The next set of peaks in the ~1600–1750 cm^{-1} region mark the presence of carbon–oxygen double bonds such as the C=O of carboxylic acids. Peak contribution in the ~1750–1800 cm^{-1} region can be ascribed to the C=O of lactones or carboxylic anhydrides, likely present in small quantities. In situ heating of the HNO_3-treated carbon has been used in an attempt to parse out the contribution of these functional groups with complementary information available from XPS and TGA-MS.

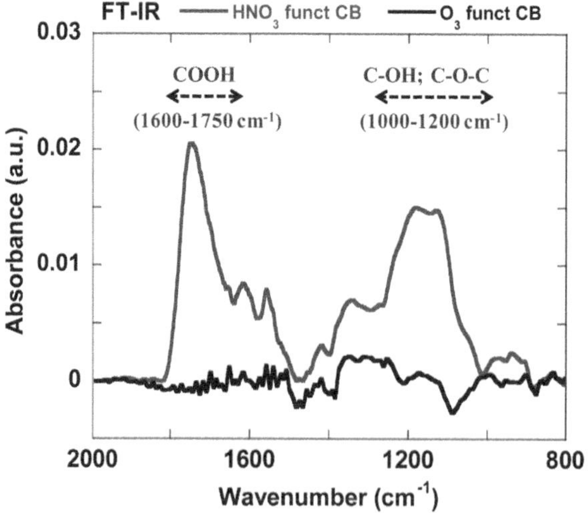

FIGURE 7.21 FT-IR spectra for HNO_3 and O_3 functionalized carbon at room temperature. HNO_3 funct CB shows two prominent peaks, while O_3 funct CB does not have significant IR-absorbance peaks.

7.7.3.1 In Situ Heating Stage FT-IR

Both O_3- and HNO_3-carbon were subjected to the following heating protocol in situ. Spectra were collected after a 20-min hold at each temperature.

In situ heating with the spectra collected at

a. 30°C: room temperature, as a baseline before temperature ramp,
b. 125°C: to remove moisture,
c. 300°C: carboxylic acid groups have been observed to leave at this temperature using XPS and TGA-MS,
d. 500°C: to explore changes in the FT-IR spectrum due to leaving functional groups, if any.

All spectra have been processed for atmospheric H_2O and CO_2 with KBr at 300°C as a reference. Heating KBr to 300°C ensures that there is no moisture in the reference, which would result in negative spectral features at 1590–1650 cm^{-1} and 3200–3600 cm^{-1}. Baseline corrected spectra are presented here. Spectra collected using the above protocol with the in situ heating cell are shown in Figure 7.22.

A consistent broad peak and overlapping C–O or C–O–H contributions from phenols, ethers, alcohols, and carboxylic acids are present in the ~1000–1400 cm^{-1} region. Peak contributions in the ~1500–1800 cm^{-1} region show a consistent drop in absorbance, especially in the 1600–1800 cm^{-1} range. Given the visible decrease, the relative change in area under the curve for this region, along with complementary information from XPS and TGA–MS, is used in an attempt to quantify the carboxylic groups. Percentage values are computed as a function of the total area under the curve in the 800–1900 cm^{-1} range. These values are presented in Table 7.8.

Given the lack of peaks observed for O_3-treated carbon black, no such quantification was attempted.

A consistent decrease in the area under the curve for the region that contains the COOH absorbance signature as the temperature is increased agrees with observations from XPS and TGA–MS. The presence of small peaks and some remaining intensity after in situ heat treatment to 300 and 500°C is probably due to the presence of small quantities of anhydrides or lactone groups, which have been observed to evolve and leave as CO_2 at temperatures >500°C [43].

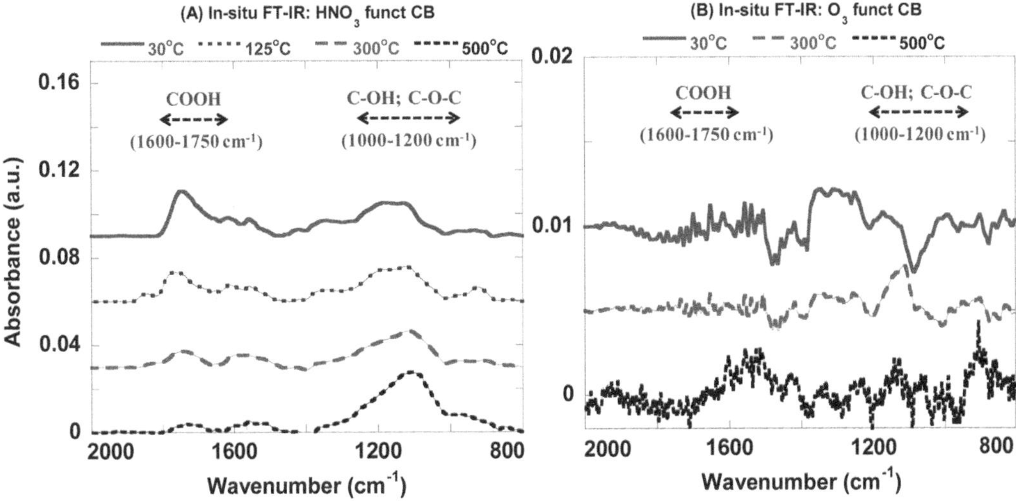

FIGURE 7.22 Effect of temperature on FT-IR spectra of the (B) O_3-treated and (A) HNO_3-treated carbon subjected to in situ heating (note different *Y*-axis scales). Temperature increases top down.

TABLE 7.8
Relative % of COOH Using Area Under the Curve of the FT-IR Spectrum

Area Under the Curve/Temperature (°C)	COOH (%) (1600–1800 cm^{-1})
30	20
200	14
300	11
500	5

7.7.3.2 Ex Situ Heating and FT-IR

A possible concern is the effect of in situ vacuum heat treatment on the evolution of functional groups and their associated temperature ranges as reported by in situ FT-IR analysis when compared with the same material heated ex situ to drive off functionalization and subsequently analyzed by FT-IR. To compare observations of such in situ to ex situ heating, HNO_3-treated carbon is heated outside of the FT-IR spectrometer (ex situ) to temperatures of 300 and 500°C under controlled and isothermal conditions in a TGA until a near-stable weight reading was attained. The heating protocol is as follows.

Ex situ heating isothermally until a near stable weight loss at

a. 300°C: to remove COOH before FT-IR analysis,
b. 500°C: to explore changes, if any, including loss of other functionality.

The ex situ heated carbon is then transferred to the FT-IR spectrometer for spectra acquisition at room temperature (~30°C) and after moisture removal at ~125°C. Thus, spectra from samples heated ex situ are compared to the FT-IR spectra of the sample heated in situ as a check for consistency in functional group evolution with treatment conditions. This comparison is shown in Figure 7.23.

Spectra for both temperature treatments are consistent for both in situ and ex situ heating, although there may be some minor contribution in the 1400–1800 cm^{-1} range from leftover functional groups

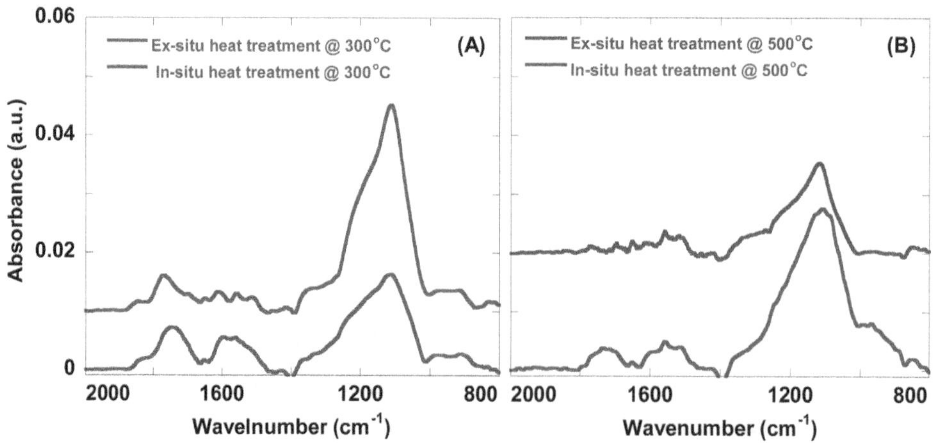

FIGURE 7.23 Comparison of FT-IR spectra from HNO_3-treated carbon heated to (A) 300°C and (B) 500°C, in situ (bottom-curve) and ex situ (top-curve).

in the sample heated in situ, given their vacuum conditions and limited time at temperature. This contribution is higher for the 300°C treatment as compared to 500°C, likely due to leftover carboxylic anhydrides, with their presence evidenced by their signature peak around the ~900 cm^{-1} mark [36], visible in the spectra of carbons heated in situ.

7.8 STRENGTHS AND LIMITATIONS OF THE CHARACTERIZATION TECHNIQUES

To summarize, the presence of functional groups works toward altering molecular arrangements and layer plane structure of the carbon black, observed by the differences in lattice parameters and d_{002} measured using XRD and the resultant change in vibrational modes seen by Raman. Visual differences in nascent and functionalized materials are shown by the TEM micrographs, with elemental carbon and oxygen quantified by EDS. Elemental mapping using EDS in a TEM typically has nanometer resolution and gives an approximate analysis of the volumetric elemental composition of the sample, given that the interaction of beam is with the entire thickness of an electron-transparent sample such as that used for TEM.

Compositional analysis of the material is done using XPS providing for an accurate and precise relative measure of the elements with an accuracy of <1.5% for quantified carbon and oxygen with ~1% standard deviation. XPS also provides information on how carbon may be bonded to oxygen. Bonding information is extracted by XPS spectra curve fitting because of known changes in the BE of carbon when singly or doubly bonded to oxygen. XPS can inform the user about the oxidation state of the element, making it a very useful technique for elemental as well as functional group quantification. However, XPS probes the top 1–10 nm of the sample and is most sensitive to the top 1–3 nm of the material, classifying it as a surface-sensitive technique.

In addition to XPS, FT-IR is used to identify signature peaks from the same functional groups seen by XPS. The signal detected by the FT-IR spectrometer is considered to be from the bulk of the sample given that the sample is mixed and ground with KBr before analysis and the depth of signal detection is of the order of a few microns into the sample. The presence of carboxylic acid and phenolic groups is confirmed by prominent peaks in the 1600–1800 cm^{-1} and 1000–1400 cm^{-1} regions, respectively. Given the IR-absorbing nature of the carbon material and the instrument's signal-to-noise ratio, quantification of the functional groups is a challenge, and reliable relative quantification will require a uniformly functionalized and externally quantified carbon as a calibration standard.

Thermal treatment of the functionalized carbon black drives away functional groups over a range of temperatures. With a resolution of <0.1 μg, weight loss from a bulk sample (~10 s of mg) at different temperatures can be attributed to the presence of different functional groups. Based on the reactivity of the functional group on the carbon black, these are observed to leave at the ~300 and ~500°C mark, with loss of the carbon backbone thereafter. Typical sample amounts required for analysis make it a bulk analysis technique. The associated MS analyses of the evolved off-gases for molecular mass and the release of H_2O, CO, and CO_2 molecules, for instance, can be correlated with observed weight loss for a range of temperatures. Thus, the TGA records the response of the sample with temperature, and the MS can help identify the presence of elements with prior knowledge about the sample. It cannot, however, be used for elemental quantification but can provide a relative measure of the degree of functionalization by comparing the weight loss of untreated carbon black to functionalized carbon black in the same temperature range. For carboxylic acid groups, this assumes minimal oxidation of the carbon matrix at the ~300°C mark with weight loss then attributed to the evolution of carboxylic acid as CO_2, in agreement with the literature [31,36].

Correlating functional group evolution with temperature has been further resolved by in situ XPS and FT-IR measurements that show a consistent reduction in the intensity of the peak attributed to the presence of carboxylic acid, indicative of this being the leaving group at 300°C. Complimenting this is the significant weight loss as CO_2 and H_2O observed by TGA–MS at the same temperature.

TABLE 7.9
Comparison of Measurement Parameters Across Analytical Techniques

Analytical Technique	Signal Detected From	Depth Probed	Pressure Conditions	Resolution	Accuracy	Precision
XPS	Surface	<10 nm	Vacuum	<1 eV	<1.5 at.%	<5%
TGA	Bulk	~1 mm	Ambient	<0.1 μg	1°C	0.1°C 0.01 wt.%
FT-IR	Bulk	<5 μm	Ambient/ Vacuum	<4 cm^{-1}	<0.2%	0.1%

The presence of carboxylic acid groups at room temperature and their subsequent evolution with thermal treatment in TGA, XPS, and FT-IR show consistency in the temperature–functional group relationship across the techniques. Given the overlapping temperature ranges of the comparatively less reactive carbonyl and phenolic groups with the carbon backbone oxidation at temperatures >500°C, delineating these by TGA–MS is a challenge. Specific peak assignments and quantification of these groups by XPS and FT-IR can result in erroneous values given their overlapping peak contributions and have, therefore, not been calculated.

Table 7.9 summarizes key parameters across analytical techniques. Each technique has its advantages and limitations, and complementary information from different techniques can potentially help with better material characterization and elemental quantification based on the functionalization technique used, which also influences the type and quantity of functional groups introduced. A discussion on the process feasibility and amount of oxygen introduced for both wet and dry chemical treatment follows.

7.9 ADVANTAGES AND LIMITATIONS OF WET AND DRY CHEMICAL TREATMENT

As seen from the microscopic and spectroscopic material characterization, a clear distinction is observed when carbon black is functionalized using a wet method versus a dry chemical treatment.

7.9.1 Ease of Functionalization

Modifying the chemistry of carbon with a dry functionalization method such as treatment with ozone gas is likely to be the least challenging approach regarding sample preparation, retrieval postfunctionalization, and processing time. The setup does require ozone gas and UV light to excite and dissociate ozone molecules into the reactive O radical so that it reacts with the surface of the carbon.

Comparatively, HNO_3-treated carbon black is an involved process. It requires the use of an extremely strong acid, which can become a concern for the safety of the personnel handling the acids. Touching concentrated HNO_3 can burn one's skin, and therefore, extreme caution is required when functionalizing using HNO_3 or any other strong acid. The material to be functionalized is mixed with the acid and allowed to reflux for a few hours. Apart from being time intensive, this process requires a reliable reflux apparatus set up within a fume hood to ensure no acid, in liquid or vapor form, leaks to the surroundings. Sample retrieval can be challenging because of the introduction of highly acidic groups with the material soaked in the acid. The functionalized material requires repeated washing with distilled water to dilute any acid yet wetting the carbon black surface to a near neutral pH of 7 from an initial pH of 0.1–1. If required for use in a dry form thereafter, separating this material from distilled water into which it will now dissolve given the presence of hydrophilic oxygen functional groups is a task unto itself. An ultra-high-speed centrifuge may be used to separate most of the material from its solution in distilled water. However, there may still be sufficient amounts of carbon dissolved, and complete separation will require more labor- and time-intensive

TABLE 7.10
Relative Amounts of COOH Calculated Using XPS, TGA, and FT-IR

Relative Amount of COOH (%)	Analytical Techniques		
Temperature (°C)	XPS (at.%)	TGA (wt.%)	FT-IR (Funct.%)
30	24	35	20
200	—	—	14
300	19	—	11
500	14	—	5

processing. Even more problematic than water removal is the inevitable clumping and random agglomeration that will challenge achieving a "dry" and flowable powder and render any subsequent aqueous re-dispersal difficult.

7.9.2 Amount of Oxygen Introduced

Wet chemical treatment being more rigorous with a strong oxidizing agent like HNO_3 introduces a lot more atomic oxygen as compared to dry O_3 treatment, which only lightly modifies the surface of the carbon. XPS, TGA–MS, and FT-IR are used to quantify this difference as complementary analytical techniques. A summary of the amount of oxygen introduced as per the results of these analytical techniques is shown in Table 7.10.

It is important to note that these values are calculated from the results, wherein the measurements have an inherently different physical basis, restricting direct comparisons. XPS is a surface analysis technique, probing the top 5 nm of the surface, while TGA–MS registers bulk weight changes based on leaving groups at various temperatures. FT-IR spectroscopy is typically used as a qualitative measure to detect signature peaks and confirms the presence (or absence) of COOH at different temperature conditions. However, relative quantities have been computed here to give the reader a sense of the change in the COOH signature peak intensity with temperature.

7.9.3 Distribution of Functional Groups

Because it is a strong acid, HNO_3 imparts a number of carboxylic acid groups to the carbon black. Carboxylic acid accounts for almost a third of all functional groups on the surface, quantified by XPS, and a relatively similar proportion in bulk, quantified by TGA–MS and FT-IR. Phenolic groups are observed to be quantitatively approximately half of the carboxylic groups present. Carbonyls and phenols are challenging to differentiate using TGA given that these groups and carbon begin to oxidize and leave in a similar temperature range, observed to be typically $<500°C$. The XPS and TGA–MS curves of the two functionalizations and the missing prominent carboxylic acid peak for O_3-treated carbon imply the introduction of phenolic and maybe some carbonyl groups with this dry functionalization. In this functionalization, carboxyls are rare or present in insignificant quantities.

7.9.4 Process Scalability

Given the ease of O_3 functionalization, a dry gaseous process such as this is more scalable with fewer requirements for bulk functionalization of carbon black. However, the amount of oxygen introduced is limited to $\sim$10 at.%, with phenols and carbonyls dominating the functional group makeup. Strong acid treatment, when scaled for bulk functionalization, will result in a strong oxidation treatment and an abundance of acidic functional groups but requires apparatus and multiple steps to retrieve the material in the desired form for further applications. We note the availability of industrial-scale ozone generators.

7.10 APPLICATIONS OF FUNCTIONALIZED CARBON

Carbon and its forms have been functionalized and used as a reinforcing filler [21], as an adsorbent [23,45], or as cathodes for lithium-ion batteries due to their high-power densities postfunctionalization [20], to name a few applications. Oxygen and nitrogen functionalized carbons are used as catalysts, with the functional groups being active sites where chemisorption of the reactants can occur [43]. They have been employed in direct methanol fuel cells as an electrocatalyst [39] and as catalyst support for platinum nanoparticles with enhanced properties after nitrogen functionalization, relative to an amorphous carbon as catalyst support [35]. The presence of functional groups imparts cation exchange properties to the carbon when these groups are acidic, while basic functionality promotes anion exchange [25]. The carbon black work function, for instance, is a key parameter defining charge exchange at the material's interface and is a function of its surface chemistry and may be modified with appropriate treatment [46]. Surface functionalized CNTs, given their unique end-capped container-like structure, are being explored and developed for use in biomedical applications such as drug delivery, vaccinations, and oncology with functionalization increasing their biocompatibility [6]. These are also being used as fillers in nanocomposite materials given their enhanced dispersion and interfacial interaction capability after functionalization [47]. Controlled functionalization of 2D graphene sheets, which may otherwise be aggregated and have inaccessible interlayer channels, and doping heteroatom groups on their surface have shown promise for storage, making it favorable for supercapacitors [37].

Oxygen functionalization on carbon black, CNTs [26,48,49], and nano-onions [50] has largely been used as a gateway to introducing other groups tailored to one's requirement. Functionalities such as the homogeneous introduction of the diazonium cation to an oxygenated carbon black have been studied as a function of reaction and pretreatment conditions, where oxygenated sites were used to introduce diazonium cations [51]. Ozonolysis has been used to pretreat single-walled CNTs forming unstable ozonides, which can then be replaced by desired functional groups and skew the distribution away from carbonyl groups and in favor of other desired functional groups, based on the reagent used (nitric acid, hydrogen peroxide, piranha, etc.) [38]. A variety of chemical oxidation treatments on CNTs using nitric acid, hydrogen peroxide, sulfuric acid, ammonium hydroxide, hydrochloric acid, or a combination of these have been studied systematically for their functionalization versus degradation effect on nanotubes [27]. Reflux in nitric acid has been observed to degrade CNTs relative to other reagents. However, CNTs, when treated with nitric acid vapor, showed an intact morphology and functionalized to a greater extent as compared to wet chemical reflux [29]. Efficient methods of functionalizing CNTs such as microwave radiation under a concentrated acid mixture can functionalize single-walled and multi-walled CNTs in ~5 min. Aniline functional groups have been introduced through amidation after microwave treatment [28]. Electrochemical modification of CNTs is used as an alternate clean and efficient route for functional group addition [52]. Inductively coupled radiofrequency plasma has been used to introduce oxygen, fluorine, and nitrogen groups on CNTs [30]. Nitrenes and carbenes have been used to functionalize the sidewalls of single-walled CNTs using a microwave-driven hydrogen-atom glow discharge method [15].

7.10.1 COMMERCIAL CARBON BLACK AS A SURROGATE FOR SOOT

Soot structure has widely been observed to be a function of its formation conditions, which may lead to a variety of functional groups being introduced onto its surface given its formation temperature and its reactivity with atmospheric gaseous species such as sulfur oxides (SOx), oxides of nitrogen (NOx), and/or ozone (O_3) [39,53]. Specific health effects of soot particles and the functional groups adsorbed on its surface are being studied combining research and knowledge from combustion science, material surface chemistry, atmospheric pollution, and epidemiology research [54]. Different particle characteristics are associated with the different health-related aspects, with an added dependence on exposure time and concentration, making this a nontrivial and compounded problem to tackle [54]. Ultrafine particles have been reported to contain a higher proportion of organic and elemental carbon that may be the cause for oxidative stress when it contacts human lung cells [55]. In the same study, the effect of redox-active species such as quinones has also been reported to be medically detrimental.

Over decades, carbon black has been widely used as a proxy for studies on soot oxidation [56–60] given the challenges involved in sample acquisition and tailored chemistry for controlled analysis. Soot is an ancillary combustion product, while carbon black is a purposefully manufactured product made from hydrocarbons. It consists of dangling bonds at the edges of the carbon lamellae, primarily satisfied by hydrogen [25]. Thus, carbon black surfaces can be modified to closely replicate the surface of soot and thereby perform systematic analysis on its chemistry and structure as a surrogate to soot, to better infer the effects of soot with such controlled experiments. However, there are numerous grades of carbon blacks, all tailored for specific applications, and therefore, it is important to use one that resembles the soot type of interest in morphology, nanostructure, and decomposition characteristics.

7.11 SUMMARY

A commercially manufactured carbon black has been subject to oxygen functionalization using a dry and wet chemical treatment with ozone and nitric acid, respectively. Subsequent characterization of the material and functional groups introduced is done using complementary analytical techniques. Wet functionalization introduces a significant amount of oxygen functional groups volumetrically relative to the dry, gaseous treatment for which surface modification is observed. Acid functionalized carbon shows prominent carboxylic acid groups, while ozone introduces relatively mild phenolic and carbonyl groups. TEM, EDS, Raman, and XRD are used to infer changes in the material's structure and elemental content after chemical treatment, while XPS, TGA, and FT-IR characterize and quantify major functional groups. Effect of treatment time shows a threshold duration above which functionalization reaches a plateau or a saturation point. In situ heating of functionalized carbon black analyzed with XPS and FT-IR shows a consistent decline in oxygen content as the temperature is increased. Evolution of the type of functional groups is tracked and complimented by TGA–MS. Characterization and quantification of functional groups are shown to be consistent across the analytical techniques, with results complimenting one another to better comprehend the degree and type of functionalization along with its effect on particle nanostructure.

7.12 AUTHOR CONTRIBUTIONS

RVW conceptualized the outline. MS conceived and planned experiments and material characterization. JS performed XPS and analyzed the data. JLG performed and helped analyze EDS. TJZ performed FT-IR experiments and helped with related analyses. MS wrote the manuscript in consultation with RVW, TJZ, JLG, and JS. All authors provided critical feedback and helped shape the analysis and manuscript.

7.13 ACKNOWLEDGMENTS

Characterizations (TEM, EDS, XRD, Raman, XPS, TGA, and FT-IR) were performed using the facilities of the Materials Research Institute at The Pennsylvania State University. HRTEM guidance from Dr. Jennifer Gray and Dr. Ke Wang is gratefully acknowledged. Help from Dr. Ekaterina Bazilevskaya for TGA and Maxwell Wetherington for Raman at the Materials Characterization Laboratory in measurement and guidance with the analysis is much appreciated.

REFERENCES

1. C. Harvey. Air pollution in India is so bad that it kills half a million people every year. *The Washington Post* (2016).
2. India, China account for over half of global deaths due to air pollution: Report. *The Times of India* (2018).
3. G. Anand. India's Air Pollution Now Rivals China's as Deadliest in the World. *New York Times* (2017).
4. Three million people dying yearly due to outdoor air pollution, says WHO. *PR Newswire* (2016).
5. Anonymous. Breathe uneasy; Air pollution in India. *The Economist.* **414,** 38 (2015).
6. M. Prato, K. Kostarelos, and A. Bianco. Functionalized carbon nanotubes in drug design and discovery. *Acc. Chem. Res.* **41,** 60–68 (2008).
7. P. J. F. Harris. *Carbon nanotube science: synthesis, properties and applications.* (Cambridge University Press, 2009).
8. K. P. Loh, Q. Bao, G. Eda, and M. Chhowalla. Graphene oxide as a chemically tunable platform for optical applications. *Nat. Chem.* **2,** 1015–1024 (2010).
9. D.-W. Wang and D. Su. Heterogeneous nanocarbon materials for oxygen reduction reaction. *Energy Environ. Sci.* **7,** 576–591 (2014).
10. R. S. Rajeev and S. K. De. Crosslinking of rubbers by fillers. *Rubber Chem. Technol.* **75,** 475–510 (2002).
11. K. Lawrence et al. Functionalized carbon nanoparticles, blacks and soots as electron-transfer building blocks and conduits. *Chem. – An Asian J.* **9,** 1226–1241 (2014).
12. J. Liu, Y. Ye, Y. Xue, X. Xie, and Y.-W. Mai. Recent advances in covalent functionalization of carbon nanomaterials with polymers: Strategies and perspectives. *J. Polym. Sci. Part A Polym. Chem.* **55,** 622–631 (2016).
13. L. P. Biro, C. A. Bernardo, G. G. Tibbetts, and P. Lambin, Eds. *Carbon Filaments and Nanotubes: Common Origins, Differing Applications?* (Springer). doi:10.1007/978-94-010-0777-1
14. A. Hirsch and O. Vostrowsky. Functionalization of carbon nanotubes BT-functional molecular nanostructures (ed. Schlüter, A. D.) 193–237 (Springer Berlin Heidelberg, 2005). doi:10.1007/b98169
15. M. Holzinger et al. Sidewall functionalization of carbon nanotubes. *Angew. Chemie Int. Ed.* **40,** 4002–4005 (2001).
16. G. Goncalves et al. Surface modification of graphene nanosheets with gold nanoparticles: the role of oxygen moieties at graphene surface on gold nucleation and growth. *Chem. Mater.* **21,** 4796–4802 (2009).
17. A. Hirsch. Functionalization of single-walled carbon nanotubes. *Angew. Chemie Int. Ed.* **41,** 1853–1859 (2002).
18. S. Banerjee, T. Hemraj-Benny, and S. S. Wong. Covalent surface chemistry of single-walled carbon nanotubes. *Adv. Mater.* **17,** 17–29 (2005).
19. I. Sutherland, E. Sheng, R. H. Bradley, and P. K. Freakley. Effects of ozone oxidation on carbon black surfaces. *J. Mater. Sci.* **31,** 5651–5655 (1996).
20. A. Jaffe, A. Saldivar Valdes, and H. I. Karunadasa. Quinone-functionalized carbon black cathodes for lithium batteries with high power densities. *Chem. Mater.* **27,** 3568–3571 (2015).
21. N. Tsubokawa. Functionalization of carbon black by surface grafting of polymers. *Prog. Polym. Sci.* **17,** 417–470 (1992).
22. T. Takada, M. Nakahara, H. Kumagai, and Y. Sanada. Surface modification and characterization of carbon black with oxygen plasma. *Carbon* **34,** 1087–1091 (1996).
23. C. Moreno-Castilla et al. Activated carbon surface modifications by nitric acid, hydrogen peroxide, and ammonium peroxydisulfate treatments. *Langmuir* **11,** 4386–4392 (1995).
24. Y. Otake and R. G. Jenkins. Characterization of oxygen-containing surface complexes created on a microporous carbon by air and nitric acid treatment. *Carbon* **31,** 109–121 (1993).
25. H. P. Boehm. Some aspects of the surface chemistry of carbon blacks and other carbons. *Carbon* **32,** 759–769 (1994).
26. Y. Wang et al. An integrated route for purification, cutting and dispersion of single-walled carbon nanotubes. *Chem. Phys. Lett.* **432,** 205–208 (2006).
27. V. Datsyuk et al. Chemical oxidation of multiwalled carbon nanotubes. *Carbon* **46,** 833–840 (2008).
28. B. A. Kakade and V. K. Pillai. An efficient route towards the covalent functionalization of single walled carbon nanotubes. *Appl. Surf. Sci.* **254,** 4936–4943 (2008).
29. W. Xia, C. Jin, S. Kundu, and M. Muhler. A highly efficient gas-phase route for the oxygen functionalization of carbon nanotubes based on nitric acid vapor. *Carbon* **47,** 919–922 (2009).

30. A. Felten, C. Bittencourt, J. J. Pireaux, G. Van Lier, and J. C. Charlier. Radio-frequency plasma functionalization of carbon nanotubes surface O2, NH3, and CF4 treatments. *J. Appl. Phys.* **98,** 74308 (2005).
31. H. P. Boehm. Surface oxides on carbon and their analysis: a critical assessment. *Carbon* **40,** 145–149 (2002).
32. T. I. T. Okpalugo, P. Papakonstantinou, H. Murphy, J. McLaughlin, and N. M. D. Brown. High resolution XPS characterization of chemical functionalised MWCNTs and SWCNTs. *Carbon* **43,** 153–161 (2005).
33. D. B. Mawhinney et al. Infrared spectral evidence for the etching of carbon nanotubes: ozone oxidation at 298 K. *J. Am. Chem. Soc.* **122,** 2383–2384 (2000).
34. B. N. Khare, M. Meyyappan, A. M. Cassell, C. V Nguyen, and J. Han. Functionalization of carbon nanotubes using atomic hydrogen from a glow discharge. *Nano Lett.* **2,** 73–77 (2002).
35. S. S. Rich et al. Nitrogen functionalized carbon black: a support for Pt nanoparticle catalysts with narrow size dispersion and high surface area. *Carbon* **81,** 115–123 (2015).
36. J. L. Figueiredo, M. F. R. Pereira, M. M. A. Freitas, and J. J. M. Órfão. Modification of the surface chemistry of activated carbons. *Carbon* **37,** 1379–1389 (1999).
37. W. Tian et al. Three-dimensional functionalized graphenes with systematical control over the interconnected pores and surface functional groups for high energy performance supercapacitors. *Carbon* **85,** 351–362 (2015).
38. S. Banerjee and S. S. Wong. Rational sidewall functionalization and purification of single-walled carbon nanotubes by solution-phase ozonolysis. *J. Phys. Chem. B* **106,** 12144–12151 (2002).
39. J.-O. Müller, D. S. Su, U. Wild, and R. Schlögl. Bulk and surface structural investigations of diesel engine soot and carbon black. *Phys. Chem. Chem. Phys.* **9,** 4018–4025 (2007).
40. A. C. Ferrari and D. M. Basko. Raman spectroscopy as a versatile tool for studying the properties of graphene. *Nat. Nanotechnol.* **8,** 235 (2013).
41. A. L. Patterson. The Scherrer formula for X-ray particle size determination. *Phys. Rev.* **56,** 978–982 (1939).
42. M. A. Short and P. L. Walker. Measurement of interlayer spacings and crystal sizes in turbostratic carbons. *Carbon* **1,** 3–9 (1963).
43. J. L. Figueiredo and M. F. R. Pereira. The role of surface chemistry in catalysis with carbons. *Catal. Today* **150,** 2–7 (2010).
44. H. P. Boehm, E. Diehl, W. Heck, and R. Sappok. Surface oxides of carbon. *Angew. Chemie Int. Ed. English* **3,** 669–677 (1964).
45. C. Stoquart, P. Servais, P. R. Bérubé, and B. Barbeau. Hybrid membrane processes using activated carbon treatment for drinking water: a review. *J. Memb. Sci.* **411–412,** 1–12 (2012).
46. T. J. Fabish and D. E. Schleifer. Surface chemistry and the carbon black work function. *Carbon* **22,** 19–38 (1984).
47. C. A. Dyke and J. M. Tour. Covalent functionalization of single-walled carbon nanotubes for materials applications. *J. Phys. Chem. A* **108,** 11151–11159 (2004).
48. H. Kong, C. Gao, and D. Yan. Functionalization of multiwalled carbon nanotubes by atom transfer radical polymerization and defunctionalization of the products. *Macromolecules* **37,** 4022–4030 (2004).
49. J.-K. Lee, H. M. Pathan, K.-D. Jung, and O.-S. Joo. Electrochemical capacitance of nanocomposite films formed by loading carbon nanotubes with ruthenium oxide. *J. Power Sources* **159,** 1527–1531 (2006).
50. V. Georgakilas, D. M. Guldi, R. Signorini, R. Bozio, and M. Prato. Organic functionalization and optical properties of carbon onions. *J. Am. Chem. Soc.* **125,** 14268–14269 (2003).
51. M. Toupin and D. Bélanger. Spontaneous functionalization of carbon black by reaction with 4-nitrophenyldiazonium cations. *Langmuir* **24,** 1910–1917 (2008).
52. K. Balasubramanian and M. Burghard. Chemically functionalized carbon nanotubes. *Small* **1,** 180–192 (2004).
53. J.-O. Muller et al. Morphology-controlled reactivity of carbonaceous materials towards oxidation. *Catal. Today* **102–103,** 259–265 (2005).
54. J. S. Lighty, J. M. Veranth, and A. F. Sarofim. Combustion aerosols: factors governing their size and composition and implications to human health. *J. Air Waste Manage. Assoc.* **50,** 1565–1618 (2000).
55. N. Li et al. Ultrafine particulate pollutants induce oxidative stress and mitochondrial damage. *Environ. Health Perspect.* **111,** 455–460 (2003).

56. M. D. Hays and R. L. Vander Wal. Heterogeneous soot nanostructure in atmospheric and combustion source aerosols. *Energy & Fuels* **21,** 801–811 (2007).
57. B. R. Stanmore, V. Tschamber, and J.-F. Brilhac. Oxidation of carbon by NO_x, with particular reference to NO_2 and N_2O. *Fuel* **87,** 131–146 (2008).
58. C. K. Gaddam, R. L. Vander Wal, X. Chen, A. Yezerets, and K. Kamasamudram. Reconciliation of carbon oxidation rates and activation energies based on changing nanostructure. *Carbon* **98,** 545–556 (2016).
59. R. L. Vander Wal, A. Yezerets, N. W. Currier, D. H. Kim, and C. M. Wang. HRTEM study of diesel soot collected from diesel particulate filters. *Carbon* **45,** 70–77 (2007).
60. J.-O. Müller, D. S. Su, R. E. Jentoft, U. Wild, and R. Schlögl. Diesel engine exhaust emission: oxidative behavior and microstructure of black smoke soot particulate. *Environ. Sci. Technol.* **40,** 1231–1236 (2006).

8 Fabrication of Conductive Nanoporous Membranes Using Self-Assembling Block Copolymers

Ece Isenbike Ozalp[a], *Sungho Kim*[a], *Vignesh Sundar,*[a] *Jian-Gang (Jimmy) Zhu*[a] *and Jeffrey A. Weldon*[b]

[a]Department of Electrical & Computer Engineering, Carnegie Mellon University, Pittsburgh, Pennsylvania, USA
[b]Department of Electrical Engineering, University of Hawaii at Manoa, Honolulu, Hawaii, USA

CONTENTS

8.1 INTRODUCTION

Nanoporous membranes with sub-100 nm pore sizes have been widely investigated due to high demand for biological applications such as biosensing, filtration, and drug delivery.[1–3] Their properties such as small pore size and high surface-to-volume ratio make them applicable for nanofluidic applications via controlled diffusion due to the comparable sizes between the charged molecules and the nanochannels.[4] Controlling the transport of charged molecules is essential for smart drug delivery systems, considering drug molecules are typically charged.[5,6] Implantable drug delivery devices with controllable drug delivery rates can reduce side effects and maximize efficacy of the drug by keeping the drug concentration relatively constant at a targeted concentration. Nanoporous membranes that can control the diffusion of charged molecules could be used for drug delivery systems.

In the application of the majority of previously reported nanoporous membranes, the diffusion rate is primarily determined by the pore size.[7–9] As a result, the diffusion rate is fixed for a given fabrication process. The pore diameter of the nanoporous membranes typically lies within the range of 1–100 nm. As the pore size approaches the electrical double layer (EDL) thickness, also known as the Debye length, a phenomenon called the exclusion-enrichment effect impacts the flow characteristics.[10] The EDL is a specific charge distribution at the liquid–solid interface and is caused by the fixed surface charges on the membrane and compensating counterions in the solution. When the diameter of the nanochannel is comparable to the EDL thickness, the formation of an overlapped EDL prevents coions with the same polarity as the surface charge from entering the channel. The exclusion of coions and the enrichment of counterions are the mechanisms of the exclusion-enrichment effect that allows the control of diffusion through the membrane. It has been shown that the channel electrical potential can modify the diffusion rate of charged molecules.[11] This technique is applicable to drug delivery because many molecules of interest, including drug molecules and proteins, are charged and thus lend themselves to control through electrical gating.[5,6] Therefore, if the electrical potential in the nanochannel can be externally manipulated, the transport of drug molecules through the membrane can be actively controlled.[12]

Karnik et al.[13] developed a nanofluidic transistor structure that uses a gate electrode on nanofluidic channels to control the surface charge and therefore modulate the flow using field-effect control. However, these devices were fabricated as planar structures, and the fabrication process is not suitable for membrane structures. Nuclear track-etched membranes have attracted attention to understand the electrokinetic flow in nanoporous channels.[14,15] Martin et al.[14] demonstrated ion transport selectivity using polycarbonate filtration membranes (nuclear track-etched membranes) as templates to form metal nanotubules via electroless plating. However, the nuclear track-etched membranes have low pore density and high aspect ratios, which can limit the flow rate.

Conductive nanoporous membranes could potentially allow for external electrical control of the transport behavior through the nanochannels via the exclusion-enrichment effect when the diameter of the channel is comparable to the EDL thickness.[10,12,16] For the fabrication of such membranes, e-beam lithography and focused ion beam–based processes can allow a great deal of control when it comes to the pore sizes, shapes, and position of the pores; however, they cannot be used for large areas due to the high cost and long fabrication time.[17] Anodic aluminum oxide (AAO) membranes have been widely used in the fabrication of nanoporous membranes. These membranes have many desirable properties, some of which are biocompatibility, small pore size, and uniformity.[7,18] The pore diameter of the AAO membranes can vary from 2 to 300 nm, and the pore density is in the range of 10^9–10^{11} pores/cm^2. The pore size can be controlled by varying the applied voltage, reaction time, etc.[19,20] However, the anodization process converts the aluminum to alumina, which is electrically insulating. Wu et al.[21] transferred the pattern of nanoporous AAO membrane into a tungsten thin film and then oxidized the tungsten surface to form a dielectric layer. However, the pattern transfer of the AAO membrane into a tungsten film significantly reduced the pore density. Pardon et al.[22] used atomic layer deposition techniques to deposit platinum and aluminum oxide dual layer on top of AAO membranes to create a metal-insulator stack inside the nanopore. However, to electrically gate with minimal gate leakage current, a pinhole-free oxide layer is necessary, which is difficult to achieve unless the insulating layer is sufficiently thick. The deposition of the insulating layer is further complicated by the nanoporous structure of these membranes.

Our fabrication process is fundamentally different; the membrane is intrinsically conductive. Block copolymers (BCPs) are used to fabricate the initial pattern of our membrane structure. BCPs can be used to fabricate patterns with higher pore densities ($>10^{11}$ per cm^2) with uniform, narrow pore sizes, pore shapes, and size tunability by changing the molecular weights and composition, making them a very desirable material to use.[2,23] The nanochannels with 25 nm pore diameters were fabricated into an amorphous silicon thin film sheathed by a native oxide in order to enable gating. The pore size (~25 nm) and pore density ($\sim 10^{11}$pores cm^{-2}) of the membrane are inherent and do not change throughout the process. Therefore, the pore density of the final membrane is still as high as that of BCP patterns. Further, the optimal channel size required to provide a high diffusion rate and low-voltage electrical control, as indicated in simulations, can be achieved using BCPs, while still maintaining a high pore density. Amorphous silicon was chosen as a membrane material due to its properties such as biocompatibilty, conductivity, amenability to thin film deposition, well-understood etch conditions, and surface chemistry.[13] For the electrical gating of the membrane, a conformal insulating layer is necessary. The insulating layer in this case does not

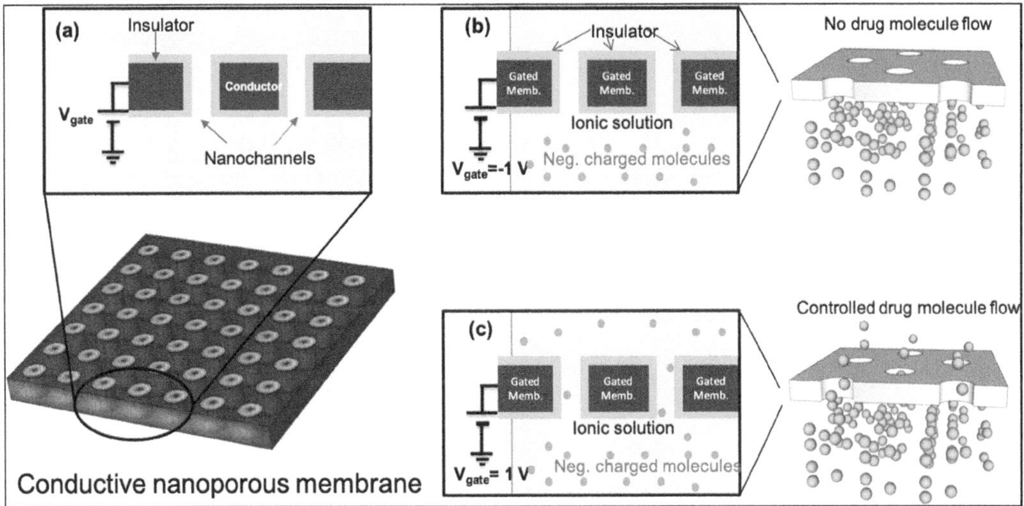

FIGURE 8.1 Illustration of the proposed electrically gated nanoporous membrane. (a) Schematic of the gated conductive nanoporous membrane structure with a thin insulating layer that prevents current leakage. (b) When a negative voltage is applied to the gate, negatively charged molecules are excluded from the channel due to the electrostatic repulsion. (c) When a positive voltage is applied, negatively charged molecules are enriched inside the channel due to the electrostatic attraction.

require an additional deposition, because amorphous silicon can be oxidized to form a pinhole-free insulating layer.[24] Furthermore, any damage to the oxide layer will self-heal due to the oxidation of the underlying amorphous silicon structure. Figure 8.1(a) shows a conceptual diagram of the conductive nanoporous membrane where each pore is effectively gated by an applied gate voltage and a thin oxide layer prevents current leakage. Figure 8.1(b and c) shows the fundamental physical mechanism. An applied gate voltage controls the potential in the nanochannel, resulting in either enrichment or exclusion of charged molecules. When the target molecule is negatively charged and a negative gate voltage is applied, the diffusive transport is blocked due to the electrostatic repulsion in the nanochannel as shown in Figure 8.1(b). When the target molecule is negatively charged and a positive gate voltage is applied, the diffusive transport is enhanced due to the electrostatic attraction in the nanochannel as shown in Figure 8.1(c).[12] Therefore, the diffusive flow of the charged molecules can be controlled by a single-gate voltage without the need for high driving voltage.

8.2 SIMULATIONS

We modeled the effect of the membrane potential on the diffusion of charged molecules through silicon/silica nanochannels in our previous work.[12] We demonstrated through simulations that cylindrical nanochannel structures with chemically treated surfaces have the ability to control the diffusive transport with a small applied DC gate voltage.[12] The gated nanochannel structures were modeled in the finite element software COMSOL Multiphysics. If the diameter of the nanochannel is too small and comparable to the drug molecule, single file diffusion occurs, which limits the overall diffusion rate. Conversely, the channel diameters larger than 100 nm require an unrealistically high control voltage. From our previous study, channels with diameters of about 20 nm balance the trade-off between controllability and diffusion rate.

To predict the controllability of the fabricated nanoporous membrane structure, we analyzed the gated molecular diffusion via simulations in nanochannels with 25 nm in diameter and 50 nm in length between two chambers with asymmetric concentrations of charged drug molecules. The

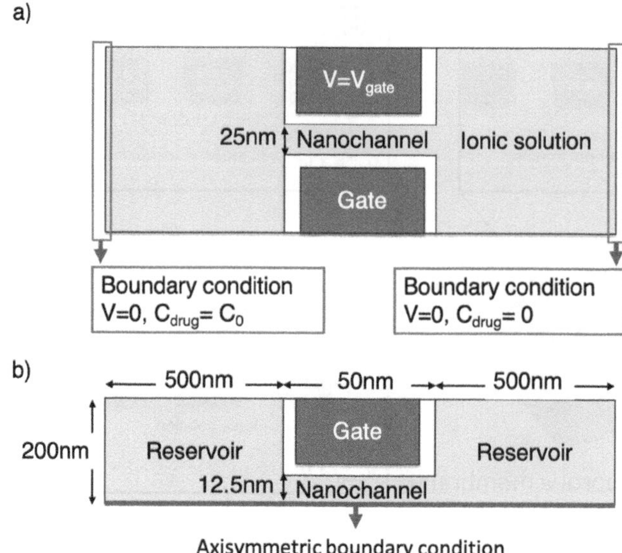

FIGURE 8.2 (a) Schematic of the nanochannel and simulation conditions. (b) Geometry of the simulation structure. Axisymmetric boundary condition was applied to simulate the cylindrical channel structure.

geometry of the modeled nanochannel is shown in Figure 8.2. Two reservoirs connect the conductive nanochannel with gate-insulator structure. Both reservoirs were filled with the electrolyte solution. The background electrolyte was modeled as sodium chloride (NaCl) aqueous solution, and the ionic concentration was varied from 10^{-4} M to 10^{-2} M. The modeled insulator was a 5-nm-thick silicon dioxide (SiO_2) layer, and the gate voltage was applied to the gate electrode that surrounds the nanochannel. The reservoir wall is grounded as shown in Figure 8.2(a). In order to maximize the controllability, the surface charge needs to be removed. Karnik et al.[13] determined that the surface charge of SiO_2 can be removed by chemically treating the surface with 3-glycidoxypropyl trimethoxysilane followed by ethanolamine. Therefore, the surface charge of the SiO_2 nanochannel wall was neglected and set to 0 in the simulations. The boundary conditions of the simulations are shown in Figure 8.2(a) to test the gating ability of the gated nanochannel. Two oppositely charged representative drugs were chosen for simulation purposes. Both interferon alpha-2b (INFα-2b) and leuprolide are used for the cancer treatment and have a net charge of −2q (-3.204×10^{-19} C) and +1q ($+1.602 \times 10^{-19}$ C), respectively.[25] Here, we assumed that the sizes of the drug molecules are considerably smaller than the nanochannel size so we ignored single-file diffusion behavior and modeled the drug molecules as point charges.

The simulation results are shown in Figure 8.3. The simulations were done with various background ionic strengths. The EDL thickness is inversely proportional to the background ionic concentration. The results showed that because the EDL thickness was at its thickest, the gating was maximized when the background ionic concentration was at the lowest value. For the positively charged drug molecules, the flux was increased with the negative gate voltage and decreased by the positive gate voltage [Figure 8.3(a)] as expected. For the negatively charged molecules, the opposite behavior was observed [Figure 8.3(b)]. The results shown in Figure 8.3 demonstrate that with 10^{-4} M background ionic concentration, the flux was altered more than 10-fold compared to the simulated flux without applied gate voltage. Even at a higher background ionic concentration (10^{-2} M), the electrical gating of the nanochannel was achieved. The flux at +1 V and −1 V showed 2.3-fold of difference for leuprolide and fivefold of difference for INFα-2b. This result indicates that a nanochannel with 25 nm diameter can alter the diffusion rate by many orders of magnitude under certain conditions with a ±1 V gate voltage. Hence, from the simulation, we can conclude that the

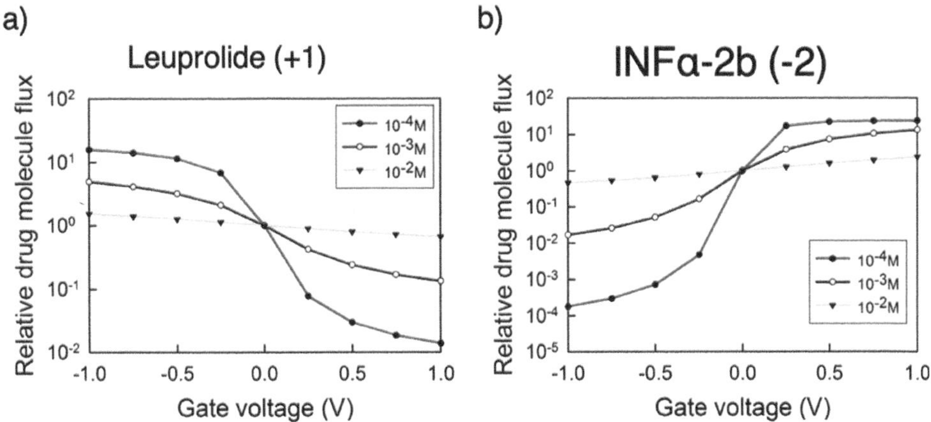

FIGURE 8.3 Simulated diffusive transport flux of (a) leuprolide (formal charge: +1q) and (b) INFα-2b (formal charge: −2q) measured at the end of the nanochannel. Note that the y-axis is log scale.

fabricated conductive nanoporous membrane has the ability to control the drug molecule flux with applied gate voltage.

8.3 FABRICATION METHODS

A 50-nm-thin film of amorphous silicon (am-Si) was sputter deposited with an argon pressure of 5 mT at the rate of 0.1 nm/s. The resistivity of the resulting thin film was measured using the four point probe method to be 40 Ωm. A thin layer of native SiO_2 was used to enable the gating of the device. While the conductivity of amorphous silicon is lower than metals, it can be patterned using fluoride-based plasmas in a reactive ion etching (RIE) system. Using a bias voltage, anisotropic etch profiles can be achieved, enabling high-aspect ratio patterning into the membrane layer.

BCPs, which are comprised of two blocks of different polymers connected by a covalent bond, have attracted extensive research over the last few years due to their self-assembly characteristics. These have been used to form both periodic and aperiodic nanostructures for semiconductor and data storage applications.[26] Diblock copolymer poly styrene-b-dimethyl siloxane (PS-b-PDMS) was used in this work. The two blocks in this polymer have a high enthalpy of mixing, as indicated by its relatively high Flory–Huggins interaction parameter (χ) of 0.26.[27] Based on the high Flory–Huggins interaction parameter, PS-b-PDMS is more promising for small-scale fabrication compared to other polymers such as polystyrene poly(methyl methacrylate) (χ ~0.04–0.06).[27] Using PS-b-PDMS with a molecular weight of 51 kg/mol and 17 vol.% PDMS, a regular array of PDMS spheres in a polystyrene matrix can be obtained, with a center-to-center distance of 35 nm between the PDMS spheres as shown in the plane-view scanning electron micrograph (SEM), Figure 8.4(b). Further, the polystyrene block (which consists of a carbon-like backbone) etches much faster in an oxygen plasma in a RIE process compared to PDMS (which consists of a silicon-like backbone), allowing pattern transfer into the underlying mask structure.[27]

Figure 8.4 illustrates the process flow for the fabrication of amorphous silicon membranes using BCP nanostructures as the initial pattern. An array of PDMS dots (from the PS-b-PDMS BCP thin film) was used as an etch mask to pattern an array of pillars into a suitable mask structure. The pattern was then reversed using a spin-on-glass and subsequently transferred into the am-Si layer. The initial film stack [Figure 8.4(a)] consists of the mask structure (for carbon pillar fabrication and pattern reversal), the am-Si membrane layer, and an etch stop layer of thermal SiO_2 grown onto a single crystal silicon substrate.

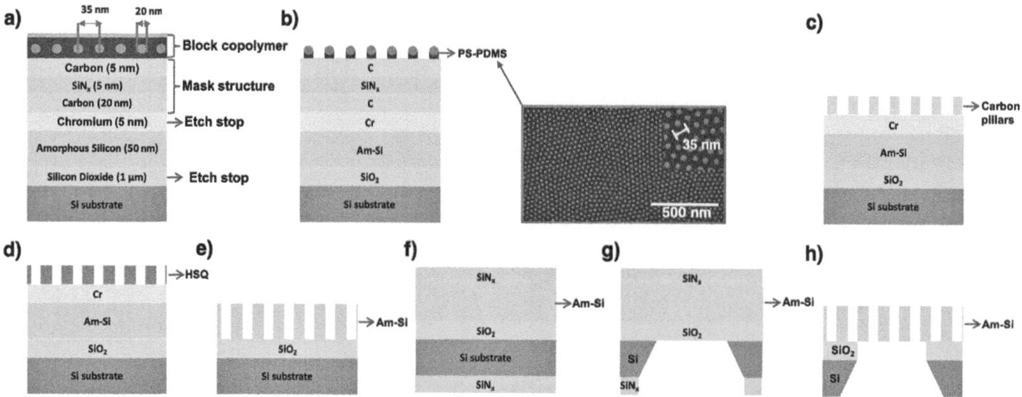

FIGURE 8.4 Schematic process flow diagram of the fabrication of the conductive nanoporous membrane at the wafer scale. (a) Initial film stack. (b) BCP etch and SEM of the BCP pattern. (c) Amorphous carbon pillar fabrication. (d) Reverse patterning using a spin-on-glass (HSQ). (e) Pattern transfer into the 50-nm-thick amorphous silicon layer. (f) Low-pressure chemical vapor deposition of low-stress silicon nitride for through-wafer etch in KOH. (g) Etching of the silicon nitride and KOH through-wafer etch after patterning the backside of the chip. (h) The membrane is deprotected by silicon nitride removal using a phosphoric acid etch and is released by using a buffered hydrofluoric acid etch to remove the silicon dioxide layer.

PS-b-PDMS, as purchased from Polymer Source Inc., was diluted (0.08 gm in 10 ml) in a toluene solution. A thin film was deposited onto the film stack using a Solitec spin-coating system using the dynamic dispense technique (spread parameters: 600 rpm, 6 s; spin parameters: 4000 rpm, 60 s). The concentration and spin conditions were optimized so that one monolayer of PDMS spheres in a PS matrix was obtained. Thermal annealing for 24 h at 175°C was performed to facilitate the micro-phase separation process [Figure 8.4(a)]. A carbon tetrafluoride (CF_4) plasma was first utilized to etch the PDMS film that segregated to the polymer–air interface in a PlasmaTherm 790 RIE system.[28] A low-bias oxygen (O_2) plasma was then used to selectively etch the PS matrix exposing an array of PDMS spheres [Figure 8.4(b)]. This pattern can then be transferred using a second oxygen plasma into the underlying mask structure.

The mask structure is composed of layers of an amorphous carbon (am-C) (5 nm), silicon nitride (5 nm), and a thicker amorphous carbon layer (20 nm). Alternating thin layers of am-C and silicon nitride are necessary to obtain a pattern with desired height and vertical side-walls due to their contrasting etch chemistries. Alternating O_2 and CF_4 plasmas were utilized to transfer the pattern from the PDMS dots to the final thick am-C layer [Figure 8.4(c)]. Without these intermediate layers, the pattern transfer into the 20-nm am-C layer is incomplete. This is because the physical erosion component during the RIE process affects the PDMS spheres more than it affects the silicon nitride.

In order to reverse the pattern, hydrogen silsesquioxane (HSQ), which is a commonly used negative e-beam resist and a spin-on-glass, was utilized. The HSQ was dispensed onto the sample by spin coating (spread parameters: 600 rpm, 4 s; spin parameters: 4000 rpm, 60 s) then baked at 180°C for 1 min. Upon baking, the HSQ molecules cross link, forming a network structure similar to SiO_2.[29] To reverse the pattern, the thickness of the HSQ layer was needed to be much higher than the height of the fabricated pillars, resulting in a nearly planar surface, as shown by the SEM image in Figure 8.5(b). HSQ has been used for planarization in the semiconductor industry,[29] and since it is spin coated from a solution, it fills the gaps between the pillars as well. Since the properties of the baked HSQ are similar to SiO_2, it can be etched using a CF_4 plasma in RIE. This was done until the tops of the am-C pillars were exposed, as shown in Figure 8.5(c). Then, using an O_2 plasma (which does not etch the HSQ), the am-C pillars were etched away, and thus a negative of the previous pattern was created, as seen from the plane-view SEM in Figure 8.5(d). The HSQ alone is insufficient for the

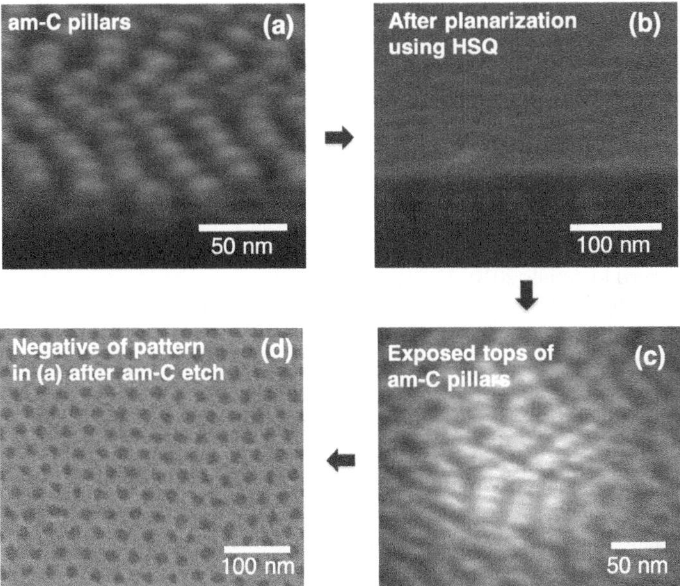

FIGURE 8.5 SEM images. (a) Tilt-view image of amorphous carbon pillars after pattern transfer. (b) Tilt-view image after spin-coating HSQ. (c) Tilt-view image of etch back to expose the tops of amorphous carbon pillars. (d) After the removal of amorphous carbon pillars.

final pattern transfer step into the am-Si membrane layer since their etch rates in a fluoride plasma are similar. Thus, the HSQ pattern was first transferred to a thin 5-nm chromium layer using a chlorine plasma in a Plasma Therm Versaline ICP RIE system, and then finally to the 50-nm-thick am-Si membrane layer using a CF_4 plasma [Figure 8.4(e)].

To release the membrane, the porous amorphous silicon layer was protected by deposition of a 100-nm-thick low-stress stoichiometric silicon nitride layer on both sides [Figure 8.4(f)]. The backside of the substrate was patterned using photolithography and plasma etching of silicon nitride using CF_4 and O_2 plasmas. The silicon substrate was etched through with 40% potassium hydroxide (KOH) at 80°C. Once the etch was complete, silicon nitride was removed using a phosphoric acid, and then a buffered hydrofluric acid etch was used to strip the oxide layer to release the porous membrane, as shown in Figure 8.4(h).

8.4 CONCLUSION

We have fabricated a gateable, highly porous (10^{11} pores cm^{-2}), intrinsically conductive nanoporous membrane structures with ~25-nm pore diameters. The channel dimension was determined by modeling the gated diffusion through the nanochannel. The simulation results showed that the nanochannel with 25-nm diameter can significantly alter the molecular diffusion rate of drug molecules with a ±1 V gate voltage. To fabricate our intrinsically conductive highly nanoporous membrane structure, PS-b-PDMS was used to pattern the nanochannels. The pattern of BCP then was transferred to a mask structure. Once the pillar structure is achieved, the sample is covered with spin-on-glass for reverse patterning. Amorphous silicon is etched to get the porous structure, and then the membrane is released using KOH. Overall, this novel structure allows for a self-healing, gateable, high pore density, intrinsically conductive nanoporous membrane that enables the controlled diffusion of the charged molecules. This membrane could be used for drug delivery systems to reduce side effects and maximize efficacy of the drug by controlling the drug concentration and keeping it relatively constant at a targeted location.

8.5 ACKNOWLEDGMENTS

This project was funded by the CMU-SYSU Collaborative Innovation Research Center (CIRC) of Carnegie Mellon University. The authors would like to thank Dr. Matthew Moneck, executive manager of the Carnegie Mellon University Nanofabrication Facility.

REFERENCES

1. Bernards, D.A.; Desai, T.A. Nanotemplating of biodegradable polymer membranes for constant-rate drug delivery. Advanced Materials 2010, 22(21), 2358–2362.
2. Jackson, E.A.; Hillmyer, M.A. Nanoporous membranes derived from block copolymers: From drug delivery to water filtration. ACS Nano 2010, 4(7), 3548–3553.
3. Desai, T.A.; Hansford, D.J.; Leoni, L.; Essenpreis, M.; Ferrari, M. Nanoporous anti-fouling silicon membranes for biosensor applications. Biosensors and Bioelectronics 2000, 15(9–10), 453–462.
4. Schoch, R.B.; Han, J.; Renaud, P. Transport phenomena in nanofluidics. Reviews of Modern Physics 2008, 80(3), 839–883.
5. Ahuja, S. Overview of capillary electrophoresis in pharmaceutical analysis. Separation Science and Technology. Separation Science and Technology 2008, 9, 1–8.
6. Kim, K.; Lee, J.W.; Shin, K.S. Polyethylenimine-capped Ag nanoparticle film as a platform for detecting charged dye molecules by surface-enhanced Raman scattering and metal-enhanced fluorescence. ACS Applied Materials and Interfaces 2012, 4(10), 5498–5504.
7. Boss, C.; Meurville, E.; Sallese, J.M.; Ryser, P. Size-selective diffusion in nanoporous alumina membranes for a glucose affinity sensor. Journal of Membrane Science 2012, 401–402, 217–221.
8. Leoni, L.; Boiarski, A.; Desai, T.A. Characterization of nanoporous membranes for immunoisolation: Diffusion properties and tissue effects. Biomedical Microdevices 2002, 4(2), 131–139.
9. Montagne, F.; Blondiaux, N.; Bojko, A.; Pugin, R. Molecular transport through nanoporous silicon nitride membranes produced from self-assembling block copolymers. Nanoscale 2012, 4(19), 5880.
10. Plecis, A.; Schoch, R.B.; Renaud, P. Ionic transport phenomena in nanofluidics: Experimental and theoretical study of the exclusion-enrichment effect on a chip. Nano Letters 2005, 5(6), 1147–1155.
11. Daiguji, H.; Yang, P.; Majumdar, A. Ion transport in nanofluidic circuits. Nano Letters 2004, 4(1), 137–142.
12. Kim, S.; Ozalp, E.I.; Sundar, V.; Zhu, J.-G.; Weldon, J.A. Modeling of electrically controlled molecular diffusion in a nanofluidic channel. Journal of Applied Physics 2015, 118(7), 74301.
13. Karnik, R.; Castelino, K.; Majumdar, A. Field-effect control of protein transport in a nanofluidic transistor circuit. Applied Physics Letters 2006, 88(12), 1–4.
14. Nishizawa, M.; Menon, V.P.; Martin, C.R. Metal nanotubule membranes with electrochemically switchable ion-transport selectivity. Science 1995, 268(5211), 700–702.
15. Kemery, P.J.; Steehler, J.K.; Bohn, P.W. Electric field mediated transport in nanometer diameter channels. Langmuir 1998, 14(10), 2884–2889.
16. Stein, D.; Kruithof, M.; Dekker, C. Surface-charge-governed ion transport in nanofluidic channels. Physical Review Letters 2004, 93(3), 35901.
17. Nabar, B.P.; Çelik-Butler, Z.; Dennis, B.H.; Billo, R.E. A nanoporous silicon nitride membrane using a two-step lift-off pattern transfer with thermal nanoimprint lithography. Journal of Micromechanics and Microengineering 2012, 22(4), 45012.
18. Gultepe, E.; Nagesha, D.; Sridhar, S.; Amiji, M. Nanoporous inorganic membranes or coatings for sustained drug delivery in implantable devices. Advanced Drug Delivery Reviews 2010, 62(3), 305–315.
19. Kim, K.; Kim, M.; Cho, S.M. Side wall anodization of aluminum thin film on silicon substrate. Korean Journal of Chemical Engineering 2005, 22(5), 789–792.
20. Stroeve, P.; Ileri, N. Biotechnical and other applications of nanoporous membranes. Trends in Biotechnology 2011, 29(6), 259–266.
21. Wu, S.; Wildhaber, F.; Bertsch, A.; Brugger, J.; Renaud, P. Field effect modulated nanofluidic diode membrane based on Al 2O3/W heterogeneous nanopore arrays. Applied Physics Letters 2013, 102(21), 1–5.
22. Pardon, G.; Gatty, H.K.; Stemme, G.; van der Wijngaart, W.; Roxhed, N. Pt-Al2O3 dual layer atomic layer deposition coating in high aspect ratio nanopores. Nanotechnology 2013, 24(1), 15602.

23. Yang, S.Y.; Yang, J.A.; Kim, E.S.; Jeon, G.; Oh, E.J.; Choi, K.Y.; Hahn, S.K.; Kim, J.K. Single-file diffusion of protein drugs through cylindrical nanochannels. ACS Nano 2010, 4(7), 3817–3822.
24. Gaspard, F.; Halimaoui, A.; Sarrabayrouse, G. Electrical properties of thin anodic silicon dioxide layers grown in pure water. Revue de Physique Appliquée 1987, 22(1), 65–69.
25. Grattoni, A.; Shen, H.; Fine, D.; Ziemys, A.; Gill, J.S.; Hudson, L.; Hosali, S.; Goodall, R.; Liu, X.; Ferrari, M. Nanochannel technology for constant delivery of chemotherapeutics: Beyond metronomic administration. Pharmaceutical Research 2011, 28(2), 292–300.
26. Stoykovich, M.P.; Nealey, P.F. Block copolymers and conventional lithography. Materials Today 2006, 9(9), 20–29.
27. Cummins, C.; Gangnaik, A.; Kelly, R.A.; Borah, D.; Connell, J.O.; Petkov, N.; Georgiev, Y.M.; Holmes, J.D.; Morris, M.A. Aligned silicon nanofins via the directed self-assembly of PS-b-P4VP block copolymer and metal oxide enhanced pattern transfer. Royal Society of Chemistry 2015, 6712–6721.
28. Jung, Y.S.; Ross, C.A. Orientation-controlled self-assembled nanolithography using a polystyrene – polydimethylsiloxane block copolymer. Nano Letters 2007, 7(7), 2046–2050.
29. Choi, S.; Word, M.J.; Kumar, V.; Adesida, I. Comparative study of thermally cured and electron-beam-exposed hydrogen silsesquioxane resists. Journal of Vacuum Science & Technology B: Microelectronics and Nanometer Structures 2008, 26(5), 1654.

9 Theranostics with Near-Infrared Sensitive Rare-Earth Doped Lumino-Magnetic Nanoplatforms

G. A. Kumar[a], *G. C. Dannangoda*[b], *and K. S. Martirosyan*[b]

[a]Department of Physics and Astronomy, University of Texas at San Antonio, San Antonio, Texas, USA

[b]Department of Physics and Astronomy, University of Texas at Rio Grande Valley, Brownsville, Texas, USA

CONTENTS

9.1 INTRODUCTION

Theranostics is a novel and innovative field of nanomedicine that combines both targeted specific therapy and imaging [1, 2]. The theranostics approach uses multifunctional nanostructured systems specifically designed on a single delivery platform enabled by nanotechnology [3]. These nanoplatforms can perform multiple functions, such as tracking and imaging the infected cells and delivering drug very precisely at the cellular level, and can perform various other therapeutic procedures such as photodynamic therapy (PDT) [4], photothermal therapy (PTT) [5], and magnetic field–based hyperthermia [6].

Several types of nanoparticles (NPs) have been approved for cancer treatment, and many other nanotechnology-enabled therapeutic modalities are under clinical investigation, including chemotherapy, hyperthermia, radiation therapy, gene or RNA interference therapy, and immunotherapy [7, 8]. Due to the altered anatomy of tumor vessels, NPs can be easily escaped from the blood into tumor tissues and be retained through the process of enhanced permeability and retention [9]. Additionally, due to the high surface area-to-volume ratios, NPs can be loaded with a high dose of drugs or other carriers for targeted delivery. A single NP can be properly designed with very

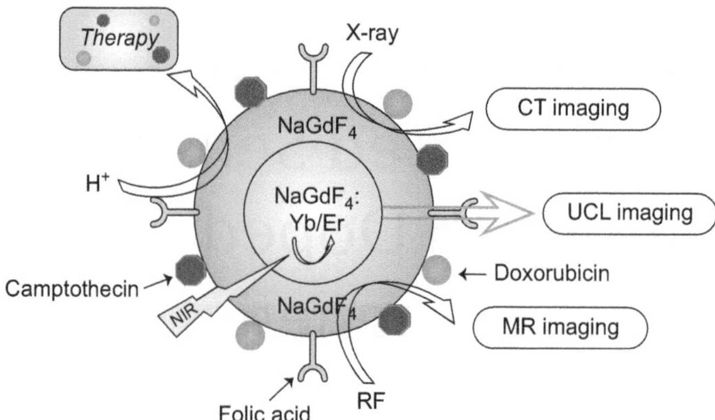

FIGURE 9.1 Design concepts of an Ln-doped multifunctional theranostic NP. The core of the NPs contains the Yb:Er pair, which gives strong green and red fluorescence by UC luminescence (UCL) under near-infrared (NIR) excitation, and the emission can be used in NIR imaging. The Gd atom present in the shell provides X-ray computed tomography (CT) and magnetic resonance (MR) imaging features. The surface of the NPs is covered with the anticancer drugs doxorubicin (DOX) and camptothecin [10].

complex architecture to be used as an imaging or therapeutic agent. For example, in optical imaging modality, the core of the NPs is designed as a fluorescent light emitter under proper excitation wavelength, which enables the optical imaging feature of the NPs. The core material can also be chemically modified with a magnetic species to enable the magnetic imaging features, and this can be done either in the core of NPs or as core-shell structure. This complex structure can be further coated with therapeutic drugs such as chemotherapy and PDT agents, and finally, multifunctional NPs are surface-functionalized with targeting ligands and cloaking agents such as poly(ethylene glycol). Introduction of targeting ligands may help to increase the target-to-background contrast in imaging and improve the selective delivery at the target of interest. The schematic architecture of such a theranostic NPs is shown in Figure 9.1.

9.2 INTRACELLULAR IMAGING

Intracellular imaging allows the visualization, tracking, and quantification of molecular interactions in living cells as well as in entire organisms. Understanding the intracellular molecular mechanisms and pathways is the only way to learn the normal cell physiology and the origin of many diseases. To reveal the cellular level origin of a disease, comprehensive, microscopic molecular mapping of individual cells is required, which is done through intracellular imaging. A number of advanced analytical techniques have been developed in the past for single-cell molecular imaging such as mass spectrometry [11–13], mass cytometry [14–16], surface-enhanced Raman spectroscopy (SERS) [17–22], and optical imaging using organic fluorophores [23–25], carbon nanotubes [26], and quantum dots (QDs) [27–31]. Compared to optical imaging, all other analytical methods have their limitations, such as high cost and low resolution. Optical imaging with fluorescent dyes and QD is so versatile in the biomedical industry because of its simplicity and low cost. However, there are significant limitations with these current imaging agents; these are as follows. (1) *Photobleaching and instability:* Organic dyes and fluorescent proteins have been used successfully for in vivo imaging but suffer from a high photobleaching rate when used in high-intensity cell imaging studies, thus making long-term experiments unfeasible. While single fluorophore detection has been demonstrated in proof-of-concept experiments, organic fluorescent dyes still do not have sufficient long-term brightness for many biomedical applications. Recently, organic dyes have been replaced by QD, but the luminescence efficiency of QD strongly depends on the environment and can create signal instability.

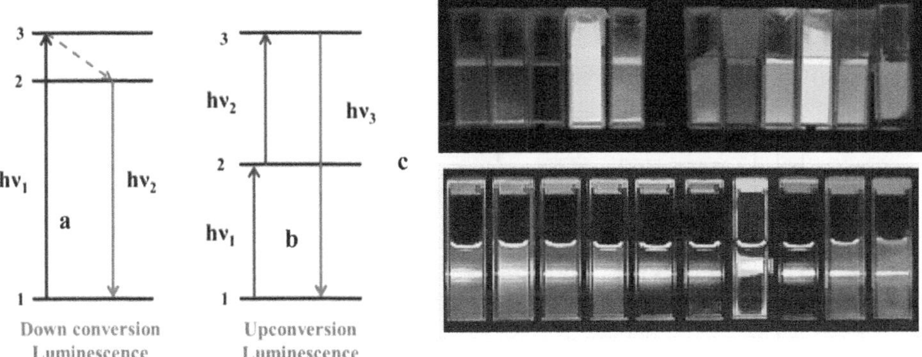

FIGURE 9.2 Scheme of excitation and emission in Ln-doped materials: (a) Downconversion process where a high-energy photon excites the upper energy level and emission photon appears in the low-energy longer wavelength; (b) UC process where a low-energy photon excites the lower energy state and, through multiphoton absorption, emission occurs in the high-energy lower wavelength; (c) photograph of the downconversion (top) and UC (bottom) emission in various Ln-doped phosphors.

In addition, QD shows fluorescence blinking, which is not desirable in imaging. (2) *Toxicity:* The toxicity inherent in QD and radioactive materials precludes use in human medical applications. (3) *Autofluorescence:* Short-wavelength excitation of organic fluorophores often produces background autofluorescence from tissue that limits image contrast. (4) *Information density:* The greatest limitation of organic fluorophores is the relatively broad emission spectra that limit the number of independent phenomena that can be analyzed in a single experiment, and this prevents high information density. (5) *Technology cost:* Both QD and fluorophore-based technology are expensive in several aspects. Fluorophore-based imaging technology requires expensive detection systems to eliminate the autofluorescence noise level. Also, QD-based technology requires size-tunable materials for adjusting the excitation/emission, which requires precise and expensive synthesis procedures.

In order to overcome the above limitations, lanthanide (Ln)-doped nanophosphors/NPs (LnNPs) are being proposed as the next-generation optical imaging agents. Due to the large number of NIR excitation channels, LnNPs can be utilized in two ways, as shown in Figure 9.2a and b: (1) upconversion (UC) where the low-energy NIR excitation (longer wavelength) yields high-energy (shorter wavelength) photons, and (2) downconversion where the shorter wavelength excitation gives longer wavelength emission. A typical photograph of the visible emitting colors under downconversion and UC is shown in Figure 9.2c. In the figure, for the UC emission, all materials are excited with the well-known 980-nm band from the Ln Yb^{3+}, and the multiple colors in UC and downconversion processes are obtained through different dopants from the Ln series as well by adjusting their concentrations and selecting various host materials.

For biological applications, since most of the tissues are transparent in the 700–1300 nm region (II biological window), it is considered as the optimum window for background-free imaging, and so far, Ln-doped phosphors are the only candidates that show strong up- and downconversion emission under NIR excitation window in the range of 800–980 nm. Ln-doped phosphors have been well known in the photonics area for a long time because of their several technological applications. In these phosphors, fractions of the host cations are replaced with either single or multiple Ln atoms in proper valence state. Normally, the trivalent oxidation states of Ln-s are stable and show a wide absorption and emission spectral range extending from UV–VIS-IR.

Figure 9.3 shows the energy level schemes of all trivalent Ln-s, which clearly indicate the wide range of excitations from UV to infrared range. Because of the multitude of closely spaced energy levels, many of them could be excited with several IR wavelengths and emit in the NIR–VIS–UV

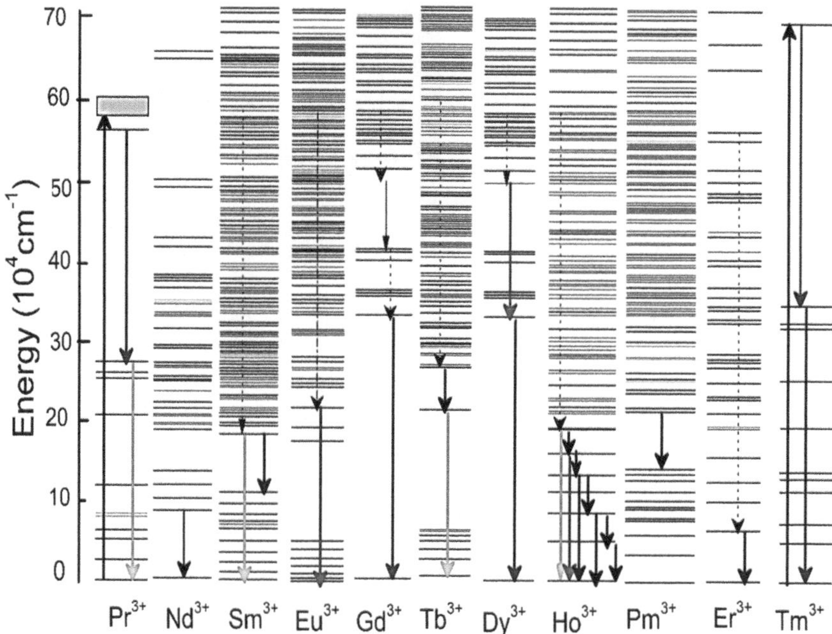

FIGURE 9.3 Theoretically computed energy levels of various trivalent Ln ions. The arrows show some of the excitation and emission channel.

region. As opposed to the organic nonlinear materials, where the multiphoton absorption occurs through the virtual levels, the UC mechanism in Ln happens through real intermediate levels, either sequential excitation of the same atom or excitation of two atoms and subsequent energy transfer [12]. Further, because of the shielding of the *f* orbital by the *d* orbital, the absorption and emission in Ln are sharp lines characteristic of atomic transitions in a well-ordered matrix. By using different rare-earth dopants, a large number of distinctive emission spectra can be obtained [32, 33] in the VIS-NIR region, as shown in Figures 9.4a and b.

Compared to organic fluorophores and QDs, LNPs have several advantages in medical imaging. Unlike QDs, they are not likely to be toxic. The LD50 for Ln oxides is on the order of 1000 mg/kg, while the LD50 value for many selenium oxide QDs is on the order of 1 mg/kg [34]. The other advantage is the extremely low-power excitation needed for the emission processes in LnNP. Normally, in organic fluorophores, nonlinear UC can be seen under intense pulsed (ns or fs) laser excitation. However, in Ln, because of the presence of the real intermediate excited states, comparatively less power is needed to initiate the UC process. For example, recently we invented a composition where the green UC could be seen by the naked eye even with 15 μW excitation [35]. In Ln-doped UC phosphors, the excitation intensities needed are 10^7 times less than the intensities needed for 2-photon excitation of typical organic dyes and are easily achievable using very low-cost IR continuous-wave diode lasers.

Another obvious advantage of LnNPs is their photostability under external conditions. Unlike organic fluorophores, LnNPs do not photobleach, and their emission intensity remains stable for years, making them favorable for long-term use. Other advantages of the technology are: (1) high signal-to-noise ratio (since the excitation is in the NIR biological transparent window, tissue does not show fluorescence, which allows UC technology to get background-free images with high signal-to-noise ratio), (2) high penetration depth (because of the higher penetration depth of NIR radiation to tissues, in vivo imaging is possible), (3) high resolution and information density (the sharp level continuously tunable emission spectral features enable high information density and resolution as

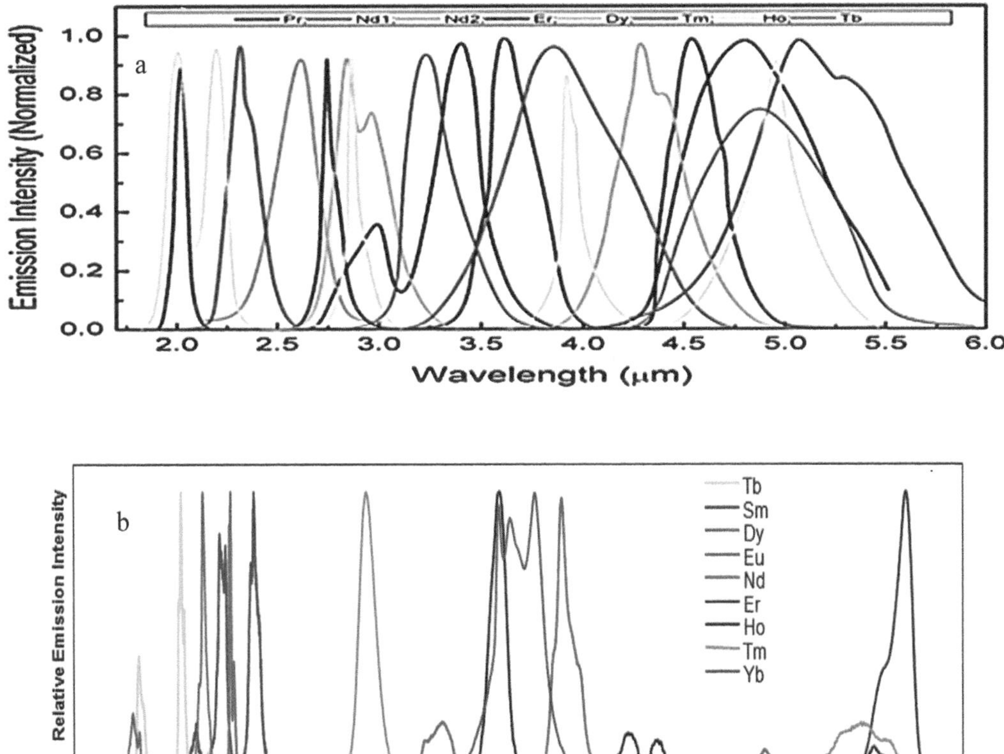

FIGURE 9.4 Emission spectra of various trivalent Ln-doped phosphors. Top: UV–VIS-NIR emission spectra of selected Ln-s; bottom: NIR-mid-infrared spectra of selected trivalent Ln-doped phosphors.

low as 10 nm), (4) multimodal imaging (multifunctional features can easily be incorporated in this material by properly adjusting the material structure; for example, by replacing some of the cations with metals such as Gd, Fe, Ni, and Pt, strong magnetic features can be added that result in a lumino-magnetic phosphor), and (5) low-cost technology (in LNP, since the color tunability is achieved by the dopant selection and composition, size tunability is not needed as in QD, which makes this technology cheaper). In addition, LnNPs can also be imaged in scanning electron microscopy (SEM) due to their cathodoluminescence, so that ultra-high spatial resolution multiphoton imaging is possible using SEM [36].

A web of science survey showed more than 136 articles published in 2018 on the application of various types of Ln-doped UC phosphors for biomedical imaging with multiple functionalities such as optical, magnetic, and X-ray. The importance of LnNP in the medical imaging area is clearly indicated by the exponential growth of the number of publications every year. However, the majority of these publications are concentrated on NIR excited UC imaging where the phosphors are excited under 800/980 nm radiation, and the emission was recorded in the visible region, mainly green and red [37–63]. Though in principle this type of imaging is feasible for in vitro applications, it is limited for in vivo imaging because of the excess tissue scattering in the UV–VIS region by the inhomogeneities in the tissue structure, as illustrated in Figure 9.5. To overcome the limitation in penetration depth, LNPs should work within a "biological window" of human tissues where tissues are partially transparent. The biological windows (BW) are classified as I, II, III, and IV according to the wavelength window of 700–950 nm (I window), 1000–1400 nm (II window), and 1500–1750 nm

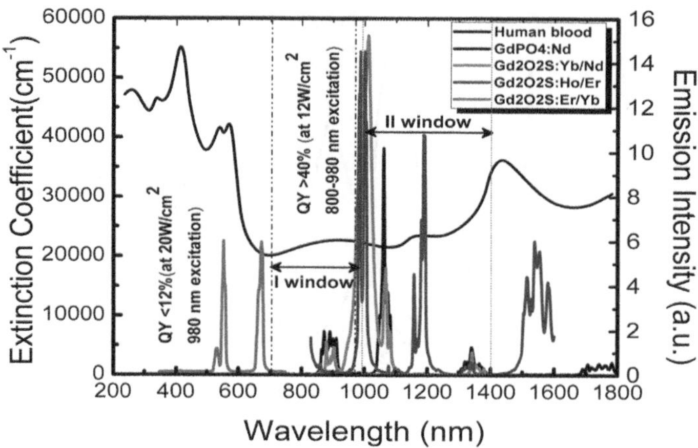

FIGURE 9.5 Tissue absorption and overlap of several NIR emissions from selected LNPs in I and II biological windows.

(III window). For the UC imaging, although the I-BW is good for good excitation penetration depth and low water absorption, the visible light is highly attenuated by the tissue scattering. Though there are few LnNPs that show downconversion in the I-BW, their efficiency is limited for an imaging application.

Compared to I-BW, in the II-BW, the scattering is further reduced due to the longer wavelength that would benefit for improvement in resolution of sub-tissue images. Though this spectral range is limited by two absorption bands of water allocated at about 980 and about 1500 nm, the reduction in the scattering cross-section could overcome the slight increment in the absorption coefficient due to water, thus leading to longer penetration depths compared to the I-BW. Because of these expected improvements in both image resolution and penetration depths, there is now great interest in the development of Ln downconversion phosphors working in the II-BW.

A closer observation of Figure 9.5 shows various emission bands from several Ln-s, such as Yb, Nd, and Ho, that overlap well with the II-BW. Among all trivalent Ln-s, Nd^{3+} is one of the most potential NIR emitters, with reported 100% quantum yield (QY) at 1064 nm in several low phonon energy hosts such as halides and chalcogenides. It has a strong excitation peak at 808 nm that is almost not affected by the water absorption. The higher QY and longer excitation and emission wavelengths make Nd^{3+}-doped low phonon hosts desirable candidates for sub-tissue imaging. Nd^{3+} can also be co-doped with Yb^{+} and Ho^{3+} as potential NIR emitters in the II-BW.

9.3 MULTIMODAL IMAGING WITH LnNPs

NPs can serve as a sensitive and adaptable mode of tracking cells in animal experiments. Particles that can be tracked by both optical and nonoptical imaging such as magnetic, X-ray, and other methods offer the advantages of multimodal imaging features. Optical imaging allows very sensitive localization of very small numbers of cells, even single cells, and allows clear visualization of the cells in relationship to their surrounding tissues. On the other hand, nonoptical imaging, such as MR imaging, can be used to a depth of tissue that is not currently usable with optical imaging. It too can be used down to the single-cell level. Similarly, X-ray imaging can also give information at a far deeper level that is not possible with optical imaging. Therefore, the ideal type of NPs is one that can be used for multiple forms of imaging in the same experiment.

Compared to other materials such as QD and metallic NPs, multifunctional features can be easily added to LnNP with using a very simple procedure, which makes LnNPs versatile multimodal imaging agents. Bimodal [49–53], trimodal [54–63], and hexa-modal imaging [64] have been recently reported using nLNP. In general, since the LNP can be properly engineered to have customized NIR excited emissions as well as strong magnetic properties, both of which are favorable for in-depth imaging, they are considered to be the ideal candidates for many kinds of medical imaging applications. The biomedical applications of Ln-doped NPs were first explored by the Prasad group [65]. They reported the use of Yb/Er/Tm-doped Y_2O_3 UC NPs of 25 nm size as a platform for cancer imaging, therapy, and drug delivery. Though several Ln oxide materials were tried for this purpose, their performance was not superior compared to the halide-based host $NaYF_4$, which was first reported by Yi et al. [66]. Today, Ln-doped $NaYF_4$ is one of the most explored UC phosphors for various biomedical applications such as imaging, therapy, and drug delivery.

Zeng et al. [67] proposed and demonstrated a strategy for simultaneous phase, size manipulation, color tuning, and remarkably enhanced red UCL in $NaLnF_4$:Yb/Er (Ln: Lu, Gd, and Yb), simply through transition metal Mn^{2+} doping. They demonstrated that 980 nm excited the UC bioimaging with high sensitivity, high contrast, and deeper tissue penetration with no autofluorescence. In addition to NIR imaging, they used X-ray radiation to image the blood vessels in the lung of a mouse. Ln-ion-doped $NaGdF_4$ NPs have been widely exploited as new-generation magnetic and optical probes. Therefore, the chlorotoxin-conjugated NaGdF4:Ho^{3+} NPs could potentially serve as an MRI and optical probe for the detection of tiny brain cancer [68]. Wang et al. [69] reported magnetic and fluorescent bifunctional $YbPO_4$:Er, Dy microspheres with UC and magnetic properties that can be used for UC optical and magnetic imaging. The microspheres successfully entered the human cancer cells with low toxicity. They also demonstrated infrared PDT (IRPDT) with the proposed phosphor by conjugating them with the PDT drug merocyanine 540 (MC540) with absorption maximum at 540 nm. Since the UC fluorescence emitting from the microspheres could be absorbed by MC540, $YbPO_4$:Er, Dy-MC540 could destroy the cancer cells via IRPDT mechanism effectively. A chemotherapy drug, DOX, was further loaded onto a nanosphere to achieve enhanced antitumor effect based on chemotherapy. NaYF4:Yb^{3+}, Tm^{3+}, and Co^{2+} nanorods are proposed as high-performance dual contrast agents for both T2-weighted MR and NIR to NIR UC bimodal imaging. Compared to the visible UC, the NIR UC from Tm^{3+} at 800 nm has better depth of imaging and better performance compared to green and red UC [70]. Kim et al. [71] reported the bimodal imaging and drug delivery applications of Eu^{3+}-doped $GdVO^4$ NPs. NPs exhibit strong red photoluminescence, and the Gd^{3+} in $GdVO_4$ can be used as a T-1 contrast agent for MRI. MR contrast enhancement, as well as intracellular drug delivery, can be achieved by the mesoporous silica layer coating onto the NP. $GdVO_4$:Eu^{3+}@$mSiO_2$ NPs, a new type of theragnostic agent, can provide new opportunities in cancer treatment. Polyethylenimine (PEI) coated, fully biocompatible $NaGdF_4$:Yb/Er-NPs are demonstrated as potential platform for both efficient gene therapy and MRI/CT/UCL trimodal imaging both in vitro and in vivo. In vitro studies revealed that the gene carriers are able to transfer the enhanced green fluorescence protein plasmid DNA into HeLa cells [72].

Radio isotopically labeled nanomaterials have recently attracted much attention as theranostic agents. Paik et al. [73] report size- and shape-controlled synthesis of Ln fluoride nanocrystals such as YF3 and GdF3 doped with the beta ray–emitting radioisotope yttrium-90 (Y-90) and demonstrated Cerenkov radioluminescence and MR imaging features, which offer unique opportunities as a promising platform for multimodal imaging and targeted therapy. Lei et al. [74] reported water-soluble Yb^{3+}/Er^{3+} co-doped Bi_2O_3 UC nanospheres with uniform morphology and the possibility of using these nanoprobes with red UC emission for optical imaging in vivo. Furthermore, Bi^{3+} and Yb^{3+} containing nanospheres also exhibited significant enhancement of contrast efficacy compared with iodine-based contrast agent via X-ray CT imaging.

Recently, several articles have been published on the application of using 800 nm excitation as the better wavelength for biological applications due to the decreased water heating effect. Among the trivalent Ln-s, both Nd and Tm show strong absorption around 800 nm, and by using these two dopants, several up- and downconversion phosphors can be synthesized for various biomedical imaging and therapeutic applications. Some of the recent work in this area includes the $NaYF_4$:Yb,

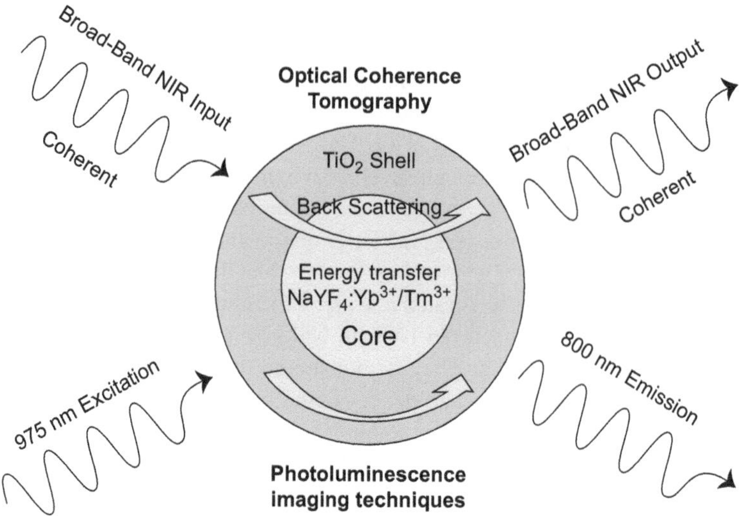

FIGURE 9.6 Schematic illustration of TiO_2-coated (NaYF4:Yb 60%, Tm 0.5%)/$NaYF_4$ core-shell NPs for multimodal optical imaging with OCT [77].

Nd@CaF_2 core/shell NPs that showed deep tissue penetration depth (similar to 10 mm) in in vitro imaging and high spatial resolution of blood vessels (similar to 0.19 mm) in in vivo imaging [75]. Feng et al. [76] used paramagnetic $NaHoF_4$ hosts doped with Nd^{3+} for NIR optical and MR imaging. The NP shows strong 1200 nm emission from Ho^{3+}, and the $r(2)$ value of $NaHoF_4$: Nd^{3+} NPs reaches 143.7 s^{-1} mM^{-1} at a high magnetic field of 11.7 T. Braz et al. [77] introduced a novel $NaYF_4$:Yb 60%, Tm (0.5%)/$NaYF_4$ core (51 nm), and TiO_2 shell (15 nm) nanostructure design that serves as a multimodal optical imaging contrast agent with optical coherence tomography (OCT) features for dental adhesion evaluation. The NPs architecture is shown in Figure 9.6.

The TiO_2 shell provides enhanced contrast for OCT, while the rare-earth-doped core upconverts excitation light from 975 nm to an emission peaked at 800 nm for photoluminescence imaging, and using this, the OCT and the photoluminescence wide-field images of the human tooth were collected. In addition, the core-shell NPs were dispersed in a commercially available dental bonding system, allowing clear identification of dental adhesive layers with OCT. They evaluated that the presence of the NP in the adhesive induced an enhancement of 67% scattering coefficient to significantly increase the OCT contrast. In addition to the OCT features, the UC photoluminescence in the near-infrared spectral region is suitable for imaging of deep dental tissue.

Optical imaging in the NIR II window presents an unprecedented improvement in imaging sensitivity and spatial resolution over the traditional visible and near-infrared light. Recently, there have been several reports on the development of several multifunctional contrast agents with optical imaging features in the NIR-II window. Li et al. [78] reported a new NIR II-emitting probe based on Nd-diethylenetriamine pentaacetate acid (DTPA) complex. The designed probe with bright narrow-band emission at 1330 nm was successfully used for highly sensitive in vivo bioimaging with high biocompatibility and optical-guided tiny tumor detection. In addition, the Nd-DPTA complex also holds great promise for CT imaging.

9.4 MULTIMODAL IMAGING WITH PDT

By properly engineering the composition and design, NPs can be used for various therapeutic applications along with their various imaging modalities. Light-activated imaging agents are more favorable for these applications due to the fact that various photochemical processes can happen inside the material under various excitation wavelengths. PDT is one such photochemical process wherein some organic molecules under UV or red wavelength excitation can generate highly active singlet oxygen, which can effectively kill the malignant cells. This process, called PDT or photochemotherapy, has been effectively used in the medical field for treating various cancers. Porfimer sodium (Photofrin) is the most widely used and studied photosensitizer, and in clinical practice, the procedure is done with a red laser. Conventional PDT done for cancer treatment has major drawback due to the low penetration depth of the red laser in the human body and the fact that the treatment cannot be done for deeply located tumors. To circumvent this issue, X-ray PDT and infrared light-activated PDT were recently proposed by several researchers and was successfully verified in various animal models.

The fundamental mechanism involved in X-ray PDT (XPDT) and IRPDT is purely energy transfer-induced spectroscopic process. In XPDT, the PDT agent can be conjugated with a high X-ray absorbing material such as another Ln phosphor that can convert the X-ray to UV and green fluorescence that can be absorbed by the photofrin. The absorption cross-section of photofrin in the UV and green region is stronger than that of red, which makes XPDT more efficient than normal PDT [79]. Li et al. [80] demonstrated XPDT in a scintillator based on $NaYF_4$:Gd nanorods doped with different concentrations of terbium (Tb) to be used as a multifunctional material for both imaging and PDT application. While XPDT is good for treating deeply located tumors, the ionizing nature of the X-ray radiation can also cause damage to healthy cells.

As an alternative method to solve this issue, IRPDT was proposed wherein the PDT was activated through infrared radiations, which have more penetration depth in tissues and cause no cellular damage. The schematic spectrophotochemical process that occurs in IRPDT is shown in Figure 9.7. The major component of this process is the infrared-sensitive phosphor that converts infrared light to red, blue, or green fluorescence that can be readily absorbed by the PDT agent. Among the various infrared-sensitive phosphors, Ln-doped UC phosphors are the best choice for these applications.

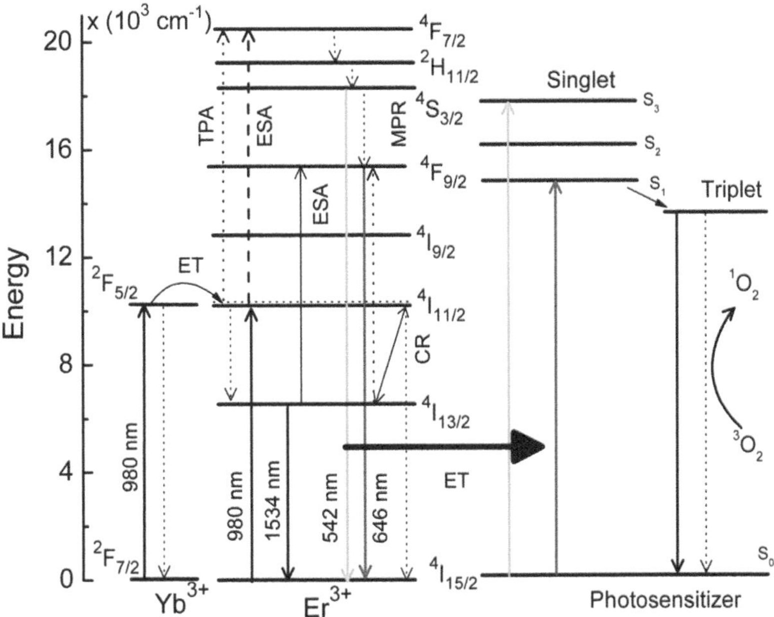

FIGURE 9.7 Mechanism of IRPDT with Ln-doped infrared-sensitive phosphors. The green and red fluorescence coming from the phosphor under two-photon excitations with 980 nm light is absorbed by the PDT agent and converts the singlet oxygen to highly reactive triplet oxygen.

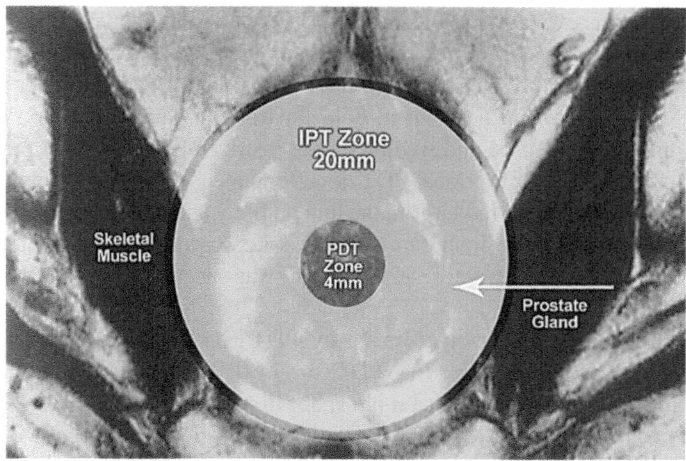

FIGURE 9.8 Comparison of the depth of penetration of normal PDT and IRPDT. Normal PDT zone is shown in brown color at the center, and the blue zone indicates the IRPDT zone.

Since various Ln-doped infrared UC nanophosphors have already been proved as efficient multifunctional contrast agents in biomedical imaging, the same phosphors can be suitably modified to enable the IRPDT activity. Thus, a single Ln-doped UC nanophosphor can be suitably designed to perform both imaging and therapy, which opens up a myriad of theranostic applications with Ln-doped NPs.

Wang et al. [81] demonstrated magnetic and fluorescent bifunctional $YbPO_4$:Er, Dy microspheres with UC and magnetic properties that can be used for UC and magnetic imaging with low toxicity. They also validated IRPDT with the proposed phosphor by conjugating them with the PDT drug merocyanine 540 (MC540), which has absorption at 540 nm. Since the UC fluorescence emitting from the microspheres could be absorbed by MC540, $YbPO_4$:Er,Dy-MC540 could kill the hepatocellular carcinoma cells via IRPDT mechanism effectively. The advantage of IRPDT compared to conventional PDT is that the IR light has more penetration depth (approximately five times) compared to the red light used in normal PDT, as shown in Figure 9.8.

A core-shell-shell nanoplatform of the form $NaYbF_4$:Nd@$NaGdF_4$:Yb/Er@$NaGdF_4$ loaded with Chlorin e6 and folic acid for simultaneous imaging and PDT was developed by Ai et al. [82]. They claim that the nanoplatform effectively destroys cancer cells in concentration-, time-, and receptor-dependent manners beneath 15-mm thickness of muscle tissue. Highly efficient in vitro and in vivo 915 nm light-triggered photodynamic therapies are realized for the first time by utilizing Na0.52$YbF_3$52:Er@SrF_2 NPs as multifunctional nanoplatforms. Unlike the 980 nm excitation, which induces undesirable heating effect due to water molecules, the 915 nm excitation offers dramatically diminished overheating yet similar therapeutic effects in comparison to those triggered by 980 nm light. Along with IRPDT activity, the material also offers good MR and X-ray CT imaging features [83]. Kalluru et al. [84] demonstrated an all-in-one Ln-doped mesoporous silica frameworks (EuGdOx@MSF) loaded with the anticancer drug DOX to facilitate simultaneous bimodal MR imaging with approximately twofold higher T1 contrast as compared to the commercial Gd(III)-DTPA complex and fluorescence imaging with excellent photostability. Upon a very low dose (130 mW cm^{-2}) of 980 nm irradiation, the NPs not only induce PDT by itself but also can phototrigger the release of the DOX effectively to impart combined chemo-photodynamic therapeutic effects and destroy solid tumors completely. They also found that the EuGdOx@MSF-mediated PDT effect can suppress the level of the key drug-resistant protein, i.e., p-glycoprotein (p-gp), and help alleviate the problem of drug resistance commonly associated with many cancers.

A new theranostic scintillator NP composite in a core-shell-shell arrangement, of the form $NaLuF_4$:Gd(35%), Eu(15%)@$NaLuF_4$:Gd(40%)@$NaLuF_4$:Gd(35%), Tb (15%), which is capable of being excited by a single X-ray radiation source to allow deep tissue PDT and optical imaging with low cytotoxicity and effective photocytotoxicity, was reported by Hsu et al. [85]. With the X-ray excitation, the nanocomposite can emit visible light at 543 nm from Tb^{3+} to stimulate the loaded RB photosensitizer and cause the death of cancer cells. The nanocomposite can also emit light at 614 and 695 nm from Eu^{3+} for luminescence imaging. The middle shell in the core-shell-shell is unique to separate the Eu^{3+} in the core and the Tb^{3+} in the outer shell to prevent resonance quenching between them and to result in good PDT efficiency. Luminescence/MR dual-modal imaging and IRPDT were also reported in Gd_2O_3:Yb^{3+}/Ln [86]. In addition, 800 nm-driven imaging and PDT were reported in core-shell $NaErF_4$@$NaLuF_4$ NPs by Li et al. [87]. Their results confirmed that the optimal 5 nm shell thickness can well balance the enhancement of UCL and the attenuation of energy transfer efficiency from Er^{3+} toward a photosensitizer, to achieve efficient production of singlet oxygen for PDT. In addition to strong UC, the obtained NPs showed X-ray CT and T-2 MRI features, thus making this nanomaterial an excellent theranostic agent for multimodal imaging and image-guided therapy. $NaYF_4$:Yb, Er@$NaGdF_4$:Nd@SiO_2-RB core-shell-shell structured NPs that integrate luminescence imaging in biological window II, MR imaging, and IRPDT were recently reported by Xu et al. [88].

9.5 NANOPLATFORMS FOR PTT

PTT is another cure technique for the treatment of various medical conditions, including cancer. Unlike in PDT where UV–VIS light radiations are used to generate reactive oxygen species through photosensitization with the supply of tissue oxygen content, in PTT, the treatment was done through infrared radiations and no oxygen was needed for the procedure. The PTT agent absorbs the IR radiation and generates a heating effect, which readily kills the malignant cells. Further, the infrared radiation is less energetic with high tissue penetration depth and is less harmful to other cells and tissues. A schematic of the PTT procedure is shown in Figure 9.9 [89]. The photothermal agent was initially injected into the affected cells followed by excitation with NIR light to create a heating effect. Several PTT agents are reported in the literature, among which the most common agents are metallic NPs, nanoshells, graphene, graphene oxide, and LnNPs [90].

Lv et al. [91] designed a nanophosphor of the structure TiO_2@$Y_2Ti_2O_7$@YOF:Yb, Tm, which could be used for optical imaging, CT imaging, and therapeutic effects from PDT and PTT. Under 800 nm laser irradiation, reactive oxygen species can be generated due to the energy transfer from YOF:Yb, Tm to the $Y_2Ti_2O_7$ photocatalyst, which is responsive to blue emission, and the thermal effect can be simultaneously produced due to the nonradiative transition and the recombination of

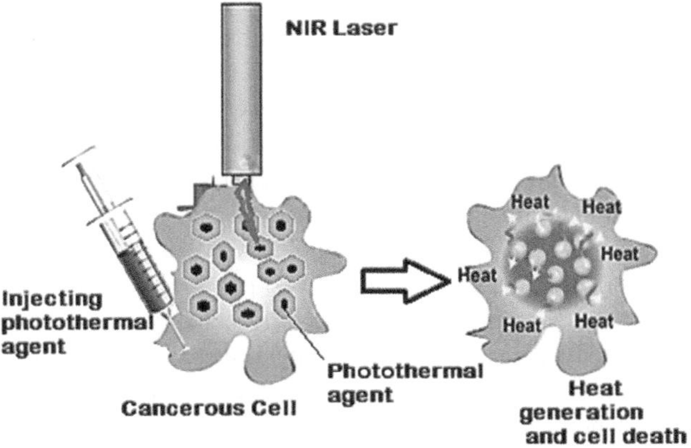

FIGURE 9.9 Mechanism of PTT. The photothermal agent was initially injected into the affected cells followed by excitation with NIR light to create a heating effect. The local heating from the photothermal agent destroys the affected cells [89].

electron-hole pairs. The IRPDT and PTT can efficiently suppress tumor growth, which was evidenced by both in vitro and in vivo results. PDT with PTT and chemotherapy for enhanced antitumor efficiency was reported by Lv et al. [92] in GdOF:Ln@SiO_2 (Ln = 10%Yb/1%Er/4%Mn) mesoporous capsules with a carbon dots (CDs) outside shell. This phosphor showed remarkably strong red emission under 980 nm excitation, which could trigger IRPDT activities in the surrounding photodynamic agents ZnPc, and at the same time, the incorporated CDs outside the shell can generate obvious thermal effect under 980 nm laser excitation. At the same time, the thermal effect can enhance the DOX release, which greatly improves the chemotherapy, resulting in a simultaneous therapeutic effect. The system shows superior therapeutic efficiency against tumor growth, as demonstrated both in vitro and in vivo. In addition to these therapeutic effects, the phosphor core particles show optical, magnetic, and CT imaging properties.

An UC imaging agent with CT imaging and PTT features was designed using Ln-doped oxide NPs by Yang et al. [93]. They prepared Y_2O_3:Yb, Er@mSiO_2-CuxS double-shelled hollow spheres where the inner core is YbEr-doped Y_2O_3, which is surrounded by mesoporous silica shell; 2.5 nm-sized CuxS NPs served as photothermal agents, and the chemotherapeutic agent DOX was then attached to the surface of mesoporous silica, forming a DOX-DSHS-CuxS composite. The composite exhibits high anticancer efficacy due to the PTT induced by the CuxS NPs and enhanced chemotherapy promoted by the heat from the CuxS-based PTT when irradiated by 980 nm light. Moreover, the composite shows excellent in vitro and in vivo X-ray CT and upconversion fluorescence imaging properties due to the doped Ln ions, thus making it possible to achieve imaging-guided synergistic therapy. Xu et al. [94] reported $NaGdF_4$:Yb,Ce,Ho@$NaGdF_4$@mSiO_2-ZnPc-CDs-P(NIPAm-MAA) nanocomposites capable of pH/thermo-coupling sensitive drug release; multimodal imaging with CT, UC, and MRI features; and synergetic antitumor therapy through photothermal therapy, photodynamic therapy, and chemotherapy. Due to the tri-doped Yb/Ce/Ho ions in the core and the inert shell coating, the NPs show intense red emission under 980 nm laser excitation, the emitted red light triggers the PDT agent zinc(II)-phthalocyanine (ZnPc) to produce singlet oxygen, and the CDs generate photothermal effect upon 980 nm laser irradiation as well as avoiding ZnPc leakage. The thermal effect together with the acidic environment in the cancer sites induces the shrinkage of the pH/thermal sensitive polymer P(NIPAm-MAA), enabling targeted and controllable release of DOX.

Self-assembled gold nanostar-$NaYF_4$:Yb/Er clusters with optimized plasmon resonance provide both in vitro imaging contrast and combination cell killing through simultaneous PTT and PDT therapy. Here, the photothermal effect was achieved through the heating effect of the gold nanostar under 980 nm excitation [95]. Prussian blue (PB) functionalized $CaMoO_4$:Eu@SiO_2@ Au nanorod hybrid NPs (HNPs), with multimodal capabilities such as fluorescence imaging, SERS detection, PTT, and good photostability and biocompatibility, were reported by Parchur et al. [96]. Under 808 nm laser excitation, HNPs show hyperthermia temperature. PB-functionalized NPs can be used in clinical trials for the treatment of radioactive exposure, and PB acts as a Raman reporter molecule with good biosafety and stability in the human body. DOX and indocyanine green (ICG)-loaded Gd_2O_3:Eu^{3+}@P(NIPAm-co-MAA)@THA@cRGD nanocomposites were designed and successfully applied in multimodal imaging diagnosis and therapy by Shen et al. [97]. The designed nanocomposites display a versatile, multifunctional platform that includes (1) simultaneous targeting with cRGD; (2) multimodal imaging, including two-photon luminescence, MRI, CT, and photothermal imaging(PTI); and (3) stimuli-responsive drug delivery and highly efficient synergistic chemo-/photothermal/photodynamic anticancer therapy. Liu et al. [98] used a $CsLu_2F_7$: Yb, Er, Tm-based visual therapeutic platform for imaging-guided cancer therapy. Due to the presence of cesium (Cs) in host lattice, the nanomaterial can provide higher resolution X-ray CT imaging than many other reported Ln-based CT contrast agents, and the presence of Yb:Er Ln provides UC imaging features under 980 nm excitation. By using the chemotherapy drug alpha-tocopheryl succinate and photothermal coupling agent ICG, this nanoplatform can provide targeted chemo-PTT.

Zhang et al. [99] reported a core/shell structured nanotheranostic (CuSgcopolymer) consisting of hydrophilic CuS NPs, styrene (St), *N*-isopropylacrylamide (NIPAm), methacrylic acid, and polymerizable Ln complex (Gd(AA)3phen) for MRI-guided chemo-PTT. The synthesized complex with excellent biocompatibility showed high loading capacity for DOX and HCl and excellent drug release under low pH and high temperature. The photosensitive CuS cores, which can absorb NIR light and convert to fatal heat, leading to a simultaneous therapeutic effect, combined PTT with chemotherapy. Moreover, the temperature-sensitive copolymer attached to the CuS NPs was affected by the thermal effect and gives rise to a highly controllable DOX release. The CuS@ copolymer/DOX showed an enhanced therapeutic efficacy against 4T1 cancer cells compared with separate PTT or chemotherapy, and the drug delivery procedure could be visualized by in vivo MR images.

Core-shell nanostructures consisting of plasmonic materials and Ln-doped UC NPs show many promising applications in theranostics. Wang et al. [100] report a novel type of thermal fluorescent core-shell hybrid nanocomposite incorporating rare-earth Yb^{3+} and Er^{3+} ion-doped GdOF as the shell and gold nanorods (GNRs) as the core. Under 980 nm light irradiation, the green and red emissions of GdOF: Yb^{3+}, Er^{3+} generally overlap with the visible absorbance of GNRs, thereby creating a photothermal effect, and the PTT efficiency can be easily controlled by altering the contents of Yb^{3+} and Er^{3+} ions. This multifunctional nanocomposite, which possesses UCL and photothermal and biocompatibility properties, shows strong potential for imaging and PTT [100]. Lv et al. [101] proposed a fully biocompatible organic-inorganic hybrid core-shell theranostic nanoplatform, based on mesoporous silica-coated UC core-shell NPs ($NaGdF_4$:Yb,Er@$NaGdF_4$:Yb@$mSiO_2$-Dopa) that incorporates dopamine (Dopa) in the silica layer. Silica-attaching PEI makes the Dopa transform into an active form that strongly absorbs 980 nm light. They showed that the activated nanoplatform is able to produce a pronounced photothermal effect that elevates water temperature from room temperature to 41.8°C within 2 min while emitting strong upconverted luminescence for optical imaging. The same material also shows MRI and X-ray CT features due to the presence of gadolinium (Gd) element.

Several polymer-coated Ln nanophosphors have been increasingly used recently in the theranostic application. The biocompatibility of nanocomposites and the efficiency of tumor ablation are the two important factors when constructing a nanotheranostic probe. Fully biocompatible polydopamine (PDA)-coated $NaYF_4$:Nd^{3+}@$NaLuF_4$ was reported as a multifunctional theranostic agent by Dai et al. [102], where they report that the photothermal conversion efficiency of the nanocomposites was optimized and maximized by the increase of the PDA shell thickness. The nanocomposite also shows better CT imaging and NIR imaging feature in the NIR-II window. A novel NIR-sensitive PDA core-shell-shell multifunctional core/satellite nanotheranostic platform, $NaYF_4$:20%Yb^{3+},2%Er^{3+}/$NaYF_4$:10%Yb^{3+},10%Nd^{3+}/$NaYF^4$:20%Nd^{3+}, for in vivo imaging-guided PTT under 808 nm excitation was reported by Ding et al. [103]. The design enabled outstanding UCL properties and strong X-ray absorption, thereby making the nanocomposites potential candidates for bimodal imaging. In addition, the PDA core provides high photothermal conversion efficiency and outstanding antitumor effect. Imaging-guided and temperature-monitored photothermal treatments using single core-shell $NaGdF_4$:YbEr-based UCNPs with a highly Yb^{3+} or Nd^{3+}-doped shell were reported by Hao et al. [104], where the spatial separation between the emission-sensing core and the heating shell was able to tailor the competition between the light and heat generation processes, and hence higher UCL efficiency and enhanced heating capability were achieved by introducing the core-shell design. The system was also able to measure the temperature, thus making the NPs a multifunctional platform for imaging and temperature-guided therapy.

Ln nanocomposites and core-shell structures with metallic nanostructures such as rods, spheres, and shells have been recently explored as potential theranostic agents. A multifunctional 980 nm excited GNR GNR@SiO_2@U NaGdF4:Er^{3+}, Yb^{3+} nanocomposite with attached photosensitizer (ZnPc) was reported by Huang et al. [105] for luminescence imaging, photothermal generation, nano thermometry, and photodynamic effects. The presence of gold nanorod not only increases the

UC process but at the same time induces photothermal effect. A unique kind of fluorescent thermal-MR core-shell UC nanostructure of GNRs@$NaGdF_4$:Yb^{3+}, Er^{3+} has been designed and fabricated to simultaneously achieve PTT and multimodal imaging. The presence of cetyltrimethylammonium bromide on the surface of GNRs offers the benefits of reducing toxicity and increasing biocompatibility. More significantly, the red and green UC emission appropriately overlaps with absorbance of GNRs, which improves the photothermal conversion efficiency. An 808 nm excited and T1-MRI signal is found in one single-shell NP $NaErF_4$@$NaGdF_4$ (Er@Gd), which is used as a dual-modality imaging contrast agent in vivo to accurately determine the position of tumors. With the guidance of dual-modality imaging, PTT is effectively used to ablate tumors in a mouse model.

Recently, there has been a lot of interest in Ln-doped apatite (hydroxy and fluoro) materials for various biomedical imaging and therapeutic applications. Hu et al. [106] reported the applications of dextran conjugated Yb^{3+} and Ho^{3+} co-doped fluorapatite (FA:Yb^{3+}/Ho^{3+}) as suitable for tracking and monitoring cartilage development in bone marrow mesenchymal stem cells in vitro and in vivo. Villa et al. [107] show how the 1340 nm emission band of Nd^{3+}-doped SrF_2 NPs can be used to produce autofluorescence-free, high-contrast in vivo fluorescence images. They also demonstrated that the complete removal of the food-related infrared autofluorescence is imperative for the development of reliable biodistribution studies. Core-shell structured $NaLuF_4$:Gd/Yb/Er@ $NaLuF_4$:Yb@ $NaLuF_4$:Nd/Yb@ $NaLuF_4$:NPs was reported as a viable candidate for optical and CT imaging [108]. The proposed material can be used both in up- and downconversion optical imaging under proper NIR excitation. Our recent work on the Nd^{3+}-doped fluorapatite (FAP)-coated Fe_3O_4 NPs shows that these NPs show better NIR emission in the second biological window and the NPs also show room temperature superparamagnetic nature with saturation magnetization value up to 7.8 emu/g [109]. To further improve the optical imaging features, we also explored several trivalent Ln combinations such as Nd/Yb,Yb/Er in FAP and found that such co-doped systems show better luminescence imaging platform in the second and third biological windows [110, 111].

9.6 MAGNETIC LUMINESCENT NANOPARTICLES

There is significant interest in magnetic nanoparticles in biomedical applications due to their unique magnetic properties. The advantages of specific properties of magnetic nanoparticles can be used in magnetic responsiveness and magnetic imaging devices. The main challenge in the synthesis of magnetic-fluorescent nanoparticles is quenching of the luminescence of the fluorophore when it is in direct contact with the magnetic particle. Since the magnetic fields are permeable to human tissues, magnetic particles are utilized as a drug delivery tool as well as for therapy. Magnetic particle imaging is a new imaging method intended for use in medical diagnosis, where it is subjected to a magnetic field similar to that used in conventional MR imaging, but it allows the direct quantitative mapping of the spatial distribution of magnetic nanoparticles. A new type of multifunctional nanoparticles based on magnetic and luminescent response was synthesized by covalently linking multiple carboxyl-functionalized superparamagnetic Fe_3O_4 nanoparticles and individual amino-functionalized silica-coated fluorescent $NaYF_4$: Yb, Er UC nanoparticles [112]. These nanocomposite particles may emit visible light in response to the irradiation by NIR light, allowing the application of the bioimaging. The Fe_3O_4/$NaYF_4$: Yb, Er magnetic/luminescent nanocomposites with the presence of active functional groups on the surface were successfully conjugated with transferrin specifically recognized on the receptors overexpressed on HeLa cells and can be employed for bio-labeling and fluorescent imaging of HeLa cells. The presence of both NIR-responsive UC nanoparticles and superparamagnetic properties will enable the practical application of the nanocomposites in bio-imaging and magnetic hyperthermia.

Magnetic hyperthermia, on the other hand, is a heat treatment method that uses magnetic particles to generate heat to destroy the local cancer cell by heating it to the predetermined temperature without damaging the healthy cells and tissues. For this method, the temperature of the treatment area has to be kept within the exact temperature range for a certain period of time to work effectively.

Magnetic hyperthermia has been studied and used for many years as an adjunct to cancer radiotherapy or chemotherapy [113, 114]. Its use is based on the fact that tumor cells are more sensitive to temperature in the range of 42–45°C (which yields necrosis, coagulation, or carbonization) than normal tissue cells up to 56°C [115]. Therefore, the temperature range of 45–56°C has become critical for cancer treatment. The hyperthermia has seen more clinical usage in Europe and Asia. In the United States, this method still remains largely unpopular due to possible overheating and necrosis of normal tissue. For most nanoparticles suggested so far for cancer hyperthermia treatment, uniform controlled induction heating and selectivity remain major challenges. Development of nanoparticles that are not affected by alternating magnetic fields above 50°C has become essential in order to prevent overheating of normal cells. A significant limitation for application of the existing magnetic nanoparticles is their nonselective and excessive heating of tumors and surrounding normal tissues, especially the fat tissue, which is unacceptable in case of breast cancer considering abundance of fat in mammary glands and close proximity of tumor to vital organs located in thoracic cavity. Basically, hyperthermia increases perfusion and oxygenation of neoplastic hypoxic cells, which are more resistant to ionizing radiation than normal cells [116]. Moreover, increased tumor tissue perfusion facilitates the absorption of chemotherapeutic drugs through cell membrane without being more toxic [117–119]. As a result, the action of combination of hyperthermia with radiotherapy or chemotherapy becomes more efficient. Consequently, hyperthermia allows reducing of tumors resistant to various chemotherapeutic drugs such as DOX, cisplatin, bleomycin, nitrosoureas, and cyclophosphamide. It has been demonstrated that hyperthermia also has an antiangiogenic action and an immunotherapeutic role, due to thermal shock proteins, which are produced by stressed tumor cells [120, 121]. The main challenges of the magnetic-induced hyperthermia cancer therapy are the enhancement of heating power of such nanoparticles and the control of the local tumoral temperature [122, 123].

In order to be efficient in biological applications, magnetic particles must fulfill essential properties such as biocompatibility, long-term stability, nontoxicity, and high responsivity to an external field, which targets the particles to the preferred location. Biocompatibility and long-term stability of magnetic particles are ensured by using magnetic core/shell and shell as biocompatible polymers or surfactants that are consistent with the carrier fluid. Special attention has been given to magnetite (Fe_3O_4) nanoparticles for their properties such as strong ferromagnetic/superparamagnetic behavior with high saturation magnetization, lower sensitivity to oxidation, and relatively low toxicity, which are more applicable for biomedical applications.

Although there are several forms of magnetism that exist in nature, superparamagnetism seems to be preferred for biomedical applications [124]. First, due to lower remnant magnetization (magnetic moment when no external field), attractive forces between particles are minimized, which lessens the particle aggregation. Second, higher magnetic susceptibility and saturation magnetization can be achieved with a lower external field. The superparamagnetic state can be attained when the particle size decreases to a sufficiently small value (<50 nm), where it is favorable to exist as particles with a single domain. This single-domain magnetic moment will randomly flip between the two opposite directions, separated by an energy barrier ($\Delta E = KV$, where K is magnetic anisotropy constant, and V is the volume of the particle), along with the so-called easy axis, where the energy state for these atoms is lowest. The time between two magnetization flips, τ_N (Néel relaxation time), is given by the following Néel–Arrhenius equation:

$$\tau_N = \tau_0 \exp\left(\frac{\Delta E}{k_B T}\right)$$

where τ_0 is a length of time, characteristic of the material, k_B is the Boltzmann constant, and T is the temperature. If the time between two magnetization flips is smaller than the time used to measure the magnetization of the nanoparticles, the magnetization appears to be average zero for the nanoparticles. This state of the particles is called superparamagnetic. The temperature transition between superparamagnetism and the so-called blocked state, where measured time is greater than relaxation

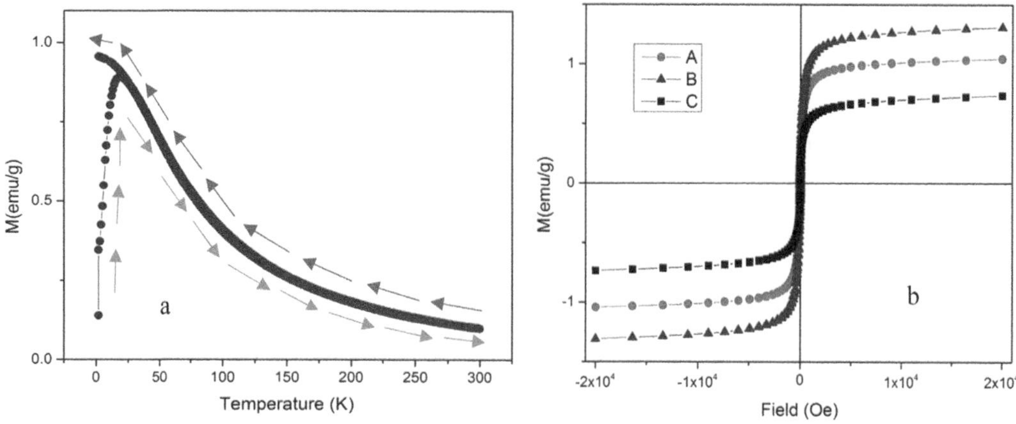

FIGURE 9.10 (a) The ZFC and FC curves of the Fe_3O_4 sample at magnetic field $H = 100$ Oe; (b) hysteresis loops of the surface-modified Fe_3O_4 samples at 300 K.

time, is called the blocking temperature (T_B). The zero-field-cooled (ZFC) and field-cooled (FC) curves obtained from temperature-dependent magnetization measurement are usually used to obtain the information of the energy barriers (ΔE) or the blocking temperature of superparamagnetic particles. ZFC-FC measurements were performed by cooling down the sample from room temperature or well above T_B to a temperature well below the T_B with no external magnetic field. After reaching lower temperature, by applying a small magnetic field (typically around 100 Oe), sample was heated up to a higher temperature (~300 K) while measuring the magnetization to obtain the ZFC curve. The FC curve was obtained by cooling down the sample to the same lower temperature while keeping the same external magnetic field.

Figure 9.10a shows a temperature dependence of the magnetization curve of a Fe_3O_4 sample with average particle size around 10 nm. The nanoparticles were prepared by using the co-precipitation method. Their surfaces were to be coated with fluorapatite and Nd^{3+}-doped fluorapatite for the luminomagnetic properties [109]. The peak in the ZFC curve suggests that the blocking temperature is around 25 K, which confirms that particles at room temperature should be in the superparamagnetic state. Also, qualitative information about particle size distribution can be obtained by analyzing the ZFC peak shape. Since the blocking temperature depends on the size of the crystallites, each crystallite with specific volume distribution blocked at a different temperature, creating a broad peak in the ZFC curve. The narrow peak corroborates the narrow size distribution of the as-prepared Fe_3O_4 nanoparticles [109]. Figure 9.10b shows the hysteresis results of surface-modified Fe_3O_4 samples at room temperature. The samples A, B, and C referred to the 2.1 mol% Nd^{3+}-doped fluorapatite and pure fluorapatite-coated Fe_3O_4, respectively. All samples exhibit negligible coercivity, which confirms the superparamagnetic nature at room temperature. The saturation magnetization values of samples are lower than the values of Fe_3O_4 nanoparticles due to the diamagnetic contribution of the fluorapatite surrounding the Fe_3O_4 nanoparticles. This result indicates that the superparamagnetic particles can be modified to accomplish luminomagnetic properties, which benefits diverse biomedical applications such as efficient magnetic separation, MRI, and hyperthermia.

Figure 9.11 shows the time dependence of the temperature of water-dispersed Fe_3O_4 nanoparticles as well as dextran-coated Fe_3O_4 nanoparticles measured applying an AC magnetic field amplitude up to 300 Gauss and frequency of 604 kHz [125]. The specific power absorption (SPA) is the key parameter describing the heating efficiency of magnetic nanoparticles. The SPA represents the power released by the magnetic colloid per unit mass of magnetic nanoparticles.

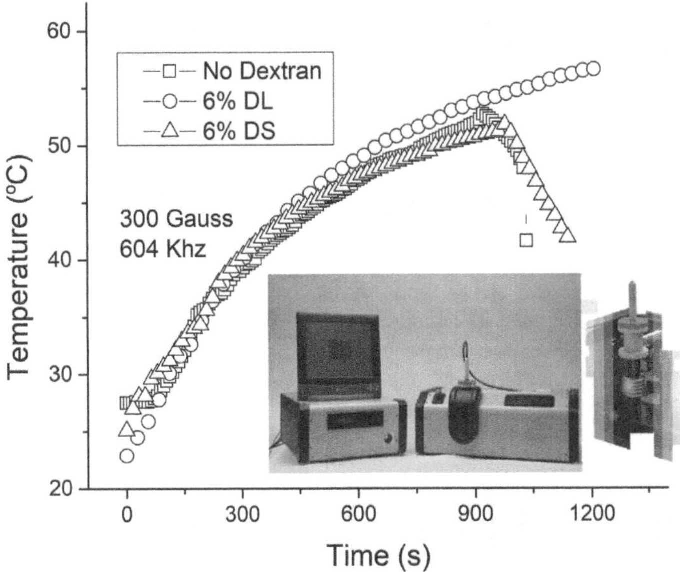

FIGURE 9.11 Magnetic-induced hyperthermia experiments conducted with an alternating magnetic field (DL, dextran from *Leuconostoc* spp. MW ~6000; DS, dextran with sodium sulfate MW ~500,000). Inset: DM100 series (nB nanoScale Biomagnetics) magnetic hyperthermia device with optical fiber temperature sensor.

The SPA is influenced by applied magnetic field amplitude, frequency, nanoparticle size, and magnetic saturation [126, 127], and based on the relationship:

$$\mathrm{SPA} = \left[\frac{m_{LIQ}c_{LIQ} + m_{NP}c_{NP}}{m_{NP}}\right]\left(\frac{\Delta T}{\Delta t}\right)$$

where m_{LIQ} and c_{LIQ} represent the mass and the heat capacity of the dispersion medium, which is mainly in a liquid state; m_{NP} and c_{NP} represent the mass and the heat capacity of the nanoparticles; and $(\Delta T/\Delta t)$ represents the temperature elevation rate. The SPA values were extracted from the initial slope of the corresponding temperature-versus-time curves. The SPA of most nanoparticles was estimated to be ~130 W/g.

9.7 SUMMARY

The recent developments in theranostics with NIR-sensitive rare-earth-doped luminomagnetic nanomaterials for bioimaging and therapy applications are introduced in this chapter. Several novel compositions and key technologies are described in this review. Owing to rapid technological developments over the last 10 years, the luminomagnetic nanomaterials have progressively advanced. A new imaging modality for photodynamic and photothermal therapies can be used both in diagnostics and therapeutic tools for the prospective routine clinical practice. Core-shell-shell rare-earth-doped luminomagnetic nanoplatforms have been extensively explored for PDT and imaging due to their representative large anti-Stokes shifts, deep penetration into living tissues, narrow emission bands, and high spatial-temporal resolution. Stable luminomagnetic properties can be potentially used to make rare-earth-doped nanostructured materials for wide use in biological imaging; however, the low luminescence efficiency and unknown toxicity may limit its further applications.

REFERENCES

1. J. Xie, S. Lee, X. Chen. Nanoparticle-based theranostic agents. *Adv. Drug Deliv., Rev.*, 62, 1064–1079, 2010.
2. D.E. Lee, H. Koo, I.C. Sun, J.H. Ryu, K. Kim, I.C. Kwon. Multifunctional nanoparticles for multimodal imaging and theragnosis. *Chem. Soc. Rev.* 41 (7), 2656–2672, 2012.
3. M.M. Yallapu, S.F. Othman, E.T. Curtis, B.K. Gupta, M. Jaggi, and S.C. Chauhan. Multi-functional magnetic nanoparticles for magnetic resonance imaging and cancer therapy. *Biomaterials* 32 (7), 1890–1905, 2011.
4. Z. Huang. A review of progress in clinical photodynamic therapy. *Technol. Cancer Res. Treat.*, 4(3):283–93, 2005.
5. L. Zou, H. Wang, B. He, et al. Current approaches of photothermal therapy in treating cancer metastasis with nanotherapeutics. *Theranostics*, 6(6):762–72, 2016.
6. R. Di Corato, G. Béalle, J. Kolosnjaj-Tabi, A. Espinosa, O. Clément, A. K. A. Silva, C. Ménager, and C. Wilhelm. Combining magnetic hyperthermia and photodynamic therapy for tumor ablation with photoresponsive magnetic liposomes. *ACS Nano*, 9 (3), 2904–2916, 2015.
7. J. Shi, P. W. Kantoff, R. Wooster, and O. C. Farokhzad. Cancer nanomedicine: progress, challenges and opportunities. *Nat. Rev. Cancer*, 17 (1): 20–37, 2017.
8. K. Park, S. Lee, E. Kang, K. Kim, K. Choi, and I.C. Kwon. New generation of multifunctional nanoparticles for cancer imaging and therapy. *Adv. Funct. Mater.* 19 (10), 1553–1566, 2009.
9. J. W. Nichols, B.Y. Han. EPR: Evidence and fallacy. *J. Control. Release*. 190: 451–464, 2014.
10. G. Tian, W. Yin, J. Jin, X. Zhang, G. Xing, S. Li, Z. Gu and Y. Zhao. Engineered design of theranostic UC NPs for tri-modal UC luminescence/magnetic resonance/X-ray computed tomography imaging and targeted delivery of combined anticancer drugs. *J. Mater. Chem. B*, 2, 1379–1389, 2014.
11. K. Schwamborn and R.M. Caprioli. Molecular imaging by mass spectrometry-looking beyond classical histology. *Nat. Rev. Cancer* 10, 639–646, 2010.
12. J. Pol, M. Strohalm, V. Havlıcek, M. Volny, Molecular mass spectrometry imaging in biomedical and life science research. *Histochem. Cell Biol.*, 134, 423–443, 2010.
13. J.C. Vickerman. Molecular imaging and depth profiling by mass spectrometry – SIMS, MALDI or DESI? *Analyst*, 136, 2199–217, 2011.
14. C. Giesen, H.A.O. Wang, D. Schapiro, N. Zivanovic, A. Jacobs, B. Hattendorf, P. J. Schüffler, D. Grolimund, J. M. Buhmann, S. Brandt, Z. Varga, P. J. Wild, D. Günther, and B. Bodenmiller. Highly multiplexed imaging of tumor tissues with subcellular resolution by mass cytometry. *Nat. Methods*, 11, 417–422, 2014.
15. A.D. Amir el, K.L. Davis, M.D. Tadmor, E.F. Simonds, J.H. Levine, S.C. Bendall, D.K. Shenfeld, S. Krishnaswamy, G.P. Nolan, and D. Pe'er. viSNE enables visualization of high dimensional single cell data and reveals phenotypic heterogeneity of leukemia. *Nat. Biotechnol.*, 31, 545–552, 2013.
16. E.F. Simonds, P. Qiu, E.D. Amir, P.O. Krutzik, R. Finck, R.V. Bruggner, R. Melamed, A. Trejo, O.I. Ornatsky, R.S. Balderas, S.K. Plevritis, K. Sachs, D. Pe'er, S.D. Tanner, and G.P. Nolan. Single cell mass cytometry of differential immune and drug responses across a human hematopoietic continuum. *Science*, 332, 687–696, 2011.
17. A.F. Palonpon, J. Ando, H. Yamakoshi, K. Dodo, M. Sodeoka, S. Kawata, and K. Fujita. Raman and SERS microscopy for molecular imaging of live cells. *Nat. Protoc.*, 8, 677–692, 2013.
18. A.J.K. Fujita, N.I. Smith, S. Kawata. Dynamic SERS imaging of cellular transport pathways with endocytosed gold NPs. *Nano Lett.* 11, 5344–5348, 2011.
19. D.C. Kennedy, K.A. Hoop, L.-L. Tay, and J.P. Pezacki. Development of NP probes for multiplex SERS imaging of cell surface proteins. *Nanoscale*, 2, 1413–1416, 2010.
20. M. Xiao, L. Lin, Z. Li, J. Liu, S. Hong, Y. Li, M. Zheng, X. Duan, and X. Chen. SERS imaging of cell-surface biomolecules metabolically labeled with bioorthogonal Raman reporters. *Chemistry*, 9, 2040–2044, 2014.
21. S. Lee, H. Chon, J. Lee, J. Ko, B. H. Chung, D.W. Lim, and J. Choo. Rapid and sensitive phenotypic marker detection of breast cancer cells using surface-enhanced Raman scattering (SERS) imaging. *Biosens. Bioelectron.*, 51,15 238–243, 2014.
22. S. McAughtrie, K. Lau, K. Faulds, and D. Graham. 3D optical imaging of multiple SERS nanotags in cells. *Chem. Sci.,* 4, 3566–3572, 2013.

23. M. Heilemann, S. van de Linde, A. Mukherjee, and M. Sauer. Super-resolution imaging with small organic fluorophores. *Angew. Chem. Int. Ed.*, 48, 6903–6908, 2009.
24. M. Fernández-Suárez and A. Y. Ting. Fluorescent probes for super-resolution imaging in living cells. *Nat. Rev. Mol. Cell Biol.*, 9, 929–943, 2008.
25. G.T Dempsey, J.C Vaughan, K.H. Chen, M. Bates, and X. Zhuang. Evaluation of fluorophores for optimal performance in localization-based super-resolution imaging. *Nat. Methods*, 8, 1027–1036, 2011.
26. H. Gong, R. Peng, and Z. Liu. Carbon nanotubes for biomedical imaging: the recent advances. *Adv. Drug Deliv. Rev.*, 65 (15):1951–63, 2013.
27. P. Zrazhevskiy and X. Gao. Quantum dot imaging platform for single-cell molecular profiling. *Nat. Commun.*, 4, 1–12, 2013.
28. X. Michalet, F.F. Pinaud, L.A. Bentolila, J.M. Tsay, S. Doose, J.J. Li, G. Sundaresan, A.M. Wu, S.S. Gambhir, and S. Weiss. Quantum dots for live cells, in vivo imaging, and diagnostics. *Science*, 307, 538–544, 2005.
29. J. Li and J.-J. Zhu. Quantum dots for fluorescent biosensing and bio-imaging applications. *Analyst*, 138, 2506–2515, 2013.
30. M.A. Walling, J.A. Novak, and J.R.E. Shepard. Quantum dots for live cell and in vivo imaging. *Int. J. Mol. Sci.*, 10, 441–491, 2009.
31. B.A. Kairdolf, A.M. Smith, T.H. Stokes, M.D. Wang, A.N. Young, and S. Nie. Semiconductor quantum dots for bioimaging and biodiagnostic applications. *Annu. Rev. Anal. Chem.*, 6,143–162, 2013.
32. J. Gallo, I.S. Alam, J. Jin, Y.-J. Gu, E.O. Aboagye, W.-T. Wong, and N.J. Long. PET imaging with multimodal UC NPs. *Dalton Trans.*, 43, 5535–5545, 2014.
33. L. Zhou, X. Zheng, Z. Gu, W. Yin, X. Zhang, L. Ruan, Y. Yang, Z. Hu, and Y. Zhao. Mesoporous NaYbF4@NaGdF4 core-shell up-conversion NPs for targeted drug delivery and multimodal imaging. *Biomaterials* 35, 7666–7678, 2014.
34. Y.J. Bao, J.J. Li, Y.T. Wang, L. Yu, L. Lou, W.J. Du, Z.Q. Zhu, H. Peng, and J.Z. Zhu. Probing cytotoxicity of CdSe and CdSe/CdS quantum dots. *Chin. Chem. Lett.*, 22, 843–846, 2011.
35. G.A. Kumar, M. Pokhrel, and D.K. Sardar. Synthesis and up conversion spectroscopy of YbEr doped M_2O_2S (M=La, Gd, Y) phosphors. *Sci. Adv. Mat.*, 4, 623–630, 2012.
36. S.F. Lim, R. Riehn, W.S. Ryu, N. Khanarian, C.K. Tung, D. Tank, and R.H. Austin. In vivo and scanning electron microscopy imaging of up converting NPs in *Caenorhabditis elegans*. *Nano Lett.*, 6, 169–174, 2006.
37. L. Cheng, K. Yang, Y. Li, J. Chen, C. Wang, M. Shao, S.-T. Lee, and Z. Liu. Facile preparation of multifunctional UC nanoprobes for multimodal imaging and dual-targeted photothermal therapy. *Angew. Chem. Int. Ed.*, 50, 7385–7390, 2011.
38. J. Gallo, I.S. Alam, J. Jin, Y.-J. Gu, E.O. Aboagye, W.-T. Wong, and N.J. Long. PET imaging with multimodal UC NPs. Dalton Trans. 43, 5535–5545, 2014.
39. S.P. Tiwari, S.K. Maurya, R.S. Yadav, A. Kumar, V. Kumar, M.F. Joubert, and H.C. Swart. Future prospects of fluoride based upconversion nanoparticles for emerging applications in biomedical and energy harvesting. *J. Vacuum Sci. Technol. B*, 36, 06080, 2018.
40. W. Fan, B. Shen, W. Bu, F. Chen, Q. He, K. Zhao, S. Zhang, L. Zhou, W. Peng, Q. Xiao, D. Ni, J. Liu, J. Shi. A smart UC-based mesoporous silica nanotheranostic system for synergetic chemo-/radio-/photodynamic therapy and simultaneous MR/UCL imaging. *Biomaterials*, 35, 2014, 8992–9002, 2014.
41. Q. Lu, D. Wei, J. Cheng, J. Xu, J. Zhu. A novel contrast agent with Ln-doped up-conversion luminescence and Gd-DTPA magnetic resonance properties. *J. Solid State Chem.*, 192, 75–80, 2012.
42. L. Cheng, K. Yang, Y. Li, X. Zeng, M. Shao, S. T. Lee, and Z. Liu. Multifunctional NPs for UC luminescence/MR multimodal imaging and magnetically targeted photothermal therapy. *Biomaterials*, 33, 2215–2222, 2012.
43. D. Yang, Y. Dai, J. Liu, Y. Zhou, Y. Chen, C. Li, P. Ma, and J. Lin. Ultra-small $BaGdF_5$-based UC NPs as drug carriers and multimodal imaging probes. *Biomaterials*, 35, 2011–2023, 2014.
44. L. He, L. Feng, L. Cheng, Y. Liu, Z. Li, R. Peng, Y. Li, L. Guo, and Z. Liu. Multilayer dual-polymer-coated UC NPs for multimodal imaging and serum-enhanced gene delivery. *ACS Appl. Mater. Interfaces*, 5, 10381–10388, 2013.
45. Y. Xing, J. Zhao, P.S. Conti, and K. Chen. Radiolabeled NPs for multimodality tumor imaging. *Theranostics*, 4, 290–306, 2014.

46. Y. Min, J. Li, F. Liu, P. Padmanabhan, E.K.L. Yeow, and B. Xing. Recent advance of biological molecular imaging based on Ln-doped UC-luminescent nanomaterials. *Nanomaterials*, 4, 129–154, 2014.
47. Z. Yi, W. Lu, Y. Xu, J. Yang, L. Deng, C. Qian, T. Zeng, H. Wang, L. Rao, H. Liu, and S. Zeng. PEGylated NaLuF4: Yb/Er UC NPs for in vivo synergistic fluorescence/X-ray bioimaging and long-lasting, real-time tracking. *Biomaterials*, 35, 9689–9697, 2014.
48. A. Xia, X. Zhang, J. Zhang, Y. Deng, Q. Chen, S. Wu, X. Huang, and J. Shen. Enhanced dual contrast agent, Co^{2+}-doped NaYF4:Yb^{3+}, Tm^{3+} nanorods, for near infrared-to-near infrared UC luminescence and magnetic resonance imaging. *Biomaterials*, 35, 9167–9176, 2014.
49. W. Fan, B. Shen, W. Bu, F. Chen, Q. He, K. Zhao, S. Zhang, L. Zhou, W. Peng, Q. Xiao, D. Ni, J. Liu, and J. Shi. A smart UC-based mesoporous silica nanotheranostic system for synergetic chemo-/radio-/photodynamic therapy and simultaneous MR/UCL imaging. *Biomaterials*, 35, 8992–9002, 2014.
50. Y. Zhang, W. Wei, G.K. Das, T.T.Y. Tan. Engineering Ln-based materials for nanomedicine. *J. Photochem. Photobiol. C Photochem. Rev.*, 20, 71–96, 2014.
51. J. Niu, X. Wang, J. Lv, Y. Li, and B. Tang. Luminescent nanoprobes for in-vivo bioimaging. *Trends Anal. Chem.*, 58, 112–119, 2014.
52. Y. Wu, Y. Sun, X. Zhu, Q. Liu, T. Cao, J. Peng, Y. Yang, W. Feng, F. Li. Ln-based nanocrystals as dual-modal probes for SPECT and X-ray CT imaging. *Biomaterials*, 35, 4699–4705, 2014.
53. S. Zeng, H. Wang, W. Lu, Z. Yi, L. Rao, H. Liu, J. Hao. Dual-modal UC fluorescent/X-ray imaging using ligand-free hexagonal phase NaLuF4:Gd/Yb/Er nanorods for blood vessel visualization. *Biomaterials*, 35, 2934–2941, 2014.
54. A. Gnach, K. Prorok, M. Misiak, B. Cichy, and A. Bednarkiewicz. Up-converting NaYF4:0.1%Tm3+, 20%Yb3+ NPs as luminescent labels for deep-tissue optical imaging. *J. Lns*, 32, 207–212, 2014.
55. D. Yang, Y. Dai, J. Liu, Y. Zhou, Y. Chen, C. Li, P. Ma, and J. Lin. Ultra-small BaGdF5-based UC NPs as drug carriers and multimodal imaging probes. *Biomaterials*, 35, 2011–2023, 2014.
56. J. Zhou, Z. Lu, G. Shan, S. Wang, and Y. Liao. Gadolinium complex and phosphorescent probe-modified NaDyF4 nanorods for T1- and T2-weighted MRI/CT/phosphorescence multimodality imaging. *Biomaterials*, 35, 368–377, 2014.
57. Z. Liu, E. Ju, J. Liu, Y. Du, Z. Li, Q. Yuan, J. Ren, and X. Qu. Direct visualization of gastrointestinal tract with Ln-doped BaYbF5 UC nanoprobes. *Biomaterials*, 34, 7444–7452, 2013.
58. M.-K. Tsang, S. Zeng, H.L.W. Chan, and J. Hao. Surface ligand-mediated phase and UC luminescence tuning of multifunctional NaGdF4: Yb/Er materials with paramagnetic and cathodoluminescent characteristics. *Opt. Mater.* 35, 2691–2697, 2013.
59. F. Liu, X. He, L. Liu, H. You, H. Zhang, Z. Wang. Conjugation of NaGdF4 upconverting NPs on silica nanospheres as contrast agents for multi-modality imaging. *Biomaterials*, 34, 5218–5225, 2013.
60. J. Shen, L. Zhao, and G. Han. Ln-doped upconverting luminescent NP platforms for optical imaging-guided drug delivery and therapy. *Adv. Drug Deliv. Rev.*, 65, 744–755, 2013.
61. D. Chen, Y. Chen, H. Lu, and Z. Ji. A bifunctional Cr/Yb/Tm: $Ca_3Ga_2Ge_3O_{12}$ phosphor with near-infrared long-lasting phosphorescence and UC luminescence. *Inorg. Chem.*, 53, 8638–8645, 2014.
62. J.A. Damasco, G. Chen, W. Shao, H. Ågren, H. Huang, W. Song, J.F. Lovell, and P.N. Prasad. Size-tunable and monodisperse Tm3+/Gd3+-doped hexagonal NaYbF4 NPs with engineered efficient near infrared-to-near infrared UC for in vivo imaging. *ACS Appl. Mater. Interfaces*, 6, 13884–13893, 2014.
63. Y. Cen, Y.-M. Wu, X.-J. Kong, S. Wu, R.-Q. Yu, and X. Chu. Phospholipid-modified UC nanoprobe for ratiometric fluorescence detection and imaging of phospholipase D in cell lysate and in living cells. *Anal. Chem.*, 86, 7119–7127, 2014.
64. J. Rieffel, F. Chen, J. Kim, G. Chen, W. Shao, S. Shao, U. Chitgupi, R. Hernandez, S. A Graves, R. J. Nickles, P. N. Prasad, C. Kim, W. Cai, and J. F Lovell. Hexamodal imaging with porphyrin-phospholipid-coated UC NPs. *Adv. Mater.*, 27, 1785–1790, 2015.
65. B.A. Holm, E.J. Bergey, T. De Rrasad et al. Nanotechnology in biomedical applications. *Mol. Cryst. Liquid Cryst.*, 374, 589–598, 2002.
66. G.S. Yi and G.M. Chow. Synthesis of hexagonal-phase NaYF4: Yb,Er and NaYF4: Yb,Tm nanocrystals with efficient up-conversion fluorescence. *Adv. Funct. Mater.*, 16(18), 2324–2329, 2006.
67. S.J. Zeng, et al. Simultaneous realization of phase/size manipulation, UC luminescence enhancement, and blood vessel imaging in multifunctional nanoprobes through transition metal Mn^{2+} doping. *Adv. Funct. Mater.*, 24(26), 4051–4059, 2014.

68. Y. Deng, H. Wang, W. Gu, S. Li, N. Xiao, C. Shao, Q. Xu, and L. Ye. Ho3+ doped NaGdF4 NPs as MRI/optical probes for brain glioma imaging. *J. Mater. Chem. B*, 2(11), 1521–1529, 2014.
69. W. Wang, et al. Magnetic-luminescent $YbPO_4$:Er,Dy microspheres designed for tumor theranostics with synergistic effect of photodynamic therapy and chemotherapy. *Int. J. Nanomed.*, 9, 4879–4891, 2014.
70. A. Xia, et al. Enhanced dual contrast agent, Co^{2+}-doped NaYF4:Yb^{3+},Tm^{3+} nanorods, for near infrared-to-near infrared UC luminescence and magnetic resonance imaging. *Biomaterials*, 35, (33), 9167–9176, 2014.
71. T. Kim, et al. Mesoporous silica-coated luminescent Eu3+ doped GdVO4 NPs for multimodal imaging and drug delivery. *RSC Adv.*, 4(86), 45687–45695, 2014.
72. L. Wang, et al. Efficient gene delivery and multimodal imaging by Ln-based UC NPs. *Langmuir*, 30, (43), 13042–13051, 2014.
73. T. Paik, et al. Shape-controlled synthesis of isotopic yttrium-90-labeled Ln fluoride nanocrystals for multimodal imaging. *ACS Nano*, 9(9), 8718–8728, 2015.
74. P.P. Lei, et al. Yb3+/Er3+-Co doped Bi_2O_3 nanospheres: probe for UC luminescence imaging and binary contrast agent for computed tomography imaging. *ACS Appl. Mater. Interfaces*, 7(47), 26346–26354, 2015.
75. C. Cao, et al. Energy transfer highway in Nd^{3+}-sensitized NPs for efficient near-infrared bioimaging. *ACS Appl. Mater. Interfaces*, 9(22), 18540–18548, 2017.
76. Y.M. Feng, et al. Neodymium-doped NaHoF4 NPs as near-infrared luminescent/T-2-weighted MR dual-modal imaging agents in vivo. *J. Mater. Chem. B*, 5(3), 504–510, 2017.
77. A.K.S. Braz, et al. TiO_2-coated fluoride NPs for dental multimodal optical imaging. *J. Biophoton.*, 11, (4), 2018.
78. Y.B. Li, et al. Second near-infrared emissive Ln complex for fast renal-clearable in vivo optical bioimaging and tiny tumor detection. *Biomaterials,* 169, 35–44, 2018.
79. E. Abliz, J.E. Collins, J.S. Friedberg, A. Kumar, H. Bell, R.W. Waynant, and D.B. Tata. Novel applications of diagnostic X-rays in activating photo-agents through X-ray induced visible luminescence from rare-earth particles: an in-vitro study. *Proc. SPIE.*, Vol. 7565 75650B-1-13, 2010.
80. X.L. Li, et al. Soft X-ray activated NaYF4:Gd/Tb scintillating nanorods for in vivo dual-modal X-ray/X-ray-induced optical bioimaging. *Nanoscale*, 10(1), 342–350, 2018.
81. W. Wang, et al. Magnetic-luminescent YbPO4:Er,Dy microspheres designed for tumor theranostics with synergistic effect of photodynamic therapy and chemotherapy. *Int. J. Nanomed.* 9, 4879–4891, 2014.
82. F.J. Ai, et al. A core-shell-shell nanoplatform upconverting near-infrared light at 808 nm for luminescence imaging and photodynamic therapy of cancer. *Sci. Rep.*, 5, 10785, 2015.
83. Y.A. Huang, et al. 915 nm light-triggered photodynamic therapy and MR/CT dual-modal imaging of tumor based on the nonstoichiometric Na0.52YbF3.52:Er UC nanoprobes. *Small*, 12(31), 4200–4210, 2016.
84. P. Kalluru, et al. Unprecedented "all-in-one" ln-doped mesoporous silica frameworks for fluorescence/MR imaging and combination of NIR light triggered chemo-photodynamic therapy of tumors. *Adv. Funct. Mater.*, 26(43), 7908–7920, 2016.
85. C.C. Hsu, et al. Ln-doped core-shell-shell nanocomposite for dual photodynamic therapy and luminescence imaging by a single X-ray excitation source. *ACS Appl. Mater. Interfaces* 10(9), 7859–7870, 2018.
86. J. Liu, L. Huang, X. Tian, X. Chen, Y. Shao, F. Xie, D. Chen, and L. Li. Magnetic and fluorescent Gd2O3:Yb3+/Ln(3+) NPs for simultaneous UC luminescence/MR dual modal imaging and NIR-induced photodynamic therapy. *Int. J. Nanomed.*, 12, 1–14, 2016.
87. Q.Q. Li, et al. An 800 nm driven NaErF4@NaLuF4 UC platform for multimodality imaging and photodynamic therapy. *Nanoscale*, 10(26), 12356–12363, 2018.
88. F.Y. Xu, et al. Ln-doped core-shell NPs as a multimodality platform for imaging and photodynamic therapy. *Chem. Commun.*, 54(68), 9525–9528, 2018.
89. J. Estelrich and M.A. Busquets. Iron oxide nanoparticles in photothermal therapy. *Molecules*, 23(7), 1567, 2018.
90. D. Jaque, L. Martínez Maestro, B. del Rosal, P. Haro-Gonzalez, A. Benayas, J. L. Plaza, E. Martín Rodríguezand, and J. García Solé. Nanoparticles for photothermal therapies. *Nanoscale*, 6, 9494–9530, 2014.
91. R.C. Lv, et al. Multifunctional anticancer platform for multimodal imaging and visible light driven photodynamic/photothermal therapy. *Chem. Mater.*, 27(5), 1751–1763, 2015.

92. R. Lv, P.P. Yang, F. He, S.L. Gai, C.X. Li, Y.L. Dai, G.X. Yang, and J. Lin. A yolk-like multifunctional platform for multimodal imaging and synergistic therapy triggered by a single near-infrared light. *ACS Nano*, 9, 2, 1630–1647, 2015.
93. D. Yang, et al. Y2O3:Yb, Er@mSiO(2)-CuxS double-shelled hollow spheres for enhanced chemo-photothermal anti-cancer therapy and dual-modal imaging. *Nanoscale*, 7(28), 12180–12191, 2015.
94. J.T. Xu, et al. Design, fabrication, luminescence and biomedical applications of UCNPs@mSiO(2)-ZnPc-CDs-P(NIPAm-MAA) nanocomposites. *J. Mater. Chem. B*, 4(35), 5883–5894, 2016.
95. L.C. He, et al. Self-assembled gold nanostar-NaYF4:Yb/Er clusters for multimodal imaging, photothermal and photodynamic therapy. *J. Mater. Chem. B*, 4(25), 4455–4461, 2016.
96. A.K. Parchur, et al. Near-infrared photothermal therapy of Prussian-blue-functionalized Ln-ion-doped inorganic/plasmonic multifunctional nanostructures for the selective targeting of HER2-expressing breast cancer cells. *Biomater. Sci.*, 4(12), 1781–1791, 2016.
97. T.T. Shen, et al. Versatile rare-earth oxide nanocomposites: enhanced chemo/photothermal/photodynamic anticancer therapy and multimodal imaging. *J. Mater. Chem. B*, 4(48), 7832–7844, 2016.
98. Y. Liu, L. Li, Q. Guo, L. Wang, D. Liu, Z. Wei, and J. Zhou. Novel Cs-based upconversion nanoparticles as dual-modal CT and UCL imaging agents for chemo-photothermal synergistic therapy. *Theranostics*, 6, (10), 1491–505, 2016.
99. L. Zhang, et al. Dual-stimuli-responsive, polymer-microsphere-encapsulated CuS NPs for magnetic resonance imaging guided synergistic chemo-photothermal therapy. *ACS Biomater. Sci. Eng.*, 3(8), 1690–1701, 2017.
100. C. Wang, et al. Multimodal imaging and photothermal therapy were simultaneously achieved in the core-shell UCNR structure by using single near-infrared light. *Dalton Trans.*, 46(36), 12147–12157, 2017.
101. R.C. Lv, et al. Dopamine-mediated photothermal theranostics combined with up-conversion platform under near infrared light. *Sci. Rep.*, 7.1, 13562, 2017.
102. Y. Dai, et al. Mussel-inspired polydopamine-coated Ln NPs for NIR-II/CT dual imaging and photothermal therapy. *ACS Appl. Mater. Interfaces*, 9(32), 26674–26683, 2017.
103. X. Ding, et al. Multifunctional core/satellite polydopamine@Nd3+-sensitized UC nanocomposite: a single 808 nm near-infrared light-triggered theranostic platform for in vivo imaging-guided photothermal therapy. *Nano Res.*, 10(10), 3434–3446, 2017.
104. Q.Y. Hao, et al. Multifunctional Ln-doped core/shell NPs: integration of UC luminescence, temperature sensing, and photothermal conversion properties. ACS Omega 3(1), 188–197, 2018.
105. Y. Huang, et al. Upconverting nanocomposites with combined photothermal and photodynamic effects. *Nanoscale*, 10(2), 791–799, 2018.
106. X.Q. Hu, et al. Dextran-coated fluorapatite crystals doped with Yb3+/Ho3+ for labeling and tracking chondrogenic differentiation of bone marrow mesenchymal stem cells in vitro and in vivo. *Biomaterials*, 52, 441–451, 2015.
107. I. Villa, et al. 1.3 mu m emitting SrF2:Nd3+ NPs for high contrast in vivo imaging in the second biological window. *Nano Res.*, 8(2), 649–665, 2015.
108. Z. Wang, et al. Nd^{3+}-sensitized $NaLuF_4$ luminescent NPs for multimodal imaging and temperature sensing under 808 nm excitation. *Nanoscale*, 7(42), 17861–17870, 2015.
109. S. Karthi, G.A. Kumar, D.K. Sardar, C. Dannangoda, K.S. Martirosyan, and E.K. Girija. Luminomagnetic Nd^{3+} doped fluorapatite coated Fe_3O_4 nanostructures for biomedical applications. *J. Am. Ceram. Soc.*, 102.5, 1–11, 2018.
110. S. Karthi, G. Suresh Kumar, G.A. Kumar, D.K. Sardar, S. Chidangil, and E.K. Girija. Microwave assisted synthesis and characterizations of near infrared emitting Yb/Er doped fluorapatite NPs. *J. Alloys. Compd.*, 689, 525–532, 2016.
111. S. Karthi, G.A. Kumar, D.K. Sardar, C. Santhosh, & E.K. Girija. Synthesis and characterization of Nd^{3+}:Yb^{3+} co-doped near infrared sensitive fluorapatite NPs as a bioimaging probe. *Opt. Mater.*, 77, 39–47, 2018.
112. C. Mi, J. Zhang, H. Gao, X. Wu, M. Wang, Y. Wu, Y. Di, Z. Xu, C. Mao, and S. Xu. Multifunctional nanocomposites of superparamagnetic (Fe_3O_4) and NIR-responsive rare earth-doped up-conversion fluorescent (NaYF4:Yb, Er) nanoparticles and their applications in biolabeling and fluorescent imaging of cancer cells. *Nanoscale*, 2, 1141, 2010.
113. P. Moroz, S.K. Jones, and B.N. Gray. Magnetically mediated hyperthermia: current status and future directions. *Int. J. Hyperthermia,* 18, 267–284, 2002.

114. N. Seppa. Microwavable cancers: heat plus radiation shrinks some tumors. *Sci. News*, 167, 294, 2005.
115. K.S. Martirosyan. Thermosensitive magnetic nanoparticles for self-controlled hyperthermia cancer treatment. *J. Nanomed. Nanotechol.*, 3, 6, 2012.
116. P. Burgman, A. Nusenzweig, G.C. Li, C.C. Vernon, S.K. Sahu, et al. Thermoradiotherapy and thermochemotherapy. *Biol. Physiol. Phys.*, Vol. 1. New York: Springer, 1995.
117. H.H. Kampinga. Cell biological effects of hyperthermia alone or combined with radiation or drugs: a short introduction to newcomers in the field. *Int. J Hyperthermia*, 22, 191–196, 2006.
118. B.E. Dayanc, S.H. Beachy, J.R. Ostberg, and E.A. Repasky. Dissecting the role of hyperthermia in natural killer cell mediated anti-tumor responses. *Int. J. Hyperthermia*, 24, 41–56, 2008.
119. G. Van der Heijden, L.A. Kiemeney, O.N. Gofrit, O. Nativ, A. Sidi, et al. Preliminary European results of local microwave hyperthermia and chemotherapy treatment in intermediate or high risk superficial transitional cell carcinoma of the bladder. *Eur. Urol.*, 46, 65–71, 2004.
120. H. Nakano, K. Kurihara, M. Okamoto, S. Toné, and K. Shinohara. Heat-induced apoptosis and p53 in cultured mammalian cells. *Int. J. Radiat. Biol.*, 71, 519–529, 1997.
121. A. Ito, H. Saito, K. Mitobe, Y. Minamiya, N. Takahashi, et al. Inhibition of heat shock protein 90 sensitizes melanoma cells to thermosensitive ferromagnetic particle-mediated hyperthermia with low Curie temperature. *Cancer Sci.*, 100, 558–564, 2009.
122. Z. Hedayatnasab, F. Abnisa, and W.M.A.W. Daud. Review on magnetic nanoparticles for magnetic nanofluid hyperthermia application. *Mater. Design*, 123(5), 174–196, 2017.
123. M. Bañobre-Lópeza, A. Teijeirob, and J. Rivasa. Magnetic nanoparticle-based hyperthermia for cancer treatment. *Rep. Pract. Oncol. Radiother.*, 18(6), 397–400, 2013.
124. A.J. Cole, V.C. Yang, and A.E. David. Cancer theranostics: the rise of targeted magnetic nanoparticles. *Trends Biotechnol.* 29(7), 323–332, 2011.
125. N. Ohannesian, C.T. De Leo, and K.S. Martirosyan. Dextran coated superparamagnetic iron oxide nanoparticles produced by microfluidic process. *Mater. Today Proc.*, 13, 2, 397–403, 2019.
126. G.F. Goya, E. Lima Jr, A.D. Arelaro, T. Torres, H.R. Rechenberg, L. Rossi, C. Marquina, and M. Ricardo Ibarra. Magnetic hyperthermia with Fe_3O_4 nanoparticles: the influence of particle size on energy absorption. *IEEE Trans. Magn.*, 44(11), 4444–4447, 2008.
127. E. Garaio, J.M. Collantes, J.A. Garcia, F. Plazaola, S. Mornet, F. Couillaud, and O. Sandre. A wide-frequency range AC magnetometer to measure the specific absorption rate in nanoparticles for magnetic hyperthermia. *J. Magn. Magn. Mater.*, 368, 432–437, 2014.

10 Comparison of Optical, Electrical, and Magnetic Properties of Co- and Fe-Doped Nanosturctured ZnO Media

Parameswar Hari[a], *Amrit Kaphle*[b], *and Ganga R. Neupane*[b]

[a]The University of Tulsa and the Oklahoma Photovoltaic Research Institute, Department of Physics, Tulsa, Oklahoma, USA

[b]The University of Tulsa Department of Physics, Tulsa, Oklahoma, USA

CONTENTS

10.1 INTRODUCTION

Nanostructured ZnO is the focus of intense investigation due to its application in fabricating optoelectronic devices and sensors.[1] In addition, ZnO is a material of interest in designing nanostructured photovoltaic cells due to its high chemical, thermal, and physical stability; large exciton-binding energy; and high electrical conductivity.[2] Doping of nanostructures fabricated from ZnO is another area of study. In order to investigate doping in ZnO nanostructures, Group III components such as B,[3] Al,[4] and Ga[5] can be utilized to enhance the optical properties of ZnO nanoparticles (NPs). Doping can be used to tune the band gap of ZnO, which in turn influences the optical properties of ZnO NPs.[6,7] ZnO NPs can be grown utilizing a wide range of experimental techniques, such as chemical vapor deposition,[8] hydrothermal methods, and chemical bath deposition methods.[8]

Cobalt (Co) is an important dopant which has been studied extensively[9,10] due to the magnetic and electrical properties of bulk and thin film ZnO. However, there are only a few reported studies on the structural, optical, magnetic, and optoelectronic properties of Co-doped nanostructured ZnO. There are even fewer studies on the optoelectronic parameters, the index of refraction, and the dielectric constant of Co-doped ZnO nanostructured films.[11] In addition, there are conflicting reports in the literature on the effect of Co doping on the band gap of ZnO.[12,13] This may be due to structural defects in the ZnO crystal lattice such as oxygen vacancies or zinc interstitials, which are crucially dependent on the method of deposition.[14] The impurities that arise while growing the ZnO affect the bond length, charge transfer, and band structure of ZnO nanostructures in general. Incorporation of impurities during deposition makes it very difficult to obtain reproducible and reliable performance from devices made of nanostructured materials.

Iron (Fe) is another important dopant that has been studied widely due to its structural, optical, electrical, and magnetic properties in bulk and thin film Fe-doped ZnO.[15] Doping with Fe enhances electron–hole pair separation by decreasing the band gap, which in turn helps to shift the absorption to the visible light spectrum. Fe-doped ZnO exploits the spin of an electron associated with its charge for potential spintronic applications. Due to the observation of room temperature ferromagnetism (RTF), several groups have devoted their research activities in this transition metal-doped ZnO. However, there is still controversy about the origin of RTF in Fe-doped ZnO. There are reports on RTF arising from the formation of secondary $ZnFe_2O_4$ phase, whereas other studies indicated that the presence of defects such as vacancies might be responsible for getting magnetic order in Fe-doped ZnO.[16] Materials with high dielectric constants have also received special attention because of their application in telecommunication, satellite broadcasting, and low loss substrate for microwave-integrated circuits. There are various ways to modify the dielectric properties of ZnO using several additives such as Fe. When ZnO is doped with Fe, the dielectric properties are changed by extrinsic defects.[17]

ZnO can be fabricated as NPs of varying morphologies. Like other semiconductor NPs, ZnO has size-dependent properties. ZnO NPs have attracted lots of interest in photovoltaic applications and medicine due to their tunability. ZnO NPs have different properties than those of bulk structures such as thin films. These unique properties are of interest in numerous applications such as light-emitting devices, fluorescent probes for cellular imaging, and lasers. Semiconductor materials with large band gap composed of *n*-type semiconductor have drawn lots of attention in third-generation photovoltaic cell design.

One of the most interesting properties of NPs of few nanometers (typically <50 nm) in size is that their physical properties are dominated by the confinement effect of excitons. ZnO NPs <50 nm are widely studied because they possess significant conductivity and exhibit quantum confinement. Size-dependent variation in color is another important property of ZnO NPs <50 nm.

The inhomogeneous structure of the NPs affects ensemble-averaged properties in different ways.[18] Defects in NPs caused by uneven cation–anion proportion and surface vacancies significantly influence the optical characteristics of NPs in general. Recently, studies have been performed on the physical, electrical, and optical properties of transition-metals-doped NPs.[19] The significant advantages of ZnO NPs are due to their size-dependent optical properties, and their size can be altered by changing the deposition time and annealing temperature during preparation.[20]

There are several methods for synthesizing ZnO NPs, including the aqueous solution method, sol–gel method, hydrothermal method, vapor deposition method, precipitation in microemulsions, and mechanical–chemical processes. In this chapter, we focus mainly on the synthesis of ZnO by a precipitation method and microwave-assisted method. In the following sections, we will outline the synthesis and characterization of Co- and Fe-doped ZnO NPs prepared by precipitation and the microwave-assisted method.

10.2 SYNTHESIS OF Co-DOPED ZnO NPs

ZnO is a semiconductor with a wide band gap (3.37eV) and a large exciton-binding energy (60 meV).[21] It was found that doping ZnO with Co strongly enhances structural, optical, electrical, and magnetic properties.[22] Several methods have been used for the synthesis of ZnO with Co. Hays et al.,[23] Arshad et al.,[24] and Park et al.[25] used the sol–gel technique for synthesizing Co-doped ZnO. Vanaja et al.[26] also followed the same sol–gel technique for synthesizing Co-doped ZnO. They added a 0.5-M concentration of Co acetate into the zinc solution before adding a potassium hydroxide solution for obtaining Co-doped ZnO NPs. Gandhi et al.[27] used the coprecipitation method to synthesize Co-doped ZnO. They mixed Co acetate in water with zinc acetate solution separately; then, the required amount of aqueous NaOH was added dropwise to the mixture to get white precipitate with a pale pink color and repeated the procedure for obtaining Co-doped ZnO NPs. Kripal et al.[28] and Nair et al.[29] also used the co-precipitation method for synthesizing Co-doped ZnO NPs. Carrero et al.[30] used an auto combustion technique, and Elilarassi et al.[31] used the sol-gel auto combustion technique for the synthesis of Co-doped ZnO NPs. Martinez et al.[32] used the vaporization-condensation method in a solar reactor. Samples of Co-doped ZnO (2%, 5%, and 10%) have been prepared with different pressure conditions (low pressure of 10 Torr and high pressure of 70–100 Torr) inside the evaporation chamber. The pressure inside the chamber was controlled by pumping with a rotatory pump. Other techniques such as the polymerizable precursor method by Maensiri et al.[33] were also reported. They used nitrate salts of zinc and Co and a mixed solution of citric acid and ethylene glycol as a chelating agent and reaction medium, respectively. Microwave solvothermal method by Wojnarowicz et al.[34] and a simple solution route synthesis technique followed by Singh et al.[35] for the synthesis of Co-doped ZnO were also reported. We will discuss the precipitation method and the microwave method in detail in the following sections.

10.3 SYNTHESIS OF Fe-DOPED ZnO NPs

Doping ZnO with Fe has also shown a strong influence on structural, optical, electrical, and magnetic properties.[36,37] A wide range of techniques was used to synthesize Fe–ZnO NPs. Yu et al.,[38] Zhang et al.,[40] and Ashraf et al.[39] used a sol–gel method for synthesis of Fe-doped ZnO NPs. Sharma et al.[40] used the co-precipitation method to synthesize Fe–ZnO NPs. In this study, zinc nitrate and Fe nitrate were dissolved in 100 ml of methanol with continuous stirring for 2 h at room temperature. They also separately prepared 140 mmol KOH solution and mixed them with the prepared solution by continuous stirring for 2 h. Finally, after cooling the solution to room temperature, the solution was centrifuged and washed several times with ethanol, and the final product was placed in a vacuum oven for 24 h at 50°C to obtain powders. Shu et al.[41] also used similar co-precipitation methods. They used zinc nitrate, ferric nitrate, and urea as precursors. In their process, they separately dissolved nitrates and urea in de-ionized and adjusted the pH equal to 9 for the mixture. Finally, after heating at 100°C for 2 h, precipitates were centrifuged and washed with de-ionized water and ethanol several times and dried in air at 100°C to get powder. Srinivasulu et al.[42] used chemical spray pyrolysis. Glaspell et al.[43] used microwave method to prepare Fe-doped ZnO. Elilarassi et al.[44] used ball milling with different milling time to prepare Fe-doped ZnO. Table 10.1 summarizes various deposition techniques in fabricating Fe- and Co-doped ZnO NPs.

10.4 PRECIPITATION METHOD

Co-precipitation is one of the widely used methods for synthesizing ZnO NPs. For the synthesis of Co- and Fe-doped ZnO NPs, the precipitation method is widely used because it is cost effective and requires relatively low temperatures for processing. With this method, a high degree of solubility can be achieved.[58] This method is widely used for controlling the shape and size of NPs. He et al.[59] used this synthesis technique for doping Co (up to 5%). The materials used

TABLE 10.1
Techniques for Synthesis of Co- and Fe-Doped ZnO NPs

Method	References	Summary of Findings
Sol–gel technique	23, 25, 26, 45	Co-doped ZnO NPs
	46–48	Fe-doped ZnO NPs
Co-precipitation	27, 49, 50	Co-doped ZnO NPs
	51, 52	Fe-doped ZnO NPs
Auto combustion	53	Co-doped ZnO NPs
	54	Fe-doped ZnO NPs
Sol-gel auto combustion	55	Co-doped ZnO NPs
	54	Fe-doped ZnO NPs
Vaporization condensation	32	Co-doped NPs
	56	Fe-doped NPs
Polymerizable precursor	33	Co-doped NPs
Microwave solvothermal	34	Co-doped NPs
Simple solution route	57	Fe-doped NPs
	35	Co-doped NPs

were Zn $(CH_3COO)_2.H_2O$, Na_2CO_3, and $CoCl_2.6H_2O$. They first prepared a mixed solution of Zn $(CH_3COO)_2.H_2O$ and $CoCl_2.6H_2O$ in distilled water, and the mixed solution was added into the Na_2CO_3 solution dropwise. After aging, they washed slurries several times with distilled water using centrifuge until the salinity of the supernatant became <100 ppm. Samples were dried and heated at 350°C to decompose $Zn_xCO_{1-x}CO_3$ into $Zn_xCO_{1-x}O$. After structural analysis, they found NPs of size <50 nm in diameter. Nair et al.[50] found the seed like morphology of Co-doped ZnO NPs using this technique. They prepared a precursor solution by the reaction of Zn^{2+}, CO^{2+}, and OH^- in an alcoholic medium (methanol). They found an increase in crystallite size as the concentration increased compared to undoped ZnO. A similar result was obtained by Udayakumar et al.,[60] where the size of NPs increased from 18 to 34 nm. Chithra et al.[61] observed the formation of NPs with Co doping by the preparation method and found an increase in particle size on doping. They estimated the size of particles to be 29.1 nm for 1% Co doping and 41.6 nm for 3% doping. This increase in size reveals the presence of Co in their prepared samples. However, Devi et al.[62] used the precipitation technique and found spherical-shaped NPs of size 52 nm. Compared to the size of pure ZnO NPs, Co-doped ZnO NPs had a smaller size. Chemicals such as zinc nitrate hexahydrate, Co nitrate hexahydrate, sodium hydroxide, and polyethylene glycol were used as the precursors in the precipitation method. They explained that this reduction in size is due to the distortion in ZnO host by Co ion. In addition, with this precipitation method, Godavarti et al.[63] observed the formation of spherical NPs with a reduction in particle size (around 20 nm) compared to pure ZnO NPs (around 30 nm). They also explained that this reduction in size was because of the incorporation of Co ion in the ZnO host matrix. Rath et al.[64] found Co-doped NPs with a reduction of crystallite size using the same technique. Uses of metal nitrate solutions and NaOH as precipitant were reported by this group. Similarly, Gandhi et al.[27] found a decrease in average crystallite size as the Co content increased using this preparation method. This reduction was mainly attributed to distortion in the host of ZnO lattice by CO^{2+} ion. Bhatt et al.[65] synthesized NPs having a grain size of 17 nm when doped with Co using a chemical co-precipitation method. Physical structure and optical properties of Co-doped ZnO NPs were studied by He et al.[66] using a similar co-precipitation method. The concentration up to 5% was found to incorporate Co ion in ZnO structure. Pure and Co-doped ZnO NPs were also prepared by Kalpana et al.[67] by mixing zinc sulfate heptahydrate (0.6 M) and sodium hydrogen carbonate (0.6 M) as starting material with the co-precipitation method. Also, a direct precipitation method using wet chemical reaction was used by Rana et al.[68] to synthesize pure and Co-doped ZnO NPs. This study used zinc

TABLE 10.2
Precipitation Method with the Summary of Findings

Method	References	Summary of Findings
Co-precipitation	59	NPs of size <50 nm in diameter
	71	NPs with particles morphology of different size
Co-precipitation	50	Seed like morphology. Crystallite size increases as Co concentration increases
	72	Crystallite size decreases as Fe concentration increases
Precipitation	60	NPs have a size ranging from 18 to 34 nm
		NPs have a size ranging from 25 to 36 nm
Co-precipitation	62	Spherical-shaped nanoparticles of size 52 nm
	40	Fe-ZnO NPs of mean size 3-10 nm
Co-precipitation	63, 64	Spherical Co-doped ZnO nanoparticles with size around 20 nm
Precipitation	61	Increase in particle size. 29.1 and 41.6 nm for 1% and 3% Co doping
	69	Increase in crystallite size to 90 nm for 5% Fe doping, and then size decreases to 37 nm for 20% Fe doping
Co-precipitation	27	Decrease in crystalline size with increase in Co content. 33, 31, and 29 nm for 5%, 10%, and 15% doping
Co-precipitation	65	NPs having grain size of 17 nm
Co-precipitation	66	Nanoparticles size ranging from 10 to 50 nm
Co-precipitation	67	Nanostrips at a lower concentration of Co dopant and changes cauliflower like ZnO
	126	NPs with decreasing crystallite size as Fe concentration increases up to 15%
Direct precipitation	68	NPs calcinated at 500°C were homogeneous and agglomerated with grain size 80 nm and particles calcinated at 700°C were more than 100 nm

nitrate and sodium hydroxide as precursors. Pure metal was dissolved in concentrated nitric acid to obtain aqueous of metal nitrates. A known quantity of NaOH was mixed with deionized water, and the solution was mixed with the zinc nitrate solution to form a white precipitate. This precipitate was then washed several times with deionized water and ethyl alcohol to remove impurities. The final powder was then dried in an oven at 100°C for 15 h, and the powder was grounded and calcinated at different temperatures. The obtained powder was then analyzed, and it was observed that the dopant element 'Co' influenced the particle size.

Fabbiyola et al.[69] used the co-precipitation method to prepare Fe-doped ZnO NPs. Fe concentration was used up to 2%, and they found remarkable morphologies. Chemicals such as zinc nitrate, Fe nitrate, and sodium hydroxide were used. Raja et al.[69] found that particle size varied from 26 to 37 nm with spherical shape with large agglomeration when ZnO was doped with Fe up to 9% using the co-precipitation method. They observed a decrease in NP size with increasing Fe concentration. Kumar et al.[70] also used the same technique to prepare samples and studied ferromagnetic behavior. They used zinc nitrate and Fe nitrate dissolved in de-ionized water to obtain a 0.6 M solution. The solution was stirred for 1 h at 25°C, and sodium hydroxide was added drop wise to get a pH value of 9. The solution was continuously stirred for 3 h at room temperature and filtered. Finally, the mixture was dried at 800°C for 15 h and annealed at 500°C for 3 h to get the samples. In our study, we prepared Fe-doped ZnO NPs using the co-precipitation method. We found that particle size decreased as Fe concentration increased by up to 15% and ranged from 87 to 55 nm. Zinc acetate, Fe nitrate, and sodium hydroxide were used as a precursor solution. Zinc acetate was mixed with de-ionized water and stirred continuously with a magnetic stirrer. Sodium hydroxide was then added drop wise to make the pH of the solution 11. The solution was heated at 70°C for 2 h, and it was centrifuged, washed several times, and then dried in an oven to get powder. Table 10.2 summarizes various precipitation methods for synthesizing Co- and Fe-doped ZnO NPs.

10.5 MICROWAVE METHOD

Among other deposition methods for fabricating metallic-doped ZnO, the microwave method is the fastest synthesis method. The microwave oven had already been widely adopted in the 1980s for the synthesis of the organic compound.[34] In conventional heating, heat is transferred to the material by convection process, which leads the non-uniform temperature distribution between bulk material and surface. In contrast, microwave heating results in a uniform distribution of temperature and fast formation of NPs. Glaspell et al.[43] used the conventional microwave to synthesize Co-doped ZnO NPs. They set the power of microwave to 33% of 650 W and operated in 30-s cycles (on for 10 s and off for 20 s) for 10 min. Badhusha[73] also prepared Co-doped ZnO (1%, 3%, and 5%) NPs using microwave synthesis. They put the solution in LG microwave oven at a power of 300 W for 30 min and left the solution to cool at room temperature, and the resulting precipitate was centrifuged, washed, and kept in the oven to dry. Finally, the obtained powder was analyzed and found to be spherical NPs with decreasing grain size as the Co concentration increases. Wojnarowicz et al.[34] also used this method for the synthesis of Co-ZnO (1%, 5%, 10%, and 15%). The reaction was conducted in a Teflon vessel in a Magnum II reactor (Ertec, Poland) at 220°C. The reaction duration time for all experiments was 25 min under the constant pressure of 0.1 MPa at a microwave power of 600 W. They observed the formation of NPs, which was confirmed by X-ray diffractometry (XRD) analysis and scanning electron microscopy (SEM) images. Co-doped ZnO (5%, 10%, and 15%) NPs synthesis was also done by Bhatti et al.[74] using the microwave method. The reaction was not heated continuously in the microwave; rather, it was heated nonspontaneously with gaps of 10 s until the reaction was completed and found the NPs had porous morphology. Varadhaseshan et al.[75] synthesized Co-doped ZnO NPs for the study of structural and magnetic properties. They found that the average crystallite size of NPs was nearly 25 nm. Good crystalline quality and good surface morphology $Zn_{1-x}Co_xO$ nanopowders were obtained by Hoang et al.[76] as the Co was doped with ZnO and showed that Co doping up to 5%, Co^{2+} ion is successfully incorporated into ZnO host matrix. Mary et al.[77] synthesized pure and Co-doped ZnO by microwave combustion technique using urea as fuel and obtained NPs having an average size of 47–23 nm. Basu et al.[78] used microwave-assisted nonaqueous sol–gel route for synthesis and obtained Co-doped ZnO NPs with crystal sizes of 8–9 nm.

Kwong et al.[79] synthesized undoped and Fe-doped ZnO by a microwave-assisted method without any thermal calcination treatment. They used a similar protocol reported by Han et al.[80] This study used zinc acetate and Fe nitrate in deionized water to make the solution and heated the solution in the microwave at a fixed temperature of 80°C with a constant microwave power of 400 W for 30 min. Bhatti et al.[81] also prepared Fe–ZnO NPs using the microwave method. They observed NPs of uniform size, shape, and morphology. This study used zinc nitrate, citric acid, and ferric nitrate to prepare Fe-doped ZnO NPs. One molar zinc nitrate was dissolved in 25 ml distilled water and 10% 1 M ferric nitrate was dissolved in other 25 ml water with citric acid. After a few minutes of stirring, the mixture was put into the microwave at a power of 650 W. It was heated in cycles of 10-s gaps until the solution became brownish color. Our group[126] also synthesized Fe–ZnO NPs using the microwave method. Zinc acetate, Fe nitrate, and sodium hydroxide were used as precursors. Zinc acetate and Fe nitrate were dissolved in ethanol separately and stirred with magnetic stirred for 10 min. We then mixed the solution, and sodium hydroxide was added dropwise to make pH 11 of the solution. The solution was then kept in the microwave for heating. After heating solution in the microwave with a constant power of 210 W for 2 min, the solution was centrifuged and washed several times with deionized water and finally precipitated solution was dried in oven to obtain Fe-doped ZnO powder. The obtained powder was analyzed using transmission electron microscopy, and NPs with a size range from 11 to 17 nm were found. Table 10.3 summarizes the various microwave studies to synthesize Co- and Fe-doped ZnO NPs.

TABLE 10.3
Microwave Method with a Summary of Result

Method	Reference	Summary of Findings
Microwave	43	Single-phase ZnO NPs without any indication of dopant
Microwave	73	Spherical Co-doped NPs having size 20–30 nm
	82	Triangular-shaped Fe-doped NPs having size 9 nm
Microwave solvothermal	34	Co-doped NPs had an average diameter ranging from 30 to 70 nm before annealing, and after annealing at 800°C, particles size ranged from 30 nm to 2 μm
Microwave	74	Morphology of Co-doped particles was porous. Pores in 15% doped sample were large in comparison to 5% doped sample
Microwave	75	The average crystallite size of NPs was nearly 25 nm
Microwave	76	Co-doped NPs
	83	ZnO nanostructures
Microwave	77	Co-doped ZnO NPs having an average size of 47–23 nm
Microwave	78	Co-doped ZnO NPs with crystal sizes of 8–9 nm

10.6 XRD ANALYSIS

Co- and Fe-doped ZnO NPs can be characterized by X-ray diffraction. This analysis is used for finding crystallite size, dislocation density, stress, and strain. Various studies have found hexagonal wurtzite structure for Co- and Fe-doped ZnO NPs. Chithra et al.[61] investigated Co-doped ZnO NPs by XRD pattern and found well-pronounced diffraction peaks corresponding to (100), (002), (101), (102), (110), (103), (200), (112), and (201) planes indicating wurtzite structure. All the peaks matched well with the JCPDS data card No. 36-1451. It was also noticed that the intensity of (100) and (101) peak increases with doping, whereas the intensity of (002) peak decreases with doping. This is indicative of Co is incorporated into the ZnO structure. Particles sizes were found to be increasing as the doping concentration increases. The decreasing trend of dislocation density shows that the higher concentration of Co leads to a reduction in imperfection. Also, microstrain was found to decrease as the Co content increased.

El Ghoul et al.[84] investigated prepared NPs with XRD and found wurtzite single crystallite phase without a secondary phase, which indicates that Co substituted Zn ions. Lima et al.[85] found the Co-doped ZnO NPs had hexagonal wurtzite structure with 10 mol% dopant with no evidence of any secondary phase. The average crystallite size of the samples was in the range of 25–50 nm. Pal et al.[86] also doped 3% and 5% Co with ZnO using a mechanical ball milling process, and from the XRD analysis, they confirmed wurtzite-type structure with no secondary phase. They also found one intense peak of around 44°, but with Co doped, there was no peak, indicating that Co ion successfully was incorporated into the lattice site rather than as an interstitial state. Co doping was observed to change the XRD pattern in doped samples because of size reduction and induced lattice strain. XRD analysis was also done by Castro et al.,[87] who synthesized Co-doped ZnO (0%, 0.5%, 3%, 5%, and 10%).

This study also observed a ZnO wurtzite structure from the XRD analysis. For higher Co-content (5% and 10%), a weak extra feature was observed. This peak was assigned to a secondary Co_3O_4 phase. Doan et al.[88] also observed XRD spectra for all Co-doped samples as shown in Figure 10.1. They observed no shift in peaks among the samples. In addition, they observed the appearance of a peak that corresponds to Co_3O_4 phase besides the peaks related to the wurtzite phase. The width of the peaks was enhanced for 4% Co-doped ZnO sample as shown in Figure 10.2. Many other groups, including Muthukumaran et al.,[89] Dinesha et al.,[90] Rath et al.,[64] and Maensiri et al.,[33] observed a single-phase wurtzite structure with no other secondary peaks from the XRD analysis, confirming the incorporation of Co ion in ZnO. Bhattacharyya et al.[91] characterized Co-doped ZnO using XRD

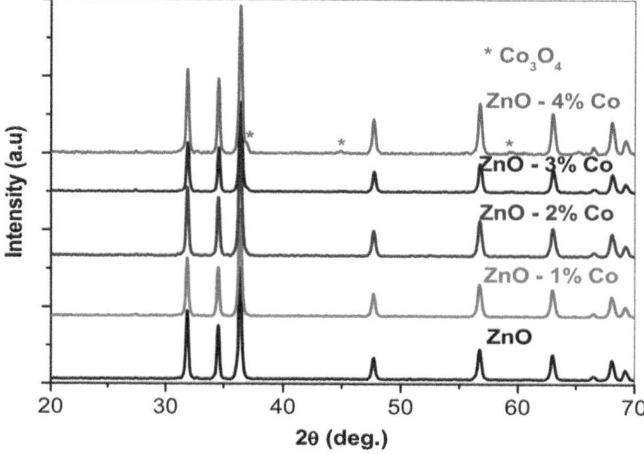

FIGURE 10.1 XRD analysis of Co-doped ZnO. *Source*: "Reprinted from Ref.[88], Copyright (2014), with the permission of IOP Publishing."

and observed hexagonal ZnO phase, which matched well with JCPDS card No. 89-0510. They did not find any other secondary impurity phase. In addition, they found the relative intensity of XRD peak varied with the increase in Co concentration, which confirms the incorporation of Co ion in the ZnO host matrix.

Ashraf et al.[39] investigated Fe-doped ZnO NPs with XRD and found hexagonal wurtzite structure of ZnO in all samples. They observed a shift in peak position to higher angles corresponding to the plane (101) and explained due to the difference in ionic radius of zinc and Fe ions. They also calculated the crystallite size of NPs using the Debye-Scherer formula. This study observed a decrease in crystallite size from 19 to 29 nm as the Fe concentration increased up to 10%. They also observed an increase in full width at half maxima and dislocation density from the XRD analysis with increasing Fe concentration. Vidhya et al.[92] also studied Fe-doped ZnO quantum dots structural properties by XRD analysis. This study observed cubic-hexagonal crystal nature with fine crystallinity. XRD

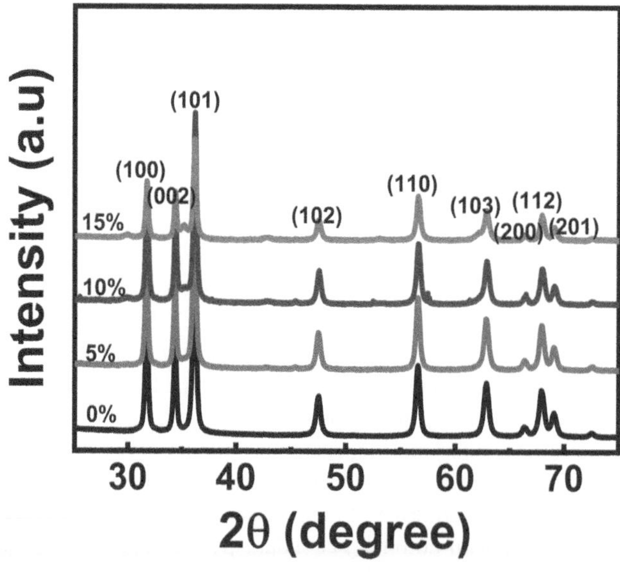

FIGURE 10.2 XRD spectra of Fe–ZnO NPs prepared by the precipitation method.

TABLE 10.4
Synthesis Techniques and XRD Measurement Results

Method	References	Results
Precipitation	61	Hexagonal wurtzite structure of NPs and reduction in dislocation density and microstrain as Co content increases
Sol–gel	84	Wurtzite single crystallite phase with no secondary phase
	93	Single phase with a hexagonal wurtzite structure with no undesired peaks
Sol–gel	85	Hexagonal wurtzite structure. The average crystallite size was in the range 25–50 nm
Co-precipitation	89	Different microstructure without changing a hexagonal wurtzite structure. Average crystallite size decreases from 22.24 to 15.93 nm for $x = 0$–0.04 then reaches 26.54 nm for $x = 0.06$
Co-precipitation	64	Single-phase wurtzite structure with no secondary phase
	71	Hexagonal wurtzite structure with good crystalline formation
Microwave	76	Co doping up to 5%; Co ion is successfully incorporated into the ZnO host matrix. Single-phase wurtzite structure
	94	Single phase of the cubic structure with no clear change in peak's position
Microwave solvothermal	34	Wurtzite structure; the density of powder after synthesis was nearly 5.2 g/cm^3; the average particle diameter was 31 nm; crystalline particle size was 26 nm

peak broadening was clearly observed in this study and confirmed the formation of small size particles. They also used Debye-Scherer formula to estimate the size of the crystal and found 10 nm. We have also studied Fe-doped ZnO NPs prepared by both precipitation and microwave method through XRD analysis.[126] Figure 10.2 shows the XRD spectra of Fe-doped ZnO NPs. Single-phase hexagonal wurtzite structure was observed in all samples. No secondary phases and impurity within the detection limit of XRD were observed. It was also observed that the peak (101) of doped samples prepared by both precipitation and microwave method shifts toward lower diffraction angle with respect to undoped ZnO, confirming the change in lattice parameters and cell volume. We also measured the crystallite size of NPs using the Debye-Scherer formula and observed a decrease in crystallite size with increasing Fe concentration up to 15%. Table 10.4 summarizes various synthesis methods of Co- and Fe-doped ZnO and XRD results of doped ZnO nanostructures.

10.7 PHOTOLUMINESCENCE STUDIES

For the optical properties of Co- and Fe-doped ZnO NPs, photoluminescence (PL) analysis was carried out at an excitation wavelength of 325 nm. Room temperature PL studies of Co-doped ZnO (0–4%) under 325 nm line excitation of a neon laser were done by Doan et al.[88] and found two emissions in the ultraviolet and visible regions. PL intensity of doped samples was found to decrease with increasing concentration from 0% to 3%, whereas the 4% sample was considerably enhanced in comparison to pure ZnO. The enhancement of PL intensity in 4% was explained by the introduction of Co_3O_4 phase. Chitra et al.[61] also compared undoped ZnO and Co-doped ZnO (1% and 3%) NPs samples and found that there was no trace of UV and violet emission in doped samples. It was also found that, due to Co doping, oxygen vacancies were created that produce blue and green emissions. In addition, there was a decrease in PL intensity with an increase in Co concentration, which was attributed to the nonradiative recombination process taking place. It was also confirmed that Co substitution in ZnO results in the absence of characteristic UV emission at 390 nm and the enhancement of the intensity of blue emission in 484–490 nm range.

Hammad et al.[95] studied PL of Co-doped ZnO (0%, 1%, 3%, 5%, and 7%), as shown in Figure 10.3. Co-doped ZnO UV emission peak exhibits a large blue shift from 410 to 390 nm compared to the pure ZnO emission peak. This blue shift was attributed to the shift in the optical band gap

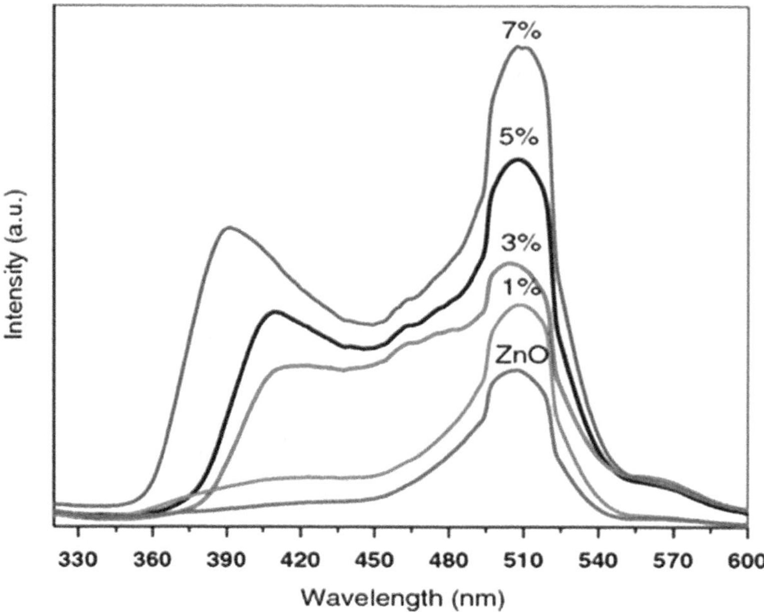

FIGURE 10.3 PL spectra of Co-doped ZnO NPs at room temperature. *Source*: "Reprinted from Ref.[95], Copyright (2013), with the permission of Springer."

in samples. This result confirmed the Co incorporation in the lattice site in ZnO. Devi et al.[62] found a red shift for Co-doped ZnO compared to pure ZnO, which was attributed to electron–phonon coupling, lattice distortion, and localization of charge carriers due to interface effects and points defects. Anand et al.[96] use excitation wavelength of 239 nm for 0.01 M% of Co-doped and found different peaks in PL spectra around 468 nm belonging to blue emission and the peak around 437 nm belongs to blue–green emission. In addition, there was another peak around 519 nm, which belongs to green emission. The bands at 340 and 396 nm correspond to defect-related emission and are believed to be caused by the transition from a level of ionized oxygen vacancies to the valence band. Room temperature PL spectra were also studied by Basith et al.[97] of undoped and Co-doped ZnO NPs (0.5 wt% Co, 1.0 wt% Co, 1.5 wt% Co, and 2.0 wt% Co) using 325 nm excitation wavelength. They observed two main peaks, a UV emission peak at 400 nm and a weak broad green emission peak in the visible region at 536 nm. When the Co was doped, they found the 400 nm UV peak diminished while the green peak around 536 nm became prominent.

Sathya et al.[47] studied PL properties of Fe-doped ZnO NPs at room temperature. They observed a dominant peak at 389 and 390 nm. These peaks in the UV region correspond to the near band edge (NBE) emission peak. Kumar et al.[52] also studied optical properties of Fe-doped ZnO NPs with PL analysis using an excitation wavelength of 325 nm. They observed a strong and wide PL signal for pure ZnO sample ranging from 423 to 850 nm and centered at 706 nm. For all other Fe-doped samples, they observed broad emission spectra ranging from 520 to 850 nm. In their observation, they found that the PL intensity gradually decreased with increasing Fe content. Wu et al.[98] also studied the optical properties of Fe-doped ZnO NPs at room temperature. They observed six main peaks, including three obvious broad bands centered at 390, 440, and 490 nm and three comparatively weak emission peaks at 420, 460, and 525 nm. In addition, they also observed the emission peak position of doped samples other than 1% Fe-doped ZnO sample exhibits a slight blue shift with increasing Fe content.

In our work,[126] we have also investigated optical properties of Fe-doped ZnO NPs prepared by both precipitation and the microwave method with PL characterization. All the PL spectra were

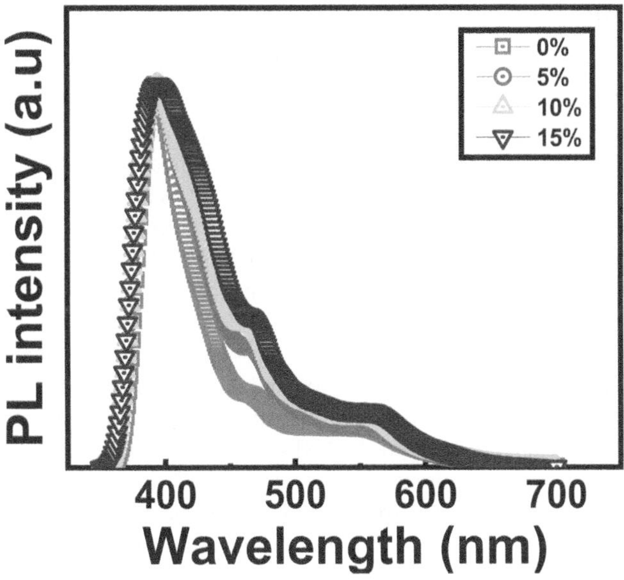

FIGURE 10.4 PL spectra of Fe-ZnO NPs prepared by microwave method.

recorded at room temperature for all samples. We used 325 nm as the excitation wavelength for our study. We observed three prominent peaks, one around 395 nm and the other two around 467 and 564 nm, as shown in Figure 10.4. The first peak was related to NBE emission peak, and the other two were defect-related peaks (oxygen vacancies or Zn interstitials, respectively). We also observed a blue shift in NBE peak for all doped samples with respect to pure ZnO NPs. For the samples prepared by the precipitation method, NBE peak shifted from 393.4 to 389.1 nm, and for the samples prepared from the microwave method, NBE peak shifted from 396.78 to 395.04 nm.

Table 10.5 summarizes PL results on Co- and Fe-doped ZnO prepared by the microwave and precipitation method.

TABLE 10.5
Different Synthesis Techniques with PL Study's Findings

Method	References	Findings
Precipitation	61	No trace of UV and violet emission results in blue and green emission due to oxygen vacancies, decrease in PL intensity with an increase in Co concentration
	99	The spectra of all the samples show well-defined sharp peaks both in the UV region and in visible region. PL study clears Fe doping favors for the violet emission in ZnO NPs
Simple chemical	95	The large blue shift from 410 to 390 nm, a weak UV emission at 405 nm, and very strong and broad emission at 508 nm
Co-precipitation	62	The red shift of PL peaks for Co-doped samples in comparison with pure ZnO
	69	A very strong and intense violet emission at 418 nm is seen for all undoped and doped Fe–ZnO
Chemical precipitation	96	Different PL spectra around 468 nm belonging to blue emission and peaks around 437 nm belonging to blue–green emission
	100	A relatively sharp, weak UV emission band at 3.23 eV and a broad stronger emission band in the green part of the visible spectrum are observed
Microwave-assisted	97	Two main peaks, UV emission peak at 400 nm and a weak broad green emission peak at 536 nm

10.8 SEM STUDIES

The morphology of ZnO nanostructure was studied by SEM. Size and distribution of Co- and Fe-doped ZnO NPs can be analyzed by this technique. Nair et al.[50] found that for Co-doped ZnO, all of the samples were nanometer in size, but there was no any significant change in grain size as the Co concentration increased. Particle size and morphology changes were also investigated by Basith et al.[97] for undoped ZnO, 0.5 wt%, and 2.0 wt% Co-doped ZnO samples. Large spherical aggregates of smaller individual NPs of various size were obtained. The average particles size was found to be in the range of 30–35 nm up to 0.5 wt% Co-doped ZnO and 37–42 nm up to 2.0 wt% Co-doped ZnO with a hexagonal shape. Rana et al.[101] observed that SEM images of Co-doped ZnO samples were homogeneous with dopant substituting Zn site in ZnO compound. High porosity in the material was confirmed by SEM image by Dinesha et al.[90] For 0.01 M% concentration of Co-doped ZnO, NPs were observed with small agglomeration and the distribution of about 6.2 nm, indicating the moderate distribution of NPs by Anand et al.[96]. And Ahmed et al.[102] used field emission SEM result and found that as prepared Co-doped ZnO NPs are nearly in spherical in shape with a particle size <50 nm. The morphologies of pure and 7% Co-doped ZnO (Figure 10.5a and b) were analyzed by Hammad et al.,[95] who found nearly spherical NPs with the homogeneous size distribution of ZnO nanocrystals. For the 7%, Co-doped ZnO microsphere was seen from the SEM image. These microspheres were porous with vacancies at their center. Vanaja et al.[26] reported cauliflower-like ZnO on the surface of ZnO and almost spherical shape morphology in Co-doped ZnO using SEM images.

Elilarassi et al.[44] reported SEM images of Fe-doped ZnO NPs as randomly distributed spherical NPs. Fabbiyola et al.[69] also studied structural properties of Fe-doped ZnO and found that the average

FIGURE 10.5 SEM micrographs of Co-doped ZnO NPs. (a) 0% and (b) 7% of Co. *Source*: "Reprinted from Ref.[95], Copyright (2013), with the permission of Springer."

FIGURE 10.6 SEM image of Fe-doped ZnO. *Source*: "Reprinted from Ref.[47], Copyright (2017), with the permission of Magnolithe GmbH."

size of particles in nanometer size. First, they observed spheroid like structure for pure ZnO and found transformation into nanosheet and nanoplates for 5% and 10% Fe doping, and with further doping, they observed the transformation to nanosphere. Shu et al.[41] also studied morphological features of undoped and Fe-doped ZnO NPs. They observed the irregular shape and lateral dimension in the range of 100–350 nm for undoped ZnO. When Fe was doped with ZnO, they found a significant change in morphology of ZnO. Sathya et al.[47] also did morphological studies of 1%, 3%, 5%, and 7% Fe-doped ZnO. They observed the morphology of NPs with spherical structure in the surface without any isolated grains or larger agglomerates, with nano crystallites revealing the polycrystalline nature as shown in Figure 10.6. A similar result was also observed by Raja et al.[103] Morphology of pure and Fe-doped ZnO NPs was studied by SEM. They observed particles in a spherical shape with large agglomeration for pure ZnO and particles size varying from 26 to 37 nm. Agglomeration was almost restricted in Fe-doped ZnO particles. Uniform shape and narrow size distribution with some porosity in the background portion were observed. Summary of findings on SEM studies of Co- and Fe-doped ZnO is provided in Table 10.6.

TABLE 10.6
Different Synthesis Techniques and Summary of Findings of SEM Analysis

Method	References	Summary of Results
Co-precipitation	50	Undoped ZnO grain size was slightly less than Co-doped ZnO
	104	Fe-doped ZnO NPs with strong agglomeration
Microwave-assisted	97	NPs had average particles size in the range of 30–35 nm up to 0.5% and 37–42 nm up to 2% with a hexagonal shape
Direct precipitation	101	Homogeneous NPs with an average particle size of about 40 nm
Chemical precipitation	96	NPs with small agglomeration having distribution of about 6.2 nm
Co-precipitation	21	Nano hollow rods and nanosheets transformed into nano spherical shape with random size distribution as Co content increases
	105	The average diameter of NPs varies from 80 to 230 nm with increasing annealing temperature

10.9 OPTICAL ABSORPTION STUDIES

Optical properties of Co- and Fe-doped ZnO NPs were studied using UV–vis absorption spectroscopy. This technique can be utilized to estimate the optical band gap of the material with various doping levels. Hammad et al.[95] studied absorption spectra for undoped and Co-doped ZnO and found an obvious blue shift of absorption edges from 361 to 340 nm as shown in Figure 10.7. The band gap energy for the sample was also investigated using the Tauc plot, and an increase in value from 3.33 eV (pure ZnO) to 4.13 eV (7% Co) was found. This increase in band gap or blue shift can be explained by the Burstein–Moss effect.[106] Kumar et al.[107] studied the absorption spectra, and using the Tauc plot, and they showed an increase in band gap due to Co doping. However, after irradiation, they found a decrease in the band gap. This may be due to the fact that ion irradiation has created some new localized energy states above the valence band. Optical absorption spectra were also analyzed by Chithra et al.[61] for 1% and 3% Co-doped ZnO NPs.

The absorption peaks at 345, 359, and 383 nm have been recorded for undoped, 1% doped, and 3% doped ZnO, respectively, with a corresponding energy gap of 3.59, 3.45, and 3.24 eV. This reduction in band gap with an increase in Co concentration was explained by s-d and p-d exchange interaction. Maswanganye et al.[108] also used UV–vis spectroscopy and analyzed the absorption spectra in Co-doped ZnO. For Co-doped ZnO, three absorption peaks were reported between 550 and 700 nm. The band gap energy was found to be 2.48 eV for 5% Co-doped ZnO. A red shift was observed between the undoped and doped ZnO NPs. Pal et al.[109] reported a blue shift in the band gap energy of Co-doped ZnO compared to the undoped ZnO. The shift in absorption peak was explained by the sp–d exchange between the ZnO band electrons and localized d electrons of doped Co^{2+} cation. For the Co-doped ZnO (2%, 4%, 8%, and 17%), Palacios et al.[110] observed shift and deformation of the peak of absorption. Band gap calculation was done using first principles ab initio calculations of the electronic band diagram and density of states. This calculation found a reduction in the electronic gap due to hybridization between the sp-d electrons of the host semiconductor and d electrons from

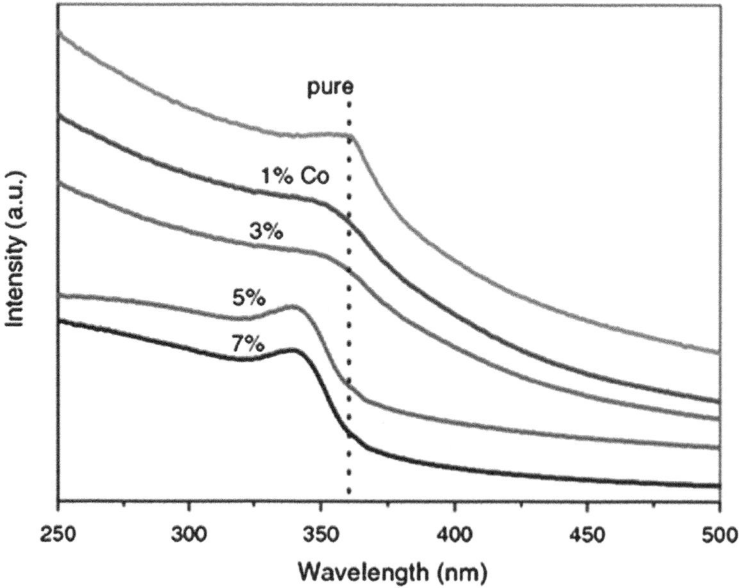

FIGURE 10.7 UV–vis spectra of Co-doped ZnO NPs at room temperature. *Source*: "Reprinted from Ref.[95], Copyright (2013), with the permission of Springer."

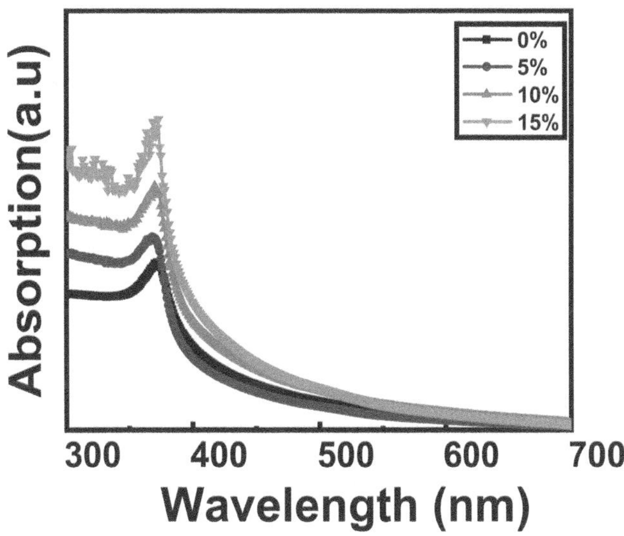

FIGURE 10.8 UV–vis absorption spectra of Fe-doped ZnO NPs.

the Co. This study also observed a shift and deformation in the features of optical absorption of Co-doped ZnO.

Elilarassi et al.[44] studied optical absorption properties of undoped and Fe-doped samples using UV–vis near-infrared spectroscopy. With different milling temperature, they observed a shift in the absorption edge toward higher wavelength (lower energy) for Fe-doped samples. They explained this shift as an indication of the incorporation of the Fe ion into the ZnO lattice. In addition, they also calculated the optical band gap of NPs by fitting the Tauc model. They found a band gap of 3.25 eV for pure ZnO NPs and band gap decrease to 3.05 eV for sample milled for 24 h. Hammad et al.[111] studied optical properties of pure and Fe-ZnO NPs by UV–vis absorption spectroscopy with different level of Fe concentration. They observed an obvious blue shift in absorption edges from 354 to 335 nm with an Fe doping level from 0 to 5%. They reported an increase in absorption in the region of 3.5–3.71 eV is due to the d–d transition of Fe ions. They also explained the blue shift on the basis of the Moss–Burstein theory. Since Fe-doped ZnO NPs are *n*-type semiconductors, the Fermi level will be inside of the conduction band. As electrons occupy the states below the Fermi level in the conduction band, the absorption edge should shift toward higher energy.

We have also studied the optical properties of Fe-doped ZnO (0–15%) NPs through absorption spectra using UV–vis spectroscopy at room temperature.[126] Prepared NP was dissolved in deionized water and sonicated for 10 min. The solution was kept in the sample holder and was measured in the wavelength between 300 and 700 nm. All absorption spectra are shown in Figure 10.8. We have observed a blue shift from 373 to 368 nm in the absorption edge with an increasing Fe concentration in all samples prepared from the precipitation method as shown in Figure 10.8. For the samples prepared from microwave method, we again observed a blue shift in the absorption edge from 356 to 349 nm. This blue shift can be attributed to the Moss–Burstein band filling effect. We have calculated band gap energy using the Tauc plot equation. From the Tauc plot, we found an increase in band gap energy from 3.13 to 3.19 eV in the samples (Fe-doped ZnO) prepared by the precipitation method and increase in band gap from 3.21 to 3.25 eV in the samples (Fe-doped ZnO) prepared by the microwave method. Table 10.7 summarizes the UV spectroscopy results of Co- and Fe-doped NPs by the microwave and precipitation methods.

TABLE 10.7
Different Synthesis Techniques and Results of UV Spectroscopy Analysis

Method	References	Summary of Findings
Co-precipitation	66	Incorporation of Co^{2+} created a new emission band at 1.85 eV but quenched the NBE luminescence
	103	Strong absorbance is found for wavelength below 364.21 nm of Fe-doped ZnO nanomaterials, while a very low absorbance in the visible region, the band gap, decreases with Fe concentration
Precipitation	61	The absorption peak at 345, 359, and 383 nm has been recorded for undoped, 1% doped, and 3% doped ZnO, respectively, with a corresponding energy gap of 3.59, 3.45, and 3.24 eV
Microwave assisted	76	Several absorption peaks at 660, 611, and 565 nm. Band edge energy red shifts with Co-doping
	81	Blue shift of absorption edge with Fe-doping. Band gap energy has a higher value than the pure ZnO particles
Precipitation	112	A systematic increase in band gap from 3.40 to 3.69 eV. Blue shift in the absorption peaks as the Co concentration increases from 0 to 15%

10.10 IMPEDANCE SPECTROSCOPY

Electrical properties of Co- and Fe-doped ZnO were studied by impedance analysis. Dielectric behavior of the particles can be analyzed over a wide range of frequencies. Generally, resistive and capacitive components of the electrical parameter are separated using this technique. Arshad et al.[113] studied Co-doped ZnO (0%, 1%, 3%, and 5%) using impedance spectroscopy. They observed a double semicircle for pure ZnO and a single semicircle for Co-doped ZnO in the cole–cole plots, which suggests the dominance of grain boundary resistance over grain resistance in the doped samples. It was also observed that the resistivity increases as Co concentration increases in Co-doped ZnO. In pure ZnO, grain size was maximum, and the size decreases as Co concentration increases. Ansari et al.[114] also used impedance spectroscopy technique to characterize the electrical properties of Co-doped ZnO (0%, 1%, 3%, 5%, and 10%). They studied the variation of the resistive part of impedance (Z') with frequency and found that Z' decreases with an increase in frequency for all compositions. It was also observed that Z' has a strong frequency dependence in the lower region and has no dependence on the higher frequency region. They observed a single circular arc corresponding to pure and Co-doped ZnO, suggesting the dominance of grain boundary resistance in all samples. Rostami et al.[115] also performed electrochemical studies of ZnO and Co-doped ZnO NPs. They immersed steel samples in solutions containing ZnO and Co-ZnO NPs for different time (2, 4, and 24 h). The samples dipped in the solution containing Co-ZnO extract showed semicircles in Nyquist plots. In addition, two relaxation times were observed for these samples. Our previous study[112] on Co-doped ZnO NPs (0%, 5%, 10%, and 15%) found resistance decreases as the Co content increases, as shown in Figure 10.9.

Denisha et al.[116] studied dielectric properties of Fe-doped ZnO samples at room temperature using an impedance analyzer with a Kelvin fixture over the frequency range of 1 to 5 MHz. They observed that dielectric loss decreased substantially with increasing frequency and reached nearly a constant value for each sample. The dielectric constant, dielectric loss, and dielectric loss factor values were found to decrease with the increase in Fe concentration and also with an increase in frequency. This behavior was attributed to different hopping mechanism and defects formed during synthesis. Cherifi et al.[116] also studied electrical properties of Fe-doped ZnO samples using complex impedance analysis. They observed single circular arc typical behavior in all samples suggesting the dominance of grain resistance. In addition, they studied the temperature dependence of complex

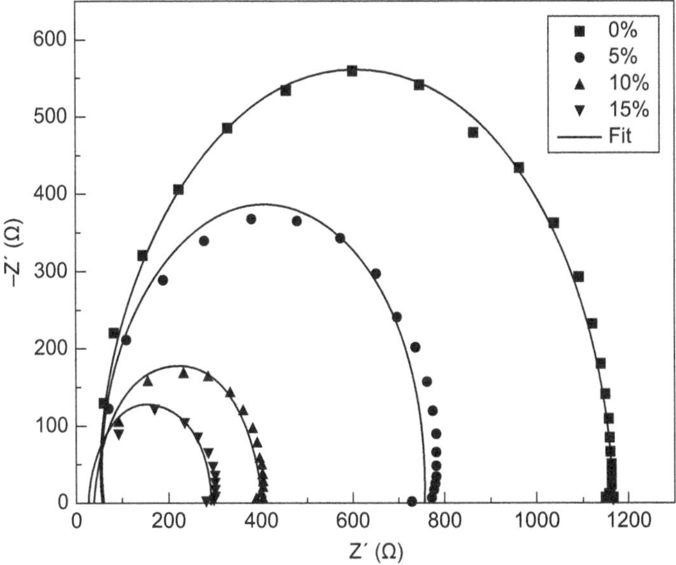

FIGURE 10.9 Cole–cole plot of Co-doped ZnO QDs. *Source*: "Reprinted from Ref.[112], Copyright (2016), with the permission of One Central Press."

impedance spectra (Nyquist plot) of Fe-doped ZnO over a wide range of frequency from 1 kHz to 1 MHz. They found a distinct effect on the impedance characteristic spectra with temperature. At high temperature (100°C), they observed the second semicircle due to bulk and grain boundary conduction.

In our work, we have also studied the electrical properties of Fe-doped ZnO NPs prepared by the precipitation and microwave method through impedance spectroscopy.[126] We studied grain boundary resistance and grain boundary capacitance using the cole–cole plot technique. The cole–cole plot was analyzed by an RC parallel equivalent circuit. For Fe-doped ZnO (0–15%) samples, prepared by the precipitation method as shown in Figure 10.10, we observed that all the samples exhibit single

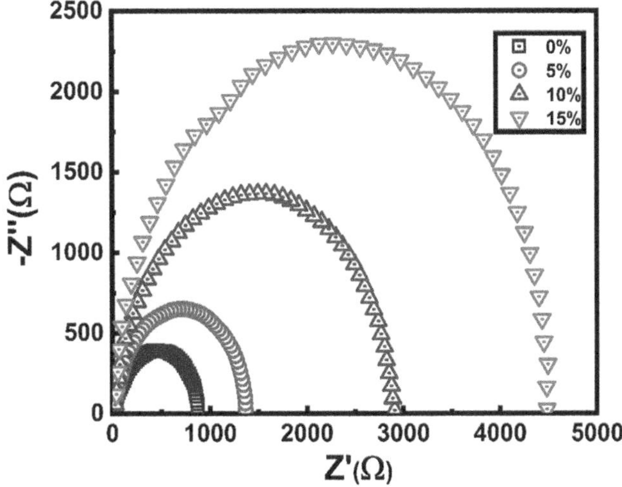

FIGURE 10.10 Impedance spectra of Fe–ZnO NPs prepared by precipitation method.

TABLE 10.8
Precipitation and Wet Chemical Route Method and Analysis of Impedance Spectroscopy Results

Method	References	Results
Precipitation	112	The decrease in diameter of a semicircle with an increase of Co content in cole–cole plot fitting. Series resistance and effective capacitance decrease with an increase in doping level
	41	Doping of Fe content greatly improves the dielectric properties of ZnO
Wet chemical route	117	Single semicircular arc appeared in all Co-doped samples. Semicircular arc is modeled by an equivalent circuit consisting of resistor and capacitor connected in parallel. The diameter of semicircular arc increases as the content increases

semicircular behavior, suggesting the dominance of grain boundary resistance over the grain resistance. Similar behavior was also observed in Fe-doped ZnO NPs prepared by microwave method. The diameter of the semicircle decreased as Fe content increased for samples prepared by both precipitation and microwave methods. Table 10.8 summarizes impedance studies on Fe- and Co-doped ZnO NPs prepared by precipitation and microwave methods.

10.11 MAGNETIC PROPERTIES

There have been several studies on the magnetic properties of Co- and Fe-doped ZnO NPs, but the explanation of room temperature ferromagnetism (RTFM) is still controversial. The origin of RTFM has been attributed to both intrinsic and extrinsic properties, while some have reported pure paramagnetic behavior at room temperature. Franco et al.[118] investigated magnetic properties of Co-doped ZnO (0%, 1%, 3%, 5%, 7%, and 9%) in the temperature range from 5 to 750 K for magnetic fields up to 85 KOe and found both ferromagnetic and paramagnetic behavior at room temperature. Also, the antiferromagnetic phase was observed for temperatures below 260 K. Castro et al.[119] also studied magnetic properties of Co-doped ZnO (0%, 0.5%, 1%, 3%, 5%, 7%, and 10%) and found the presence of long-range magnetic ordering in these samples. They found evidence of ferromagnetism in samples by analyzing magnetization as a function of magnetic field and temperature. Pure and Co-doped ZnO (5%, 10%, and 15%) were studied by Gandhi et al.,[27] who found a significant change in M–H loop resulting in change in diamagnetic behavior of pure ZnO to ferromagnetic behavior in Co-doped ZnO, as shown in Figure 10.11. Figure 10.11 clearly demonstrates the diamagnetic nature of pure ZnO (A). Ferromagnetic nature was observed when ZnO was doped with Co (5%, 10%, and 15%). Magnetization tends to increase with increasing Co concentration, as shown in the figure.

Hays et al.[23] also studied magnetic behavior of Co-doped ZnO (0%, 1%, 3%, 8%, and 12%) and found paramagnetic properties for all samples and weak ferromagnetic behavior for 12% Co-doped sample. Martinez et al.[32] prepared Co-doped ZnO NPs at low pressure (nearly 10 Torr) and high pressure (70–100 Torr) and studied magnetic properties. They found that sample prepared at low pressure exhibits ferromagnetic behavior at low temperature, but samples prepared with the high-pressure condition are paramagnetic and no ferromagnetism was induced as Co content increased. Iqbal et al.[120] studied Co-doped ZnO (0%, 2%, 4%, 6%, 8%, and 10%) NPs and found a typical diamagnetic behavior for undoped ZnO, which was attributed to the presence of a paired electron in a d orbital. For the doped ZnO, up to 6% paramagnetic nature was observed, which is attributed to the local moment induced by Co ion in the host matrix. For 8% and 10%, Co-doped ZnO ferromagnetism was observed. This was attributed to the sp–d and d–d exchange interaction. Ashraf et al.[121] also found room temperature ferromagnetism in Co-doped ZnO NPs. This room temperature ferromagnetism was due to bound magnetic polarons because of the successful doping of Co ions

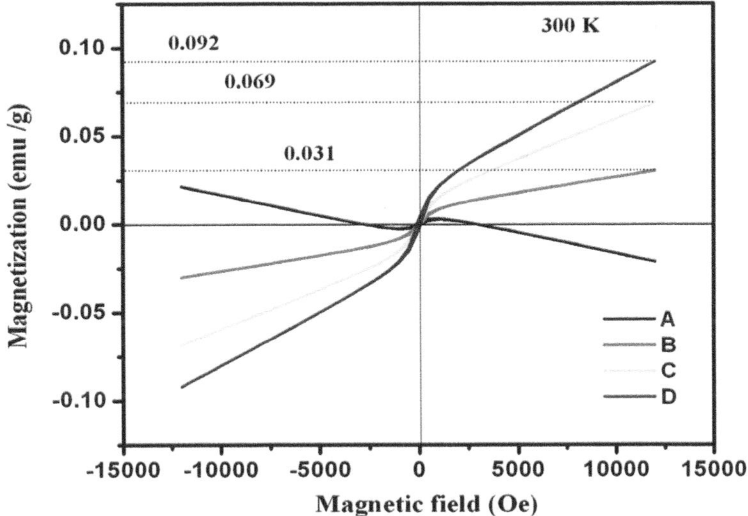

FIGURE 10.11 Magnetization vs. magnetic field of Co-doped (A) ZnO, (B) 5% Co-doped ZnO, (C) 10% Co-doped ZnO, and (D) 15% Co-doped ZnO. *Source*: "Reprinted with the permission from Ref.[27], Copyright (2016 American Chemical Society."

in Zn sites. For low concentration (0% and 1%) of Co in ZnO diamagnetic behavior and for higher concentration (3% and 5%) of Co, both paramagnetic and ferromagnetic behavior was reported by El Ghoul et al.[122] at room temperature. Table 10.9 shows a summary of experimental studies in the synthesis of Co-doped ZnO NPs with different preparation methods.

Wu et al.[41] studied the magnetic properties of Fe-doped ZnO powder using vibrating sample magnetometer (VSM). From the magnetization curve, they observed ferromagnetic behavior at room temperature. They observed paramagnetic behavior at room temperature for samples with lower doping of Fe (1% and 5%), whereas obvious ferromagnetic behavior was observed in higher doped samples such as 10% and 20% Fe-doped ZnO. The saturation magnetization and coercive force values were 0.74 emu/g, 1.74 emu/g, 90 and 78 Oe, respectively, for 10% and 20% Fe doping samples.

Ashraf et al.[39] also studied magnetic properties of Fe-doped ZnO NPs using VSM. They observed high saturation magnetization for Fe-doped ZnO NPs compared to pure ZnO NPs. They observed

TABLE 10.9
Overview of the Reported Experimental Result of the Magnetic Ground State of Co–ZnO NPs

Preparation Method	Obtained Morphology	Magnetic State
Polymerizable precursor method	Agglomerated irregular NPs (20–100 nm)	Room temperature ferromagnet
Hydrothermal method	Spherical NPs (50–100 nm)	Room temperature ferromagnet
Thermal hydrolysis	10 nm NPs	Room temperature ferromagnet
Sol–gel	140 nm NPs	Paramagnetic behavior and low temperature (150 K) ferromagnet
Solvothermal synthesis	NPs	Room temperature ferromagnet
Autocombustion method	NPs (15 or 40 nm)	Paramagnetic or room temperature
Alkaline-activated hydrolysis and condensation	20 nm spherical NPs	Paramagnetic behavior

Source: Adapted from Ref.[123] with permission of The Royal Society of Chemistry.

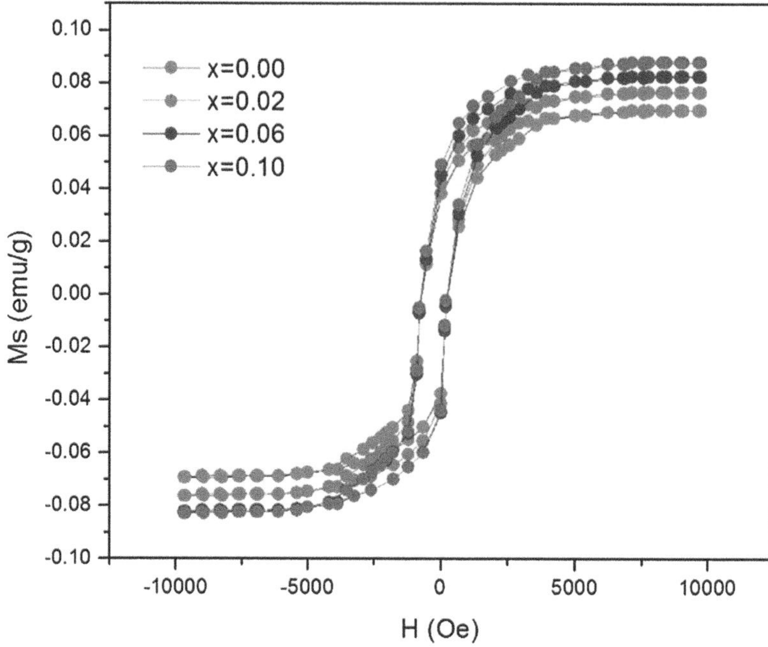

FIGURE 10.12 M–H curve of Fe-doped ZnO NPs with different Fe concentration. *Source*: "Reprinted from Ref.[39], Copyright (2015), with the permission of Elsevier."

ferromagnetic behavior even for pure ZnO, as shown in Figure 10.12. This effect may be due to the presence of Zn vacancies that result in the spin polarization of oxygen 2p orbitals. With Fe doping, an increase in saturation magnetization was attributed to RKKY interactions. They found that saturation magnetization increased from 0.065 to 0.088 emu/g with an increase in doping concentration. Table 10.10 summarizes the magnetic studies on Fe and Co-doped ZnO nanoparticles prepared by precipitation and microwave method.

TABLE 10.10
Synthesis Techniques and the Results of the Magnetic Study

Method	References	Results
Co-precipitation	21	M–H curve shows the presence of two components: paramagnetic and ferromagnetic
	69	Magnetic properties are dependent upon the variation of particles' size, shape, and composition; 20% of Fe-doped NPs have the highest saturation magnetization of 1.627 memu/g
Microwave	78	All the samples show paramagnetic behavior
Co-precipitation	27	The diamagnetic behavior of pure ZnO changes to ferromagnetic nature when doping with Co
	70	Magnetization measurement performed at 5 and 300 K represents that all Fe-doped samples exhibit room temperature ferromagnetism
Chemical precipitation	22	6% doping of Co shows weak ferromagnetic behavior at very low temperature, i.e., 5 K
Microwave assisted	97	All compositions show room temperature ferromagnetic behavior. Magnetization increases with increasing Co content
	43	Hydrogen reduction at 573 K for the Fe-doped samples transforms the magnetic nature from paramagnetic to ferromagnetic

10.12 SUMMARY

In this chapter, we summarized the results of Co- and Fe-doped ZnO NPs. We discussed two different techniques in fabricating ZnO NPs in this chapter—precipitation from chemical bath and microwave-assisted deposition methods—because of the simplicity, scalability, and reproducibility of NPs. We have summarized the structural, optical, electrical, and magnetic properties of Co- and Fe-doped ZnO NPs deposited by wet chemical methods. From the XRD analysis, we found a hexagonal wurtzite structure for all Fe- and Co-doped ZnO NPs. PL studies by different researchers showed that both red shift and blue shift of UV emission peak were observed with both Co and Fe doping in ZnO. SEM results showed the morphology of NPs when ZnO was doped with Fe and Co by both the precipitation and microwave method. Different authors have reported both increases and decreases in band gap energy when doped with Co and Fe in optical absorption studies by UV–vis spectroscopy. Electrical properties of Fe- and Co-doped ZnO were studied with impedance spectroscopy, and the dominance of grain boundary resistance over grain resistance was found. Magnetic properties were studied, and room temperature ferromagnetism behavior was observed in both Fe- and Co-doped ZnO NPs prepared by both the precipitation and microwave method.

10.13 ACKNOWLEDGMENTS

This work was supported by NASA EPSCoR (grant nos. NNX13AN01A and NNX15AM75A). The authors would like to acknowledge Dr. Alexie Grigoriev for helping with XRD measurements.

REFERENCES

1. Nam, G., Yoon, H., Kim, B., Lee, D.-Y., Kim, J. S., Leem, J.-Y. Effect of Co Doping Concentration on Structural Properties and Optical Parameters of Co-Doped ZnO Thin Films by Sol–Gel Dip-Coating Method. *J. Nanosci. Nanotechnol.* **2014**, *14* (11), 8544–8548.
2. Bedia, A., Bedia, F. Z., Aillerie, M., Maloufi, N., Benyoucef, B. Morphological and Optical Properties of ZnO Thin Films Prepared by Spray Pyrolysis on Glass Substrates at Various Temperatures for Integration in Solar Cell. *Energy Procedia* **2015**, *74*, 529–538.
3. Yu, Q., Li, L., Li, H., Gao, S., Sang, D., Yuan, J., Zhu, P. Synthesis and Properties of Boron Doped ZnO Nanorods on Silicon Substrate by Low-Temperature Hydrothermal Reaction. *Appl. Surf. Sci.* **2011**, *257* (14), 5984–5988.
4. Kim, S., Kim, M. S., Nam, G., Leem, J.-Y. Structural and Blue Emission Properties of Al-Doped ZnO Nanorod Array Thin Films Grown by Hydrothermal Method. *Electron. Mater. Lett.* **2012**, *8* (4), 445–450.
5. Escobedo-Morales, A., Pal, U. Defect Annihilation and Morphological Improvement of Hydrothermally Grown ZnO Nanorods by Ga Doping. *Appl. Phys. Lett.* **2008**, *93* (19), 193120.
6. Jie, J., Wang, G., Han, X., Hou, J. G. Synthesis and Characterization of ZnO:In Nanowires with Superlattice Structure. *J. Phys. Chem. B* **2004**, *108* (44), 17027–17031.
7. Xu, L., Su, Y., Chen, Y., Xiao, H., Zhu, L. A., Zhou, Q., Li, S. Synthesis and Characterization of Indium-Doped ZnO Nanowires with Periodical Single-Twin Structures. *J. Phys. Chem. B* **2006**, *110* (13), 6637–6642.
8. Kim, S., Nam, G., Park, H., Yoon, H., Lee, S.-H., Kim, J. S., Kim, J. S., Kim, D. Y., Kim, S.-O., Leem, J.-Y. Effects of Doping with Al, Ga, and In on Structural and Optical Properties of ZnO Nanorods Grown by Hydrothermal Method. *Bull. Korean Chem. Soc.* **2013**, *34* (4), 1205–1211.
9. Kim, K.-C., Kim, E., Kim, Y.-S. Growth and Physical Properties of Sol–gel Derived Co Doped ZnO Thin Film. *Superlattices Microstruct.* **2007**, *42* (1–6), 246–250.
10. Prellier, W., Fouchet, A., Simon, C., Mercey, B. Ferromagnetic Co-Doped ZnO Thin Films Grown Using Pulsed Laser Deposition from Zn and Co Metallic Targets. *Mater. Sci. Eng. B* **2004**, *109* (1–3), 192–195.
11. Nath, S. K., Chowdhury, N., Gafur, M. A. Effect of Co Doping on Crystallographic and Optoelectronic Properties of ZnO Thin Films. *J. Supercond. Nov. Magn.* **2015**, *28* (1), 117–123.

12. Kim, K. J., Park, Y. R. Spectroscopic Ellipsometry Study of Optical Transitions in Zn1−xCoxO Alloys. *Appl. Phys. Lett.* **2002**, *81* (8), 1420–1422.
13. Bhat, S. V., Deepak, F. L. Tuning the Bandgap of ZnO by Substitution with Mn2+, Co2+ and Ni2+. *Solid State Commun.* **2005**, *135* (6), 345–347.
14. Ivill, M., Pearton, S. J., Rawal, S., Leu, L., Sadik, P., Das, R., Hebard, A. F., Chisholm, M., Budai, J. D., Norton, D. P. Structure and Magnetism of Cobalt-Doped ZnO Thin Films. *New J. Phys.* **2008**, *10* (6), 065002.
15. Cherifi, Y., Chaouchi, A., Lorgoilloux, Y., Rguiti, M., Kadri, A., Courtois, C. Electrical, Dielectric and Photocatalytic Properties of Fe-Doped ZnO Nanomaterials Synthesized by Sol Gel Method. *Process. Appl. Ceram.* **2016**, *10* (3), 125–135.
16. Mishra, A. K., Das, D. Investigation on Fe-Doped ZnO Nanostructures Prepared by a Chemical Route. *Mater. Sci. Eng. B Solid-State Mater. Adv. Technol.* **2010**, *171* (1–3), 5–10.
17. Mehedi Hassan, M., Ahmed, A. S., Chaman, M., Khan, W., Naqvi, A. H., Azam, A. Structural and Frequency Dependent Dielectric Properties of Fe3+doped ZnO Nanoparticles. *Mater. Res. Bull.* **2012**, *47* (12), 3952–3958.
18. Li, Z. H., Truhlar, D. G. Nanothermodynamics of Metal Nanoparticles. *Chem. Sci.* **2014**, *5* (7), 2605–2624.
19. Brintha, S. R., Ajitha, M. Synthesis and Characterization of ZnO Nanoparticles via Aqueous Solution, Sol-Gel and Hydrothermal Methods. *IOSR J. Appl. Chem.* **2015**, *8* (11), 66–72.
20. Zhang, L., Yin, L., Wang, C., Qi, Y., Xiang, D. Origin of Visible Photoluminescence of ZnO Quantum Dots : Defect-Dependent And. **2010**, 9651–9658.
21. Fabbiyola, S., Kennedy, L. J., Aruldoss, U., Bououdina, M., Dakhel, A. A., Judith Vijaya, J. Synthesis of Co-Doped ZnO Nanoparticles via Co-Precipitation: Structural, Optical and Magnetic Properties. *Powder Technol.* **2015**, *286*, 757–765.
22. Samanta, A., Goswami, M. N., Mahapatra, P. K. Structural, Optical, Dielectric, Magnetic and Magnetoelectric Properties of Co-Doped ZnO Nanoparticles. *J. Mater. Sci. Mater. Electron.* **2016**, *27* (11), 12271–12278.
23. Hays, J., Reddy, K. M., Graces, N. Y., Engelhard, M. H., Shutthanandan, V., Luo, M., Xu, C., Giles, N. C., Wang, C., Thevuthasan, S., Punnoose, A. Effect of Co Doping on the Structural, Optical and Magnetic Properties of ZnO Nanoparticles. *J. Phys. Condens. Matter* **2007**, *19* (26).
24. Arshad, M., Azam, A., Ahmed, A. S., Mollah, S., Naqvi, A. H. Effect of Co Substitution on the Structural and Optical Properties of ZnO Nanoparticles Synthesized by Sol-Gel Route. *J. Alloys Compd.* **2011**, *509* (33), 8378–8381.
25. Park, J. H., Lee, Y. J., Bae, J.-S., Kim, B.-S., Cho, Y. C., Moriyoshi, C., Kuroiwa, Y., Lee, S., Jeong, S.-Y. Analysis of Oxygen Vacancy in Co-Doped ZnO Using the Electron Density Distribution Obtained Using MEM. *Nanoscale Res. Lett.* **2015**, *10* (1), 186.
26. Vanaja, A., Rao, K. S. Effect of Co Doping on Structural and Optical Properties of Zinc Oxide Nanoparticles Synthesized by Sol-Gel Method. *Adv. Nanoparticles* **2016**, *5* (February), 83–89.
27. Gandhi, V., Ganesan, R., Abdulrahman Syedahamed, H. H., Thaiyan, M. Effect of Cobalt Doping on Structural, Optical, and Magnetic Properties of ZnO Nanoparticles Synthesized by Coprecipitation Method. *J. Phys. Chem. C* **2014**, *118* (18), 9715–9725.
28. Kripal, R., Gupta, A. K., Srivastava, R. K., Mishra, S. K. Photoconductivity and Photoluminescence of ZnO Nanoparticles Synthesized via Co-Precipitation Method. *Spectrochim. Acta - Part A Mol. Biomol. Spectrosc.* **2011**, *79* (5), 1605–1612.
29. Nair, M. G., Nirmala, M., Rekha, K., Anukaliani, A. Structural, Optical, Photo Catalytic and Antibacterial Activity of ZnO and Co Doped ZnO Nanoparticles. *Mater. Lett.* **2011**, *65* (12), 1797–1800.
30. Carrero, A., Sagredo, V., Pernechele, C., Rossi, F. Synthesis and Characterization of Co-Doped ZnO Nanocompound. *IEEE Trans. Magn.* **2013**, *49* (8), 4614–4617.
31. Elilarassi, R., Chandrasekaran, G. Influence of Co-Doping on the Structural, Optical and Magnetic Properties of ZnO Nanoparticles Synthesized Using Auto-Combustion Method. *J. Mater. Sci. Mater. Electron.* **2013**, *24* (1), 96–105.
32. Martínez, B., Sandiumenge, F., Balcells, L., Arbiol, J., Sibieude, F., Monty, C. Structure and Magnetic Properties of Co-Doped ZnO Nanoparticles. *Phys. Rev. B* **2005**, *72* (16), 165202.

33. Maensiri, S., Sreesongmuang, J., Thomas, C., Klinkaewnarong, J. Magnetic Behavior of Nanocrystalline Powders of Co-Doped ZnO Diluted Magnetic Semiconductors Synthesized by Polymerizable Precursor Method. *J. Magn. Magn. Mater.* **2006**, *301* (2), 422–432.
34. Wojnarowicz, J., Kusnieruk, S., Chudoba, T., Gierlotka, S., Lojkowski, W., Knoff, W., Lukasiewicz, M. I., Witkowski, B. S., Wolska, A., Klepka, M. T., Story, T., Godlewski, M. Paramagnetism of Cobalt-Doped ZnO Nanoparticles Obtained by Microwave Solvothermal Synthesis. *Beilstein J. Nanotechnol.* **2015**, *6*, 1957–1969.
35. Singh, J., Chanda, A., Gupta, S., Shukla, P., Chandra, V. Effect of Cobalt Doping on Structural and Optical Properties of ZnO Nanoparticles. **2016**, *1* (1), 050091.
36. Parra-Palomino, A., Perales–Perez, O., Singhal, R., Tomar, M., Hwang, J., Voyles, P. M. Structural, Optical, and Magnetic Characterization of Monodisperse Fe-Doped ZnO Nanocrystals. *J. Appl. Phys.* **2008**, *103* (7), 07D121.
37. Dhiman, P., Rani, R., Singh, M. Structural and Electrical Properties of Fe Doped ZnO Nanoparticles Synthesized by Solution Combustion Method. In *AIP Conference Proceedings*; 2012; Vol. 1447, pp 307–308.
38. Yu, X., Meng, D., Liu, C., Xu, K., Chen, J., Lu, C., Wang, Y. Enhanced Photocatalytic Activity of Fe-Doped ZnO Nanoparticles Synthesized via a Two-Step Sol-Gel Method. *J. Mater. Sci. Mater. Electron.* **2014**, *25* (9), 3920–3923.
39. Ashraf, R., Riaz, S., Kayani, Z. N., Naseem, S. Structural and Magnetic Properties of Iron Doped ZnO Nanoparticles. *Mater. Today Proc.* **2015**, *2* (10), 5384–5389.
40. Sharma, P. K., Dutta, R. K., Pandey, A. C., Layek, S., Verma, H. C. Effect of Iron Doping Concentration on Magnetic Properties of ZnO Nanoparticles. *J. Magn. Magn. Mater.* **2009**, *321* (17), 2587–2591.
41. Shu, R., Wang, X., Yang, Y., Tang, X., Zhou, X., Cheng, Y. Coprecipitation Synthesis of Fe-Doped ZnO Powders with Enhanced Microwave Absorption Properties. *Nano* **2016**, *11* (12), 1650136.
42. Srinivasulu, T., Saritha, K., Reddy, K. T. R. Synthesis and Characterization of Fe-Doped ZnO Thin Films Deposited by Chemical Spray Pyrolysis. *Mod. Electron. Mater.* **2017**, *3* (2), 76–85.
43. Glaspell, G., Dutta, P., Manivannan, A. A Room-Temperature and Microwave Synthesis of M-Doped ZnO (M = Co, Cr, Fe, Mn & Ni). *J. Clust. Sci.* **2005**, *16* (4), 523–536.
44. Elilarassi, R., Chandrasekaran, G. Synthesis and Characterization of Ball Milled Fe-Doped ZnO Diluted Magnetic Semiconductor. *Optoelectron. Lett.* **2012**, *8* (2), 109–112.
45. Arshad, M., Azam, A., Ahmed, A. S., Mollah, S., Naqvi, A. H. Effect of Co Substitution on the Structural and Optical Properties of ZnO Nanoparticles Synthesized by Sol-Gel Route. *J. Alloys Compd.* **2011**, *509* (33), 8378–8381.
46. Ciciliati, M. A., Silva, M. F., Fernandes, D. M., de Melo, M. A. C., Hechenleitner, A. A. W., Pineda, E. A. G. Fe-Doped ZnO Nanoparticles: Synthesis by a Modified Sol–gel Method and Characterization. *Mater. Lett.* **2015**, *159*, 84–86.
47. Sathya, B., Porkalai, V., Anburaj, D. B., Nedunchezhian, G. Low Temperature Ferromagnetism and Optical Properties of Fe Doped ZnO Nanoparticles Synthesized by Sol-Gel Method 7. *MMSE J.* **2017**, No. April, 38–43.
48. Liu, H., Yang, J., Zhang, Y., Yang, L., Wei, M., Ding, X. Structure and Magnetic Properties of Fe-Doped ZnO Prepared by the Sol–gel Method. *J. Phys. Condens. Matter* **2009**, *21* (14), 145803.
49. Kripal, R., Gupta, A. K., Srivastava, R. K., Mishra, S. K. Photoconductivity and Photoluminescence of ZnO Nanoparticles Synthesized via Co-Precipitation Method. *Spectrochim. Acta - Part A Mol. Biomol. Spectrosc.* **2011**, *79* (5), 1605–1612.
50. Nair, M. G., Nirmala, M., Rekha, K., Anukaliani, A. Structural, Optical, Photo Catalytic and Antibacterial Activity of ZnO and Co Doped ZnO Nanoparticles. *Mater. Lett.* **2011**, *65* (12), 1797–1800.
51. Saleh, R., Djaja, N. F. UV Light Photocatalytic Degradation of Organic Dyes with Fe-Doped ZnO Nanoparticles. *Superlattices Microstruct.* **2014**, *74*, 217–233.
52. Kumar, K., Chitkara, M., Sandhu, I. S., Mehta, D., Kumar, S. Photocatalytic, Optical and Magnetic Properties of Fe-Doped ZnO Nanoparticles Prepared by Chemical Route. *J. Alloys Compd.* **2014**, *588*, 681–689.
53. Carrero, A., Sagredo, V., Pernechele, C., Rossi, F. Synthesis and Characterization of Co-Doped ZnO Nanocompound. *IEEE Trans. Magn.* **2013**, *49* (8), 4614–4617.
54. Irshad, K., Khan, M. T., Murtaza, A. Synthesis and Characterization of Transition-Metals-Doped ZnO Nanoparticles by Sol-Gel Auto-Combustion Method. *Phys. B Condens. Matter* **2018**, *543* (April), 1–6.

55. Elilarassi, R., Chandrasekaran, G. Influence of Co-Doping on the Structural, Optical and Magnetic Properties of ZnO Nanoparticles Synthesized Using Auto-Combustion Method. *J. Mater. Sci. Mater. Electron.* **2013**, *24* (1), 96–105.
56. Silva, L. G., Solis-Pomar, F., Gutiérrez-Lazos, C. D., Meléndrez, M. F., Martinez, E., Fundora, A., Pérez-Tijerina, E. Synthesis of Fe Nanoparticles Functionalized with Oleic Acid Synthesized by Inert Gas Condensation. *J. Nanomater.* **2014**, *2014*.
57. Kwong, T.-L., Yung, K.-F. Surfactant-Free Microwave-Assisted Synthesis of Fe-Doped ZnO Nanostars as Photocatalyst for Degradation of Tropaeolin O in Water under Visible Light. *J. Nanomater.* **2015**, *2015*, 1–9.
58. Saleh, R., Djaja, N. F., Prakoso, S. P. The Correlation between Magnetic and Structural Properties of Nanocrystalline Transition Metal-Doped ZnO Particles Prepared by the Co-Precipitation Method. *J. Alloys Compd.* **2013**, *546*, 48–56.
59. He, R., Hocking, R. K., Tsuzuki, T. Co-Doped ZnO Nanopowders: Location of Cobalt and Reduction in Photocatalytic Activity. *Mater. Chem. Phys.* **2012**, *132* (2–3), 1035–1040.
60. Udayakumar, S., Renuka, V., Kavitha, K. Structural Optical and Thermal Studies of Cobalt Doped Hexagonal ZnO by Simple Chemical Precipitation Method. **2012**, *4* (2), 1271–1280.
61. Chithra, M. J., Pushpanathan, K., Loganathan, M. Structural and Optical Properties of Co-Doped ZnO Nanoparticles Synthesized by Precipitation Method. *Mater. Manuf. Process.* **2014**, *29* (7), 771–779.
62. Devi, P. G., Velu, A. S. Synthesis, Structural and Optical Properties of Pure ZnO and Co Doped ZnO Nanoparticles Prepared by the Co-Precipitation Method. *J. Theor. Appl. Phys.* **2016**, *10* (3), 233–240.
63. Godavarti, U., Mote, V. D., Dasari, M. Role of Cobalt Doping on the Electrical Conductivity of ZnO Nanoparticles. *J. Asian Ceram. Soc.* **2017**, *5* (4), 391–396.
64. Rath, C., Singh, S., Mallick, P., Pandey, D., Lalla, N. P., Mishra, N. C. Effect of Cobalt Substitution on Microstructure and Magnetic Properties in ZnO Nanoparticles. *Indian J. Phys.* **2009**, *83*, 415–421.
65. Bhatt, R. C., Kar, M., Arora, M., Singh, N. Synthesis and Characterization of Pure and Doped Nanocrystalline Zinc Oxide. *J. Optoelectron. Adv. Mater.* **2010**, *1* (2), 231–238.
66. He, R., Tang, B., Ton-That, C., Phillips, M., Tsuzuki, T. Physical Structure and Optical Properties of Co-Doped ZnO Nanoparticles Prepared by Co-Precipitation. *J. Nanoparticle Res.* **2013**, *15* (11).
67. Kalpana, S., Krishnan, S. S., Senthil, T. S., Elangovan, S. V. Cobalt Doped Zinc Oxide Nanoparticles For Photocatalytic Applications. *J. Ovonic Res.* **2017**, *13* (5), 263–269.
68. Rana, S. B., Singh, P., Sharma, A. K., Carbonari, A. W., Dogra, R. Synthesis and Characterization of Pure and Doped ZnO Nanoparticles. *J. Optoelectron. Adv. Mater.* **2010**, *12* (2), 257–261.
69. Fabbiyola, S., Kennedy, L. J., Ratnaji, T., Vijaya, J. J., Aruldoss, U., Bououdina, M. Effect of Fe-Doping on the Structural, Optical and Magnetic Properties of ZnO Nanostructures Synthesised by Co-Precipitation Method. *Ceram. Int.* **2016**, *42* (1), 1588–1596.
70. Kumar, S., Kim, Y. J., Gautam, S., Chae, K. H., Koo, B. H., Lee, C. G. Ferromagnetism in Fe Doped ZnO Synthesized by Co-Precipitation Method. *J. Ceram. Soc. Japan* **2009**, *117* (1365), 616–618.
71. Pal Singh, R. P., Hudiara, I. S., Bhushan Rana, S. Effect of Calcination Temperature on the Structural, Optical and Magnetic Properties of Pure and Fe-Doped ZnO Nanoparticles. *Mater. Sci.* **2016**, *34* (2), 451–459.
72. Moussa, D., El-Said Bakeer, D., Awad, R., Abdel-Gaber, A. M. Physical Properties of ZnO Nanoparticles Doped with Mn and Fe. *J. Phys. Conf. Ser.* **2017**, *869* (1), 012021.
73. Sheik Muhideen Badhusha, M. Microwave Assisted Synthesis of ZnO and Co Doped ZnO Nanoparticles and Their Antibacterial Activity. *Pharma Chem.* **2016**, *8* (20), 78–84.
74. Bhatti, S., Surve, S., Shukla, V. N. Structure and Magnetic Characteristics of Co & Cr-Doped ZnO Nanoparticles, Synthesized by Using Microwave Method. *Mater. Today Proc.* **2017**, *4* (2), 3825–3831.
75. Varadhaseshan, R., Meenakshi Sundar, S., Prema, C. On the Preparation, Structural and Magnetic Properties of ZnO:Co Nanoparticles. *Eur. Phys. J. Appl. Phys.* **2014**, *66* (1), 10602.
76. Hoang, L. H., Hai, P. Van, Hai, N. H., Vinh, P. Van, Chen, X. B., Yang, I. S. The Microwave-Assisted Synthesis and Characterization of $Zn_{1-x}Co_xO$ Nanopowders. *Mater. Lett.* **2010**, *64* (8), 962–965.
77. Mary, J. A., Vijaya, J. J. Influence of ZnO Co-Doping On the Structural, Optical, and Antibacterial Properties of ($Zn_{1-2x}Ce_xFe_x$)O Nanoparticles. *Int. J. Innov. Res. Sci. Eng. Technol.* **2015**, *4* (1), 9–14.
78. Basu, S., Inamdar, D. Y., Mahamuni, S., Chakrabarti, A., Kamal, C., Kumar, G. R., Jha, S. N., Bhattacharyya, D. Local Structure Investigation of Cobalt and Manganese Doped ZnO Nanocrystals and Its Correlation with Magnetic Properties. *J. Phys. Chem. C* **2014**, *118* (17), 9154–9164.

79. Kwong, T.-L., Yung, K.-F. Surfactant-Free Microwave-Assisted Synthesis of Fe-Doped ZnO Nanostars as Photocatalyst for Degradation of Tropaeolin O in Water under Visible Light. *J. Nanomater.* **2015**, *2015*, 1–9.
80. Han, L., Wang, D., Lu, Y., Jiang, T., Chen, L., Xie, T., Lin, Y. Influence of Annealing Temperature on the Photoelectric Gas Sensing of Fe-Doped ZnO under Visible Light Irradiation. *Sens. Actuat. B Chem.* **2013**, *177*, 34–40.
81. Bhatti, S., Surve, S., Shukla, V. N. Structural Characterization and Optical Properties of Fe & Ni-Doped Zinc Oxide Nanopored Particles , Synthesized Using Microwave Method. *Int. J. Adv. Technol. Eng. Sci.* **2015**, *3* (08), 123–128.
82. Bilecka, I., Luo, L., Djerdj, I., Rossell, M. D., Jagodič, M., Jagličić, Z., Masubuchi, Y., Kikkawa, S., Niederberger, M. Microwave-Assisted Nonaqueous Sol–Gel Chemistry for Highly Concentrated ZnO-Based Magnetic Semiconductor Nanocrystals. *J. Phys. Chem. C* **2011**, *115* (5), 1484–1495.
83. Rana, A. ul H. S., Kang, M., Kim, H.-S. Microwave-Assisted Facile and Ultrafast Growth of ZnO Nanostructures and Proposition of Alternative Microwave-Assisted Methods to Address Growth Stoppage. *Sci. Rep.* **2016**, *6* (1), 24870.
84. El Ghoul, J., Kraini, M., El Mir, L. Synthesis of Co-Doped ZnO Nanoparticles by Sol–Gel Method and Its Characterization. *J. Mater. Sci. Mater. Electron.* **2015**, *26* (4), 2555–2562.
85. Lima, M. K., Fernandes, D. M., Silva, M. F., Baesso, M. L., Neto, A. M., de Morais, G. R., Nakamura, C. V., de Oliveira Caleare, A., Hechenleitner, A. A. W., Pineda, E. A. G. Co-Doped ZnO Nanoparticles Synthesized by an Adapted Sol–gel Method: Effects on the Structural, Optical, Photocatalytic and Antibacterial Properties. *J. Sol-Gel Sci. Technol.* **2014**, *72* (2), 301–309.
86. Pal, B., Giri, P. K. High Temperature Ferromagnetism and Optical Properties of Co Doped ZnO Nanoparticles. *J. Appl. Phys.* **2010**, *108* (8).
87. Castro, T. J., Rodrigues, P. A. M., Oliveira, A. C., Nakagomi, F., Mantilla, J., Coaquira, J. A. H., Franco, A., Pessoni, H. V. S., Morais, P. C., Da Silva, S. W. Optical and Magnetic Properties of Co-Doped ZnO Nanoparticles and the Onset of Ferromagnetic Order. *J. Appl. Phys.* **2017**, *121* (1).
88. Thuy Doan, M., Vinh Ho, X., Nguyen, T., Nguyen, V. N. Influence of Doping Co to Characterization of ZnO Nanostructures. *Adv. Nat. Sci. Nanosci. Nanotechnol.* **2014**, *5* (2), 025011.
89. Muthukumaran, S., Gopalakrishnan, R. Structural, FTIR and Photoluminescence Studies of Cu Doped ZnO Nanopowders by Co-Precipitation Method. *Opt. Mater. (Amst).* **2012**, *34* (11), 1946–1953.
90. Dinesha, M. L., Jayanna, H. S., Mohanty, S., Ravi, S. Structural, Electrical and Magnetic Properties of Co and Fe Co-Doped ZnO Nanoparticles Prepared by Solution Combustion Method. *J. Alloys Compd.* **2010**, *490* (1–2), 618–623.
91. Bhattacharyya, S., Gedanken, A. Synthesis, Characterization, and Room-Temperature Ferromagnetism in Cobalt-Doped Zinc Oxide (ZnO:Co2+) Nanocrystals Encapsulated in Carbon. *J. Phys. Chem. C* **2008**, *112* (12), 4517–4523.
92. Vidhya, K., Bhoopathi, G., Devarajan, V. P., Saravanan, M. Green Synthesis of Glucose Capped ZnO:Fe Quantum Dots. A Study on Structural, Optical Properties and Application. *Res. J. Recent Sci.* **2014**, *3* (ISC-2013), 4,238-241.
93. Bousslama, W., Elhouichet, H., Férid, M. Enhanced Photocatalytic Activity of Fe Doped ZnO Nanocrystals under Sunlight Irradiation. *Optik (Stuttg).* **2017**, *134*, 88–98.
94. Manikandan, A., Kennedy, L. J., Bououdina, M., Vijaya, J. J. Synthesis, Optical and Magnetic Properties of Pure and Co-Doped ZnFe2O4nanoparticles by Microwave Combustion Method. *J. Magn. Magn. Mater.* **2014**, *349*, 249–258.
95. Hammad, T. M., Salem, J. K., Harrison, R. G. Structure, Optical Properties and Synthesis of Co-Doped ZnO Superstructures. *Appl. Nanosci.* **2013**, *3* (2), 133–139.
96. Anand, B., Muthuvel, A. Synthesis and Characterization of ZnO Nanoparticles Using Effective of Co Doped. *Int. J. Res. Appl. Sci. Eng. Technol.* **2017**, *5* (Viii), 2321–9653.
97. Basith, N. M., Vijaya, J. J., Kennedy, L. J., Bououdina, M., Jenefar, S., Kaviyarasan, V. Co-Doped ZnO Nanoparticles: Structural, Morphological, Optical, Magnetic and Antibacterial Studies. *J. Mater. Sci. Technol.* **2014**, *30* (11), 1108–1117.
98. Wu, X., Wei, Z., Zhang, L., Wang, X., Yang, H., Jiang, J. Optical and Magnetic Properties of Fe Doped ZnO Nanoparticles Obtained by Hydrothermal Synthesis. *J. Nanomater.* **2014**, *2014*, 1–6.
99. Kanchana, S., Chithra, M. J., Ernest, S., Pushpanathan, K. Violet Emission from Fe Doped ZnO Nanoparticles Synthesized by Precipitation Method. *J. Lumin.* **2016**, *176*, 6–14.

100. Padmavathy, N., Vijayaraghavan, R. Enhanced Bioactivity of ZnO Nanoparticles—An Antimicrobial Study. *Sci. Technol. Adv. Mater.* **2008**, *9* (3), 035004.
101. Rana, S. B., Singh, P., Sharma, A. K., Carbonari, A. W., Dogra, R. Synthesis and Characterization of Pure and Doped ZnO Nanoparticles. *Int. J. Eng. Manuf.* **2010**, *12* (2), 257–261.
102. Ahmed, F., Kumar, S., Arshi, N., Anwar, M. S., Koo, B. H., Lee, C. G. Doping Effects of Co2+ Ions on Structural and Magnetic Properties of ZnO Nanoparticles. *Microelectron. Eng.* **2012**, *89* (1), 129–132.
103. Raja, K., Ramesh, P. S., Geetha, D. Structural, FTIR and Photoluminescence Studies of Fe Doped ZnO Nanopowder by Co-Precipitation Method. *Spectrochim. Acta – Part A Mol. Biomol. Spectrosc.* **2014**, *131*, 183–188.
104. Grotel, J., Pikula, T., Siedliska, K., Ruchomski, L., Panek, R., Wiertel, M., Jartych, E. Structure and Hyperfine Interactions of Fe-Doped ZnO Powder Prepared by Co-Precipitation Method. *Acta Phys. Pol. A* **2018**, *134* (5), 1048–1052.
105. Gao, D., Zhang, Z., fu, J., xu, Y., qi, J., Xue, D. Room Temperature Ferromagnetism of Pure ZnO Nanoparticles. *J. Appl. Phys.* **2009**, *105* (11), 119.
106. Nirmala, M. Synthesis of Pure And Transition Metals (TM=Mn, Co) Doped ZnO Nanoparticles And Their Photocatalytic, Antibacterial And Ethanol Sensing Applications. *PhD diss.*Bharathiar University, **2011**.
107. Kumar, S., Asokan, K., Singh, R. K., Chatterjee, S., Kanjilal, D., Ghosh, A. K. Investigations on Structural and Optical Properties of ZnO and ZnO:Co Nanoparticles under Dense Electronic Excitations. *RSC Adv.* **2014**, *4* (107), 62123–62131.
108. Maswanganye, M. W., Rammutla, K. E., Mosuang, T. E., Mwakikunga, B. W. Synthesis , Structural and Optical Characterisation of Cobalt and Indium Co-Doped ZnO Nanoparticles. *Proceedings of SAIP*, **2015**, 1, 61–66.
109. Pal, B., Giri, P. K. High Temperature Ferromagnetism and Optical Properties of Co Doped ZnO Nanoparticles. *J. Appl. Phys.* **2010**, *108* (8), 1–8.
110. Palacios, P., Aguilera, I., Wahnón, P. Electronic Structure and Optical Properties in ZnO:M(Co, Cd). *Thin Solid Films* **2010**, *518* (16), 4568–4571.
111. Hammad, T. M., Griesing, S., Wotocek, M., Kuhn, S., Hempelmann, R., Hartmann, U., Salem, J. K. Optical and Magnetic Properties of Fe-Doped ZnO Nanoparticles Prepared by the Sol-Gel Method. *Int. J. Nanoparticles* **2013**, *6* (4), 324.
112. Tiwari, R. C. Structural, Optical and Electrical Properties of Cobalt Doped ZnO Quantum Dots. *Funct. Nanostructures* **2016**, *1* (2), 60–66.
113. Arshad, M., Ahmed, A. S., Azam, A., Naqvi, A. H. Exploring the Dielectric Behavior of Co Doped ZnO Nanoparticles Synthesized by Wet Chemical Route Using Impedance Spectroscopy. *J. Alloys Compd.* **2013**, *577*, 469–474.
114. Ansari, S. A., Nisar, A., Fatma, B., Khan, W., Naqvi, A. H. Investigation on Structural, Optical and Dielectric Properties of Co Doped ZnO Nanoparticles Synthesized by Gel-Combustion Route. *Mater. Sci. Eng. B* **2012**, *177* (5), 428–435.
115. Rostami, M., Rasouli, S., Ramezanzadeh, B., Askari, A. Electrochemical Investigation of the Properties of Co Doped ZnO Nanoparticle as a Corrosion Inhibitive Pigment for Modifying Corrosion Resistance of the Epoxy Coating. *Corros. Sci.* **2014**, *88*, 387–399.
116. Dinesha, M. L., Prasanna, G. D., Naveen, C. S., Jayanna, H. S. Structural and Dielectric Properties of Fe Doped ZnO Nanoparticles. *Indian J. Phys.* **2013**, *87* (2), 147–153.
117. Oves, M., Arshad, M., Khan, M. S., Ahmed, A. S., Azam, A., Ismail, I. M. I. Anti-Microbial Activity of Cobalt Doped Zinc Oxide Nanoparticles: Targeting Water Borne Bacteria. *J. Saudi Chem. Soc.* **2015**, *19* (5), 581–588.
118. Franco, A., Pessoni, H. V. S., Ribeiro, P. R. T., Machado, F. L. A. Magnetic Properties of Co-Doped ZnO Nanoparticles. *J. Magn. Magn. Mater.* **2017**, *426* (August 2016), 347–350.
119. Castro, T. J., Rodrigues, P. A. M., Oliveira, A. C., Nakagomi, F., Mantilla, J., Coaquira, J. A. H., Franco Júnior, A., Pessoni, H. V. S., Morais, P. C., da Silva, S. W. Optical and Magnetic Properties of Co-Doped ZnO Nanoparticles and the Onset of Ferromagnetic Order. *J. Appl. Phys.* **2017**, *121* (1), 013904.
120. Iqbal, J., Janjua, R. A., Jan, T. Structural, Optical and Magnetic Properties of Co-Doped ZnO Nanoparticles Prepared via a Wet Chemical Route. *Int. J. Mod. Phys. B* **2014**, *28* (25), 1450158.

121. Ashraf, R., Riaz, S., Bashir, M., Naseem, S. Magnetic and Structural Properties of Co Doped ZnO Nanoparticles. *ANBRE* **2013**, 1, 346–352.
122. El Ghoul, J., Kraini, M., Lemine, O. M., El Mir, L. Sol–Gel Synthesis, Structural, Optical and Magnetic Properties of Co-Doped ZnO Nanoparticles. *J. Mater. Sci. Mater. Electron.* **2015**, *26* (4), 2614–2621.
123. Djerdj, I., Jagliić, Z., Aron, D., Niederberger, M. Co-Doped ZnO Nanoparticles: Minireview. *Nanoscale.* **2010**, July 7, 1096–1104.

11 First-Order Reversal Curve Study of Nanomagnetic Materials

Brad C. Dodrill

Lake Shore Cryotronics, Westerville, Ohio, USA

CONTENTS

11.1 INTRODUCTION

The measurement most commonly performed to characterize a material's magnetic properties is that of a major hysteresis loop. More complex magnetization curves covering states with field and magnetization values located inside the major hysteresis loop, such as minor hysteresis loops and first-order reversal curves (FORC), can provide additional information that can be used to characterize magnetic interactions and coercivity distributions in magnetic materials. In this chapter, we will discuss the FORC measurement and analysis protocol and present results for several relevant nanomagnetic materials including exchange-coupled hard/soft nanocomposite permanent magnets, nanomagnet arrays, core/shell granular magnetic heteronanostructures, doped magnetic nanoparticles, exchange bias magnetic multilayer thin films, and nanocrystalline amorphous nanocomposites.

11.2 FIRST-ORDER REVERSAL CURVES (FORCS)

The most common measurement used to characterize a material's magnetic properties is measurement of the hysteresis or $M(H)$ loop as illustrated in Figure 11.1. The main parameters extracted from the hysteresis loop, which are used to characterize the properties of magnetic materials include the saturation magnetization M_s (the magnetization at maximum applied field), the remanence M_r (the magnetization at zero applied field after applying a saturating field), and the coercivity H_c (the field required to demagnetize the sample).

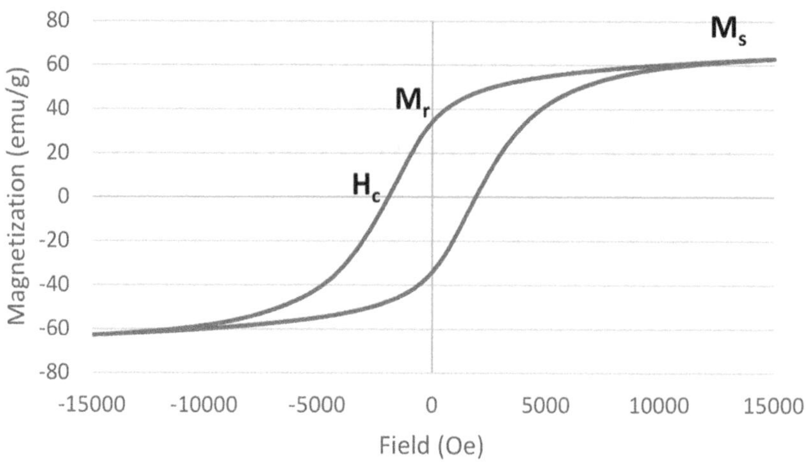

FIGURE 11.1 A typical hysteresis loop showing the extracted parameters M_s, M_r, and H_c.

More complex magnetization curves covering states with field and magnetization values located inside the major hysteresis loop, such as minor hysteresis loops and FORCs, can give additional information that can be used for characterization of magnetic interactions [1].

FORC is relevant to any hysteretic magnetic material that is comprised of fine micron- or nano-scale particles, grains, and so on. It has been extensively used by earth and planetary scientists studying the magnetic properties of natural samples (rocks, soils, sediments, etc.) because FORC can distinguish between single domain (SD), multidomain (MD), and pseudo-SD (PSD) behavior, and because it can distinguish between different magnetic mineral species [2, 3]. FORC has proven to be useful in better understanding the nature of magnetization reversal and interactions in magnetic nanowires [4, 5, 6, 7, 8], nanomagnet arrays [9, 10, 11, 12], thin film magnetic recording media [13, 14, 15], exchange bias magnetic nanodot arrays [16], thin film magnetic multilayers [17, 18], nanostructured permanent magnet materials [19, 20], soft magnetic bilayers [21], and magneto-caloric effect materials [22]. It has also been used to differentiate between phases in multiphase magnetic materials because it is very difficult to unravel the complex magnetic signatures of such materials from a hysteresis loop measurement alone [23, 24].

An FORC is measured by saturating a sample in a field H_{sat}, decreasing the field to a reversal field H_a, then measuring moment versus field H_b as the field is swept back to H_{sat}. This process is repeated for many values of H_a, yielding a series of FORCs as shown in Figure 11.2 for a ferrite permanent magnet sample. The FORC distribution $\rho(H_a, H_b)$ is the mixed second derivative:

$$\rho\left(H_a, H_b\right) = -\left(\frac{1}{2}\right)\frac{\partial^2 M\left(H_a, H_b\right)}{\partial H_a \partial H_b}$$

The FORC diagram is a 2D or 3D contour plot of $\rho(H_a, H_b)$. It is common to change the coordinates from (H_a, H_b) to:

$$H_c = \frac{H_b - H_a}{2} \text{ and } H_u = \frac{H_b + H_a}{2}$$

H_u represents the distribution of interaction or reversal fields, and H_c represents the distribution of switching or coercive fields of the hysterons. The 2D FORC diagram for the ferrite permanent magnet sample is shown in Figure 11.3.

An FORC diagram not only provides information regarding the distribution of interaction and switching fields, but also serves as a "fingerprint" that gives insight into the domain state and nature of interactions occurring in magnetic materials. In an FORC diagram, entirely closed contours

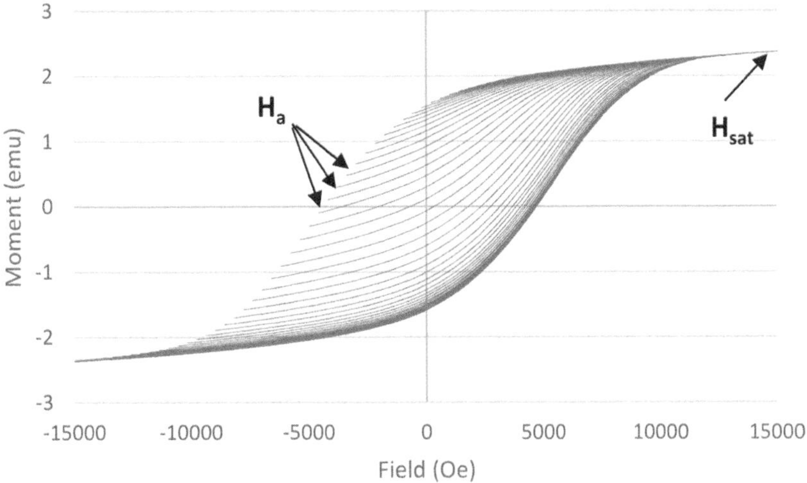

FIGURE 11.2 Measured first-order reversal curves for a ferrite permanent magnet.

are usually associated with SD behavior, while open contours that diverge toward the H_u axis are associated with MD, and open and closed contours together are associated with PSD. The peak in the FORC distribution is usually centered at a switching field H_c that correlates with the coercivity as determined from a hysteresis loop measurement. If the peak in the FORC distribution is centered at an interaction field $H_u = 0$, this means that interactions between particles, grains, and so on, are weak. Conversely, if the peak is shifted toward positive H_u, they are strong. Multiple peaks in an FORC diagram mean that there are multiple magnetic phases in a material. The very shape of the FORC distribution provides insight into the nature of interactions (dipolar and exchange) that are occurring in a magnetic material.

There are a number of open source FORC analysis software packages such as FORCinel [25] and VARIFORC [26]. In the results that follow, a Lake Shore Cryotronics vibrating sample magnetometer (Model 8600 VSM) was used to measure the FORCs, and all results are presented in cgs-units, i.e., magnetic field H(Oe) versus magnetic moment m(emu). FORCinel was used to calculate the FORC distributions and plot the FORC diagrams.

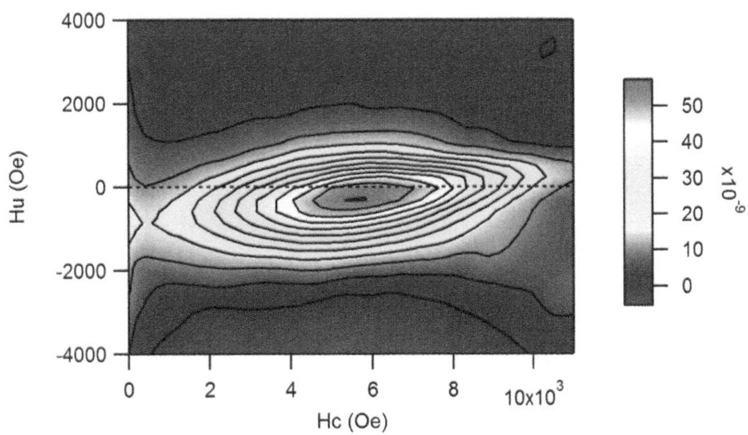

FIGURE 11.3 The 2D FORC diagram for a ferrite permanent magnet.

11.3 FORC DISTRIBUTIONS OF TYPICAL NANOSTRUCTURED MAGNETS

In this section, we present FORC measurement and analysis results for a variety of families of nanostructured materials, ranging from soft magnetic nanocomposites to magnetic recording media. These include nanocomposite permanent magnet, nickel (Ni) nanowire array, high-density patterned media consisting of CoPt nanomagnet array, core/shell granular magnetic heteronanostructure, doped magnetic nanoparticles, exchange-biased magnetic multilayer thin film, and a nanocrystalline nanocomposite magnet.

11.3.1 Nanocomposite Permanent Magnets

Rare-earth permanent magnet materials are indispensable elements in many devices such as electrical motors, hybrid vehicles, and portable communications devices. The magnets have major influence on the size, efficiency, stability, and cost of these devices and systems. Owing to the increasing price of rare-earth materials, there is keen interest in developing strong permanent magnet materials that do not rely as heavily on rare-earth constituents, such as nanostructured magnets, and exchange-coupled nanocomposite alloys with co-existing soft and hard phases because of the coercivity enhancement that is obtained at the SD size (nanometer scale).

To demonstrate the utility of FORC analysis for differentiating phases in exchange-coupled nanocomposites, FORC data were acquired at $T = 300$ K on nanometer-sized (~60 nm) barium hexaferrite $BaFe_{12}O_{19}$.

Figures 11.4 and 11.5 show the measured hysteresis *M*(*H*) loop and FORCs, respectively. Figure 11.6 shows the resultant FORC diagram. There is a subtle "kink" in the *M*(*H*) loop (Figure 11.4) at low fields, suggesting the presence of a low and high coercivity phase. The FORC diagram (Figure 11.6) shows two peaks corresponding to the low and high coercivity components, and the region between the two peaks is related to the exchange coupling between the two phases [27].

11.3.2 Nanomagnet Arrays

Magnetic nanowires, nanodots, and nanoparticles are an important class of nanostructured magnetic materials. At least one of the dimensions of these structures is in the nanometer (nm) range, and thus, new phenomena arise in these materials due to size confinement. These structures are ideal candidates for important technological applications in spintronics, high-density recording media, microwave electronics, and permanent magnets, and for medical diagnostics and targeted drug delivery applications. In addition to technological applications, these materials represent an experimental

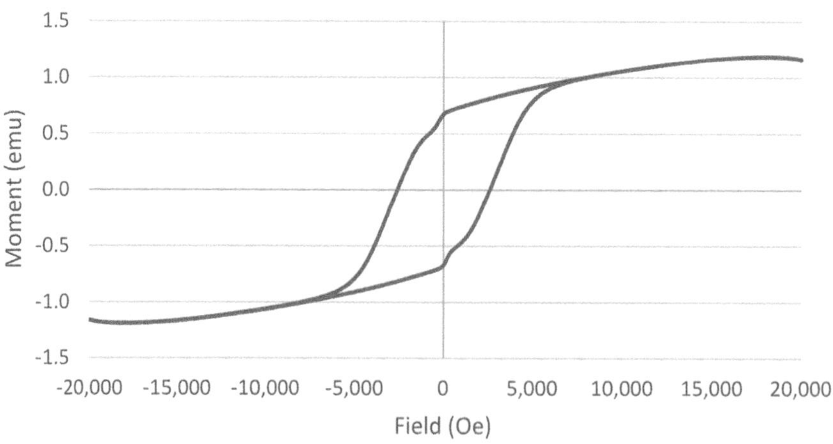

FIGURE 11.4 Hysteresis *M*(*H*) loop for $BaFe_{12}O_{19}$ nanoparticles.

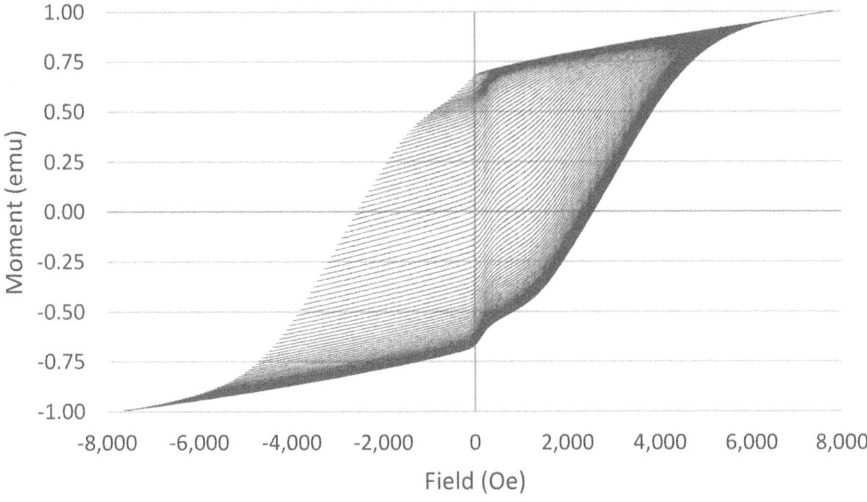

FIGURE 11.5 FORCs for $BaFe_{12}O_{19}$ nanoparticles.

playground for fundamental studies of magnetic interactions and magnetization mechanisms at the nanoscale level. When investigating the magnetic interactions in these materials, one of the most interesting configurations is a periodic array of magnetic nanowires, because both the size of the wires and their arrangement with respect to one another can be controlled. Interwire coupling is one of the most important effects in nanowire arrays because it significantly affects magnetization switching and microwave and magneto-transport properties. Experimentally, FORCs are used to investigate the effect and strength of these interactions.

Ni nanowire samples were fabricated by electro-deposition using anodic aluminum oxide (AAO) membrane as a template. Figure 11.7 shows scanning electron microscopy (SEM) images of a series of alumina membranes prepared with different pore diameters.

Figure 11.8 shows a series of FORCs measured at $T = 300$ K for a periodic array of Ni nanowires with a mean diameter of 70 nm and interwire spacing of 250 nm. Figure 11.9 shows the FORC diagram and shows the distribution of coercive and interaction fields resulting from coupling between adjacent nanowires [7]. The long tail centered at $H_u = 0$ has been theoretically shown to be due to interwire interactions occurring between higher coercivity wires within the array [28].

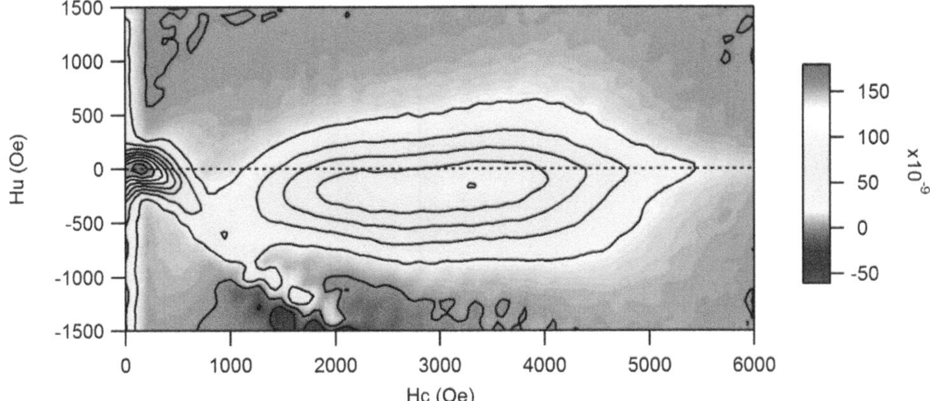

FIGURE 11.6 FORC diagram for $BaFe_{12}O_{19}$ nanoparticles; the low and high coercivity phases are clearly differentiated.

FIGURE 11.7 SEM images of the top surfaces of AAO templates with pore diameters of (a) 40, (b) 60, and (c) 80 nm, and interpore distances of 100 nm.

Figure 11.10 shows a series of FORCs measured at $T = 300$ K for an array of sub-100 nm CoPt nanomagnets in a high-density patterned magnetic recording media, and Figure 11.11 shows the resultant FORC diagram. Note that the peak in the FORC distribution is shifted toward negative interaction fields (H_u) and that the distribution has a "boomerang" shape. These features are usually associated with exchange interactions [29], suggesting that exchange interactions are occurring between adjacent CoPt nanomagnets within the array.

11.3.3 Core/Shell Granular Heteronanostructures

A wide range of multicomponent nanostructures can be realized through the colloidal synthesis of inorganic nanocrystals. It enables the combination of different phases and provides a powerful approach for the creation of smart materials with novel properties.

Figure 11.12 shows transmission electron microscopy (TEM) images of a magnetically contrasted granular heteronanostructure consisting of two oxide phases with different magnetic order: a core consisting of 10 nm ferrimagnetic partially oxidized iron oxide ($Fe_{3-x}O_4$) nanoparticles in an antiferromagnetic (AFM) cobalt oxide (CoO) shell.

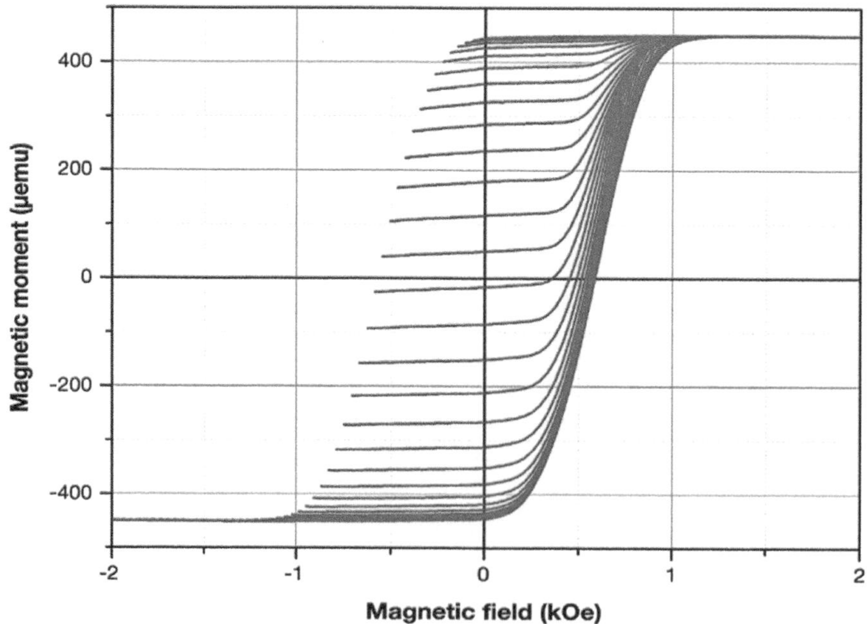

FIGURE 11.8 FORCs for an array of Ni nanowires.

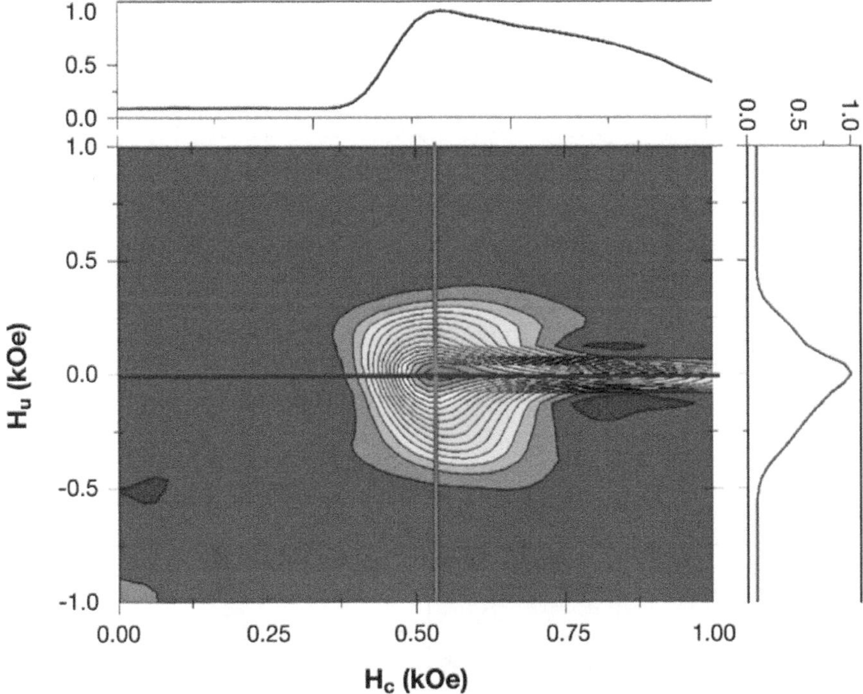

FIGURE 11.9 FORC diagram for an array of Ni nanowires. The upper and right curves correspond to the profile of the distribution along the horizontal and vertical marked lines, respectively.

Figure 11.13 shows a series of FORCs measured at T = 120 K, and Figure 11.14 shows the resultant FORC diagram for the 10-nm $Fe_{3-x}O_4$ nanoparticles core/CoO shell [30]. The FORC diagram consists of open contours and is consistent with a dispersed collection of predominantly MD particles. The peak in the FORC distribution is shifted toward positive interaction fields (H_u), which suggests that interactions are occurring between the iron oxide nanoparticles.

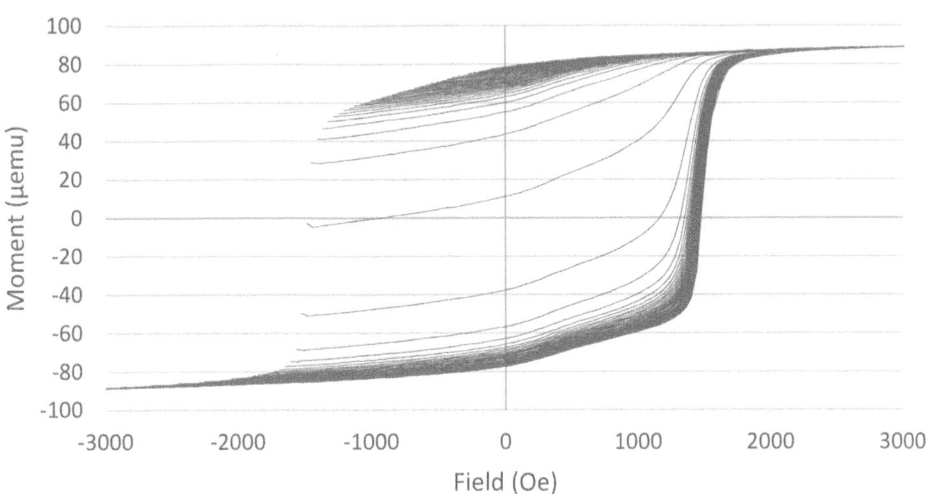

FIGURE 11.10 FORCs for an array of sub-100 nm CoPt nanomagnets.

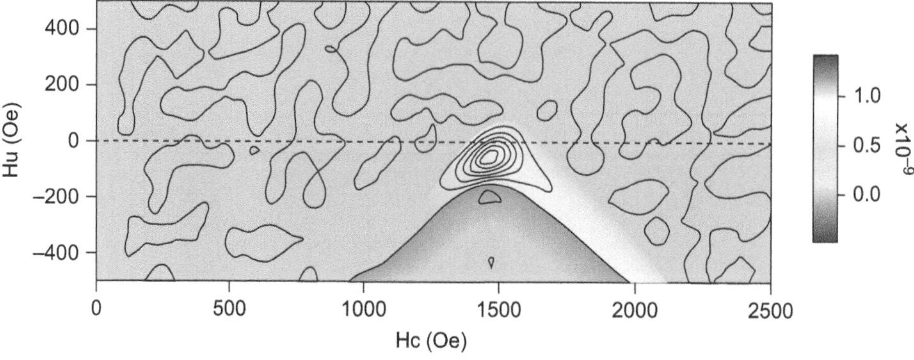

FIGURE 11.11 FORC diagram for an array of sub-100 nm CoPt nanomagnets.

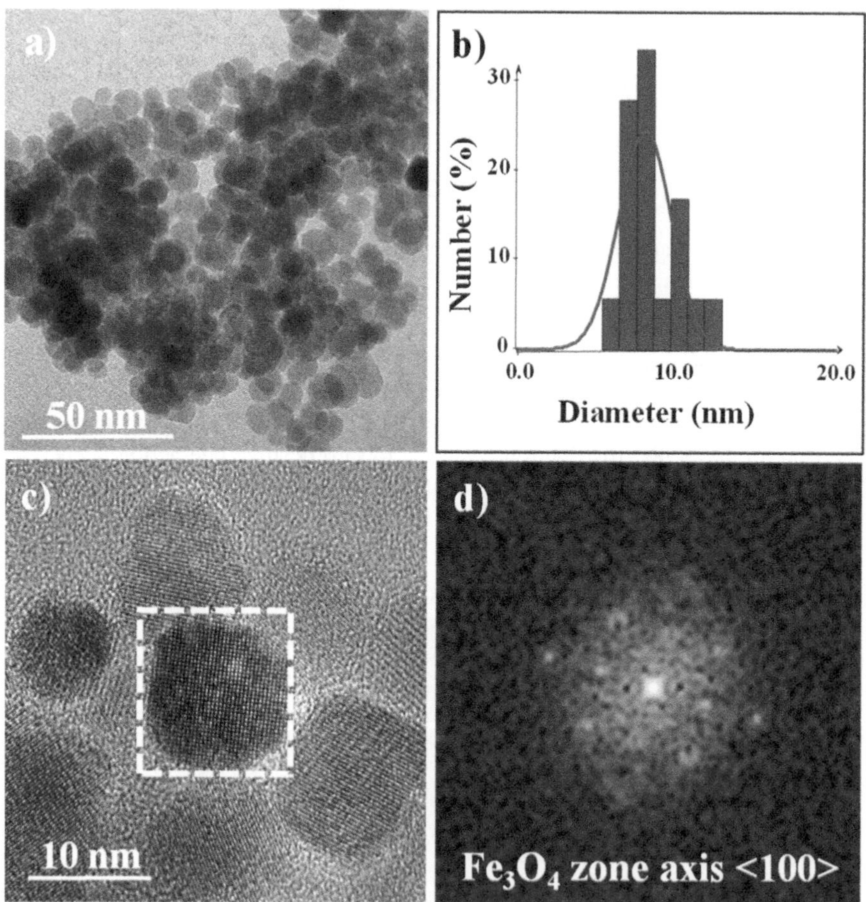

FIGURE 11.12 (a) TEM image of an assembly of 10-nm-sized polyol-made iron oxide particles, their size distribution (b), and one representative HRTEM micrograph (c) with the FFT pattern (d) calculated for the selected area.

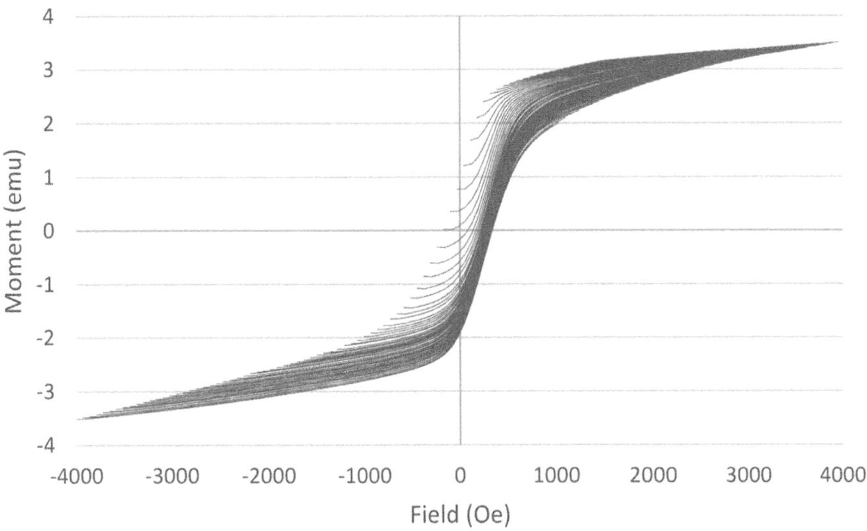

FIGURE 11.13 FORCs at T = 120 K for 10-nm $Fe_{3-x}O_4$ nanoparticles core/CoO shell.

11.3.4 Doped Magnetic Nanoparticles (DMNP)

The magnetic properties of DMNP-based nanomagnets could impact targeted drug delivery as well as become basic building blocks for future permanent magnets and other electromagnetic devices. By incorporating a magnetic impurity ion, Fe^{2+} in a paramagnetic manganese oxide Mn_3O_4 nanoparticle host, the paramagnetic phase of Mn_3O_4 is converted to a ferromagnetic (FM) phase when the FM nanoparticle is in the range of 5 to 10 nm.

Figure 11.15 shows a series of FORCs measured at T = 300 K for 0.5% Fe-doped Mn_3O_4 nanoparticles, and Figure 11.16 shows the resultant FORC diagram. The FORC diagram consists of both closed and open contours and thus suggests that the DMNPs are behaving as a collection of PSD particles. The peak in the diagram correlates with the coercivity of the sample as determined from the hysteresis loop measurement, and it is centered at $H_u = 0$, which means that there are negligible interactions between Fe-ions, which is reasonable given the dilution of the Fe^{2+} in the Mn_3O_4 paramagnetic nanoparticles host [31].

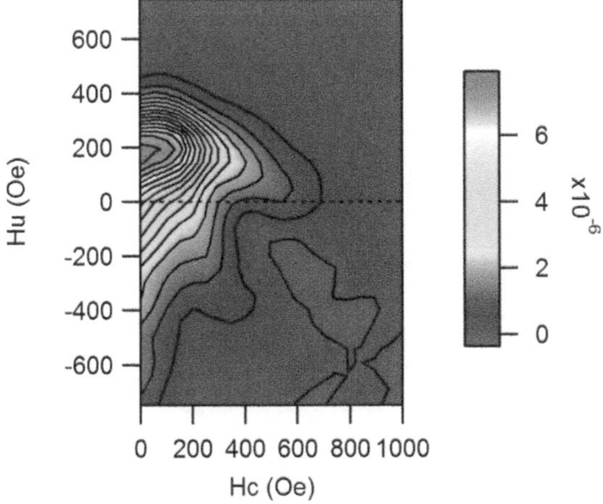

FIGURE 11.14 FORC diagram at T = 120 K for 10-nm $Fe_{3-x}O_4$ nanoparticles core/CoO shell.

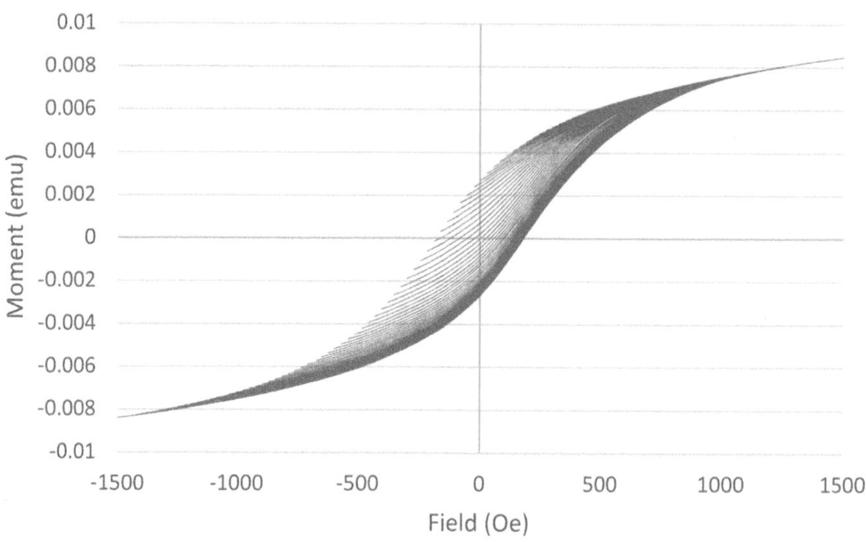

FIGURE 11.15 FORCs for 0.5% Fe^{2+}-doped Mn_3O_4 nanoparticles.

11.3.5 Exchange Bias Magnetic Multilayer Films

Exchange bias magnetic multilayer films are technologically important materials for applications such as spin-valve read heads for hard disk drives and gigahertz-range microwave devices. In these materials, at least one anti-ferromagnetic AFM layer is intercalated between FM layers. In addition to their technological applications, they are also useful for fundamental studies of magnetic interactions and magnetization reversal processes in magnetic nanostructures because both the number (n) of AFM/FM interfaces and the thickness of the FM and AFM layers can be controlled.

Figure 11.17 shows the $M(H)$ loop measured at $T = 300$ K, with the applied field oriented in-plane and parallel to the easy axis for a multilayer film [32] of composition [FeNi (60 nm)/IrMn (20 nm)]$_n$, where FeNi represents Ni (80%) Fe (20%) and the number of layers is $n = 5$.

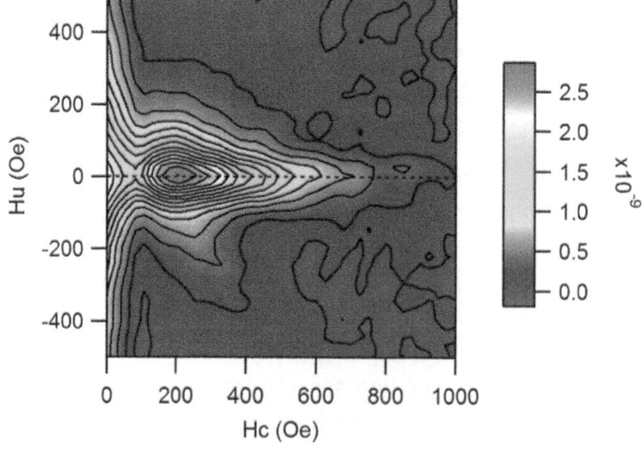

FIGURE 11.16 FORC diagram for 0.5% Fe^{2+}-doped Mn_3O_4 nanoparticles.

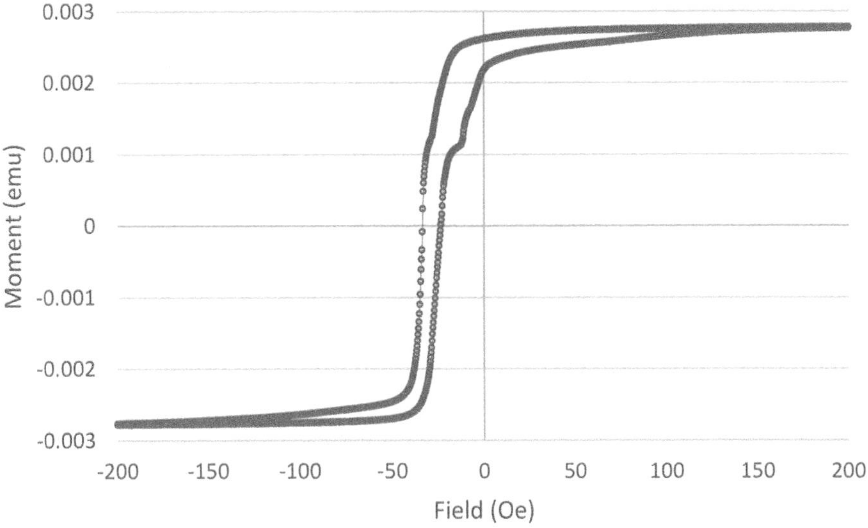

FIGURE 11.17 Hysteresis loop for [FeNi (60 nm)/IrMn (20 nm)]$_5$ for the applied field parallel to the easy axis.

When the magnetic field is applied parallel to the exchange bias field, the loop is shifted toward the left (negative field values), and the exchange bias and coercivity fields are: $H_{ex} = -30$ Oe and $H_c = 4$ Oe. The extra steps in the ascending and descending branches of the magnetization curve in the second quadrant are related to microstructural defects/roughness of the AFM/FM interfaces.

Figure 11.18 shows a series of FORCs measured at $T = 300$ K for the field oriented parallel to the exchange bias field, and Figure 11.19 shows the corresponding FORC diagram [33]. The diagram indicates that there is a main peak of the distribution that is centered around H_c (4 Oe); however, note that the distribution of switching fields extends over several Oe. The peak of the distribution in the H_u direction corresponds to the exchange bias field H_{ex} (−30 Oe). The spread of the distribution in

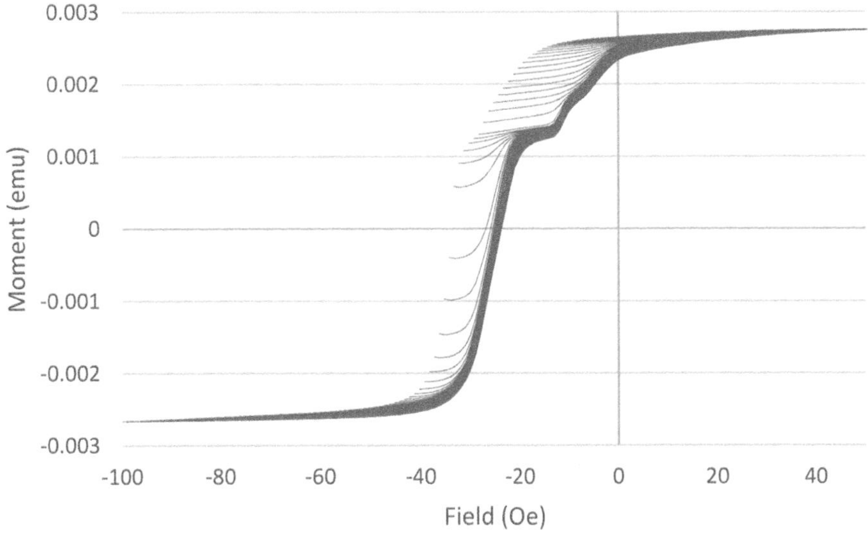

FIGURE 11.18 FORCs for [FeNi (60 nm)/IrMn (20 nm)]$_5$ for the applied field parallel to the easy axis.

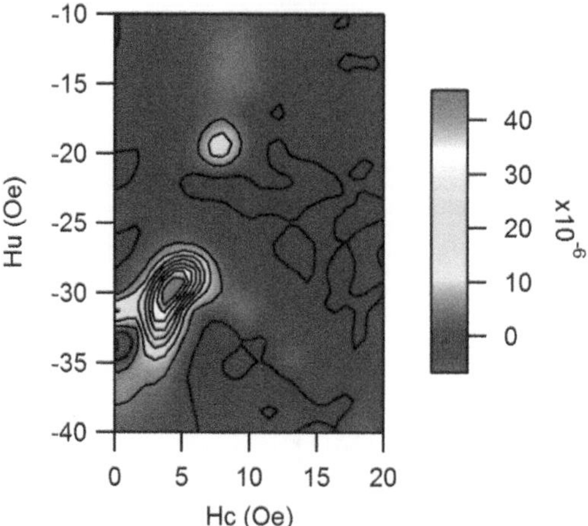

FIGURE 11.19 FORC diagram for [FeNi (60 nm)/IrMn (20 nm)]$_5$ for the applied field parallel to the easy axis.

the H_u direction is related to interactions between the AFM and FM layers. The satellite distribution centered at $H_u = -20$ Oe and $H_c = 7$ Oe is related to structural inhomogeneities at the AFM/FM interface and is more pronounced the higher the number of layer repetitions or, equivalently, the higher the number of AFM/FM interface inhomogeneities. The FORC measurement and analysis protocol provide additional information that cannot be obtained from the standard hysteresis loop measurement alone.

11.3.6 Nanocrystalline Nanocomposites

Soft magnetic materials are important for high-frequency power electronic components and systems for power conditioning and grid integration. Weight and size reductions are possible in these applications through operation at increased frequencies, and FeCo-based nanocrystalline nanocomposites can be used for frequency operation at 100 kHz and above. Figure 11.20 shows a series of FORCs

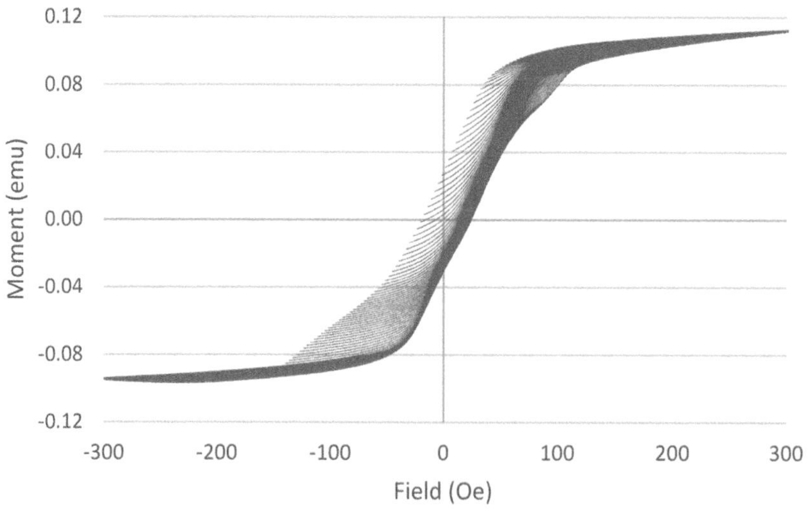

FIGURE 11.20 FORCs at $T = 780°C$ for a $Fe_{56}Co_{24}Si_2B_{13}Nb_4Cu_1$ nanocrystalline amorphous/nanocomposite melt spun ribbon.

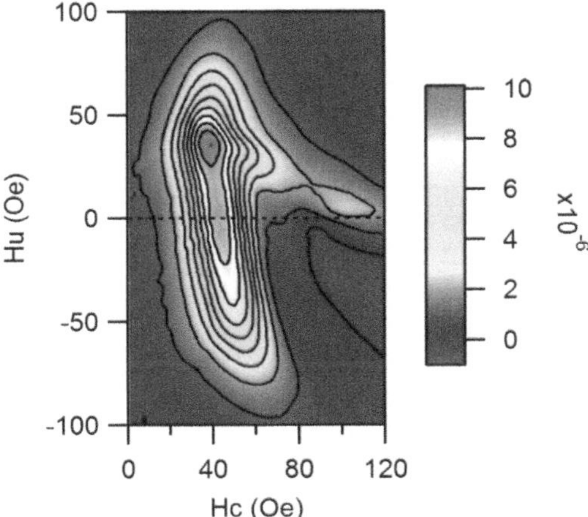

FIGURE 11.21 FORC diagram at $T = 780°C$ for a $Fe_{56}Co_{24}Si_2B_{13}Nb_4Cu_1$ nanocrystalline amorphous/nanocomposite melt spun ribbon.

measured at $T = 780°C$ (1053 K), which is above the Curie temperature (520°C) of the amorphous phase, for a $Fe_{56}Co_{24}Si_2B_{13}Nb_4Cu_1$ nanocrystalline (<10 nm) amorphous/nanocomposite melt spun ribbon. Figure 11.21 shows the corresponding FORC diagram [34] and shows a "wishbone-like" feature. Wishbone FORCs typically mean that there are dipolar (or magnetostatic) interactions between particles. The peak in the FORC distribution is shifted toward positive interaction fields (H_u), which suggests that the dipolar interactions are strong.

11.4 CONCLUSION

FORC analysis is indispensable and serves as a "fingerprint" for understanding the nature of and characterizing interactions and coercivity distributions in a wide array of magnetic materials, including natural magnets, magnetic recording media, nanowire and nanomagnet arrays, exchange-coupled permanent magnets, exchanged-biased magnetic multilayers, core/shell magnetic hetero-nanostructures, doped magnetic nanoparticles, and nanocrystalline nanocomposites. In this chapter, we discussed the FORC measurement technique and subsequent analysis that leads to the FORC diagram and presented results for several relevant nanomagnetic materials.

REFERENCES

1. I. D. Mayergoyz, Mathematical Models of Hysteresis and their Applications, 2nd Ed.; Academic Press, 2003.
2. C. R. Pike, A. P. Roberts, K. L. Verosub, Characterizing Interactions in Fine Magnetic Particle Systems Using First Order Reversal Curves, *J. Appl. Phys.* 85, 6660,1999.
3. A. P. Roberts, C. R. Pike, K. L. Verosub, First-Order Reversal Curve Diagrams: A New Tool for Characterizing the Magnetic Properties of Natural Samples, *J. Geophys. Res.*, 105, 461, 2000.
4. A. Rotaru, J. Lim, D. Lenormand, A. Diaconu, J. Wiley, P. Postolache, A. Stancu, L. Spinu, Interactions and Reversal Field Memory in Complex Magnetic Nanowire Arrays, *Phys. Rev. B*, 84, 13, 134431, 2011.
5. O. Trusca, D. Cimpoesu, J. Lim, X, Zhang, J. Wiley, A. Diaconu, I. Dumitru, A. Stancu, L. Spinu, Interaction Effects in Ni Nanowire Arrays, *IEEE Trans. Mag.*, 44, 11, 2730, 2008.
6. A. Arefpour, M. Almasi-Kashi, A. Ramazani, E. Golafshan, The Investigation Of Perpendicular Anisotropy of Ternary Alloy Magnetic Nanowire Arrays Using First Order Reversal Curves, *J. Alloys and Comp.*, 583, 340, 2014.
7. B. C. Dodrill, L. Spinu, First-Order-Reversal-Curve Analysis of Nanoscale Magnetic Materials, Technical Proceedings of the 2014 NSTI Nanotechnology Conference and Exposition, CRC Press, 2014.

8. A. Sharma, M. DiVito, D. Shore, A. Block, K. Pollock, P. Solheid, J. Feinberg, J. Modiano, C. Lam, A, Hubel, B. Stadler, Alignment of Collagen Matrices Using Magnetic Nanowires and Magnetic Barcode Readout Using First Order Reversal Curves (FORC), *J. Mag. Mag. Mat.*, 459, 176, 2018.
9. F. Beron, L. Carignan, D. Menard, A. Yelon, Extracting Individual Properties from Global Behavior: First Order Reversal Curve Method Applied to Magnetic Nanowire Arrays, Electrodeposited Nanowires and Their Applications, edited by N. Lupu, 228, 2010, INTECH, Croatia.
10. R. Dumas, C. Li, I. Roshchin, I. Schuller, K. Liu, Magnetic Fingerprints of sub-100 nm Fe Nanodots, *Phys. Rev. B*, 75,134405, 2007.
11. D. Gilbert, G. Zimanyi, R. Dumas, M. Winklhofer, A. Gomez, N. Eibagi, J. Vincent, K. Liu, Quantitative Decoding of Interactions in Tunable Nanomagnet Arrays, *Sci. Rep.*, 4, 4204, 2014.
12. J. Graffe, M. Weigand, C. Stahl, N. Trager, M. Kopp, G. Schutz, E. Goering, Combined First Order Reversal Curve and X-ray Microscopy Investigation of Magnetization Reversal Mechanisms in Hexagonal Antidot Arrays, *Phys. Rev. B*, 93, 014406, 2016.
13. B. Valcu, D. Gilbert, K. Liu, Fingerprinting Inhomegeneities in Recording Media Using First Order Reversal Curves, *IEEE Trans. Mag.*, 47, 2988, 2011.
14. A. Stancu, E. Macsim, Interaction Field Distribution in Longitudinal and Perpendicular Structured Particulate Media, *IEEE Trans. Mag.*, 42, 10, 3162, 2006.
15. M. Winklhofer, R. K. Dumas, K. Liu, Identifying Reversible and Irreversible Magnetization Changes in Prototype Patterned Media Using First- and Second-Order Reversal Curves, *J. Appl. Phys.* 103, 07C518, 2008.
16. R. Dumas, C. Li, L. Roshchin, I. Schuller, K. Liu, Deconvoluting Reversal Modes in Exchange Biased Nanodots, *Phys. Rev. B*, 144410, 2012.
17. R. Gallardo, S. Khanai, J. Vargas, L. Spinu, C. Ross, C. Garcia, Angular Dependent FORC and FMR of Exchange Biased NiFe Multilayer Films, *J. Phys. D: Appl. Phys.*, 50, 075002, 2017.
18. N. Siadou, M. Androutsopoulos, I. Panagiotopoulos, L. Stoleriu, A. Stancu, T. Bakas, V. Alexandrakis, Magnetization Reversal in [Ni/Pt](6)/Pt(x)/[Co/Pt](6) Multilayers, *J. Mag. Mag. Mat.*, 323, 12, 1671, 2011.
19. T. Schrefl, T. Shoji, M. Winklhofer, H. Oezeit, M. Yano, G. Zimanyi, First Order Reversal Curve Studies of Permanent Magnets, *J. Appl. Phys.*, 111, 07A728, 2012.
20. M. Pan, P. Shang, H. Ge, N. Yu, Q. Wu, First Order Reversal Curve Analysis of Exchange Coupled SmCo/NdFeB Nanocomposite Alloys, *J. Mag. Mag. Mat.*, 361, 219, 2014.
21. M. Rivas, J. Garcia, I. Skorvanek, J. Marcin. P. Svec, P. Gorria, Magnetostatic Interaction in Soft Magnetic Bilayer Ribbons Unambiguously Identified by First Order Reversal Curve Analysis, *Appl. Phys. Lett.*, 107, 132403, 2015.
22. V. Franco, F. Beron. K. Pirota, M. Knobel, M. Willard, Characterization of Magnetic Interactions of Multiphase Magnetocaloric Materials using First Order Reversal Curve Analysis, *J. Appl. Phys.*, 117, 17C124, 2015.
23. B. C. Dodrill, First-Order-Reversal-Curve Analysis of Nanocomposite Permanent Magnets, Technical Proceedings of the 2015 TechConnect World Innovation Conference and Expo, CRC Press, 2015.
24. C. Carvallo, A. R. Muxworthy, D. J. Dunlop, First-Order-Reversal-Curve (FORC) Diagrams of Magnetic Mixtures: Micromagnetic Models and Measurements, *Phys. Earth Planet. Inter.*, 154, 308, 2006.
25. R. J. Harrison, J. M. Feinberg, FORCinel: An Improved Algorithm for Calculating First-Order Reversal Curve Distributions Using Locally Weighted Regression Smoothing, Geochemistry, Geophysics, Geosystems. 9, 11, 2008. FORCinel may be downloaded from: https://wserv4.esc.cam.ac.uk/nanopaleomag/?page_id=31.
26. R. Egli, VARIFORC: An Optimized Protocol for Calculating Non-regular First-Order Reversal Curve (FORC) Diagrams, *Global Planetary Change*, 203, 110, 203, 2013.
27. Y. Cao, M. Ahmadzadeh, K. Xe, B. Dodrill, J. McCloy, Multiphase Magnetic Systems: Measurement and Simulation, *J. Appl. Phys.* 123(2), 023902, 2018.
28. C. Dubrota, A. Stancu, What Does a First Order Reversal Curve Really Mean: A Case Study: Array of Ferromagnetic Nanowires, *J. Appl. Phys.* 113, 043928, 2013.
29. D. Roy, P. S. A. Kumar, Exchange Spring Behaviour in $SrFe_{12}O_{19}$-$CoFe_2O_4$ Nanocomposites, *AIP Advances*, 5, 2015.

30. G. Franceschin, T. Gaudisson, N. Menguy, B. C. Dodrill, N. Yaacoub, J. Greneche, R. Valenzuela, S. Ammar, Exchange-Biased $Fe_{3-x}O_4$ Granular Composites of Different Morphologies Prepared by Seed-Mediated Growth in Polyol: From Core-Shell to Multicore Embedded Structures, *Part. Part. Syst. Charact.*, 1800104, 2018.
31. R. Bhargava, C. Michel, V. Parashar, A. Mehta, B. Dodrill, A. Goswami, J. Chen, T. Thundat, Doped Magnetic Nanoparticles (DMNP) – A New Class of Ferromagnetic Nanoparticles, submitted to *Phys. Rev. Mater.*, 2018.
32. Sample courtesy of C. Garcia, Massachusetts Institute of Technology (MIT).
33. B. C. Dodrill, Magnetometry and First-Order-Reversal-Curve (FORC) Studies of Nanomagnetic Materials, Dekker Encyclopedia of Nanoscience and Nanotechnology, Taylor & Francis, 2016.
34. B. Dodrill, P. Ohodnicki, M. McHenry, A. Leary, High-Temperature First-Order-Reversal-Curve (FORC) Study of Magnetic Nanoparticle Based Nanocomposite Materials, *MRS Adv.*, 2017.

12 Spin-Dependent Scattering of Electrons on Magnetic Heterostructure with Noncoplanar Element

A. S. Sahakyan,[1] *A. V. Poghosyan,*[1] *R. M. Movsesyan,*[1] *and A. N. Kocharian*[2]

[1]The State Engineering University of Armenia, Yerevan, Armenia
[2]California State University, Los Angeles, California

CONTENTS

12.1 INTRODUCTION

The main objective of spintronics is emerged from the research areas on spin-dependent electron transport phenomena, spin-polarized electric current and its control after the research discoveries done by A. Fermit and P. Grünberg and their groups in the ferromagnet/superconductor tunneling experiments. The researchers have considered a variety of magnetic systems including multilayer systems, or granular materials consisting of separated regions of magnetic material with independent magnetization orientations can be controlled by externally applied magnetic fields. The spin-dependent elastic scattering through materials controlled by changing the magnetic configuration led to the discovery of giant magnetoresistance, magnetic tunnel junctions, spin valve effects, and related phenomena, which are of important fundamental and applied significance [1,2]. The spin-dependent scattering is encountered in the systems when an initial antiparallel orientation of internal magnetic field is changed into a parallel configuration by application of an external magnetic field. The scattering probability depends upon the number of available quantum states for the electron to magnetic moment of nanoelement, which depends strongly on the relative orientation of the electron's spin and the magnetic field of nanoelement. The more states are available, the higher the probability of scattering, and the lower the electron transmittance. If the spin and magnetic field are parallel, fewer states are available for electron scattering, so the transmittance is larger than if the spin and the magnetic field are anti-parallel. This is the basic idea of spin-dependent scattering. The observation result strongly depends on geometric parameters (the thicknesses of the layers); intrinsic parameters (magnetization, and effective masses); bulk sample properties (conductivity relaxation times) and interface scattering properties. In general, the magnetic structure on the interface might cause a considerable asymmetry of scattering at two different electron spin orientations. We consider the tunneling through a magnetic "defect" layer (nanoelement) having internal magnetic field which orientation is noncoplanar to directions of internal magnetic fields of other layers.

Many publications with interesting results have appeared both experimental and theoretical [3–7] provide the rapid progress of this field of science. Highly popular objects of research are systems with so-called magnetic barriers, as well as films containing magnetic nanolayers. For example, in [8–13] transport properties of systems with magnetic barriers were investigated. Under certain conditions it is possible to carry out theoretical studies through the use of Lippmann-Schwinger equation. As perspective materials for ferromagnetic layers, it is convenient to choose semiconductors with a wide band gap, for example ZnO doped with Mn impurities, which ensure the ferromagnetic properties of the nanolayer: very strong internal magnetic fields and high Curie temperature [14–17]. The energy profiles of films containing the described nanolayers contain potential barriers with high degrees of magnetization separated from each other by nonmagnetic quantum wells. Either non-magnetic semiconductors with a narrow band gap or normal metals can be used as non-magnetic layers in the film. Hereinafter we assume that the thicknesses of the nonmagnetic layers are large enough so that there is no exchange interaction between two neighboring barriers. Systems possessing such an energy profile can be used as spin filters in which various transport characteristics can be controlled by varying the spatial orientations of the internal magnetic fields' vectors of the nanolayers [18–20].

Here we consider one-dimensional elastic scattering of polarized and nonpolarized quantum waves (electrons) on the systems described earlier, which contain potential barriers with noncoplanar vectors of internal magnetic fields. In the case of a single (isolated) barrier, noncoplanarity is due to the mutual spatial orientations of the vectors of the incident wave spin polarization and the internal field of the barrier. The multi-barrier system has its "own" noncoplanarity. In the context of the quantum theory of scattering, the existence of noncoplanar picked out directions characterizing the system provides the conditions for scattering having a two-channel character: without and with a spin-flip scattering of the electron. These selected directions can be characterized by certain angular degrees of freedom and both the transmission and reflection coefficients and the degrees of spin polarization of the scattered waves also depend on these parameters. Thus, by varying these degrees of freedom, the transport characteristics can be controllably varied. Obviously, in the case when the incident wave is polarized, this problem is reduced to question about possible transformations of the spin polarization of the scattered waves. In the case of a nonpolarized incident wave, the question can be posed as follows: under what conditions can the spin polarization of at least one of the scattered waves reach a maximum value. For observer, the wave scattered forward (the transmitted wave) is actual. Of course, the magnitude of the transmission coefficient is also essential. Therefore, the novel theories are in demand to understand the origin of the spin dependence of the experimentally measured parameters, and how these parameters relate to properties of magnetic nanoelement.

A number of effects that are presented here are similar to well-known effects from the existing literature and, in fact, are their analogs. We note separately the basic approximation, which is used here: the subbarrier (under-barrier) motion of a particle excludes orbital quantization—only spin-dependent scattering is possible. In addition, the estimates show that the scattering time is much less (100-1000 times) than the cyclotron period. These circumstances make it possible to omit the term $\frac{e}{c}\vec{A}$ in the expression for the canonical momentum of the electron. The electron interacts with the electrostatic field of the barrier, and its spin with its magnetic field. Therefore, a situation here becomes similar to the case of neutron optics [21–24].

12.2 TRANSFER MATRIX METHOD

For the study of spin-dependent scattering of electron on different structures one can use a very effective transfer matrix method developed in optics and acoustics for wave propagation, so here we present the construction of it. When referring to the spin-dependent scattering we mean the scattering of particles with half spin on a magnetized barrier, characterized by the induction vector of the internal magnetic field, which is assumed to be homogeneous. An electron with the spin $\vec{s}$ has a magnetic moment $\vec{\mu} = -g\mu_B\vec{s}$, where $\mu_B = \frac{|e|\hbar}{2mc} \approx 0,9 \cdot 10^{-20}$ Erg/G is the Bohr magneton, m – the mass of the electron, g – g-factor which is equal to 2 in vacuum (in crystals, its value is somewhat different from 2).

The most general form of the Hamiltonian for the interaction of the spin of an electron with the internal magnetic field is [25]

$$H_{\text{int}}^{(2)} = -\frac{1}{2}\mu_B g_{\alpha\beta}\sigma_\alpha B_\beta, \tag{12.1}$$

where μ_B – is the Bohr magneton, σ_α – are the Pauli matrices, B_β – the components of the induction of the internal magnetic field, $g_{\alpha\beta}$ – the tensor of g-factors. The summation over repeating indices is carried. Later, we will neglect anisotropy, by assuming $g_{\alpha\beta} = g\delta_{\alpha\beta}$, and then

$$H_{\text{int}}^{(2)} = -\gamma\vec{\sigma}\vec{B}, \quad \gamma = \frac{1}{2}\mu_B g. \tag{12.2}$$

Thus, the total Hamiltonian of electron – magnetized barrier interaction has the following form:

$$H_{\text{int}} = H_{\text{int}}^{(1)} + H_{\text{int}}^{(2)} = (V_0 - \gamma\vec{B}\vec{\sigma})\varphi(y), \tag{12.3}$$

where V_0 is the “amplitude” of the potential barrier, $\varphi(y)$ is a finite function on the y coordinate. The coordinate axes are chosen as follows: axis x and z lie in the film plane, and the axis y is perpendicular to it.

The components k_x, k_z of the momentum are conserved. Thus, the electron scattering processes are one-dimensional in nature, and occur along the y axis.

The corresponding Schrödinger equation that describes the electron scattering is

$$(\hat{H}_0 + \hat{H}_{\text{int}})\hat{\psi} = E\hat{\psi}, \tag{12.4}$$

where $\hat{H}_0$ is the Hamiltonian of a free particle, $\hat{H}_{\text{int}}$ is defined by the expression Eq. 12.3, $\hat{\psi} = \begin{pmatrix}\psi_1\\ \psi_2\end{pmatrix}$ is the spinor wave function of the electron, $E = \varepsilon - \frac{\hbar^2 k_\parallel^2}{2m}$, ε is the electron energy, E – the scattering energy or the electron energy corresponding to the y – degree of freedom of the electron, $k_\parallel$ – the parallel to the interface component of the wave vector. We get Equation 12.4 by means of the separation of variables that has a one-dimensional character.

Here we consider the case when the operator of the spin-magnetic field interaction in the Hamiltonian (Eq. 12.3) has the form $-\gamma B\sigma_z\varphi(y)$, where the z axis is selected as the quantization axis. Let's search for solution of the asymptotic wave function in the form

$$\begin{aligned} \hat{\psi}_I &= \hat{I}e^{iky} + \hat{r}e^{-iky}, \quad y \to -\infty \\ \hat{\psi}_{III} &= \hat{t}e^{iky}, \quad y \to +\infty, \end{aligned} \tag{12.5}$$

where $\hat{I} = \begin{pmatrix}I_1\\ I_2\end{pmatrix}$ is the spinor amplitude of the incident wave, $\hat{r} = \begin{pmatrix}r_1\\ r_2\end{pmatrix}$, $\hat{t} = \begin{pmatrix}t_1\\ t_2\end{pmatrix}$ – the amplitudes of the reflected and transmitted waves, respectively. Hereinafter we'll call r_ℓ and $t_\ell(\ell = 1,2)$ partial amplitudes. By the choice of the quantization axis components $\psi_{1,2}$ in the Equation 12.4 isn't entangled and as a result we conclude that $r_\ell = r(E, V_0 \pm \gamma B), t_\ell = t(E, V_0 \pm \gamma B)(\ell = 1,2)$, where

$r(E, V_0)$ and $t(E, V_0)$ are the corresponding amplitudes of the scalar scattering problem, when there is no interaction between the spin and the internal magnetic field.

In this simplest case when the quantization axis is chosen on the analogy of the scalar scattering problem, the transfer matrix can be presented in the following block-diagonal form:

$$\tilde{S} = \begin{pmatrix} \tilde{S}_{01} & 0 \\ 0 & \tilde{S}_{02} \end{pmatrix}, \quad \tilde{S}_\ell = \begin{pmatrix} \frac{1}{t_\ell^*}, & -\frac{r_\ell^*}{t_\ell^*} \\ -\frac{r_\ell}{t_\ell}, & \frac{1}{t_\ell} \end{pmatrix}. \tag{12.6}$$

It is easy to obtain expressions for the corresponding amplitudes of scattering on a system consisting of two consequent identical barriers of the same height and with the same internal magnetic fields.

The calculated amplitudes of transmission and reflection correspondingly are equal to

$$T_{0\ell} = \frac{t_\ell^2}{1 - r_\ell^2 e^{-2ikc}}, \quad R_{0\ell} = -\frac{r_\ell\left[1 + e^{2i(\delta_\ell + kc)}\right]}{1 - r_\ell^2 e^{-2ikc}}, \tag{12.7}$$

where c is the distance between the barriers, δ_ℓ–the phase of the amplitude of transmission: $t_\ell = |t_\ell|e^{i\delta_\ell}$. In the case of a rectangular barrier asymptotic wave function (Eq. 12.5), coincides with the exact ones in the areas I and III, and then for the partial amplitudes we obtain the following expressions:

$$t_\ell = \frac{m_2}{m_1}\frac{4ikaq_\ell e^{-ika}}{\left(q_\ell + i\frac{m_2}{m_1}k\right)^2 e^{-q_\ell a} - \left(q_\ell - i\frac{m_2}{m_1}k\right)^2 e^{q_\ell a}}, \quad r_\ell = \frac{2\left(q_\ell^2 + \frac{m_2^2}{m_1^2}k^2\right)^2 sh(q_\ell a)}{\left(q_\ell + i\frac{m_2}{m_1}k\right)^2 e^{-q_\ell a} - \left(q_\ell - i\frac{m_2}{m_1}k\right)^2 e^{q_\ell a}}$$
$$k = \frac{1}{\hbar}\sqrt{2m_2 E}, \quad q_\ell = \frac{1}{\hbar}\sqrt{2m_1(V_0 \mp \gamma B - E)}, \tag{12.8}$$

where m_1 and m_2 are the effective masses of an electron in the regions of the quantum well and the magnetic barrier, respectively.

From Equation 12.8 we can see that for electrons with a spin oriented along the magnetic field of the barrier, its effective height is $(V_0 - \gamma B)$, and for electrons whose spins are directed opposite to the magnetic field - $(V_0 + \gamma B)$.

In Equation 12.7, δ_ℓ are the phases of the transmission amplitudes: $t_\ell = |t_\ell|e^{i\delta_\ell}$, for a rectangular barrier, they are as follows:

$$\delta_\ell = \frac{\pi}{2} - 2ka + \arctan\left\{\frac{2q_\ell k\, cth(q_\ell a)}{\frac{m_2}{m_1}k^2 - \frac{m_1}{m_2}q_\ell^2}\right\}. \tag{12.9}$$

At values of the scattering energy determined from one of the equations

$$\delta_\ell = -kc + \pi\left(n + {}^1\!/_2\right) \tag{12.10}$$

A resonant transmission occurs when one of the coefficients of transmission turns to unity. This phenomenon is the quantum analogue of the well-known Fabry-Perot resonance in optics [26]. Further, it is more convenient to reintroduce the transfer matrix (Eq. 12.7) as follows:

$$S_0\begin{pmatrix} \hat{I} \\ \hat{r} \end{pmatrix} = \begin{pmatrix} \hat{t} \\ 0 \end{pmatrix}. \tag{12.11}$$

It is easy to show that S_0 and $\tilde{S}_0$ are related as follows:

$$S_0 = A\tilde{S}_0 A \tag{12.12}$$

where

$$A = \begin{pmatrix} 1 & 0 & 0 & 0 \\ 0 & 0 & 1 & 0 \\ 0 & 1 & 0 & 0 \\ 0 & 0 & 0 & 1 \end{pmatrix}, \quad A^{-1} = A \tag{12.13}$$

Then

$$S_0 = \begin{pmatrix} \alpha^*, & -\beta^* \\ -\beta, & \alpha \end{pmatrix}, \quad \alpha = \begin{pmatrix} 1/t_1 & 0 \\ 0 & 1/t_2 \end{pmatrix}, \quad \beta = \begin{pmatrix} r_1/t_1 & 0 \\ 0 & r_2/t_2 \end{pmatrix}. \tag{12.14}$$

We also can choose the interaction operator $H_{\text{int}}^{(2)}$ in the form (Eq. 12.2): where the z axis is no longer the quantization axis, and its matrix is nondiagonal in contrast to the previously discussed case. The Hamiltonian (Eq. 12.2) and, consequently, the total Hamiltonian can be diagonalized by an appropriately chosen unitary transformation, *i.e.*, $\hat{H}' = U^{+}HU$. Then the amplitudes of transmission and reflection of the unitary-transformed problem are reduced to the following $\hat{r}' = U^{+}\hat{r}$ and $\hat{t}' = U^{+}\hat{t}$, where $\hat{r}$ and $\hat{t}$- are the amplitudes of the original problem. Then the transfer matrix of the unitary transformed problem, consistent with the definition (Eq. 12.11), is

$$S' \begin{pmatrix} \hat{I}' \\ \hat{r}' \end{pmatrix} = \begin{pmatrix} \hat{t}' \\ 0 \end{pmatrix}, \tag{12.15}$$

Or equivalently,

$$S'\hat{U}^{+} \begin{pmatrix} \hat{I} \\ \hat{r} \end{pmatrix} = \hat{U}^{+} \begin{pmatrix} \hat{t} \\ 0 \end{pmatrix}, \quad \hat{U} = \begin{pmatrix} U & 0 \\ 0 & U \end{pmatrix}, \tag{12.16}$$

So finally

$$\hat{U}S'\hat{U}^{+} \begin{pmatrix} \hat{I} \\ \hat{r} \end{pmatrix} = \begin{pmatrix} \hat{t} \\ 0 \end{pmatrix}. \tag{12.17}$$

Thus, we obtain the following expression for the transfer matrix [18–20], consistent with Equations 12.14 and 12.17:

$$S = \hat{U}S'\hat{U}^{+} = \begin{pmatrix} U\alpha^* U^{+} & -U\beta^* U^{+} \\ -U\beta U^{+} & U\alpha U^{+} \end{pmatrix} \tag{12.18}$$

To complete the task, it is necessary to construct a matrix U that diagonalizes the Hamiltonian $H_{\text{int}}^{(2)}$ of the spin interaction with the magnetic field of the barrier. We assume that the induction vector of the internal field is oriented arbitrarily and is given by the angular degrees of freedom θ, ξ, φ (Figure 12.1), where ξ – is the angle between the total induction $\vec{B}_0$ and the z axis, φ – is the angle between the x axis and the projection of $\vec{B}_0$ on the plane $x0z$, θ – is the noncollinearity angle, or the angle between $\vec{B}'$ and the z axis. In the case when the transverse component $\vec{B}_{\perp}$ is zero, the vector $\vec{B}_0$ lies in the plane of the magnetic layer $x0z$.

As can be seen from Figure 12.1,

$$\vec{B}_0 = \vec{i}B_x + \vec{j}B_y + \vec{k}B_z (B_y = B_{\perp}). \quad B_{0x} = B_x = B\sin\theta, \quad B_{0y} = B_{\perp}, \quad B_{0z} = B_z = B\cos\theta \tag{12.19}$$

and finally we have:

$$B_{0x} = B_0 \cos\varphi \sin\xi, \quad B_{0y} = B_0 \sin\varphi \sin\xi, \quad B_{0z} = B_0 \cos\xi,$$

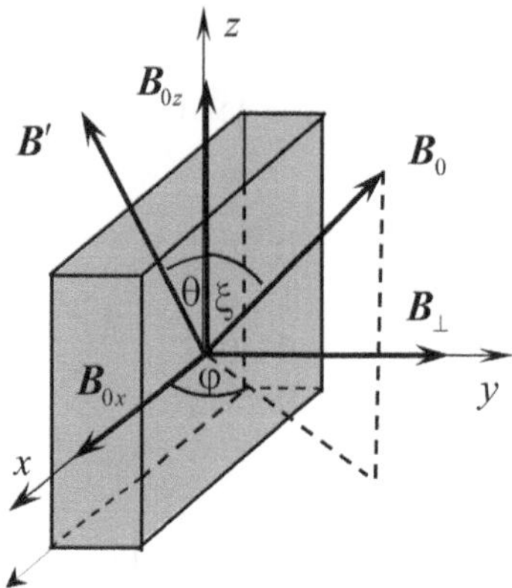

FIGURE 12.1 The projections of the magnetic layer internal field induction $\vec{B}_0$ on the corresponding coordinate axes and planes.

where φ and ξ are the angles in the spherical frame. They are related to the field components, $\vec{B}_0$, $\vec{B}'$, $\vec{B}_\perp$, and the nonlinearity angle θ, in the following way:

$$\sin\xi = \sqrt{1-\sin^2\alpha\cos^2\theta}, \quad \sin\varphi = \frac{\cos\alpha}{\sqrt{1-\sin^2\alpha\cos^2\theta}}, \\ \sin\alpha = \frac{1}{\sqrt{1+\eta^2}}, \quad \eta = \frac{B_\perp}{B}, \tag{12.20}$$

and the dimensionless parameter, η, can be either less than or more than the unit.

Then the Schrödinger equation in the variables ξ and φ takes the form:

$$-\frac{\hbar^2}{2m}\begin{pmatrix}\psi_1''\\ \psi_2''\end{pmatrix} - \mu B_0\begin{pmatrix}\cos\xi & \sin\xi e^{-i\varphi}\\ \sin\xi e^{i\varphi} & -\cos\xi\end{pmatrix}\begin{pmatrix}\psi_1\\ \psi_2\end{pmatrix} + V(y)\begin{pmatrix}\psi_1\\ \psi_2\end{pmatrix} = E\begin{pmatrix}\psi_1\\ \psi_2\end{pmatrix}, \tag{12.21}$$

It is easy to show that diagonalization is carried out using unitary transformation $H' = U^{+}HU$, $\psi = U\chi$

$$U = U_0U_1 = \begin{pmatrix}e^{-i\varphi/2} & 0\\ 0 & e^{i\varphi/2}\end{pmatrix}\begin{pmatrix}\cos{}^{\xi}\!/_2 & -\sin{}^{\xi}\!/_2\\ \sin{}^{\xi}\!/_2 & \cos{}^{\xi}\!/_2\end{pmatrix}, \tag{12.22}$$

Thus unitary-transformed Hamiltonian takes the form

$$H' = -\frac{\hbar^2}{2m}\frac{d^2}{dy^2} + V(y) - \gamma B_0\sigma_z, \tag{12.23}$$

where $V(y)$ is the scalar potential of an electrostatic interaction.

We note that in the case when the transverse component of the internal field is absent ($B_\perp = 0$), $\varphi = 0$ and $\xi = \theta$.

12.3 SCATTERING ON A SINGLE-POTENTIAL BARRIER

Consequently, the transmission and reflection amplitudes are connected with the partial amplitudes, $\hat{T}$ and $\hat{R}$, by the same unitary transformation:

$$\hat{T} = U_0U\hat{t}, \quad \hat{R} = U_0U\hat{r}, \quad \hat{t} = \begin{pmatrix}t_1\\ t_2\end{pmatrix}, \quad \hat{r} = \begin{pmatrix}r_1\\ r_2\end{pmatrix}. \tag{12.24}$$

As a result, we obtain the following expressions for the transmission and reflection amplitudes:

$$\begin{aligned} T_{\uparrow\uparrow} &= t_1 \cos^2\frac{\xi}{2} + t_2 \sin^2\frac{\xi}{2}, & T_{\downarrow\downarrow} &= t_1 \sin^2\frac{\xi}{2} + t_2 \cos^2\frac{\xi}{2}, \\ T_{\uparrow\downarrow} &= \frac{1}{2}e^{-i\varphi}(t_1 - t_2)\sin\xi, & T_{\downarrow\uparrow} &= \tfrac{1}{2}e^{+i\varphi}(t_1 - t_2)\sin\xi. \end{aligned} \tag{12.25}$$

$$\begin{aligned} R_{\uparrow\uparrow} &= r_1 \cos^2\frac{\xi}{2} + r_2 \sin^2\frac{\xi}{2}, & R_{\downarrow\downarrow} &= r_1 \sin^2\tfrac{\xi}{2} + r_2 \cos^2\tfrac{\xi}{2}, \\ R_{\uparrow\downarrow} &= \frac{1}{2}e^{-i\varphi}(r_1 - r_2)\sin\xi, & R_{\downarrow\uparrow} &= e^{+i\varphi}(r_1 - r_2)\sin\xi. \end{aligned} \tag{12.26}$$

This implies a transition to the case, $B_\perp = 0$, that is, the absence of the transverse component of the magnetic field. The indices in Equations 12.25 and 12.26 indicate the spin states in the scattered and incident waves, respectively.

Using the transfer matrix method, one can show that:

$$\hat{T} = \tau I, \quad \hat{R} = \rho I, \tag{12.27}$$

where I is the spinor amplitude of the incident wave and it defines the spin polarization of this wave:

$$\tau = \begin{pmatrix} T_{\uparrow\uparrow} & T_{\downarrow\uparrow} \\ T_{\uparrow\downarrow} & T_{\downarrow\downarrow} \end{pmatrix}, \quad \rho = \begin{pmatrix} R_{\uparrow\uparrow} & R_{\downarrow\uparrow} \\ R_{\uparrow\downarrow} & R_{\downarrow\downarrow} \end{pmatrix}, \tag{12.28}$$

and $T_{i,j}, R_{i,j}(i,j = \uparrow,\downarrow)$ are defined by the expressions Equations 12.49 and 12.50, $\hat{T}$ and $\hat{R}$ are the spinor amplitudes of the waves scattered forward and backward, respectively.

We take an incident wave amplitude be $\begin{pmatrix} 1 \\ 0 \end{pmatrix}$. Then we obtain for the transmission and reflection amplitudes:

$$\begin{aligned} D &= |t_1|^2 \cos^2\frac{\xi}{2} + |t_2|^2 \sin^2\frac{\xi}{2}, \\ R &= |r_1|^2 \cos^2\frac{\xi}{2} + |r_2|^2 \sin^2\frac{\xi}{2} \end{aligned} \tag{12.29}$$

These amplitudes contribute into both channels of scattering, forward and backward, with and without the spin overturn:

$$D = |T_{\uparrow\uparrow}|^2 + |T_{\uparrow\downarrow}|^2, \quad R = |R_{\uparrow\uparrow}|^2 + |R_{\uparrow\downarrow}|^2. \tag{12.30}$$

As it is known, the polarization vectors are defined as follows:

$$\vec{P}_T = \frac{T^+\vec{\sigma}T}{T^+T}, \quad \vec{P}_R = \frac{R^+\vec{\sigma}R}{R^+R}, \tag{12.31}$$

where T and R are the forward and backward scattering amplitudes and they are defined by the following spinors:

$$P_{Tx} = \frac{T_{\uparrow\uparrow}T^*_{\uparrow\downarrow} + T^*_{\uparrow\uparrow}T_{\uparrow\downarrow}}{D}, \quad P_{Ty} = i\frac{T_{\uparrow\uparrow}T^*_{\uparrow\downarrow} - T^*_{\uparrow\uparrow}T_{\uparrow\downarrow}}{D}, \quad P_{Tz} = \frac{|T_{\uparrow\uparrow}|^2 - |T_{\uparrow\downarrow}|^2}{D}. \tag{12.32}$$

$$P_{Rx} = \frac{R_{\uparrow\uparrow}R^*_{\uparrow\downarrow} + R^*_{\uparrow\uparrow}R_{\uparrow\downarrow}}{R}, \quad P_{Ry} = i\frac{R_{\uparrow\uparrow}R^*_{\uparrow\downarrow} - R^*_{\uparrow\uparrow}R_{\uparrow\downarrow}}{R}, \quad P_{Rz} = \frac{|R_{\uparrow\uparrow}|^2 - |R_{\uparrow\downarrow}|^2}{R}. \tag{12.33}$$

We do not present here the explicit expressions for these components, because they are very cumbersome. Anyhow, they are expressed by the partial amplitudes, t_ℓ and r_ℓ $(\ell = 1,2)$, and the variable noncoplanarity angles, ξ and φ.

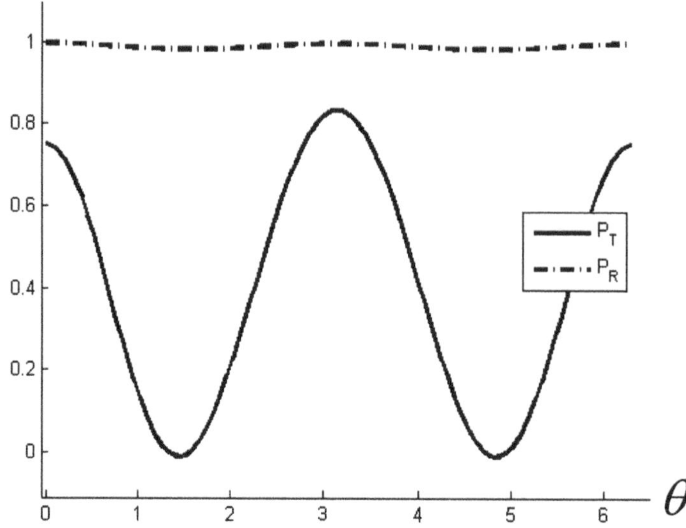

FIGURE 12.2 The dependence of P_T and P_R (P_{Tz} and P_{Rz}) on the angle θ in the under-barrier region ($x = 0.5$); $\eta = 0.5$.

12.4 POLARIZATION VECTORS OF SCATTERED WAVES

The components of the polarization vectors, (Eq. 12.32) and (Eq. 12.33), are macroscopic quantities and their number is three. They are functions of three variables, namely θ, η and x. We constructed the dependences of physical parameters on these variables as follows. We fixed two of them and found the dependences of the components on the third one. For instance, we choose $\eta = 0,5$ and 5 (weak and strong transverse magnetic fields $B_{\perp}$correspondingly), leaving parameter B to stay constant. Besides, the value, $x = 0,5$ was fixed, which corresponds to the under-barrier scattering. Combining these four values, we constructed four variants of dependences of the components on the nonlinearity angle, θ. Thus, Figures 12.2 and 12.3 correspond to the under-barrier scattering; Figures 12.4

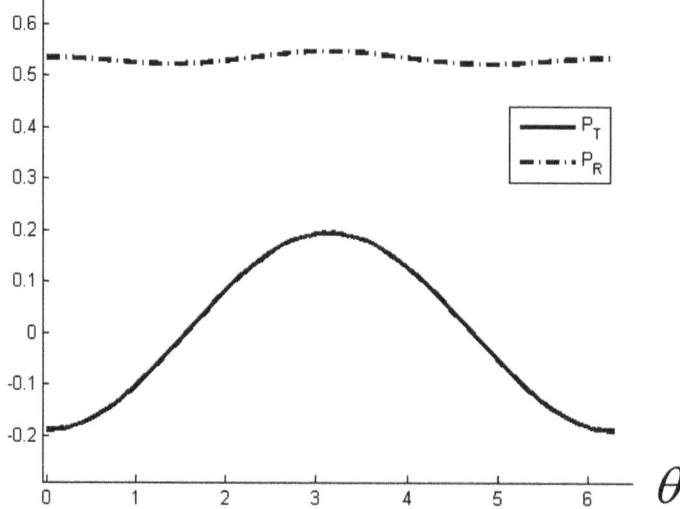

FIGURE 12.3 The dependence of P_T and P_R on the angle θ in the under-barrier region ($x = 0.5$); $\eta = 5$.

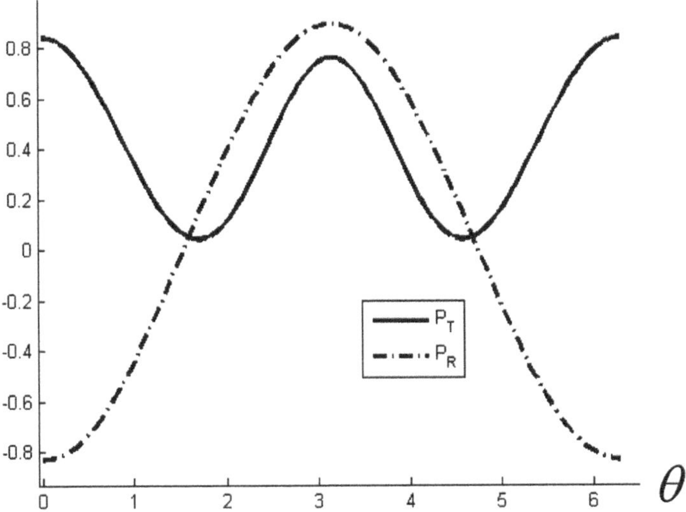

FIGURE 12.4 The dependence of P_T and P_R on the angle θ.

and 12.5 are for the over-barrier scattering. It is seen in Figures 12.2 and 12.3 that the polarization degree P_R of the reflected wave depends very weakly on θ both in weak and strong transverse fields correspondingly. We find also that the tenfold increase of η leads to the twofold decrease of parameter P_R. In weak fields, P_T definitely is a periodic function of θ with an amplitude equal to 0,8.

The tenfold increase of the field leads to retardation (increase) of the periodicity (it increases about two times), and the amplitude decreases two times, and a weak z-repolarisation is emerged.

Consequently, one can conclude that the increase of the transverse field leads to suppression of oscillations of the z-polarization of the transmitted wave. The polarization degree of the reflected wave is more sensitive with respect to the transverse field and the oscillations practically are suppressed even in weak fields; the increase of η leads to a decrease of the z-polarization about two times. In the over-barrier region of the scattering the suppression of the z-repolarization is observed, too, if η increases (see Figures 12.4 and 12.5).

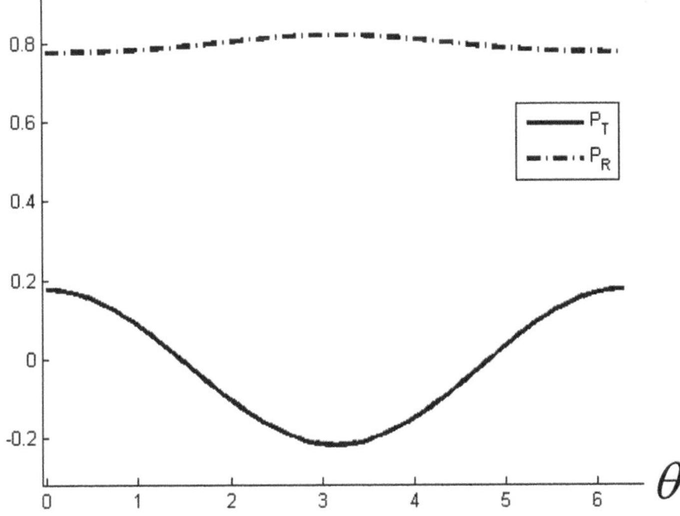

FIGURE 12.5 The dependence of P_T and P_R on the angle θ in the over-barrier region ($x = 1,5$); $\eta = 5$.

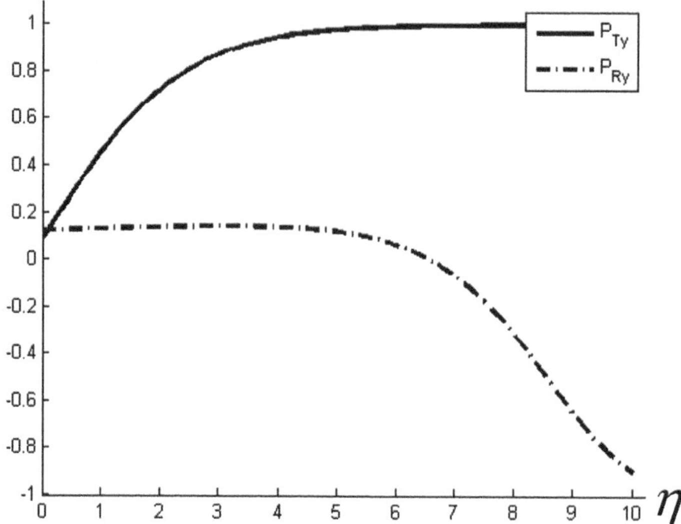

FIGURE 12.6 The dependence of $y-$components of the polarization vectors, P_{Ty}, P_{Ry}, on the parameter η for the following fixed values: $x = 0, 5$; $\theta = \pi/4$.

Now we consider the dependence of the polarization degree on the parameter, η, for the fixed angle: $\theta = \pi/4$ and $x = 0.5; 1.5$.

As it is seen in Figures 12.6 and 12.7, the polarization degree of the transmitted wave is decreased if η increases, then it asymptotically vanishes, i.e., polarization vanishes for sufficiently large η and z both in the under-barrier and over-barrier regions. The polarization degree of the reflected wave has non-monotonous behavior, and in certain regions of η z-repolarisation is observed. Now we consider the behavior of the y-components of the polarization vectors. It is interesting, because the transverse magnetic field is directed along the y-axis.

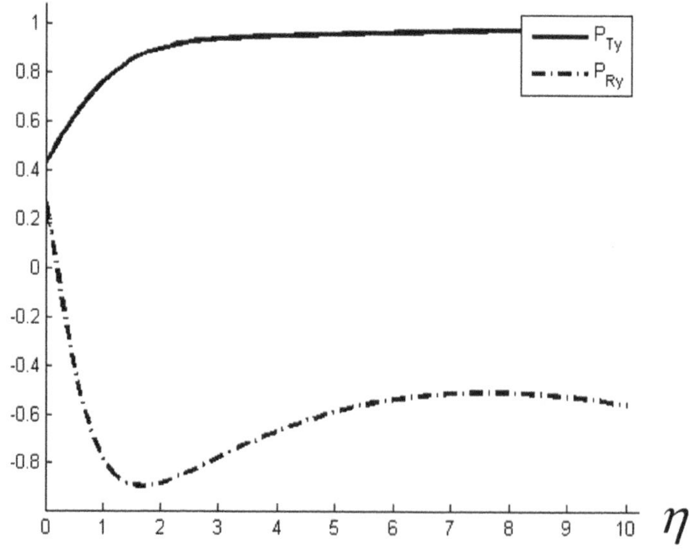

FIGURE 12.7 The dependence of $y-$components of the polarization vectors, P_{Ty}, P_{Ry}, on the parameter η for the following fixed values: $x = 1, 5$; $\theta = \pi/4$.

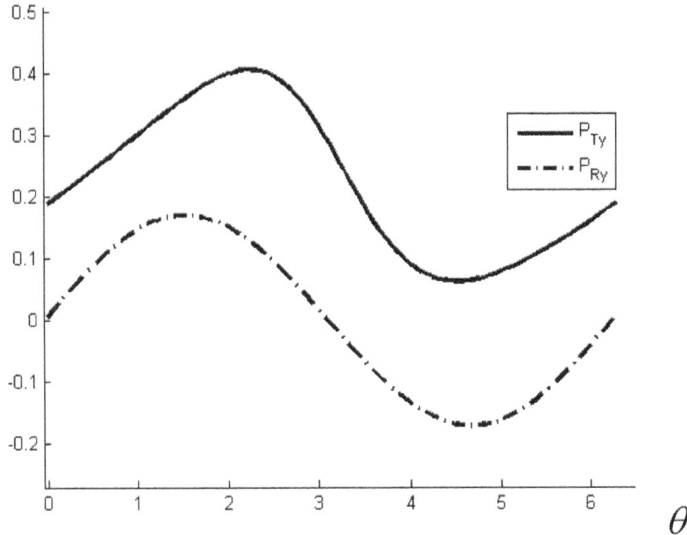

FIGURE 12.8 The dependence of P_{Ty}, P_{Ry} on θ for the following fixed values: $x = 0,5$; $\eta = 0,5$.

It is seen in Figures 12.6 and 12.7 that complete y-polarization is formed in the transmitted wave if η increases, because $P_{Ty} \sim 1$ both in the under-barrier and over-barrier regions of energy. In the reflected wave, P_{Ry} has a plateau if $\eta \approx 6$ and it begins to decrease in the under-barrier region. The behavior in the over-barrier region is reversed; if η increases, y-polarization decreases and y-repolarization occurs; then, P_{Ry} increases and reaches its plateau. Now we consider the dependences of P_{Ty} and P_{Ry} on θ. It is seen in Figures 12.8 and 12.9 that these dependences are oscillating functions of θ in weak transverse fields in an under-barrier region; also, these oscillations can be suppressed by a strong field.

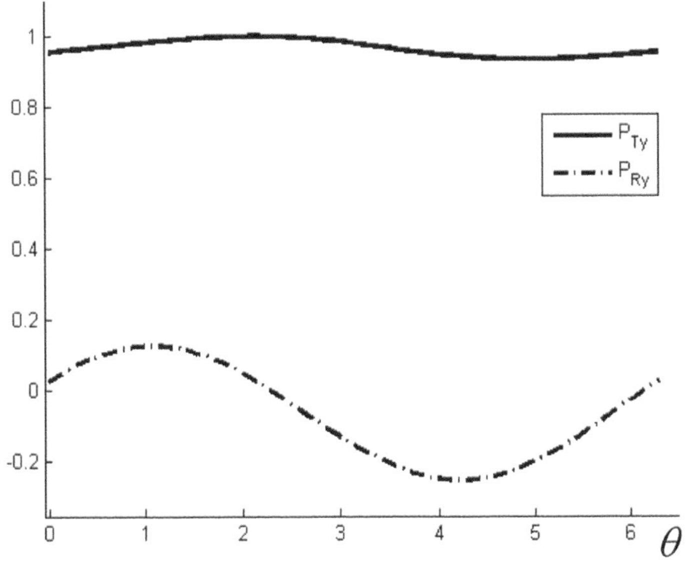

FIGURE 12.9 The dependence of P_{Ty}, P_{Ry} on θ for the following fixed values: $x = 0,5$; $\eta = 5$.

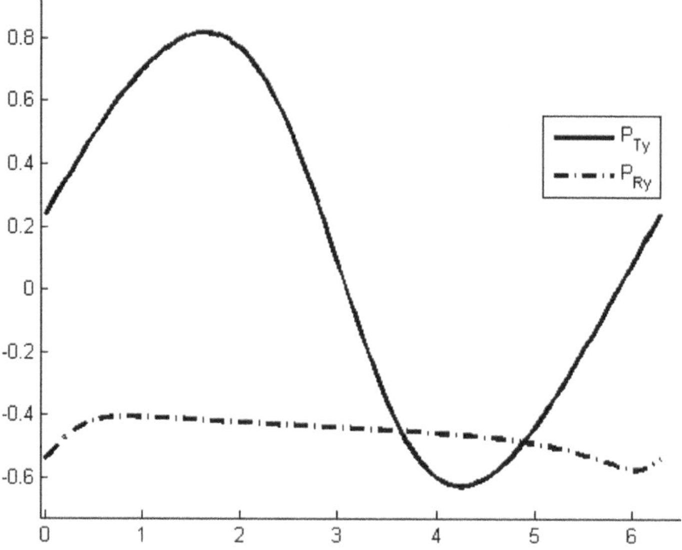

FIGURE 12.10 The dependence of P_{Ty}, P_{Ry} on θ for the following fixed values: $x = 1, 5$, $\eta = 0, 5$.

This suppression is clearly seen in the over-barrier region on curves in Figures 12.10 and Fig. 12.11. For instance, the oscillations with the amplitude, 1.4, of P_{Ty} become weak, changing in the approximate region ~ 0.2 and, in addition, the reorientation of the y – polarization vanishes.

Moreover, the oscillations of P_{Ry} vanish in the over-barrier region even in weak transverse fields. Taking the above-said into account, one can say that if the transverse magnetic field increases, the tendency of suppression of the polarization vector component oscillations – polarized with respect to the nonlinearity angle θ – increases too. In some cases this suppression can be manifested even in weak transverse fields.

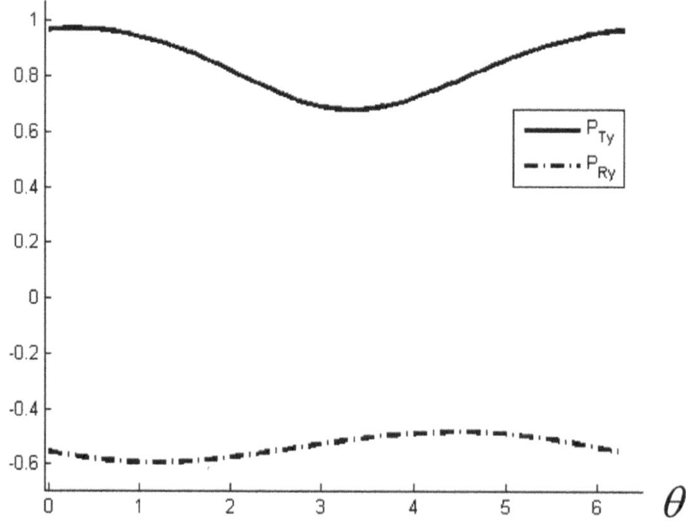

FIGURE 12.11 The dependence of P_{Ty}, P_{Ry} on θ for the following fixed values: $x = 1, 5$, $\eta = 5$.

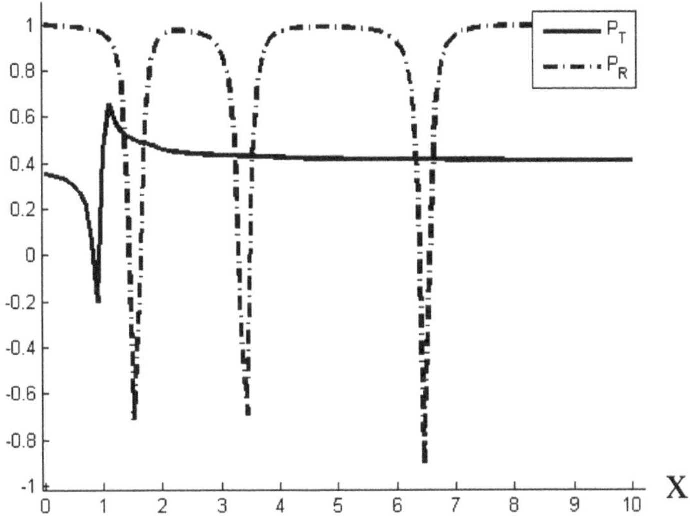

FIGURE 12.12 The dependence of P_T and P_R on the parameter x for the following fixed values: $\eta = 0,5$; $\theta = \pi/4$.

In connection with this, it is interesting to follow this suppression in the dependences of P_{Ti} and $P_{Ri}(i = z, y)$ on the dimensionless energy of scattering. It is seen in Fig. 12.12 and Fig. 12.13 that when η is small, P_T depends weakly on parameter x except in the small vicinity of the point, $x = 1$, where it has a sufficiently large jump. For the tenfold increase of η this jump is significantly suppressed (Figure 12.13) and the z-polarization practically vanishes. The degree of the polarization of the reflected wave consists of quasi periodically repeated narrow and deep dips and maxima in the form of plateau, but as η increases, the plateaus vanish, turning into local minima, and the dips are widened. The polarization vectors depend on three variables, the noncollinearity angle θ, the

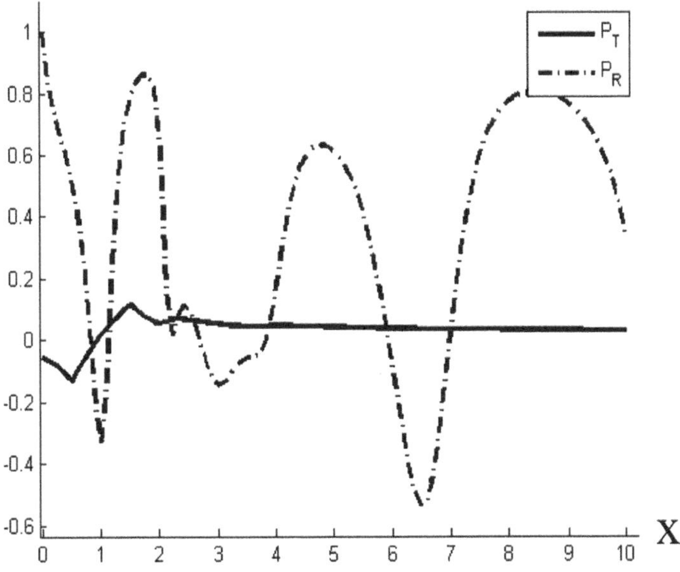

FIGURE 12.13 The dependence of P_T and P_R on the parameter x for the following fixed values: $\eta = 5$; $\theta = \pi/4$.

dimensionless energy x and the ratio of two components of the internal magnetic field induction. Thus the investigation was carried out as follows. Two of the above-said variables were fixed and the dependences of the vectors, $\vec{P}_T$ and $\vec{P}_R$ on the third one was constructed. All the results indicate an increase of the transverse field component leading to suppression of the polarization characteristics of the oscillations. Thus, decreases of amplitudes, as well as an increase of periods take place. Additionally, suppression of reorientation of polarization vectors takes place. On the other hand, an increase of $B_\perp$ leads to a significant orientation of the polarization vectors, a tendency in orientations of $\vec{P}_T$ and $\vec{P}_R$ along one of the coordinate axis, meanwhile, the other components practically vanish. This occurs for very large transverse fields ($B_\perp >> B$, where B is the longitudinal component of the field). In connection with this it is interesting to discuss the case when the longitudinal component vanishes ($B = 0$) and only the transverse field ($B_\perp$) is present. For instance, if the polarization of the incident wave is $\begin{pmatrix} 1 \\ 0 \end{pmatrix}$, then the transmission and reflection coefficients respectively are: $D = \frac{1}{2}\left|t_1 - t_2\right|^2$ and $R = \frac{1}{2}\left|r_1 - r_2\right|^2$, and only the y-components of the polarization vectors do not vanish: $P_{Ty} = P_{Ry} = \pm 1$. In the case of the following incident wave, $\begin{pmatrix} 0 \\ 1 \end{pmatrix}$, we have: $D = \frac{1}{2}\left|t_1 + t_2\right|^2$, $R = \frac{1}{2}\left|r_1 + r_2\right|^2$ and $P_{Ty} = P_{Ry} = -1$. Both scattered waves become completely y-polarized.

12.5 SCATTERING ON A TWO-BARRIER SYSTEM

Here we consider the scattering of an unpolarized wave on a system consisting of two identical barriers. We assume that the inductions of the barrier internal fields are noncoplanar. Thus, in contrast to the single-barrier system, the system possesses its "own" noncoplanarity.

The Hamiltonian of the system is chosen as follows:

$$V(y) = (V_0 - \gamma B\sigma_z)\varphi(y) + (V_0 - \gamma\vec{B}'\vec{\sigma})\varphi(y - c - a) \tag{12.34}$$

where $\vec{B}$ and $\vec{B}'$ are the inductions of the internal fields of the left and right barriers, respectively. $\vec{B}$ is directed along the z axis, and the orientation of $\vec{B}'$- is determined by the angular variables introduced in the previous paragraph (noncoplanar degrees of freedom); $\varphi(y)$ and $\varphi(y - c)$ are finite functions describing the shape of barriers and are localized near $y = 0$ and $y = c$, respectively, a is the width of the localization of barriers.

Then the transmission and reflection amplitudes are determined from the equation

$$\begin{pmatrix} I \\ R \end{pmatrix} = S^{-1}(0)\, V(c)\, S^{-1}(\varphi, \xi)\, V^{-1}(c) \begin{pmatrix} T \\ 0 \end{pmatrix}, \tag{12.35}$$

where $S^{-1}(0)$ and $S^{-1}(\varphi, \xi)$ are the transfer matrix of the scattering problem on the left and right barriers, respectively. The corresponding expressions for them are given in the previous paragraphs; $V(c)$-is the displacement matrix over distance c:

$$V(c) = \begin{pmatrix} e^{-i\hat{k}c} & 0 \\ 0 & e^{i\hat{k}c} \end{pmatrix}, \qquad e^{-i\hat{k}c} = e^{ikc}\begin{pmatrix} 1 & 0 \\ 0 & 1 \end{pmatrix}. \tag{12.36}$$

The solutions of Equation 12.35 for spinor amplitudes T and R are as follows

$$\begin{aligned} T &= \hat{T}I = U(\alpha U\alpha + \beta^* U\beta e^{-2ikc})^{-1} I \\ R &= \hat{R}I = (\beta U\alpha + \alpha^* U\beta e^{-2ikc})(\alpha U\alpha + \beta^* U\beta e^{-2ikc})^{-1} I \end{aligned} \tag{12.37}$$

$\hat{T}$ and $\hat{R}$ are 2×2 matrices composed of the transmission and reflection amplitudes, respectively, I – spinor amplitude of the incident wave.

Writing the solutions of Equation 12.37 in an explicit form, we get [27,28]:

$$T_{\uparrow\uparrow} = \frac{1}{d}\left(\frac{\cos^2 \xi/2}{T_{01}} + \frac{\sin^2 \xi/2}{T_\pi}\right), \quad T_{\downarrow\downarrow} = \frac{1}{d}\left(\frac{\sin^2 \xi/2}{T_{02}} + \frac{\cos^2 \xi/2}{T_\pi}\right),$$
$$T_{\downarrow\uparrow} = \frac{1}{2d}e^{i\varphi}\left(\frac{1}{T_\pi} - \frac{1}{T_{01}}\right)\sin\xi, \quad T_{\uparrow\downarrow} = -\frac{1}{2d}e^{-i\varphi}\left(\frac{1}{T_\pi} - \frac{1}{T_{01}}\right)\sin\xi, \tag{12.38}$$
$$d = \frac{\cos^2 \xi/2}{T_{01}T_{02}} + \frac{\sin^2 \xi/2}{T_\pi^2}, \quad T_\pi = \frac{t_1 t_2}{1 - r_1 r_2 e^{rikc}}.$$

Here $T_{\pi 1} = T_{\pi 2} = T_\pi$ is the amplitude of a wave that has transmitted through a system of two barriers with anti-parallel internal fields ($\theta = \pi, \varphi = 0$), and $T_{0\ell}$ are partial amplitudes of a wave that has transmitted through a double-barrier system, with parallel internal fields, where t_ℓ and r_ℓ are partial amplitudes for an isolated barrier.

The most interesting is the situation when the incident wave is not spin-polarized, *i.e.*, when its degree of polarization is zero. In this case, for the transmission coefficient, we get:

$$D_T = \frac{1}{2}\left(|T_{\uparrow\uparrow}|^2 + |T_{\downarrow\downarrow}|^2 + |T_{\uparrow\downarrow}|^2 + |T_{\downarrow\uparrow}|^2\right). \tag{12.39}$$

Considering that the interaction energy of the spin and the magnetic field is small compared with the height of the barrier $(\gamma << V_0)$, Equation 12.39 can be represented in the form:

$$D_T = \frac{1}{2}\left(|T_{01}|^2 + |T_{02}|^2\right)\cos^2\frac{\xi}{2} + |T_{01}||T_{02}|\sin^2\frac{\xi}{2}. \tag{12.40}$$

In the case when the transverse component of the internal field of the right barrier is zero ($\eta = 0$), and when the condition of partial resonance of transmission is satisfied (for example, $|T_{01}| = 1$), at angles of noncollinearity $\theta = 0, \pi$, then the transmission coefficient $D_{T\,\mathrm{rez}} = |T_{02}|$, *i.e.*, it is an exponentially small quantity. This is analogous to the effect of giant magnetoresistance. The presence of a sufficiently strong transverse field does not affect the smallness of the transmission coefficient. When the condition of partial resonance is fulfilled, with an accuracy of exponentially small terms, we have

$$D_{T\,\mathrm{rez}} = \frac{1}{2}\cos^2\frac{\xi}{2}. \tag{12.41}$$

And the reflection coefficient in the same conditions is as follows:

$$D_{R\,\mathrm{rez}} = \frac{1}{2}\left(1 + \sin^2\frac{\xi}{2}\right). \tag{12.42}$$

$D_{T\mathrm{rez}}$ is an oscillating function of the noncollinearity angle θ with extreme values

$$D_{T\mathrm{rez}}^{(1,2)} = \frac{1}{4}(1 \pm \sin\alpha), \tag{12.43}$$

which are achieved when $\theta = 0$ or $\theta = \pi$.

For small transverse fields ($\eta << 1$)

$$D_{T\,\mathrm{rez}} = \frac{1}{2}(\cos^2\theta/2 + \eta^2), \tag{12.44}$$

where the first term describes the effect, which is a spin analogue of the well-known in optics so called Malus law [29]. With large transverse fields ($\eta >> 1$)

$$D_{T\,\mathrm{rez}} = \frac{1}{2}\left(1 + \frac{1}{\eta}\cos\theta\right), \tag{12.45}$$

that is, the dependence of the transmission coefficient on η and θ practically disappears.

For the three components of the polarization vector of the transmitted wave, we have:

$$P_{Tx} = \frac{\sin\xi}{4D_T|d|^2}\left\{2\left(\frac{1}{|T_{01}|^2}-\frac{1}{|T_{02}|^2}\right)\cos^2\frac{\xi}{2}\cos\xi + \right.$$
$$\left. + \left[\left\{\left(\frac{1}{T_\pi T_{01}^*}-\frac{1}{T_\pi^* T_{01}}\right)\cos^2\frac{\xi}{2}+\left(\frac{1}{T_\pi T_{02}^*}-\frac{1}{T_\pi^* T_{02}}\right)\sin^2\frac{\xi}{2}\right\}e^{-i\varphi}+\text{c.c.}\right]\right\},$$
$$P_{Ty} = i\frac{\sin\xi}{4D_T|d|^2}\left[\frac{1}{T_\pi}\left(\frac{1}{T_{01}^*}+\frac{1}{T_{02}^*}\right)-\text{c.c.}\right], \tag{12.46}$$
$$P_{Tz} = \frac{1}{4D_T|d|^2}\left\{\left(\frac{1}{|T_{02}|^2}-\frac{1}{|T_{02}|^2}\right)\cos^2\frac{\xi}{2}\cos\xi+\frac{1}{2}\left[\left(\frac{1}{T_\pi T_{02}^*}-\frac{1}{T_\pi^* T_{01}^*}\right)+\text{c.c.}\right]\right\},$$

where D_T is determined by the Equation 12.39.

In conditions of partial resonance of transmissions

$$P_T = \frac{\cos^2\xi/2}{\cos^2\xi/2+2|T_{02}|^2}, \tag{12.47}$$

where the degree of polarization of the transmitted wave $P_T = \sqrt{\overline{P_{Tx}}^2+\overline{P_{Ty}}^2+\overline{P_{Tz}}^2}$ and averaging of the P_{Ti} components is carried out over an ensemble of incident particles. Thus, the degree of polarization of the transmitted wave under certain conditions may be close to unity, but it never turns into unity due to the two-channel nature of scattering. In the situation considered, $|T_{02}|^2$ is an exponentially small quantity that can be easily estimated $|T_{02}| \approx (V_0/\gamma B)\,|t_2|^2$, where t_2 is one of the transmittance amplitudes of the single barrier. Under these conditions, the transmission coefficient weakly depends on the angle θ. The minimum value of the polarization degree of the transmitted wave $P_{\min}$ is a function of the parameter η. It is easy to show that in the case of weak transverse fields, when ($\eta << 1$)

$$P_{\min} \approx \eta^2, \tag{12.48}$$

and in the case of strong transverse fields ($\eta >> 1$)

$$P_{\min} \approx 1-4|T_{02}|^2, \tag{12.49}$$

i.e., practically equal unity. Thus, with increasing η valve effect gradually disappears. This result is consistent with the statement that the presence of transverse field components suppresses oscillation effects in the scattering of a polarized wave on a single barrier.

The polarization of the reflected wave is investigated in a similar way. Using the Equation 12.5, for the corresponding amplitudes we get:

$$R_{\uparrow\uparrow} = \left(\frac{R_{01}}{T_{01}}\cos^2\xi/2+\frac{R_{\pi1}}{T_\pi}\sin^2\xi/2\right)T_{\uparrow\uparrow}+\frac{1}{2}\left(\frac{R_{01}}{T_{01}}-\frac{R_{\pi1}}{T_\pi}\right)e^{-i\varphi}T_{\downarrow\uparrow}\sin\xi,$$
$$R_{\downarrow\downarrow} = \left(\frac{R_{02}}{T_{02}}\cos^2\xi/2+\frac{R_{\pi2}}{T_\pi}\sin^2\xi/2\right)T_{\downarrow\downarrow}-\frac{1}{2}\left(\frac{R_{02}}{T_{02}}-\frac{R_{\pi2}}{T_\pi}\right)e^{i\varphi}T_{\uparrow\downarrow}\sin\xi,$$
$$R_{\uparrow\downarrow} = \left(\frac{R_{01}}{T_{01}}\cos^2\xi/2+\frac{R_{\pi1}}{T_\pi}\sin^2\xi/2\right)T_{\uparrow\downarrow}+\frac{1}{2}\left(\frac{R_{01}}{T_{01}}-\frac{R_{\pi1}}{T_\pi}\right)e^{-i\varphi}T_{\downarrow\downarrow}\sin\xi, \tag{12.50}$$
$$R_{\downarrow\uparrow} = \left(\frac{R_{02}}{T_{02}}\cos^2\xi/2+\frac{R_{\pi2}}{T_\pi}\sin^2\xi/2\right)T_{\downarrow\uparrow}-\frac{1}{2}\left(\frac{R_{02}}{T_{02}}-\frac{R_{\pi2}}{T_\pi}\right)e^{i\varphi}T_{\uparrow\uparrow}\sin\xi.$$

where

$$\frac{R_{0\ell}}{T_{0\ell}} = \frac{r_\ell}{t_\ell^2}+\frac{r_\ell}{|t_\ell|^2}e^{-2ikc}(\ell=1,2),\quad \frac{R_{\pi1}}{T_\pi}=\frac{r_1}{t_1t_2}+\frac{r_2}{t_1^*t_2}e^{-2ikc},\quad \frac{R_{\pi2}}{T_\pi}=\frac{r_2}{t_1t_2}+\frac{r_1}{t_1t_2^*}e^{-2ikc}, \tag{12.51}$$

where $R_{\pi\ell}(\ell = 1, 2)$ are reflection amplitudes for a double-barrier system with anti-parallel internal fields.

Thus, the reflection amplitudes are expressed in terms of the quantities that are the corresponding amplitudes of the quantum resistances. They, in turn, represent the ratio of the corresponding reflection and transmission partial amplitudes. Here, the components of the polarization vector of the reflected wave are expressed in terms of amplitudes (Eq. 12.50) and are as follows:

$$\begin{aligned} P_{Rx} &= \frac{1}{2D_R}(R_{\uparrow\uparrow}R^*_{\uparrow\downarrow} + \text{c.c.} + R^*_{\downarrow\downarrow}R_{\downarrow\uparrow} + \text{c.c.}), \\ P_{Ry} &= \frac{i}{2D_R}(R_{\uparrow\uparrow}R^*_{\uparrow\downarrow} - \text{c.c.} + R^*_{\downarrow\downarrow}R_{\downarrow\uparrow} - \text{c.c.}), \\ P_{Rz} &= \frac{1}{2D_R}\left(|R_{\uparrow\uparrow}|^2 - |R_{\downarrow\downarrow}|^2 + |R_{\downarrow\uparrow}|^2 - |R_{\uparrow\downarrow}|^2\right), \end{aligned} \tag{12.52}$$

where $D_R = \frac{1}{2}(|R_{\uparrow\uparrow}|^2 + |R_{\downarrow\downarrow}|^2 + |R_{\uparrow\downarrow}|^2 + |R_{\downarrow\uparrow}|^2)$ – is the reflection coefficient when the incident wave is not spin-polarized. Then in the mode of partial transmission resonance ($|T_{01}| = 1, |T_{02}| << 1$) we will come to the following result: the reflected wave turns out to be strongly polarized with a degree of polarization close to unity. Interestingly, this result is true for any value of the transverse component of the internal field, and hence for its zero value. Consequently, the result obtained is valid not only for noncoplanar, but also for non-collinear systems, which is easy to check out.

These results allow us to conclude that the considered double-barrier noncoplanar structure has strong spin-polarizing properties. The growth of the transverse component of the internal field of the second barrier leads to the suppression of the phenomenon characteristic of double-barrier systems – an analogue of the well-known phenomenon of the valve effect. It is important that when the condition of transmission partial resonance is fulfilled, the degree of polarization of both the transmitted and reflected waves turns out to be close to unity. Besides, while in the absence of a transverse component or its small value the transmission coefficient satisfies the quantum analogue of the Malus law, in the case of strong transverse fields the resonance transmission coefficient reaches a plateau, that is, the characteristic oscillations on noncollinearity angle θ disappear. Along with this, in the polarization of the transmitted wave the deep dip to zero at $\theta = \pi$ disappears. Note that earlier [see reference 26] we have shown that the suppression of the effects of oscillation in the angle θ also is taking place in a single-barrier noncoplanar system, both in transmission and reflection coefficients, and in degrees of polarization of forward and backward scattered waves. (see previous paragraph).

12.6 MULTI-BARRIER SYSTEM WITH A "NONCOPLANAR DEFECT"

Here we consider a more complicated system consisting of N periodically located barriers and discuss the appearance of specific features which are consequences of the introduced noncoplanarity into this multilayered system. The barriers are grouped into two "ferromagnetic domains" with parallel inductions of internal fields and are separated from each other by a "noncoplanar defect" – a single barrier, the internal field induction of which is noncoplanar with respect to the inductions of the other barriers (Figure 12.14).

The first (left) domain contains $(n - 1)$ barriers and the second (right) – $(N - n)$ barriers. The spatial orientation of the induction of the n-th barrier's internal field, as in the previous paragraphs, is described by two degrees of freedom θ and η, and the magnitudes of the inductions of all barriers are equal to each other. We are interested in the dependence of the spin polarization degree of the transmitted wave on these degrees of freedom and on the momentum variable, and it is assumed that the wave incident on the system is not spin-polarized.

As it is well known, when scattering on an isolated domain, a spin-flip doesn't happen – due to the parallel character of the internal fields of the barriers, however, when scattering on the "defect", a spin-flip occurs with a certain probability, which significantly affects the spin polarization of the

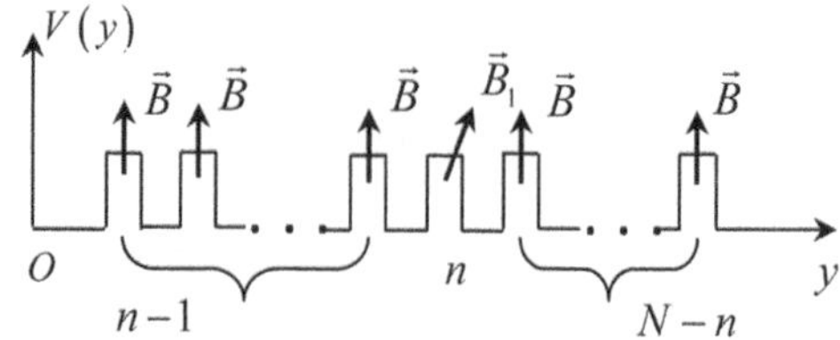

FIGURE 12.14 The potential relief of a film containing magnetic layers.

transmitted wave. Here we give only a scheme for calculating the transmission amplitude, without writing out the final results due to their cumbersome nature.

The transfer-matrix of the considered scattering problem can be represented as follows:

$$\tilde{S} = \tilde{S}_0^{n-1}(U\tilde{S}_0U^+)\tilde{S}_0^{N-n}, \tag{12.53}$$

where $\tilde{S}_0^{n-1}$ and $\tilde{S}_0^{N-n}$ are the transfer matrices of the scattering problem on domains containing $(n-1)$ and $(N-n)$ barriers, respectively, and $U\tilde{S}_0U^+$ on the "defect".

We represent these degrees (**степени**) as follows:

$$\tilde{S}_0^m = \begin{pmatrix} \alpha_m, & \beta_m^* \\ \beta_m, & \alpha_m^* \end{pmatrix}, \quad m = n-1,\ N-n \tag{12.54}$$

where α_m, β_m are 2×2 diagonal matrices with elements

$$\alpha_{m\ell} = T_{m\ell}^{-1},\ \beta_{m\ell} = R_{m\ell}T_{m\ell}^{-1},\ \ell = 1,2 \tag{12.55}$$

And the amplitudes $T_{m\ell}$ and $R_{m\ell}$ are the transmission and reflection amplitudes for the m – barrier "domain":

$$\begin{aligned} T_{m\ell} &= \frac{e^{imk}}{P_{m\ell} - e^{2i\delta_\ell}P_{(m-1)\ell}}, \qquad R_{m\ell} = \frac{r_\ell P_{m\ell}}{t_\ell P_{m\ell} - e^{2i\delta_\ell}P_{(m-1)\ell}} \\ P_{m\ell} &= \frac{\sin[(m+1)\gamma_\ell]}{\sin\gamma_\ell}, \qquad \cos\gamma_\ell = \frac{\cos(\delta_\ell - 2k)}{|t_\ell|}, \end{aligned} \tag{12.56}$$

δ_ℓ are the phases of the partial amplitudes of transmission on a solitary barrier, and $|t_\ell|$-their magnitudes (module).

The transmission amplitude is defined from the relation

$$T_N = S^{-1}I, \tag{12.57}$$

where I is the amplitude of the incident wave,

$$S = (\alpha_{n-1}U\alpha U^+ + \beta_{n-1}^*U\beta U^+)\alpha_{N-n} + (\alpha_{n-1}U\beta^*U^+ + \beta_{n-1}^*U\alpha^*U^+)\beta_{N-n}, \tag{12.58}$$

and the calculation of the polarization degree is described in the previous sections.

12.7 RESULTS AND DISCUSSIONS

Here we present some of the most characteristic 3D- and 2D-plots for the dependences of the spin polarization degree of the transmitted wave on independent variables k, θ, η. The exact functional expression for P is very cumbersome and difficult to analyze analytically. The plot constructions were carried out with the help of the "Mathematics" program for a particular case $N = 7$ and the fixed values of the parameters k, θ, η – are given on each of the plots.

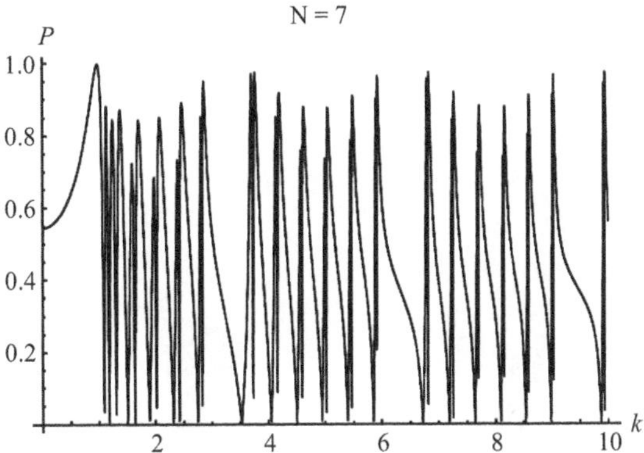

FIGURE 12.15 The dependence of the degree of polarization of transmitted wave on the electron momentum, $P(k, \theta = 0; \eta = 0)$.

Let us consider first the scattering of an unpolarized wave on a collinear system ($\theta = 0$, $\eta = 0$) consisting of $N = 7$ identical barriers. Figure 12.15 shows the dependence of the degree of polarization of the wave transmitted through a collinear system, provided that the incident wave is not spin-polarized. As seen from the figure, it is a sequence of peaks whose maxima appear in the region $(0, 85 \div 1)$. This has a simple explanation: in this case, the degree of polarization is defined by the following expression

$$P_{coll} = \frac{|T_{N1}|^2 - |T_{N2}|^2}{|T_{N1}|^2 + |T_{N2}|^2}, \tag{12.59}$$

where $T_{N\ell}(\ell = 1, 2)$ are Zeeman-split partial transmission amplitudes (Eq. 12.56). In a collinear system, P_{coll} reaches a maximum at a partial resonance of transmission, i.e., when for example, $|T_{N1}| = 1$, then $|T_{N2}|$ is an exponentially small quantity. Scattering on a "noncoplanar defect" leads to the fact that the narrow resonances on Figure 12.15 acquire a certain width Γ depending on the electron momentum and inversely proportional to the time of the forward scattering on the "defect". These results in the picture shown in Figure 12.16 [30]: there appears a sequence of sufficiently wide plateaus at relatively large $k(k > 1, 5)$ values separated by deep dips to zero polarization. The regions of these drastic changes are very narrow and are determined from the inequality condition $\cos(\delta_\ell - k) > |t_\ell|$. In the region of small values $k(k < 1, 5)$ we have a sequence of peaks as in Figure 12.16. Such a behavior of $P(k)$ is easily explained on the basis of the uncertainty principle

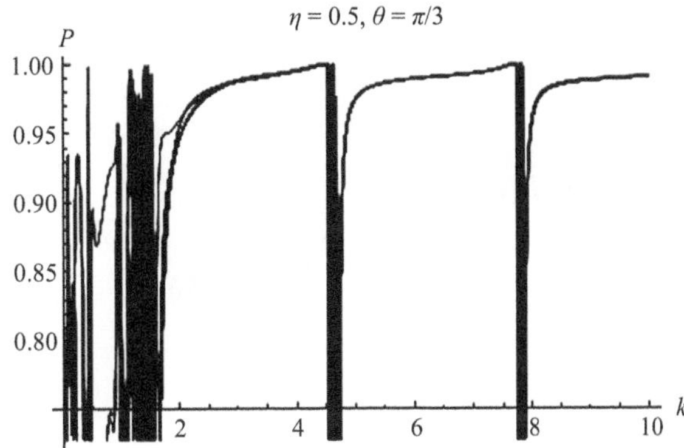

FIGURE 12.16 The momentum dependence of $P(k)$ in the presence of a "noncoplanar defect".

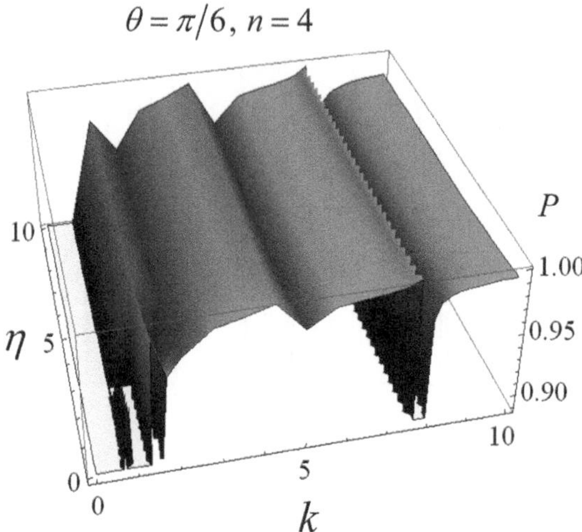

FIGURE 12.17 The dependence of the degree of polarization $P(k, \eta)$ on the variables k and η with fixed values of $\theta = \pi/6$ and $n = 4$.

for energy and time. Also note that the dependence of $P(k)$ very weakly depends on the coordinate of the defect n and the noncoplanarity parameter η [30].

Fig. 12.17 is a 3D-plot of the degree of polarization of the transmitted wave from the electron momentum k and the parameter η. A comparison of the plots on Figures 12.16 and 12.17 indicates their coherence.

Figure 12.18 is a 3D-plot of the dependence of the degree of polarization $P(\theta, \eta)$ on the variables θ and η with fixed $k = 1$, $n = 1$. $P(\theta, \eta)$ oscillates over the angle θ; however, as the parameter η grows, the amplitude of these oscillations significantly decreases.

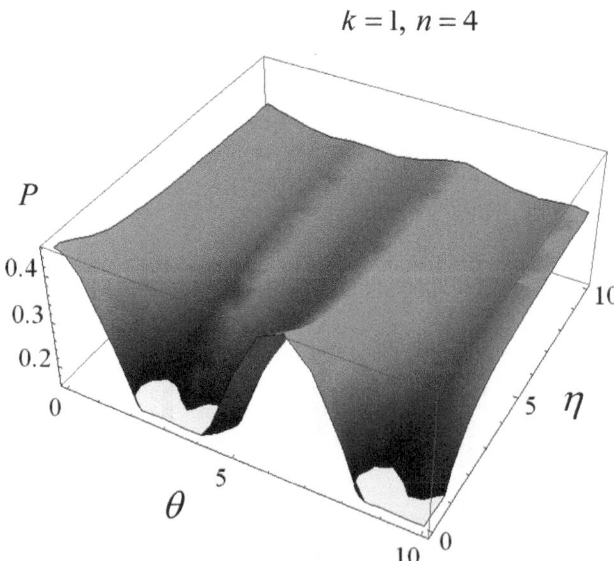

FIGURE 12.18 The dependence of the degree of polarization $P(\theta, \eta)$ on the variables θ and η with fixed values of $k = 1$ and $n = 1$.

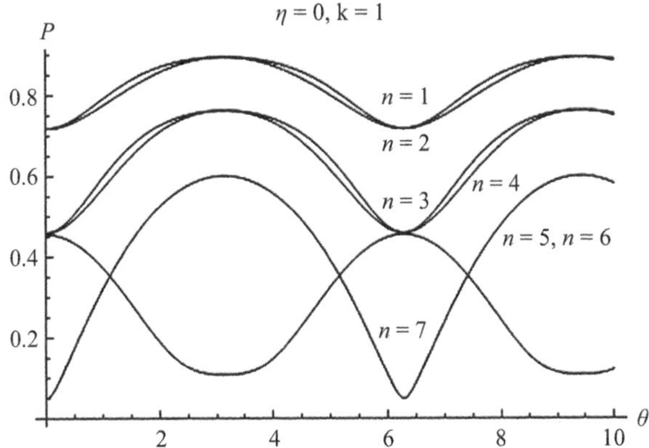

FIGURE 12.19 The dependence of the degree of polarization of the transmitted wave on the angle θ for a non-collinear system ($\eta = 0$).

Figures 12.19 and 12.20 show the dependences of the degree of polarization of the transmitted wave for a non-collinear system ($\eta = 0$) and a noncoplanar system with a large value of parameter η.

$P(\theta)$ is an oscillating function of angle θ with a rather large value of the amplitude of oscillations (Figure 12.19). It is important that $P(\theta)$ substantially depends on the coordinate of the "defect". As the parameter η increases, the amplitude of oscillations decreases, however, there still remains the dependence on the "defect" coordinate.

In Figure 12.21 it can be seen that with the growth of the parameter η, the $P(\eta)$ curves are asymptotically flattened and turned into a plateau: the polarization saturates. However, as can be seen from the plot, the saturated value essentially depends on the coordinate of the "defect". Depending on the coordinate, partial or full y−polarization of the transmitted wave occurs as η increases.

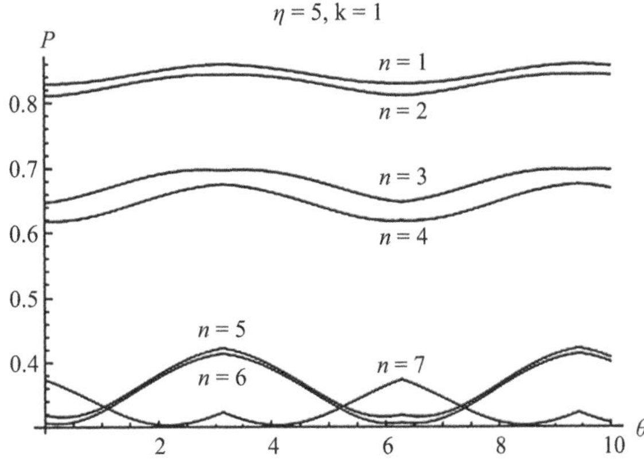

FIGURE 12.20 The dependence of the degree of polarization of the transmitted wave on the angle θ for a strongly non-coplanar system ($\eta = 5$).

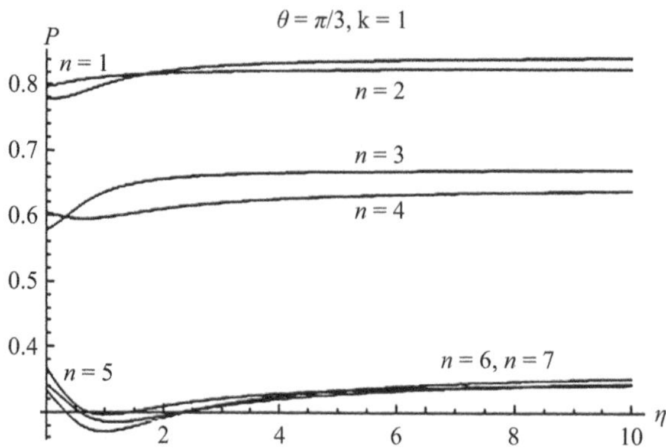

FIGURE 12.21 The dependence of the degree of polarization $P(\eta)$ on the parameter η with fixed $\theta = \pi/6$ and $k = 1$.

12.8 CONCLUSION

Here we have proposed a system consisting of magnetic nanolayers alternating with non-magnetic layers. The potential relief of the system is a sequence of magnetic barriers separated by non-magnetic quantum wells. The system possesses noncoplanarity: the induction of the internal magnetic field of one of the layers (barriers) is noncoplanar with respect to the inductions of the internal fields of the other barriers, and its spatial orientation is determined by certain (noncoplanar) degrees of freedom. The existence of the latter contributes to spin-dependent two-channel elastic scattering of electrons during propagation through this system. As a result, the transmission and reflection amplitudes are dependent on the "noncoplanar" degrees of freedom, and hence the transport characteristics, for example, the transmission coefficients and the degree of polarization of the electron wave, also strongly depend on these degrees of freedom. These quantities in noncoplanar layered system have several interesting properties. In the case, when the system is noncollinear, *i.e.*, when the induction vector of the internal field is parallel to the plane of separation of the layers, these characteristics are oscillating values of the noncollinearity angle, and the amplitude of these oscillations is sufficiently large. The noncoplanarity degree increases with the transverse component of the internal field in one of the interface barrier (separation plane). With increasing the degree of noncoplanarity, the amplitude of oscillations decreases, which significantly affects the behavior of the transport characteristics of the system. These features are set forth in detail in the previous sections. Under certain conditions, the considered systems possess sufficiently strong spin-polarizing properties, which, in general, are varied due to the presence of "noncoplanar" degrees of freedom. The ability to control the transport characteristics makes these systems with noncoplanar nanoelelements perspective from the point of view of designing various nanoelectronic devices: it is important not only to obtain a spin-polarized current, but also be able to control it by local impact on a noncoplanar barrier, for example, by means of an external magnetic field. To implement this in practice, it is very important that the substance of the layer corresponding to the noncoplanar barrier be magnetically soft with respect to substances of other magnetic layers (magnetic-hard layers) and, therefore, the noncoplanar barrier both in profile and in size will differ from the rest. In terms of our use of the transfer-matrix method, this implementation does not lead to difficulties in solving the corresponding scattering problem even for complex noncoplanarity structures. Generally speaking the scattering data for the amplitude and degrees of polarization of scattered waves depend on the direction of the initial wave which can be incident from left or right.

The assumed absence in this paper of other possible scattering channels, leading to spin relaxation, provides the system with the sufficient ballistic transport. However, there are several issues that we didn't consider in this article. In particular, this is concerned to the current-voltage characteristics in these systems and their dependencies on noncoplanar degrees of freedom, or studies of diagrams linking the magnitudes of the spin-polarization degree of the transmitted wave and the transmission coefficient. These problems will be considered somewhere else in future.

12.9 ACKNOWLEDGMENT

This work at California State University Los Angeles was supported by the NSF Centers of Research Excellence in Science and Technology (CREST) and NSF Partnerships for Research and Education in Materials (PREM) under Grant DMR-1523588.

REFERENCES

1. Fert A., Nobel Lecture: Origin, development, and future of spintronics, Rev. Mod. Phys. 80 1517 (2008).
2. Grünberg P., Nobel lecture: From spin waves to giant magnetoresistance and beyond. Rev. Mod. Phys. 80 1531 (2008).
3. Fert A., The present and the future of spintronics, Thin solid film. 517 2–5 (2008).
4. Zutic I., Pas Sarma S., Spintronics: fundamentals and applications. Rev. Mod. Phys. 76 323 (2004).
5. Miyazak T., Tezuka N., Giant magnetic tunneling effect in Fe/Al_2O_3/Fe junction JMMM. 139 L 231–234 (1995).
6. Moodera J.S., Mathon G., Spin-polarized tunneling in ferromagnet junctions, JMMM. 200 248 (1999).
7. Papp G., Peeters F.M., Magneto conductance for tunneling through double magnetic barriers, Physica E. 25 339 (2005).
8. Lu J.-D., Li Y.-B., Dai H.-M., Electron transport properties in a nanostructure consisting of periodic magnetic-electric barriers, Physica E. 40 457 (2008).
9. Zhang H.-j., Tang W.-h., Zhao X., Spin transmission in a series of magnetic barrier structure, Physica B. 389 281 (2007).
10. Lu M., Zhang L., Jin Y., Yan X., Spin-dependent tunneling in nanostructures consisting of magnetic barriers, Eur Phys. J. B 27 4 565–570 (2002).
11. Li S., Lu M.-W., Jiang Y.-Q., Chen S.-Y., Spin filtering in a δ-doped magnetic-electric-barrier nanostructure, Phys. Lett. A 378 3189 (2014).
12. Brataas A., Bauer G.E., Kelly P.J., Non-collinear magnetoelectronics. Phys. Rep. 427 157 (2006).
13. Lu M.-W., Zhang G.-L., Kong Y.-H., Bias-tunable electron-spin polarization in an antiparralel double delta-magnetic barrier nanostructure, Microelectron J 38(3) 401–405 (2007).
14. Radovanovich P.V., Gamelin D.R., High-temperature ferromagnetism in Ni^{2+}-Doped ZnO aggregates prepared from colloidal diluted magnetic semiconductor quantum dots, Phys. Rev. Lett. 91 157202 (2003).
15. Dhar S., Brandt O., Ramsteiner M., Sapega V.F., Ploog K.H., Colossal magnetic moment of Gd in GaN, Phys. Rev. Lett. 94 037205 (2005).
16. Le Clair P., Ha J.K., Swagten J.M., Kohlhepp J.T., vande Vin C.H., de Longew J.M., Large magnetoresistance using hybrid spin filter devices, Appl. Phys Lett. 80 625 (2002).
17. Ramos A.V., Guittet M.J., Moussy J.B., Room temperature spin filtering in epitaxial cobalt-ferrite tunnel barriers, Appl. Phys Lett. 91 122107 (2007).
18. Kocharian A.N., Sahakyan A.S., Movsesyan R.M., Spin-polarizing properties of heterostructures with magnetic nano elements, JMMM. 322 19 (2010).
19. Sahakyan A.S., Movsesyan R.M., Kocharian A.N., The manifestation of one-dimensional noncollinear spin-dependent scattering on magnetized nanoelement structures, J. Nanopart. Res. 16 2198 (2014).
20. Sahakyan A.S., Movsesyan R.M., Kocharian A.N., Heterostructure spin-dependent transport: non-collinearity in magnetic barrier nanoinclusions. In: Dekker Encyclopedia of Nanoscience and Nanotechnology. Third Edition. Taylor & Francis (2017).
21. Auker J.F., Felcher G.P., Polarized-neutron reflectometry, JMMM. 200 741 (1999).
22. Felcher G.P., Adenwalla S., De Haan V.O., Observation of the Zeeman splitting for neutrons reflected by magnetic layers. Physica B. 221 1–4 494–499 (1996).
23. Utsuro M., Ignatovich V.K., Handbook of Neutron Optics. WILEW-VCH (2010).
24. Ignatovich V.K., The physics of ultracold neutrons, Clarendon Press, Oxford (1990).

25. Bir G.L., Pikus G.E., Shelnitz P., Louvich D., Symmetry and strain effects in semiconductors, Wiley, New York (1974).
26. Sahakyan A.S., Poghosyan A.V., Movsesyan R.M., Kocharian A.N., Noncoplanarity effects in the one-dimensional spin-dependent scattering. J. Contemp. Phys. (Armenian Academy of Sciences) 52 pp 49–57 (2017).
27. Poghosyan A.V., Sahakyan A.S., Movsesyan R.M., Kocharian A.N., On scattering of the spin-non-polarized wave on the two-barrier noncoplanar system, J. Contemp. Phys. 52 3 pp 209–215 (2017).
28. Sahakyan A.S., Poghosyan A.V., Movsesyan R.M., Kocharian A.N., Multilayered nanostructures with controlled magnetic configurations, Tech.Connect Briefs 2017 Tech.Connect Org. ISBN 978-0-9975117-8-9 Advanced Materials. pp 180–183.
29. Ditchburn R.W., Light, Blackie & Son Ltd. 1952.
30. Sahakyan A.S., Poghosyan A.V., Movsesyan R.M., Kocharian A.N., Magnetic heterostructures with tunable spin-dependent transport properties, Tech. Connect Briefs 2018, Tech. Connect Org. ISBN 978-0-9988782-2-5 Advanced Materials pp 208–211.

13 Graphene-Layered Structures for Next-Generation Energy Storage

Tereza M. Paronyan
Hexalayer, LLC, Lousville, Kentucky, 40206

CONTENTS

13.1 INTRODUCTION

The interest in the intercalation of alkali metals within graphitic carbon started in the 1950s [1], and it has been found that, along with other alkali metals, lithium has the most efficient intercalation into graphite due to its smaller atomic radius. In 1991, Sony reported about commercializing the concept of lithium reversible intercalation into graphite as efficient energy storage in the form of lithium-ion battery (LIB) devices. An LIB is a rechargeable battery in which lithium ions move from the negative electrode to the positive electrode during discharge and back when charging, where the active material is responsible for the amount of stored lithium.

Since then, various forms of carbon materials, including natural and synthetic graphite, carbon blacks, active carbons, carbon fibers, cokes, carbon nanotubes, spherical carbon nanostructures, exfoliated graphite, and graphene in many different forms, have been evaluated as candidates for lithium intercalation, and significant improvement in LIB devices has been achieved. LIBs have become an important part of our lifestyle as a portable mobile power source of unnamed devices. However, they suffer from a short lifetime due to the low capacity of storing energy (carried by lithium) and long recharge time. Many efforts have been made to improve the capacity of those devices, including replacing the active material (referred to as an anode) to achieve an enhanced lithium absorption.

To maximize the stored energy per unit mass in these batteries, it is crucial to determine which types of carbon may react reversibly with the largest amount of lithium per unit mass of the carbon. Battery scientists measure the number of electrons transferred through the external circuit per gram of carbon; this quantity is the specific capacity of the carbon in milliampere-hours per gram (mAh/g). As one lithium ion is transferred for each electron, the specific capacity can be directly related to the stoichiometry of the hosting carbon electrode. If one lithium atom is transferred per six carbon atoms, the maximum limit for graphite under ambient conditions [2], a specific capacity of 372 mAh/g results. The effective capacity of graphite is considerably lower than this theoretical capacity because of the limited penetration of lithium within graphite interlayer spaces, which require a larger amount of anode material for an efficient store of energy. Therefore, the efficient intercalation of lithium into interplanar spaces of graphitic anode can significantly reduce the weight of efficient batteries.

Silicon is considered another promising candidate for a high-energy density LIB anode with a theoretical capacity of 4200 mAh/g. However, silicon anode faces two challenges: poor conductivity and volumetric expansion. During lithiation/delithiation, Si anodes exhibit significant volume change, resulting in pulverization of the particles, loss of the electrical contact, rupture of the solid-electrolyte interphase (SEI), and consequently, rapidly faded storage performance [3]. The challenges of silicon devices immediately lead scientists and engineers to search for other competitive anode materials, and current efforts mainly focus on developing new materials with higher capacities [4,5].

However, reducing the battery weight with existing capacities or developing new lightweight material with higher capacities is also important. Advanced lightweight materials are the key to the development of future flexible batteries.

Graphene, a carbon monolayer packed into a 2D honeycomb lattice, is considered to be a building block for carbonaceous materials such as graphite, fullerenes, and carbon nanotubes. Since 2004, when Novoselov and coworkers isolated a single atomic layer of graphene from graphite [6], the interest in graphene has grown constantly, giving rise to the perspectives for next-generation electronic devices and energy storage. Graphene-layered materials in general exhibit rich physics and application potential owing to their exceptional electronic properties, which arise from the intricate π-orbital coupling and the symmetry breaking in twisted multilayer systems [7–21].

In this review, we discuss some synthesis routes of 2D- and 3D-layered graphene, key distinguishing characteristics of graphene in a single-layer or multilayered form, and their perspectives in energy storage. We discuss the features of newly developed incommensurate multilayer graphene (IMLG), the feasibility of IMLG production, and the extraordinary features of IMLG intercalating large amounts of lithium as an anode in LIBs. These results provide a new approach to the quick development of advanced rechargeable lightweight and flexible batteries.

13.2 EXPERIMENTAL AND DISCUSSIONS

13.2.1 Graphene Synthesis: Two- and Three-Dimensional Graphene

In recent years, a large number of publications have discussed the preparation techniques of graphene for various applications. The focus has recently shifted to the epitaxial growth of graphene as being a more advanced process to control the graphene geometry that would be beneficial for its application in electronics [22,23]. Indeed, the preparation route affects the quality, structural architecture (Figure 13.1), and physical, mechanical, and electrical properties of the final product.

Graphene, a *defect-free flat carbon monolayer*, is the only basic member of a large family of 2D carbon forms. Single-layer graphene can be obtained by either exfoliation (mechanically or chemically) of graphite [6, 24] or directly growing it on catalytical substrates at high temperatures in a result of decomposition hydrocarbon [25]. Mechanically exfoliated graphene is usually flat and high quality but of small size (micrometers), while the chemically exfoliated graphene is randomly oriented and has enriched defects due to the oxidation–reduction process of graphite (Figure 13.2a).

Large area graphene (centimeters) can be easily produced by chemical vapor deposition (CVD) technique using many transitional metals (Cu, Ni, Pt, Pd, Ir, etc.) as catalytical substrates [25–30]. It can be easily isolated from the metallic substrate and used for numerous applications [31].

Copper foil has been found to be a good catalytic substrate for developing two-dimensional (2D) large-scale graphene (single or very few) sheets by CVD and transfer to desired substrates

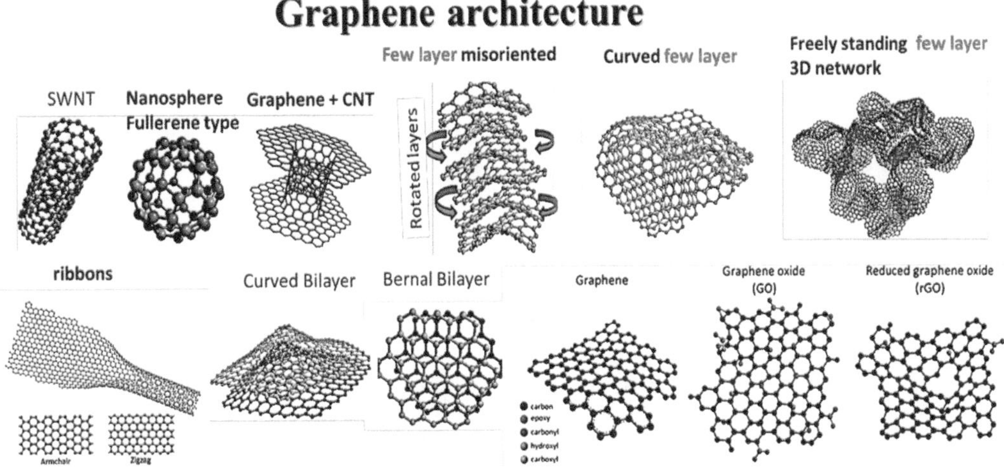

FIGURE 13.1 Graphene-based structures.

(Figure 13.2b). Many factors affect the graphene quality grown on copper by CVD, including the purity, thickness, and surface morphology of the catalyst, hydrocarbon source, growth temperature, pressure, and mechanical stresses [32].

Multilayer graphene can also be easily grown on nickel catalyst due to its high solubility of carbon into nickel [26]. The shape of nickel substrate and synthesis conditions can be crucial defining the internal and external structural architecture of graphene. Nickel foil is an excellent substrate to prepare flattened, large area, few-layer graphene [33], whereas nickel powder or particles can be used to prepare 3D porosity graphene [34–36] (Figure 13.3b). Usually, graphene grown on nickel catalyst results preferably commensurately AB and ABA stacking layers (CMLG) like graphite's stacking order. In some cases of curved-like substrates at certain growing conditions, graphene may misplace the commensurate ordering and result in misoriented, rotated, or twisted layers. As we mentioned in layered structures, the electronic properties can be manipulated through variation of layers and their orientation [22,23], and in this point of view, multilayer graphene structures become particularly interesting.

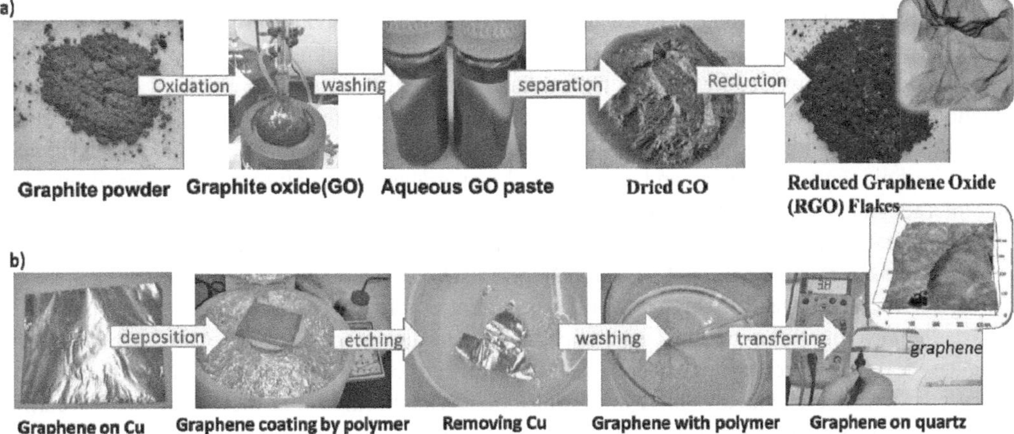

FIGURE 13.2 (a) Graphite exfoliation into graphene flakes using the Hummer method. (b) Large area CVD, flat, single-layer graphene synthesis on copper foil.

FIGURE 13.3 (a) Flattened multilayer graphene growth on nickel foil. (b) Multilayer 3D graphene preparation using nickel powder. The insets in the top show high-resolution SEM images.

Usually, the layered materials can be assembled to form heterostructures held together by van der Waals interlayer interactions. For a given assembly, the relative orientation of the individual layers can significantly change the optical and electronic properties of those layered structures [37–40]. In multilayer graphene, for a given number of layers (*N*), a wide range of properties is accessible by changing the relative orientation of the individual layers [41]. In general, the graphene superstructure is incommensurate with the layered lattice leading to a strong suppression of the interlayer coupling, particularly for the rotation angles >10°. Consequently, the single-layer electronic spectrum reappears for twisted graphene systems, especially in the undoped multilayer system [17–20].

It has been observed by several research groups that the AA-stacked graphene epitaxial systems experience a misorientation of ~2–5° [17,22,23] between the layers and exhibit the electronic properties of a monolayer graphene because of decoupling of the layers, which also occurs in graphene obtained by CVD [42,43] and ultrasonicate graphite [44]. This behavior is different from that known for the AB-stacked graphene. The absence of commensurately stacking order results in weaker van der Waals forces between adjacent layers bringing a change in electronic properties. An increase in rotation angle between layers decreases interplanar interaction so that the IMLG may be considered as a single layer (with modified electronic structure) [19,20,38–40]. In fact, the incommensurately stacked infinite layer graphene can be considered new graphenic structure with weakened interplanar interaction exhibiting novel physical and electronic properties. From this point of view, IMLG would be a new form of graphene structure with another perspective of energy storage.

Supercapacitors have emerged as a highly promising device technology for temporary energy storage, thanks to intrinsically high power densities that enable fast charging and discharging as well as excellent cycling behavior. Ideal supercapacitor electrode materials should exhibit high electrical conductivity, high specific surface area, and area-specific capacitance, as well as good mechanical strength, yet be chemically inert to ensure long device lifetime. For these reasons, sp^2 carbon materials have attracted a great deal of attention [45–49].

The calculations have shown that the capacitance of graphene-derived supercapacitors may improve by modifying the local structure and internal morphology of the graphene electrode rather than the global surface area and pore geometry [47, 50–52]. Also, it has been demonstrated that

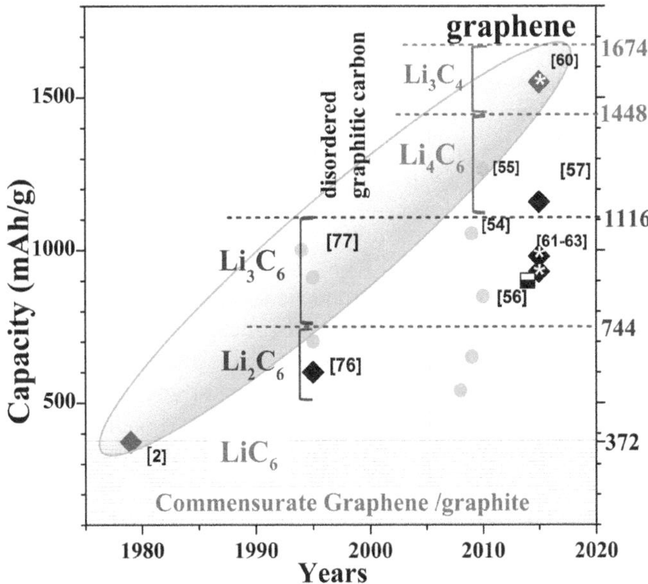

FIGURE 13.4 Historically evaluated data of carbon-based LIBs. The solid diagonals show the reversible stable data; half-solid square, the increased capacity after hundreds of cycles; solid bright circles, irreversible high capacities; and highlighted asterisks, data with IMLG.

multilayer regions are far more effective at screening the double-layer potential than are single-layer regions, which leads to higher area-specific capacitance. Hence, the theoretical quantum capacitance of multilayer graphene can be significantly improved by altering its local structure and morphological features [53].

There have been several publications on testing graphene anodes in which researchers registered unusual higher capacities [54–57] (Figure 13.4). High capacity of Li-ion absorption was measured for chemically derived defective graphene (~1050 mAh/g) and high-quality few-layer graphene (FLG) (850 mAh/g), but these cells were unstable for high discharge rates and for long-term cycling [54,55]. A 900 mAh/g capacity was measured for defective porous graphene network after a 300 charge–discharge cycling process, which proved to be stable over many hundreds of additional cycles [56]. The authors devised a new mechanism in which defects in the graphene lattice act as seed points that initiate plating of lithium metal within the interior of the porous graphene structure. Theoretical calculations have confirmed that topological defects in graphene plane increase lithium absorption capacity [58,59], which would explain the capacity enhancement with highly defective graphene. However, there was no explanation of the capacity increase during the cycling as no additional defects would be created by the lithium insertion/deinsertion process as lithium binds to carbon physically not chemically. The mechanically strong solvated graphene foam has been prepared by chemical exfoliation of graphite and used as an anode without adding any other binders or conductive additives. It has delivered a very high reversible capacity of 1158 mAh/g at a charge–discharge rate of 100 mAh/g [57]. This capacity was even larger than LiC_2 (1116 mAh/g) composition, which was the highest capacity suggested for sp^2 carbon structures. A new explanation needs to explain lithium intercalation into graphenic structures. The authors suggested that enhanced capacity could be largely attributed to the uniqueness of the solvated porous network, which leads to a higher SSA and faster Li-ion diffusion. However, the authors also mentioned unusual stacking order of graphene observed in broadening of the (002) peak of the X-ray diffraction pattern.

Recently, our group has measured significantly higher (up to 1540–1600 mAh/g) reversible capacities at the same charge–discharge rate based on as-grown incommensurately stacked layered graphene (IMLG) in the 3D form [60–63]. Our studies revealed capacity increased with the decrease of commensurate stacking order of layered graphene structures, which led us to develop a completely new adequate lithium intercalation model that would explain any capacity up to 1674 mAh/g for multilayer graphene.

13.2.2 Multilayer Graphene: Commensurate Versus Incommensurate Stacking

CMLG is a form of graphite where the individual layers are called graphene and, in each layer, the carbon atoms are arranged in a honeycomb lattice with separation of 0.142 nm, and the distance between planes is 0.335 nm. When multiple graphene sheets are layered on top of each other, van der Walls bonding occurs, and the three-dimensional structure of graphite is formed with a lattice spacing between sheets $c = 6.71$ Å (Figure 13.5). The sheets align such that their two-dimensional hexagonal lattices are staggered, either in an ABAB pattern or an ABCABC pattern. Atoms in the plane are bonded covalently, with only three of the four potential bonding sites satisfied. The fourth electron is free to migrate in the plane, making graphite electrically conductive.

Incommensurate graphene is also multilayer graphene (IMLG), consisting of single graphene sheets that are closely parallel packed together similar to graphite but incommensurately, such as rotated, twisted, or randomly misoriented (Figure 13.6a), unlike graphite. Twisted graphene is characterized by a rotational mismatch between adjacent sheets that results in a Moiré superstructure [64].

As we learned, both copper and nickel metals have been found to be good catalytical substrates to develop 2D [25, 26] and 3D multilayer graphene for various applications [34–36]. The synthesis conditions such as temperature, pressure, and hydrocarbon source and the choice of catalyst (metal type, shape, and size) as well the quality, the number of layers, and physical, mechanical, and structural properties of graphene define the architecture and application of graphene.

Commensurate flattened graphene can be easily grown at high temperature under ambient pressure using nickel foil or deposited nickel films [26]. Due to the high solubility of the carbon into nickel, it is difficult to control the thickness of graphene (number of layers) precipitated on the nickel surface. The shape and size of the catalyst are very important to govern the inner and outer structural architecture of the graphene, such as its curvature, stacking orientation, and geometrical shape [22,34,35,60].

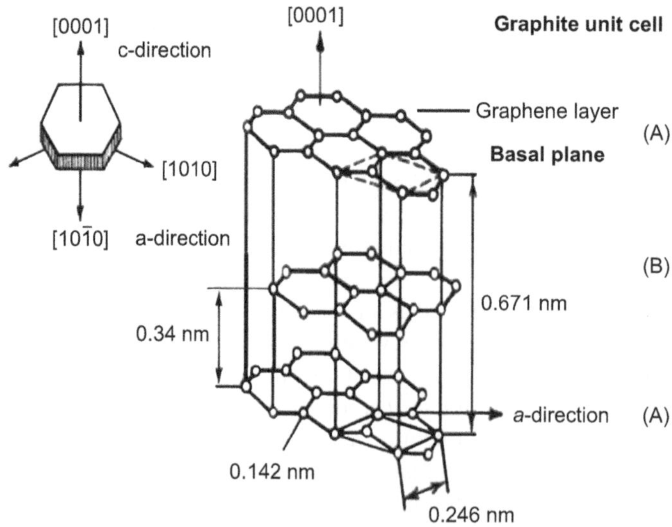

FIGURE 13.5 ABAB-stacked graphite unit cell.

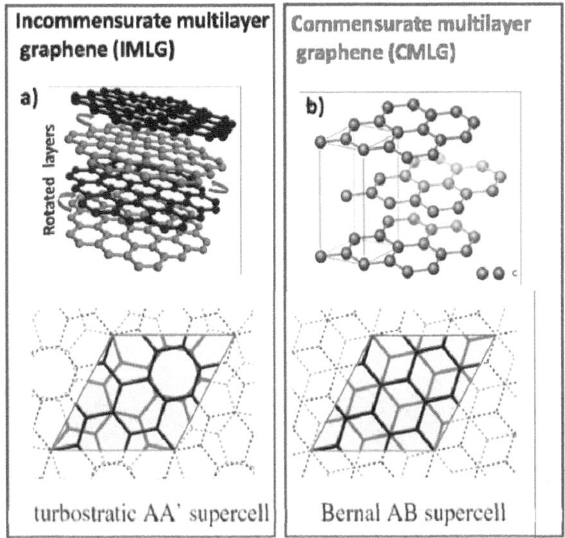

FIGURE 13.6 A schematic illustration of (a) incommensurate and (b) commensurate (graphite) multilayer graphene structures.

The interconnection between crystallites of electrode improves the charge transfer in batteries since the electrical conductivity arises in part from the hopping of carriers between crystallites. Therefore, direct synthesis of well-interconnected crystallites of graphenic nanostructures would be an advance for efficient collection of the electric charges.

For this purpose, a nickel 3D template was built by sintering of nickel particles together at high temperatures, similar to that shown in Figure 13.3b [60]. Creating high porosity within layers and increasing of surface area, a better LIB performance was expected by replacing the graphite anode with 3D graphene structures. However, first efforts with using 3D graphene have not demonstrated significant improvement in the battery's performance compared to the graphite anode (Chen) even when the surface area has been significantly increased. The reason was pointed out to be the larger pore size (hundreds of microns) of the 3D network, which depleted the electrons hopping for charge transfer.

Here, we discuss unique features of incommensurate-layered graphenic structures that demonstrate significant improvement of lithium intercalation (up to 600%) used as an anode in LIBs by replacing graphite [60–63]. IMLG structures were grown on few-micron-sized nickel particles by decomposition of the methane at high temperatures under low pressure (Figure 13.3b). After removing the nickel, a high-purity and -quality graphene network, as a foam, was analyzed by energy dispersive spectroscopy (EDS) (Figure 13.7b).

3D IMLG graphene structures consist of micron-sized curved graphene sheets connected to each other without significant damage and agglomeration and that are separated from each other by a few micropores (Figure 13.8). The curvature of the graphene sheets that is originated of curved shape of catalyst seems the main reason for incommensurate stacking.

X-ray diffraction (XRD) pattern shows a broadened (002) peak with the maximum at $2\theta = 26.45°$ corresponding to the average layer spacing of d = 3.36 Å (Figure 13.9), which includes several Gaussian peaks with d spacing ranging from 3.32 to 3.48 Å. The absence of the other oblique diffraction peaks such as (101), (102), and (103) indicates the absence of commensurate ordering within layers. The (100) and (110) peaks originate from in-plane crystallinity of graphene. The broadened Gaussian peaks are the evidence of the reduced number of graphene layers, which seems to be varied as several Gaussian peaks are fitted. We observed that decreased intensity of XRD (002) peak with nearly

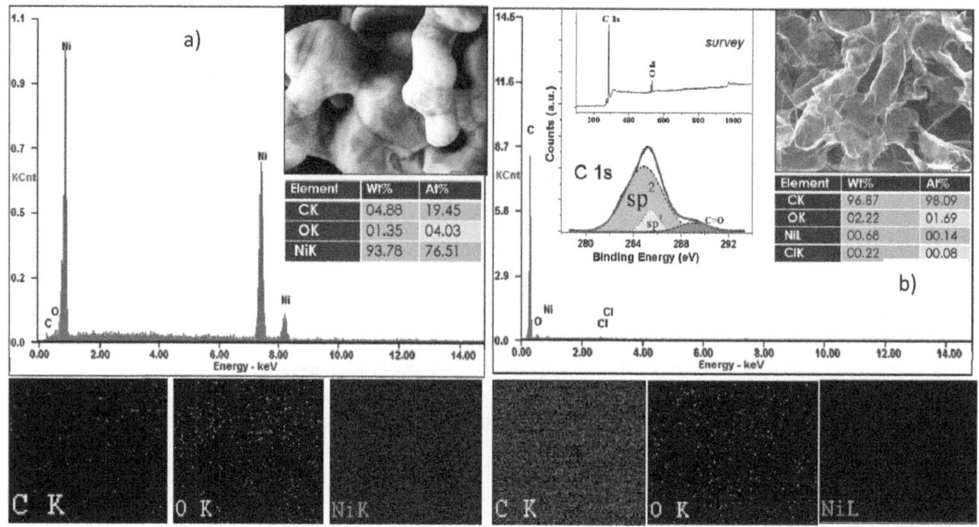

FIGURE 13.7 EDS analysis of graphene/Ni network (a) with and (b) without Ni.

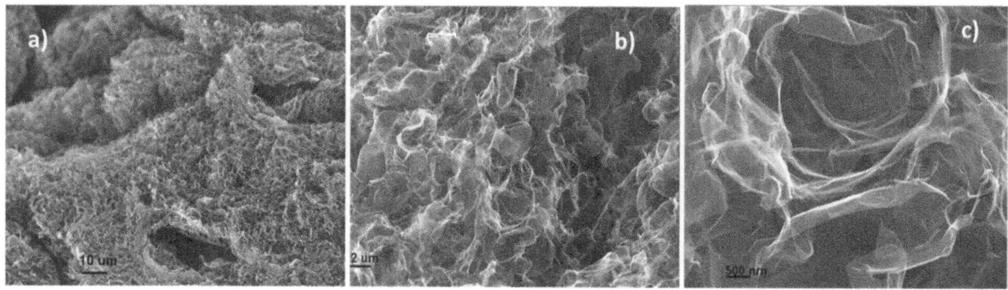

FIGURE 13.8 SEM images of IMLG foam in variety magnifications: (a) scale bar 10 μm, (b) 2 μm, and (c) 500 nm.

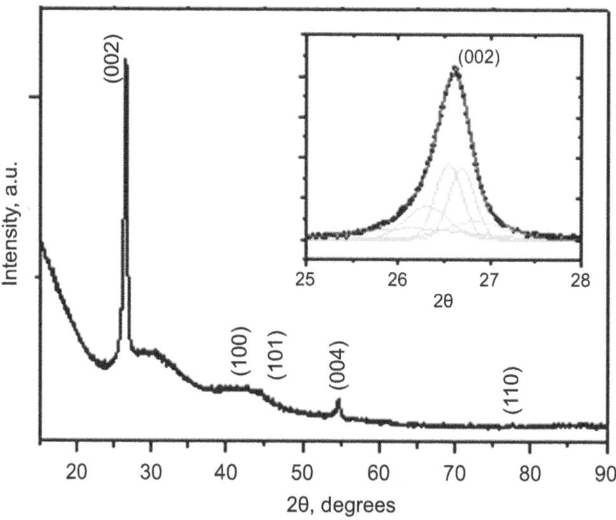

FIGURE 13.9 XRD pattern of as-grown IMLG graphene foam. The inset shows Gaussian fit of (002) peak (zoom in of the (002) peak area).

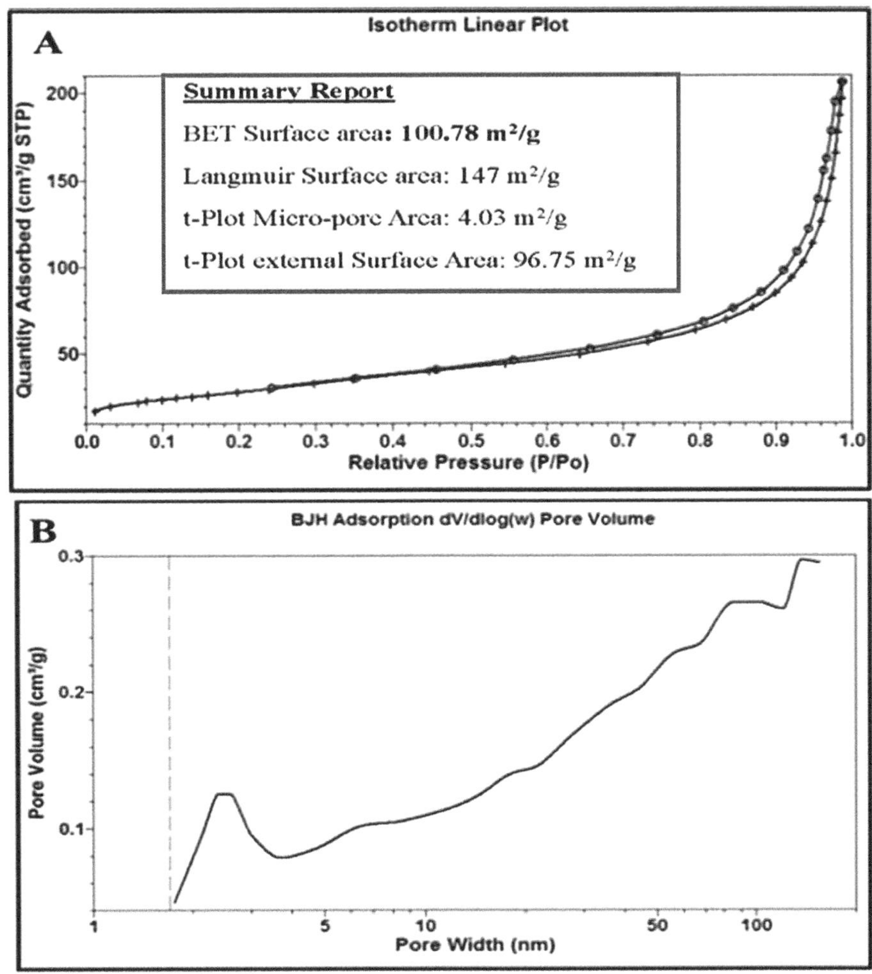

FIGURE 13.10 BET results of IMLG foam with 93% incommensurateness (Sample 1). (A) N_2 molecules adsorption/desorption isotherms. (B) N_2 molecules adsorption.

noticeable (004) peak (sometimes even absent) is correlated to the increased lithium intercalation capacity.

The SSA of IMLG foams measured by Bruner–Emmett–Teller method is 90–100 m^2/g (Figure 13.10), which is ~10 times higher than graphite, assuming these graphene samples consist of 26–30 layers on average, as the theoretical surface area of SLG is 2630 m^2/g.

Raman spectroscopy is a powerful tool to characterize the structural and electronic properties of carbon materials. It teaches us about the stacking order, the number of layers, quantity, and nature of defects, doping [65,66], etc. The shape, width, and intensity of the 2D band distinguish the stacking [67,68] and relative orientations [69]. Raman signature of graphite, SLG (Figure 13.11b), and IMLG (Figure 13.11a) show that both SLG and IMLG have a single Lorentzian fit of the 2D band, whereas graphite has two Lorentzian fits. The number of CMLG layers (two to five layers) can be recognized by the *n* umber of Lorentzian fits into the 2D band.

Multilayer graphene 2D band of either single- or multi-Lorentzian fits discriminates between *commensurate and incommensurate stacking*. In addition to Lorentzian fit, two values of the spectrum—*2D bandwidth* (presented as the full width of half maxima, FWHM) and I_{2D}/I_G (here, I_{2D} and I_G are the heights of 2D and G bands, respectively)—can classify the range of *incommensurate* and

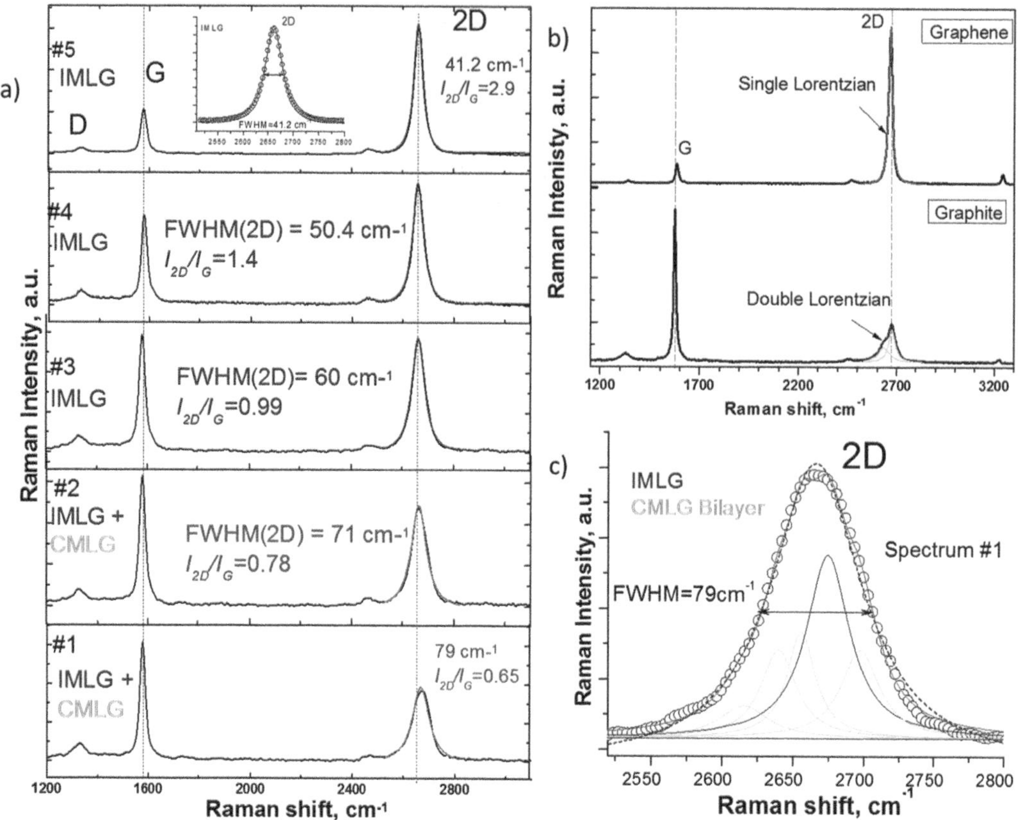

FIGURE 13.11 Raman analysis of commensurate to incommensurate transition. (a) Raman spectra of five individual spots by 5 μm (X:Y) mapping steps of laser (λ = 638 nm) showing the transition of commensurate (spectrum #1) to fully incommensurate (spectrum #5). (b) Details of 2D peak Lorentzian fit of spectrum #5. (c) Details of multi-Lorentzian fit of 2D band of spectrum #1.

commensurate stacking. The higher the ratio I_{2D}/I_G and the narrower the 2D band, the weaker is the interplanar coupling, thus closer resembling spectrum of SLG (#5 graph in Figure 13.11a). A direct determination of FWHM and I_{2D}/I_G bulk values can be acquired by the generation of Raman maps (Figure 13.12). The batch of 2D and G bands was processed and analyzed from several maps to estimate the fraction of incommensurate stacking in graphene material. Aberration of these values of the incommensurate stacking area is likely caused by the variety of interplanar rotation angles of adjacent layers as noted in Ref. 69. The rotation angle between layers can be defined from selected area electron diffraction (SAED) patterns (Figure 13.13b–d). Raman 2D band becomes broader and consists of a mixture of multi-Lorentzian of IMLG and CMLG due to their overlapping exposure to laser [70], which still can be counted as CMLG (Figure 13.10c). Commensurate to incommensurate transition occurs at the set range of FWHM (2D) $\geq$56 cm^{-1} with $I_{2D}/I_G \geq 0.94$. The overlapping areas can be either IMLG or CMLG, where 2D band has a similar fit of either single- or multi-Lorentzian. More than 45 individual spectra were analyzed to find the set of FMHM and I_{2D}/I_G values that determine CMLG to IMLG transition [60].

High-resolution transmission electron microscopy (HRTEM) revealed that graphene contains 2–30 layers of curved sheets, and single-layer sheets are seldom observed (Figure 13.13d). HRTEM images of graphite or as-grown CMLG (Figure 13.12e) differ from IMLG (Figure 13.13g and h) as the twist or rotation of layers results in Moiré patterns [64]. Those patterns were observed on IMLG

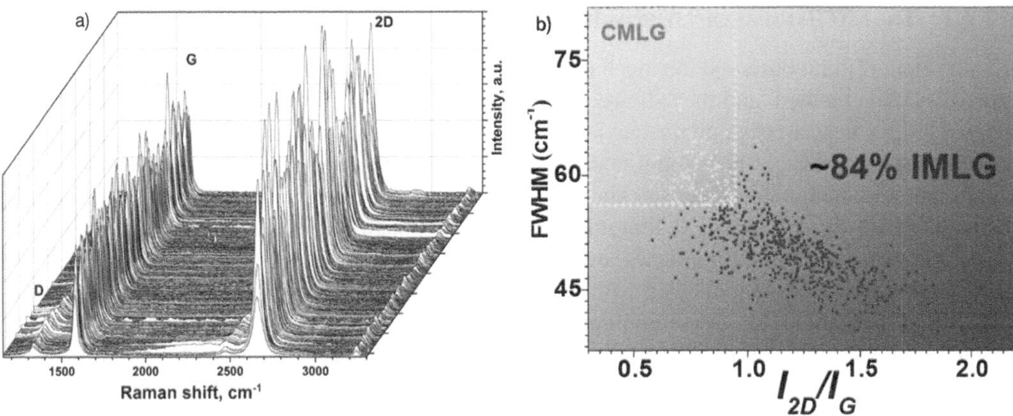

FIGURE 13.12 (a) Scattering of Raman spectrum by mapping of X;Y-5x5 using 638nm wavelength of Laser, (b) FWHM of 2D vs. I2D/IG of hundreds of individual Raman spectrum showing the boundary of IMLG and CMLG.

sheets after fast Fourier transform (FFT) filtering of the HRTEM image (the inset of Figure 13.13h) when layers are rotated or twisted. SAED measurements are another way to reveal multilayer graphene is either *commensurate or incommensurate*. SAED pattern of single-layer graphene is similar to the graphite, as shown in Figure 13.12a, while IMLG has two or more hexagonal patterns rotated to each other at 5–30° angles (Figure 13.12b–d), indicating the misorientation of adjacent layers. Many numbers of rotated layers exhibit a circular pattern, as shown in Figure 13.13d. In fact, the signal of single-Lorentzian fit of Raman 2D band associated with rotated SAED multihexagon patterns is the key characteristic to recognize whether graphene is *commensurate* or *incommensurate*. In the case of larger rotation angles, XRD (002) peak may disappear due to weakened interplanar interaction becoming similar to the XRD of single-layer graphene.

Thus, Raman analysis is the most powerful tool to estimate multilayer graphene stacking order.

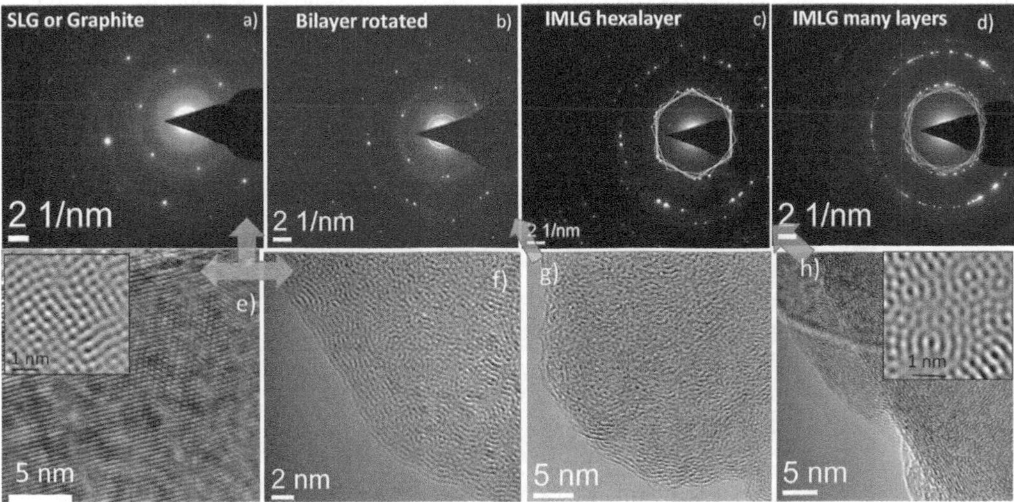

FIGURE 13.13 SAED patterns: (a) SLG or graphite, (b) bilayer rotated, (c) IMLG hexalayer, (d) IMLG many layers. HRTEM patterns of (e) graphite, (f) single-curved layer, (g) IMLG bilayer, and (h) IMLG six layers. The insets of (e) and (h) show FFT filtering of masked areas of the images.

13.2.3 The Mechanisms of Lithium Intercalation into Graphitic/Graphenic Carbon

Kambe et al. [2] first observed the mechanism of lithium intercalation into the graphite by the electron diffraction method, and they showed the formation of LiC_6. They found that, at low temperatures, Li^+ ions in successive intercalate layers are commensurate with the graphite layers in a C6Li configuration and that, at near room temperature, Li ions remained in a C_6Li commensurate configuration even though interlayer stacking order of graphite was absent. The authors suggested that lattice constant of metallic *b.c.c.* lithium (a = 3.51 Å) is such that sequential (111) planes sandwiched between *AAA* graphite planes are nearly commensurate with the LiC_6 structure. Li atoms occupy next nearest-neighbor sites within the van der Waals spaces between every pair of carbon sheets (Figure 13.14). Those findings suggested that lithium tends to intercalate into graphite in LiC_6 configuration, and it was significant input for the further development of lithium intercalation into graphite at ambient conditions using electrochemical reactions in LIB cells assembled as an anode/electrolyte/separator/cathode [71].

As the measured number of electrons transferred through the external circuit per gram of carbon is the specific capacity, then stoichiometry of the carbon electrode with lithium defines this value. The theoretical capacity of lithium intercalation into graphite can be calculated based on LiC_6 compound formation when one lithium ion can be absorbed on every other hexagon of carbon by resulting LiC_6 stoichiometry by Equation 13.1.

$$\mathrm{Li} + x\bar{e} + 6\mathrm{C} \Leftrightarrow \mathrm{LiC_6} \tag{13.1}$$

Specific capacity ($C_{s,th}$) = 372 mAh/g based on Faraday's well-known equation.

$$C_{s,th} = \frac{nF}{M} \tag{13.2}$$

where n is the number of moles of electrons transferred per mole, F is Faraday's constant, $F = 9.6485 \times \frac{10^4\mathrm{C}}{\mathrm{mol}\,\bar{e}}$, and M is the molar mass of active (carbon) material – 72 g/mol.

Nearly, this capacity is expected to be achieved with highly crystalline graphite considering LiC_6 formation. However, it appears that lithium reversible intercalation into graphite is considerably lower than this value. The limit of graphite's capacity is due to excluded lithium intercalation within interlayer spaces, which consists of commensurately stacked infinite layers of graphene. The strength of repulsive interactions that arise due to the orthogonality of interplanar π orbitals of sp^2 carbon excludes lithium diffusion between graphene layers, and as a result, the predominate intercalation occurs at prismatic surfaces (arm-chair and zig-zag faces), and no significant lithium reversibly intercalates within commensurately stacked graphene layers.

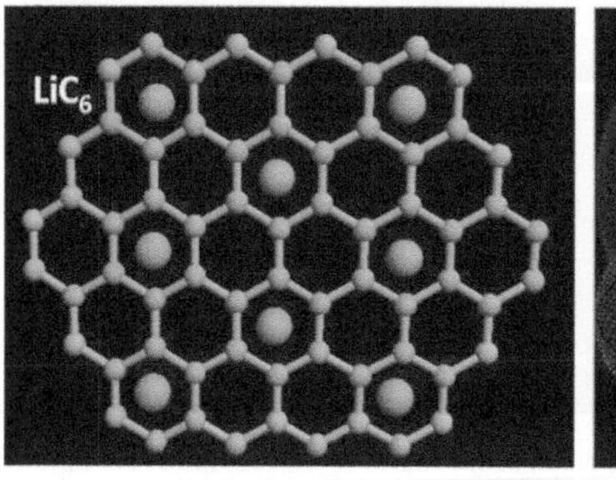

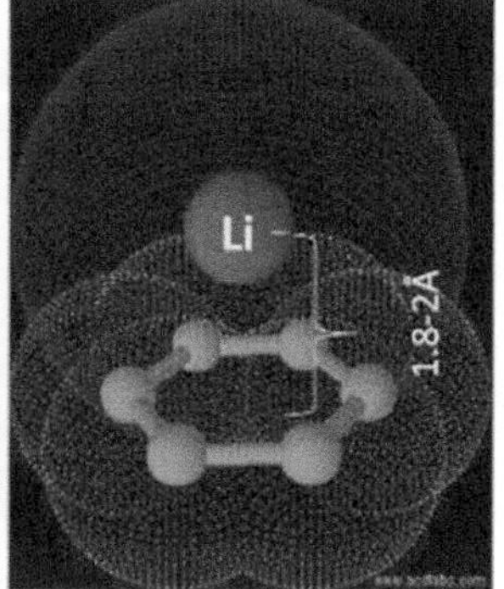

FIGURE 13.14 Molecular illustration of lithium intercalation mechanism of LiC_6.

Many other carbonaceous materials have been studied as active materials for lithium intercalation, and it appears that lithium can reversibly intercalate within most carbonaceous materials with various capacities. Besides graphitic carbon, lithium also intercalates into non-graphitizable hard carbons and also reversibly binds near hydrogen atoms in carbonaceous materials containing substantial hydrogen. It has been found that the efficiency of lithium absorption strongly depends on multiple factors including a crystalline structure, surface area, and stacking geometry [72–75].

In the 1990s, the increased capacities were measured by Dahn's group [76] and Sato's group [77] using hydrogen-bonded and disordered carbonaceous materials. Dahn's group offered a new mechanism from which they inferred that there is a double-site insertion of lithium into single-layer graphene (SLG) that can result in a maximum capacity of 744 mAh/g (Li_2C_6). Sato et al. proposed another mechanism suggesting covalent arrangements of Li2 that occupy nearest neighbor sites between pairs of carbon sheets that would have a capacity of 1116 mAh/g (LiC_2) [77]. Later, Binda et al. prepared and studied LiC_2 under low pressures, but it was unstable at ambient conditions [78]. Thus, achieving a reversible stable absorption of lithium that would provide either Li_2C_6 or LiC_2 composition has remained a challenge.

Recently, graphene became the most studied material for energy storage due to its fundamental properties such as superior electrical conductivity, extraordinarily high surface area, and stability. It has great potential for the significant change of the energy-storage landscape. Specifically, graphene could present several new features for energy-storage devices, such as lightweight and fast-charging batteries, powerful ultralight capacitors, flexible and transparent batteries, and high-capacity and fast-charging mobile devices.

As we mentioned, the electronic properties of multilayer graphene can be varied with the number of graphene layers and their stacking orientation [37–39]. In particular, the rotation of adjacent layers results in the incommensurate stacking order that weakens interlayer interactions within adjacent layers. An increase in rotation angle between layers decreases the interplanar interaction so that the incommensurately stacked multilayers can be considered as a single graphene layer with a modified electronic structure [39].

As-grown multilayer graphene films on SiC(000) (*C*-face), typically 10–20 graphene layers thick, exhibit high carrier mobility with characteristics of highly doped (single-layer graphene) [79,80]. Subsequent calculations confirmed that those stacking faults decouple adjacent graphene sheets so that their band structure is nearly identical to isolated graphene. Santos et al. [19] found that a small angle rotation destroys the particle-hole symmetry of an AB-stacked bilayer and a small stacking defect such as rotation can have a profound effect on the low-energy properties of the bilayer. Unlike the case of the AB-stacked bilayer, a potential difference between layers does not open a gap in the spectrum. Therefore, we can assume that the rotated many layer graphene can be considered as a new form of graphitic structure with weakened interplanar interaction that exhibits similar physical properties of isolated SLG with newly modified electronic configuration. The supercell of rotated layers differs from supercell of graphite (Figure 13.6), and new phenomena of quantum physics may occur. Recently, a quantum mechanical model of lithium storage in multilayer graphene has shown dramatic changes in quantum capacity as local structure and morphology are altered [53].

Here, we demonstrate the preparing of free-standing incommensurate graphene structures and their potential application as anodes in LIB cells by replacing the graphite. We discuss the features of this graphene and dramatic enhancement of lithium intercalation capacity with finite incommensurate layers synthesized in the form of a 3D network.

13.2.4 IMLG-Based Batteries: New Mechanism of Lithium Insertion into Multilayer Graphene

The charge collection generated in the electrodes of batteries is governed by the size and nature of crystallites as the electrical conductivity arises in part from the hopping of carriers between them.

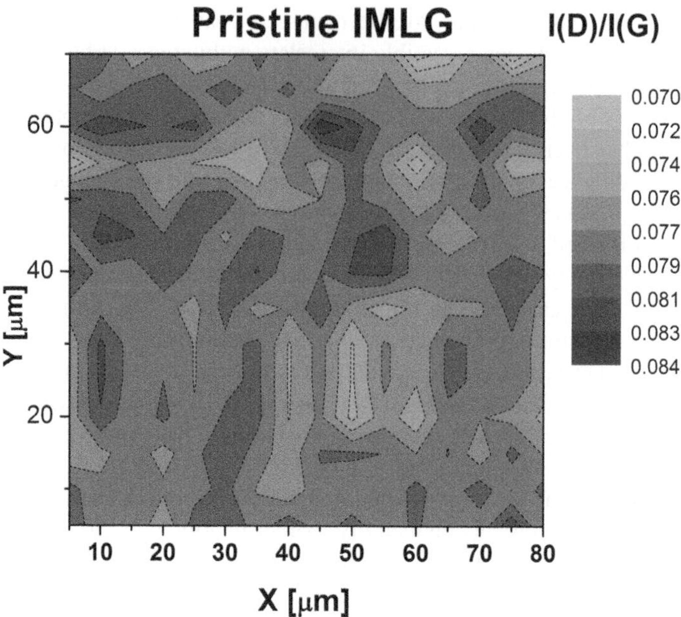

FIGURE 13.15 Raman mapping analysis of pristine graphene foam performed by 5 × 5 (X, Y) μm step (λ = 638 nm laser wavelength). Showing I_D/I_G distribution.

L_a crystallite size and defect concentration of IMLG samples were estimated by Raman I_D/I_G (here, I_D is the height of D band) [81,82], and a low rate of I_D/I_G distribution was revealed.

Figure 13.15 shows higher crystallite size of L_a = 450–580 nm, confirming overall high crystallinity in-plane. Overall, a low defect concentration (<0.02% of carbon) neglects any impact of original defects on the battery performance. Several samples of pristine graphene synthesized at various conditions were estimated that vary from 19% to 85% of incommensurateness by generating Raman maps (as shown in Figure 13.12). High-quality graphene foams with various incommensurateness were acquired (~1 mg for each coin cell) and tested as anodes with up to 100 cycles of electrochemical cells (ECs) at 100 mAh/g current density (Figure 13.16a).

Discharge and charge measurements were carried out at a voltage range of 3.0–0.005 V with different current densities of 0.1–5.0 A/g. The cyclic voltammetry (CV) measurements were performed at the voltage range of 3.0–0.005 V with a scan speed of 1 mV/s. The initial discharge (Li-insertion) capacity profiles (Figure 13.16b) are different from the subsequent cycles of charge or discharge caused by the formation of a passivating film (referred to as the SEI). Two cathodic peaks appeared during the initial lithiation process that were measured by CV (Figure 13.16d). The first current peak, which appeared in the voltage range of 0.7 V, is likely related to *SEI* formation as it disappeared at subsequent cycles. The second peak, located at 0.15 V, appears for high incommensurate (93%) samples. After two to three cycles, the current peaks became stable, which indicates that the insertion–extraction of Li^+ produces good reversibility with low hysteresis. The anodic peak reaches a maximum after the second cycle at +0.34 V. The anodic peak was minimized from 1 to +0.34 V by an increase in incommensurateness degree indicating larger lithium insertion. It appears that higher incommensurateness creates easy accessibility for the lithium atoms as weakened interlayer interaction occurs with higher incommensurateness. The first charge capacity increases by an increase of incommensurateness degree, confirming that larger lithium diffusion occurs due to interplanar accessibility. The tendency of capacity increase occurs also after the 100th cycle from 410 mAh/g (~20% IMLG) up to 1600 mAh/g (93% IMLG) by an increase in incommensurateness

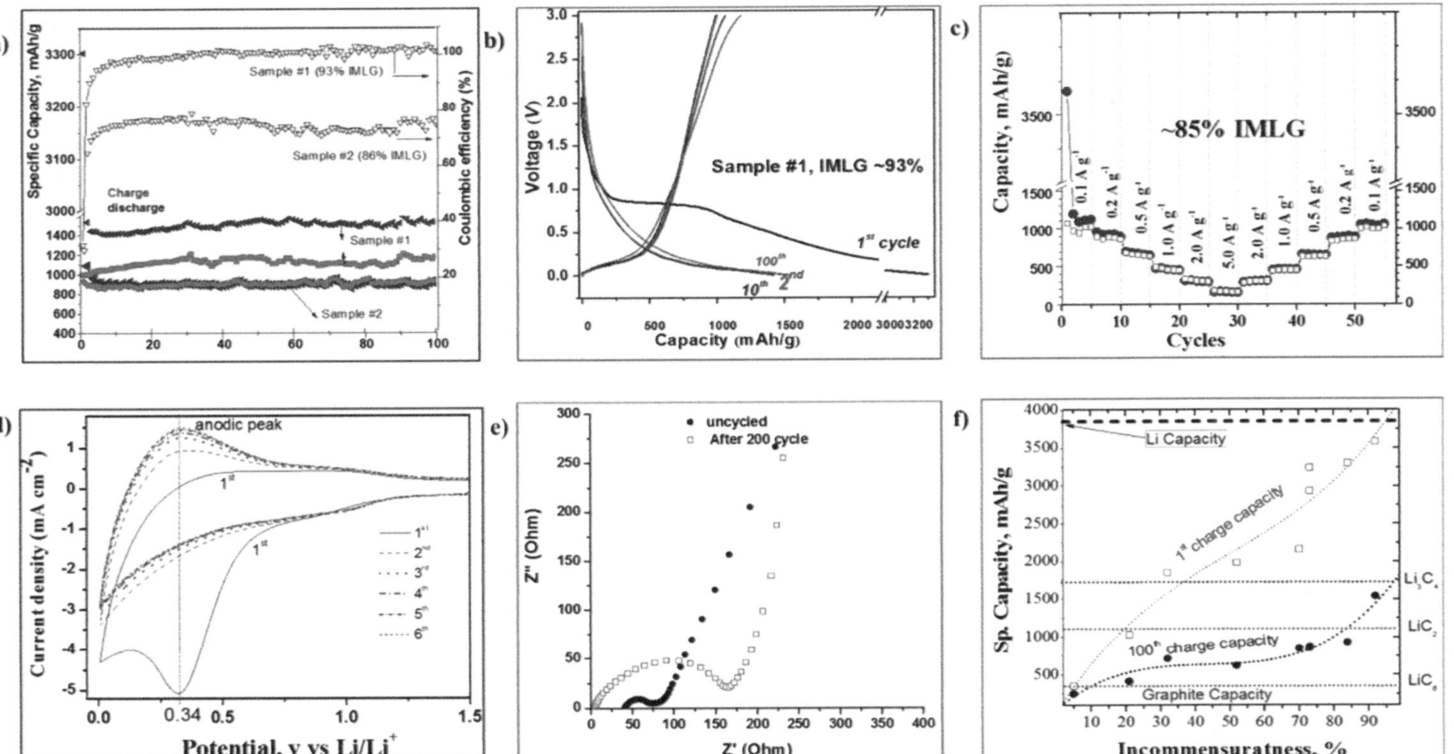

FIGURE 13.16 Electrochemical characterization of IMLG-based coin cells. (a) Capacity cycling for Samples 1 and 2 tested at 100 mA/g: solid triangles – charge (lithiation), solid dots – discharge capacity, open triangles – coulombic efficiencies (axis in right). (b) Charge–discharge voltage vs. specific capacity at 100 mA/g: 1st, 2nd, 10th, and 100th cycle for Sample 1. (c) *C*-rate testing for Sample 2 at different current densities: solid dots – charge, open squares – discharge capacities. (d) CV curves of Sample 2 at 3.0–0.005 V with a scan speed of 1 mV/s. (e) Impedance measurement of IMLG Sample 2 electrodes before (solid cycles) and after 200th cycle (open squares). (f) The plot of charge capacities versus incommensurateness of pristine graphene samples; open squares – first cycle, solid dots – 100th cycle. The curve presents polynomial fit of 100th capacity.

(Figure 13.16f). Commercial graphite was also tested, which performs at ~250 mAh/g capacity at the same current density.

Overall, the cells that exhibit 1030–1100 mAh/g at the second charge cycle (as Sample 2 with 85% IMLG) demonstrate higher coulombic efficiency up to 99.9% throughout 100 cycles. The cells that exhibit a higher capacity (as Sample 1) of 1100 mAh/g in the second charge cycle demonstrate lower (~74–76%) coulombic efficiency (as sample with 93% IMLG) even though the retained capacity remains over 95% after the second cycle. *C*-rate testing (Figure 13.16c) exhibits stability of the cells under high-current densities, which makes these cells feasible. Impedance measurements (Figure 13.16e) before and after 200 cycles show no significant damage in conductivity of the electrodes, which show efficient charge–discharge of IMLG electrodes without further accumulation of metallic lithium.

The specific charge capacity shows nonlinear dependence versus incommensurateness (Figure 13.16f), which allows us to consider other factors such as the layer–layer rotation angle, and the number of layers can also affect charge collection.

Due to obtained high values of reversible capacities we considered a new lithium insertion model which would explain any LiC_{2-x} $(0 \leq x \leq 2)$ formation by Equation 13.3.

$$LixC_{2-x} \rightarrow xLi^{+} + x\bar{e} + C_{2-x} \tag{13.3}$$

XRD analysis of highly lithiated and delithiated anodes prepared from 85% to 93% IMLG does not reveal any Bragg peaks at the range of $2\theta = 9 \div 35°$, whereas the samples with ≤80% IMLG exhibit a low-intensity broader (002) peak that has a tendency to be increased with lower capacity and incommensurateness degree (Figure 13.17a). The absence of the (002) peak in highly intercalated graphene electrodes is due to delamination of graphene into single sheets, with distances $d \geq 3.6$ Å

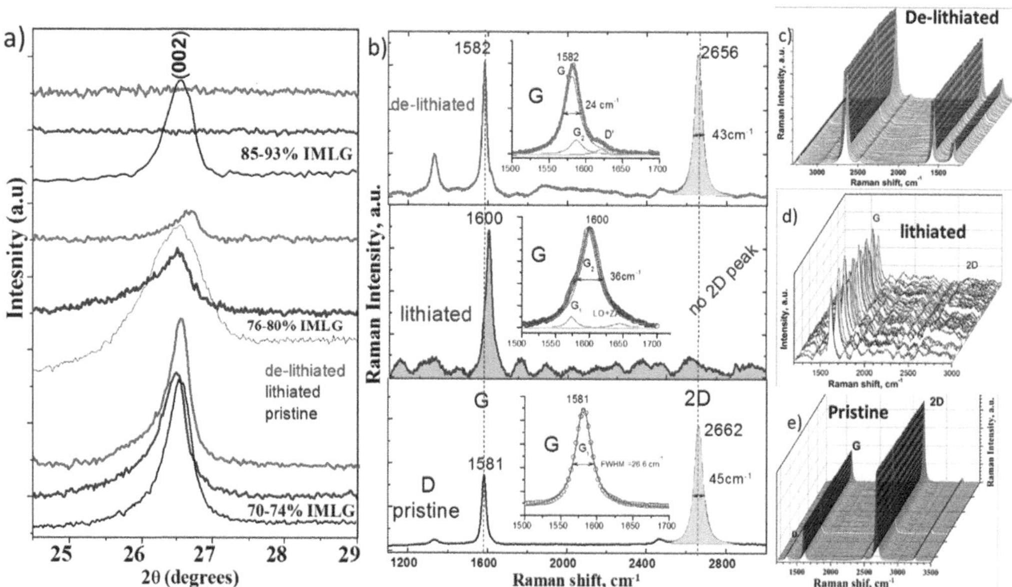

FIGURE 13.17 (a) The performance of XRD (002) peak of lithiated and delithiated electrodes with variety of IMLG fractions: (bottom) pristine graphene, (middle) lithiated, and (top) delithiated. No other Bragg peaks have been observed for highly lithiated samples at the range of 8–35°. (b) The behavior of Raman spectra of graphene electrodes (bottom) pristine, (middle) lithiated, and (top) delithiated. The insets show Lorentzian fits of G band. 3D images of Raman mapping (c) delithiated, (d) lithiated, and (c) pristine IMLG graphene films.

than from in-plane defects, as low D and high 2D bands of Raman spectra of deinserted samples support no significant damage after lithium desorption (Figure 13.11b-top graph). The absence of highly lithiated electrodes is due to the sandwiched lithium layers, while lower capacity samples still have some stronger interacted graphene layers where the lithium has no diffusion at all as (002) peak position has not changed to the left.

SAED analysis revealed d-spacing peaks at $d_{\text{lithiated}} = 3.9–4.06$ Å and $d_{\text{delithiated}} = 3.65–3.8$ Å, indicating $A\acute{A}$ stacking for both lithiated and delithiated anodes (Figure 13.12f). The lithium has adsorbed between two incommensurate graphene layers at the distance of 1.85–2.03 Å = $1/2\ d_{\text{lithiated}}$, which correlates with the theoretically calculated values of 1.84–2.02 Å [83].

Thus, X-ray probing confirms that all galleries of IMLG layers are sandwiched by lithium layers that are reassembled graphene layers into AA' stacking ($d = 3.9–4.06$ Å). Once the sheets were adjusted to AA′ stacking during the insertion, they stayed in that position even after deinsertion, which, in fact, is the key to governing the further reversible cycling. This assembly is different from graphite intercalation, where the sheets return back to AB initial stacking during the initial deinsertion and prevent further lithium penetration within interlayer spacing.

Structural and electronic changes in the graphene were obvious also by Raman analysis of electrodes (Figure 13.17b). The shift and splitting of G band by Lorentzian fit was observed: G related to interior carbon layer with no lithium intercalated, and G1 and G2 related to bonded carbon layer with partially and fully bonded lithium, respectively [84,85]. Due to the transfer of electron charge, the G band position is changed by 19 cm^{-1} with the significant increase of G2 band height ratio to G band. The height ratio I_G/I_{G2}, estimated at 0.48, is significantly increased up to 8.07 when G peak position is shifted from 1581 to 1600 cm^{-1}. The peak at 1645 cm^{-1} was revealed for highly lithiated samples, which are associated with the out-of-plane LO+ZA phonon mode [86] and here is likely caused by bonding of lithium as it was not observed for delithiated samples. The 2D and D bands nearly disappeared due to lithium insertion, which is similar to highly intercalated graphite [87]. Raman mapping analysis (Figure 13.17c and d) confirms homogeneity of the signals through the electrode that occurs due to the lithium insertion and deinsertion. Surprisingly, delithiated graphene performs high-quality graphene Raman signal, confirming that lithium adsorption–desorption does not affect to the in-plane crystallinity of graphene, and this graphene can handle longer cycling of lithium batteries.

HRTEM of highly lithiated graphene nanosheets revealed well-aligned periodic patterns with respect to graphene planes (Figure 13.18a). The arrangement and periodicity of those patterns led us to assign it to lithium atoms in a manner that every lithium is associated with each hexagonal ring of graphene (Figure 13.18e). FFT filtering of HRTEM image results in Moiré patterns, either lithiated (Figure 13.18c) or delithiated graphene (Figure 13.18d), as a sign of incommensurateness; furthermore, lithium atoms have a commensurate ordering on graphene. A similar construction was observed earlier under in situ HRTEM study by Kambe et al. [2] by changing the temperature. In fact, graphene layers are not completely constructed as AA stacking during lithium intercalation besides commensurately sandwiched lithium layers.

X-ray photoemission spectroscopy (XPS) analysis of highly lithiated/delithiated anodes revealed no metallic lithium or Li_2O peaks have been found after 100 cycles (Figure 13.19a), confirming that all lithium atoms are participating in the charge transfer as Li^+. Often Li_2CO_3 peak was found as a result of Li^+ reaction to CO2 due to expose of electrodes during the transfer to XPS chamber. In general, the Li 1s XPS peak tends to the decrease toward the metallic lithium peak direction with an increase of Li concentration into the carbon [88]. Sample 1 performs single-Gaussian 1s peak of lithium at 55.5 eV (Figure 13.19), which can be assigned to LiC_{2-x} ($0 \leq x \leq 2$). To confirm consistency of the capacity values obtained by EC measurements, we estimated Li:C ratio of another piece of Sample 1 by appropriate interpretation of Li 1s and C 1s peak areas (dividing by sensitivity factors). $LiC_{1.4}$ (capacity of 1594 mAh/g) was acquired after stabilized capacity at the 5th cycle, which demonstrated capacity of 1607 mAh/g at the 5th cycle of EC measurement. It is important to note that

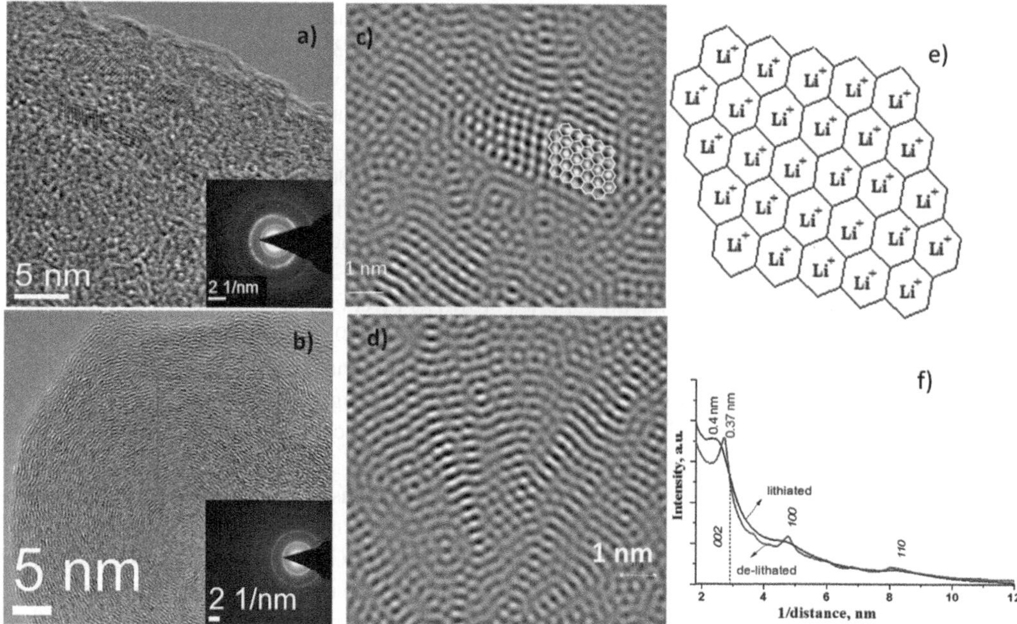

FIGURE 13.18 (a) HRTEM images of Li-inserted sheets (inset) – SAED pattern of the area. (b) FFT filtering of masked area in (a). (c) HRTEM image of deinserted graphene sheets. (inset) SAED pattern of the area. (d) FFT filtering of the masked area in image (c). (e) Schematic illustration of lithium insertion on graphene plane. (f) The plots of intensity vs. inverse distance based on SAED analysis.

no metallic Li (XPS-52.5 eV), Li_2O, or Li_2O_2 XPS peaks were found even after 200 cycles which is a good sign for safe battery charge discharge. Also, no significant electrolyte degradation was found such as C1S the peak at 289 eV is absent, which usually is originated by R-CH_2OCO_2Li formation as result of electrolyte degradation. Impedance measurements for the IMLG electrode before the charge–discharge process and after 200 cycles (Figure 13.16e) confirms no significant degradation of the electrode's conductivity. Usually, electrolyte degradation causes significant conductivity damage by the formation of isolated compounds inside of the anode that causes fade of the capacity cells. XPS and impedance measurements confirm that this graphene actively transfers lithium charges and

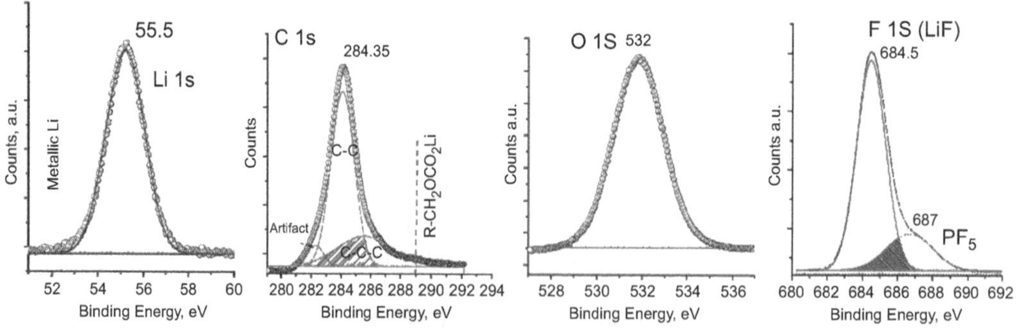

FIGURE 13.19 XPS analysis of charged Sample 2 after 100th cycle of charge–discharge. Non-monochromatized Mg Kα X-ray ($h\nu \approx 1253.6$ eV) was used for the electrode surface excitation.

would avoid further lithium plating. Lithium plating is deleterious to the performance, reliability, and safety of lithium-ion cells.

13.2.5 Discussion of The Mechanism

The study of lithiated and delithiated electrodes led us to build a model that would explain such a high capacity possible variation as well as depend on the number of graphene layers in the same insertion mechanism. This model is based on the multilayer graphene where lithium atoms adsorb above and below each hexagon sandwiched between graphene layers as ordered layer (Figure 13.20). Then Li:C ratio would vary by N number of layers. The specific capacity of graphite LiC_6 is known to be 372 mAh/g. We determined the Sp.C. for other possible stoichiometry following the same logic considering each hexagonal ring can host one lithium atom. We consider the ideal case with single-layer graphene, and Li intercalates on both sides above and below each hexagonal ring of carbon. Equation 13.2 of specific capacity could be expressed also in terms of N number of graphene layers by Equation 13.4. We consider $(N + 1)$ valence electrons (atoms) of Li per molar mass of 1/3 carbon; then for each hexagon, it would be $1/3 \times 6 \times 12.01 \times N$ depending on the number of carbon layers N. Therefore,

$$C_{s,th,N} = \left(\frac{N+1}{N} \cdot \frac{1 \text{ mol } \bar{e}}{12.01 \cdot 6 \cdot \frac{1}{3}[g]}\right) \cdot \left(9.6485 \cdot 10^4 \frac{\text{coulomb}}{\text{mol } \bar{e}}\right) \cdot 1 \cdot \frac{\text{Amp} \times \text{sec}}{\text{coulomb}} \cdot \frac{1 \text{ hour}}{3600 \, s} \text{or}$$

$$C_{s,th,N} = 1116 \cdot \left(\frac{\boldsymbol{N+1}}{\boldsymbol{N}}\right) \left[\frac{\text{mAh}}{\text{g}}\right] \tag{13.4}$$

For example, for a bilayer graphene, $N = 2$, Sp.C. $= 1116 \cdot (\frac{2+1}{2}) \frac{[\text{mAh}]}{\text{g}} = 1674 \cdot \frac{[\text{mAh}]}{\text{g}}$ corresponds to three lithium atoms for each four carbon atoms, giving Li_3C_4 formula. Continuing the process of adding a layer of graphene and layer of lithium gives the overall stoichiometry of $Li_{N+1}C_{2N}$, where N is the number of graphene layers in the stack. Thus, an infinite layer of a stack would approach the stoichiometry of LiC_2 (capacity of 1116 mAh/g). Thus, we can use Equation 13.4 and define the capacity with the variable number of graphene (Figure 13.21) when Li intercalates on both sides above and below each hexagonal ring of carbon.

Another new finding of bilayer graphene has just recently been recorded confirming the lithium storage capacity exceeds that expected from the formation of LiC6 [89]. The authors refer as densest configuration known under normal conditions for lithium intercalation within bulk graphitic carbon point to the possible existence of distinct storage arrangements of ions in two-dimensional-layered materials as compared to their bulk parent compounds.

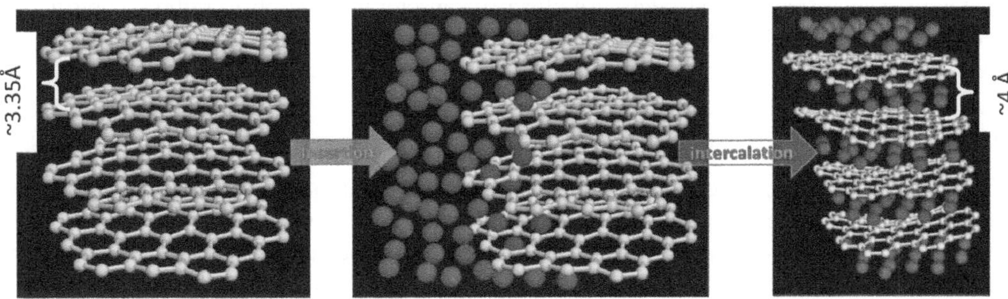

FIGURE 13.20 Schematic illustration of lithium insertion into IMLG layers.

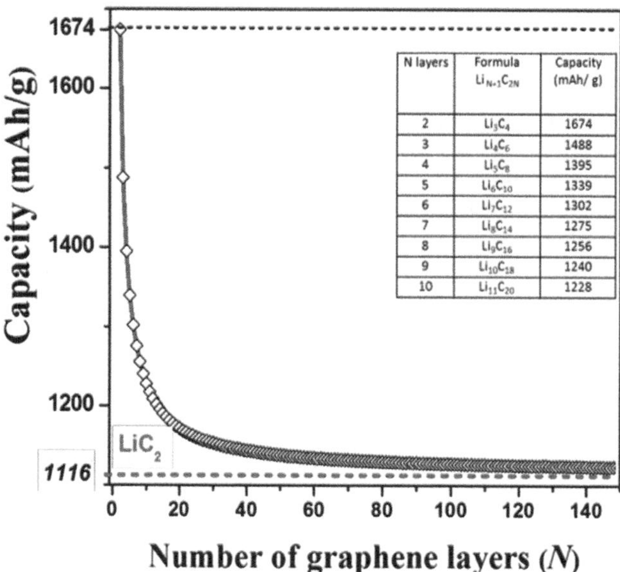

FIGURE 13.21 The dependency of theoretical capacity values versus on graphene layers based on $Li_{N+1}C_{2N}$ stoichiometric formula.

13.3 CONCLUSION

We analyze the unique properties of as-grown, free-standing, incommensurate graphene network as an exceptional new candidate for high-capacity LIB anodes. We reveal that weakened interplanar interaction as it occurs within IMLG layers enables easy and full penetration of lithium atoms, followed by flexible adjustment of the layers for stable reversible cycling. We also found the set of Raman 2D bandwidth and associated I_{2D}/I_G can be defining parameters to estimate incommensurate fraction in bulk material.

We reveal that a high amount of lithium can be commensurately intercalated/deintercalated into incommensurately stacked graphene layers due to physical absorption without any damage of the graphene plane. We achieved up to 1600 mAh/g reversible capacity for finite layers, which is stable in long-term cycling and high current loads. We propose a new model of lithium insertion mechanism when capacity can be depended on N numbers of graphene layers that participate in the charge transfer as $Li_{N+1}C_{2N}$ compound resulting 1674 mAh/g maximal capacity for the bilayer.

These findings offer a new class of high-capacity graphitic material for use as an advanced anode in high-energy-density LIBs. IMLG graphene provides an effective charge transfer, promising long-term cycling stability and safety. An effective capacity increase of over 400% compared to commercial graphite cells proves the feasibility of this graphene for new-generation lightweight and powerful LIBs.

Acknowledgments

The major experiments related to this project have been done at the University of Louisville. We gratefully acknowledge my colleagues at the University of Louisville, Dr. Arjun Thapa, Dr. A Sherihy, Dr. J. B. Jasinski, and Dr. J. S. D. Jangam, who supported the experiments. We also thank Prof. F. Zamborini for providing Raman spectrometer and Prof. B. Alphenaar for providing unlimited time on CVD equipment.

REFERENCES

1. Hérold, A., Recherches sur les composés d'insertion du graphite. *Bull. Soc. Chim. Fr.* 187, 999–1012 (1955).
2. Kambe, N., Dresselhaus, M.S., Dresselhaus, G., Basu, S., McGhie, A.R., Fischer, J.E. Intercalate ordering in first stage graphite-lithium. *Material Science Engineering.* 40, 1–4 (1979).
3. Wu, H., Cui, Y. Designing nanostructured Si anodes for high energy lithium ion batteries. *Nano Today* 7, 414–429 (2012).
4. Zhao, Y., Feng, J., Liu, X., Wang, Y., Wang, L., Shi, C., Huang, L., Freng, X., Chen, X., Xu, L., Yan, M., Zhang, Q., Bai, X., Wu, H., Mai, L. Self-adaptive strain-relaxation optimization for high-energy lithium storage material through crumpling of graphene. *Nature Communications* 5 Article 4565 (2014), doi: 10.1038/ncomms5565.
5. Wang A., Kadam S., Li H., Shi S., Qi Y. Review on modeling of the anode solid electrolyte interphase (SEI) for lithium-ion batteries. *Npj Computational Materials* 4, 1–26 (2018).
6. Novoselov K.S., Geim A.K., Morozov S.V., Jiang D., Zhang Y., Dubonos S.V., Grigieva I.V., Firsov A.A., Electric field effect in atomically thin carbon films, *Science* 22 V 306, Issue 5696, 666–669 (2004).
7. Geim, A. K. Graphene: status and prospects. *Science* 324, 1530–1534 (2009).
8. Ponomarenko, L.A. et al. Cloning of Dirac fermions in graphene superlattices. *Nature* 497, 594–597 (2013).
9. Yankowitz, M. et al. Emergence of superlattice Dirac points in graphene on hexagonal boron nitride. *Nat. Phys.* 8, 382–386 (2012).
10. Dean, C.R. et al. Hofstadter's butterfly and the fractal quantum hall effect in Moiré superlattices. *Nature* 497, 598–602 (2013).
11. Hunt, B. et al. Massive Dirac fermions and Hofstadter butterfly in a van der Waals heterostructure. *Science* 340, 1427–1430 (2013).
12. Schmidt, H., Rode, J.C., Smirnov, D., Haug, R.J. Superlattice structures in twisted bilayers of folded graphene. *Nat. Commun.* 5, 5742 (2014).
13. Yan, W., He, W.-Y., Chu, Z.-D., Liu, M., Meng, L., Dou, R.-F., Zhang, Y., Liu, Z., Nie, J.-C., He, L. Strain and curvature induced evolution of electronic band structures in twisted graphene bilayer, *Nat. Commun.* 4, Article 2159 (2013).
14. Kim, K.S. et al. Coexisting massive and massless Dirac fermions in symmetry broken bilayer graphene. *Nat. Mater.* 12, 887–892 (2013).
15. Trambly de Laissardière, G., Mayou, D., Magaud, L. Localization of Dirac electrons in rotated graphene bilayers. *Nano Lett.* 10, 804–808 (2010).
16. Lee, D. S. et al. Quantum Hall effect in twisted bilayer graphene. *Phys. Rev. Lett.* 107, 216602 (2011).
17. Hass, J. et al. Why multilayer graphene on 4 H – SiC (0001–) behaves like a single sheet of graphene. *Phys. Rev. Lett.* 100, 125504 (2008).
18. Luican, A. et al. Single-layer behavior and its breakdown in twisted graphene layers. *Phys. Rev. Lett.* 106, 126802 (2011).
19. Lopes dos Santos, J.M.B., Peres, N.M R., Castro Neto, A.H. Graphene bilayer with a twist: electronic structure. *Phys. Rev. Lett.* 99, 256802 (1–4) (2007).
20. Shallcross, S., Sharma, S., Kandelaki, E., Pankratov, O.A. Electronic structure of turbostratic graphene. *Phys. Rev. B* 81, 165105 (2010)
21. Koren, E., Leven, I., Lörtscher, E., Knoll, O.H., Duerig, U. Coherent commensurate electronic states at the interface between misoriented graphene layers. *Nat. Nanotechnol.* 11, 752 (2016), www.nature.com/naturenanotechnology.
22. Varchon, F., Mallet, P., Magaud, L., Veuillen, J.-Y., Rotational disorder in few-layer graphene films on 6H-SiC(000-1): a scanning tunneling microscopy study. *Phys. Rev. B* 77, 165415 (2008).
23. Li, G., Luican, A., Lopes dos Santos, J.M.B., Castro Neto, A.H., Reina, A., Kong, J., Andrei, E.Y., Observation of Van Hove singularities in twisted graphene layers, *Nature Phys.* 6, 109 (2009).
24. Marcano, D.C., Kosynkin, D.V., Berlin, J.M., Sinitskii, A., Sun, Z., Slesarev, A., Alemany, L.B., Lu, W., Tour, J.M. Improved synthesis of graphene oxide, *ACS Nano* 4, 8, 4806–4814 (2010).
25. Li, X., Cai, W., An, J., Kim, S., Nah, S., Yang, W., Piner, R. Large-area synthesis of high-quality and uniform graphene films on copper foils. *Science* 324, 1312–1314 (2009).

26. Yu, Q., Lian, J., Siriponglert, S., Li, H., Chen, Y.P., Pei, S.-S. Graphene segregated on Ni surfaces and transferred to insulators. *Appl. Phys. Lett.* 93, 113103(1–3) (2008).
27. Peter, S., Jerzy, T. S., Eli, S. Graphene on Pt (111): growth and substrate interaction. *Phys. Rev. B* 80, 245411 (2009).
28. An, X., Liu, F., Yung, J.J., Kar, S. Large-area synthesis of graphene on palladium and their Raman spectroscopy. *J. Phys. Chem. C 116* (31), 16412–1642 (2012).
29. Coraux, J., N'Diaye, A.T., Engler, M., Busse, C., Wall, D., Buckanie, N., Meyer zu Heringdorf, F.-J., Gastel, R., Poelsema, B., Michely, T. Growth of graphene on Ir(111). *New J. Phys.* 11 (2009).
30. Paronyan, T.M., Pigos, E.M., Chen, G., Harutyunyan, A.R. The formation of ripples in graphene as a result of interfacial instabilities. *ACS Nano* 5(12), 9619–9627 (2011).
31. Wang, Z., Puls, C.P., Staley, N.E., Zhang, Y., Todd, A., Xu, J., Casey, A.H, Hollander, J.M., Robinson, A.J., Liu, Y. Technology ready use of single layer graphene as a transparent electrode for hybrid photovoltaic devices. *Physica E* 44(2), 521–524 (2011).
32. Mattevi, C., Kim, H., Chhowalla, M. A review of chemical vapour deposition of graphene on copper, *J. Mater. Chem.* 21, 3324–3334 (2011).
33. Reina, A., Jia, X., Ho, J., Nezich, D., Son, H., Bulovic, V., Dresselhaus, M.S., Kong, J. Large area, few-layer graphene films on arbitrary substrates by chemical vapor deposition. *Nano Lett.* 9(1), 30–35 (2009).
34. Chen, Z., Ren, W., Gao, L., Liu, B., Pei, S., Cheng, H.-M. Three-dimensional flexible and conductive interconnected graphene networks grown by chemical vapour deposition. *Nature Mater.* 10, 424–428 (2011).
35. Zhou, S., Xu, J., Xiao, Y., Zhao, N., Wong, C.-P., Low-temperature Ni particle-templated chemical vapor deposition growth of curved graphene for supercapacitor applications. *Nano Energy* 13, 458–466 (2015).
36. Paronyan, T.M., Harutyunyan, A.R. Graphene originated 3D structures grown on the assembled nickel particles. *APS March Meeting* 58, B B6.00005 (2013).
37. McCann, E. Asymmetry gap in the electronic band structure of bilayer graphene. *Phys. Rev. B* 74, 161403(1–4) (2006).
38. Latil, S., Menuier, V., Henrard, L. Massless fermions in multilayer graphitic systems with misoriented layers: ab initio calculations and experimental fingerprints. *Phys. Rev. B* 76, 201402(1–4) (2007).
39. Berashevich, J., Chakraborty, T. Interlayer repulsion and decoupling effects in stacked turbostratic graphene flakes. *Phys. Rev. B* 84, 033403(1–4) (2011).
40. Nam, B.L., Tran, D.H., Lilia, M.W. Interlayer interactions in van der Waals heterostructures: Electron and phonon properties, *ACS Appl. Mater. Interfaces* 8(9), 6286–6292 (2016).
41. Koshino, M., Ando, T. Electronic structures and optical absorption of multilayer graphenes. *Solid State Commun.* 149, 1123–1127 (2009).
42. Lu, C.-C., Lin, Y.-C., Liu, Z., Yeh, C.-H., Suenaga, K., Chiu, P.-W. Twisting bilayer graphene superlattices, *ACS Nano* 7(3), 2587–2594 (2013).
43. Fang, W., Hsu, A.L., Song, Y., Birdwell, A.G., Amani, M., Dubey, M., Dresselhaus, M.S., Palacios, T., Kong, J. Asymmetric growth of bilayer graphene on copper enclosures using low-pressure chemical vapor deposition. *ACS Nano*, 8(6), 6491–6499 (2014).
44. Swain, A.K., Bahadur, D. Facile synthesis of twisted graphene solution from graphite-KCl. *RSC Adv.* 3, 19243–19246 (2013).
45. Chen, X., Paul, R., Dai, L. Carbon-based supercapacitors for efficient energy storage *Natl. Sci. Rev.* 4(3), 453–489 (2017).
46. Brownson, D.A.C., Kampouris, D.K., Banks, C.E. Graphene electrochemistry: fundamental concepts through to prominent applications. *Chem. Soc. Rev.* 41, 6944–6976 (2012).
47. Candelaria, S.L., Shao, Y., Zhou, W., Li, X., Xiao, J., Zhang, J.-G., Wang, Y., Liu, J., Li, J., Cao, G. Nanostructured carbon for energy storage and conversion. *Nano Energy* 1, 195–220 (2012).
48. Chen, D., Tang, L., Li, J. Graphene-based materials in electrochemistry. *Chem. Soc. Rev.* 39, 3157–3180 (2010).
49. Frackowiak, E. Carbon materials for supercapacitor application. *Phys. Chem. Chem. Phys.* 9, 1774–1785 (2007).
50. Pandolfo, A., Hollenkamp, A. Carbon properties and their role in supercapacitors. *J. Power Sources* 157, 11–27 (2006).
51. Chmiola, J., Yushin, G., Gogotsi, Y., Portet, C., Simon, P., Taberna, P.L. Anomalous increase in carbon capacitance at pore sizes less than 1 nanometer. *Science* 313, 1760–1763 (2006).

52. Kalluri, R.K., Biener, M.M., Suss, M.E., Merrill, M.D., Stadermann, M., Santiago, J.G., Baumann, T.F., Biener, J., Striolo, A. Unraveling the potential and pore-size dependent capacitance of slit-shaped graphitic carbon pores in aqueous electrolytes. *Phys. Chem. Chem. Phys.* 15, 2309–2320 (2013).
53. Wood, B.V., Ogitsu, T., Otani. M., Biener, J. First-principles-inspired design strategies for graphene-based supercapacitor electrodes. *J. Phys. Chem. C* 118, 4–15 (2014).
54. Pan, D., Wang, S., Zhao, B., Wu, M., Zhang, H., Wang, Y., Jiao, Zh. Li storage properties of disordered graphene nanosheets. *Chem. Mater.* 21, 3136–3142 (2009).
55. Lian, P., Zhu, X., Liang, Sh., Li, Zh., Yang, W., Wang, H. Large reversible capacity of high quality graphene sheets as an anode material for lithium-ion batteries. *Electrochim. Acta* 55, 3909–3914 (2010).
56. Mukharajee, R., Thomas, A.V., Datta, D., Singh, E., Li, J., Eksik, O., Shenoy, V.B., Koratkar, N. Defect-induced plating of lithium metal within porous graphene networks. *Nat. Commun.* 5, 3710 (1–10) (2014).
57. Xu, Y., Lin, Z., Zhong, X., Papandrea, B., Huang, Y., Duan X. Solvated graphene frameworks as high-performance anodes for lithium-ion batteries. *Angew. Chem.* 54(18), 5345–5350 (2015).
58. Liu, Y., Wang, Y.M., Yakobson, B.I., Wood, B.C. Assessing carbon-based anodes for lithium-ion batteries: a universal description of charge-transfer binding. *Phys. Rev. Lett.* 113, 028304(1–5) (2014).
59. Datta, D., Li, J., Koratkar, N., Shenoy, V.B. Enhanced lithiation in defective graphene. *Carbon* 80, 305–310 (2014).
60. Paronyan, T.M., Thapa, A.K., Sherehiy, A., Jasinski, J.B., Dilip, J.S. Incommensurate graphene foam as a high capacity lithium intercalation anode. *Nat. Sci. Rep.* 7, 39944 (2017), doi: 10.1038/srep39944.
61. Paronyan, T.M., Thapa, A.K., Sherehiy, A., Jasinski, J.B., Dilip, J.S. Exceptional lithium intercalation capacity of incommensurate graphene foam in rechargeable batteries. *ECST* 77, 311–320 (2017).
62. Paronyan, T.M., Thapa, A.K., Sherehiy, A., Jasinski, J.B., Dilip, J.S. Novel graphene structures: substantial capacity enhancement in lithium batteries. *Advanced Materials TechConnect Briefs Graphene & 2D-Materials* Chapter 3, 95–98 (2017).
63. Paronyan, T.M. "Advanced graphene anode for the Li-ion batteries": review for Dekker Encyclopedia of Nanoscience and Nanotechnology (2018), doi: 10.1081/E-ENN3.
64. Warner, H., Rümmeli, M.H., Gemming, T., Bücher, B., Broggs, G.A.D. Direct imaging of rotational stacking faults in few layer graphene. *Nano Lett.* 9, 102–106 (2009).
65. Ferrari, A.C., Basko, D.M. Raman spectroscopy as a versatile tool for studying the properties of graphene. *Nat. Nanotechnol.* 8, 235–246 (2013).
66. Malard, L.M., Pimenta, M.A., Dresselhaus, G., Dresselhaus, M.S. Raman spectroscopy in graphene. *Phys. Rep.* 473, 51–87 (2009).
67. Ponchara, P., Ayari, A., Michel, T., Sauvajol, J.-L. Raman spectra of misoriented bilayer graphene. *Phys. Rev. B* 78, 113407 (1–4) (2008).
68. Lui, C.H., Li, Z., Chen, Z., Klimov, P.V., Brus, L.B., Heinz, T.F. Imaging stacking order in few-layer graphene. *Nano Lett.* 11, 164–169 (2011).
69. Kim, K., Coh, S., Tan, L.Z., Regan, W., Yuk, J.M., Chatterjee, J.M., Crommie, M.F., Cohen, M.L., Louie, S.G., Zettl, A. Raman spectroscopy study of rotated double-layer graphene: misorientation-angle dependence of electronic structure. *Phys. Rev. Lett.* 108, 246103 (1–6) (2012).
70. Wu, J.-B., Zhang, X. Ijäs, M., Han, M.-P., Qiao, X.-F., Li, X.-L., Jiang, D.-S., Ferrari, A.C., Tan, P.-H. Resonant Raman spectroscopy of twisted multilayer graphene. *Nat. Commun.* 5, 5309(1–10) (2014).
71. Bernier, P., Fischer, J.E., Roth, S., Solin, S.A. Eds. Chemical Physics of Intercalation, Plenum, New York, pp. 59–78 (1987).
72. Zheng, T., Liu, Y., Fuller, E.W., Tseng, S., Sacken, U., von and Dahn, J.R. Lithium insertion in high capacity carbonaceous materials. *J. Electrochem. Soc.* 142(8), 2581–2592 (1995).
73. Winter, M., Besenhard, J., Spahr, M.E., Novak, P. Insertion electrode materials for rechargeable lithium batteries. *Adv. Mater.* 10, 725–763 (1998).
74. Dahn, J.R. Lithium Batteries: New Materials, Developments and Perspectives edited by G. Pistoia, Amsterdam-London-New York-Tokyo (1994).
75. Zheng, T., Reimers, J. N., Dahn, J.R. Effect of turbostratic disorder in graphitic carbon hosts on the intercalation of lithium. *Phys. Rev B* 51, 734–741 (1995).
76. Dahn, J.R., Zheng, T., Liu, Y., Xue, J.S. Mechanisms for lithium insertion in carbonaceous materials. *Science* 270, 590–593 (1995).
77. Sato, L., Noguchi, M., Demanchi, A., Oki, N., Endo, M. A mechanism of lithium storage in disordered carbons. *Science* 264, 556–558 (1994).

78. Bindra, C., Nalimova, V.A., Sklovsky, D.E., Benes, Z., Fischer J.E. Super dense LiC2 as a high capacity li intercalation anode. *J. Electrochem. Soc.* 145, 2377–2380 (1998).
79. Berger, C., Song, Zh., Li, Xl., Wu, X., Brown, N., Naud, C., Mayou, D. Electronic confinement and coherence in patterned epitaxial graphene. *Science*, 312, 1191 (2006).
80. Wu, X., Li, X., Song, Zh., Berger, C., de Heer, W.A. Weak antilocalization in epitaxial graphene: evidence for chiral electrons. *Phys. Rev. Lett.* 98, 136801 (2007).
81. Cançado, L.G., Takai, K., Enoki, T. General equation for the determination of the crystallite size La of nanographite by Raman spectroscopy. *Appl. Phys. Lett.* 88, 163106(1–3) (2006).
82. Cançado, L.G., Jorio, A., Ferreira, E.H.M., Stavale, F., Achete, C.A., Capaz, R.B., Moutinho, M.V.O., Lombardo, A., Kulmala, T.S., Ferrari, A.C. Quantifying defects in graphene via Raman spectroscopy at different excitation energies. *Nano Lett.* 11, 3190–3196 (2011).
83. Rytkönen, K., Akola, J., Manninen, M. Density functional study of alkali-metal atoms and monolayers on graphite (0001). *Phys. Rev. B* 75, 075401(1–9) (2007).
84. Inaba, M., Yoshida, H., Ogumi, Z., Abe, T., Mizutani, Y., Asano, M. In situ Raman study on electrochemical Li intercalation into graphite. *J. Electrochem. Soc.*, 142 (1), 20–26 (1995).
85. Pollak, E., Baisong Geng, B., Ki-Joon Jeon, K.-J., Lucas, I.T., Richardson, T.J., Wang, F., Kostecki, R. The interaction of Li+ with single-layer and few-layer graphene. *Nano Lett.* 10, 3386–3388 (2010).
86. Sato, K., Park, J.S., Saito, R., Cong, C., Yu, T., Lui, C.H., Heinz, T.F., Dresselhaus, G., Dresselhaus, M.S. Raman spectra of out-of-plane phonons in bilayer graphene. *Phys. Rev. B* 84, 035419(1–5) (2011).
87. Sole, C., Drewett, N.E., Hardwick, L.J. In situ Raman study of lithium-ion intercalation into microcrystalline graphite. *Faraday Discuss.* 172, 223–237 (2014).
88. Mordkovich, V.Z. Synthesis and XPS investigation of superdense lithium-graphite intercalation compound, LiC2. *Syn. Mater.* 80, 243–247 (1996).
89. Kühne, M., Börrnert, F., Fecher, S., Ghorbani-Asl, M., Biskupek, J., Samuelis, D., Krasheninnikov, A.V., Kaiser, U., Smet, J.H. Reversible superdense ordering of lithium between two graphene sheets. *Nature* 564, 234–239 (2018).

14 Self-Assembled Peptide Nanostructures Doped with Electroluminescent Compounds for Energy Conversion

Tatiana Duque Martins, Geovany Albino de Souza, Diéricon Sousa Cordeiro, Ramon Silva Miranda, Lucas Fernandes Aguiar, Adão Marcos Ferreira Costa
Chemistry Institute, Federal University of Goiás, Goiânia, Goiás, Brazil

Antonio Carlos Chaves Ribeiro
Goianian Federal Institute of Education, Science and Technology, Morrinhos, Goiás, Brazil

CONTENTS

14.1 INTRODUCTION

Several distinct nanoscale structures formed of self-assembled peptides can be obtained by simple environmental and procedures control. They can be easily designed for a given application and have been used as alternative materials in several nanotechnology applications. For instance, Reches and Gazit showed that they can be used as templates to nanowires fabrication,[1] while their morphology can be controlled for drug delivery and diagnoses applications.[2]

Since they can be chemically modified or physically combined to several other compounds, they find applications in bionanotechnology and bioengineering,[3,4] and due to their intrinsic biocompatibility and stability, they are more suitable for industrial application than proteins, once they can be synthesized on large scale.[2]

Among all possible peptides, diphenylalanine (Phe-Phe) and its derivatives have been extensively investigated due to their capability to form distinct self-assembled structures, such as nanotubes, nanowires, fibrils, spheres, and so on, by simple procedure adjustments[5,6,7]; therefore, they find application as materials for light harvesting and energy conversion, as fluorescent probes, and as compounds for treatment of several diseases, due to their ability of generate hydrogels.[8,9,10] In the nanotube form, they are considered as substitutes for the carbon nanotubes, since they can also act as conductors, with the advantage of biocompatibility and hazard safety. Also, Phe-Phe nanotubes are sensitive to external electric fields, which confer on them the optical nonlinearity needed for applications in electro-optical devices.[11] As shown by Gilboa et al.,[12] Phe-Phe nanotubes present effective electro-optic coefficients comparable to those of inorganic crystals such as lithium niobate and potassium titanyl phosphate, which gives rise to the possibility of application of Phe-Phe structures to generate nonlinear photonic crystals.

14.2 ELECTRONIC EXCITED STATE COMPLEX FORMATION

Electronic energy transfers in nanomaterials are ruled by the spontaneous activation and deactivation processes[13] that occur in the HOMO and LUMO orbitals, with probabilities governed by their energy separation. In peptide supramolecular structures, it is expected that a number of chromophores interact with each other, resulting in a new mixed electronic HOMO, which can be related to a LUMO accessed upon the right excitation energy. These mixed orbitals now enable new processes that describe the energy quanta needed for a given transition and are characteristic of the structure formed.

When light is absorbed by the structure, it reaches the excited electronic state, and from there, it can interact with ground state structures and form complexes of energy transfer known as exciplexes. They exist only when one of the components of the complex is at the electronic excited state; therefore, they are short-lived and occur for a given propose. To understand the photophysical properties of organic compounds that are responsible for the energy transfer is of major importance to design a material for energy conversion application. These properties can be understood in terms of the transition moment from a lower energy level (lm) to a higher one (un), which is given by the solution of the Schrodinger equation[14]:

$$\mathrm{M}_{\mathrm{lm,un}} = \int \psi_{\mathrm{un}} \mu \psi_{\mathrm{lm}} \mathrm{d}\tau \tag{14.1}$$

where ψ_{un} is the excited electronic state wavefunction, and ψ_{lm} is the ground electronic state wavefunction.

The transition rate is given as

$$\Delta N_{\mathrm{lm,un}} = N_{\mathrm{lm}} B_{\mathrm{lm,un}}\, \rho(\nu_{\mathrm{lm}\to\mathrm{un}}) \tag{14.2}$$

where $B_{\mathrm{lm,un}}$ is the Einstein coefficient for the absorption process, and it is related to the transition moment by $B_{\mathrm{lm,un}} = \frac{8\pi^3 e^2}{3h^2 c} M^2_{\mathrm{lm,un}}$; N_{lm} is the population of the lowest energy level, and $\rho(\nu_{\mathrm{lm}\to\mathrm{un}})$ is the radiation density of frequency ν given by Planck's black body radiation law.[15]

The emission process, which comprehends the electronic transition from a high-energy electronic level to the ground electronic state, is given as

$$\Delta N_{\mathrm{un,lm}} = N_{\mathrm{un}} A_{\mathrm{un.lm}}\, \rho(\nu_{\mathrm{un}\to\mathrm{lm}}) \tag{14.3}$$

where $A_{\mathrm{un,lm}}$ is the Einstein coefficient for the spontaneous emission process and is related to the probability of transition.

Peptides that present aromatic rings often can self-assemble into larger structures with fluorescence and absorption spectra that are very distinct from the isolated peptide. It is not unusual that self-assembled structures behave as organic semiconductors, due to the communication between the HOMOs of the isolated units that form new and highly conjugated structures, which are responsible for the charge delocalization and charge generation, separation, and transport. The charge transport in conjugated molecules is governed by electronic and electron–vibration interactions, so-called phonons, which are described by the Hamiltonian for non-interacting electrons and phonons[16]:

$$H = \sum_{m} \varepsilon_m a_m^+ a_m + \sum_{mn} t_{mn} a_m^+ a_n + \sum_{Q} \hbar\omega_Q \left(b_Q^+ b_Q + 1/2\right) \tag{14.4}$$

where $a_m{}^+$ and a_m are the creation and annihilation operators, respectively, for an electron in the molecular lattice (m); $b_Q{}^+$ and b_Q are the creation and annihilation operators of a phonon with a wavevector Q and frequency ω_Q; ε_m is the electron energy site; and t_{mn} is the electron coupling integral that describes the electronic coupling and is dependent on the vibrational and phonon coordinates.

Therefore, energy transport in such systems is dependent on the extent of the electronic coupling, governed by the conjugation length and electron mobility. Several mechanisms can be proposed for the charge transport; generally, they involve the initial step of charge hop, which can be interpreted in terms of Marcus theory:

$$k_{ET} = \frac{4\pi^2}{h} t^2 \frac{1}{\sqrt{4\pi\lambda kT}} exp\left[-\frac{\left(\lambda + \Delta G^0\right)^2}{4\lambda kT}\right] \tag{14.5}$$

where t is the transfer integral; λ is the reorganization energy that needs to be minimized to promote a more efficient energy transfer, and ΔG^o is the driving force, which, for a self-exchange reaction, is zero.

Therefore, the transfer rate depends on the maximum electronic coupling, representing the interaction strength, and on the minimal reorganization energy with no geometry changes. The mechanisms of energy transfer can be described by the excited electronic state characteristics. In steady-state conditions, the fluorescence rate is k_{FM} and the quantum yield (q_{FM}) is:

$$q_{\text{FM}} = \frac{k_{\text{FM}}[^1M^*]}{I_0} = \frac{k_{\text{FM}}}{k_{\text{FM}} + k_{\text{IM}}} \tag{14.6}$$

where I_0 is the incident electromagnetic radiation; k_{IM} is the internal quenching rate; and $[^1M^*]$ is the population of the molecules in the singlet electronic state.

When the electronic excitation produces an initial quantity $[^1M^*]_0$, the molecular lifetime (τ_M) is given as

$$\frac{1}{k_{\text{FM}} + k_{\text{IM}}} = \tau_M \tag{14.7}$$

Which enables the energy transfer mechanism experimental description in terms of time-resolved fluorescence spectroscopy, which gives fluorescence decay curves that reveal the nature of the electronic excited state deactivation.

Using time-correlated single photon counting technique, the compound's resulting decay curve is deconvoluted from the instrument response to result in the sum of all exponential decays from the compound (N_{tk}), which is given as[17]

$$N(t_k) = \sum_{t=0}^{t-t_k} L(t_k)I(t - t_k)\Delta t \tag{14.8}$$

With the convolution being:

$$I_k(t) = L(t_k)I(t - t_k)\Delta t \qquad (t > t_k) \tag{14.9}$$

The term $(t - t_k)$ is necessary because the pulse is initiated at $t = t_k$, since there is no emission prior to excitation. The data evaluated by the least-square method must give an adjustment parameter that is able to express the data reliability. Thus, equivalence between experimental and theoretical data is given by χ^2 parameter:

$$\chi^2 = \sum_{k=1}^{n} \frac{1}{\sigma_k^2}[N(t_k) - N_c(t_k)]^2 = \sum_{k=1}^{n} \frac{[N(t_k) - N_c(t_k)]^2}{N(t_k)} \tag{14.10}$$

where ${\sigma^2}_k$ is the standard deviation of each of n channels used in the analysis. Thus, χ^2 represents the sum of the standard deviations of the experimental data $N(t_k)$ and the expected data $N_c(t_k)$.

The most prominent mechanism through which the deactivation of energy transfer complexes of organic compounds occurs is through a bimolecular process, such as the Forster resonant energy transfer (FRET).[18] In FRET, the electric field around the donor created upon excitation is perceived by an acceptor that is at an optimal distance from the donor. This proximity induces the acceptor to oscillate with the donor electric field and energy from the donor is transferred. The FRET rate is given as

$$k_{DA} = \frac{9000k^2 \ln 10}{128\pi^5 n^4 N_A \tau_{DA} r^6} \int \frac{F_D(\tilde{\nu})\varepsilon_A(\tilde{\nu})}{\tilde{\nu}^4} d\tilde{\nu} \tag{14.11}$$

where k^2 gives the relative orientation of the donor and acceptor dipoles, F_D is the fluorescence intensity of the donor, ε_A is the molar extinction coefficient of the acceptor, τ_{DA} is the fluorescence lifetime of the donor in the presence of the acceptor, r is the distance between donor and acceptor, and n is the refractive index. It requires that the electronic transitions of the donor and the acceptor are allowed and that coulombic interactions between donor and acceptor are effective.

FRET can occur between similar components, and it is assigned as homoFRET or, most commonly, involving distinct compounds or structures, which is assigned as heteroFRET. In heteroFRET, fluorescence intensity of the donor decreases when in the presence of the acceptor, as well as the fluorescence lifetime, while anisotropy is increased, whereas the only effect of homoFRET is the decrease in anisotropy, with no changes in fluorescence intensity or fluorescence lifetime. Therefore, it is not a property of identity, but rather a property of photophysical characteristics summed to environmental conditions. This means that FRET can be modulated and, therefore, exploited for technological applications. Figure 14.1 shows the scheme of FRET occurrence.

In this perspective, considering the characteristics of the peptides and the mechanisms of formation of each supramolecular structure that they can originate, applications are proposed, depending on the control exerted on their assembling mechanisms, as well as on the interaction they will present with additives. By performing an accurate control of these conditions, they can be designed to present the desired properties.

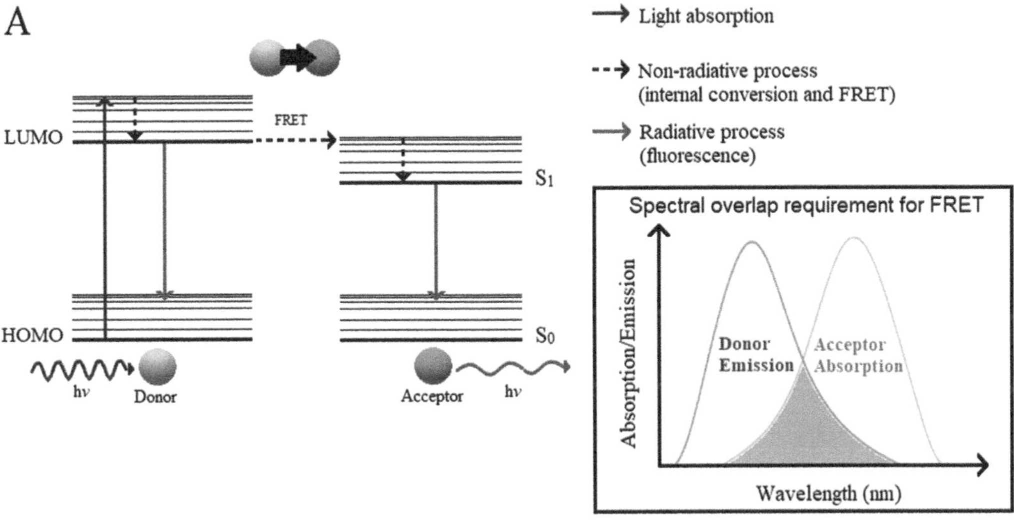

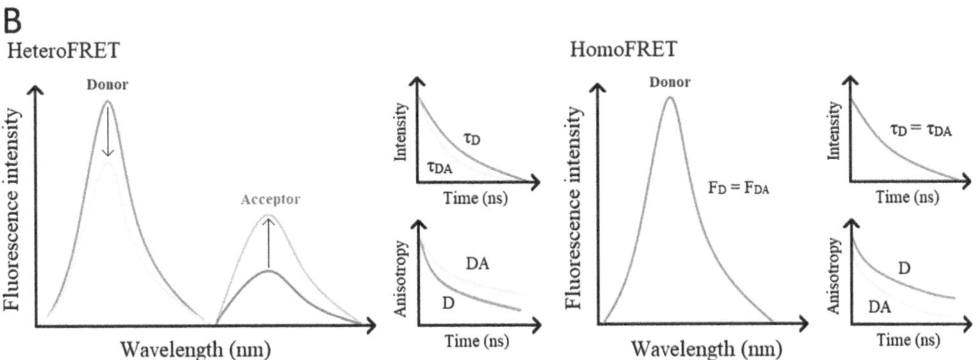

FIGURE 14.1 Scheme of the FRET process. (A) Electronic processes that occur during FRET upon the requirement of spectral overlap between the donor emission spectrum and the acceptor absorption spectrum. (B) Effects on fluorescence intensity, fluorescence lifetime, and anisotropy when FRET involves fluorophores with different spectroscopic characteristics (heteroFRET) and involving fluorophores of similar spectroscopic characteristics (homoFRET).

14.3 PEPTIDE SELF-ASSEMBLING

Self-assembling of peptides is driven by the balance of noncovalent interaction forces that are in dynamic fluctuation during the process. During the self-assembling, molecules are spontaneously organized into ordered structures by a balance of several physical interactions between groups and structures, such as Van der Waals and electrostatic bonding, hydrogen bonding, and π–π stacking interactions.[19] The accepted general mechanism of self-assembling of peptides considers the balance of attractive and repulsive forces between molecules, affecting their arrangement. It is described by the way that molecular composition, molecular orientation, and diffusion rates influence the directional noncovalent interactions and give rise to the energetically favorable structure. These parameters are strongly dependent on environmental conditions such as pH, temperature, solvent structure, concentration, and the presence of impurities[20] and can be exploited to give multi-component supramolecular structures.[21,22,23,24] Scheme 14.1 shows the supramolecular structures that can be obtained by controlling peptide self-assembling.

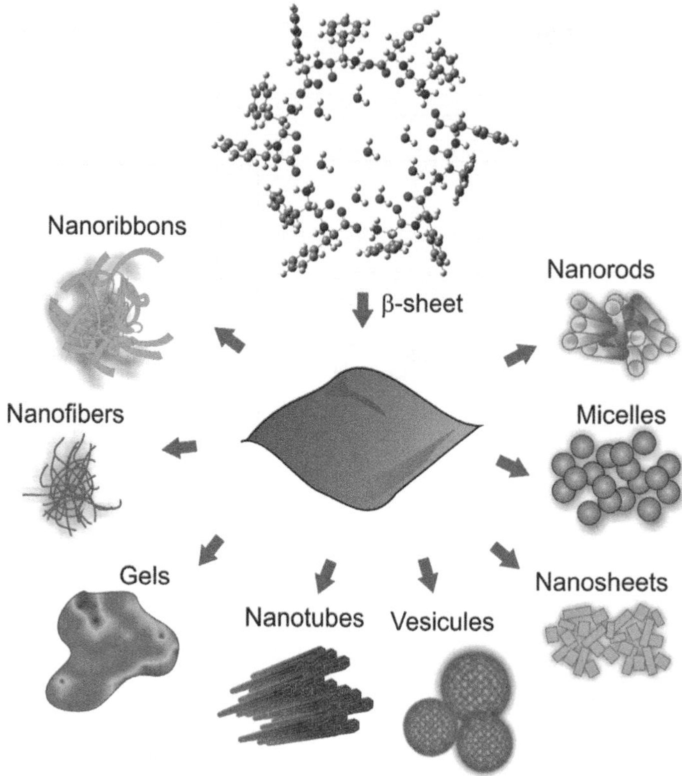

SCHEME 14.1 Variety of nanostructures obtained by exerting peptide self-assembling control.

Control of peptide self-assembling is usually carried out by supramolecular strategies that confer an adaptable behavior to the synthesis that are very difficult to address in conventional covalent protocols.[18,25] This synthesis control is crucial for aiming applications.

14.4 RECENT TECHNOLOGICAL APPLICATION

Application in strategic fields of science and technology such as sensing, diagnosis, targeting, photodynamic therapy, drug delivery, and vaccines technology has been the driving force that has fostered the fast development of such supramolecular structures for technological applications.[26] Therefore, by the control of environmental characteristics such as pH, temperature, solvent, and balance of molecular interactions, which consist of the crucial characteristics that enable supramolecular bottom-up strategies, it is possible to design the peptide structures with outstanding characteristics for uses in chemical encapsulation and develop new mechanisms of drug delivery, which have been used for medical applications, for instance, for vaccine delivery and drug targeting.

Although there are several successful systems used for drug and vaccine delivery systems, such as liposomes and nanoparticles, peptide self-assembled nanostructures are thought to be advantageous because some of them are amphiphilic, which confers on them high drug-loading capability, low drug leakage, and high permeability to the membranes of living cells.[27,28,29,30]

Another interesting characteristic of such structures is their ability of structure interconversion driven by environmental control. They are interesting because they enable the release of an encapsulated drug or vaccine with specific environmental characteristics, which can result in a more efficient delivery system with less side effects. Another use of the interconversion is the possibility of producing structures such nanotubes that are easily prepared in water but very difficult to be obtained in organic solvents. By interconversion, water-free nanotubes can be produced for application in electronic devices and solar cells, which require a humidity-free environment.

Also, peptide molecules that self-assemble in water can form hydrogels, by the three-dimensional arrangement of peptide polymers that absorb water due to hydrophilic groups in their networks. They are easily produced and are responsive to temperature, pH, ionic strength, light, etc., which can find application as delivery systems, for instance, and perform controlled drug release.[31,32] In another perspective, structures can be obtained from cyclic peptide, which are rigid, and, due to this, can easily bind to the receptor in the cell. In the work by Shirazi et al.,[33] a tryptophan-based peptide was produced to result in vesicle-like structures for phosphopeptide delivery, but a nanofiber that was degraded with time was also produced for Mazza drug transport into the central nervous system.[34] For drug delivery purposes, nanomicelles and dendrimers also have been proposed,[35] aimed at distinct tissues and organs; for instance, short repetitive sequences of aspartic acid are used for drug delivery to bone tissues.[36]

Seeking a variety of applications, self-assembled peptides can be doped with selected additives to present the desired properties by forming complexes with them; for instance, the use of luminescent additives can result in excited electronic state complexes that find application in photodynamic therapy.[37,38,39] This strategy can also be employed to generate photosensitizers, as shown in the work of Wu et al.[40] and Kirchherr et al.[41] in which poly(ethylene glycol)-*b*-poly(L-lysine)-*b*-poly(L-leucine) micelles were self-assembled in the presence of a photosensitizer; it generated a more stable system than the free ICG and with enhanced cellular internalization when compared to the free photosensitizer.

It is interesting that the luminescence of the complexes formed with these compounds is distinct from the isolated photosensitizers. On some of these systems, the photophysical responses are restored when the photosensitizer is delivered to the target cell or tissue, which indicates that the new materials are formed by noncovalent interactions. This characteristic, along with the prime peptide characteristic of biocompatibility, can be extensively exploited to generate materials for several applications. For instance, Ejiofor and Webster[36] evaluated the benefits of developing nanomaterials for implants, arguing that they present distinct bonding potential, mechanical characteristics, high surface energy, increased surface roughness of the bulk material, strength, ductility, and electrical conductivity, specific properties that are important for the synthesis, efficacy and biocompatibility,[42] adhesion to alloy surfaces in implants, and antimicrobial properties.[43,44,45]

Many self-assembled peptide structures present strong electrical-optic activity that is modulated by the monomers' distribution on the structure, which influences the balance of interaction forces to keep the unities together during assembly. Specially, diphenylalanine derivatives present intense electro-optic properties originated on the porous triclinic structure of their nanotubes.[12] The possibility of modulation of this property is exploited in applications such as imaging, sensing, and energy conversion. For instance, Moini et al.[46] constructed Lab-on-a-Chip devices using self-assembled vertical diphenylalanine nanotubes with electro-optical properties.

Chemical modifications to introduce a desired property to the structures are often taken as a strategy for new materials proposal. This is the case for optoelectronic or probing applications, in which it is important to present luminescence. Although some self-assembled peptides present intrinsic electro-optical properties, this strategy can provide a secure pathway to obtain the yields required for an application. Nevertheless, the strategies based on the exploitation of the balance of interactions that are involved in the self-assembling process, specially π–π stacking, as shown by Takahashi et al.,[47] are the most prominent due to the simple procedures they involve. Liang et al.[48] combined peptide nanotubes with chromophores for light harvesting, and Ryu et al.[49] incorporated luminescent lanthanides and phenanthroline to peptide nanotubes to give them them tunable luminescent properties. In our work,[50,51,52] we showed that self-assembled diphenylalanine can be doped with small fluorescent molecules as well as with electroluminescent polymers to generate energy transfer complexes with distinct photophysical properties that can find several applications, from electroluminescent devices to solar cells and sensors. Moreover, we showed that doping these structures with these molecules can be a way to control the self-assembling. This was shown in Ribeiro et al.[51] in which poly[2-methoxy-5-(2-ethylhexyloxy)-1,4-phenylenevinylene] (MEH-PPV) was the electron donor doping agent that, due to the interference it causes in the balance of interaction forces of diphenylalanine self-assembling, resulted in spheres, while with poly(vinyl-carbazole) (PVK), an electron acceptor as doping agent, nanotubes are obtained.

Also, several other classes of additives can be thought to result in optical active peptides. Dias[52] obtained diphenylalanine nanotubes with structural differences upon doping with maghemite, a magnetic iron nanoparticle, and cadmium and tellurium–quantum dot (QD) (Sigma-Aldrich, CdTe core-type quantum dots, used as received), resulting in nanotubes with distinct photophysical properties from the initial materials. Diphenylalanine stock solution is obtained by the dissolution of 20 mg of diphenylalanine (Sigma-Aldrich), used as received, into 0.2 mL of 1,1,1,3,3,3-hexafluoro-2-propano (Sigma-Aldrich, ≥99% pure, used as received), as suggested by Ghadiri[53] resulting in 100 mg mL^{-1}. This solution is diluted to 2 mg mL^{-1} for preparing QD solution. QD solution at correct concentration is obtained first by the dilution of 50 μL of the QD stock solution to 10 mL with deionized water, and second by the dilution of previous solution by 200 times. The sample is obtained by the mixture of 0.01 mL of Phe-Phe solution, with 0.1 mL of QD solution and 0.4 mL of deionized water, resulting in a 2-mg mL^{-1} Phe-Phe solution.[54]

In their results, fluorescence emission increased when nanotubes were doped with QD or maghemite, because a charge transfer complex was readily formed with more efficient fluorescence than the simple nanotube. Moreover, upon doping with QD and maghemite, the resulting material presents fluorescence even more intense and occurring at a wider wavelength range, which is an effect of the relaxed electronic states of the energy transfer complex modulation and which is why these complexes are be able to perform energy transfer and conversion and find application in electroluminescent or photovoltaic devices.

Many self-assembled structures can be chemically modified to contain a chemical group with a desired property, or a compound presenting the specific property can be physically adhered to the structures in a way that adds new properties to the self-assembled peptide. This is the case when they are used for optoelectronic or probing applications, in which it is important to present luminescence. In fact, some self-assembled peptide structures can present intrinsic electro-optical properties due to the balance of interactions that are involved in the self-assembling process, especially π–π stacking, as shown by Takahashi et al.[47] In their work, Liang et al.[48] combined peptide nanotubes with chromophores for light harvesting, and Ryu et al.[49] incorporated luminescent compounds such as lanthanides and phenanthroline to peptide nanotubes to give them tunable luminescent properties. In our work,[50,54,55,56] we showed that self-assembled diphenylalanine can be doped with a variety of fluorescent compounds, from small molecules to polymers, and that energy transfer complexes are the result of these combinations, promoted by weak interaction forces. These complexes present distinct photophysical properties, and due to the nature of the interactions that maintain the structures together, they can find several applications, from electroluminescent devices to sensors. Moreover, doping can be a tool to promote the control of the self-assembling. For instance, Ribeiro[51] showed that using poly[2-methoxy-5-(2-ethylhexyloxy)-1,4-phenylenevinylene] (MEH-PPV) (Sigma-Aldrich, Mn average ≈125,000 molar weight), an electron donor, as doping agent interferes in the interaction forces balance between diphenylalanine units during the self-assembling process, which results in spheres instead of nanotubes, whereas doping it with PVK, an electron acceptor, results in nanotubes (Figure 14.2).

Indeed, doping with organic materials has been widely used in a number of systems, with the objective, among others, of creating new materials with interesting properties for an application. Nevertheless, it is not often easy to predict which would be the properties presented by the new

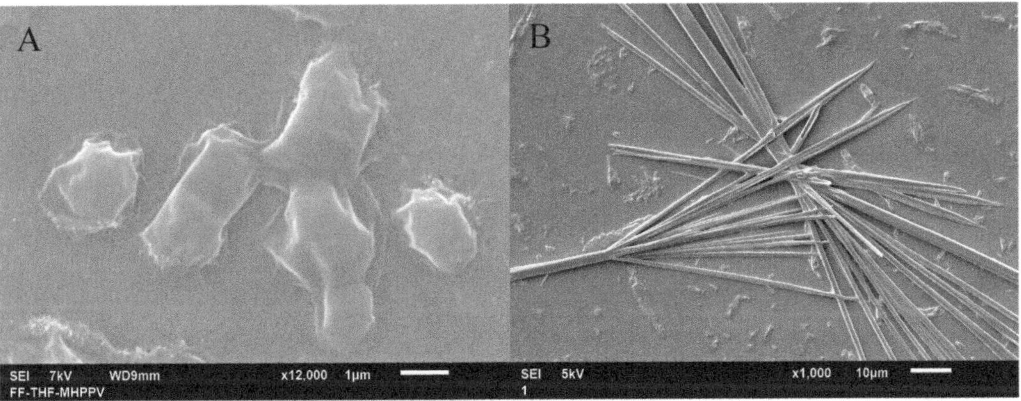

FIGURE 14.2 SEM images for Phe-Phe self-ensembled supramolecular structures: (A) with MEH-PPV as additive, resulting in spheres; (B) with PVK as additive, resulting in nanotubes.

materials, because the ways in which components will interact depend on very sensitive balances of forces. The products are usually complex, and their properties are unique and frequently distinct from the initial materials. In a previous work, Dias[45] showed that diphenylalanine nanotubes with structural differences are obtained when doped with maghemite, a magnetic iron nanoparticle, and cadmium and tellurium–QD, and the photophysical properties are quite distinct from the initial materials.

The results of Dias showed that fluorescence emission increases when nanotubes are doped with QD or maghemite, due to the nature of interaction between the components that originate charge transfer complexes with more efficient fluorescence. In addition, when both doping agents are added simultaneously to diphenylalanine during the self-assembling process, the resulting material presents fluorescence even more intense and occurring at a wider wavelength range. This effect is a result of luminescent processes occurring from relaxed electronic states of the energy transfer complex that is formed when the material is doped with the magnetic particle and QD, simultaneously. These findings evidenced that peptide nanostructures such as diphenylalanine nanotubes can efficiently be doped with distinct agents to form complexes able to perform energy transfer and conversion and find application in electroluminescent or photovoltaic devices. It is noteworthy that the self-assembled structures are formed in aqueous media, due to the balance of interactions between the peptide residues and the doping agents, i.e., metallic nanoparticle QDs or organic polymers. Distinct interactions are dominant depending on the nature of the components. In these preparations, no further chemical reactions were employed, and the final structure is subject of conversion into another structure, simply by controlling the environmental conditions of preparation. They assemble based on physical interactions rather than chemical modifications.

Another application that can benefit from the energy transfer complex formation with peptide nanotube is given in Souza's work.[54] By performing the diphenylalanine self-assembling with a (2-benzothiazolyl)-7-dietilamino-coumarin, a complex very sensitive to small concentrations of oxygen (O_2) in water was obtained, even though the specific coumarin used as doping agent was not sensitive to O_2, resulting in a system for O_2 sensor with success.[57] Figure 14.3 shows (A) the steady-state fluorescence spectroscopic study of this system in the presence of distinct concentrations of O_2 dissolved in water, and (B) the nanotube structure used in this sensor.

Souza also suggests that, even not being part of the self-assembled structure, C6 contributes to self-assembly process, by favoring π–stacking interactions between peptide aromatic rings.

More efficient solar cells are being inspired by the light harvesting machinery of natural organisms. Among the most recent organic self-assembled materials proposed for light harvesting

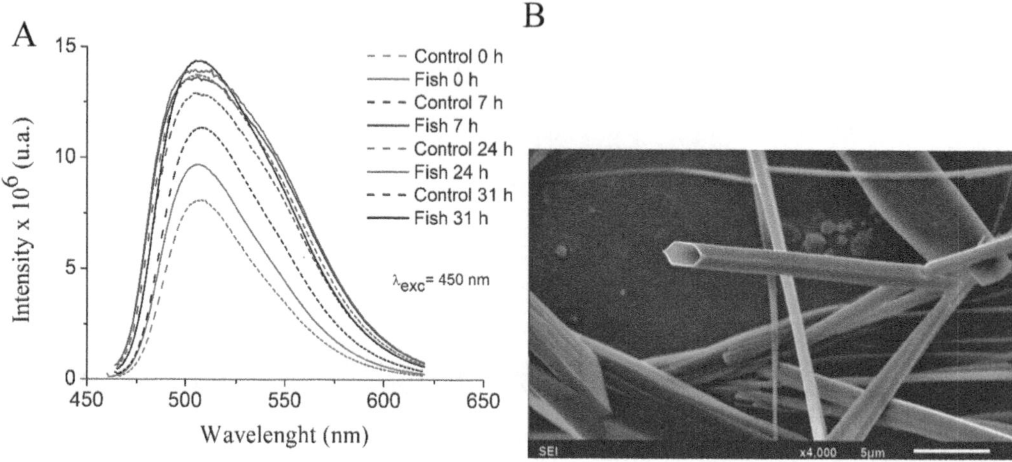

FIGURE 14.3 (A) Steady-state fluorescence spectra of the sensor, at distinct times of $O_{2(g)}$ consumption in water sample. Solid lines refer to fluorescence of the water with $O_{2(g)}$ consumption by fish; dashed lines refer to control water, with no changes in $O_{2(g)}$ concentration. (B) SEM images of Phe-Phe/C6 nanotubes used in the $O_{2(g)}$ sensors.

application are water-soluble porphyrin functionalized Phe-Phe nanostructures, which behave as biomimetic antenna systems. According to Charalambidis et al.,[58] the major drawbacks of these artificial light-harvesting devices are their irreversibility and low stability, which limits their application. In a successful attempt to overcome such limitations, they prepared chiral Fmoc-Phe-Phe-meso-tetraphenylporphyrin (TPP) conjugates, in which the Phe-Phe fibrils behave as quenched antennas, exhibiting efficient excitonic coupling as identified in its circular dichroism spectra. Simply by increasing the solvent polarity by a small extent, the fibers convert into fluorescent spherical assemblies with no exciton couplets. These spheres, in turn, can be reversed into the former configuration by diluting further with nonpolar solvent, making this reversible system a switchable photovoltaic device that showed long-term stability under many activation/deactivation cycles.[57] Photovoltaic devices based on isolated porphyrin derivatives still suffer from photobleaching and low stability of electron–hole pairs for efficient generation of photocurrent. In order to optimize the performance of porphyrin-based photovoltaic devices, Feng et al.[59] combined a tetraphenyl-porphine zinc (TPP-Zn) with aromatic peptide Fmoc-Leucine-OMe hydrogel. In their system, the porphyrin-derived TPP-Zn dopant acts as a conversion antenna, whereas the hydrogel matrix plays the role of charge acceptor and electron–hole transfer matrix. In this way, π-stacking interactions between the fluorenyl moieties and aromatic rings of the porphyrin promote photostability.[59] In addition, Tao et al.[60] produced Phe-Phe multiporous microspheres containing 5-mono(4-carboxyphenyl)-10,15,20-triphenyl porphine as additive, in which extended π–π interactions resulted in a broad-spectrum sensitivity and photoelectron transfer. Distinctly, they had demonstrated that these Phe-Phe-Porphyrine micro-assemblies are useful in biocatalytic redox reactions, because they were able to regenerate the NADH co-enzyme used in these reactions.[57]

These organic self-assembling peptides are not limited to light harvesting applications. Wu et al.[61] showed that photosensitive peptide hydrogels have been developed for smart materials that respond to a variety of external stimuli.[61] For instance, Xiong et al.[62] recently prepared a low-molecular-weight gelator that is responsive to a variety of environment stimuli, composed of an amino acid, an azobenzene linker, and a short oligoethylene glycol tail (AA-Azo-EG_6). In this material, the hydrogen bonding and salt bridges reversibly respond to pH change, and the azo unit provides rigidity through π–π interactions and is responsive toward UV light, which triggers trans–cis isomerization. The oligoethylene glycol tail offers flexibility and hydrophilicity and is thermoresponsive.[62]

Self-assembling peptides with quantum confinement properties have been investigated to replace the conventional inorganic quantum dots for biological applications due to their high cytotoxicity. Fan et al.[9] prepared quantum dots based on dimers of aromatic cyclo-peptides, the cyclo-phenylalanine-tryptophan (cyclo-FW), and cyclo-WW, which can be further self-assembled into larger supramolecular structures of various morphologies and tunable photoluminescence properties. Self-assemblies of cyclo-dihistidine, cyclo-diphenylalanine, and cyclo-dityrosine were studied, which showed distinct morphologies and photoluminescent properties. According to the authors, the pivotal role of the aromatic side chains on self-assembly is the basis for modulating the photoluminescent and morphological properties of the bulk material, which can be accomplished by changing the aromatic moieties. The luminescence of the assemblies can also be controlled by doping with metal ions. For example, nanospheres of cyclo-WW coordinated with Zn(II) presented remarkable luminescence in the visible and near-infrared region. These spheres were successfully tested for their biocompatibility and stability for bio-imaging in cancer cell lines and in mice. An LED prototype based on the system was also demonstrated.[63]

Ely et al.[64] explored the intrinsic quantum confinement properties of self-assembled peptides to construct a hybrid field-effect transistor of Phe-Phe self-assembled dipeptide and poly(3-hexylthiophene)-P3HT, an organic semiconductor. The resulting material benefitted from a synergistic effect that showed improved charge transport through the polymer and represented an environmentally friendly bio-organic device.[59]

Among the intrinsic properties of Phe-Phe nanostructures are their piezoelectricity that arises from their noncentrosymmetric crystals; however, uniform and stable polarization has been difficult to achieve. Innovation in fabrication methods and doping with additives have been the strategies to overcome this limitation. Nguyen et al.[65] have attained controlled polarization in vertically grown Phe-Phe microrods as well as improved piezoelectric strength, by applying an electric field during the self-assembling process. They were able to grow microrods from a substrate coated with Phe-Phe seed film in contact with an aqueous solution of the dipeptide. The electrical dipoles of free dipeptide molecules in solution align parallel to the applied electric field, and so do the molecules, carrying out the self-assembly process. A power generator based on these microrods produced an open-circuit voltage of 1.4 V and a power density of 3.3 nW cm^{-2} that showed remarkable stability. A device made of three layers of this power generator was able to light up a small liquid crystal by pressing the device with a human finger.[63] In a different approach, Ivanov et al.[66] prepared Phe-Phe microtubes covered by graphene oxide decorated with nickel nanoparticles, which showed a 100-fold increase in electrical conductivity, a twofold increase in mechanical stiffness, and significant piezoelectricity.[60]

14.5 CONCLUSION

Modified peptide self-assembled structures have provided great advances in several areas of knowledge and technology, and therefore, they have found application in distinct fields. Due to their ease of preparation and the wide variety of shapes they can assume, these self-assembled structures have been developed for applications in which inorganic materials have dominated for decades. In particular, self-assembled peptides conjugated to luminescent compounds find application in strategical scientific and technological fields, such as energy conversion and diagnosis and treatment of diseases; nevertheless, there still is an urgent need for scrutiny of their characteristics and properties to guarantee the efficient control of their properties upon application. Many studies have proved that noncovalent interaction forces are dominant on self-assembling mechanisms, independently of the peptide identity, and that these interactions rule the final structures and the nature of the energy transfer complexes generated by these modifications. Therefore, these interactions are responsible for the properties these materials present. At this point, great advances will be only possible if the nature of these charge transfer complex formations and their characteristics are well known. This is the breakthrough we must face now.

REFERENCES

1. Reches, M., Gazit, E., Casting metal wires within discrete self-assembled peptide nanotubes. *Science*. 2003, 300, 625-627.
2. Gazit, E. Self-assembled peptide nanostructures: the design of molecular building blocks and their technological utilization. *Chem. Soc. Rev.* 2007, 36, 1263–1269.
3. Gao, X., Matsui, H. Peptide-based nanotubes and their applications in bionanotechnology. *Adv. Mater.*, 2005, 17, 2037–2050.
4. Zhang, S., Marini, D. M., Hwang W., Santoso, S. Design of nanostructured biological materials through self-assembly of peptides and proteins. *Curr. Opin. Chem. Biol.*, 2002, 6, 865–871.
5. Carny, O., Shalev, D. E., Gazit, E. Fabrication of coaxial metal nanocables using a self-assembled peptide nanotube scaffold. *Nano Lett.* 2006, 6(8), 1594-1597.
6. Kasotakis, E., Mossou, E., Adler-Abramovich, L., Mitchell, E. P., Forsyth, V. T., Gazit, E., Mitraki, A. Design of metal-binding sites onto self-assembled peptide fibrils. *Biopolymers*. 2009, 92(3), 164-172.
7. Mason, T. O., Chirgadze, D. Y., Levin, A., Adler-Abramovich, L. Gazit, E., Knowles, T. P. J., Buell, A. K. Expanding the solvent chemical space for self-assembly of dipeptide nanostructures. *ACS Nano* 2014, 8(2), 1243-1253.
8. Raeburn, J., Pont, G., Chen, L., Cesbron, Y., Levy, R., Adams, D. J. Fmoc-diphenylalanine hydrogels: understanding the variability in reported mechanical properties *Soft Matter* 2012, 8, 1168–1174.
9. Fan, Z., Sun, L., Huang, Y., Wang, Y., Zhang, M. Bioinspired fluorescent dipeptide nanoparticles for targeted cancer cell imaging and real-time monitoring of drug release. *Nat. Nanotechnol.* 2016, 11(4), 388-394.
10. A. Handelman, N. Kuritz, A. Natan, G. Rosenman, Reconstructive phase transition in ultrashort peptide nanostructures and induced visible photoluminescence. *Langmuir*. 2016, 32 (12), 2847-2862.
11. Handelman, A., Lavrov, S., Kudryavtsev, A., Khatchatouriants, A., Rosenberg, Y., Mishina, E., Rosenman, G. Nonlinear optical bioinspired peptide nanostructures. *Adv. Opt. Mater.* 2013, 1 (11), 875-884.
12. Gilboa, B., Lafargue, C., Handelman, A., Shimon, L. J. W., Rosenman, G., Zyss, J., Ellenbogen, T. Strong electro-optic effect and spontaneous domain formation in self-assembled peptide structures. *Adv. Sci.*, 2017, 4, 1700042 (7p).
13. Yariv, A. Quantum electronics. 3rd ed. Wiley & Sons. New York. 1989.
14. Birks, J. B. Photophysics of organic molecules. John Wiley & Sons, London, United Kingdom, 1970.
15. Suzuki, H Absorption spectra of organic molecules. Academic Press, New York, NY, 1967.
16. Beljonne, D., Cornil., J. Coropceanu, V., Silva-Filho, D. A., Geskin, V., Lazzaroni, R., Leclere, P., Bredas, J. L. On the transport, optical, and self-assembly properties of π-conjugated materials: a combined theoretical/experimental insight. In Handbook of conjugated polymers. 3rd ed. Rev. Conjugated polymers, theory, synthesis, properties, and characterization. Eds. Sktheim, T. A., Reynolds, J. R. CRC Press, Taylor and Francis Group. Boca Raton, 2007.
17. Lakowicz, J. R. Principles of fluorescence. 2nd ed. Kluwer Academic/Plenum Publishers. New York, NY, 1999.
18. Forster, T. 10th Spiers memorial lecture. Transfer mechanism of electronic excitation. *Discuss. Faraday Soc.* 1959, 27, 7-17.
19. Mandal, D., Shirazi, A. N., Parang, K. Self-assembly of peptides to nanostructures, *Org. Biomol. Chem.* 2014, 12, 3544–3561.
20. Ahlers, P., Fischer, K., Spitzer, D., Besenius, P. Dynamic light scattering investigation of the kinetics and fidelity of supramolecular copolymerizations in water . Macromolecules 2017, 50, 7712–7720.
21. Gilroy, J. B., Gädt, T., Whittell, G. R., Chabanne, L., Mitchels, J. M., Richardson, R. M., Winnik, M. A., Manners, I. Monodisperse cylindrical micelles by crystallization-driven living self-assembly. *Nat. Chem.* 2010, 2, 566–570.
22. Zhang, W., Jin, W., Fukushima, T., Saeki, A., Seki, S., Aida, T. Supramolecular linear heterojunction composed of graphite-like semiconducting nanotubular segments. *Science*. 2011, 334, 340–343.

23. Gorl, D., Zhang, X., Stepanenko, V., Würthner, F. Supramolecular block copolymers by kinetically controlled co-self-assembly of planar and core-twisted perylene bisimides. *Nat. Commun.* 2015, 6, 7009 (7p).).
24. Besenius, P. Controlling supramolecular polymerization through multicomponent self-assembly. *J. Polym. Sci., Part A: Polym. Chem.* 2017, 55, 34–78.
25. Valery, C., Artzner, F., Paternostre, M. Peptide nanotubes: molecular organisations, self-assembly mechanisms and Applications. *Soft Matter*, 2011, 7, 9583-9594.
26. Eskandari, S., Guerin, T., Toth, I., Stephenson, R. J. Recent advances in self-assembled peptides: Implications for targeted drug delivery and vaccine engineering. *Adv. Drug Deliv. Rev.* 2017, 110–111, 169–187.
27. Cui, H., Webber, M. J., Stupp, S. I. Self-assembly of peptide amphiphiles: from molecules to nanostructures to biomaterials. *Pept. Sci.* 2010, 94, 1–18.
28. Keyes-Baig, C., Duhamel, J., Fung, S.-Y., Bezaire, J., Chen, P. Self-assembling peptide as a potential carrier of hydrophobic compounds. *J. Am. Chem. Soc.* 2004, 126, 7522–7532.
29. Fung, S., Yang, H., Chen, P. Formation of colloidal suspension of hydrophobic compounds with an amphiphilic self-assembling peptide. *Colloids Surf. B* 2007, 55, 200–211.
30. Samad, M. B. , Chhonker, Y. S., Contreras, J. I., McCarthy, A., McClanahan, M. M., Murry, D. J., Conda-Sheridan, M. Developing polyamine-based peptide amphiphiles with tunable morphology and physicochemical properties . *Macromol. Biosci.* 2017, 17, 1700096.
31. Fletcher, N. L., Lockett, C. V., Dexter, A. F. A pH-responsive coiled-coil peptide hydrogel. *Soft Matter.* 2011, 7, 10210–10218.
32. Xing, R., Li, S., Zhang, N., Shen, G., Möhwald, H., Yan, X. Self-assembled injectable peptide hydrogels capable of triggering antitumor immune response. *Biomacromolecules* 2017, 18, 3514-3523.
33. Shirazi, A. N., Tiwari, R. K., Oh, D., Banerjee, A., Yadav, A., Parang, K. Efficient delivery of cell impermeable phosphopeptides by a cyclic peptide amphiphile containing tryptophan and arginine. *Mol. Pharm.* 2013, 10, 2008–2020.
34. Mazza, M., Patel, A., Pons, R., Bussy, C., Kostarelos, K. Peptide nanofibres as molecular transporters: from self-assembly to in vivo degradation. *Faraday Discuss.* 2013, 166, 181–194.
35. Xu, X.-D., Liang, L., Chen, C.-S., Lu, B., Wang, N., Jiang, F.-G., Zhang, X.-Z., Zhuo, R.-X. Peptide hydrogel as an intraocular drug delivery system for inhibition of postoperative scarring formation. *ACS Appl. Mater. Interfaces* 2010, 2, 2663–2671.
36. Jiang, T., Yu, X., Carbone, E. J., Nelson, C., Kan, H. M., Lo, K. W. -H. Poly aspartic acid peptide-linked PLGA based nanoscale particles: potential for bone-targeting drug delivery applications. *Int. J. Pharm.* 2014, 475, 547–557.
37. Abbas, M., Zou, Q., Li S., Yan, X. Self-assembled peptide- and protein-based nanomaterials for antitumor photodynamic and photothermal therapy. *Adv. Mater.* 2017, 29, 1605021.
38. Wu, S., Butt, H.-J. Near-infrared-sensitive materials based on upconverting nanoparticles. *Adv. Mater.* 2016, 28, 1208-1226.
39. Song, X. J., Liang, C., Gong, H., Chen, Q., Wang, C., Liu, Z. Photosensitizer-conjugated albumin-polypyrrole. nanoparticles for imaging-guided in vivo photodynamic/photothermal therapy. *Small* 2015, 11(32), 3932-3941.
40. Wu, L., Fang, S. T., Shi, S., Deng, J. Z., Liu, B., Cai, L. T. Hybrid polypeptide micelles loading indocyanine green for tumor imaging and photothermal effect study. *Biomacromolecules* 2013, 14, 3027-3033.
41. Kirchherr, A. K. , Briel, A., Mader, K. Stabilization of indocyanine green by encapsulation within micellar systems. *Mol. Pharmaceut.* 2009, 6(2), 480-491.
42. Ejiofor, J., Webster, T. J. Biomedical applications: implants, pp 327-338, in: Dekker Encyclopedia of Nanoscience and Nanotechnology, 2nd Ed., Taylor & Francis, 2009.
43. Yucesoy, D. T., Hnilova, M., Boone, K., Arnold, P. M., Snead, M. L., Tamerler, C. Chimeric peptides as implant functionalization agents for titanium alloy implants with antimicrobial properties. *JOM* 2015, 67(4), 754-766.
44. Sánchez-Gómez, S., Martínez-de-Tejada, G. Antimicrobial peptides as anti-biofilm agents in medical implants. *Curr. Top. Med. Chem.* 2017, 17(5), 590-603.
45. Yazici, H., O'Neill, M. O., Kacar, T., Wilson, B. R., Oren, E. E., Sarikaya, M., Tamerler, C. Engineered chimeric peptides as antimicrobial surface coating agents toward infection-free implants. *ACS Appl. Mater. Interfaces* 2016, 8 (8), 5070–5081.

46. Moini, E., Dadkhah, A. A., Allafchian, A., Habibi, N. Patterning protein conjugates into organized microarrays with diphenylalanine peptide nanotubes self-assembled on graphite and gold electrode. *J. Mater. Sci: Mater. Electron.* 2017, 28, 16910–16920.
47. Takahashi, R., Wang, H., Lewis, J. P. Electronic structures and conductivity in peptide nanotubes. *J. Phys. Chem. B* 2007, 111 (30), 9093–9098.
48. Liang, Y., Guo, P., Pingali, S. V., Pabit, S., Thiyagarajan, P., Berland, K. M., Lynn, D. G. Light harvesting antenna on an amyloid scaffold. *Chem. Commun. (Camb.)* 2008, 48, 6522–6524.
49. Ryu, J., Lim, S. Y., Park, C. B. Photoluminescent peptide nanotubes. *Adv. Mater.* 2009, 21 (16), 1577–1581.
50. Souza, G. A. Photophysical and morphological characterization of peptide nanostructures containing fluorescent compound for environmental application. Universidade Federal de Goiás, Goiânia-Brazil, 2014, Masters Dissertation.
51. Ribeiro, A. C. C., Camargo, H. S., Pereira, D. H., Custodio, R., Martins, T. D. Photoluminescence of Solvent-selected Fluorescence Moieties in MEH-PPV Solutions and Films. *J. Braz. Chem. Soc.* 2018, 29, 543-559.
52. Dias, D. L. Fluorescence spextroscopy applied to the study of the interactions of formation of peptide nanostructures containing maghemite. Universidade Federal de Goias, Goiânia-Brazil, 2014. Masters Dissertation.
53. Ghadiri, M. R. et al. Self-assembling organic nanotubes based on a cyclic peptide architecture. *Nature* 1993, 366 (6453), 324–327.
54. Ribeiro, A. C. C., Souza, G. A., Pereira, D. H., Cordeiro, D. S., Miranda, R. S., Custódio, R., Martins, T. D. Phe-Phe di-peptide nanostructures self-assembling modulated by luminescent additive. *ACS Omega.* 2019, 4 (1), 603-619.
55. Ribeiro, A. C. C. Influence of diphenylalanine (FF) self-assembling in the photophysics of systems FF/fluorophore: a theoretical-practical approach. Universidade Federal de Goiás, Goiânia-Brazil. 2017. PhD Thesis.
56. Dias, D. L. Fluorescence spectroscopy applied to the study of the interactions of formation of peptide nanostructures containing maghemite. Universidade Federal de Goias, Goiânia-Brazil, 2014. Masters Dissertation.
57. Martins, T. D., Souza G. A., Costa-Filho, P. A., Ribeiro, A. C. C., Universidade Federal de Goiás. Sistema Fluorescente para sensor óptico de oxigênio dissolvido, Método para preparar dispositivo de sensor óptico e uso de sistema fluorescente, 2015. Brazil Patent BR1020150206658. August 27 2015.
58. Charalambidis, G., Georgilis, E., Panda, M. K., Anson, C.. E., Powell, A. K., et al. A switchable self-assembling and disassembling chiral system based on a porphyrin-substituted phenylalanine-phenylalanine motif. *Nat. Commun* 2016, 7, 12657. Doi https://doi-org.ez49.periodicos.capes.gov.br/10.1038/ncomms12657
59. Feng, L., Wang, A., Ren, P., Wang, M., Dong, Q., Li, J., Bai, S. Self-assembled peptide hydrogel with porphyrin as a dopant for enchanced photocurrent generation. *Colloid Interface Sci. Commun.* 2018, 23, 29. doi: 10.1016/j.colcom.2018.01.006
60. Tao, K., Xue, B., Frere, S., Slutsky, I., Cao, Y., Wang, W., Gazit, E. Multiporous supramolecular microspheres for artificial photosynthesis. *Chem. Mater.* 2017, 4454. doi: 10.1021/acs.chemmater.7b00966
61. Wu, D., Xie, X., Kadi, A. A., Zhang, Y. Photosensitive peptide hydrogels as smart materials for applications. *Chin. Chem. Lett.* 2018, 29, 1104. doi https://doi.org/10.1016/j.cclet.2018.04.030
62. Xiong, W., Zhou, H., Zhang, C., Lu, H. An amino acid-based gelator for injectable and multi-responsive hydrogel. *Chin. Chem. Lett.* 2017, 28, 2128. doi: 10.1016/j.cclet.2017.09.019.
63. Tao, K., Fan, Z., Sun, L., Makam, P., Tian, Z., et al. Quantum confined peptide assemblies with tunable visible to near-infrared spectral range. *Nat. Commun.* 2018, 9, 3217. doi: 10.1038/s41467-018-05568-9
64. Ely, F., Cipriano, T. C., da Silva, M. O., Peressinotto, V. S. T., Alves, W. A. Semiconducting polymer-dipeptide nanostructures by ultrasonically-assisted self-assembling. *RSC Adv.* 2016, 6, 32171. doi: 10.1039/c6ra03013k

65. Nguyen, V., Zhu, R., Jenkins, K., Yang, R. Self-assembly of diphenylalanine peptide with controlled polarization for power generation. *Nat. Commun.* 2016, 7, 13566. doi: 10.1038/ncomms13566
66. Ivanov, M. S., Khomchenko, V. A., Salimian, M., Nikitin, T., Kopyl, S., Buryakov, A. M., Mishina, E. D., Salehli, F., Marques, P. A. A. P., Goncalves, G., Fausto, R., Paixão J. A., Kholkin, A. L. Self-assembled diphenylalanine peptide microtubes covered by reduced graphene oxide/spiky nickel nanocomposite: an integrated nanobiomaterial for multifunctional applications. *Mater. Des.* 2018, 142, 149. doi: 10.1016/j.matdes.2018.01.018

15 Nanoscale Characterization of Bone: Mechanical, Ultrasound, Electrical, and Spectroscopic Methods

Efstathios I. Meletis, Farad Sarker, Meet Shah, Harry F. Tibbals, Suni Yadev

Department of Materials Science and Engineering, The University of Texas at Arlington, Arlington, Texas, USA

CONTENTS

15.1 INTRODUCTION

Bone is organized and structured on multiple levels.[1] Knowledge of the skeletal system has historically progressed from the macro to the micro, and then to the cellular, chemical, and molecular level as science has developed, along with medical and surgical capacities to deal with diseases, injuries, and disorders.[2] Developments in nanoscale materials science and engineering are providing knowledge in science and guidance in medicine regarding the organization of osseous tissue at the nanoscale.[3] Nanoscale research provides an understanding of how structure and activity at this level affect the strength, resilience, and toughness of bones.[4,5] This chapter reviews selected nano-based techniques applied to the study of osseous tissues: bone and cartilage. Each instrumental approach has its strength and weaknesses.[6] New characterization options are enhanced by emerging nanotechnologies, with potential for medical diagnosis and monitoring.

15.2 FUNDAMENTALS OF BONE CHARACTERIZATION

Mechanical characterization deals with forces acting on a material and its internal structures. Materials science characterizes the properties described as strength, hardness, brittleness, and so on.[7,8] For clinical purposes, an index is needed to assess the state of bone strength, in order to predict fracture risk. Ideally, such a measure should be obtainable noninvasively, with minimal radiation risk and low cost, and should be easy to conduct.

Small forces acting on a confined object can produce microscopic localized deformations. A force that produces changes in shape or structure is termed a *stress* (or load). The resulting deformation is termed a *strain*. The strength of a material is defined in terms of the yield stress (or yield strength) and ultimate stress (ultimate strength). The yield strength is the stress at which the material begins to deform. The ultimate strength is the stress at which the material breaks. For ductile or plastic materials, there is generally a poorly reproducible divergence between the yield point and the ultimate strength. For brittle materials, the two points converge: the more brittle the material, the less difference between the yield and rupture points. In materials science, toughness is the ability of a material to absorb stress energy without rupture. Resilience is the ability of material to absorb energy without permanent deformation, that is, to behave elastically. The fatigue strength is defined in two indices: the fatigue limit is the maximum stress that a material can undergo in a given number of repeated deformations before fracturing; the fatigue life measure is the number of stress repetitions the material can withstand with a given fixed load before failure.

Different types of forces are distinguished, based on direction relative to the object: *tension*, *compression*, *bending*, *shearing*, and *twisting* or *torque*. Tension forces pull on an object, producing a stretch or extension deformation; compression forces push. Shear forces simultaneously act on a body in different directions, tending to pull it apart. Torque forces act to rotate portions of a body around an axis. Compression forces can act in one direction, as along a length axis, or simultaneously in all directions, as when compressing a gas or liquid.

Deformations generally have complex shapes; their characterization involves quantitative measurement of the solid geometry. Small symmetrical deformations such as those produced by a test probe can be characterized by depth alone, which is a sufficiently accurate approximation for comparison purposes.

Most solids are in a stable equilibrium state; when a force creates a small strain, they return to the original shape. Deformations which are reversed when the stress is removed are called elastic. Deformations that remain after the causative force is terminated are inelastic. A basic property of solids is the elastic limit: the magnitude of strain beyond which an elastic recovery does not take place.

Small strains within the elastic limit obey Hooke's law:

$$F = k\Delta L \tag{15.1}$$

where ΔL is the magnitude of the deformation (e.g., the depth of an indentation, or the angle of a bend), resulting from the force F, and k is a constant that depends on the shape of the material and the applied force. Thus, the force is not constant, but depends on the degree of deformation. Within the elastic limit, the force is directly proportional to the deformation, and vice versa. Note that in tension, since the object is fixed in place, an equal and opposite force is implied to restrain it from motion.

The length deformation ΔL for an idealized volume of cylindrical rod of material stretched elastically in the direction of its length is:

$$\Delta L = \frac{1}{Y}\frac{F}{A}L_0 \tag{15.2}$$

where A is the cross-sectional area of the cylinder, and L_0 is the original undeformed length. The constant factor Y is defined as the *elastic modulus*, or *Young's modulus*. This elastic modulus is a characteristic property of a material, giving an index of its resistance to deformation within its elastic limit when opposing forces act along an axis.

The elastic modulus summarizes the combined effects at the nano level of all the forces that hold a solid material in its fixed shape. Other moduli are defined separately for shear forces and for bulk compression, in order to characterize the behavior of a body under different types of loading and stress. The shear modulus characterizes the resistance to deformation by transverse forces, acting in opposite directions. The bulk modulus is usually defined in terms of the ratio of change in pressure, or force per unit area, to the change in volume of a material.

The various forms of moduli are useful for comparison of the strength of materials. For uniform crystalline and polymer materials, the moduli can be related theoretically to intermolecular and interdomain forces and microstructures. For complex composites, it is more difficult to precisely define these relationships, but the moduli are still useful for empirical comparisons. Strictly, deformation moduli may be anisotropic, and the properties represented by the elastic, shear, and bulk moduli depend upon the thermodynamic state of the material, e.g., the temperature, pressure, and entropy. But for most biomaterials, the thermodynamic state can generally be assumed to be constant.

The shape of a plot of stress versus strain is characteristic for a material. Strain rises linearly with stress in the region exhibiting elastic behavior. The slope of the rise is equal to the elastic modulus. Up to the elastic limit, strain is reversible. Beyond the elastic limit, the behavior of simple uniform materials typically exhibits one of two behaviors: brittle or ductile. For brittle materials, such as glasses and ceramics, the curve departs from the linear relationship, just before the failure point of ultimate strain, where the material undergoes fracture. For ductile materials, such as malleable metals, the substance exhibits less resistance to deformation in a region between the elastic limit and the ultimate failure point—a yield region. Composite materials and many complex polymers typically exhibit behavior that does not fit either of these modes. The stress versus strain plot for complex materials may have several straight line regions separated by yield regions, each corresponding to a change in internal configuration states.

Another important parameter for characterization of material deformation is the *Poisson's ratio*, which quantifies the tendency of a force applied in one direction to produce deformations in orthogonal directions. The *Poisson's ratio* is the ratio of transverse strain to longitudinal strain. Normal solid materials tend to resist a change in volume (corresponding to the bulk modulus) more strongly than a change in shape (corresponding to the shear modulus). Thus, when stretched in a longitudinal direction, normal materials shrink in cross-section, and conversely, expand in width when compressed in length. The definition of *Poisson's ratio*, η, is:

$$\eta = -\frac{e_{\text{trans}}}{e_{\text{longit}}} \tag{15.3}$$

where e_{trans} is the transverse strain, or ratio of change in width (or radius) to the original width (or radius), and e_{longit} is the ratio of the change in length (or axial extension) to the original value, e.g.:

$$e_x = \frac{\Delta x}{x} \tag{15.4}$$

where x is one of two orthogonal directions of deformation measurement.

Note that by convention, tensile (stretching) strain is considered positive, and compression is considered negative. Thus, the minus sign in the definition of η gives normal materials a positive value for Poisson's ratio.

15.3 BONE ARCHITECTURE AND THE NANOSCALE

Bone is organized in a hierarchy of composite structures from the macro to the sub-nano level (Figure 15.1).[1,4,5,9] At the cellular level, specialized cells adapt to stresses by redistributing minerals and polymers. On the gross anatomical level, bones have a hard, dense outer shell (cortical bone), which may enclose a volume of porous network (cancellous bone, also known as spongy or trabecular bone). The spaces in the trabecular bone are filled with extracellular matrix, which plays an important role in bone development. In large bones, the trabecular layer may surround a cavity, filled with matrix fluid (marrow) in which cells are stored. Bones are connected to each other and to muscles by tough, flexible ligaments and tendons composed of biopolymers (primarily collagen).

The cortical skeleton supports muscles, nerves, skin, blood vessels, and lymphatic network and protects soft organ systems. Parts of the skeleton form protective cages around critical organs such as the brain, heart, and lungs. Portions of the skeleton are integral parts of the circulatory, sensory, and other organ systems. Bone channels carry blood and lymph vessels. Eye sockets and ear canals are shaped by bony enclosures, which protect sensitive neural systems and sensors from interference. The ear depends on three tiny bones (malleus, staples, incus = hammer, stirrup, and anvil) to transmit and amplify sound with sensitivity to nanoscale vibrational displacements.

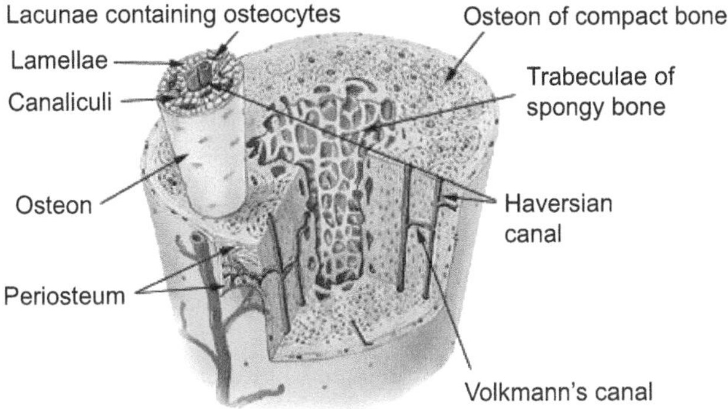

FIGURE 15.1 Hierarchical levels of bone structure. (*Source*: NCI SEER Training, adapted image licensed under a Creative Commons Attribution 4.0 international license.)

15.3.1 Cortical Bone

Cortical bone encloses specialized cells (osteocytes, osteoblasts, and osteoclasts). Mature osteocytes are buried in the cortical matrix and interconnected by dendritic processes similar to those in nerve cells, forming a signaling network that is responsive to mechanical forces and other influences. The osteocytes regulate the activity of osteoblasts, which deposit materials in an organized pattern, and osteoclasts, which reabsorb material to break down bone. These cells coordinate to adjust density and structure in their surroundings in response to external loading forces, redistributing material for growth and repair during development and healing of damage (Figure 15.2).

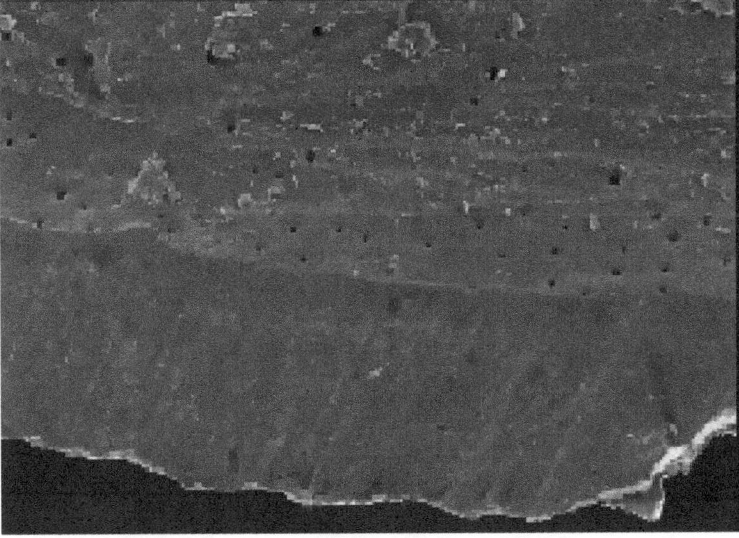

FIGURE 15.2 Scanning electron microscopy (SEM) image of mouse tibia cross-section, showing boundary and differences between laminar cortical bone (lower half of image) and immature woven bone (60 × magnification). (From UT Arlington Characterization Center for Materials and Biology, Drs. Jiang and Tibbals, on genetically modified mouse from UTSW and Texas Scottish Rite Childrens' Hospital Research Center, Dr. J. Rios, W. Pierce; sample preparation and micrographs by M. Shah and S. Yadev, UTA Materials Science and Engineering Department.)

At the micro level, cortical bone consists of microcrystalline minerals (hydroxyapatite, with many trace element components) embedded in a collagen-based matrix. The main microstructural element of cortical bone is the osteon, a column of concentric layers, typically several millimeters long and around 0.2 mm in diameter, surrounding a central canal (Haversian canal). Volkmann channels extending within the cortical bone carry blood vessels that connect the central canal to the outer circulatory system. In much of the skeleton, a dense layer of vascular tissue, the periosteum, separates the outer layer of the bone from the body.

At the nano level, mineral crystals and collagen fibrils are arranged in a matrix, with the mineral phase providing compression strength for load support and the collagen providing toughness through elasticity and resistance to crack propagation. The exact structural architecture at this level is important for the functional behavior of bone. Aspects such as crystal size and orientation, the polymer length and cross-linkage, and other factors are related to proper development and healing, diseases, and disorders, although their role is not yet fully elucidated. Thus, analysis and characterization of cortical bone at the nano level is an important area for ongoing research.

15.3.2 Trabecular Bone

The inner porous mass found enclosed in larger bones is called trabecular bone (also known as spongy or cancellous bone). Unlike the compact-layered architecture of cortical bone, the trabecular microstructure is a highly porous, honeycomb-like lattice. The network of connecting trabeculae consists of rods (or filaments) connecting pore walls, or plates, enclosing fluid-filled spaces (Figure 15.3).

The trabecular bone has several functions:

1. *Structural*: The resilient internal support network distributes loads. Stresses and strains are dispersed among multiple load-bearing paths, diluting the effect of forces and lessening the probability that the deformation exceeds the elastic limit in any one path.

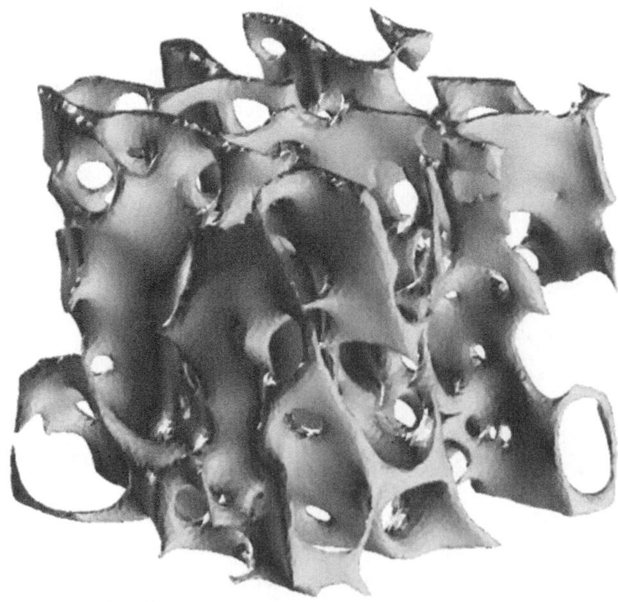

FIGURE 15.3 Three-dimensional image of trabecular bone from computed tomography scan (section ~4 mm^3; pore sizes = ~1 mm).

2. *Weight and mass reduction*: The spongeous network, surrounded by a dense outer cortical shell, reduces the weight and amount of material required to maintain the load-bearing structure, while providing the same or better strength and ruggedness.
3. *A protected cell building area*: The trabecular network provides a large surface area to which stem cells are anchored in spaces filled with fluids circulating the proteins, minerals, and other building blocks for cell growth, differentiation, and maturation. Within the relatively secure protected chambers, shielded from mutagenic radiation and invasive pathogens, progenitor stem cells are nurtured into specialized cells: red blood cells, white blood cells forming the complex adaptive immune system, platelets, and others. The surfaces are attachment points for signaling molecules creating a guiding environment for stem cells. Thus, bone is not merely a supportive tissue but also an active organ with important roles in the immune and blood systems.

15.3.3 Collagen and Mineral

Collagen polymer underlies the structure of both cortical and trabecular bone. Collagens constitute the largest nonmineral component in bone. The mineral component is predominately hydroxyapatite [$Ca_{10}(PO_4)_6(OH)_2$], with some other salts and trace elements with important functions in the crystal lattice. During development and repair, bone formation begins with growth of a collagen structure, into which mineral crystals are precipitated to form hard bone. In mature cortical bone, mineralized collagen fibers are organized in parallel layers (lamellar bone). In newly formed skeleton, collagen fibers are visible under the microscope arranged in random directions (woven bone). Woven bone has less mineral and a higher density of osteocytes than mature lamellar tissue. During development, some tissues remain less mineralized, forming ligaments and tendons. Genetically expressed proteins selectively promote and block the mineralization of collagen. Their distribution during development guides the process of mineralization. Characterization at the nano level is helping the understanding of how these processes work, in order to provide guidance for improved healing and regeneration.[3–5,8–10]

15.4 CHARACTERIZATION TECHNIQUES

The present standards for bone characterization in biomedicine are X-ray diffraction and transmission electron microscopy.[6] In clinical monitoring, bone mineral density (BMD) is the gold standard for monitoring osteoporosis and similar conditions, and dual-energy X-ray absorptiometry (DXA) is the accepted measurement technique. However, there is a growing recognition of the shortcomings, both of DXA for obtaining meaningful densitometry data and of BMD as a reliable predictor of bone strength. The following sections review some selected promising areas for improved nano-level characterization of bone.

15.4.1 Mathematical and Physical Simulation

Computational simulation is increasingly used in investigation, to guide the framing of physical analysis. Numerical digital simulation has become an essential tool in biomedical research for modeling and analysis of bone at all levels of organization.[11–13]

Bone research and medicine benefit from computer simulation models. Digital models aid in visualization and comprehension of results from imaging and are useful in planning of experiments and surgical procedures and for studying mechanical behavior for cortical and trabecular bone.[14,15] In one such study, a model of trabecular bone was used to evaluate the relationship between structure and strength.[16,17] This type of model is relevant to clinical management of fracture risks. The degree to which bone loss occurs with age and disease is important in medicine. Bone weakening can lead to fractures in the spine, the proximal femur, and the distal radius. These sites are prioritized for bone monitoring in patients at risk of disease and degeneration, by measuring BMD with micro-computed

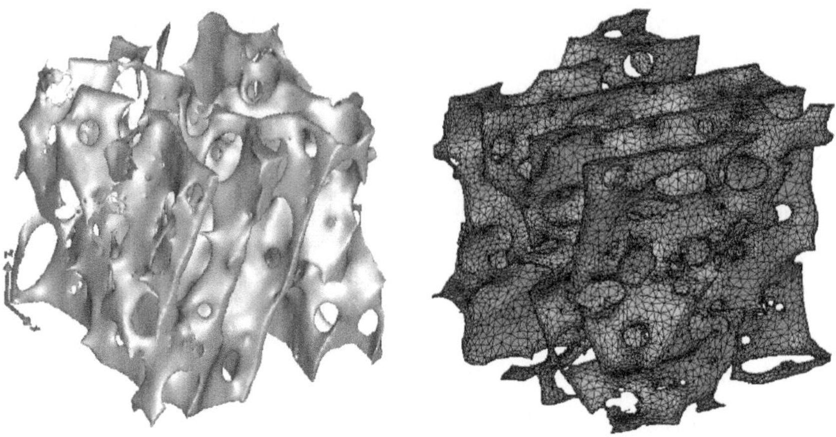

FIGURE 15.4 3D solid and meshed models of human femoral neck trabecular bone (sections ~4 mm^3; pore sizes = ~1 mm) (from Ref. 16. Sarker).

tomography (CT) radiology imaging. Three-dimensional digital models are made by capturing data from 3D CT bone images, using triangular surface grids to make a digital model to any desired resolution. With these digital models and data about the material composition, multiphysics simulations can be conducted to study strength, elasticity, and other characteristics of the trabecular network. The digital model can be modified to project the effects of future bone loss and predict the risk of fracture.

15.4.1.1 Morphological Indices

In the clinical management and research of bone disease, a number of morphometric indices are used to assess the degree of bone loss. Idealized plate and rod models are used to quantify and interpret data from radiological images of trabecular bone, producing a measure called the structure model index.[18] The index is a rough measure of properties such as trabecular thickness, spacing, and porosity.[19] To compute the index, the complex trabecular network is calculated by a differential analysis from a triangulated surface and compared against the simplified plate and rod structure. A recent study evaluated more complex models than the conventional plates and rods, with models based on geometrical elements including triangular prism, cubic, hexagonal, rhombic dodecahedron, and truncated octahedron open unit cells.[16] Digital computer simulations of stress and strain response were carried out for different materials assumptions and varying thicknesses of the structure (Figures 15.4 and 15.5). Calculated relationships were determined by multiphysics simulation, between the fracture toughness, elastic modulus, and apparent density of the trabecular structures.

15.4.1.2 3D Printing of Bone Models

Rapid prototyping models, using three-dimensional printing, are fabricated to visualize and test bone models. Examples are shown in Figures 15.4 to 15.6. In the complex model study cited earlier, 3D printing was used to make molds from which solid polymer models were fabricated using different polymer formulations. The physical models were subjected to mechanical testing to evaluate which forms best matched the characteristics of natural bone.[16,17]

One formulation used epoxy resin with fillers of silicon carbide nanoparticles. The mechanical behavior of bone depends on the response of structural elements over a broad range of length scales spanning from the nanoscale collagen fibrils to the microscale trabecular rods and plates. Polymer composites for scaffolding that simulate this behavior may be more compatible with the host bone.

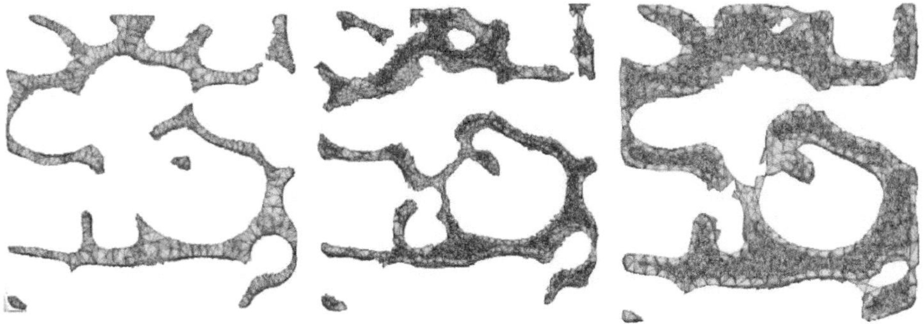

FIGURE 15.5 Simulated bone loss: three cross-sections of human trabecular bone models, with different bone densities and triangulated data points for digital modeling (section ~4 mm^2; pore sizes = ~1–2 mm).

Of the different open unit cell geometries, the rhombic dodecahedron microstructure had the micro-elastic properties most similar to human bone.[16] These models may guide the design of artificial bone implants and uncover factors relating tissue loss to risk of fracture.

15.4.2 Mechanical Properties

An essential biological property of bone is its structural strength. A comprehensive area of study and medical practice deals with the mechanical properties of various kinds of bone.[20] As one of the oldest forms of medicine is the treatment of fractures, a major area of research has been based on bone strength and how to improve it.[21,22]

Failure of bone is different depending on the rate at which load is applied. Loading may be steady or variable, cyclic at different frequencies, or rapid impulsive impact. Failure may be permanent deformation, shear, sudden fracture, or fatigue fracture with steady or repetitive/cyclic loading.[23,24]

Failure mode depends on many factors, from chemical bonding to interaction between crystalline grain domains or forces binding components of a composite material. Load forces include indentation (force applied directly) or abrasion (force applied obliquely, such as scratching or cutting). Cartilage may exhibit creep, in which deformation occurs slowly in response to stress, via slip and viscous behavior. The degree of hydration and the behavior of liquid phases within bone may affect strength. Marrow within the bone channels may behave like a non-Newtonian fluid, in which the resistance to deformation (the dynamic viscosity) depends on the rate at which a force is applied.[25]

FIGURE 15.6 3D-printed scale model of trabecular bone and reversed 3D-printed model of spaces within the bone, used as a mold to test the performance of composite polymer/nanoparticle scaffold materials (size of model = 4 cm^3).

15.4.2.1 Characterizing Materials: Mutual Contributions from Engineering and Biology

The techniques for characterization of strength in bone are based on methods developed for engineered materials such as metals and ceramics.[26,27] Biomaterials science relies on engineering techniques, but engineering owes much to biology for understanding how complex natural structures at the nano level produce strong and lightweight properties with economy of energy and materials usage. A growing area of materials science focuses on biomimetics, which is the study of natural materials and how they can be replicated to optimize artificial material properties.[28] Biomimetic approaches are especially useful in designing materials that must interface with and substitute for native tissue, as with tissue engineering scaffolding and implants.[29]

Stress and response in composite materials such as bone are more complex than for typical uniform materials. In a composite, the strength depends on the micro- and nano-level interactions between nonuniform structures in the material.[30,31] Bone is a complex composite with components composed of layers of rigid crystalline plates and deformable polymer networks, separated by interspersed water, macromolecules, and polymers. Each of these components and their interfaces affect the behavior under stress. The reaction to stress is distributed at the micro and nano level to contribute to the overall response to loads of all kinds.[32–34] It appears that the elastic modulus of bone at the bulk level is only weakly related to mineral density and bone health. Nano-level methods are needed to characterize strength-related moduli locally in order to understand the relationships between structural hierarchy and mineral crystal organization and how they affect bone resilience.[35] Nanoscale characterization supports comprehension of phenomena at the micro and bulk levels of tissue. To comprehend the micro level, one must have an understanding of mechanisms at the next lower level.

15.4.2.2 Mechanical Properties: Indentation Testing

Hardness is defined as the resistance of a material to localized mechanical deformation. Surface indentation is used to test hardness of materials. Test probes apply a measured force to a surface to form an indentation. In the classical test, hardness is estimated according to the depth of deformation after the probe is withdrawn. Methods differ in features such as test probe geometries. Methods include the Rockwell Hardness, Vickers, Knoop, and Brinell tests, for which there are defined standards; at least 12 different methods are in use.[36,37] High-resolution instrumentation allows measurement of hardness, elastic modulus, creep, residual stresses, and other properties at precise local levels.[38]

Some indentation instruments can be programmed to apply cyclic loading in order to study the transition from elastic deformation to fatigue-induced plastic deformation.[39] A related technique is dynamic mechanical analysis (DMA), in which materials are subjected to cyclic loading, measuring the response to the force cycle over time.[40] Viscoelastic properties are characterized as the stiffness modulus, viscosity, and damping, by measuring the amplitude of the deformation at the peak of the oscillating force wave and the lag between the stress and strain. Although DMA does not directly involve nanoscale instrumentation, it yields information on the nanoscale behavior of polymers and composites and complements micro- and nanoindentation. DMA has been applied in a number of studies of bone properties, where its application is still relatively new.[41] DMA is relevant to bone because it provides a unique means of measuring brittleness, which is related to the storage modulus.[42]

Stress–strain measurements and load strength have been extended down to the nano level with instrumentation for nanoindentation.[43] With high-resolution measurement of nanoscale displacements, discrete events can be studied, such as dislocation source activation, shear instability initiation, and phase transformations in very small areas, using quasi-static and dynamic force programs.[44] Testing at this scale can be useful for materials with complex and varied composite structures, provided attention is given to details such as the type and shape of probe employed. Nanoindentation with nanoscale probes and precisely controlled and measured nanoscale forces can test the local behavior of small regions of material under loading.

Recently, nanoindentation has been applied to bone, to investigate properties such as the longitudinal and transverse anisotropy of modulus and hardness in cortical bone and differences between healthy and degenerative bone tissue.[45] Nanoindentation has been combined with nanoscratch tests, and multiscale simulation models of bone nanoindentation have been developed to aid calibration and analysis.[46–48]

Most commercial indentation instruments are designed to give results for hardness and elastic modulus, using the Oliver–Pharr method to obtain these two quantities from the raw data.[49] This model has been shown to require careful interpretation or modification in order to provide accurate results for complex materials such as composites and bone.[50] The limitations and improved approaches have been examined, with special reference to bone.[51] In nanoindentation of bone, instruments and methods may differ. This applies to the type of probe used as well as the methodology applied to compute parameters.[52]

As with many other techniques that involve probing sub-micro areas of material, each measurement covers only a very small portion of sample. This can be an advantage when examining nanoscale material interactions in complex polymers, fine-grained crystalline materials, or composites. However, to relate nanoscale probe measurements to overall properties, multiple measurements must be taken in the larger scale context. Advanced instruments combine nanoindentation with atomic force microscopy (AFM) for imaging and calibration.[53] In nanoindentation and AFM, the nanoscale probe can potentially damage delicate biological samples. Instruments with direct force control minimize the risk of damage to tissue samples, as well as to the probe tip.

15.4.2.3 Mechanical Properties: Ultrasound

Bone strength can be determined through the interaction of bone with sound waves. Although the most common ultrasound imaging modalities are incapable of imaging at the nanoscale, because of limited resolution, the interaction of sound with materials can provide information on the nanoscale structure and binding forces. Ultrasound elastometry and elastography employ sound as the impulse for measurement of bulk and elastic mechanical properties. A newer technique is photoacoustic elastography, which is capable of producing information about bone and bone–tissue interfaces at the micro- and nanoscale levels, using the photoacoustic effect, by which light produces sound impulses within materials.

It is useful to introduce a few of the basic relationships and equations describing sound waves and their interaction with materials.[54,55] Sound is a form of mechanical energy, transmitted as vibrations through material media. The vibration frequency is expressed in Hertz.

$$1 \text{ Hz} = 1 \text{ vibration/s} \tag{15.5}$$

The range of sound frequencies, or acoustic spectrum, is classified into three areas: audible, infrasound, and ultrasound. Audible sound is the range perceived by most humans, defined arbitrarily as between 20 and 20,000 Hz (or 20 kHz). Infrasound is at lower frequencies, and ultrasound is at higher frequencies (>20 kHz).

The vibrations that make up sound waves consist of compressions around an equilibrium pressure in the medium. (Sound does not involve bulk transfer of matter: individual particles vibrate in place around a mean position.) Sound waves carry energy across the medium in which they travel. The energy carried by a wave is the sum of potential energy of compression plus the kinetic energy of movement. In practice, the energy delivered by a sound wave can be measured by its compression of a piezoelectric crystal in a microphone, through conversion to an electric potential.

Although sound does not involve transport of matter, sound propagates through material as waves of compression and decompression in the media. Sound waves propagate through air, water, and tissue at different velocities depending on the density and elasticity of the medium. Sound can travel in four types or modes: compression (longitudinal), shear (lateral), surface (primarily in liquids), and plate waves (in thin layers of material). In using sound to interrogate nanoscale properties of tissue, such as elasticity, we deal primarily with compression and shear waves.

The velocity of *compression sound waves* passing transversely through a fluid medium is related to the material's density and the bulk modulus:

$$c = \sqrt{\frac{K_b}{\rho}} \tag{15.6}$$

where c is the velocity of sound across the medium, ρ is the material density, and K_b is the bulk modulus (which is a measure of the volumetric elasticity, or the resistance to pressure in all directions, i.e., compressibility).

Note that by itself, increased density results in a lower velocity of sound (heavier particles move more slowly than lighter particles with the same kinetic energy). However, increased density is usually associated with much greater elasticity (due to stronger bonding between particles). Sound velocity is also dependent upon temperature and other factors. So any relationship between density and sound velocity is valid only for materials with similar elastic properties, in similar environmental conditions.

In rigid media (solids), the velocity of compression waves depends on the resistance to shear (transverse) forces as well as the compressibility. If we ignore the shear forces, as is approximated in a long thin rod of material, the velocity of a compression wave in a solid is given by an equation similar to the one for fluids, with the Young's modulus (elasticity), Y, replacing the bulk modulus:

$$c = \sqrt{\frac{Y}{\rho}} \tag{15.7}$$

However, in bulk rigid solids, a second, lower intensity wave is generated, perpendicular to the direction of transverse propagation, called the *shear wave*. The velocity of the shear wave is given as

$$s = \sqrt{\frac{G}{\rho}} \tag{15.8}$$

where s is the velocity of propagation of the shear wave, G is the *shear modulus*, or stiffness, and ρ is the material density. The shear wave arises because of strong connections between adjacent layers of particles in solids. Because of the loss of energy to shear waves, the speed of compression waves in solid volumes of material such as tissue is given by a modified form of the Equation 15.6 above for fluids, namely:

$$c_{\text{solid}} = \sqrt{\frac{K_b + \frac{4}{3}G}{\rho}} = \sqrt{\frac{E\left(1-\eta\right)}{\rho\left(1+\eta\right)\left(1-2\eta\right)}} \tag{15.9}$$

where η is the Poisson's ratio, defined as the ratio of transverse expansion (strain) to the longitudinal expansion (strain) in response to an applied force (see Equation 15.3).

In gases and liquids, no shear wave is generated, and in soft tissues, the shear wave is of low intensity. In bone, and some denser soft tissues, shear waves are significant, and their use in characterizing bone strength will be discussed later.

As shown in the above equations, since sound involves the compression of matter, its propagation depends on the elasticity and density of the medium. In general, elasticity effects outweigh density, so sound travels more rapidly in high-density elastic solids and less rapidly in lower density deformable tissues and compressible gases. Vacuum acts as a perfect sound insulator, since no matter is available to propagate mechanical motion.

Air gaps between ultrasonic transducers and tissue attenuate sound waves, mainly through reflection. In practice, specially formulated gels are used to fill the gap and ensure connectivity for sound conductance between transducers and materials for ultrasonic evaluation.

Acoustic intensity is defined as the power carried by sound waves per unit area in a direction perpendicular to that area. Acoustic intensity is important in the measurement of sound energy transfer and attenuation. Intensity, I, as power per area, is a rate of energy transfer, usually expressed in watts per area, or joules per second per meter squared:

$$I = w/m^2 \tag{15.10}$$

Relative sound intensities are expressed on the decibel logarithmic scale, in decibels (dB), which is convenient for comparing attenuated or reflected sound intensity to a source.

The resistance of a medium to the propagation of sound is called the *acoustic impedance* and is a characteristic property of a material. The acoustic impedance of a material is defined as

$$Z_m = \rho \times c \tag{15.11}$$

where c is the velocity of sound in the medium, ρ is the material density, and Z_m is the acoustic impedance of the material. The characteristic acoustic impedance of a material reflects the extent to which particles are bound in place, resisting mechanical motion. Z is related to elasticity or stiffness and other factors.

Depending upon how tightly bound particles are to each other and how many complex internal vibrational modes exist within the particles, energy can be transferred across the medium with minimal loss, conserving the direction and momentum of the vibration that originates the sound waves. However, multiple mechanisms create reflection, refraction, scattering, absorption, and dispersion of sound in matter.

Sound is reflected at interfaces between materials of different acoustic impedance (termed the *acoustic mismatch*). In general, reflection is partial, with a portion of the incident energy being reflected and the remainder transmitted into the medium. The greater the difference in acoustic impedance at a boundary, the greater is the fraction reflected. Most biological soft tissues are sufficiently similar that incident sound loses only a small portion of its energy as it passes through a body. The reflected portions at each interface can be detected and used to construct an image, while portions of the beam continue with small enough loss of energy to allow significant penetration into the body.

Reflection of sound can be specular or diffuse. Interface surfaces with a diameter larger than one wavelength of the incident sound reflect sound at the same angle as the angle of incidence. Specular reflections are mirror-like, preserving the image information; smooth surfaces provide areas of reflection that are large relative to incident wavelengths. Diffuse reflection occurs from rough surfaces, with small reflecting surfaces relative to incident wavelength. In diffuse reflection, sound is redirected in multiple directions, depending upon the exact geometry of the boundary and structures within the interface. Sound waves can be treated like light waves for some aspects of reflection. An ideal model of diffuse reflection of light is Lambertian reflectance, in which light is reflected with equal luminance or radiance in all opposing directions.

Specular reflection at small angles produces reflected echos, which can be detected at the source of a transmitted sound beam. The time lapse before detection of an echo gives depth information, and directional scanning can be performed to generate image information. If the angle of incidence is 90° (normal incidence), the energy of the beam is reflected entirely back in the opposite direction toward the source. In this special case of specular reflection, the ratio of the intensity of the incident and reflected sound is:

$$\frac{I_r}{I_i} = \frac{(Z_1 - Z_2)^2}{(Z_1 + Z_2)^2} \tag{15.12}$$

where I_r and I_i are the reflected and incident sound intensities, respectively, Z_1 is the acoustic impedance of the incident media, and Z_2 is the acoustic impedance of the reflecting media.

Scattering occurs because the binding of particles within matter is not absolute, and space can exist between individual molecules in a polymer or ions within a crystal lattice. Entropic losses can occur in the collisions that transfer energy, resulting in scattering of the direction of sound waves as well as energy loss. The degree of loss of ultrasound is one characteristic indicative of the structure of materials at the nanoscale. Scattering increases rapidly with increasing frequency, resulting in attenuation of sound at the higher ultrasound frequencies. Although potential image resolution increases with increased frequency, scattering limits the penetration of ultrasound in tissue at frequencies above around 15 MHz.

Refraction occurs when the angle of incidence at an acoustic boundary is nonzero, and the transmitted portion of a sound beam deviates from the direction of incidence. The change in direction, or refraction, is either toward or away from the normal, depending on the relative velocities of sound in the two media. Snell's law describes the relation between the angles of incidence, *I*, and angle of refraction, *R*, in both sound and optics:

$$\text{Snell's law} \equiv \frac{\sin I}{\sin R} = \frac{c_i}{c_r} \tag{15.13}$$

where c_i is the speed of sound in the incident medium and c_r is the speed in the refracting medium. Refraction can interfere with ultrasound imaging though loss of beam intensity and production of artifacts.

15.4.2.4 Sonography Overview: Modes, Resolution, and Applications

The detection and analysis of reflected ultrasound waves are the basis for ultrasound medical imaging, or sonography.[56] The most common ultrasound imaging modalities are incapable of measurements at the nanoscale, because of the limited resolution attainable through imaging with straightforward reflected and absorbed sound. In addition, the frequencies most useful for imaging of soft tissues do not penetrate bone.

However, the techniques of elastography and photoacoustics are applicable to bone and can yield information about material properties that characterize nanoscale structure; imaging using these modalities can relate measured properties in localized areas. These modes make the advantages of sound imaging available for bone. The advantages include low cost of instrumentation, penetration depth into tissues, lack of harm from ionizing radiation, few interferences, fast data acquisition, and good distinction of tissue types, among others.

The simplest diagnostic ultrasound mode uses a single transducer to measure the transversal and reflection of sound along a one-dimensional directional axis. This mode, called the A-mode, is analogous to depth-finding sonar. A more complex mode, the B-mode, uses a linear array of transducers to profile a two-dimensional section of tissue for each measurement of transmitted and reflected sound waves with the array (analogous to imaging of geological strata by seismic imaging). Multiple measurements are then used to construct a two-dimensional image. Computation produces three-dimensional images and can produce four-dimensional real-time interaction with imaged tissues.[57]

A special class of ultrasound imaging uses the Doppler effect for highly accurate measurement of motion, such as the flow of blood cells in arteries and the heart. In clinical and instrumental practice, this is referred to as M-mode Doppler ultrasound.

Sound waves in the ultrasound range (frequencies above 20 kHz, above the upper limit of most human hearing) are only partially reflected and weakly absorbed by soft tissues. Thus they propagate through layers of tissue, losing energy as they pass from layer to layer. The detection of a series of reflections from deeper and deeper layers is the basis for the simpler modes of ultrasound imaging. For highest performance in ultrasound imaging, there is a trade-off between depth of penetration and strength of reflected signal. Lower frequency sound is absorbed more weakly and penetrates to deeper tissues but has less resolution than the more scattered and absorbed higher frequencies.

The lateral and depth resolution of ultrasound imaging is limited by the wavelengths used. As discussed earlier, the wavelengths of sound in air audible to human ears range from roughly 17 m to 17 mm, corresponding to frequencies from 20 Hz to 20 kHz. Ultrasound wavelengths are shorter than 17 mm, corresponding to 20 kHz and higher. Frequencies approaching the gigahertz ranges become increasingly difficult to produce and control and are strongly absorbed and scattered, limiting propagation. Thus, the practical lateral resolution of conventional ultrasound imaging is limited to a few millimeters, far above the nanometer range for cellular and molecular structures.

Techniques are available using signal processing and modulation to increase the effective resolution of reflective ultrasound imaging. These more complex modes can increase effective resolution by signal processing and synthetic focusing arrays, down to the range of a few micrometers.[58]

Most ultrasound frequencies are almost entirely reflected by hard bony tissues, so ultrasound is not useful for analysis and diagnosis of their internal structure. Higher frequency sound, which is capable of higher imaging resolution, is the most readily reflected and scattered by the dense, hard, and cortical bone. Thus, conventional medical ultrasound frequencies are not useful for imaging the internal microstructure of bone, as most trabecular bone is shielded behind a strong cortical shell. Despite this limitation for imaging, ultrasound propagation data can provide useful information on the internal strength of tissue and bone.

15.4.2.5 Physical Interaction of Sound with Tissue

The interaction of sound waves with tissue is essentially one of mechanical force and obeys the same laws of physics as any other applied force. The effect of microstructure and nanostructure on the propagation of sound through the tissues can be related to internal strength and structure on the micro- and nanoscales, as averaged over selected volumes. The impact of sound pressure waves on bone can be considered in manner analogous to the application of mechanical force from physical probes.

The details of the interaction are complicated by the oscillating nature of sound waves and the multiple modes for their propagation, reflection, and absorption by materials. The greater complexity of the interactions with sound waves compared with a material probe gives opportunities for gathering more information about the material, including reconstruction of images from reflected waves, and determining velocities of particles in motion by means of the Doppler effect.[59]

15.4.2.6 Interactions of Sound with Bone: Micro- and Nanostructural Mechanical Effects

The interactions of sound with bone that give information at the micro- and nanoscales are generally not those most useful for morphological imaging. Some modes of sound interaction can be configured to measure basic properties that depend on internal structure at the lowest levels, such as hardness, elasticity, and toughness. In these modes, the energy of sound waves is used to probe structure in a manner analogous to use of a physical probe in nanoindentation.[60]

Sound interrogation of mechanical properties can measure longitudinal compression waves or lateral shear waves, which reflect the bulk compressibility or elasticity, respectively. In general, measurements of bulk compressibility give information on short range molecular interactions, as determined mainly by tissue molecular composition. Shear elasticity, on the other hand, is influenced by the structural properties of tissue at the nano, cellular, and higher levels of architecture.

Sound propagation, penetration, and reflection differ in cortical bone, trabecular bone, and cartilage. The propagation velocity depends on the density and the bonding strength of the components of the material. The homogeneity of the tissue will affect the degree of reflection, which is greatest at phase and domain boundaries. Microstructures, nanostructures, and molecular bonding may cause resonance absorption of sound energy.

Lamellar cortical bone is the strongest and hardest form of bone and will thus have the highest velocity and the greatest reflection at its onset boundary. Immature woven cortical bone has less mineral and more collagen, and its softer structure absorbs more sound energy. Trabecular bone is less hard and filled with matrix fluid; its interaction with sound is more like that of soft tissues, dominated by the liquid phase. Cartilage, making up collagen-rich tendons and ligaments, is similar in sound interaction to soft tissues. Sound propagation in bone is typically anisotropic, varying with the direction across inhomogeneous micro- and nanostructures. This and other complicating factors can be used to obtain information on structure.

15.4.2.7 Mechanical Characterization of Bone with Ultrasound

Since its inception in the early 1990s, the application of ultrasound to measure elastic properties of bone has used both single-axis values (quantitative ultrasound [QUS]) and imaging of strain and elasticity maps of tissue (elastography). With both approaches, techniques and instruments have been improved, and QUS and elastography have steadily gained in clinical use.[60,61]

15.4.2.8 Ultrasound Elastometry: Estimating Elasticity from Speed of Sound in Bone

The propagation of sound through a material can probe properties that characterize the structure and dynamics of the system on the micro and nano levels. Variation of ultrasound reflection data can be analyzed with models and assumptions to yield information on material properties.[61] Strains produced by sound waves and the effect of material on velocity of sound can indicate the internal strengths of the material. In typical engineering materials for which strength data have been studied, no simple relationship has been found between strength and elasticity, because strength is a complex property determined by many factors. In contrast, bone elastic modulus and ultimate strength have been shown to have a positive relationship.[62]

Interpretation of relationships between the complex microstructure of bone and ultimate strength involves complex models and assumptions. In practice, the bulk modulus is often measured to obtain an estimate of bone strength. The bulk modulus is related to the velocity of sound in bone, as discussed in the earlier section on ultrasound theory (see Equation 15.6). The velocity of ultrasound can be measured directly by timing of transmission of pulses or echoes, indirectly by interference patterns, or indirectly from the critical angle of reflection.[63]

15.4.2.9 Elastometry by Critical Angle Measurement

Critical angle elastometry is based on the refraction and reflection at the inhomogeneous boundary between bone and soft tissue (or, in vitro, bone and water or air). When a wave passes from a medium in which the speed of propagation is slower into a medium with a higher propagation speed, there is a critical angle above which total reflection will occur at the interface. Since sound travels faster in bone than in soft tissue, the critical angle phenomenon will occur for sound at the tissue–bone interface (or, in vitro, at a bone–water or bone–air interface).

The value of the critical angle depends on the relative velocities of sound in the media on either side of the interface. Thus, determination of the critical angle of sound reflection can be used to calculate the velocity of the sound in bone. The velocity of sound in a material is related to the critical angle by Snell's law:

$$V_p = \frac{c}{\sin \phi_c} \tag{15.14}$$

where V_p is the velocity of a sound compression wave in bone, c is the velocity of sound in the adjacent medium, and ϕ_c is the critical angle of reflection of sound at the interface. The velocity of sound in water and in soft tissue is known from laboratory data, so the velocity in bone can be determined by measurement of the critical angle. This method, termed *critical angle reflectometry*, has been verified experimentally by comparison with other means of sound velocity measurement.[64]

15.4.2.10 Elastometry by QUS

QUS is the term generally used for clinical measurement of ultrasound velocity and attenuation in bone. The advantage of QUS is to provide an inexpensive, portable, and easily used point-of-care modality.In the versions that have found widespread in vivo clinical use, an ultrasound pulse or beam is transmitted into in vivo anatomically accessible bone. The sound speed and attenuation are measured by either pulse reflection or transmission mode. Pulse reflection (or pulse echo) mode uses a single transducer to emit and receive the pulses and reflections. Transmission mode instruments use two ultrasound transducers, a separate transmitter and receiver.[65]

The compressive strength of a material is proportional to its bulk modulus. As described in the previous discussion of ultrasound theory, the speed of sound is proportional to the square root of the elastic modulus divided by the density of the material. This relation holds under ideal conditions, where the wavelength is large relative to the cross-sectional area of the material being measured, where the material is isotropic (the biomechanical properties are invariant with spatial direction) and where the material is not dispersive. None of these conditions apply strictly to bone, which is in general a heterogeneous, anisotropic, and dispersive composite material. Models and assumptions are used with QUS to give measures of strength estimated from sound velocity measurements.[66]

Attenuation of ultrasound intensity takes place through two primary mechanisms: scattering and absorption. Scattering redistributes the energy of a propagating wave, depending on the wavelength and the specific acoustic properties of the material. Scattering includes reflection, or backscattering, which occurs at interfaces in inhomogeneous tissues. Absorption converts a portion of the wave energy into heat. Absorption is a complex process that depends upon the molecular and nanoscale structure of the material. Relaxation mechanisms occur when density fluctuations are out of phase with pressure fluctuations as sound traverses a material, leading to wave cancellations and conversion of organized wave oscillations into random heat. Absorption increases with wave frequency. Bone can be characterized by the dependence of absorption on frequency, but the most common and simpler measurements are for broadband attenuation, averaging attenuation over a range of frequencies. But unlike the case for velocity, no theoretical relationship has been found between attenuation and elastic modulus, so specific attenuation coefficients are empirical.[66]

15.4.2.11 Ultrasound Elastography: Mapping and Imaging Elasticity of Tissue

Sonography gives a means of displaying internal organs, but imaging with ultrasound compression waves does not provide information on elastic properties and does not inform on the internal structure of bone. The bulk modulus, upon which the propagation of ultrasound in the body depends, is similar for most soft tissues and is not directly related to elasticity. A class of alternative techniques, termed elastography, has been developed employing ultrasound to map and visualize tissue elasticity.[67,68] Elastography techniques are mostly applicable to soft tissue rather than bone, although some can be used for tendons and cartilage. Recently, some newer modes of elastography have been adapted to provide elasticity and stiffness information for bone.[60]

15.4.2.12 Physical Principles of Elastography

The basic physics underlying elastography methods is the same for both soft tissue and bone. In all elastography techniques, an external force is applied to tissue and the resulting displacements are measured and mapped. The applied force can be either static (or quasi-static) or dynamic (impulse or cyclical). For harder, less compressible tissues, static measurement is less useful; impulsive or vibrational forces such as ultrasound give more information for bone stiffness.

In soft tissues with their high water content, deformations produced by sound propagation are predominately fluid in character, and thus related to the bulk modulus (see Equation 15.6). The bulk modulus is almost uniform for most healthy soft tissues. Differences in intrinsic bulk moduli can indicate differences between normal and diseased tissues. For example, tumors and fibrotic tissues are generally more resistant to deformation than normal tissue. Measurements of bulk modulus, as indicated by sound velocity, can have diagnostic value in the prostate, cervix, breast, liver, and other tissues. In practice, however, differences may be too small to be evident in a sonogram. Disease may cause uniform hardening of an entire organ, making diagnosis difficult.

The more rigid the tissue, the more sound propagation will reflect the elasticity, or Young's modulus (see Equation 15.7). But in soft tissue, the contribution of the Young's modulus to ultrasound velocity cannot be easily separated from the effect of the bulk modulus.

In rigid complex composites such as bone, the ultrasound velocity reflects mainly the elasticity, but even in bone, there are some contributions from the bulk modulus, due to semi-fluid behavior of polymer layers and the fluid matrix content in pores.

Additional information can be gained by measurement of shear waves in rigid materials. In solids, mechanical forces such as ultrasound generate shear waves, whose propagation velocity is related to elasticity and independent of the bulk modulus. For this reason, displacement produced by the shear wave can provide better discrimination of biomaterials stiffness than the compression or axial wave.

In a solid, a dynamic force will produce both compressional and shear waves. Compression waves travel rapidly, altering volume (and density) across the material in the direction of the driving force. Shear waves travel perpendicular to the compression waves, at lower speeds and amplitudes. Compression waves used in ultrasound imaging have frequencies around 10 MHz, traveling in tissue at speeds near 1500 m/s. Shear waves propagate more slowly, at frequencies lower than 2 kHz; higher frequencies are absorbed rapidly and dissipate as heat.

Shear wave velocity is proportional to the square root of the shear modulus (see Equation 15.8). The elasticity, Young's modulus, is related to the shear modulus as

$$Y = 2G\left(1 + \eta\right) \tag{15.15}$$

where Y is Young's modulus, G is the shear modulus, and η is Poisson's ratio for the material (see Equation 15.3). For water and other liquids, Poisson's ratio approaches 0.5, which is the limit of perfect incompressibility. Since biological tissues are virtually incompressible, the Poisson's ratio is usually assumed to be close to 0.5 for the purposes of approximating the Young's modulus from the shear wave velocity. This assumption weakens the rigor of evaluations of tissue by many ultrasound elasticity methods; measured estimates for Poisson's ratio in various soft tissues are typically between 0.3 and 0.4.

In heterogeneous tissue such as bone, the characteristic mechanical properties depend heavily on the sample. Reported values of Poisson's ratio measurements for cortical bone are between 0.2 and 0.5, and for cancellous bone between 0.01 and 0.35. In practice, the ratio for bone is often assumed to approximate between a multiple of two to three in diagnostic elastography.[69,70]

This assumption would be valid only for near perfectly incompressible, isotropic solid materials. But for most tissues and tumors, measured values of the elastic modulus vary over many orders of magnitude. Neither the elastic modulus nor shear modulus taken alone can give the full picture of tissue stiffness for soft tissues or bone. More research is needed to examine the effects and validity of assumptions and instrumental techniques.[70]

15.4.2.13 Photoacoustic Imaging and Measurement

Photoacoustics is an emerging technique that combines optics with ultrasound in a direct way by means of generation of sound waves from the absorption of light, or any electromagnetic radiation, such as microwaves, in materials.[71–74] Because the imaging is based on ultrasound created by absorbed photons, photoacoustic imaging is not limited by the scattering of light, which defines the optical diffusion limit, thus attaining depth without sacrifice of resolution. Because sound scatters several orders of magnitude less than light in biological tissue, photoacoustics is capable of greater imaging depth and greater spatial resolution than optical imaging based on reflected light or fluorescence. Photoacoustics can also determine the elasticity distribution of tissues by measuring stress and strain simultaneously.

15.4.2.14 The Photoacoustic Effect: Producing Sound with Light

The photoacoustic effect has been known for more than a century, in which absorption of variable pulses of light produces sound waves within a material. Although some earlier inventions were based on the phenomenon, its full potential was not realized until the development of nanopulsed lasers, high-performance ultrasonic transducers, and fast computers. It is currently emerging as an important tool for biomedical imaging and biomechanical characterization. The photoacoustic effect is one of a broader class of interactions between electromagnetic radiation and materials involving the production of acoustic waves by the interaction with matter by light, X-rays, magnetic impulses, and other electromagnetic fields. To avoid confusion with the acousto-optic effect, the term *photoacoustic* has replaced the original designation *optoacoustic* in most areas. When radiofrequency or microwave wavelength radiation is used instead of light for generation of acoustical waves, the term *thermoacoustics* is often used.

The photoacoustic effect overcomes some of the fundamental limitations of both ultrasound and optical interrogation, in a hybrid combination of both methods. Ultrasound resolution increases with frequency but so does sound wave scattering, with loss of penetration, especially for dense tissue such as bone. Optical methods such as confocal microscopy, two-photon microscopy, and optical coherence tomography can provide high resolution and specificity but lack focusing ability beyond ~1 mm depth into biological tissues because of photon scattering. Laser light at selected wavelengths can penetrate several centimeters into tissue, and ultrasound can achieve axial and lateral resolutions of a few millimeters or less, using reconstruction signal processing algorithms.

The basic physics of the photoelectric effect are described by the photoacoustic wave equation in the temporal domain:

$$\left[\nabla^2 \frac{1}{c^2}\frac{\delta^2}{\delta t^2}\right] p\left(\vec{r},t\right) = -\frac{\beta}{C_p}\frac{\delta}{\delta t}\left[\Phi\left(\vec{r}\right) I\left(t\right)\right] \tag{15.16}$$

where c is the velocity of acoustic wave propagation in the material, β is the thermal expansion coefficient, C_p is the specific heat at constant pressure, $I(t)$ is the temporal intensity profile of the laser pulse, $p\left(\vec{r},t\right)$ is the generated photoacoustic wave pressure at position $\vec{r}$ and time t, and $\Phi\left(\vec{r}\right)$ is the optical absorption energy (which is the product of the optical absorption coefficient of the material and the optical fluence = the optical energy delivered by the laser per unit area incident to the beam direction).[73] Thus, the photoacoustic ultrasound intensity increases with the intensity of the laser pulse and depends on the optical absorption of components in the material.

15.4.2.15 Photoacoustic Interactions of Light in Tissue

Photoacoustics for tissue typically uses a tunable dye laser with a wavelength range of 680 to 2500 nm, for which tissue penetration depths up to ~8 cm are possible. When a laser with a pulse width of a few nanoseconds is absorbed in tissue, a wideband ultrasound wave can be induced by thermoelastic expansion and relaxation. For efficient photoacoustic sound generation, the excitation pulses should be significantly shorter than the thermal and stress relaxation times, which in soft tissues are typically 1.7 ms and 10 ns, respectively.

When the generated photoacoustic wave pattern is acquired by an ultrasound imaging system, a map of the deposited optical energy can be computed. The degree and specificity of light absorption can be varied with dyes and markers to increase the information gained by the image mapping. Transparency or absorption varies with excitation wavelength and tissue composition and can be manipulated with optical agents and/or nanoparticles, giving flexibility in imaging targets and enabling imaging with structural and molecular selectivity. For example, lipids and collagen contents can be selectively mapped in tissue by utilizing laser excitation frequencies that correspond to their characteristic infrared absorption bands.[75]

15.4.2.16 Photoacoustic Modalities for Characterization of Bone and Cartilage

Photoacoustics can be applied to tissue characterization in the form of photoacoustic microscopy, photoacoustic tomography, photoacoustic elastometry, or photoacoustic elastography. Figure 15.7 shows the general approach used for all modalities of photoacoustic imaging and analysis in tissues. In photoacoustic microscopy, a focused ultrasonic transducer scans to obtain a two-dimensional image from photoacoustic ultrasound waves. Because the photoacoustic ultrasound can be produced by different molecules at different light absorption wavelengths, photoacoustic microscopy can be used to image anatomical, functional, molecular, flow dynamic, and metabolic contrasts in vivo.[72,73,76,77]

Photoacoustic tomography produces two- and three-dimensional imagery of internal tissue structures from the acquisition and mathematical analysis of photoacoustic ultrasound waves. This imagery can produce information about soft tissues, cartilage, and bone–tissue interfaces at the micro- and nanoscale levels. When photoacoustic tomography is used in conjunction with photoacoustic microscopy, the combined system is capable of micron-scale resolution using chromophore biomarkers.

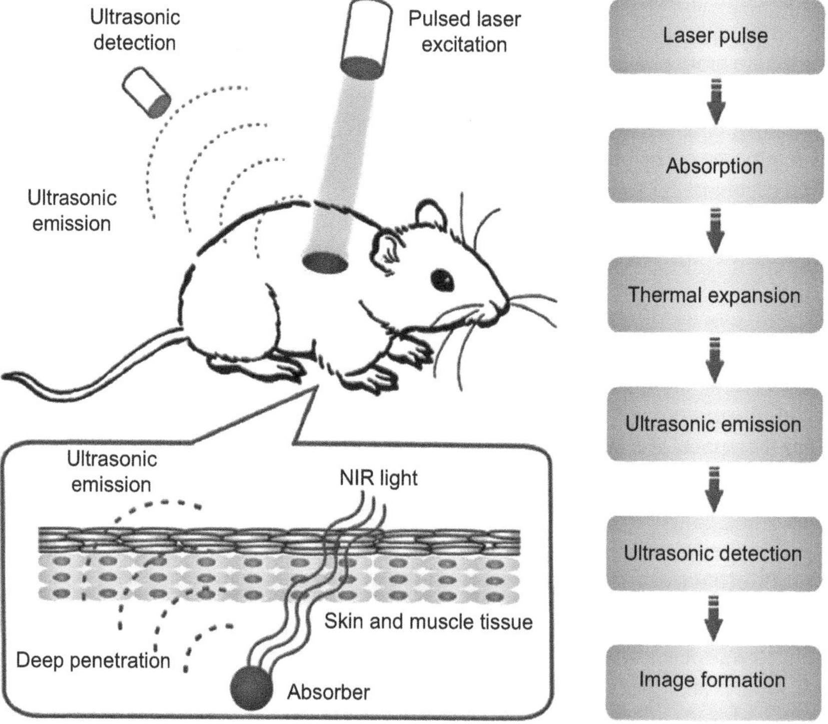

FIGURE 15.7 Schematic of photoacoustic imaging system. (1) Non-ionizing laser pulses are delivered into biological tissues; (2) the delivered energy is absorbed by endogenous or exogenous contrast agents; (3) the absorbed energy is converted into heat, leading to transient thermal expansion; (4) the thermal expansion causes the generation of ultrasonic emission; (5) the generated ultrasonic emission is detected by ultrasonic transducers; and (6) then analyzed for image formation, spectral analysis, or biomechanical properties characterization. (Reproduced with permission from Wang S, Lin J, Wang T, Chen X, Huang P. Recent Advances in Photoacoustic Imaging for Deep-Tissue Biomedical Applications. *Theranostics* 2016;6(13):2394–2413. doi:10.7150/thno.16715. Available from http://www.thno.org/v06p2394.htm)

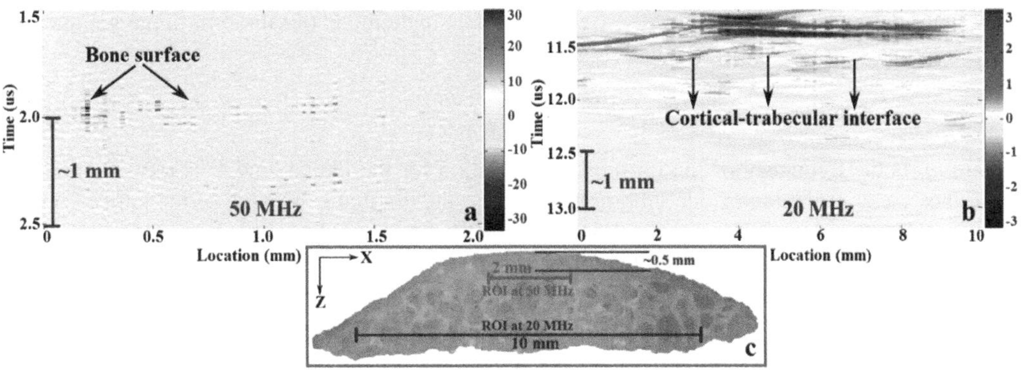

FIGURE 15.8 Cortical and trabecular interface structures resolved by photoacoustic microscopy. (a) and (b) Photoacoustic microscopy images obtained at 50 and 20 MHz, respectively, for the goat rib sample in (c). All the images share the same coordinate system shown in (c). (Reproduced with permission from: Kaiplavil, S., *et al.*, *Biomed. Opt. Express*, 5, 2488–502, 2014.)

15.4.2.17 Imaging Cortical and Trabecular Structure with Photoacoustic Microscopy

The computational process for producing images in photoacoustic microscopy is much less complex than with photoacoustic tomography. Since microscopy does not rely on mathematical reconstruction of three-dimensional images, it is less affected by multiple reflections and severe attenuation of photoacoustic ultrasound waves, such as those produced in trabecular bone, cortical-trabecular, and matrix-mineral boundaries.

Photoacoustic microscopy has been shown capable of producing images of trabecular structure and clearly resolving the cortical–trabecular interface in one study using ex vivo goat rib with some tissue overlay. Optoacoustic excitation was produced by a pulsed optical parametric oscillator laser with a wavelength of 690 nm, 8-ns pulse width, and 10-Hz pulse repetition frequency. The microscopic images were generated by raster-scanning an ultrasound transducer over the tissue sample with a step width of 0.02 mm, averaging readings over 10 excitations to produce each data point. It was demonstrated that detection of photoacoustic-generated ultrasound at 50 MHz produced poor image resolution. But using the lower frequency, less strongly scattered frequency of 20 MHz, the cortical–trabecular interface was resolved, as well as details of the matrix–trabecular interfaces (Figure 15.8). The images produced by this method were inferior in resolution to other methods studied (e.g., photothermal tomography), but the technique was used to confirm their accuracy.[78]

15.4.2.18 Thermal Photoacoustic Measurement

Another recent advance is the characterization of biomechanical properties of bone by measuring the change in photoacoustic ultrasound intensity with temperature variation. Thermal photoacoustic (TPA) measurement is an innovative modality that can give information on internal chemical and molecular properties of materials by measuring the change in photoacoustic signal intensity with temperature. The mechanism for production of photoacoustic ultrasound depends on the internal micro- and nanoscale state of bone. TPA, therefore, has high potential for providing information of relevance to bone health.

The dependence of the intensity of the photoacoustic signal on temperature is related to the Grüneisen parameter, λ, which is an dimensionless index of the effect that a change in volume of a crystal lattice has on its vibrational properties.[79] Changes in temperature and pressure affect the size and dynamics of molecular scale structures; the original microscopic formulation of the Grüneisen parameter described the relation of the frequency of a single mode of vibration to the volume of a crystal lattice:

$$\lambda_i = -\frac{\delta \ln \omega_i}{\delta \ln V} \tag{15.17}$$

where ω_i is the frequency of the ith mode of vibration for the lattice, and V is the volume. This describes the contribution of one vibrational mode to the thermal expansion of the lattice with increasing vibrational energy, corresponding to increasing frequency. The macroscopic Grüneisen parameter, Γ, applied to a material is a thermodynamic property that represents a weighted average of all vibrational modes, giving a measure of the many effects of temperature on crystal structure. Because of the equivalences between many basic thermodynamic properties, the Grüneisen parameter may be expressed in many forms; it is related to the heat capacity, thermal expansion coefficient, bulk modulus, speed of sound, and density of a material. For example:

$$\Gamma = \frac{\beta c^2}{C_p} \tag{15.18}$$

where β is the thermal coefficient of volume expansion, c is the speed of sound in the material, and C_p is the heat capacity at constant pressure.

Since c, β, and C_p are all tissue specific, the Grüneisen parameter and its change as a function of the temperature are characteristic for a material. The changes of $\Gamma$with temperature can distinguish different types of tissue and indicate changes in tissue conditions and status. Properties such as density and elasticity can be estimated from the measured effects of temperature on photoacoustic ultrasound intensity. The temperature-dependent slope of photoacoustic ultrasound signal intensity (the TPA measurement) can potentially serve as an index parameter for medical diagnosis and tissue characterization.

TPA has been tested to assess its feasibility for diagnosis of bone health. Unlike conventional photoacoustic measurements, TPA is independent of the absolute value of ultrasound intensity, relying on the relative change with temperature. Thus, TPA is less sensitive to sampling and measurement variances. Photoacoustic signal intensity increases at different rates for different tissue components. For example, the signal temperature dependence of blood and organic components is greater than for mineral-rich components. Thus, TPA can evaluate the BMD for comparison of changes with health. In one study using well-established rat models of bone mineral loss, the slopes of TPA plots correlated well with bone mineral indices, demonstrating the potential of the technique for diagnosis and monitoring of osteoporosis.[80]

15.4.2.19 Photoacoustic Measurement with Light

A drawback of ultrasound measurement with solid-state transponders is that the probes must be in good contact with the material to be imaged or measured. For conventional ultrasound and photoacoustics, this involves placement of the sensing transducer directly to tissue, usually with the application of a gel to eliminate air gaps and ensure good conductivity. Alternatively, the limb or body part to be measured is immersed in water, as is commonly the practice for elastometry using QUS of the calcaneus.

Photoacoustics makes an all-photonics mode possible, eliminating the need for contact with a transducer. Ultrasound generated by light can be detected as vibrations on the surface of material by laser interferometry. This approach allows stand-off, contactless application of photoacoustic imaging and elastometry, opening new possibilities for diagnostics. The optical system can be implemented with a small profile detector, making it feasible to incorporate all-optical photoacoustic microscopy into endoscopic probes.[81] If realized in a cost-effective form that is efficacious for in vivo tissue and bone imaging, it would have significant impact on medical diagnostics and monitoring.

15.4.3 Electrical Properties

Bone is a composite, porous, material, with fluid and solid phases and crystalline and polymer components, which contribute singly and interactively to its overall electromagnetic and electromechanical characteristics. Together, they make bone a dielectric material that exhibits intrinsic electrical and electromechanical properties, including piezoelectric effects, electrostriction, and flexoelectricity. Macro-scale measurements represent an average of all of the individual electroactive components and their interactions. The interrelationships between electrical and dielectric behavior and the underlying structural and mechanical properties are complex but may have potential as indicators of the state of nanostructures that underly bone status and health.

As a dielectric material, bone is influenced by electric fields, both static and oscillatory. For electromagnetic radiation with frequencies higher than the microwave range, the predominant interactions are at the molecular level. Lower frequency interactions will be with dielectric nano- and microscale structural components with permanent or induced dipole moments. Low-frequency electromagnetic fields will also create polarizing effects through ionic redistributions of electrolytes in matrix fluids such as blood and marrow.

15.4.3.1 Significance of Electrical Measurements of Osseous Tissue

The most precise and accurate measurements of the dielectric properties of bone can be done with samples of tissue in the laboratory, and these are significant for scientific and biomedical research. In addition, many technologies for external measurements on the body can produce results that are meaningful in a clinical context. For medical use, it is sufficient that there is a measurable contrast in properties of different tissue types and conditions and that distinctive ranges can be identified for these properties. External electromagnetic measurement technologies are low cost and minimally invasive, so they have been the focus of many research efforts. Averaged external electrical characteristics are explored because they offer the potential for a single, simple indicator of bone condition, which can be easily accessed.

The techniques and methods for electrical measurement of bone are directed toward the basic electrical properties, as well as use of those properties for spectroscopic analysis, mapping, and imaging. Measurement modalities include the following:

1. Conductance/resistance (for direct current)
2. Capacitance/reactance (static field)
3. Thermally stimulated depolarized current (for stored electrical charge)
4. Impedance (for alternating currents ranging from 1 Hz to 300 GHz [upper limit of microwaves])
5. Impedance spectroscopy
6. Microwave tomography
7. Electromechanical coupling (piezoelectricity, electrostriction, and flexoelectricity)
8. Electrochemical biosensors

Within each of these categories, a number of specific techniques have been developed and evaluated.

15.4.3.2 Dielectric Properties of Bone and Cartilage

Materials can be classified as conductors, semiconductors, and insulators, that is, dielectric materials. Bone and biological tissues in general are dielectric materials. The presence of an electric field induces dielectric polarization in the tissue, caused by displacements of charges, or polarization. In physical terms, polarization results from the reorientation of particles or molecules with electric dipole moments to align with the applied field. The applied field may also produce electronic charge redistributions within membranes, fluids, or molecules, producing an induced dipole.

To the extent that these electronic displacements are elastic, electric charge can be stored in the material and recovered by relaxation or discharge when the applied field is removed or reversed. Inelastic or dissipative losses of electrical charge may occur through a number of mechanisms analogous to friction and/or internal work done irreversibly in the material. The energy of the electrical field is not lost, but converted into heat or chemical and/or mechanical potential energy in the material. In addition to heat, in some cases, the energy may be converted and dissipated in the form of sound waves or radiation. The more complex and composite the structure of the material, the more varied the mechanisms for interaction with the electric fields become.

Since biological tissues such as bone exhibit complex interactions with electromagnetic fields, the theoretical equations describing electromagnetic behavior in simpler materials will give at best a first-order approximation. However, the basic physical models are useful for interpreting effects and characterizing tissues.

15.4.3.3 Basics of Dielectric Properties of Materials

A brief overview of the physics and electronics of dielectric materials will establish some definitions for standard measurement quantities and relationships.[82]

Polarization is defined theoretically in electromagnetic terms by the decrease in the electric field strength caused by the material. This is expressed by the equation:

$$E_{\text{effective}} = E_{\text{applied}} - E_{\text{polarization}} \tag{15.19}$$

where E is the vector of electric field strength. The electric field strength is related to the electric potential, V, which is a scalar quantity representing the amount of work required to move an electric charge upon which the field exerts electromotive force. The relation of the vector field E to the scalar potential is:

$$E = -\left(\frac{\mathrm{d}V}{\mathrm{d}x}, \frac{\mathrm{d}V}{\mathrm{d}y}, \frac{\mathrm{d}V}{\mathrm{d}z}\right) \tag{15.20}$$

Note that the vector nature of E in Equation 15.20 implies that polarizability in a material may be anisotropic: polarization under the influence of an electric potential may be different depending upon the axis along which it is measured.

For simplicity, the basic description of electrical properties is usually given first in terms of scalar, one-dimensional relationships for constant, nonvarying potentials, which can be generalized to vector descriptions for variable electric fields.

The electrical characterization of materials relies on the important interrelated material characteristics: permittivity, capacitance, and conductance.

15.4.3.4 Permittivity

The interaction of a material with an electric field is defined by the permittivity. The permittivity is the effect of a material on electric field strength, as determined by the force exerted between two charges in the material (the Coulomb force), relative to the field strength in a vacuum.

A vacuum, or free space, has a perfect permittivity: electromagnetic fields propagate freely with no losses, and a vacuum cannot absorb or store electromagnetic energy. By definition, the relative permittivity of a vacuum has the value of 1. All material substances have relative permittivity greater than 1, reflecting the degree to which electric fields affect the internal electrical charge distributions within the materials.

The interaction of dynamic electric fields with material is defined by the complex permittivity, which represents the behavior of material in absorbing energy from changing electric fields. The complex permittivity is a combination of a term measuring the ability of the material to store energy by charge separation and a term that represents the loss of energy from the field through dissipation in the material.

The complex permittivity is expressed by the equation:

$$\varepsilon(\omega) = \varepsilon' - i\varepsilon''(\omega) = \varepsilon' - j\frac{\sigma(\omega)}{\omega\varepsilon_0} \tag{15.21}$$

where $\varepsilon(\omega)$ is the permittivity of the specific material as a function of the angular cyclic frequency, ω, of the electric field; $\sigma(\omega)$ is the admittance, or frequency-dependent conductance, or the material; and ε_0 is the permittivity of a vacuum (free space). (Here, j is used as the symbol for the imaginary root of minus one: in electromagnetic formulas, j is used in place of the i used in mathematical equations by convention, so as not to confuse the imaginary unit with the symbol i used for electric current.)

The first real term, ε', is the storage, or static contribution, and the imaginary term, $i\varepsilon''(\omega)$, describes the dynamic, dissipated loss of electric field energy.

The real term, ε', is defined as the *relative permittivity* of the material and is characteristic for the interaction of the material with a static applied electric field. The relative permittivity is also commonly referred to as the *dielectric constant* of the material. A related term is the *electric susceptibility*, a measure of how easily a dielectric material is polarized in a static electric field. The electrical susceptibility, χ_e, of a dielectric material is related to the relative permittivity as

$$\chi_e = \varepsilon' - 1 \tag{15.22}$$

The dissipative term is a function of the frequency, or rate of change of the applied field. In physical terms, dissipation of electromagnetic energy is due to the limits on the rates at which polarization can occur. As the frequency of an oscillating electric field increases, dipoles are not able to orient fast enough to keep in alignment with the applied field.

15.4.3.5 Conductivity and Resistivity

Electrical conductance describes how easily electrical charge flows through a material, that is, how easily the material conducts electric current. Conversely, electrical resistance is the inverse of conductance. For historical reasons and ease of experimental measurement, resistance is usually considered the primary property in definitions.

The electric current, I, is defined in units of electric charge, q, passing through a cross-sectional area of material per unit time:

$$I = \frac{q}{t} \tag{15.23}$$

In an ideal uniform material, the electric current, or flow of electric charge, is described by Ohm's law:

$$I = \frac{V}{R} \tag{15.24}$$

where V is the electric potential measured across the material in units of volts, and R is an empirically determined constant called the resistance.

The resistivity, ρ, is defined as the intrinsic resistance of a material. In practice, resistivity is measured by the resistance exhibited by a specimen of material of a given cross-sectional area, A, and given length, L, given by the equation:

$$R = \frac{\rho L}{A} \tag{15.25}$$

The conductivity, σ, is the reciprocal of the resistivity:

$$\sigma = \frac{1}{\rho} \tag{15.26}$$

Conductivity and resistivity are intrinsic properties that depend on the composition and structure of the material. They are determined by the mobility of charge carriers. A good electrical conductor has a large number of freely delocalized charge carriers, which may be electrons, ions in solution, or electron holes (gaps in the electron structure that can be filled by adjacent electrons, thus in effect producing a reverse current). In tissues and bone, the contributions from electrolyte ions in the tissue fluids are important along with the intrinsic charge carriers and polarizable structures in the solid and colloidal components of the biomaterial.

A static electric potential across a conducting material will result in movement of charge carriers in the field, producing an electric current (a direct current). An oscillating electric potential results in the propagation of oscillations in the charge carriers, producing an alternating current. Resistance for direct current depends upon charge mobility and barriers to transport charge across the material, whereas resistance for alternating currents depends on localized mobility of charges and the transfer of electric field oscillations across the conductor. To separate the mathematical description of these two phenomena, the complete formulation of resistance to include oscillating currents is represented by a complex number, in which the real term describes the static contribution and the imaginary term quantifies the behavior of the alternating contribution.

15.4.3.6 Capacitance and Reactance

The capacitance of a material is a measure of the ability of the material to store electrical charge. Capacitance is directly related to the dielectric constant of the material. The capacitance of a specific sample of material depends on its geometry and its permittivity. In practice, the capacitance, C, is expressed in terms of the dielectric constant of material filling a gap between two parallel conducting surfaces at different electric potentials (an ideal plate capacitor device). The capacitance of a device is the ratio of the charge stored with a given material filling the space between the surfaces, to that stored by the same potential difference separated with a vacuum.

$$C = \frac{\varepsilon' \varepsilon_0 A}{d} \tag{15.27}$$

where A is the area of each face of the parallel surface, and d is the separation distance between the surfaces.

The above equation for capacitance applies to direct currents and static electric fields. For high-frequency variable electromagnetic fields, a capacitive material will exhibit resistance to the change of voltage. This opposition is termed *reactance* and is due to two phenomena, called *capacitive reactance* and *inductive reactance*.

Capacitive reactance is due to accumulated charge induced by an applied voltage. Charge buildup creates an electric field opposed to the applied field. Capacitive reactance decreases as the frequency with which an alternating current reverses polarity increases. With increasing frequency, there is less time for charges to build up before the polarity reverses. Thus, the higher the frequency, the less the capacitive reactance contributes.

Inductive reactance is the opposition to a change of current through a capacitive material, due to the creation of a magnetic field. A changing electric current creates a magnetic field, which induces a countercurrent in the opposing direction. For a sinusoidal alternating current, the inductive reactance is proportional to the frequency and the inductance, which depends on the physical shape and dimensions of the material. Any conductive material with finite dimensions has some inductance. The more intense the interactions between conduction paths, the larger is the inductance; multiple turns in an electromagnetic coil result in a large inductance. Inductive reactance is produced in analogous equivalent circuits by the effects of alternating electromagnetic fields on complex materials such as bone.

15.4.3.7 Impedance

Impedance is a generalized measure of opposition to electromagnetic field effects in a material. Impedance is the combination of resistance and reactance. When an alternating electromagnetic field acts on a material, the resistance depends on the frequency of the alternating field. The behavior of the current flow for alternating currents is described by the *complex impedance*, in which real and imaginary terms represent the static and dynamic components contributing to the interaction. Complex impedance is defined as

$$Z = R + jX \tag{15.28}$$

where the real term, R, is the resistance, and X, in the imaginary term, is the reactance.

Impedance is important in the interactions between sinusoidal and pulsed electromagnetic radiation and bone material. Impedance reflects the delay in adjustment of dipoles under the influence of a polarizing electric field, termed *dielectric relaxation*. Some adjustment leads to dissipative loss of energy. Different mechanisms come into play at different frequencies. For oscillating fields with frequencies from 1 Hz to ~300 GHz (3×10^{11} Hz), the upper frequency limit for microwave radiation, bulk dielectric effects dominate the interactions.

The upper frequency of microwaves, 300 GHz, corresponds to photon energy of 10^{-2} eV, and a wavelength of 10^{-4} m (10^{-1} mm, or 100 μm, or 10^5 nm). Above this energy range, the electromagnetic wavelength begins to approximate the scale of nanoscale structures in biomaterials. At higher levels, photon interactions increasingly begin to come into resonance with vibrational and rotational modes of large molecules, crystals, and polymer structures of biomaterials and have a high rate of absorption and dissipation into heat. Beyond this level, the interactions of electromagnetic fields with matter no longer correspond mainly to dielectric polarizations and current flows. Instead, impedance is dominated by the resonances of light with absorption and scattering modes of materials, which are described by quantum mechanical models.

Dielectric effects in composite materials such as bone are complicated by the many contributions from different components, each of which has different properties in response to different electromagnetic frequencies. Parametric models have been developed to describe the dependence of impedances on the properties of components. An expression that describes the combined contributions is

$$\varepsilon(\omega) = \varepsilon_\infty + \sum_{n=1}^{4} \frac{\Delta\varepsilon_n}{1 + \left(j\omega\tau_n\right)^{(1-\alpha_n)}} + \frac{\sigma_i}{j\omega\varepsilon_0} \tag{15.29}$$

in which, ε_∞ is the permittivity in the terahertz frequency range; σ_i is the ionic conductivity; for each dispersion region, τ is the relaxation time; and $\Delta\varepsilon$ is the drop in permittivity in the corresponding frequency range for the region.

Equation 15.29 is derived from the Cole–Cole relationship, which accounts for the delocalized "stretching" of dielectric effects due to strong interactions between polarization elements across intermolecular distances in polymers. With a choice of parameters appropriate to each tissue, Equation 15.29 may be used to estimate the dielectric behavior of a composite as a combination of its components over a desired frequency range.[83]

Bone and other tissues in general exhibit anisotropy in their electrical properties. The relative permittivity, resistance, capacitance, and other properties can vary along different directions relative to their internal structure. This adds another level of complexity to characterization and interpretation of measurements.[84]

15.4.3.8 Measurement of Electrical Properties of Bone and Cartilage

Techniques and instrumentation for dielectric measurement are well developed and generally economical because of many applications in electronics, communications, and materials industries. Applications in medical diagnostics are relatively undeveloped, but there are many studies that indicate promising potential for diagnostics using low-cost and minimally invasive dielectric characterization technologies.[85]

15.4.3.9 Basic Dielectric Measurement Techniques

Among the methods for measuring dielectric properties, several have been adapted for the characterization of bone samples and in vivo measurements.[86] These include:

1. *Coaxial probe method:* An open-ended coaxial probe is applied to the material. This method is simple, nondestructive, and easy to use. Measurements require only a network analyzer or impedance analyzer, a coaxial probe, and software.
2. *Free space method:* Antennas focus microwave energy on or through the sample. This is a noncontact method that is most useful at millimeter-wave frequencies.
3. *Resonant cavity method:* The material is placed in a high Q resonant cavity structure, altering the center resonance frequency of the cavity, which enables its permittivity to be calculated. Resonant cavities are usually of split-cylinder or split-post geometry.
4. *Parallel plate method:* The material is placed between two parallel conductive electrode plates to form a capacitor. This method works best for low-frequency measurements of uniform thin sheets of material. In principle, this is a straightforward technique, but in practice, care and measures must be taken to eliminate or compensate for the effects of stray capacitances.
5. *Inductance coil methods:* The material under test is placed inside the windings of a wire conductor coil, or the material is placed alongside within the extended field produced by an inductance coil, and the inductance is measured by its effect as an inductive core on the effective inductance of the coil. An inductance balance bridge circuit (or LCR meter for inductance, L, capacitance, C, and resistance, R) is used to measure the inductance of the coil. (This is the same principle as used in metal detectors.) This is an emerging method that can measure inductance produced by electron spin effects.
6. *Split-ring resonators:* Split-ring resonators are microfabricated antennas consisting of a pair of conducting loops with gaps in opposite ends. Rotating currents in the rings produce strong coupling with applied electric fields, making the resonator produce large coupling effects. Microfabricated split-ring resonators are used instead of high-permittivity materials to increase the bandwidth and power of antennas, while reducing the size. The bandwidth of the antenna can be tuned by adjusting the dimensions of the ring resonators with much greater precision and ease than with material substrates. Microstrip split-ring resonator antennas are built into microcircuits, which can be placed against a tissue sample. The dielectric properties of the tissue affect the resonance, with high sensitivity and penetration depth. Split-ring resonators are used in engineering measurement and have been evaluated in research as sensors for bone dielectric properties, to monitor healing in hip transplants and similar procedures.

15.4.3.10 Electrical Impedance of Bone

Impedance measurements have been applied to bone in three different ways: low-frequency impedance measurement for locating nerves in bone during surgery, impedance spectroscopy for broad characterization of osseous tissues, and impedance biosensors for detecting biomarkers and measuring forces in bone and scaffolding.

Direct measurement of impedance between two electrodes inserted during bone surgery has been explored as a safety mechanism to detect proximity between surgical instruments and important nerves embedded in the bone. Facial nerves in the temporal bone are an area of particular concern for this type of monitoring during cochlear implantation. Studies have suggested that impedance from neuromonitoring can detect nerve tissue proximity to needles and drills by monotonic decrease of impedance with distance from the facial nerve.[87]

15.4.3.11 Impedance Spectroscopy

The technique of impedance spectroscopy applies a sinusoidal voltage to a system or material to measure its effective impedance over a range of frequencies. The method is familiar for use in electronics and electrochemistry and has more recently been applied to characterize the dielectric response of ceramics, polymers, and complex composite materials. The dielectric profile can reveal information on internal structure that is complimentary to that obtainable with higher frequencies of the electromagnetic spectrum, especially for complex optically opaque materials.[88]

Impedance spectroscopy has been studied for its possible utility to characterize trabecular and subchondral bone for dielectric properties and BMD. The spectrum of effective dielectric responses over a frequency range from 100 Hz to 20 MHz contains information on bone status. The technique has the potential to differentiate between healthy and osteopenic bone. This would require that impedance spectra could be shown with high confidence to relate to bone strength. In particular, it would be valuable if electrical properties could predict early stage thinning and mineral loss associated with bone weakness. Impedance spectroscopy has potential for live patient monitoring for diagnosis and treatment of thinning and mineral loss associated with osteoporosis and other syndromes of bone weakness.

15.4.3.12 Electrochemical Biosensors

Impedance biosensors have found many biomedical applications, because of ease of implementation and measurement, and the direct response of impedance to the binding of target biomolecules and protein and DNA sequences. Impedance spectroscopy can be carried out with sensors embedded in tissues. Impedance sensors are relatively simple to integrate in microelectronics and can be screen printed onto inexpensive materials.[89]

Among the many applications of biosensors for characterization of tissues, there are a number of examples for identifying biomarkers in bone and bone-related diseases, including osteoporosis and neurodegenerative skeletal disorders.[90] There are many bone biomarkers that may have utility in biosensors, including the N-terminal propeptide of type 1 procollagen, P1NP, and sclerostin. The ability to use impedance spectroscopy with biosensors is a particularly powerful tool with potential for research and diagnostics.

15.4.3.13 Microwave Characterization

Microwave interactions have the potential to characterize tissue at energies between the modalities of ultrasound and infrared and X-ray measurements. The usual classification of the microwave region of the electromagnetic spectrum includes the span between 10^{-4} m $=10^{-1}$ mm (or 100 μm, or 10^5 nm), and frequency from 3×10^9 Hz to 3×10^{12} Hz, (300 MHz to 300 GHz). This range is between radio waves and infrared light and includes the frequencies used for radar.

The interactions between microwave electromagnetic fields and material are dominated by charge redistributions that involve multimolecular structures on the nano- to macroscale. For example, in water, these frequencies are related to the loose, fluid lattice formed by hydrogen bonding of water molecules, spread out in energy due to fluctuating hydrogen bonding. The typical absorption curve for liquid water has a broad peak in the dissipation factor, at around 3 cm, which increases in frequency with temperature, making it an important absorption region for biological architectures, proteins, cells, composite layers, and porous networks that underlie bone strength. Microwave signatures have been found to differ between healthy bone and tissue with decreased mineralization.[91]

15.4.4 Spectroscopy: Chemical and Molecular Properties

Spectroscopic methods can probe the molecular and atomic properties that underlie bone composition and structure. These methods use the electromagnetic spectrum from infrared to X-rays. Probing the internal states of molecules starts just above the energy level of microwaves, with infrared spectroscopy. At higher energies, visible and ultraviolet spectroscopy is less important for bone, although not completely inaccessible. At X-ray energies, elemental species can be identified by their inner shell electron energies, and crystal structures can be unraveled by diffraction and scattering techniques.

15.4.4.1 Infrared Vibrational Spectroscopy

Infrared light (IR radiation) spans a range from microwaves to the visible portion of the electromagnetic spectrum.[92,93] In wavelength, this is from 1 mm to 700 nm, and in frequency from 300 GHz up to 430 THz. In energy, IR ranges from 0.01 up to 2 eV. The resonances for interaction with IR frequencies correspond to dielectric alterations that occur with vibration, stretching, bending, and rotation of atoms in the molecules. If these motions involve charge separation, the resulting electric field oscillation can be in resonance with a frequency of radiation falling in the IR range. Conversely, IR radiation can excite these molecular motions through resonant interactions.

Light interacts with polarizable molecules or with dipolar groups to produce vibrational excitation of covalently bonded atoms and groups. Absorption of radiation energy excites a vibrational mode from a lower to a higher quantized energy state. The energy levels are determined by the strength of the binding and conformational barriers opposing the vibrations. The number of vibrational modes is determined by the internal degrees of freedom of the molecule. The major modes of molecular vibration are as follows:

- *Stretching*: change in the length of a bond
- *Bending*: change in the angle between two bonds
- *Rocking*: change in angle between a group of atoms
- *Wagging*: change in angle between the plane of a group of atoms
- *Twisting*: change in the angle between the planes of two groups of atoms
- *Out-of-plane*: change in angle between a bond and the plane of the rest of the molecule

A secondary type of interaction with IR light is scattering, in which the photon is re-emitted. For this interaction, the predominant mode is elastic, or Raleigh scattering, in which light is scattered without change in energy or wavelength. A much lower probability mode is inelastic, or Raman scattering, in which the scattered light has an altered wavelength due to energy exchange with the molecule. Raman scattering can result in the emitted light having a lower energy than the incident light (termed Stokes Raman scattering) or a higher energy (anti-Stokes scattering) (Figure 15.9).

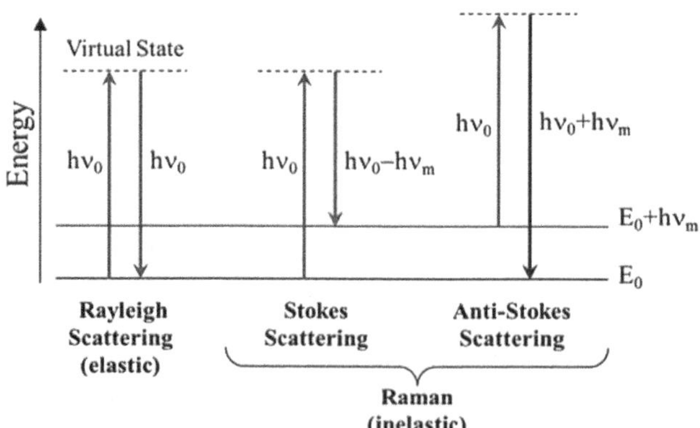

FIGURE 15.9 Energy-level representations for Rayleigh and Raman scattering ($h\nu_0$ = excitation energy, $h\nu_m$ = inelastic energy change, and E_0 = baseline energy level). (Adapted from Meet Shah, Thesis submitted for the degree of Master of Science in Materials Science and Engineering, The University of Texas at Arlington, Texas, 2018.)

An IR vibrational spectrum is produced when the absorption and/or emission is recorded across a range of wavelengths. By convention, IR spectra are described in units of reciprocal wavelength, cm^{-1}. Raman scattering spectral characteristics are complimentary to the normal IR spectrum. Raman spectra are much lower intensity than IR spectra, because of the very low probability for inelastic scattering.

15.4.4.2 FTIR Measurements in Bone

The advantages of fast transform approaches have led to Fourier transform spectroscopy becoming the standard instrumental technique for both IR (FTIR) and Raman spectroscopy.[94] For solid samples of low transmittance, such as bone, the FTIR spectrum is measured by reflectance. To amplify the absorption signal, an attenuated total reflection (ATR) technique is used to measure the attenuation of an IR beam making multiple reflections across a small surface area of the sample.[95] ATR places a smooth surface of the sample against a crystal, which is transparent to IR light with high refractive index; the IR beam enters the crystal at an angle, undergoing multiple reflections before exiting the sample area to strike a photodetector (Figure 15.10).

Figure 15.11 shows a typical FTIR spectrum taken from the cortical portion of a rat femur. The numbers above the peaks show the identification of the absorptions per Table 15.1. IR spectroscopy has been demonstrated to be useful for bone. Bone is about 80% collagen, with water and mineral components as the remainder. Both the collagen and, to a lesser extent, the mineral and fluid components have many vibrational and rotational modes that provide IR absorption bands.[96] Table 15.1 includes some of the most useful peak absorption wavenumbers for bone.

In the above figures, the hydroxy, hydrocarbon, and amide peaks correspond to collagen absorption bands, and the phosphate absorption is produced by the apatite components. The identification of the bands and assignment to functional groups is made by comparison with library spectra (Figure 15.12).

FTIR instruments with laser probes and microscopes enable sampling at selected areas of tissue in the complex heterogeneous microanatomy of bone. Since the ATR probe on an FTIR or fiber optic microprobe on a Raman instrument samples an area of a few square millimeters or less at a time, it is important to co-register the sampling for IR spectra with micrographs. Alternatively, one may

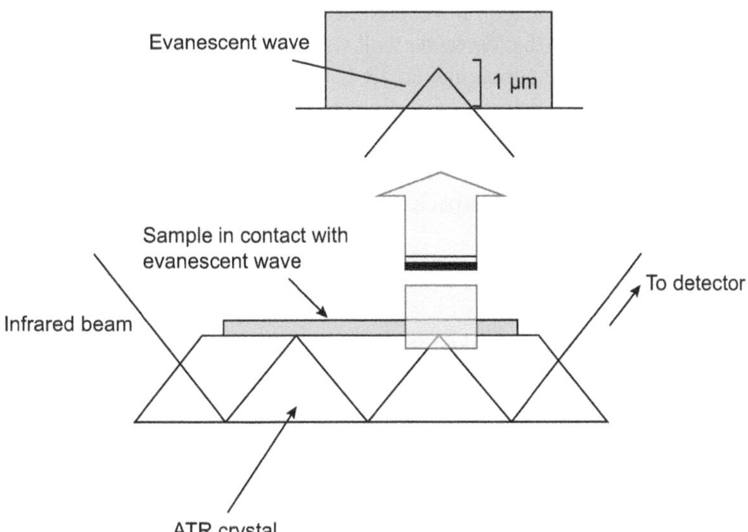

FIGURE 15.10 Surface evanescent wave technique for FTIR–ATR (Fourier Transform Infra-Red–Attenuated Total Reflectance) spectroscopy.

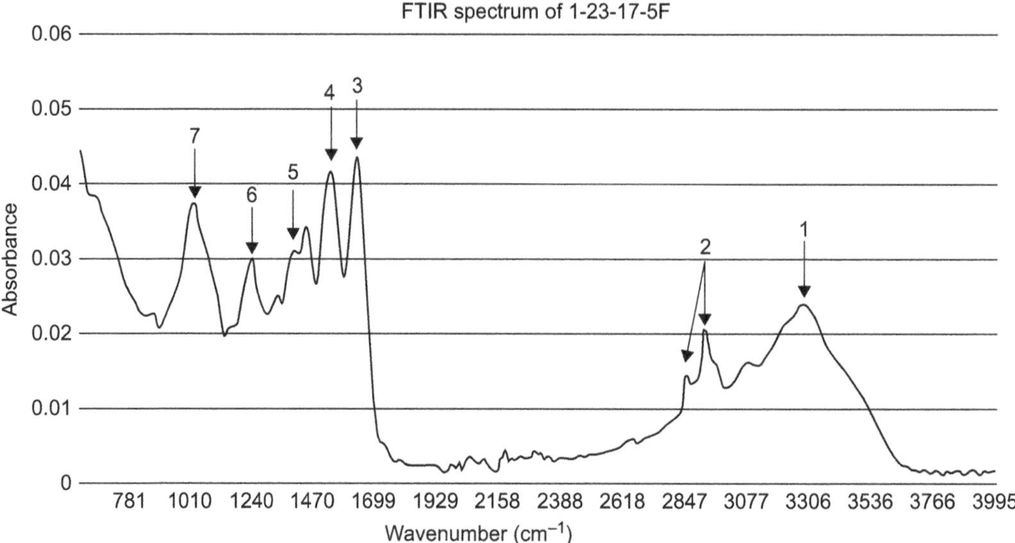

FIGURE 15.11 FTIR spectrum of bone from a fresh cross-section of cortical rat femur.

take randomized samples across the bone to ensure a representative cross-section, with pores, cells, vessels, and varied compositions.

Microscopic mapping of mineral distribution in bone has been carried out using the IR bands of carbonate and phosphate groups.[93]

15.4.4.3 Raman Spectroscopy of Bone

Raman spectroscopy is complimentary to FTIR in many respects. Raman is especially useful for information on collagen and various matrix proteins. The Raman spectrum also has information on bone mineral crystallinity and can be used to obtain information on the orientation of mineral crystallites with respect to the collagen fibril axis. Mineral and collagen matrix measurements can be assessed from Raman spectra that correlate well with bone and tissue properties.[97]

In a Raman laser microscope spectrometer, IR is beamed through a fiber optic waveguide under a microscope, enabling the selection of portions of the tissue for analysis. Reflected light from the sample is measured by a photodetector positioned with a filter to remove all but the spectral range of interest. The Raman emission for Stokes scattering is shifted to longer wavelengths than the excitation laser IR light. For Raman spectra of organic molecules, the working range is between

TABLE 15.1
Absorption Peaks for Bone Functional Groups

Sr. No.	Component	Wavenumber
1	OH^-	3280
2	CH_2	2292 and 2852
3	Amide I	1631
4	Amide II	1541
5	CO_3^{2-}	1405
6	Amide III	1235
7	PO_4^{3-}	1014

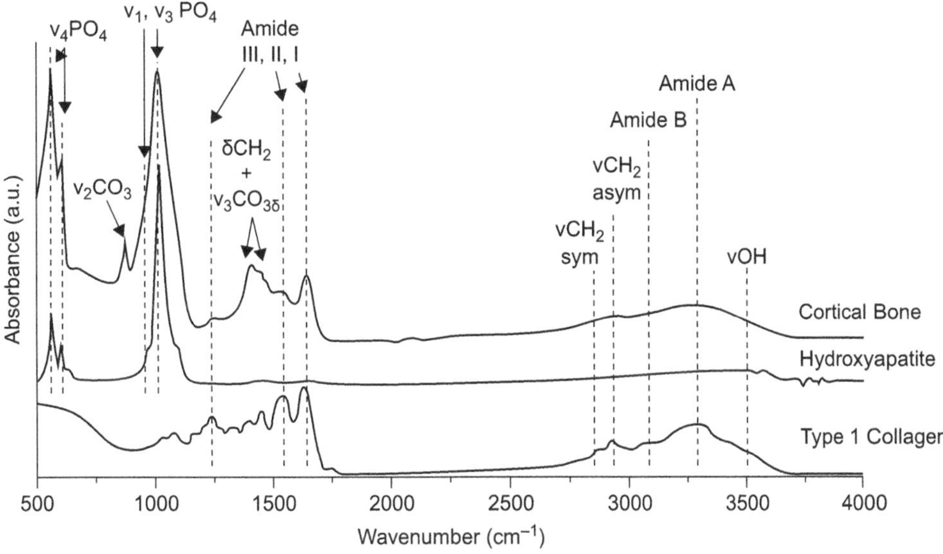

FIGURE 15.12 FTIR spectra of bone compared to major bone component compounds.

600 and 1100 nm. Excitation wavelengths are usually in the 400 to 600 nm range, but for bone, a longer wavelength gives better results, with less non-Raman scattering and deeper penetration. An excitation wavelength of 780 nm was used to obtain the spectrum in Figure 15.13. Because of the low intensity of Raman scattering modes, it is necessary to average a large number of spectra. Excitation of biological samples must be paced to avoid drying and overheating from the IR light. Noise from fluorescence can be much stronger than the Raman signal, especially in biological tissues, which necessitates special fluorescence suppression measures.[98]

Figure 15.13 shows a typical Raman shift spectrum of bone, taken from a sliced rat femur. Table 15.2 gives assignments for the important peaks shown in the spectrum. One of the differences from the FTIR spectrum is the small but clearly identifiable hydroxyproline peak at 875 cm^{-1}.

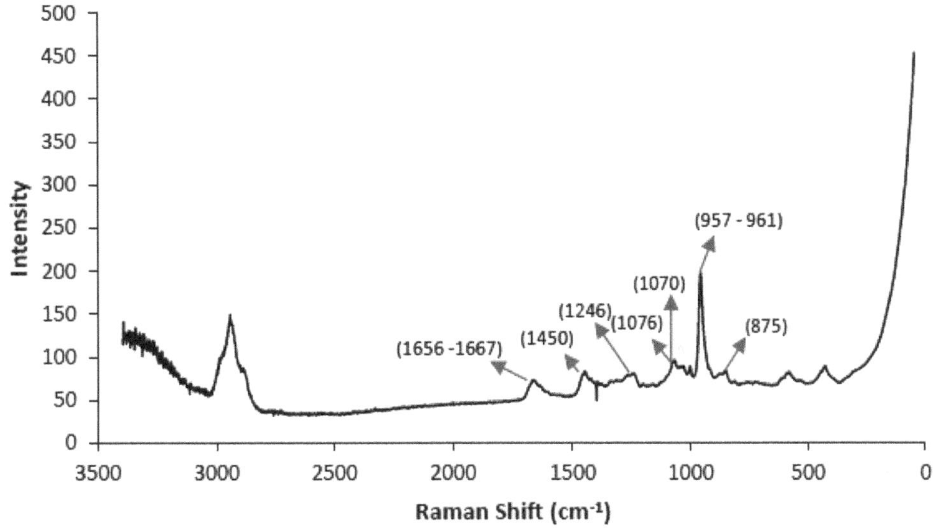

FIGURE 15.13 Raman spectrum of cortical bone from rat femur.

TABLE 15.2
Major Peaks in Raman Shift Spectrum of Rat Cortical Bone

Component	Wavenumber
Hydroxyproline	875 cm^{-1}
Phosphate (PO_4^{3-}) v1	957–961 cm^{-1}
Carbonate (CO_3^{2-})	1070 cm^{-1}
Phosphate (PO_4^{3-})	1076 cm^{-1}
Amide III	1246 cm^{-1}
CH_2 deformation	1450 cm^{-1}
Amide I	1656–1667 cm^{-1}

Hydroxyproline is a major component of collagen protein and is used as an indicator of collagen content in bone.[10]

The area under each peak can be used to estimate relative quantities of constituents from both FTIR and Raman and to compute mineral-to-collagen or -matrix ratios. A number of the peak ratios and parameter definitions that are used for assessment of mineral-to-matrix ratios from Raman spectra are shown in Table 15.3. Many of the parameters obtained from FTIR and Raman spectra of bone have been shown to correlate with the properties of bone tissue. The direction of research in this area has been to identify parameters associated with bone quality, fragility, fractures, and osteoporosis.[93,99]

15.4.4.4 Scanning Electron Spectroscopy with Energy Dispersive Spectroscopy

An emerging technology combines electron microscopy with energy dispersive X-ray spectroscopy (SEM/EDS). The electron beam used for imaging in electron microscopy can be focused with sufficiently high energy to excite inner shell electrons in the atoms of the sample under study. When an electron from another, outer, shell fills the hole, the difference in energy between the two electron states is emitted as an X-ray with the energy equal to the difference between the two electron shell levels in the atom. This precise energy difference is characteristic of each elemental atomic species. The number and energy of the emitted X-rays are measured by an energy dispersive spectrometer. The element can be identified by its characteristic X-ray emission, and an image can be constructed that maps the elemental composition on the electron micrograph.

In an application to bone, elemental composition maps were produced from electron microscopy specimens of rat femur and tibia. Figure 15.14 shows sets of micrograph images of bone samples

TABLE 15.3
Mineral-to-Matrix Ratios Derived from Raman Spectra

Most used Raman Ratios	Calculations
Mineral to matrix	(959-to-Pro, Hyp or 959-to-amide I) a measure of mineral content
Carbonate to phosphate	(1070-to-959) a measure of carbonate substitution in apatitic lattice
Crystallinity	(inverse width 959) a measure of crystal size and/or perfection
Collagen cross-link	(1685-to-1665) a measure of collagen fibril maturity

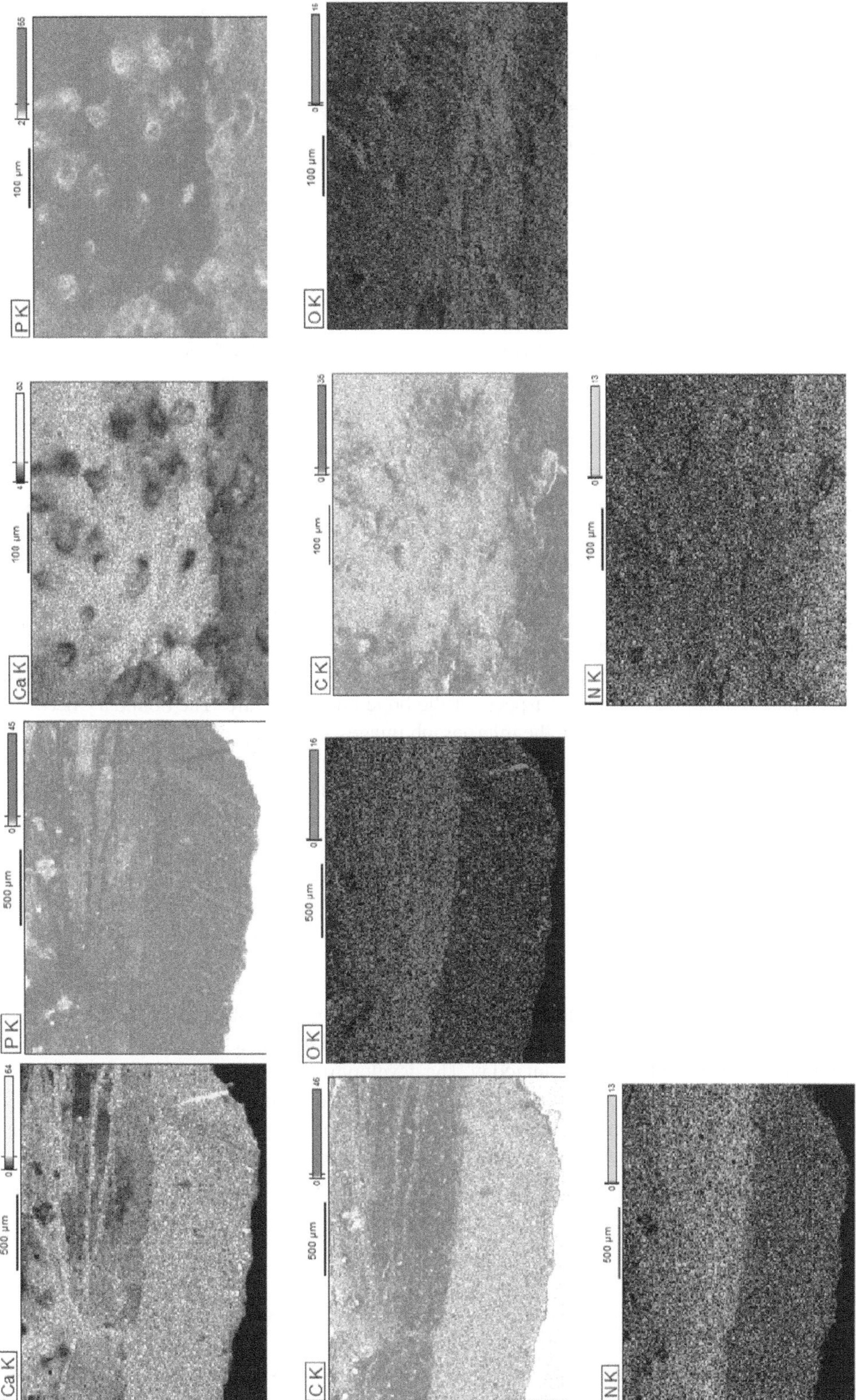

FIGURE 15.14 SEM–EDS micrographs of rat femur samples (see text).

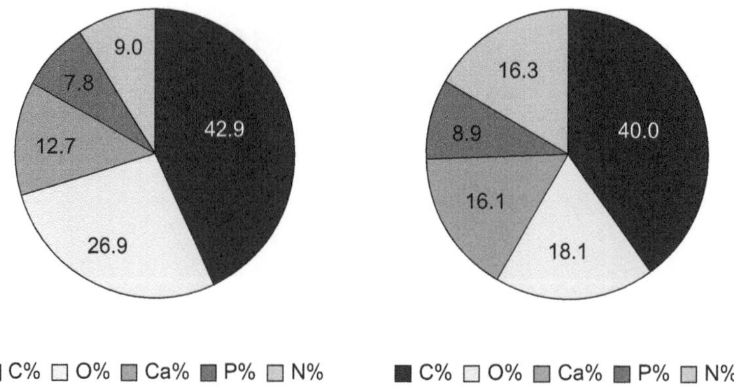

FIGURE 15.15 Relative abundances for selected elements as measured by SEM–EDS in bone samples from two different rats.

from rat femur and tibia. The images map the X-ray dispersion emission intensities for Ca (calcium), P (phosphorous), C (carbon), N (nitrogen), and O (oxygen). The elemental densities are related to the intensity of color in the images. Figure 15.15 shows the elemental ratios in pie graph form, taken from the same X-ray intensity and energy data used to construct the bone cross-section image maps.

The images are derived from the K-shell of the respective atomic species. The dimensional scale bars represent 500 μm. The image magnification is 400×. The intensity scale color bars indicate the amplitude of the emission for each element, which is roughly proportional to atomic concentration. The concentration relationship is not exact because of differences in emission probabilities, depth of atoms in the sample, and other effects, which can be computed and compensated for in the future. The different elemental mappings bring out aspects of the bone microstructure. The boundary between cortical and cancellous bone is clear in the micrograph images.

The pie charts in Figure 15.15 show differences in bone samples from two rats with different genetics and bone conditions. The potential of this method is promising, particularly if the SEM-EDS data can be correlated with measures of bone strength, development, biological and medical properties, and outcomes.

The variation in elemental content shows pores and what may be locations of cells and vessels. Further analysis is being conducted by histology experts.

15.5 CONCLUSION

Knowledge of forces that stimulate and guide bone remodeling, growth, and repair is very important for clinical guidance. Thus, improved methods for characterization of electromechanical properties and behavior of bone are highly significant today. The development of methods for deeper characterization of bone ultrastructure is an area of sustained scientific and medical interest. Nanotechnology is continuing to grow as a discipline to enhance capabilities for observing and manipulating osseous materials at the nanoscale. This chapter has surveyed a small part of the newly emerging technology applicable to bone, focusing on selected areas where progress is rapid and widespread and that appear less known to potential users in the medical community.

REFERENCES

1. Burr, D.B. and Allen, M.R., Eds., Basic and Applied Bone Biology, Academic Press, New York, NY, 2014.
2. Tibbals, H.F., Medical Nanotechnology and Nanomedicine (Perspectives in Nanotechnology), CRC Press, Boca Raton, FL, 2011.

3. Hornyak, G.L., Dutta, J., Tibbals, H.F. and Rao, A.K., Introduction to Nanoscience, CRC Press, Boca Raton, FL, 2008.
4. Hassenkam, T., Svensson, R. B., and Zalkovskij, M., Nano-science revelations in bone research, *Curr. Nanosci.*, 3, 345-51, 2007.
5. Reznikov, N., Bilton, M., Lari, L., Stevens, M.M., and Kröger R., Fractal-like hierarchical organization of bone begins at the nanoscale, *Science*, 360, pii: eaao2189, 2018.
6. Mitić, Z., Aleksandra, S., Stojanovic, S., Najman, S., Ignjatovic, N.L., Nikolić, G.M., and Trajanovic, M., Instrumental methods and techniques for structural and physicochemical characterization of biomaterials and bone tissue: a review, *Mater. Sci. Eng. C*, 79, doi: 10.1016/j.msec.2017.05.127, 2017.
7. Ratner, B.D., Hoffman, A.S., Schoen, F.J., and Lemons, J.E., Eds., Biomaterials Science: An Introduction to Materials in Medicine, 3rd edn., Academic Press, New York, 2012.
8. Wong, J.Y., Bronzino, J.D., and Peterson, D.R., Eds., Biomaterials: Principles and Practices, CRC Press, Boca Raton, FL, 2012.
9. Boskey, A.L., Biomineralization: an overview, *Connect. Tissue Res.*, 44, 5-9, 2003.
10. Boskey, A.L., Noncollagenous matrix proteins and their role in mineralization, *Bone*, 6, 111-23, 1989.
11. Zhang, M. and Fan, Y., Eds., Computational Biomechanics of the Musculoskeletal System, CRC Press, Boca Raton, FL, 2014.
12. MacArthur, B.D. and Oreffo, R.O.C., From mathematical models to clinical reality, in Principles of Tissue Engineering, 4th edn., Lanza, R., Langer, R., and Vacanti, J.P., Eds., Academic Press, San Diego, 2014, chap. 3.
13. Bhattacharya, P. and Viceconti, M., Multiscale modeling methods in biomechanics, *Wiley Interdiscip. Rev. Syst. Biol. Med.*, 9, 3, doi: 10.1002/wsbm.1375. Epub 2017.
14. Kim, H.S., Al-Hassani, S.T., A morphological model of vertebral trabecular bone, *J. Biomech*, 35, 1101–14, 2002.
15. Lin, L., Samuel, J., Zeng X., and Wang X., Contribution of extrafibrillar matrix to the mechanical behavior of bone using a novel cohesive finite element model, *J. Mech. Behav. Biomed. Mater.*, 65, 224-35, 2017.
16. Sarker, F., Degradation mechanics of bone and bone like materials via multiscale analysis, PhD Thesis, Graduate School of The University of Texas at Arlington, TX, 2014. https://rc.library.uta.edu/uta-ir/handle/10106/24768
17. _Ferdous, S., Sarker, F., and Adnan, A., Role of nanoparticle dispersion and filler-matrix interface on the matrix dominated failure of rigid C60-PE nanocomposites: a molecular dynamics simulation study, *Polymer*, 54, 2565–76, 2013.
18. Odgaard, A., Quantification of cancellous bone architecture, in Cowin, S.C., Ed., Bone Mechanics Handbook, 2nd edn., CRC Press, Boca Raton, FL, 2001, chap. 14.
19. Stauber, M., Nazarian, A. and Müller, R., Limitations of global morphometry in predicting trabecular bone failure, *J. Bone Miner. Res.*, 29, 134-41, 2014.
20. Keaveny, T.M., Morgan, E.F., and Yeh, O.C., Bone mechanics, in Biomedical Engineering and Design Handbook, 2nd edn., Kutz, M., Ed., McGraw-Hill Education, New York, 2009, chap. 9.
21. Ritchie, R.O., How does human bone resist fracture?, *Ann. N. Y. Acad. Sci.*, 1192, 72-80, 2010.
22. Fyhrie, D.P. and Christiansen, B.A., Bone material properties and skeletal fragility, *Calcif. Tissue Int.*, 97, 213-28, 2015.
23. Shackelford, J.F., Introduction to materials science for engineers, 8th edn., Pearson, New York, NY, 2014.
24. Hibbeler, R.C., Mechanics of Materials, 10th edn., Pearson, New York, NY, 2016.
25. Ali, D. and Sen, S., Permeability and fluid flow-induced wall shear stress of bone tissue scaffolds: computational fluid dynamic analysis using Newtonian and non-Newtonian blood flow models, *Comput. Biol. Med.*, 99, 201-208, 2018.
26. Temenoff, J.S. and Mikos, A.G., Biomaterials: The Intersection of Biology and Materials Science, Pearson, New York, NY, 2008.
27. Hayenga, H.N. and Aranda-Espinoza, Biomaterial Mechanics, CRC Press, Boca Raton, FL, 2017.
28. Chen, Q. and Pugno, N.M., Bio-mimetic mechanisms of natural hierarchical materials: a review, *J. Mech. Behav. Biomed. Mater.*, 19, 3-33, 2013.
29. Sprio, S. and Sandri, M., Biomimetic biomaterials: 1. Biomimetic materials in regenerative medicine, Woodhead Publishing, Cambridge, MA, 2013.
30. Carlsson, L.A., Adams, D.F., and Pipes, R.B., Experimental Characterization of Advanced Composite Materials, 4th edn., CRC Press, Boca Raton, FL, 2014.

31. Beake, B.D., Harris, A.J., and Liskiewicz, T.W., Advanced nanomechanical test techniques, in Materials Characterization: Modern Methods and Applications, Ranganathan, N.M., Ed., CRC Press , Boca Raton, FL, 2015, chap. 1, pp. 1-89.
32. Gupta, H.S., Seto, J., Wagermaier, W., Zaslansky, P., Boesecke, P., and Fratzl, P., Cooperative deformation of mineral and collagen in bone at the nanoscale, *Proceedings of the National Academy of Sciences of the United States of America.*, 103, 17741-6, 2006.
33. Bechtle, S., Ang, S.F., and Schneider, G.A., On the mechanical properties of hierarchically structured biological materials, *Biomaterials*, 25, 6378-85, 2010.
34. Hart, N.H., Nimphius, S., Rantalainen, T., Ireland, A., Siafarikas, A., and Newton, R.U., Mechanical basis of bone strength: influence of bone material, bone structure and muscle action, *J. Musculoskelet. Neuronal Interact.*, 17, 114-139, 2017.
35. Nobakhti, S. and Shefelbine, S.J., On the relation of bone mineral density and the elastic modulus in healthy and pathologic bone, *Curr. Osteoporos. Rep.*, 16, 404-10, 2018.
36. Herman, K., Hardness Testing: Principles and Applications, ASM International, Materials Park, OH, 2011.
37. Tobolski, E., Instrumented indentation testing: a draft ASTM practice, *ASTM Standardization News*, October, 2003.
38. Van Landingham, M.R., Review of instrumented indentation, *J. Res. Natl. Inst. Stand. Technol.*, 108, 249-65, 2003.
39. Kaplan, J.T., Neu, C.P., Drissi, H., Emery, N.C., and Pierce, D.M.. Cyclic loading of human articular cartilage: the transition from compaction to fatigue, *J. Mech. Behav. Biomed. Mater.*, 65, 734-742, 2017.
40. Menard, H.P., Dynamic Mechanical Analysis: A Practical Introduction, 2nd edn., CRC Press, Boca Raton, FL, 2008.
41. Yamashita, J., Li, X., Furman, B.R., Rawls, H.R., Wang, X., and Agrawal, C.M., Collagen and bone viscoelasticity: a dynamic mechanical analysis, *J. Biomed. Mater. Res.*, 63, 31-6, 2002.
42. Brostow, W., Hagg Lobland, H.E., and Narkis, M., The concept of materials brittleness and its applications, *Polym. Bull. (Berl.)*, 67, 1697-1707, 2011.
43. Schuh, C.A., Nanoindentation studies of materials, *Mater. Today*, 9, 32-40, 2006.
44. Sun, J., Ling, M. , Wang, Y., Chen, D., Zhang, S., Tong, J., and Wang, S., Quasi-static and dynamic nanoindentation of some selected biomaterials, *J. Bionic Eng.*, 11, 144-150, 2014.
45. Arnold, M., Zhao, S., Ma, S., Giuliani, F., Hansen, U., Cobb, J.P., Abel, R.L., and Boughton O., Microindentation – a tool for measuring cortical bone stiffness? A systematic review, *Bone Joint Res.*, 9, 542-549, 2017.
46. Idkaidek, A., Agarwal, V., and Jasiuk, I., Finite element simulation of reference point indentation on bone, *J. Mech. Behav. Biomed. Mater.*, 65, 574-583, 2017.
47. Zysset, P.K., Guo, X.E., Hoffler, C.E., Moore, K.E., and Goldstein, S.A., Mechanical properties of human trabecular bone lamellae quantified by nanoindentation, *Technol. Health Care.*, 6, 429-32, 1998.
48. Engels, P.S., Begau, C., Gupta, S., Schmaling, B., Ma, A., and Hartmaier, A., Multiscale modeling of nanoindentation: from atomistic to continuum models, in Nanomechanical Analysis of High Performance Materials, Tiwari, A., Ed., Springer, New York, 2014, pp. 285-322.
49. Oliver, W.C. and Pharr, G.M., An improved technique for determining hardness and elastic modulus using load and displacement sensing indentation experiments, *J. Mater. Res.* 7, 1564-83, 1992.
50. Tang, B., Ngan, A.H.W., and Lu, W.W., An improved method for the measurement of mechanical properties of bone by nanoindentation, *J. Mater. Sci. Mater. Med.* 18, 1875-81, 2007.
51. Andriollo, T., Thorborg, J., and Hattel, J., Analysis of the equivalent indenter concept used to extract Young's modulus from a nano-indentation test: some new insights into the Oliver–Pharr method, *Modell. Simul. Mater. Sci. Eng.*, 25, 045004, 2017.
52. Boughton, O.R., Ma, S., Zhao, S., Arnold, M., Lewis, A., Hansen, U., Cobb, J.P., Giuliani, F., and Abel, R.L., Measuring bone stiffness using spherical indentation, *PLoS One*. 13, e0200475, 2018.
53. Barone, A.C., Salerno, M., Patra, N., Gastaldi, D., Bertarelli, E., Carnelli, D., and Vena, P., Calibration issues for nanoindentation experiments: direct atomic force microscopy measurements and indirect methods, *Microsc. Res. Tech.*, 73, 996-1004, 2010.
54. Sidhu, P.S., Chong, W.K., and Satchithananda, K., Measurement in Ultrasound 2nd Ed., CRC Press, Boca Raton, FL, 2016.

55. Cobbold, R.S.C., Foundations of Biomedical Ultrasound, Oxford University Press, Oxford, UK, 2006.
56. Ostensen, H. and Tole, N.M., Basic Physics of Ultrasonographic Imaging, World Health Organization, Geneva, Switzerland, 2005.
57. Wells, P.N., Ultrasound imaging, *Phys. Med. Biol.*, 51, R83-98, 2006.
58. Filoux, E., Mamou, J., Aristizábal, O., and Ketterling, J.A., Characterization of the spatial resolution of different high-frequency imaging systems using a novel anechoic-sphere phantom, *IEEE Trans. Ultrason. Ferroelectr. Freq. Control*, 58, 994-1005, 2011.
59. Langton, C.M. and Njeh, C.F., The Physical Measurement of Bone, CRC Press, Boca Raton, FL, 2003.
60. Raum, K., Microelastic imaging of bone, *IEEE Trans. Ultrason. Ferroelectr. Freq. Control*, 55, 1417-31, 2008.
61. Langton, C.M. and Njeh, C.F., The measurement of broadband ultrasonic attenuation in cancellous bone—a review of the science and technology, *IEEE Trans. Ultrason. Ferroelectr. Freq. Control*, 55, 1546-54, 2008.
62. Töyräs, J., Nieminen, M.T., Kröger, H., and Jurvelin, J.S., Bone mineral density, ultrasound velocity, and broadband attenuation predict mechanical properties of trabecular bone differently, *Bone*, 31, 503-7, 2002.
63. Gluer, C.C., A new quality of bone ultrasound research, *IEEE Trans. Ultrason. Ferroelectr. Freq. Control*, 55, 1524-8, 2008.
64. Mehta, S., Antich, P., and Blomqvist, C.G., Measurement of shear-wave velocity by ultrasound critical-angle reflectometry (UCR), *Ultrasound Med. Biol.*, 23, 1123-6, 1997.
65. Laugier, P., Instrumentation for in vivo ultrasonic characterization of bone strength, *IEEE Trans. Ultrason. Ferroelectr. Freq. Control*, 55, 1179-96, 2008.
66. Wear, K.A., Nagaraja, S., Dreher, M.L., Gibson, S.L., Relationships of quantitative ultrasound parameters with cancellous bone microstructure in human calcaneus in vitro, *J. Acoust. Soc. Am.*, 131, 1605-12, 2012.
67. Sarvazyan, A., Hall, T.J., Urban, M.W., Fatemi, M., Aglyamov, S.R., and Garra, B.S., An overview of elastography – an emerging branch of medical imaging, *Curr. Med. Imaging Rev.*, 7, 255-82, 2011.
68. Gennisson, J.L., Deffieux, T., Fink, M., and Tanter, M., Ultrasound elastography: principles and techniques, *Diagn. Interv. Imaging.*, 94, 487-95, 2013.
69. Wu, D., Isaksson, P., Ferguson, S.J., and Persson C., Young's modulus of trabecular bone at the tissue level: a review, *Acta Biomater.*, 78, 1-12, 2018.
70. Sigrist, R.M.S., Liau, J., Kaffas, A.E., Chammas, M.C., and Willmann, J.K., Ultrasound elastography: review of techniques and clinical applications, *Theranostics*, 7, 1303-29, 2017.
71. Wang, L.V., Ed., Photoacoustic Imaging and Spectroscopy, CRC Press, Boca Raton, FL, 2009.
72. Elson, D.S., Li, R., Dunsby, C., Eckersley, R., and Tang, M.X., Ultrasound-mediated optical tomography: a review of current methods, *Interface Focus*, 1, 632-48, 2011.
73. Yang, S. and Xing, D., Biomedical Photoacoustics, CRC Press, Boca Raton, FL, 2019.
74. Xu, M. and Wang, L.V., Photoacoustic Tomography, CRC Press, Boca Raton, FL, 2019.
75. Wang, P., Wang, P., Wang, H.W., and Cheng, J.X., Mapping lipid and collagen by multispectral photoacoustic imaging of chemical bond vibration, *J. Biomed. Opt.*, 17, 96010-1, 2012.
76. Beard, P., Biomedical photoacoustic imaging, *Interface Focus*, 1, 602-31, 2011.
77. Cox, B., Laufer, J.G., Arridge, S.R., and Beard, P.C., Quantitative spectroscopic photoacoustic imaging: a review, *J. Biomed. Opt.*, 17, 061202, 2012.
78. Kaiplavil, S., Mandelis, A., Wang, X., and Feng, T., Photothermal tomography for the functional and structural evaluation, and early mineral loss monitoring in bones, *Biomed. Opt. Express*, 5, 2488–502, 2014.
79. Wang, L, Zhang, C, and Wang, L.V., Grueneisen relaxation photoacoustic microscopy, *Phys. Rev. Lett.*, 113, 174301, 2014.
80. Feng, T., Kozloff, K.M., Tian, C., Perosky, J.E., Hsiao, Y.S., Du, S., Yuan, J., Deng, C.X., and Wang, X., Bone assessment via thermal photo-acoustic measurements, *Opt. Lett.*, 40, 1721-4, 2015.
81. Wissmeyer, G., Pleitez, M.A., Rosenthal, A., and Ntziachristos, V., Looking at sound: optoacoustics with all-optical ultrasound detection, *Light Sci. Appl.*, 7, 53, 2018.
82. Frohlich, H., Theory of Dielectrics: Dielectric Constant and Dielectric Loss, 2nd edn., Oxford University Press, Oxford, 1987.
83. Gabriel, S., Lau, R.W., and Gabriel, C., The dielectric properties of biological tissues: III. Parametric models for the dielectric spectrum of tissues, *Phys. Med. Biol.*, 41, 2271-93, 1996.

84. Miklavcic, D., Pavselj, N., and Hart, F.X., Electric Properties of Tissues, in Wiley Encyclopedia of Biomedical Engineering, Akay, M., Ed., John Wiley & Sons, New York, 2006.
85. Venkatesh, M.S. and. Raghavan, G.S.V., An overview of dielectric properties measuring techniques, *Can. Biosyst. Eng.*, 47, 7-15, 2005.
86. D'Amore, G., Six techniques for measuring dielectric properties, *New Electron.*, 9, 15259, 2017.
87. Wyss-Balmer, T., Ansó, J., Muntane, E., Gavaghan, K., Weber, S., Stahel, A., and Büchler, P., In-vivo electrical impedance measurement in mastoid bone, *Ann. Biomed. Eng.*, 45, 1122-32, 2017.
88. Barsoukov, E. and Macdonald, J.R., Eds., Impedance spectroscopy: theory, experiment, and applications, 3rd edn., Wiley-Interscience, John Wiley & Sons, Hoboken, NJ, 2018.
89. Cosnier, S., Ed., Electrochemical Biosensors, Pan Sanford/CRC Press, Boca Raton, FL, 2013.
90. Afsarimanesh, N., Alahi, M.E.E., Mukhopadhyay, S.C., and Kruger, M., Smart sensing system for early detection of bone loss, *J. Sens. Actuator Netw.*, 7, 10, 2018.
91. Meaney, P.M., Zhou, T., Goodwin, D., Golnabi, A., Attardo, E.A., and Paulsen, K.D., Bone dielectric property variation as a function of mineralization at microwave frequencies, *Int. J. Biomed. Imaging*, 2012, 649612, 2012.
92. Paschalis, E.P., Gamsjaeger, S., and Klaushofer, K., Vibrational spectroscopic techniques to assess bone quality, *Osteoporos. Int.*, 28, 2275-91, 2017.
93. Paschalis, E.P., Mendelsohn, R., and Boskey, A.L., Infrared assessment of bone quality: a review, *Clin. Orthop. Relat. Res.*, 469, 2170-8, 2011.
94. Talari, A.C.S., Martinez, M.A.G., Movasaghi, Z., Rehman, S., and Rehman, I.U., Advances in Fourier transform infrared (FTIR) spectroscopy of biological tissues, *Appl. Spectrosc. Rev.*, 52, 456-506, 2017.
95. Kazarian, S.G. and Chan, K.L., ATR-FTIR spectroscopic imaging: recent advances and applications to biological systems, *Analyst*, 138, 1940-51, 2013.
96. Miller, L.M., Vairavamurthy, V., Chance, M.R., Mendelsohn, R., Paschalis, E.P., Betts, F., and Boskey, A.L., In situ analysis of mineral content and crystallinity in bone using infrared micro-spectroscopy of the nu(4) PO(4)(3-) vibration, *Biochim. Biophys. Acta.*, 1527, 11-9, 2001.
97. Mandai, G.S. and Morris, M.D., Contributions of Raman spectroscopy to the understanding of bone strength, *BoneKEy Rep.*, 4, 620, 2015.
98. Wei, D., Chen, S., and Liu, Q., Review of fluorescence suppression techniques in Raman spectroscopy, *Appl. Spectrosc. Rev.*, 50, 387-406, 2015.
99. Boskey, A.L., Assessment of bone mineral and matrix using backscatter electron imaging and FTIR imaging, *Curr. Osteoporos. Rep.*, 4, 71-5, 2006.

16 Flame-Retardant Polyurethane Nanocomposites

María Eugenia Mena-Navarro[a], *Luis Adolfo Torres-González*[b], *Rodolfo Fabián Estrada-Guerrero*[a]

[a]Department of Physics and Mathematics, Iberoamericana University, Mexico City, Mexico

[b]Department of Science and Engineering, Iberoamericana University, León, Mexico

CONTENTS

16.1 INTRODUCTION

Recently, polymer nanocomposites have attracted extensive attention in materials science because they exhibit quite different properties from those of their counterpart polymer microcomposites, the matrices of which contain the same inorganic components. Polyurethane (PU) foams are polymers obtained through polycondensation reaction between diisocyanates and polyols, ether, or ester reacted by catalyst system. It is a thermostable polymer according to its chemical and morphological composition, interlacing with strong covalent bonds. PU foam products do not ignite spontaneously.

Moreover, PU never ignites spontaneously under a usual foam manufacturing condition, where it is made by mixing two components: the polyol component and the isocyanate component.[1] PU is an organic material that is combustible, and this means that it will burn under extreme conditions.

In this context, it does not have the capacity to prevent the spread of fire by itself, and it needs a flame retardant on its formulations. When the flame starts on PU, it initiates to melt.[2] PUs are multiblock polymers that have been widely used because of their excellent biocompatibility, abrasion, flexion, elastic resistances, and thermal properties. The versatility of urethane chemistry allows the product to be formulated to meet the exact needs of the applications.

Nanotechnology promises cutting edge and more efficient solutions for all kinds of problems faced by humanity. The products with nanotechnology have recently been introduced worldwide. It is believed that by 2020 these new materials will be found in the market more frequently in commercial products; therefore, a medium- and long-term revolution is expected on conventional materials that are currently being used, creating a strong area of opportunity for the development of technological products.[3] Nanotechnology has been able to revolutionize materials that we currently use. The products based on nanomaterials, known as *nanocomposites*, are becoming increasingly common.

Various materials have been incorporated at the nano level, such as minerals, sheets, or fibers, to improve the mechanical performance and thermal stability of the material. For example, Mena et al. patented a PU foam for sole applications with better dielectric resistance and mechanical property using multiwall carbon nanotubes (MWCNTs).[4] In addition to these attempts, nanotechnology leads to developing the nanoscale systems and particles for various applications. Nanoparticles, according to the American Society for Testing and Materials (ASTM) standard definition, are particles with lengths that range from 1 to 100 nm in two or three dimensions.[5] However, as the size of particles decreases, their effects increase because of their larger total surface area per unit volume.[6]

The organic fibers based on flame retardants for polymers, including flexible PU foams, have the carbon-forming activity supported by foaming or intumescences, which form a protective top layer on the surface of the polymer. On the other hand, flame retardants of phosphorus remain a challenge for industrial applications.[7] Plastics are highly flammable materials, increasing their risk as fire hazards when used in practical applications. Consequently, improving the flame-retardant polymers is a major challenge for extending the polymer use to most applications.[8]

National Institute of Standards and Technology (NIST) fire research continues to be positioned in areas in which measurement science and fundamental understanding of fire can reduce the risk of fire and the total social cost of fire to the nation.[9] Some of their flame-retardant polymer nanocomposites research adds nano particles to make nanoadditives, nanocoatings, and nanopolymer matrix because they can effectively reduce polymer flammability. Flexible PU foam has been coated by nanoparticles using sodium montmorillonite, MWCNTs, or carbon nanofibers (CNFs). The CNF-coated foams showed the highest reduction in flammability.[10] Zammarano et al. worked with nanoparticles as clays and CNFs to understand the effect on rate of heat release with special attention given to melt dripping on flexible PU foam.[11,12] Other studies on polymer nanocomposites have applied carbon-based nanoparticles: single wall carbon nanotube (SWCNT), multiwall carbon nanotube (MWCNT), and CNF in a polymethyl methacrylate matrix. Carbon nanotubes can surpass nanoclays as effective flame-retardant additives if they form a jammed network structure within the polymer matrix, such that the material as a whole behaves rheologically like a gel.[13] In addition, some researchers modified the isocyanate molecules, adding nanomaterials; in this case, a novel method to covalently functionalize CNTs that bear terminated isocyanate (NCO) groups reacted covalently with urethane resins and other molecules.[14]

16.2 GENERAL CONSIDERATIONS

Otto Bayer was the first to work on synthesis of PU from the reaction between a polyester diol and a diisocyanate in 1937, and the novel process of production was patented.[15] PU are a broad class of polymers produced by the polyaddition reaction of a diisocyanate or a polymeric isocyanate with a polyol, in the presence of suitable catalysts and additives.[16] The reaction is exothermic. The reactants are mixed together; next, a polymerization expansion takes place almost simultaneously.[17]

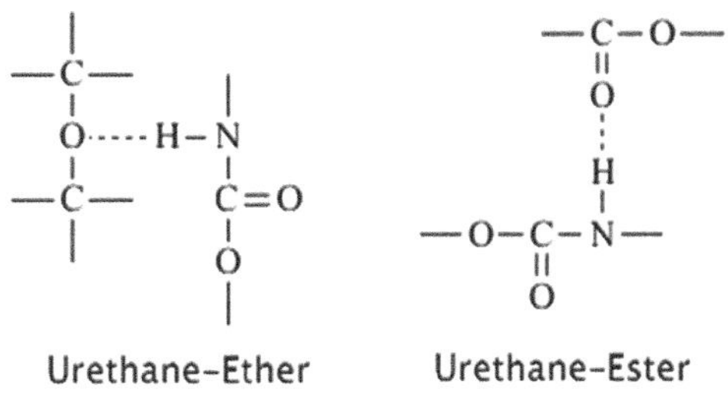

FIGURE 16.1 Schematic of urethane ether and urethane ester.

PU includes those polymers that contain a significant number of urethane groups (–HN–COO–), regardless of what the rest of the molecule might be (Figure 16.1).[18] The urethane group is usually formed by reaction between isocyanate functional group (NCO) and hydroxyl functional groups (OH), although alternative routes such as amines are used in special cases. By changing the formulation, materials can be produced with properties ranging from soft foams to relatively hard reinforced rubbers. It is common to adjust the type and content of the polyol, isocyanate, catalyst, surfactant, blowing agent, and additives in order to control the properties of the ensuing foams.

Aromatic diisocyanates, molecules that contain two isocyanate groups, are used in the production of a wide range of flexible, elastomeric, and solid PUs, as well as in adhesive and sealant formulations. Aromatic diisocyanates are more reactive than the aliphatic diisocyanates, which can only be utilized if their reactivities match the specific polymer reaction and special properties desired in the final product. PUs derived from aromatic isocyanates normally present higher glass transition, modulus, and tensile strength, but present lower elongation at break and impact resistance (Figure 16.2).[19] The degree of cross-linking is controlled by isocyanate group (–NCO) of the prepolymer and by the hydroxyl group of the polyol.

A diisocyanate and a chain extender with a more symmetrical structure increase the formation of organized structures, leading to more complete phase segregation. The chain extenders play a very important role. Without a chain extender, a PU formed by directly reacting diisocyanate and polyol generally has very poor physical properties; in this case, 1,4 butanediol ($HO-(CH_2)_4-OH$) as chain extender with prepolymer diisocyanate is used. Butanediol-1,4 is a cross-linking agent that is used because it has more structure and less flexibility because rigid foam needs hardness and stability at high temperatures.

As generally known, ether groups resist hydrolysis much better than ester groups, which make poly (ether urethanes) relatively much more stable in an aqueous environment.[20] The selection to work with polyether polyols was because the polyether system gives higher resilience with a good hydrolytic stability convenient for products that are be used in a very humid condition (i.e., construction and insulation applications). As reported in previous papers, hydrolysis decreases the thermal stability of both PUpolyether and polyesther, thus reducing their lifetimes.[21]

PU integral skin foams are the result of use of polyether resins on the polyol system matrix. The specific temperature and pressure conditions in the mold give rise to a solid outer zone and a microporous foam core. There is a gradual change in density over the foam's cross-section. This property helps to improve mechanical resistance (abrasion, tension, and tear strength) because it is

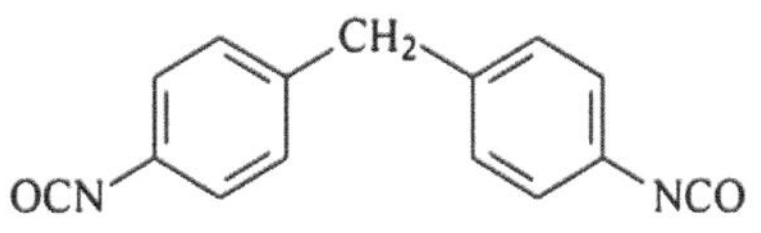

FIGURE 16.2 Formula of 4,4′-diphenylmethane–diisocyanate (MDI).

a product with less porosity on surface foam. Other advantages of PU with integral skin are that it is easy to clean, paint, finish, and polish. It is intended for use in automotive, furniture, footwear, and many other sectors.

The foaming agents are involved in the formation of the cellular structure of PU foams. There are two main types: physical foaming agents such as solvents with low boiling point (pentane, acetone, or hexane), which expand the polymer by vaporization; and chemical foaming agents such as water, which expand the polymer by the CO_2 produced.[22] Aliphatic solvent was used on the stoichiometry of nanopolyurethane for low cost.

Another essential reactant that is used in the production of PU is the surfactant. It promotes the nucleation of bubbles, stabilizes the cells, and has a significant effect on the cell size and foam air permeability. It also determines the open cell content.[23] Silicone surfactant was identified as having a significant benefit on the flammability of rigid PU foams and excellent cell structure in shoe sole applications. It can be used to reduce the amount of flame retardants required in the formulation and to provide for more consistent results when testing the foam.[24]

Catalysts are used to encourage the reaction between polyols, isocyanates, and foaming agents. Catalytic activity depends on the base and structure of the compounds. The new regulations limit the content of organic matter volatile in the products. Therefore, the current trend is toward use of catalysts that are reactive, integrated into the polymer structure, or at least lack volatility and toxicity. The organometallic catalysts improve the reaction time in the PU foam, as well as accelerate the reaction speed in the process. They play in the kinetics of the chemical reaction and also promote the formation of polymer or gelation reaction between the isocyanate and a polyol. Amines or tin catalysts are the most common catalysts, and a proper expansion is obtained through the balance between the polymerization and the gas generation. In that sense, the adjustment of catalyst type and quantity dramatically influences the expansion.[25]

16.3 BACKGROUND

PU, mainly flexible and rigid foams, is used in transportation, refrigeration, home furnishing, building and construction, marine, and business machines. For many of these products, it is necessary to add flame retardants to their chemical composition. However, since most of the end applications are internal, PU is in a critical situation and is directly subjected to increasingly stringent regulations, which on one side require high fire safety standards and on the other side limit the use of potentially toxic but extremely effective flame retardants.[26]

In previous works, Laachachi et al. commented that titanium dioxide has been used to evaluate thermal stability and reaction to fire. The thermal property was investigated after the incorporation of titanium oxide nanoparticles (TiO_2) into polymethacrylate, which is a different polymer from PU but is used on similar industrial applications.[27,28]

Research has been carried out on the thermoplastic polymer ethylene vinyl acetate (EVA), which was incorporated with clays and silica nanoparticles. However, according to the combination of the carbonized structure and the reassembly of the silicate layer, a nanocomposite of ceramic silicate was generated in layers of carbon on the surface of the material.[23–31] In the same way, the silicon-based additives also seem to provide efficient solutions to flame retardancy. However, Laoutid's research explains that the nanoparticles should be used in combination with other flame retardants to achieve the required fire performance levels. The action on the polymer of the nanoparticles alone is insufficient to ensure adequate fire resistance to meet the required standards. However, its association with other flame-retardant systems, such as phosphorus compounds, could be a very interesting approach. In several recent works, they have focused on such methods.[32] One of the most difficult problems faced while incorporating nanomaterials in nanocomposites is to evaluate their dispersability and distribution into the polymeric matrix, especially the characterization of their morphology in macroscopic systems, which involves spatial resolution and orientation distribution and requires information from the nanoscale to the macroscale.[33] The dispersion and distribution of the nanomaterials into the polymeric matrix depend not only on the compatibility between nanomaterials and the matrix, but also on the formulation and production method.[34]

Patent databases can be a part of industrial strategy to know what is new in the area of interest. Patented inventions must meet three characteristics: industrial applications, novel, and not obvious. Some industries prefer industrial secrets as a protection model rather than a patent because they do not want their know-how to be made public. However, other companies prefer the patent system because, with new technology and sophisticated equipment of measure, they can develop new materials with specific chemical compositions.

In this context, there are several interesting patents on polymer fire retardancy; some of them are on PU nanocomposites, and others are on coating applications.

For example, in 2017, Hong's patent innovated nanomodified flame-retardant PU foam using magnesium and aluminum.[35] Another innovation was made by Duan using single carbon nanotubes and graphene for a PU coating with flame resistance to reduce PU toxicity.[36] A Chinese patent by Zhenyang created PU hard foam with ignition property using nanoparticles of magnesium and nanoparticles of oxide aluminum with a particle size between 10 and 80 nm.[37]

A significant amount of research on combining flame-retardant nanocomposites with conventional flame retardants to develop more efficient materials showing improved mechanical properties has recently been reported. Analysis of nanocomposites designed to have the flammability property as the main characteristic was carried out in this chapter. PU nanocomposites have been designed with metallic nanoparticles of silver and organic nanomaterials as MWCNTs.

16.4 SYNTHESIS

16.4.1 STOICHIOMETRIC

The design of stoichiometric formulations for nanocomposite PU requires understanding the customer's requirements and the characteristics or properties of the product to be offered. Weight, density of packaging, and the resistance to high temperatures are very important terms to obtain a suitable formulation economically with the best mechanical and thermal properties. In this way, the rigid foam panels for the construction are made with little weight and a lot space volume.

The rigid PU foam was made by the follow raw materials: Polyol principal is a liquid polyether polyol; it was used as a polymer matrix in this research. It is an oxide-based triol or diol of propylene with special oxide modification of ethylene having an average molecular weight of 4000 with 800 mPas of viscosity at 25°C. Polyol graft was added because it helps to control the reaction start time; its viscosity property is 3100 mPas of at 25°C. An extra polyol component was tetrafunctional polyether polyol, which is primarily used to manufacture rigid PU foams. Hexane (C_6H_{14}) was used as a foaming agent. A chain extender was used: 1,4-butanediol ($C_4H_{10}O_2$) with a molecular weight of 90.12 and a specific gravity of 1.01 g/cm^3.

Tris 2-chloroisopropyl phosphate is an additive flame retardant with a very good hydrolytic stability used extensively in rigid PU foams, phosphorus content 9.5 wt%, and chlorine content 32.5 wt%. Its specific mass density is 1.29 g/cm^3. It was used on 8 parts by weight (p.b.w.) of the PU system because it is a cheap material. The objective was to use that additive in less percentage and mix with nanoparticles to reduce the cost on the final product because several researchers explain that organic nanomaterials are usually used as nanoscale filler for many plastics to improve their physical properties. A tertiary amine catalyst that is highly reactive and selective toward the urethane (polyol-isocyanate) reaction. Surfactant has been demonstrated to provide excellent cell control and to effectively compatibilize rigid foaming ingredients. The catalyst produces a strong catalytic activity in PU foaming reactions. Nanomaterials are used in a concentration that could provide stability on the PU system; this research proposed 0.1% and 1% of polyol system. Silver nanoparticles have <98% purity and particle sizes of 20–100 nm. MWCNTs have more than 95% purity and particle

TABLE 16.1
Stoichiometry of Flame-Retardant PU

PU System	p.b.w.	Family Name
Polyol principal	100.0	Polyol polyether
Polyol graft	11.0	Polyol polyether
Polyol tetrafunctional	3.2	Polyol polyether
Hexane	4.1	Foaming agent
1,4 butanediol	12.5	Chain extender
Flame retardant	8.0	Flame retardant
Amine	0.4	Amine catalyst
Surfactant	0.1	Surfactant
Catalyst solution	1.7	Catalyst solution
Silver nanoparticles or MWCNT	1.0	Nanomaterials
Polyol system	142.0	
Isocyanate	110.0	Prepolymer
Total PUR system	252.0	Mixing ratio 100:110

sizes of 80 nm. Both nanomaterials were from Smart Materials Company in Mexico. The specific proportion for all these compounds is described in Table 16.1. As a consideration, nanomaterials do not react with the PU system because these are not hydroxyl groups (OH) in its chemical composition. Carbon (C) or silver (Ag) adheres in the polymer chain to occupy a space and improves the physical and thermal properties in combustion reactions.

Isocyanate is an important component to produce PU foam with versatile properties. Isocyanate is a liquid modified 4,4-diphenylmethane-diisocyanate (MDI) prepolymer. It is typically used in combination with high-molecular-weight, reactive polyether polyols for manufacturing microcellular PU foams (integral skin and semi-flex). Its –NCO percentage weight was 23% at 700 mPas of viscosity at 25°C.

It is an aromatic diisocyanate; the molecules contain two isocyanate groups. It is typically used in combination with high-molecular-weight, reactive polyether polyols in the manufacture of high performance flexible elastomers and specialty foams with integral skin (Figure 16.3). Mixing ratio between both compounds is as follows: 100 p.b.w. for polyol system and 110 p.b.w. for isocyanate, because this is rigid foam formula. That means 100 k for PU system consume 110 k of isocyanate.

The chemistry of PU consists of combining resin (polyol) and isocyanate (MDI) in the correct amounts. If there is too much MDI or too much polyol, the chemical reaction will not be complete. The mixing ratio measures the specific amounts of resin and isocyanate combined in the reaction. In order to calculate the mixing ratio, the following is required:

(a) Hydroxyl content or –OH number of the polyol system, specified in the supplier's technical sheet.
(b) Percentage of water present in the resin, specified in supplier's analysis certificate.

Isocyanate Prepolymer **Polyol**

FIGURE 16.3 PU chemistry.

(c) Isocyanate content –NCO or NCO percentage of the MDI, specified in the supplier's technical sheet.

For calculating MDI amounts, use Equation 16.1:

$$\mathrm{MDI} = \frac{[(\text{polyol system OH} * 100) + (\text{water\%} * 6233.3)]}{[\text{Number of NCO} * 13.35]} \tag{16.1}$$

where 100 is the total of the resin p.b.w., 6233.3 is a constant number for H_2O, and 13.35 is a constant number for NCO.

Substituting data in Equation 16.1:

Polyol system water % = 0.21%, OH number on polyol system = 325, and NCO wt% = 23.

$$\mathrm{MDI} = \frac{[(325 \times 100) + (0.21 \times 6233)]}{23 \times 13.35}$$

$$\mathrm{MDI} = 33,808.93/307 = 110.1$$

Therefore, 100 kg for PU system (325 mg KOH/g) consume 110 k of MDI (23% NCO).

16.5 EXPERIMENTAL SETUP

The synthesized PU nanocomposites could be characterized by using the techniques listed in this section. Brief description of these techniques and methods is given in the following section.

16.5.1 Spectral Analysis

Ultraviolet and visible spectroscopy (UV–Vis) for PU nanocomposites is performed with ultraviolet visible light and near-infrared spectrophotometer. Pholnak's work proposed a UV–Vis analysis by absorption on PU coating filled with zinc oxide for economic large-scale production of composites.[38]

It is a technique of spectroscopy by molecular absorption in the UV–Vis region based on the measurement of transmittance or absorbance with defined length. Normally, the analysis of concentration of an absorbent is related linearly with the absorbance according to the Beer–Lambert law (or Beer's law). The UV–Vis spectrum will give us the characteristic band of the presence of nanoparticles.[39] The UV characteristics can be measured by UV–Vis spectrometer.

Fourier transform infrared spectroscopy has two analyses; they are reflectance and absorbance. Reflectance means that light is used in the visible and adjacent spectrum range (black light and next IR). In this region of the electromagnetic spectrum, the molecules undergo electronic transitions. The samples are mostly liquids, although the absorbance of gases and even solids can also be determined.[40]

X-ray dispersion is a nondestructive analytical technique that provides information on the crystallographic structure, chemical composition, and physical properties of materials. X-rays are electromagnetic radiations with typical photonic energies between 100 eV and 100 keV. The wavelength of X-rays is comparable to the size of atoms, and they are ideal for analyzing the structural arrangement of atoms and molecules in a wide range of materials. The energy of the X-rays can penetrate deeply into the materials and provide information about the mass structure.[40] X-ray diffraction (XRD) data indicate partial intercalation between nanomaterials and compounds.

Energy dispersive spectroscopy (EDS) allows one to identify what the particular elements are and their relative proportions (e.g., atomic %). EDS is not a surface analysis technique, but it is a qualitative analysis to understand the chemical information of each element detected by the spectrum analysis on the sample.[41] Detection area is limited by the resolution of scanning electron microscopy (SEM).

16.5.2 Thermal Analysis

Gas chromatography system–mass selective detector could be useful to identify the volatile compounds resulting in the combustion of gases on the PU nanocomposite foams and to understand if those compounds can produce toxicity. This spectroscopy technique is proposed to identify volatile compounds and their chemistry composition, which are qualitative analyses of volatile compounds.

Retention time (RT) is a measure of the time taken for a solute to pass through a chromatography column. It is calculated as the time from injection to detection—an analyte's retention time (RT) (t_R) minus the elution time of an unretained peak (t_M). Adjusted RT is also equivalent to the time the analyte spends in the stationary phase.[42] It is measured in minutes and can be calculated from the following formula

$$t'_R = t_R - t_M \tag{16.2}$$

where t'_R is adjusted retention time in minutes, t_R is analyte's RT, and t_M is elution time of an unretained peak.

The adjusted RT is not very significant by itself, but it plays an important role in another retention measurement, the retention factor (k). The retention factor is the ratio of the adjusted RT to the unretained peak time. In other words, it expresses peak retention in terms of multiples of the unretained peak time. The formula for the retention factor is shown in Equation 16.3:

$$k = \frac{t_R - t_M}{t_M} \tag{16.3}$$

Qualitative analysis relies on comparing the RT of the peaks in an unknown sample with those of known standards. If the RT of a peak in the unknown sample is the same as the standard, then a positive identification can be made. Software Agilent provides RT, percentage of area peaks, and compound names based on the NIST Mass Spectrometry Data Center, which are useful for gas chromatography (GC)–mass spectrometry (MS) analysis. Compounds were carried at 35–310°C of temperature, with temperature ramps of 4–20°C/min. This is a sampling method regularly used to analyze benzene, toluene, ethyl benzene, and xylenes elements and their quantification of volatile compounds in paints and in determination of residual solvents in polymers, among many other applications.[43]

16.5.3 Surface Morphology

Numerous studies of the PU morphological features and phase segregation behavior have been carried out by means of different imaging techniques.[44] Some of them are listed in the following section.

SEM is an analysis capable of producing high-resolution images of the surface of a sample usingelectron–matter interactions. It uses an electron beam instead of a beam of light to form an enlarged image of the surface of an object. It is an instrument that allows the observation and superficial characterization of inorganic and organic solids. It has a great depth of field and gives three-dimensional appearance to the images. This equipment can determine the particle size, morphological comparison, and chemical composition of raw materials; it can help polymer matrix to measure pore diameters of their structure and also to understand the internal nanostructure before and after combustion.

MINITAB analysis is useful in the observation and interpretation of pore diameters from porous PU, as well as from the reinforced materials mixed and supported, in order to obtain the equations of mathematical models, at the same time improving the performance of the synthesis of the different materials reinforced.[45] This technique could determine the topographic contrast between the surface and roughness.

16.5.4 Mechanical Testing

Mechanical properties can be evaluated according to international standard test methods of the ASTM or by *International Standard Organization*. The proper tests are selected depending of the material and its applications.

Hardness ASTM F1957-99 is a test method based on the penetration of a specific type of indentor when forced into the material under specified conditions. The shore hardness is applied by measurement devices known as durometers. It is determined in numbers ranging from 0 to 100 on the Shore A° scale for soft polymers or Shore D° scale for hard polymeric materials. This test method is an empirical test intended, primarily, for control purposes.[46,47] Manual operation (handheld) of a durometer will cause variations in the results attained. Improved repeatability may be obtained by using a mass securely affixed to the durometer and centered on the axis of the indentor.

Flammability test ASTM D635 is the more appropriate method for the evaluation of PU nanocomposites. The test scope is used to determine the relative rate of burning plastics. This test method was developed for polymeric materials used for parts in devices and appliances. The results are intended to serve as a preliminary indication of their acceptability with respect to flammability for a particular application. This standard is used to measure and describe the response of materials to heat and flame under controlled conditions, but it does not by itself incorporate all the factors required for fire hazards or fire risk assessment of materials, products, or assemblies under actual fire conditions.[48] The rate of burning is determined by measurements of the horizontal distance burned in relation to the time of burning and reported for each set of specimens. The rate of burning is affected by such factors as density, direction of rise, and type and amount of surface treatments. The thickness of the finished specimens must also be taken into account. These factors must be considered in order to compare materials on the same basis.[49] Calculate the burning rate from the following Equation 16.4:

$$B = \frac{D}{T}(60) \tag{16.4}$$

where B is burning rate, mm/min; D is length the flame traveled, starting from the first scribed line, mm; and T is time for the flame to travel distance, s.

This test method is intended for use as a small-scale laboratory procedure for comparing the relative horizontal burning rates of polymeric materials.

Mechanical tensile test method ASTM D412 was used to evaluate the tension properties of thermoplastic elastomers, which are the case of rigid PU nanocomposites.[50] Yield strain is the strain or elongation magnitude, where the rate of change of stress with respect to strain goes through a zero value. See Equation 16.5:

$$TS = \frac{F}{A} \tag{16.5}$$

where TS is tensile strength, the stress at rupture, MPa (N/cm^2); F is force magnitude at rupture, N; and A is cross-sectional area of unstrained specimen, cm^2.

Test method ASTM D642 describes procedures for measuring the resistance to tear strength property. Tear strength may be influenced to a large degree by stress-induced anisotropy (mechanical fibering), stress distribution, strain rate, and test piece size.[51] The tear strength T is given in kilonewtons per meter of thicknesses, by Equation 16.6:

$$T = \frac{F}{d} \tag{16.6}$$

where F is the maximum force, N; and d is the thickness of the test pieces, mm.

All these analyses explained previously (SEM, UV–Vis, XRD, and others) are complementary techniques useful for the characterization of those nanocomposites or nanomaterials. Furthermore, these techniques provide information to help derive meaningful relationships between the nanostructure and macroscale properties.

16.6 METHODOLOGY

16.6.1 Samples Prepared

PU nanocomposites were made by the synthesis of several polyether resins described previously. Nanoparticles were supplied by Smart Material; these did not synthesize because these are not the main thing in this research. The industry needs a reliable supplier chain with competitive costs. In this context, purity of more than 95% and nanoparticle size between 20 and 80 nm are considerable parameters to choose them, having a cost–benefit.

Experimental processing parameter values for in situ reaction process of PU and PU nanocomposites were as follows: low-pressure machine return speed at 6.20 min/cycle, mold temperature at 45°C, cooling time at 13 s, isocyanate at 28°C, PU system at 22°C, flow rate of 6.67 g/s, injection pressure for isocyanate of 20 pounds, and injection pressure for PU system of 40 pounds. Cooling time was 4–6 min, depending on the piece length, and it could up to 30 min. For pieces with sizes over 90 cm long or wide, high-pressure machines are recommended.

Different samples were prepared in three groups—the control PU without nanoparticles, the PU nanocomposites with silver nanoparticles (PUNAgs), and the PU nanocomposites with MWCNTs—to create a specific stoichiometry with 0.1% and 1% of the nanomaterials based on the total PU system. The industrial sector has detected an opportunity for rigid PU nanocomposite to manufacture products with applications in the construction sector; it could be useful on the roof and on walls, stones, or wood imitations.[52]

Different samples of PU rigid were studied in this work, which are products of prepolymer and polyols polymerization. The samples were cut with a measure of 2 cm in length, 2 cm in width, and 0.2 cm in thickness, for inserting on the internal plate of spectrometric equipment, as shown on Figure 16.4.

16.6.2 Sample Characterization Techniques

SEM analysis was performed with an SEM Jeol model JSM-7800F to evaluate the nanocomposite with silver nanoparticles before ignition and after combustion. SEM technique is useful to obtain the micrograph of the nanomaterial, and it is used for checking supplier specifications (length, nanoparticle size, and purity). EDS analysis was a complement to find the spectrum and chemical composition of the nanoparticles.

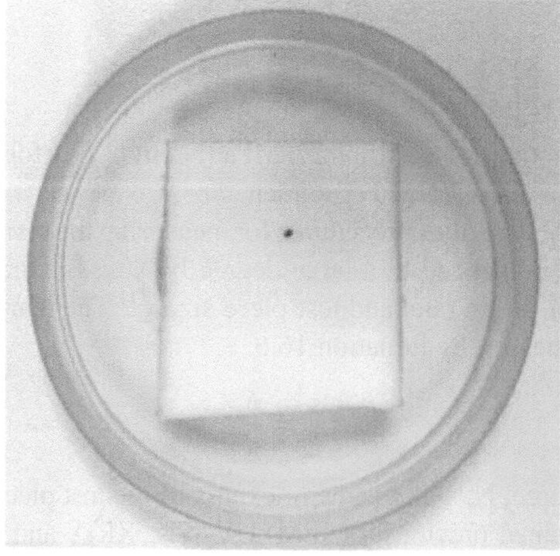

FIGURE 16.4 Sample of PU nanocomposite on the internal plate for spectrometry equipment.

GC–MDS analysis determined the volatile compounds on the samples. Sample was analyzed by a gas chromatography Agilent 7890B system coupled to a mass-selective Agilent 5977A detector with a capillary column (30 m × 250 μm ID × 0.25 μm film thickness from Agilent Technologies). Retention time is the information resulting about each component.

XRD was applied to collect information on the nanometer scale physical structure of the PU nanocomposites, performed by D2 PHASER equipment.

Infrared spectra were measured with an Agilent spectrophotometer model (Cary 670) in the wavenumber range of 400–4000 cm^{-1}. Radiation in this region can be utilized in organic structure determination by making use of the fact that it is absorbed by interatomic bonds in organic compounds. The frequencies at which there are absorptions of IR radiation ("peaks" or "signals") can be correlated directly to bonds within the compound in question.

UV–Vis analysis for PU nanocomposite and the control sample was performed with ultraviolet visible light and near-infrared spectrophotometer Cary 5000 UV–Vis NIR equipment used to measure absorbance and reflectance percentage.

16.7 RESULTS AND DISCUSSION

According to the experiment's results, the specific characteristics of nanomaterials could be used to obtain better mechanical properties and flame resistance. The results show that nanocomposite incorporated by silver nanoparticles into the polymeric matrix produces self-extinguishing flame property in the composite material.

In this context, we propose to use nanocomposites as an appropriate strategy to solve problem of fire propagation. SEM analysis shows a micrograph of top section morphology of unburned PUNAgs at 10 μm resolution; using mathematical analysis, we can determinate the pore diameter's average; for this sample, the pore diameter average is 88 μm (Figure 16.5A).

Comparative before and after tests on burns could be useful to understand how the oxidation is produced after combustion on the polymer. Figure 16.5B shows an SEM micrograph of top section morphology postburn PUNAg. The image does not have an even and regular surface as compared with the unburned sample.

EDS analysis, as explained earlier, could be applied to determine the chemical composition of the nanocomposites. Figure 16.6A shows the spectrum of PUNAg at 25 μm of resolution. In Figure 16.6B, the chemical compositions of the most important elements are shown on the graph; they are 85.8% of oxygen, 11.7% of silica, 2.4% of chlorine, and 0.1% of silver. Carbon also appears, but this element is not considered on that analysis because it is used as a substratum on the equipment. Silver nanoparticles' concentration was used at 1% based on polyol system. According to Kargarzadeh et al., nanomaterials are not completely mixed on polymer matrix because it depends on the production

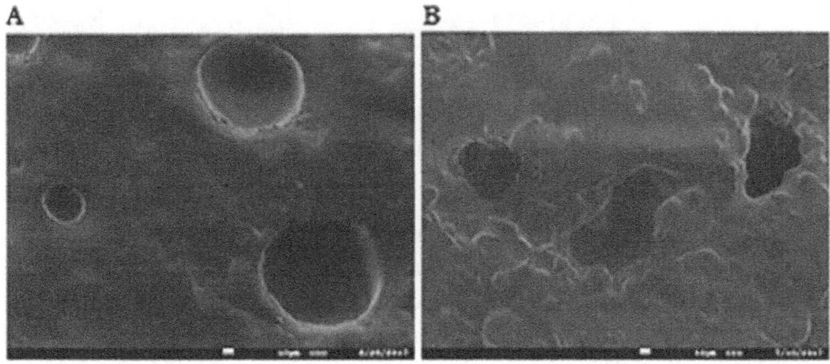

FIGURE 16.5 SEM micrographs: (A) Morphology of unburned PUNAg. (B) Morphology of postburn PUNAg.

A.

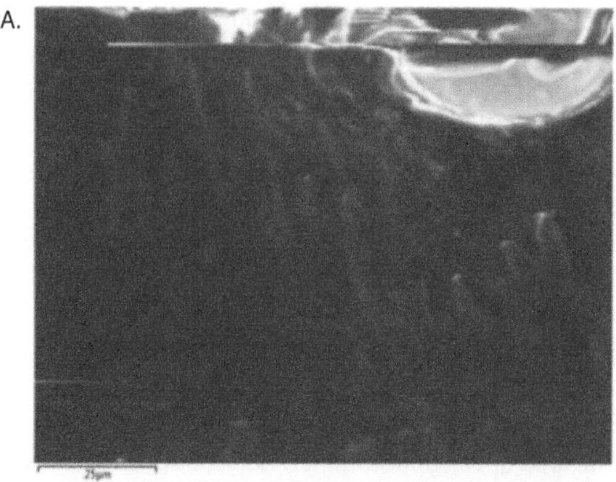

B.

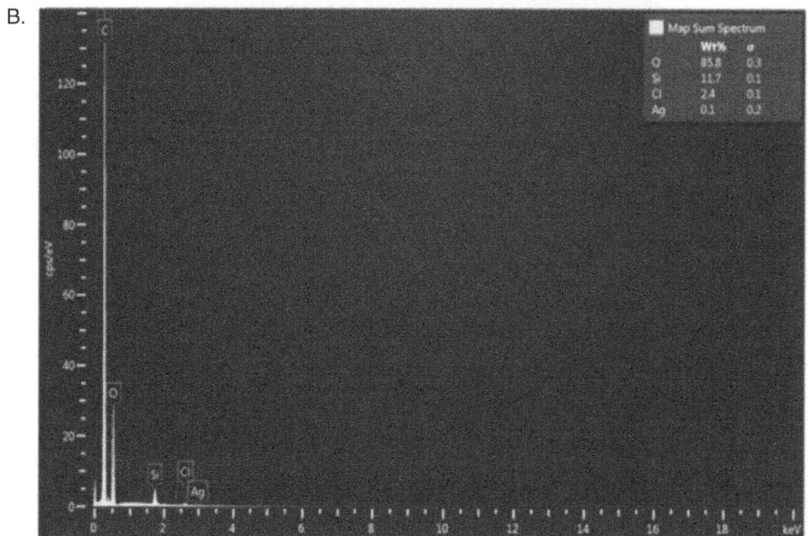

C.

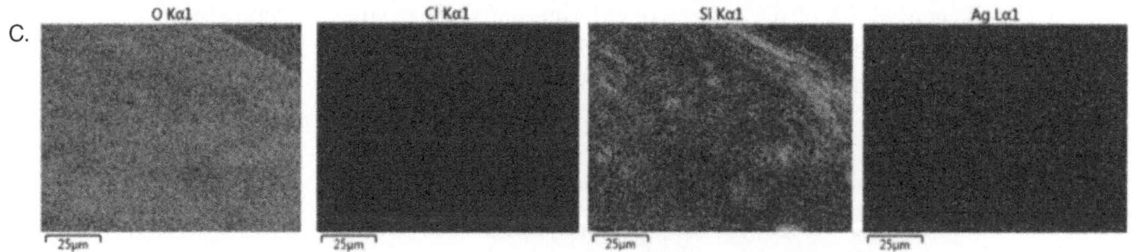

FIGURE 16.6 (A) Electron image of PUNAg at 25 μm of resolution. (B) EDS analysis and chemical composition of PUNAg. (C) EDS analysis of PUNAg at 25 μm of resolution.

method (e.g., blending).[33] Chlorine element is a part of the retardant additive composition. However, oxygen is the major element on the PU composition. Figure 16.6C shows spectrums of each element as well as their superficial distribution located on the PU nanocomposites.

GC–MDS analysis determined the volatile compounds, and it was also used to identify the chemical composition on samples. The first component to appear was the 2-(vinyloxy)-ethanol on PU control; it is not toxic because it results in 0.03% area of degradation. The second compound was the hexane on PU control, with a 0.83% area of degradation. It is the material used as foaming agent.

However, hexane on PUNAg was detected at 2.154 min. Other interesting material was the benzene, which is an important organic chemical compound composed of six carbon atoms joined in a ring with one hydrogen atom attached to each. It is classified as a hydrocarbon. This component is an essential part of prepolymer MDI; it is a human carcinogen compound with a result of 15.11% of the area of degradation on PUNAg but less percentage of area without flame retardant and without nanoparticles, which obtained 1.44% of the area. GC–MS technique was reliable and useful for the analysis.

However, at 23 min, tripropylene glycol emerged with 16.59% of the area, but on safety data sheets, it is cataloged as not hazardous on control sample. In comparison to PUNAg, tripropylene glycol appeared 1 min before at 22.9 min but with a minus percentage of only 8.12% of the total. Flame-retardant additive emerged at 28.92 min with 25.66% as the area of degradation. All compounds are shown in Figure 16.7. The American National Fire Protection Association established a time of 240 s for firemen to arrive at the fire scene.[53] In case of a fire situation, a person would inhale toxic gases produced by chloroisopropyl phosphate at 24 min of start of combustion.

Two different concentrations of silver nanoparticles were evaluated at 1% and 0.1% FTIR analysis performed in nanocomposites with silver nanoparticles is shown in Figure 16.8. It shows that the higher concentration generated lower absorbance, but when less concentration is obtained, more percentage of absorbance is generated. In zones with a wavenumber between 970 and 1755 cm^{-1}, the most abrupt changes are noted with the lower concentration percentage of silver nanoparticles, as well as another zone between 2825 and 3396 cm^{-1}. Characteristic bands of the hydroxyl and carbonyl groups were obtained in the 1726 and 3373 cm^{-1} regions. The two bands observed between

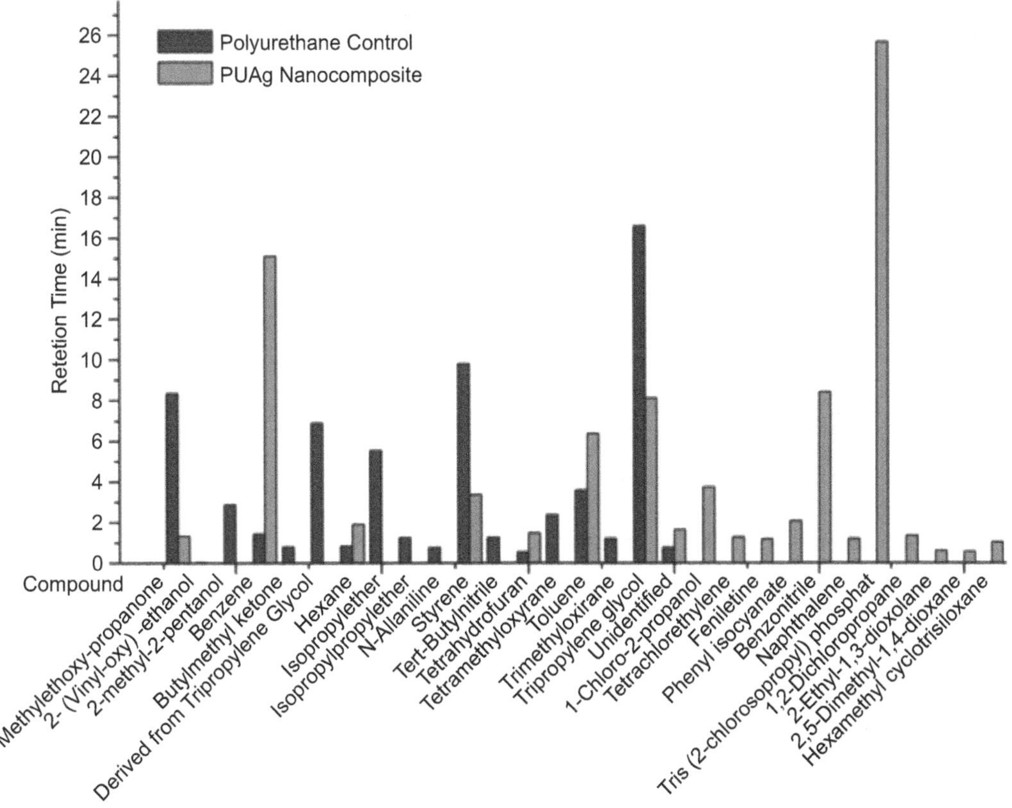

FIGURE 16.7 GC–MDS analysis of PU control and PUNAg flame retardant.

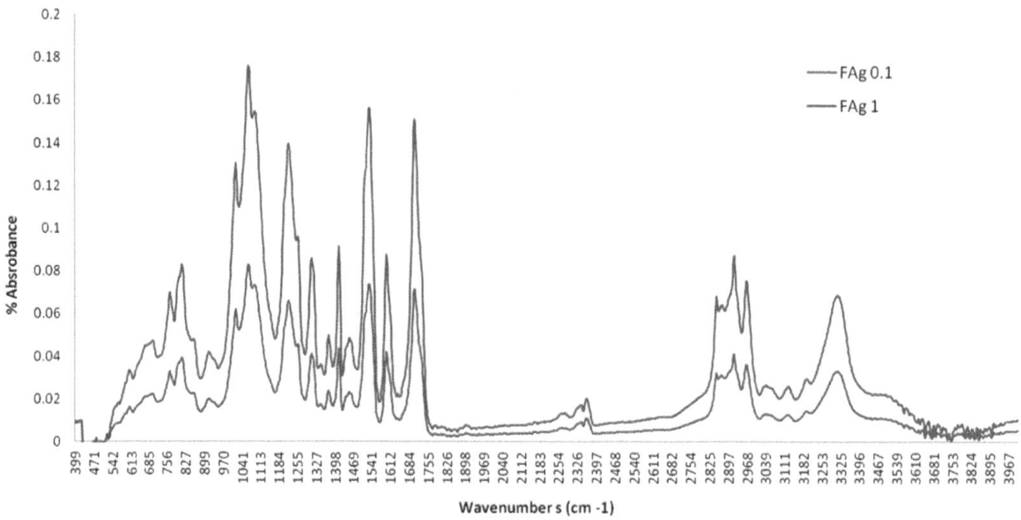

FIGURE 16.8 FTIR spectrums for PU nanocomposites with two different concentrations of silver nanoparticles.

2845 and 2938 cm^{-1} were attributed to symmetric and nonsymmetric stretching of the C=H bond with carbonyl. The two bands observed between 1630 and 1690 cm^{-1} were attributed to stretching of the C=O amide. The weak band found at 2312 cm^{-1} is assigned to isocyanates region and polymerized urethanes in 1520, 1570, 1608, and 1736 cm^{-1}. These bands are typical of the stretching of C=O and N=H bonds. Bands of amine groups were obtained in the 3325 of stretching of N=H amine. The bands observed between 3500 and 3700 cm^{-1} were attributed to stretching of N=H amide.

XRD analysis shows diffractogram of rigid PU without nanoparticles and two samples with different concentrations of 1% and 0.1% of silver nanoparticles. The diffracted intensities were recorded from 5° to 70° in 2θ. Four strong Bragg reflections were recorded at 12.5°, 18.5°, 20°, and 35°. The 2θ scan range was 5–40°, because all the intense crystalline peaks were found in this range.[54] The degree of order in these materials was established in relation of the polyol system, isocyanate, and nanomaterials mass proportions. Figure 16.9 shows the XRD patterns of the rigid PU.

PU control is observed upper than nanocomposites that are below the control sample. The greatest differences between them are observed from 13° to 23° in 2θ and also from grade 27° to 46° in 2θ for both PU silver nanocomposites. Results show that the degree of crystallinity is controlled by the ratio between isocyanate and polyol.

However, silver nanocomposites with the same prepolymer ratio had less crystallinity compared with samples. The change in crystallinity can be explained by the degree of cross-linking of the samples, which is controlled by the hydroxyl group existing in the polyol, in this case associated with the cross-linking made by silver nanoparticles.

The decrease in crystallinity of the PU samples is evidenced mainly by the disappearance of the peak located at 11°. The rigid PU sample had the most intense peaks located at 11° and 19° in 2θ. In this case, PU control sample has complete trifunctional active sites, which increase the degree of crystallinity in the structure.

According to the references cited previously, some studies used carbon nanotubes on the polymer matrix to obtain better fire resistance and mechanical property. In this context, we performed a PU nanocomposite with MWCNTs of 95% purity and 80 nm in length (Figure 16.10). Several concentrations could determine the best stoichiometry for the greatest cost–benefit of the product.

X-ray as a nondestructive technique was applied with two PU nanocomposites with different concentrations of MWCNT (0.1% and 1%) to compare with a sample of PU control. These

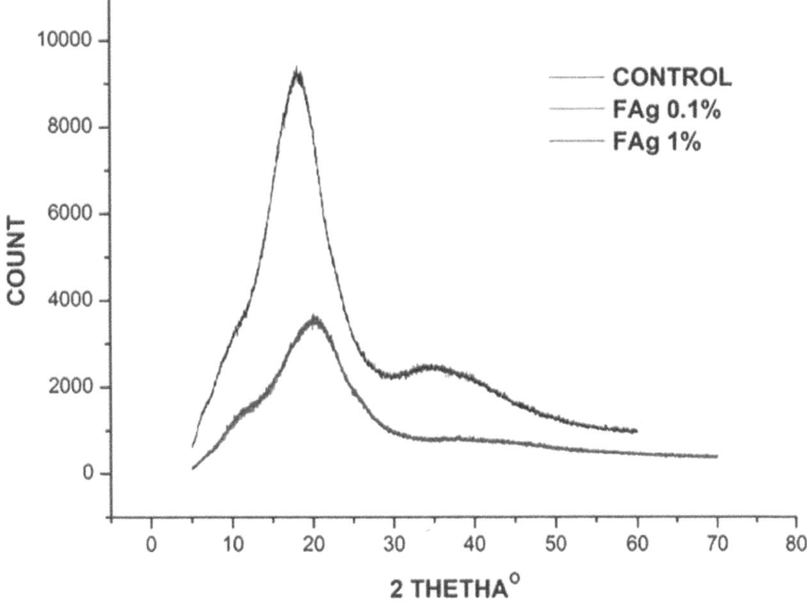

FIGURE 16.9 XRD diffractogram of PU nanocomposites with different concentration of silver nanoparticles.

diffractograms exhibit broad peaks at angles around 8°, 11°, 15°, 20°, and 33°, indicating some degree of crystallinity.

The results show that the evidence of MWCNT nanocomposites are upper than control PU on the range of 10° to 15° in 2θ and has the same results until 19° that change the crystallinity for MWCNT started to decrease. However, after that range is below the control sample, there are two

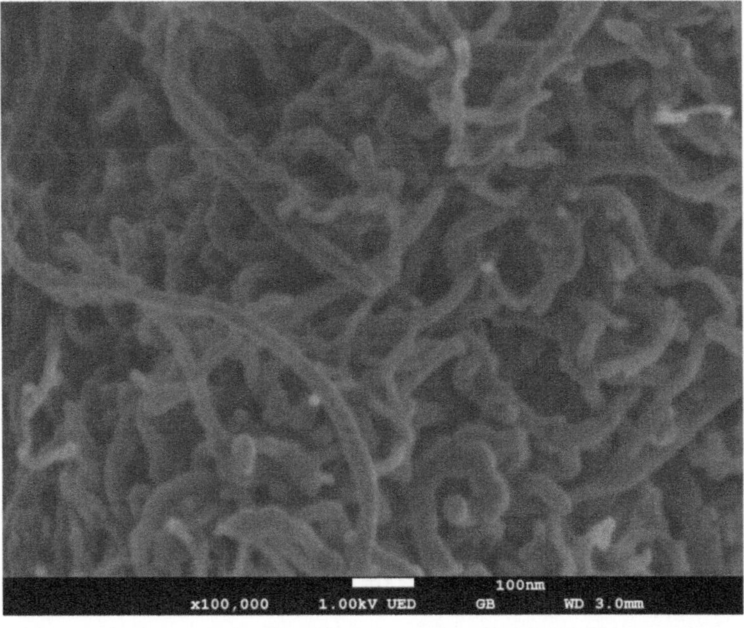

FIGURE 16.10 SEM micrograph of MWCNTs at 100 nm of resolution.

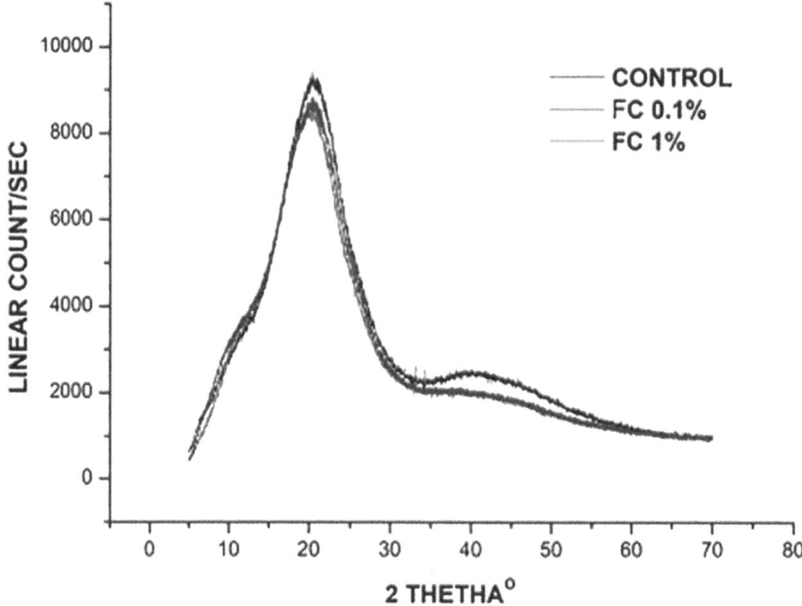

FIGURE 16.11 XRD diffractogram of PU nanocomposites with different concentration of MWCNTs.

peaks at 33° and 34° in 2θ on PU nanocomposites with 0.1% of MWCNTs (Figure 16.11). It is important to mention that MWCNT composites are more similar regarding crystallinity than silver nanocomposites compared with PU sample. Crystallinity correlated with porous carbon materials.

UV–Vis analysis is used to evaluate the absorbance and reflectance of the nanocomposites. The control curve shows less absorbance than nanocomposites; the last one has the best percentage of absorbance, which started at 1.2% and finished at 1.4% with 350 nm of wavelength (Figure 16.12). In contrast reflectance results demonstrated similar graphs on nanocomposites but totally different graphs with PU control (Figure 16.13).

Those characterizations explain the influence of nanomaterial concentration on polymer nanocomposites. Hardness ASTM D2240 tests were evaluated using PTC durometer. Control sample obtained a result of 85 Shore A°, PU silver nanocomposite resulted in 87 Shore A°, and PU nanocomposite with MWCNT resulted in 88 Shore A°.

The ASTM D635 flammability test was performed to evaluate the fire-resistant property of PU nanocomposites using Atlas horizontal flame chamber equipment. The method includes three specimens of the same sample, which need to be 125 mm in length, 130 mm in width, and 10 mm in thickness. A white line was drawn 50 mm long in order to identify whether the flame was extinguished before this displacement, resulting in a self-extinguishing material. The test specimen was supported horizontally at one end, and the free end was exposed to a gas flame for 15 s inside the Atlas horizontal flame chamber, as shown in Figure 16.14.

After the removal of the flame, the test specimen was observed for the time and extent of burning. The test was run for the three samples. After that, the distance and time of burning were measured to find out whether or not the flame in these samples is self-extinguishing (Table 16.2). PU nanocomposites of silver nanoparticles with MWCNT resulted in self-extinguishing as well as in control samples. The differences between the conventional PU and the nanostructured PU were smoke, combustion time, and melting degradation. Flame-retardant PU nanocomposite must pass stringent flammability tests in order to be approved for use.

Data from specimens were analyzed and compared in order to obtain the best rigid nanocomposite PU with fire-retardant property. After tests, rigid PU control resulted in self-extinguishing, with

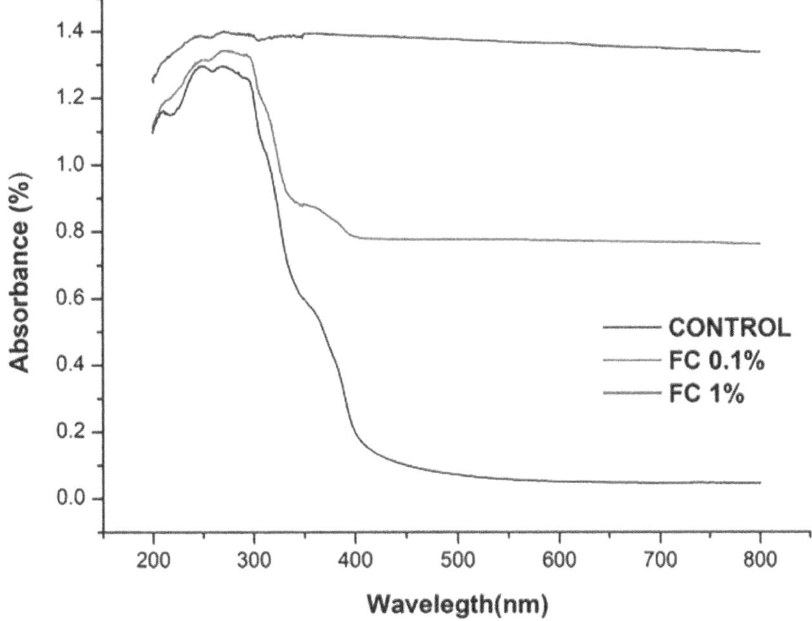

FIGURE 16.12 UV–Vis absorbance of PU nanocomposites with different concentration of MWCNTs.

a little bit of smoke but a lot of flame melt, as shown in Figure 16.15. Rigid PUNAg was self-extinguishing, with a lot of smoke and constant flame melt, as shown in Figure 16.16.

Results of horizontal burning rates of PU nanocomposites and control PU show that PU without nanomaterials had an average burning rate of 70.20 mm/min; however, PURAg obtained a burning rate of 18.28 mm/min, and PURCNT demonstrated an average burning rate of 16 mm/min, which was less than the other samples (Table 16.3).

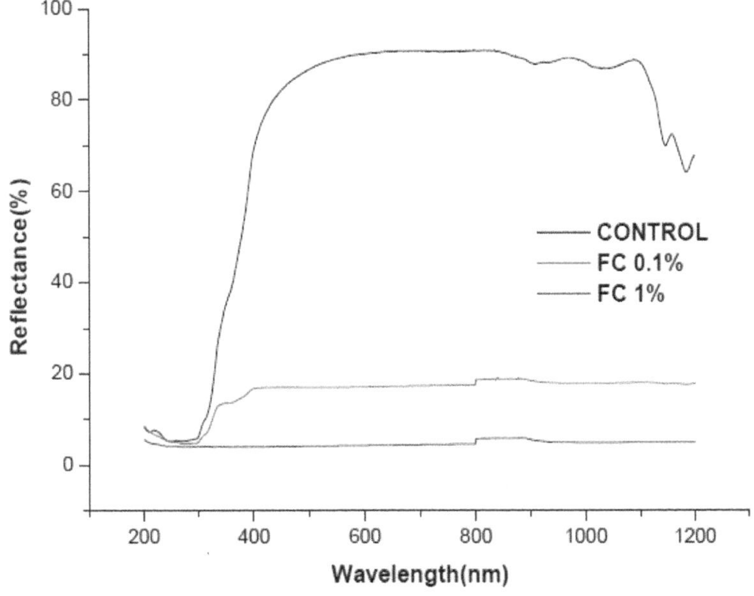

FIGURE 16.13 UV–Vis reflectance of PU nanocomposites with different concentration of MWCNTs.

FIGURE 16.14 Initiation of flammability test on PU nanocomposite with MWCNT.

TABLE 16.2
Control PU and PU Nanocomposites with Silver Nanoparticles and MWCNTs Results Determined Experimentally According to ASTM D635 Test

Sample	Time of Burning	Result	Standard ASTM D635	Visual Observation
Ctrl PUR 1	12 s	Self-extinguishing	Accepted	Flaming
Ctrl PUR 2	13 s	Self-extinguishing	Accepted	Flaming
Ctrl PUR 3	13 s	Self-extinguishing	Accepted	Flaming
PURAg 1	14 s	Self-extinguishing	Accepted	Flaming, dripping
PURAg 2	14 s	Self-extinguishing	Accepted	Flaming, dripping
PURAg 3	15 s	Self-extinguishing	Accepted	Flaming, dripping
PURCNT 1	15 s	Self-extinguishing	Accepted	Flaming
PURCNT 2	15 s	Self-extinguishing	Accepted	Flaming
PURCNT 3	15 s	Self-extinguishing	Accepted	Flaming

Ctrl PUR, control PU; PURAg, PU nanocomposites with silver nanoparticles; PURCNT, PU nanocomposites with MWCNTs.

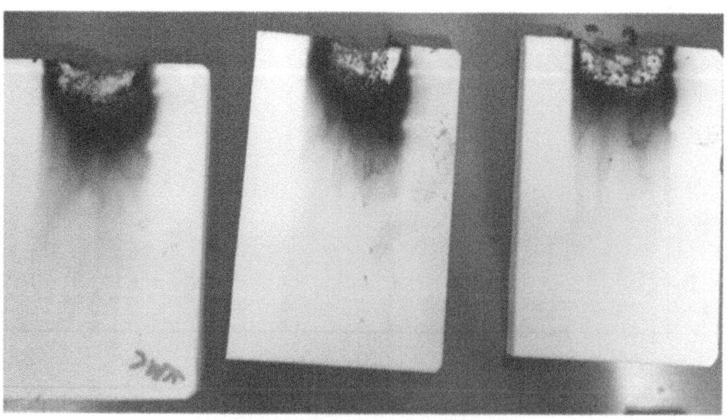

FIGURE 16.15 Rigid PU control after ASTM D635 test.

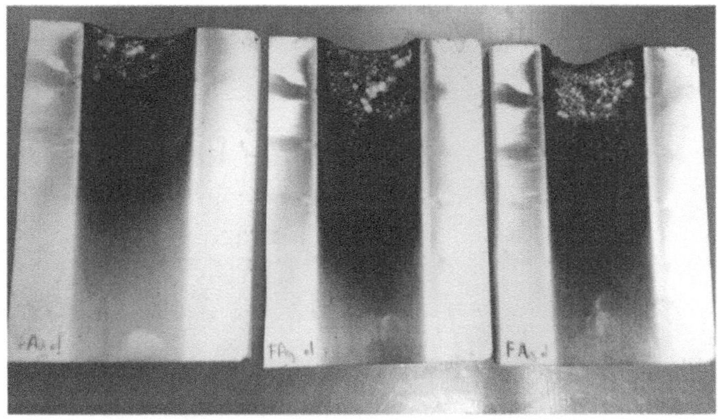

FIGURE 16.16 Rigid PU nanocomposite PURAg after ASTM D635 test.

Instron universal testing machine model 3365 was used to perform tension resistance and tear strength tests. Tensile property depends both on the material and the conditions of the test (extension rate, temperature, humidity, specimen geometry, pretest conditioning, etc.); therefore, materials should be compared only when tested under the same conditions. It is also important to note that the samples are destroyed because it is an invasive method, as shown in Figure 16.17.

Five samples of PU nanocomposite with MWCNT were evaluated for tension resistance. Results show a maximum value of 949.3 N/cm^2 and a minimum value of 736.5 N/cm^2, with an average value of 884.78 N/cm^2. In addition, the percentage of resistance to elongation for PU nanocomposite with MWCNT was ~46.78%. However, tear strength was 163 N/cm of average, as shown in Table 16.4. All of these results are satisfactory according to the evaluated standards for mechanical properties.

PU nanocomposite with MWCNT was self-extinguishing, with a lot of smoke and a few flame melts, as shown in Figure 16.18. Both of the nanocomposites could be dangerous if the material is used in high places because PU nanocomposite melts during combustion, and some drops could fall on a flammable material, spreading the flame or burning someone. The results show that the incorporation of organic and metallic nanomaterials into the polymeric matrix produces the self-extinguishing flame property in the composite PU.

The nanocomposites with silver nanoparticles and MWCNT were self-extinguishing. The main function of the nanomaterials was to reduce combustion by eliminating oxygen, preventing it from continuing to burn. In this context, we propose that the use of these nanocomposites is an appropriate strategy to solve the problem of fire propagation.

TABLE 16.3
PU Nanocomposite Burning Rate

Sample	Distance, mm	Time, s	Burning Rate, mm/min
Ctrl PUR 1	4	12	19.80
Ctrl PUR 2	3	13	13.80
Ctrl PUR 3	8	13	36.60
PURAg 1	4	14	17.43
PURAg 2	4	14	17.43
PURAg 3	5	15	20.00
PURCNT 1	3	15	12.00
PURCNT 2	4	15	16.00
PURCNT 3	5	15	20.00

FIGURE 16.17 Samples of PU nanocomposites with MWCNT for tests of tension resistance and tear strength.

TABLE 16.4
Mechanical Test for PU Nanocomposite with MWCNTs

				Results					
	Norm	Test	Unit	*Sample 1*	*Sample 2*	*Sample 3*	*Sample 4*	*Sample 5*	*Average*
PU with MWCNT	ASTM D412	Tension resistance	N/cm^2	880.7	736.5	949.3	973.8	883.6	884.78
		Percentage of resistance to elongation	%	52.89	37.21	53.42	50.82	39.55	46.78
	ASTM D624	Tear strength	N/cm	161.8	161.8	162.8	163.8	164.8	163
	ASTM D2240	Hardness	Shore A°	88	87	88	87	88	88

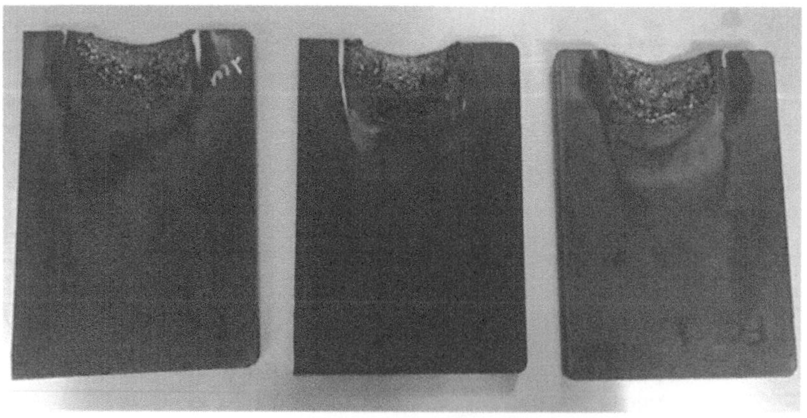

FIGURE 16.18 Rigid PU nanocomposite PURCNT after the ASTM D635 test.

Development and innovation of new materials are likely as polymer nanocomposites have been tested and demonstrated useful for safe human application.

Therefore, new materials can be used in the furniture and construction industries with the fire-resistant property to avoid the propagation of fire and can also be used to provide fire safety and improved properties for a wide range of consumer goods.

Flammability property is an area of opportunity for innovation regarding functional nanocomposites. The nanocomposite could be applied to the automotive industry or construction. It can also be used on soles for safety shoes or industrial boots for fireman, oilman, and any other industries.

16.8 ACKNOWLEDGMENTS

This research was supported by the company Development and Finished on Polyurethane (DAPSAPOL).

REFERENCES

1. Q & A on Fire and Fire Prevention of Rigid Polyurethane Foam; Japan Urethane Industry Institute (JUII), Japan, **2011**.
2. Fishback, T. The value of polyurethane foam sealants in mine ventilation. In Mine ventilation: Proceedings of the North American/Ninth US Mine Ventilation Symposium, Kingston, ON, **2002**, 85–88.
3. Mena, M.E., Torres, L.A., Martinez, M.L., Mena, F.J. Technology transfer model based on nanotechnology for the footwear sector. *Ide@s CONCYTEG* **2012**, *7* (87), 1137–1148.
4. Mena, M.E., Torres, L.A., Gonzalez, J.L., Mena, F.J. Dielectric protection sole with high shape memory. Mexican Patent WO2013012306, **2013**.
5. ASTM E 2456-06 Terminology for Nanotechnology, ASTM International, West Conshohocken, PA, **2006**.
6. Erdem, N., Cireli, A.A., Erdogan, U.H. Flame retardancy behaviors and structural properties of polypropylene/nano- SiO composite textile filaments, *J. Appl. Polym. Sci.* **2008**, *111* (4), 2085–2091.
7. Lvov, Y., Guo, B., Fakhruddin, R.F. Functional polymer composites with nanoclays. *R. Soc. Chem.* **2017**, *4*, 152-155.
8. Visakh, P.M., Arao, Y. Flame retardants, polymer blends, composites and nanocomposites. *Eng. Mater.* **2015**, *2*, 15–14.
9. Hamins, A., Averill, J., Bryner, N., Gann, R., Butry, D., Rick, R., Amon, F., Gilman, J., Maranghides, A., Mell, W., Madrzykowski, D., Manzello, S., Yang, J., Bundy, M. NIST SP 1130: Strategic Roadmap for Fire Risk in Buildings and Communities. National Institute of Standards and Technology. **2012**, 171.
10. Davis, R., Kim, Y., Harris, R., Nyden, M., Uddin, N., Gilman, J., Zammarano, M. Nanoparticles in Flame-Retardant Coatings for Flexible Polyurethane Foams: Effects on Flammability and Nanoparticle Release. Proceedings: 5th International Seminar on Modern Polymeric Materials for Environmental Applications MPM2013. **2013**, 5(1).
11. Zammarano, M., Kramer, R., Harris, R., Ohlemiller, T., Shields, J., Rahatekar, S., Lacerda, S., Gilman, J. Flammability reduction of flexible polyurethane foams via carbon nanofiber network formation. *Poly Adv. Tech.* **2008**, 19: 588–595.
12. Zammarano, M., Kramer, R., Harris, R., Ohlemiller, T., Shields, J., Rahatekar, S., Gilman, J. Effect of Carbon Nanofiber Network Formation on Flammability of Flexible Polyurethane Foams. **2017**.
13. Kashiwagi, T., Du, F., Douglas, J.F., Winey, K.I., Harris, R.H.,Shields, J.R. Nanoparticle networks reduce the flammability of polymer nanocomposites. *Nat. Mater* **2005**, 4(12) 928-933.
14. Granier, A., Nguyen, T., Eidelman, N., Martin, J. Covalent Functionalization of Carbon Nanotubes with Diisocyanate for Polyurethane Nanocomposites. Proceedings of the Adhesion Society Meeting. **2007**.
15. Bayer, O., Siefken, W., Heinrich, R., Orthner, L., Schild, H. A process for the production of polyurethanes and polyureas. German Patent DRP728981, **1937**.
16. Prisacariu, C. Polyurethane Elastomers from Morphology to Mechanical Aspects, Springer: New York, NY, USA, **2011**.
17. Ashida, K. Polyurethane and Related Foams Chemistry and Technology, Taylor & Francis Group: Boca Raton, FL, USA, **2007**.

18. Szycher, M., Szycher's Handbook of Polyurethanes, CRC Press, Boca Raton (FL), **1999**.
19. Javni, I., Zhang,W., Petrovickansaspetrovic, Z.S. Effect of different isocyanates on the properties of soy-based polyurethanes. *J. Appl. Polym. Sci.* **2003**, *88*, 2912–2916.
20. Bhowmick, A. K., Stephens, H. L. Handbook of Elastomers, 2nd ed. Marcel Dekker, New York, **2000**, Chap. 11.
21. Pegoretti, A., Fambri, L., Penati, A., Kolarik, J. Hydrolytic Resistance of Model Poly (ether urethane ureas) and Poly (ester urethane ureas). *J. Appl. Polym. Sci.* **1998**, *70*, 577–586.
22. Singh, S.N. Blowing Agents for Polyurethane Foams, Rapra Technology: Shawbury, UK, **2002**, 12.
23. Lee, S.T., Ramesh, N.S. Polymeric Foams: Mechanisms and Materials, CRC Press: New York, NY, USA, **2004**.
24. Frey, J., Grimminger, J., Stevens, R. New Silicone Surfactants for Rigid Polyurethane Foam. Conference book of paper Utech´96. International Polyurethane Industry Conference. The Netherlands. **1996**.
25. Caracciolo, P., Abraham, G. Biomedical Polyurethanes: Synthesis, Properties, Processing and Applications. Coimbra University Press. **2015**.
26. Bourbigot, S., Duquesne, S., Samyn, F., Muller, M., Lind-Say, C.I., Klein, R.A., Giannini, G. Polyurethane flameretardant formulation. Unit State of American Patent US 9499685B2, **2016**.
27. Laachachi, A., Cochez, M., Ferriol, M., Lopez-Cuesta, J.M., Leroy, E. Influence of TiO_2 and Fe2O3 fillers on the thermal properties of poly (methyl methacrylate) (PMMA). *Mater. Lett.* **2005**, 59 (1), 36–39.
28. Laachachi, A., Cochez, M., Leroy, E., Gaudon, P., Ferriol, M., Lopez-Cuesta, J.M. Effect of Al_2O_3 and TiO_2 nanoparticles and APP on thermal stability and flame retardance of PMMA. *Polym. Adv. Technol.* **2006**, 17, 327–334.
29. Zanetti, M., Kashiwagi, T., Falqui, L., Camino, G. *Chem. Mater.* **2002**, 14, 881.
30. Liu, W., Hoa, S., Pugh, M. Fracture toughness and water uptake of high-performance epoxy/nanoclay nanocomposites. *Compos. Sci. Technol.* **2005**, 65, 2364–2373.
31. Costache, M.C., David, D.J., Wilkie, C.A. Thermal degradation of ethylene-vinyl acetate copolymer nanocomposites. *Polymer.* **2005**, 46, 6947–6958.
32. Laoutid, F., Bonnaud, L., Alexandre, M., Lopez-Cuesta, J.M., Dubois, P. New prospects in flame retardant polymer materials: From fundamentals to nanocomposites. *Mater. Sci. Eng. Rep.* **2009**, 63 (3), 100–125.
33. Zammarano, M., Maupin, P.H., Sung, L.P., Gilman, J.W., McCarthy, E.D., Kim, Y.S., Fox, D.M. Revealing the interface in polymer nanocomposites. *ACS Nano*, **2011**, 5 (4), 3391–3399.
34. Kargarzadeh, H., Ahmad, I., Thomas, S., Dufresne, A. Microscopic analysis of Cellulose Nanofibril (CNF) and Cellulose Nanocrystal (CNC) based nanocomposites. In Handbook of Nanocellulose and Cellulose Nanocomposites, Wiley: Weinheim, Germany, **2017**, Vol. 1, 365–392.
35. Hong, F., Cheng, L., Jun, C., Zhiyang, B., Qun, P., Bogeng, L. Nano-modified flame retardant polyurethane foam and preparation method. Chinese Patent CN 104774311B, **2017**.
36. Duan, B., Wang, Q., Tang, Z., Ding, H., Zhang, M., Hou, L., Wang, Y., Liu, G., Qian, S. Method for preparing aqueous polyurethane coating and adhesive modified through carbon nanotubes and graphene. Chinese Patent CN 105153921B, **2017**.
37. Zhenyang, L., Ming, H., Xiaoligu, G., Yijun, S., Zhuo, L. Surface modification inorganic nanoparticle modified polyurethane rigid foam and preparation. Chinese Patent, CN101475741 B, **2010**.
38. Pholnak, C., Sirisathitkul, C., Soontaranon, S. UV-Vis absorption and small angle X-ray scattering spectra of commercial polyurethane coating filled with zinc oxide. *Natl. Acad. Sci. Lett.* **2016**, 39, 125.
39. Skoog, D.A., Holler, F.J., Crouch, S.R. Principles of Instrumental Analysis, Cengage Learning: Boston, MA, **2008**.
40. REACH for Polymers Best Available Testing Techniques and Methods, ASCAMM Private Foundation, **2011**, 94–98.
41. Moldoveanu, S.C., David, V. Basic Information Regarding the HPLC Techniques. Selection of the HPLC Method in Chemical Analysis, **2017**, 87–187.
42. Goldstein, J., Newbury, D.E., Joy, D.C., Lyman, C.E., Echlin, P., Lifshin, E., Sawyer, L., Michael, J.R. Scanning Electron Microscopy and X-Ray Microanalysis, 3rd Ed., Springer: New York, **2003**.
43. Stashenko, E., Martínez, J.R. Gas chromatography-mass spectrometry. *Adv. Gas Chromatogr.* **2014**, 1, 1–38.
44. Prisacariu, C. Polyurethane Elastomers: From Morphology to Mechanical Aspects, Springer Science & Business Media: Wien, Austria, **2011**.

45. Rangel, N., Alva, H., Romero, J., Rivera, J., Álvarez, A., Garcia, E. Synthesis and characterization of reinforced materials composite of polyurethane porous/hydroxyapatite. *Rev. Iberoam. Polim.* **2007**, 8 (2), 99–111.
46. ASTM D2240, Standard Test Method for Rubber Property-Durometer Hardness, West Conshohocken, PA, USA, **2015**.
47. ASTM F1957–99, Standard Test Method for Composite Foam Hardness-Durometer Hardness, West Conshohocken, PA, USA, **1999**.
48. ASTM D635, Standard Test Method for Rate of Burning and/or Extent and Time of Burning of Plastics in a Horizontal Position, West Conshohocken, PA, USA, **2014**.
49. ASTM D5132, Standard Test Method for Horizontal Burning Rate of Polymeric Materials Used in Occupant Compartments of Motor Vehicles, West Conshohocken, PA, USA, **2011**.
50. ASTM D412, Standard Test Methods for Vulcanized Rubber and Thermoplastic Elastomers—Tension, ASTM West Conshohocken, West Conshohocken, PA, USA, **2016**.
51. ASTM D624, Standard Test Method for Tear Strength of Conventional Vulcanized Rubber and Thermoplastic Elastomers, West Conshohocken, PA, USA, **2012**.
52. Mena, M.E., Estrada, R.F., Torres, L.A. Polyurethane nanocomposite for fireproof applications to reduce toxic gases. *Adv. Mater. TechConnect Briefs* **2017**, 5, 200–203.
53. NFPA 1710, Standard for the Organization and Deployment of Fire Suppression Operations, Emergency Medical Operations, and Special Operations to the Public by Career Fire Departments, **2016**.
54. Trovati, G., Sanches, E., Neto, S., Mascarenhas, Y., Chierice, G. Characterization of polyurethane resins by FTIR, TGA, and XRD. *J. Appl. Polym. Sci.* **2010**, 115, 263–268.

17 Neurotransmitter Release and Conformational Changes within The SNARE Protein Complex

Danko D. Georgiev
Institute for Advanced Study, Varna, Bulgaria

James F. Glazebrook
Department of Mathematics and Computer Science, Eastern Illinois University, Charleston, Illinois, USA

CONTENTS

17.1 INTRODUCTION

Neurons communicate with each other at cell-to-cell contacts, commonly referred to as *synapses*. A pioneer of modern neuroscience, du Bois-Reymond first observed the action of the synaptic potential along with the synaptic transmission at neuromuscular junctions [1]. This was further investigated by Katz [2] who unraveled the basic principle of synaptic transmission, namely how an action potential is stationed at a presynaptic nerve terminal, and how it gates Ca^{2+}-channels for functional purposes (reviewed in [3]). The traffic of protein-bearing vesicles toward the membrane surface of a cell delivers membrane-bound proteins via an energy expenditure mechanism of *exocytosis*. This vital concept has received increasing attention in both molecular (neuro)biology and neurophysiology [4], and in recent years has been linked to the chemistry of the SNARE protein complex pioneered by Südhof and coworkers [3, 5, 6] (here, SNARE stands for soluble NSF-attachment protein receptor, where NSF abbreviates N-ethylmaleimide-sensitive factor). Interestingly, in the 1990s, the concept of exocytosis gave rise to a startling perspective on a mind–brain interaction as it was hypothesized by Beck and Eccles [7](cf. [8]). They proposed a biophysical model of exocytosis that utilized quantum mechanical principles, in particular, postulating the mechanism of quantum tunneling to trigger a certain neurobiological proces. Specifically, that neurotransmitter release/exocytosis occurs postattainment of the presynaptic vesicular grid to a metastable state, through which signal transduction between neurons at a synapse is instantiated by discharge of the vesicle's neurotransmitter molecules into the synaptic cleft (reviewed in [9–12]). In [7], Beck and Eccles also proposed a macroscopic quantum entanglement between synaptic vesicles at a millisecond timescale to explain the release of only a single vesicle at an axonal bouton per axonal spike. But this particular conjecture yet remains to be verified empirically.

At the time of Beck and Eccles [7], mechanisms such as the SNARE complex [3, 5, 6], protein folding [13–15], and quasiparticle propagation through α-helical spines [16–20], were neither fully developed nor perhaps even conceived. That goes along with the relatively recent advances in understanding the nature of neurotransmitter release and synaptic transmission [21–25]. Extending earlier work [10, 11], we presented in [12] a blueprint for upgrading of the Beck–Eccles model, by incorporating the SNARE complex as the central, key mechanism. The resulting process included temperature dependent, vibrationally assisted tunneling of quasiparticles, together with the dynamics of membrane fusion and neurotransmitter release; collectively, those processes essential to understanding the nature of clinical consciousness [26].

17.2 NEURONAL COMMUNICATION AT SYNAPTIC CONTACTS

Synaptic contacts are structured in a way permitting the transmission of information in the form of electric or chemical signals from one neuron to another. At a given synapse, the plasmalemmas of two neurons are closely opposed to each other, and then a specialized molecular process permits transmission of a signal from the presynaptic (output) neuron to the postsynaptic (input) neuron. Effectively, there are two types of synapses: electrical and chemical, described as follows.

Electrical synapses provide a direct electrical coupling between two neurons through gap junctions that are pores constructed from connexin proteins [27]. The electrical synapses function rapidly, as there is no synaptic delay, but they also transmit the electric potential passively, which means that the amplitude of the signal decays exponentially with distance. Because the gap junctions are porous, the transmission is bidirectional, and signals can pass from each of the two neurons to the other one.

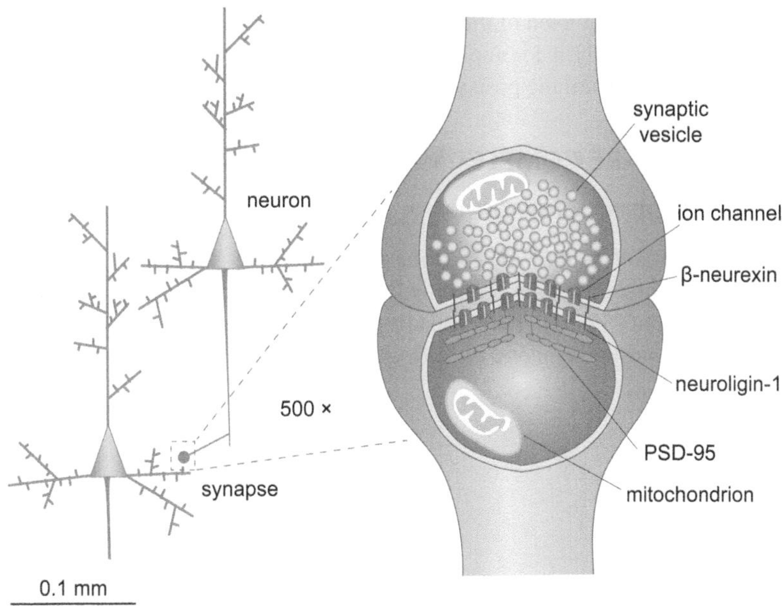

FIGURE 17.1 Communication between neurons takes place at synaptic contacts. Excitatory synapses are typically formed between presynaptic axonal boutons and postsynaptic dendritic spines. The presynaptic terminal has a large number of synaptic vesicles that contain neurotransmitter molecules. When the axonal bouton is electrically excited, the Ca^{2+} entry through presynaptic voltage-gated calcium channels leads to the fusion of a single synaptic vesicle with the plasmalemma at the active zone, thereby releasing its neurotransmitter content into the synaptic cleft. The neurotransmitter molecules then bind to postsynaptic ligand-gated ion channels and generate a postsynaptic electric potential in the target neuron. The synapse is held together by protein–protein bridges such as the β-neurexin–neuroligin-1 complexes whose anchoring in the pre- and postsynaptic plasmalemmas sets the synaptic cleft width in the range of 20–40 nm. Presynaptically, the intracellular part of β-neurexin also participates in the docking of synaptic vesicles. Postsynaptically, the intracellular part of neuroligin-1 interacts with scaffolding proteins such as postsynaptic density protein 95 (PSD-95) that organize the postsynaptic protein machinery and anchor the ligand-gated ion channels. The presence of mitochondria both in the axonal boutons and the dendritic spines ensures that the high-energy demands of the active synapses are effectively attained. Modified from Ref. [26].

Chemical synapses are the most abundant type of synaptic contacts in the brain (Figure 17.1). The main feature of the chemical synapse is that the electric signal from the presynaptic neuron is converted into a chemical signal in the form of neurotransmitter molecules released from synaptic vesicles into the synaptic cleft. The neurotransmitter molecules then bind to ligand-gated ion channels whose opening leads to the generation of a postsynaptic electric potential. The action of the chemical synapses is slow (there is $\approx$1 ms delay) due to the conversion of the signal from electrical to chemical, and back to electrical form. Crucially, postsynaptic potentials may now have a large amplitude and can be either positive or negative in sign, depending on the type of neurotransmitter released (as in fact observed in [28]).

In the brain cortex, $\approx$83% of the synapses release an excitatory neurotransmitter, such as glutamate, that binds to postsynaptic excitatory ligand-gated ion channels generating a depolarizing postsynaptic electric potential, whereas $\approx$17% of the synapses release an inhibitory neurotransmitter such as γ-aminobutyric acid (GABA), that binds to postsynaptic inhibitory ligand-gated ion channels generating a hyperpolarizing postsynaptic electric potential [29]. The transmission at chemical synapses is predominantly unidirectional from the presynaptic axonal boutons toward the postsynaptic dendrite, soma, or axon of the target neuron. At the same time, protein–protein bridges such as the β-neurexin–neuroligin-1 complexes traversing the synaptic cleft are also able to transmit information retrogressively from the postsynaptic toward the presynaptic neuron [30–34].

17.3 HOW THE SNARE COMPLEX MASTERS NEUROTRANSMITTER RELEASE

17.3.1 The Basic Protein Constituents

Synaptobrevin-2, *syntaxin-1*, and *SNAP-25* (synaptosomal nerve-associated protein of 25 kDa) are the so-called *SNARE proteins* that participate in docking synaptic vesicles to the active zone of the presynaptic plasmalemma [35–38]. These sustain the fusion pore that allows the neurotransmitter molecules to cascade into the synaptic cleft (Figure 17.2). This process is also mediated by the

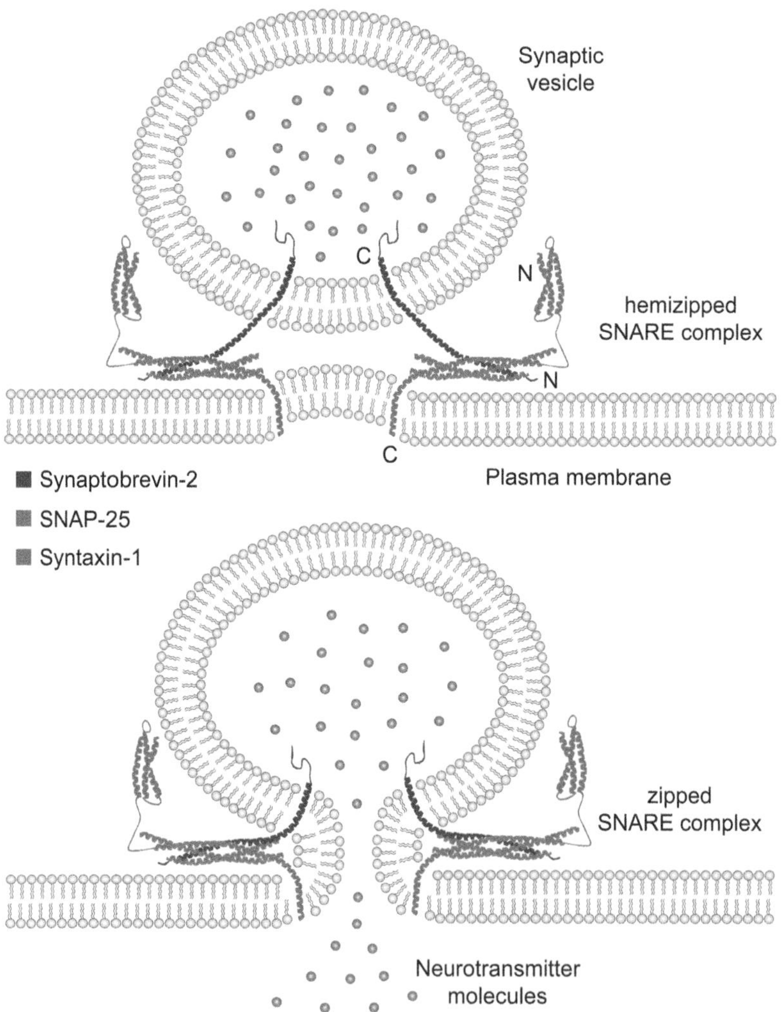

FIGURE 17.2 SNARE zipping in neurotransmitter release. Synaptic vesicles are docked at the active zone of the synapse through hemizipped SNARE complexes (top). Full zipping of the core SNARE complex into a 4-α-helix bundle proceeds from N- to C-termini and leads to a release of neurotransmitter molecules into the synaptic cleft (bottom). The neurotransmitter molecules then bind to postsynaptic ligand-gated receptors that electrically excite or inhibit the postsynaptic neuron. Modified from Ref. [26].

so-called Sec1/Munc18-like (SM) proteins. In particular, the SM protein Munc18-1 binds to the closed conformation of syntaxin-1, after which Munc13-1 opens the Munc18-1/syntaxin-1 complex, and along with other SM proteins, acts as a catalyst for SNARE complex assembly in the course of assisting synaptic vesicle fusion [3,39–44]. In the case of the neurotransmitter glutamate, Munc13-1 molecules have been observed to form multiple and discrete supramolecular self-assemblies, which on recruiting syntaxin-1, function as independent vesicular release sites, maintaining a readily releasable pool [45], and further enabling synaptic weights for robust synaptic activation [46]. Specifically, the Munc13 family comprises a large class of multidomain proteins congregating extensively at the presynaptic terminals, thus inducing conformational changes of the syntaxin-1 linker region toward formation of the SNARE complex. Instrumental are two amino acid residues (R151 and I155) in the syntaxin-1 linker region that are critical for how Munc13-1 engages priming and neurotransmitter release [44].

Synaptobrevin-2 is anchored in the membrane of the synaptic vesicle and is referred to as a *(vesicle) v-SNARE protein*, whereas syntaxin-1 and SNAP-25 are anchored in the plasmalemma and are referred to as *(target) t-SNARE proteins*. A class of v-SNAREs interacts with t-SNAREs to form a so-called *(trans) SNAREpin complex* [47–51]. Synaptobrevin-2 and syntaxin-1 possess transmembrane α-helix regions that penetrate the whole thickness of the membrane, whereas SNAP-25 binds to the membrane via palmitoyl side chains covalently bound to four cysteine amino acid residues in the middle of the molecule [52,53]. All three SNARE proteins zip together into a 4-α-helix bundle, known as the *core SNARE complex*, whose twisting applies a traction force on the opposing phospholipid bilayers of the synaptic vesicle and the plasmalemma until they merge with each other [54,55]. The generated traction force by SNARE zippering is capable of fusing the synaptic vesicle and plasmalemma, even in artificially engineered lipid-anchored synaptobrevin-2 and syntaxin-1 molecules that lack their transmembrane regions [56].

The four α-helices of the core SNARE complex in the fully zipped state are aligned parallel to each other, and close a hydrophobic gap, within which amino acid residues from each of the four α-helices are organized layer wise [57]. In the middle of the complex, there is the *zero ionic layer* formed by one arginine (R) contributed by synaptobrevin-2, along with three glutamine (Q) residues contributed by syntaxin-1 and SNAP-25 [58,59]. If SNARE complexes do indeed assemble in parallel, in relationship to their transmembrane anchors, then the outcome is a *zipping-up* of the v-SNARE and t-SNARE proteins between the two membranes [22]. With the exception of the zero ionic layer, all other layers $(-7, \ldots, -1, +\ 1, \ldots, +\ 8)$ are hydrophobic (containing mostly amino acids such as valine, leucine, and isoleucine) and form a flanking leucine zipper that tends to isolate the zero ionic layer from the solvent. This hydrophobic zipping is mediated by formation of noncovalent bonding, and thus it releases a certain amount of energy, since the bonding is associated with a transition to a lower energy conformational state.

17.3.2 N- to C-Terminal Directions

Essential for neurotransmitter release are the three SNARE proteins: synaptobrevin-2, SNAP-25, and syntaxin-1. The N-terminal coil domain of synaptobrevin-2 influences binding. It is shielded in a protein complex, but in the event the C-terminal domain is exposed. The N-terminal complex endures up to the stage of neural activity and may function to prepare synaptic vesicles for immediate exposure on Ca^{2+} entry. The initial trans-SNARE complex forms by piecing together membrane v-SNARE and t-SNARE from their N-terminal ends and progressively folds in a "zippering" fashion to the C-terminal transmembrane ends. More specifically, the SNARE complex (regulated by other SNARE master proteins) appears to start forming at the N-termini and becomes stalled there up to the point of further stimulation [60]. A consistency result is that the assembly of the core SNARE complex begins at the N-terminal domain of synaptobrevin-2 and the N-terminal domain of the H3-domain of syntaxin-1 (Figure 17.2). Observe that different SNARE motifs may not act independently, in so far that the N-terminal coil, in creating a domain of synaptobrevin-2, may then form a complex without involving the coil domain at the C-terminus. In this respect, it is sufficient that the N-terminal half of synaptobrevin-2 forms a complex with t-SNAREs [60].

17.3.3 Synaptotagmin-1 and Complexin in the Fusion Process

In neurons, the neurotransmitter release is synchronized with axonal firing and requires a Ca^{2+} influx in the presynaptic bouton [25]. In the resting state, the SNARE complex of docked vesicles is only partially zipped [61]. It is expected that on the presynaptic plasmalemma, syntaxin-1 and SNAP-25 form a 3-α-helix bundle describing a groove for the synaptobrevin-2 shoulder of the α-helix in the proximity of the synaptic vesicle. This hemizipped SNARE complex of docked vesicles is clamped by the Ca^{2+} sensor protein called *synaptotagmin-1* that represses the complete SNARE zipping, together with the spontaneous release of synaptic vesicles. In fact, all Ca^{2+}-triggered neurotransmitter release is mediated by synaptotagmin [5, 21, 34, 62–64]. Under axonal firing (depolarization), the opening of voltage-gated calcium channels leads to Ca^{2+} entry, upon which the synaptotagmin-1 molecule binds at least four Ca^{2+} ions in the five Ca^{2+} binding sites in tandem with the C2 domains [21, 65–67] and detaches from the SNARE complex, thereby both de-inhibiting the SNARE function and actively assisting membrane fusion through phospholipid interaction [6, 68].

Synaptotagmin-1 is instrumental in accompanying activation and clamping functions but works independently of each case. Synaptotagmin-1 is further assisted by the synaptic protein *complexin* that activates the SNARE/SM complex by conformational changes at the membrane-proximal C-terminus [69], and is instrumental in clamping synaptic vesicles for synaptotagmin-1 action [70–73], while in the process, conformational switches between the opened-closed states of complexin support this action toward the fusion trigerred by synaptotagmin-1 upon Ca^{2+} binding [74, 75]. This partly summarizes a dual process for complexin, since on the one hand, it controls spontaneous release at the synaptic terminal by temporarily stalling the fusion process [76–79], and on the other hand, by proceeding to co-activate with synaptotagmin-1, it continues assisting the Ca^{2+}-triggered neurotransmitter release toward completing the full fusion process [80–87]. In this process, activation of priming, clamping, and Ca^{2+}-triggering necessitates distinct complexin sequences [3]. Effectively, it is through these conformational changes that complexin-1 creates a bridge across a ternary and a binary SNARE complex, the latter consisting of syntaxin-1 and SNAP-25 (the t-SNARE complex). The apparent relative plasticity of the membrane proximal C-terminus at the end of this ternary complex is proposed to be instrumental for fusion, and the less-plastic N-terminal end is considered as supporting this process [69]. Further investigations into the role of the α-helices in the C- and N-termini of complexin in activating Ca^{2+}-triggered exocytosis are studied in [42, 43, 45, 88].

Experimental evidence reveals the fusion pores as determining Ca^{2+}-triggered exocytosis via their dynamic formation that regulates the rate of transmission of chemical signals in the form of neurotransmitters to the synaptic cleft [89]. "Full fusion" and "kiss and run" are two types of release mechanisms in which SNARE proteins reconstituted into liposomes are agents of membrane fusion *in vitro* [90] and can regulate fusion mechanisms at speeds that are necessary for secretion [91]. It is here that synaptotagmin-1 with varying concentrations of Ca^{2+} in the binding process to phospholipids in the plasmalemma implements a triggering of the fusion pore to its opening state [92]. Electrostatic interactions between the pore-lining residues of syntaxin-1 and the positively charged norepinephrine neurotransmitter molecules flowing out of the synaptic vesicle revealed the syntaxin-1 anchor to be an integral part of the fusion pore [93]. Direct interactions between t-SNAREs and neuronal Ca^{2+}-gated channels can further assist recruiting SNARE complexes near the active sites of Ca^{2+} entry in order to implement exocytosis [94].

17.3.4 The Basic Mechanism

Recall that the synaptic vesicle protein synaptobrevin-2 engages with the receptor proteins syntaxin-1 and SNAP-25 in constructing the SNARE complex. As we have pointed out in Section 17.3.2, these special proteins are anchored on the synaptic vesicle or the plasmalemma within finely structured N- and C-terminal domains. This complex comprising of four stable α-helices is firmly recognized to be instrumental in membrane/bilayer fusion reactions. The core SNARE complex consists of four α-helices: one provided by synaptobrevin-2, one by syntaxin-1, and each of the remaining two by SNAP-25 [22, 95]. The four α-helices of the core SNARE complex in the fully zipped state are aligned in parallel with each other and form a 4-α-helix bundle [96].

The present consensus is that if SNARE complexes do indeed assemble in parallel in relationship to their transmembrane anchors, then the outcome is a "zipping-up" of the v-SNARE and t-SNARE complexes between two neighboring membranes [22]. Instrumental is the role of the intermediary SNAREpins, as organized by van der Waals and certain entropic forces into groups of circular clusters that drive membranes into cohesion with each other [50]. Here, it should be noted that the rate of fusion is determined by the number of SNAREs, a quantity that is in direct proportion to the amount of cooperative effects generated via strict interactions to spontaneously organize the SNAREpins into circular clusters in the course of the membranes being forced together. Given that there are $\approx$70 v-SNAREs available per synaptic vesicle, the entropic balance between the SNAREpins may influence the sub-millisecond timescale for neurotransmitter release [50].

The structure of the SNARE complex with its C-terminal transmembrane domain extensions is known at 3.4 Å, while a propagation of the helices into the membrane suggests the passage of a bending force into the membrane that results in fusion [97]. The nature of such forces arising in the zipping mechanism of fusion has also been suggested in [10, 11]. However, in the docked vesicle formation, the SNARE complex is only partially zipped [61]. It is likely that on the presynaptic membrane, syntaxin-1 and SNAP-25 form a 3-α-helix conformation describing a groove for insertion of the synaptobrevin-2 shoulder of the α-helix in the proximity of the synaptic vesicle. Upon electric stimulation of the axonal bouton, the *trans*-SNARE complex (consisting of SNARE proteins bound on the two opposing membranes of the synaptic vesicle and plasmalemma) undergoes a conformational change which drives membrane fusion through zipping of a 4-α-helix bundle. The fully zipped SNARE complex resides on a single membrane (the plasmalemma which is now enlarged in surface) and is referred to as *cis*-SNARE complex. Multiple reuse of the SNARE protein machinery is possible at the expense of biochemical ATP energy. After completion of exocytosis, the ATP-dependent action of α-SNAP and NSF permits disassembly of the cis-SNARE complex and recycling of v-SNAREs and t-SNARES for another cycle of neurotransmitter release [98].

17.4 THE QUANTUM NATURE OF HYDROGEN BONDING, STRETCH BAND VIBRATIONS, AND TUNNELING

17.4.1 Hydrogen Bonding and Amide Vibrations

Secondary protein structure is stabilized by hydrogen bonds [99]. Amide stretch vibrations are essential for accurately predicting hydrogen-bonding dependence. Hydrogen bonding can impact the amide functional group by stabilizing its [$^{-}$O–C=N–H^{+}] structure over its [O=C–N–H] structure. A typical model of a peptide bond is N-methylacetamide. In [100], the impact of hydrogen bonding on the C=O and N–H groups of this peptide bond was studied, with several observations relating to the three amide bands (I, II, and III): (i) the frequency dependence of hydrogen bonding on the amide I vibration is derived mainly from the C=O group; (ii) the frequency dependence for that of amide II is derived from the N–H hydrogen bonding; (iii) somewhat in contrast, the hydrogen bonding dependence of the relatively conformational sensitive amide III band is equally shared by both the C=O and N–H groups, and thus is equally responsive to the hydrogen bonding at either the C=O or N–H site. The upshot is that the frequency shifts of both the amide I and the amide III bands may be effective regulators for the hydrogen bonding at the C=O and N–H sites of a peptide bond. Observe that it is essential to treat the proton in such hydrogen bonded systems as quantum. Effectively, quantum nuclear effects can weaken relatively weak hydrogen bonds but can fortify the relatively strong ones [101]. Relevant in this case is the induced fit method for quantum H-bonding, which supports the molecular interactions for inducing conformational changes in the binding sites of enzymes [102]. In such a molecular recognition event, the dynamics of proton tunneling between hydrogen bonds is highly instrumental.

17.4.2 Vibrationally Assisted Tunneling

First, let us recall that enzymes are powerful protein catalysts that operate on biochemical substrates (including other proteins) through protein sites. As powerful catalysts, they assist completing reactions in timescales that otherwise would not be biologically feasible. In fact, it is presently considered that conformational dynamics on the scale of micro-to-milliseconds may be a distinct feature of the catalytic capabilities (reviewed in [103]). Crucially, enzymes are capable of surmounting the potential energy barrier that separates reactants from products. The basic principle is that the greater the height of this energy barrier, the slower is the reaction rate. Enzymes are seen to support the passage over the barrier on reducing energy, and thus increasing the reaction rate. As pointed out in [104], the structure of the reactant at the top of the barrier is energetically unstable, thus giving rise to the *transition state*. The energy needed to pass over the barrier is the activation energy; specifically, the barrier is surmountable by the thermal excitation of the substrate. Hence, on the one hand, the transition state mechanism implements the "over," and on the other hand, quantum tunneling implements the "through" (the barrier). More formally, the overall event is determined by potential shape analysis, which for enzymatic hydrogen (quantum) tunneling has been observed for a range of proteins such as tryptophan tryptophylquinone-dependent methylamine dehydrogenase (TTQ-MADH) or tryptophan tryptophylquinone-dependent aromatic amine dehydrogenase (TTQ-AADH), in which tunneling of proton transfer dynamically induces motion of enzymes across a barrier shape that would not be realized in a fully classical region as we described earlier [101, 102, 104–107] (as reviewed in [10, 12]). Hydrogen tunneling in enzymes has been further verified by the *kinetic isotope effect*, namely the ratio between transfer rates of a biochemical reaction with substrates consisting of two different isotopes, e.g. protium, ^{1}H, or deuterium, ^{2}H [108]. In particular, quantum hydrogen tunneling has been realized for enzymatic C–H bond cleavage, thus furthering the utility of this phenomenon in the catalysis of binding and breaking processes in enzymatic dynamics [109]. The collective mechanisms, as we have described them, lead to the idea of *vibrationally assisted tunneling*. Previously, it had been verified [110] that N–H and N–D stretching accompany each other in the presence of the kinetic isotope effect of double proton transfer. It is very interesting, and significant for our purposes, that hydrogen bonding is ubiquitous for conformational changes in the SNARE-complexin-synaptotagmin complex [43, 74].

17.5 THE PROTEIN α-HELIX STRUCTURE AND CONFORMATIONAL DISTORTIONS

17.5.1 Protein α-Helices and Dynamics of the Davydov Soliton

Recall that the SNARE complex assembles by a zippering mechanism from the N- to C-terminal regions [111] instrumented by four protein α-helices: synaptobrevin-2 and syntaxin-1 contribute one α-helix each, while SNAP-25 contributes two α-helices [3, 6, 112]. Because we will focus mainly on modeling the zippering of the 4-α-helix bundle, we will hence briefly describe the secondary structure of protein α-helices.

Geometrically, the protein α-helix is a right-handed spiral with 3.6 amino acid residues per turn [113], where the N–H group of an amino acid forms a hydrogen bond with the C=O group of the amino acid four residues earlier in the polypeptide chain (Figure 17.3). Three longitudinal chains of hydrogen bonds, referred to as *α-helix spines*, run parallel to the helical axis and stabilize the α-helix structure. At the quantum level, the interaction of the amide I excitation (due to C=O bond

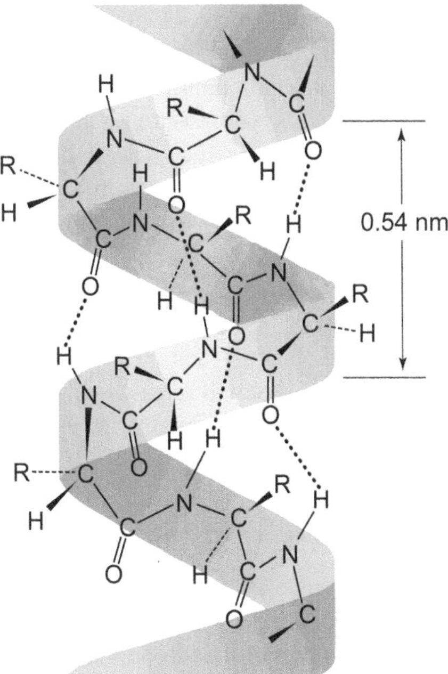

FIGURE 17.3 Structure of the protein α-helix with 3.6 amino acid residues per turn. Three α-helix spines consisting of longitudinal chains of hydrogen bonds (···) within vertically aligned –C=O···H–N–complexes thus stabilize the helical structure. Quantum quasiparticles (here modeled upon the Davydov soliton) could form, then propagate along, and conformationally twist the α-helix as a result of the interaction of the amide I (C=O) vibrations with the hydrogen bonds in the α-helical spines. The vertical distance between consecutive turns of the helix is 0.54 nm. Modified from Ref. [26].

stretching) with the vibrations of the hydrogen bonds along the α-helix spines leads to a localization of amide I excitation within a region spanning several amino acids [20].

The composite quantum state constituted by an amide I excitation, together with its associated hydrogen bond distortions, could be viewed as a quantum quasiparticle, which we have initially modeled as a self-trapped *Davydov soliton* [16,17,20,114,115]. Of the two types of Davydov soliton—symmetric, which involves all three spines of the α-helix, and asymmetric, which involves only two of the spines—we henceforth adopt the *asymmetric* Davydov soliton in which there is an associated bending of the protein α-helix in the location of the amide I excitation. Specifically, because the energy due to the propagation of an asymmetric Davydov soliton along the protein α-helix is capable of bending the α-helix structure [115,116], such a quasiparticle may be instrumental in zipping the SNARE complexes in docked synaptic vesicles, and thus triggering neurotransmitter release [10,11].

Motivated by the pioneering work of Scott and his co-workers [19, 20, 115], we described in [10, 11], and more recently in [12], the possible role of Davydov solitons as propagating within the main backbone of the polypeptide chain in relationship to the SNARE zippering mechanism. In this particular situation, the amino acid R-side chains are not taken into account in the Davydov model. Nevertheless, in the case of an *asymmetric* soliton propagation, the contortion of the α-helix through twisting and bending effectively rearranges the spatial localization of the R-side chains, and a suitably selected asymmetric Davydov soliton, as a particular polaron quasiparticle (see Section 17.8.1), may assist the leucine zippering process within the SNARE 4-α-helix bundle.

Consequently, the zippering of the core SNARE complex leads to formation of a 4-α-helix bundle as a result of synaptobrevin-2 fitting into the syntaxin-1/SNAP-25 groove. SNARE zipping leads either to a transient opening and closing of the fusion pore, which is known as the "kiss-and-run" mode of neurotransmitter release, or to complete vesicle fusion with the presynaptic membrane that requires recycling of the vesicle through clathrin-mediated endocytosis.

We propose that after detachment of synaptotagmin-1 from the core SNARE complex, the emergence and propagation of an asymmetric Davydov soliton within the synaptobrevin-2 α-helix may induce a twisting/bending of synaptobrevin-2 in such a way that it fits the groove created by the syntaxin-1/SNAP-25 complex. There is an energy barrier for fusion created by the electrostatic repulsion between the synaptic vesicle membrane and the plasmalemma. However, as a result of thermal fluctuations, the docked synaptic vesicle might be stochastically perturbed, and hence the potential barrier for fusion will be dynamically fluctuating.

If the soliton tunnels through the barrier in a way favorable for a leucine-zippered arrangement of the synaptobrevin-2 amino acid R-side chains, then the noncovalent bonding associated with the change in the core SNARE complex geometry will release energy and the Davydov soliton should dissipate at the end of the zipped 4-α-helix bundle, making the exocytotic process irreversible in the case of complete fusion of the synaptic vesicle.

17.5.2 Remarks on the Davydov Soliton and Possible Extensions

In [10, 12], we addressed several versions of the Davydov ansatz for coupling amide I vibrations (essentially the C=O stretch type) to acoustic phonons, each of which could be regarded as biologically feasible, while providing extra support of the vibrationally assisted tunneling of the Davydov soliton as instrumental for the zipping of the core SNARE complex. Though, to the best of our knowledge, there is currently no general consensus on the ideal form that the model should take on for its biological feasibility (however, see [117, 118] and remarks below concerning the N–H stretch).

Associated is the question of knowing a reliable range of values for the anharmonic, dipole–dipole interaction, and the other parameters, which characterize nonlinear exciton–phonon interactions: the phonons corresponding to the conformational oscillations of the lattice, together with the excitons describing the internal amide I excitations of the peptide groups. Significant studies by Förner [119–123] report the Davydov soliton to be stable for reasonable parameters up to temperatures of 300 K, for certain excitations at the terminal site of chains, while at higher temperatures, soliton formation can be found at both chain ends. Crucially, Förner's study, along with the amide I vibrations, takes into account the important factor of normal modes describing the N–H stretch and how this is coupled to the hydrogen bonds. Nonlinear spectroscopy methods [124, 125] have shown that this coupling of the N–H and C=O bonds is due to an anharmonic term contained in the potential energy surface of the hydrogen bond, as well as both the N–H and C=O bonds of peptide units in the α-helix (the observations of [124] verify vibrational self-trapping for crystalline acetanilide, a prototypical model system for proteins).

Thus, it is essential to keep in mind the nonlinear dynamics between those nonlinear exciton–phonon and exciton–exciton interactions, which can influence soliton formation that would require adjustments in formulating the Hamiltonian as Förner's model suggests. Further, efficiency in the phonon spectrum depends on α-helix chain length in the short-to-medium range, as was previously studied in [126, 127], showing that the coupled spectrum contains frequencies across the 200 GHz–6 THz interval, and that an increase in chain length induces an increase in amplitude of long-period oscillations. In fact, as the exciton–phonon coupling increases, the simulations of Ref. [127] predicted the dynamics as more favorable to those of a small polaron type, thus to some extent compatible with the mechanism of the earlier Fröhlich–Holstein theory [128–133] (see also Section 17.8.1).

17.5.3 Experimental Evidence for Self-Trapped States

Certain crystalline substances, such as crystalline acetanilde, can be regarded as suitable molecular models that are structurally similar to α-helices. Vibrational self-trapping for crystalline acetanilde was first observed in [134]. Related experimental evidence has been reviewed in [117, 135]. Several studies of energy states in the case of the protein myoglobin have revealed relatively short lifetimes (15 ps) for amide I vibrations [136], though these findings remain somewhat elusive, at least for myoglobin [137, 138]. On the other hand, by means of femtosecond infrared pump-probe spectroscopy, Edler and co-workers [125, 139–141] have compared the amide I band to the N–H stretch. Though the equations of motion and propagating form are formally very similar, the N–H stretch exhibits self-trapped vibrational states in a more nonlinear fashion, with greater energy when compared with the amide I band. An example is the relatively long α-helix of poly-γ-benzyl-$_L$-glutamate, where self-trapped vibrational states in the N–H stretch were observed [140]. Biologically significant is that the energy of the N–H stretch mode is high enough to approximate the energy arising from the ATP hydrolysis (as noted in [142], with analogous results to [140] in the case of β-sheets).

Further applications of two-dimensional infrared spectroscopy (2D-IR) for studying protein dynamics appear very promising in the long run. For instance, the review of [143] pinpoints amide I vibrations of the polypeptide backbone by C=O stretching and N–H bending vibrations measured at an ultra-fast picosecond timescale. Amide I vibrations have been observed between 1600 and 1700 cm^{-1}, and, for the protein glycophorin A, they are seen as delocalized in two different helices while hopping from one to the other, as influenced by the geometric structure [144]. An advantage of 2D-IR is that the spectral elements can be spread onto two frequency axes, namely excitation versus detection frequencies. This enables correlations between multiple vibrations, which associate the spectra to the ambient structure, thus further assisting the conformational behavior of α-helices via stretching-bending contortions, which are proposed as instrumental for the working dynamics of the SNARE complex.

17.6 UPGRADING THE BECK AND ECCLES MODEL

The SNARE zipping model described here provides the insight into the protein machinery involved in exocytosis that to an extent was missed in the Beck and Eccles model [7], and it also explains in detail the biomolecular conformational transitions that open the fusion pore of the synaptic vesicle. If indeed the "kiss-and-run" scenario and the "zipping" of the SNARE complex are protein folding operations, then since these involve quantum-chemical states, there is also the possibility that quantum effects can induce a neurotransmitter release into the synaptic cleft. What we have brought to the forefront in our proposal is an alternative, and a possible complementary quantum process to that of Beck and Eccles [7].

The main stages of the synaptic vesicle exocytosis are as follows:

(1) In the resting axonal bouton, the synaptic vesicle is docked in close proximity to presynaptic voltage-gated calcium ion channels that are inactive and remain in the closed channel conformation. The SNARE complex is hemizipped [145] and is reliably clamped in this conformation by complexin and the Ca^{2+}-sensor synaptotagmin-1. At rest, the free Ca^{2+} levels are very low in the axonal bouton.

(2) When the neuron fires, the electric signal reaches the synaptic bouton, and the depolarization of the presynaptic plasmalemma activates the voltage-gated calcium channels. The free Ca^{2+} levels rise up in the form of a cloud-like microdomain that persists near the docked vesicle for a time period $\Delta t = 0.3$ ms [146, 147]. Synaptotagmin-1 binds four Ca^{2+} ions and rapidly releases the clamp exerted by complexin on the SNARE complex [3]. We propose that the energy released by the synaptotagmin-1/complexin detachment induces a Davydov soliton with energy E_0 that propagates along the unzipped α-helix portion of synaptobrevin-2 until it hits a potential energy barrier parameterized by width a and height $V_0 > E_0$. As shown in [12], it can be seen that thermal fluctuations of the height of the potential energy barrier may account for temperature sensitivity of exocytosis via a quantum vibrationally assisted tunneling mechanism.

(3) If the Davydov soliton is reflected by the potential barrier V_0 (for details on soliton reflection see simulation videos in [148]), then it will propagate back to the point where the protein synaptobrevin-2 is anchored in the synaptic vesicle membrane and undergo a further reflection, again aiming toward the potential barrier V_0. The frequency n_0, with which it undertakes to cross the barrier, could be much less than the angular frequency ω_0 of the Davydov soliton, $n_0 \ll \omega_0$. The extended life of the Davydov soliton in the cavity formed by SNARE proteins is in line with experimental research confirming that enclosure in a cavity can largely insulate a quantum system from decoherence within its environment, thus allowing it to maintain quantum coherence over dynamically important timescales [149–151].
(4) If the Davydov soliton tunnels through the potential barrier V_0, then the full zipping of the SNARE complex exerts a traction force that merges the synaptic vesicle with the plasmalemma and so opens the fusion pore, thereby releasing neurotransmitter molecules into the synaptic cleft.

17.7 CONFORMATIONAL ALTERATIONS, PROTEIN FOLDING, AND THE DAVYDOV MODEL

17.7.1 Linking and Protein Folding

Intrinsically disordered proteins (IDPs) through cellular control mechanisms and signaling induce conformational states following which many types of IDP domains fold upon binding to their targets. In this way, conformational states may play a role in the binding of an IDP. As an example, in the transcription factor p53, there are IDP regulatory regions near the C-terminus, which fold into α-helical or β-strand structures on binding to different protein partners. Following the approach of [15]:

(1) A protein associates with its binding partner in a fully disordered state, and subsequently folds in association with the target protein.
(2) Folding is induced by association with the target.

It had been hypothesized that the free energy landscape is funnel shaped, with the native state situated at the base of the funnel, and folding takes place via free energy minimization. As regards the actual process, [117] proposes: (i) an initial kinetic shape in which a specific funnel is selected (often corresponding to the native state) and (ii) the structure relaxes as free energy is minimized within the funnel selected. Instrumental here is the role of vibrationally excited states in protein (reviewed in [118, 152]). A more recent extension of these ideas suggests that the distribution of protein conformations actually tends to a multi-peak distribution with individual peaks differing in structure from the native state, though seen as stable as this latter state, in which features (i) and (ii) above apply [153].

17.7.2 Protein Folding and the SNARE Complex

Rothman and co-workers [154] have shown that SNAREpins assemble between cellular membranes from cognate v- and t-SNAREs to drive lipid bilayer fusion and regulate its inherent specificity. Fusion then results from a protein folding reaction that is mechanistically coupled to two closely adherent lipid bilayers held together by folding SNAREpins. As pointed out in [95], the role of conformational selection patterns and binding procedures may be instrumental in the folding process. It is advantageous to have a preformed N-terminal helix, and it is possible that the C-terminal portion of the SNARE motif may signal a cessation to folding. It is also possible that extra triggering may be provided by the binding of synaptobrevin-2 (or by its unbinding) with the possibility of complexin dissociation to contribute to the necessary energy for facilitating, and completing the fusion process

[95, 155]. It remains to be seen if the "zippering" itself is a continuous or a piece-wise continuous (segmented) process [95].

17.7.3 Conformational Alterations and Zippering

Protein conformational transitions as microscopic biochemical processes *a fortiori* entail quantum states. Protein misfolding can lead to neurological disorders such as Huntington's and Creutzfeldt–Jakob diseases. A significant observation is that the amide I energy of certain peptide groups approximates the energy of the bending mode of water molecules, and these two modes convolute with each other [118]. Among the list of amino acids, glutamine and asparagine have a special status for the purpose of the Davydov model because each admits amide I excitations in their side chains, and they are the only two more capable of influencing "solitonic" amide I propagation occurring along the backbone for all engaging such conformational changes, that we propose as instrumental for the "zippering" process during synaptic vesicle fusion. To this extent, Cruzeiro [118] introduces a generalization of the Davydov model that incorporates an atomic interaction Hamiltonian that partially accounts for the nonlinear dynamics of the interaction between an enhanced transfer of energy from the water molecules relative to the amide I energy. As a consequence of conformational changes, protein misfolding has to be considered in the criteria for malfunctioning of the fusion process generated by the SNARE mechanism, which in turn would evoke a neurological disorder, possibly as the result of localized impairment and corruption of synaptic receptors by subsequent collapse of adhesion in the neuroligin and neurexin molecular networks [34]. For instance, a 3-α-helical (target) t-SNARE complex in a rate-limiting step can readily mis-fold. Long-range conformational alteration of t-SNAREs and how this effects the zippering process have been observed in [156].

The Davydov model for protein folding is indirectly applied, via an approximation by the nonlinear Schrödinger equation, in a more deterministic approach described in [157]. The biological setting can be related to that of [118] in terms of the amide I excitations, but the hypotheses are different. On commencing from a ϕ^4 double-well potential, the method determines the folding dynamics directly in terms of protein conformational angles. According to [157], for as long as sufficient (local) folding energy is made available, the energy of the propagating soliton can be transferred into the conformational field, thus permitting it to surpass energy barriers and eventually attain to a ground state. A reciprocal energy exchange back to the soliton enables balancing out its energy loss in the steady state, thus facilitating an efficient and regulated conformational transition.

17.7.4 Mechanical Stress Stored within the SNARE Complex

Molecular mechanical stress influences the degree of self-organization at the fusion site. Both the mechanical stress stored within the SNARE complex and the consequent membrane penetration of the transmembrane domain C-termini are functional for the opening of the fusion pore [158]. Syntaxin-1 is already formed as an α-helix prior to the SNARE complex formation, whereas synaptobrevin-2 is relatively unstructured. In order for an adequate force to be exerted on the membrane, each semi-flexible linker of the two SNARE molecules needs to be sufficiently stiff for effectiveness. Atomistic studies have provided estimates for stiffness of the syntaxin-1 linker as 1.7–50 cal mol^{-1} deg^{-2}, whereby falling within that range are the estimates 11 ± 0.2 cal mol^{-1} deg^{-2} [54] (cf. 10 ± 0.1 cal mol^{-1} deg^{-2} [159]). Consequently, transmission of force in the direction of the membrane can be impeded owing to the stress that is otherwise stored in syntaxin-1. The latter then, in an alternate fashion, relaxes by extra kinking of synaptobrevin-2, which is more flexible. In this way, the α-helical nucleation inside synaptobrevin-2 precedes the SNARE zipping procedure.

Related is how a pulling force is applied to bring two membranes together, and this force works to overcome the energy barrier for membrane fusion, as well as the necessary energy that the protein mechanism provides for fusion of synaptic vesicles with the plasmalemma. In this instance, an alteration occurs of the synaptobrevin-2 transmembrane domain location within the membrane under piconewton effects [160]. In the fusion mechanism, the C-terminus of synaptobrevin-2 detaches from an inner leaflet head group to a continuously helical state by the SNARE transition mechanism, whereby the zippering acting on the synaptobrevin-2 helix will pull upon the synaptobrevin-2 transmembrane domain, and then cause the latter to tilt in some fashion. Evidence for conformational flexibility of synaptobrevin-2 transmembrane domain α-helices has been shown by intermittent unfolding of the α-helical backbones for vibrational motions (exciton–phonon interactions) that may effect ambient liposome structures. Specifically, structures in vesicle outer leaflet mixing are claimed for Ca^{2+} triggered exocytosis, membrane fusion, and fusion pore expansion [161, 162].

17.7.5 Enhancement by Stochastic Resonance

It is commonly accepted in current neurophysiology that for finite assemblies of ion channels situated within a noisy environment, the cumulative effect of stochastically driven firing and spiking can, at some optimal noise level, enhance signal transduction—a phenomenon called *stochastic resonance* [163, 164]. Many, if not all, processes of neuronal signal transduction are stochastically driven, and in this way, they eclipse purely deterministic processes in neurophysiology [165]. Following [164], stochastic resonance is dependent upon several factors:

(1) the source of the background noise;
(2) a generally weak coherent input;
(3) a characteristic sensory barrier or threshold that the system needs to overcome in order to perform a useful task.

This is applicable to several components of the model under discussion. We are working in a biologically noisy environment in which there is constant firing of periodically stimulated neurons for which the peaks (as function of noise strength) determine a signal-to-noise ratio, which is a salient property of stochastic resonance. This is best illustrated by how NMDA (*N*-methyl-$_D$-aspartate) receptors permit an influx of Ca^{2+} ions that trigger a complex, stochastically driven chain reaction once a discharged neurotransmitter clings to a postsynaptic membrane. Since the complete triggering process occurs within noisy conduits, the situation is conducive to effects of stochastic resonance. In this way, depending on the duration of its conformational states, the NMDA receptor may take advantage of stochastic resonance for signal enhancement [166].

In our case, vibrationally assisted tunneling involves a nonlinear interplay between quantum and thermal fluctuations. As postulated in [167, 168] (and references therein), the quantum tunneling escape rate and the presence of stochastic resonance are significantly influenced by the temperature of an ambient thermal bath, and by how the system is coupled to its environment. These factors depend upon a crossover temperature T_0, above which thermal fluctuations dominate over the quantum tunneling of the quasiparticle as it is hops around within a sequence of double potential wells, whereas below T_0 the tunneling transitions can endure. There are other factors dependent on the barrier height and the prevailing signal-to-noise ratio, both of which can of course vary. In the second case, where we consider the neurotransmitter release, stochastic resonance can be induced by noisy, random fluctuations of the stochastic gating mechanism of individual fusion channels, which, as proposed, may be beneficial to the mode of signal transduction.

17.7.6 Evidence for Single Vesicle Release and Exocytosis at Central Synapses

In our previous work [12], the modification and upgrading of the Beck and Eccles hypothesis [7] did not require macroscopic quantum entanglement between synaptic vesicles at a millisecond timescale, in order to account for the release of only a single vesicle at the axonal bouton per axonal spike. There are two ways of approaching such an event, as based upon experimental evidence. First, since the calculated tunneling probability of the Davydov soliton is extremely sensitive to the height of the potential energy barrier that opposes the full SNARE zipping (see the Appendix in [12]), the molecular machinery in the axonal bouton could ensure that only a single preselected vesicle is released by raising the height of the individual potential energy barriers of all other synaptic vesicles. This is supported by the fact that SNARE master proteins such as complexins, SM proteins, Rab3/27, RIM, and RIM-BP [37,39,41,69,169] can reduce the potential energy barrier height for vesicle fusion by forming protein–protein bridges that counteract the electrostatic repulsion between the merging synaptic vesicle and plasmalemma. Indirect support for the variability in barrier heights of different synaptic vesicles is provided by the observation that a second electric spike, timed within 20 ms following a spike with successful neurotransmitter release, has a reduced probability for release of another synaptic vesicle [23]. Thus, it is plausible that the molecular machinery inside the active axonal bouton requires some time to preselect another synaptic vesicle, and then to prime it for release by adjusting its potential energy barrier height to be within the functional range for quantum tunneling.

Second, there is a readily releasable pool of synaptic vesicles consisting of vesicles that are immediately available for release. The rate at which the vesicle pool refills after depletion can vary from neuron to neuron. This rate can be increased by an accumulation of intracellular calcium in the course of the activation of synapses by action potential mediation. In this way, the rate of refill assists strengthening of the synapses when exposed to elevated bursting of the potential [24]. This contributes to how the number of vesicles contained in the readily releasable pool can be a significant factor in determining the release probability. Since in the course of an action potential, the influx of Ca^{2+} variably triggers neurotransmitter release from a presynaptic terminal, usually with a single release occurring within 5–10 Ca^{2+} signals [170–173]. Such evidence points toward only a single synaptic vesicle undergoing exocytosis at a time, with the remaining vesicles in the readily releasable pool lined up for subsequent fusion, so that the restriction of exocytosis amounting to a single vesicle, can be seen as a probabilistic event; specifically, as caused by the low release probability of each individual vesicle [171] (cf. [172]). Although varying from neuron to neuron, the rate of refilling of the pool, and an ensuing time-lag thereof, may also influence a low release probability [24].

17.8 APPENDIX: THE DAVYDOV–SCOTT QUASIPARTICLE MODEL AND GENERALIZATIONS

17.8.1 The Davydov Hamiltonian

Landau [174] studied the process of electron self-trapping and was the first to propose the concept of a polaron, namely a quasiparticle created by the coupling of an electron to a vibrating lattice, out of which a phonon is created via the electron's induction of localized distortions of the vibrational modes of the lattice. Under certain conditions, when there is an electromagnetic interaction between the electron and the lattice, the distortions arising from the phonon can reduce the potential well for the electron, and hence trapping it. Some years later, Fröhlich [128] and Holstein [132,133] modified and refined Landau's original study, in what became known as the Fröhlich–Holstein polaron (cf. [130, 131]). This latter concept inspired Davydov's work toward postulating how an intramolecular oscillator can interact with a peptide chain, seemingly analogous to how an electron interacts with a crystal lattice. The resulting quasiparticle, viewed as a large polaron, became known as the Davydov soliton. Davydov further proposed that such an interaction can induce a localized propagation of vibrational energy in α-helices (reviewed in [175]). Scott [19,20] further modified Davydov's model toward biological feasibility, and this led to the acclaimed Davydov–Scott model, which has received considerable interest for at least 30 years, and about which much has been written.

To place matters in context, and to give a basic working description, we recall that the Davydov model describes the interaction of the amide I vibrations (i.e., C=O stretch) with the hydrogen bonds that stabilize the α-helix. Davydov's Hamiltonian is formally similar to the Fröhlich–Holstein

Hamiltonian for the interaction of electrons with a polarizable lattice. Thus, the total Hamiltonian of the energy operator $\hat{H}$ is given as

$$\hat{H} = \hat{H}_{\text{ex}} + \hat{H}_{\text{ph}} + \hat{H}_{\text{int}} \tag{17.1}$$

where $\hat{H}_{\text{ex}}$ is the exciton Hamiltonian, which describes the motion of the amide I excitations between adjacent sites; $\hat{H}_{\text{ph}}$ is the phonon Hamiltonian, which describes the vibrations of the lattice; and $\hat{H}_{\text{int}}$ is the interaction Hamiltonian, which describes the interaction of the amide I excitation with the lattice [16–18, 176]. Below we will specify each of these terms for the purpose of such a working representation of the Davydov model, along with a judicial choice of parameters. As outlined in [10, 148, 177], this follows mainly from the basic structural mechanisms described by Davydov and collaborators [18–20, 116, 117, 178–180], each of which exhibits certain differences in the formulation. At a later stage, we will remark upon generalizations of the model.

First, the exciton Hamiltonian $\hat{H}_{\text{ex}}$ is given as

$$\hat{H}_{\text{ex}} = \sum_{n,\alpha} \left[E_0 \hat{a}^{\dagger}_{n,\alpha}\hat{a}_{n,\alpha} - J\left(\hat{a}^{\dagger}_{n,\alpha}\hat{a}_{n+1,\alpha} + \hat{a}^{\dagger}_{n,\alpha}\hat{a}_{n-1,\alpha}\right) + L\left(\hat{a}^{\dagger}_{n,\alpha}\hat{a}_{n,\alpha+1} + \hat{a}^{\dagger}_{n,\alpha}\hat{a}_{n,\alpha-1}\right)\right] \tag{17.2}$$

where $\hat{a}^{\dagger}_{n,\alpha}$ and $\hat{a}_{n,\alpha}$ are, respectively, the boson creation and annihilation operators for an amide I excitation at the peptide group (n, α); the index $n = 1, 2, \ldots, N$ counts the peptide groups along the α-helix spine; the index $\alpha = 1, 2, 3$ counts each α-helix spine; $E_0 = 3.28 \times 10^{-20}$ J (1650 cm^{-1}) [181] is the energy of the amide I vibration (C=O stretching); $J = 1.55 \times 10^{-22}$ J (7.8 cm^{-1}) [181] is the dipole–dipole coupling energy between a particular amide I bond (and those ahead and behind along the same spine); and $L = 2.46 \times 10^{-22}$ J (12.4 cm^{-1}) [181] is the dipole–dipole coupling energy between a particular amide I bond and those on adjacent spines in the same unit cell of the protein α-helix.

Second, the phonon Hamiltonian $\hat{H}_{\text{ph}}$ is

$$\hat{H}_{\text{ph}} = \frac{1}{2}\sum_{n,\alpha}\left[\frac{\hat{p}^2_{n,\alpha}}{M} + w(\hat{u}_{n+1,\alpha} - \hat{u}_{n,\alpha})^2\right] \tag{17.3}$$

where $\hat{p}_{n,\alpha}$ is the momentum operator and $\hat{u}_{n,\alpha}$ is the displacement operator from the equilibrium position of the peptide group (n, α), M is the mass of the peptide group, and w=13–19.5 N m^{-1} [20] is an effective elasticity coefficient of the lattice (the spring constant of a hydrogen bond).

Third, the interaction Hamiltonian $\hat{H}_{\text{int}}$ is

$$\hat{H}_{\text{int}} = \chi \sum_{n,\alpha}\left[(\hat{u}_{n+1,\alpha} - \hat{u}_{n,\alpha})\hat{a}^{\dagger}_{n,\alpha}\hat{a}_{n,\alpha}\right] \tag{17.4}$$

where χ is an anharmonic parameter arising from the coupling between the amide I excitation (exciton) and the lattice displacements (phonon) and indicates how the amide I energy depends on the hydrogen bond length. It is measured in relationship to the displacement differences $(\hat{u}_{n+1,\alpha} - \hat{u}_{n,\alpha})$ in (17.4) basically as $\chi = \mathrm{d}E_0/\mathrm{d}R$ where R denotes the length of the hydrogen bond that is directly adjacent to a particular amide I oscillator [182]. For $\chi > 0$, when the hydrogen bond length decreases, then so does the energy of amide I vibration decrease, and vice versa. When $\chi = 0$, the amide I energy is independent of relative positions of the peptide groups, while for $\chi \neq 0$, an excitation that is initially located at one peptide group will create distortion of the associated hydrogen bond, compressing it for $\chi > 0$ and expanding it for $\chi < 0$, and thus decreasing the energy of the corresponding amide I state [20, 117].

In this way, χ effectively parameterizes the strength of the exciton–phonon interaction. Various values for χ have been proposed: $\chi = 62$ pN [134], $\chi = 34$ pN [19, 116], and $\chi = -30$ pN [117, 178]. The sign of χ is related to the type of the supported soliton, namely *compressional* soliton for $\chi > 0$ or *dilatational* soliton for $\chi < 0$. We observe that both parameters J and L represent dipole–dipole coupling energies and the approximation of J and L (as constants) fails for dilatational solitons that stretch the α-helix decreasing the amplitude and width, and this indicates that the latter type of solitons are unstable when compared to the prospect of increasing stability in the compressional case [183]. Thus for the sake of relative stability, we favor the $\chi > 0$ scenario. Moreover, computer simulations of peptide bonds aimed at studying the effects of hydrogen bonding upon amide I oscillations strongly suggest that in a realistic case the energy of the amide I oscillator will decrease as the hydrogen bond length decreases [100]; hence $\chi > 0$.

17.8.2 Remarks on the Generalized Davydov–Scott Model

Since Scott's original investigations into the Davydov model [19,20], there have been a number of extensions of the model and/or a hybrid form of the model, as when it is merged with others, particularly those pertaining to the Fröhlich and Schrödinger theories, in which the original Davydov soliton may actually revert to a form more befitting a polaron [116–118,157,175,184–190]. Recently, [175] formulated a generalized Davydov–Scott model for studying polarons in a linear peptide chain with stochastic environmental interaction. To see this with respect to the basic model in Section 17.8.1, terms such as $\hat{H}_{ex}$ and $\hat{H}_{ph}$ are formally similar, but in the formalism of [175], $\hat{H}_{int}$ is generalized to:

$$\begin{aligned}\hat{H}_{\text{int}} &= \sum_{n,\alpha} \bar{\chi}[(1+\beta)\hat{u}_{n+1,\alpha} - 2\beta\hat{u}_{n,\alpha} - (1-\beta)\hat{u}_{n-1,\alpha}]\hat{a}^{\dagger}_{n,\alpha}\hat{a}_{n,\alpha} \\ &= \sum_{n,\alpha} \chi(\hat{u}_{n+1,\alpha} + (\xi - 1)\hat{u}_{n,\alpha} - \xi\hat{u}_{n-1,\alpha})\hat{a}^{\dagger}_{n,\alpha}\hat{a}_{n,\alpha} \end{aligned} \tag{17.5}$$

where $\bar{\chi} = \frac{\chi_r+\chi_l}{2}$ is the arithmetic mean of the right χ_r and left χ_l coupling parameters, $\beta = \frac{\chi_r-\chi_l}{\chi_r+\chi_l}$ is the asymmetry parameter of the interaction Hamiltonian, $\xi = \frac{1-\beta}{1+\beta}$ is the symmetry parameter of the interaction Hamiltonian, and $\chi = \bar{\chi}\frac{2}{1+\xi}$. On setting $\beta = 0$ recovers the interaction Hamiltonian of the original symmetric Davydov model, whereas on setting $\beta = 1$ recovers the interaction Hamiltonian of Scott's antisymmetric model [20].

17.8.3 Quantum Equations of Motion for the Generalized Davydov–Scott Model

In [177], we used the $|D_2\rangle$ ansatz to derive a system of quantum equations of motion from the Davydov Hamiltonian with the use of the Schrödinger equation and the generalized Ehrenfest theorem [191]. To keep the derivations concise, here we will work with a single α-helix spine. After fixing the spine index α, the total Hamiltonian becomes

$$\begin{aligned}\hat{H} = \sum_n \Bigg\{ & \left[E_0\hat{a}^{\dagger}_n\hat{a}_n - J\left(\hat{a}^{\dagger}_n\hat{a}_{n+1} + \hat{a}^{\dagger}_n\hat{a}_{n-1}\right)\right] + \frac{1}{2}\left[\frac{\hat{p}_n^2}{M} + w(\hat{u}_{n+1} - \hat{u}_n)^2\right] \\ & + \chi(\hat{u}_{n+1} + (\xi - 1)\hat{u}_n - \xi\hat{u}_{n-1})\hat{a}^{\dagger}_n\hat{a}_n \Bigg\}\end{aligned} \tag{17.6}$$

Next, we introduce an ansatz state vector that approximates the exact state vector solving the Schrödinger equation. In the literature, Davydov introduced two different possible state vectors called $|D_1\rangle$ or $|D_2\rangle$ ansätze. Here, we will again use the second of Davydov's ansatz state vectors, which has the form [192]:

$$|D_2(t)\rangle = |a\rangle|b\rangle; \quad |a\rangle = \sum_n a_n(t)\hat{a}^{\dagger}_n|0_{\text{ex}}\rangle; \quad |b\rangle = e^{-\frac{i}{\hbar}\sum_j (b_j(t)\hat{p}_j - c_j(t)\hat{u}_j)}|0_{\text{ph}}\rangle \tag{17.7}$$

Normalization of the $|D_2\rangle$ ansatz implies $\langle D_2|D_2\rangle = \sum_n |a_n|^2 = 1$, where $|a_n|^2$ is the probability for finding the amide I quantum exciton at the nth site along the α-helix spine (Figure 17.3).

With the use of the Hadamard lemma

$$e^{\hat{A}}\hat{B}e^{-\hat{A}} = \exp(\mathrm{ad}_{\hat{A}})(\hat{B}) = \sum_{k=0}^{\infty}\frac{1}{k!}(\mathrm{ad}_{\hat{A}})^k(\hat{B})$$
$$= \hat{B} + [\hat{A},\hat{B}] + \frac{1}{2!}[\hat{A},[\hat{A},\hat{B}]] + \frac{1}{3!}[\hat{A},[\hat{A},[\hat{A},\hat{B}]]] + \ldots \quad (17.8)$$

where $\mathrm{ad}_{\hat{A}}(\hat{B}) \equiv [\hat{A},\hat{B}]$ is the adjoint operator, the expectation values for $\hat{u}_n$ and $\hat{p}_n$ are found to be (for details see Section 17.8.4)

$$b_n(t) = \langle D_2(t)|\hat{u}_n|D_2(t)\rangle; \qquad c_n(t) = \langle D_2(t)|\hat{p}_n|D_2(t)\rangle \quad (17.9)$$

Provided that the $|D_2(t)\rangle$ ansatz approximates well the exact solution of the Schrödinger equation [177], its temporal evolution obeys the Schrödinger equation

$$\imath\hbar\frac{\mathrm{d}}{\mathrm{d}t}|D_2(t)\rangle = \hat{H}|D_2(t)\rangle \quad (17.10)$$

and we can use the generalized Ehrenfest theorem for the time dynamics of the expectation values (17.9), namely

$$\frac{\mathrm{d}}{\mathrm{d}t}b_n = \frac{1}{\imath\hbar}\langle[\hat{u}_n,\hat{H}]\rangle; \qquad \frac{\mathrm{d}}{\mathrm{d}t}c_n = \frac{1}{\imath\hbar}\langle[\hat{p}_n,\hat{H}]\rangle \quad (17.11)$$

For the above commutators, we obtain

$$[\hat{u}_n,\hat{H}] = \imath\hbar\frac{\hat{p}_n}{M} \quad (17.12)$$

$$[\hat{p}_n,\hat{H}] = \imath\hbar w(\hat{u}_{n+1} - 2\hat{u}_n + \hat{u}_{n-1}) - \imath\hbar\chi\left(\hat{a}^\dagger_{n-1}\hat{a}_{n-1} + (\xi-1)\hat{a}^\dagger_n\hat{a}_n - \xi\hat{a}^\dagger_{n+1}\hat{a}_{n+1}\right) \quad (17.13)$$

From equations 17.11 and 17.12, we also have

$$\frac{\mathrm{d}}{\mathrm{d}t}b_n = \frac{c_n}{M}; \qquad \frac{\mathrm{d}}{\mathrm{d}t}c_n = M\frac{\mathrm{d}^2}{\mathrm{d}t^2}b_n \quad (17.14)$$

From equations 17.11 and 17.13, together with 17.9 and 17.14, we obtain one of Davydov's equations for the phonon displacements $b_n(t)$ from the corresponding equilibrium positions

$$M\frac{\mathrm{d}^2}{\mathrm{d}t^2}b_n = w(b_{n+1} - 2b_n + b_{n-1}) + \chi\left(\xi|a_{n+1}|^2 + (1-\xi)|a_n|^2 - |a_{n-1}|^2\right) \quad (17.15)$$

The equation for the amide I probability amplitudes $a_n(t)$ can be derived by differentiating the $|D_2(t)\rangle$ ansatz. After application of the special case of the Baker–Campbell–Hausdorff formula

$$e^{\hat{A}+\hat{B}} = e^{-\frac{1}{2}[\hat{A},\hat{B}]}e^{\hat{A}}e^{\hat{B}} \quad (17.16)$$

which is valid for two operators $\hat{A}$ and $\hat{B}$ that commute with their commutator $[\hat{A},\hat{B}]$, the total time derivative of the Davydov ansatz $|D_2(t)\rangle$ is found to be [191]

$$\imath\hbar\frac{\mathrm{d}}{\mathrm{d}t}|D_2(t)\rangle = \imath\hbar\sum_n\frac{\mathrm{d}a_n}{\mathrm{d}t}\hat{a}^\dagger_n|0_{\mathrm{ex}}\rangle|b\rangle + |a\rangle\sum_j\left(\frac{\mathrm{d}b_j}{\mathrm{d}t}\hat{p}_j - \frac{\mathrm{d}c_j}{\mathrm{d}t}\hat{u}_j + \frac{1}{2}\left(b_j\frac{\mathrm{d}c_j}{\mathrm{d}t} - \frac{\mathrm{d}b_j}{\mathrm{d}t}c_j\right)\right)|b\rangle \quad (17.17)$$

Next, we calculate the terms on right-hand side of the Schrödinger equation as follows:

$$\hat{H}_{\mathrm{ex}}|D_2(t)\rangle = \sum_n[E_0a_n - J(a_{n+1} + a_{n-1})]\hat{a}^\dagger_n|0_{\mathrm{ex}}\rangle|b\rangle \quad (17.18)$$

$$\hat{H}_{\mathrm{ph}}|D_2(t)\rangle = \sum_n\hat{H}_{\mathrm{ph}}\hat{a}^\dagger_n|0_{\mathrm{ex}}\rangle|b\rangle \quad (17.19)$$

$$\hat{H}_{\mathrm{int}}|D_2(t)\rangle = \chi\sum_n a_n(t)(\hat{u}_{n+1} + (\xi-1)\hat{u}_n - \xi\hat{u}_{n-1})\hat{a}^\dagger_n|0_{\mathrm{ex}}\rangle|b\rangle \quad (17.20)$$

Then we use the Schrödinger equation to combine 17.17, 17.18, 17.19, and 17.20, and after taking the inner product with $\langle b|\langle 0_{\text{ex}}|\hat{a}_n$, we obtain

$$\imath\hbar\frac{\mathrm{d}a_n}{\mathrm{d}t} = \left[E_0 + W(t) - \frac{1}{2}\sum_j\left(\frac{\mathrm{d}b_j}{\mathrm{d}t}c_j - b_j\frac{\mathrm{d}c_j}{\mathrm{d}t}\right) + \chi(b_{n+1} + (\xi-1)b_n - \xi b_{n-1})\right]a_n - J(a_{n+1} + a_{n-1}) \tag{17.21}$$

where the expectation value of the phonon energy has been written as $W(t) = \langle D_2|\hat{H}_{\text{ph}}|D_2\rangle$.

The first three terms in 17.21 are global for all a_n, namely

$$\forall a_n: \qquad \gamma(t) = E_0 + W(t) - \frac{1}{2}\sum_j\left(\frac{\mathrm{d}b_j}{\mathrm{d}t}c_j - b_j\frac{\mathrm{d}c_j}{\mathrm{d}t}\right) \tag{17.22}$$

Furthermore, all terms in $\gamma(t)$ are real-valued and a global phase change on the quantum probability amplitudes, namely $a_n \rightarrow \bar{a}_n e^{-\frac{\imath}{\hbar}\int\gamma(t)\mathrm{d}t}$, will not change the quantum probabilities for the amide I oscillators

$$|a_n|^2 = e^{+\frac{\imath}{\hbar}\int\gamma(t)\mathrm{d}t}\bar{a}_n^*\bar{a}_n e^{-\frac{\imath}{\hbar}\int\gamma(t)\mathrm{d}t} = |\bar{a}_n|^2 \tag{17.23}$$

After the transformation and re-labeling $\bar{a}_n$ to a_n, the system of Davydov equations for a single α-helix spine becomes:

$$\imath\hbar\frac{\mathrm{d}a_n}{\mathrm{d}t} = \chi[b_{n+1} + (\xi-1)b_n - \xi b_{n-1}]a_n - J(a_{n+1} + a_{n-1}) \tag{17.24}$$

$$M\frac{\mathrm{d}^2}{\mathrm{d}t^2}b_n = w(b_{n-1} - 2b_n + b_{n+1}) - \chi\left(|a_{n-1}|^2 + (\xi-1)|a_n|^2 - \xi|a_{n+1}|^2\right) \tag{17.25}$$

where a_n is the quantum probability amplitude for finding the amide I quantum exciton at the nth site and b_n is the expectation value of the position operator $\hat{u}_n$ for longitudinal displacement of the nth peptide group.

The Davydov soliton obtained with the use of the $|D_2\rangle$ ansatz (Figure 17.4) is a very good approximation to the full quantum dynamics in the absence of thermal noise. To quantify the deviation from the Schrödinger equation, we follow the method by Sun *et al.* [193] and calculate the amplitude of the deviation vector $|\delta(t)\rangle$, which is defined as

$$\Delta(t) \equiv \sqrt{\langle\delta(t)|\delta(t)\rangle} \tag{17.26}$$

where

$$|\delta(t)\rangle \equiv \imath\hbar\frac{\partial}{\partial t}|D_2(t)\rangle - \hat{H}|D_2(t)\rangle \tag{17.27}$$

If we denote by W_0 the zero-point energy of the lattice vibrations, we can define a gauge transformed soliton energy $E = \langle D_2(t)|\hat{H}|D_2(t)\rangle - E_0 - W_0$, which is

$$E = -\sum_n\left\{2J[\mathrm{Re}(a_n)\mathrm{Re}(a_{n+1}) + \mathrm{Im}(a_n)\mathrm{Im}(a_{n+1})] - \chi|a_n|^2(b_{n+1} + (\xi-1)b_n - \xi b_{n-1}) - \frac{1}{2}M\left(\frac{\mathrm{d}b_n}{\mathrm{d}t}\right)^2 - \frac{1}{2}w(b_{n+1} - b_n)^2\right\} \tag{17.28}$$

In comparison with this gauge transformed soliton energy, the computed deviation from the Schrödinger equation is negligible, $\Delta(t) < 10^{-5}|E|$ (cf. [177]). Note that because the subtracted sum of the two constant energies $E_0 + W_0$ is two orders of magnitude larger than $|E|$, comparison with the total energy of the system will give $\Delta(t) < 10^{-7}\langle D_2(t)|\hat{H}|D_2(t)\rangle$.

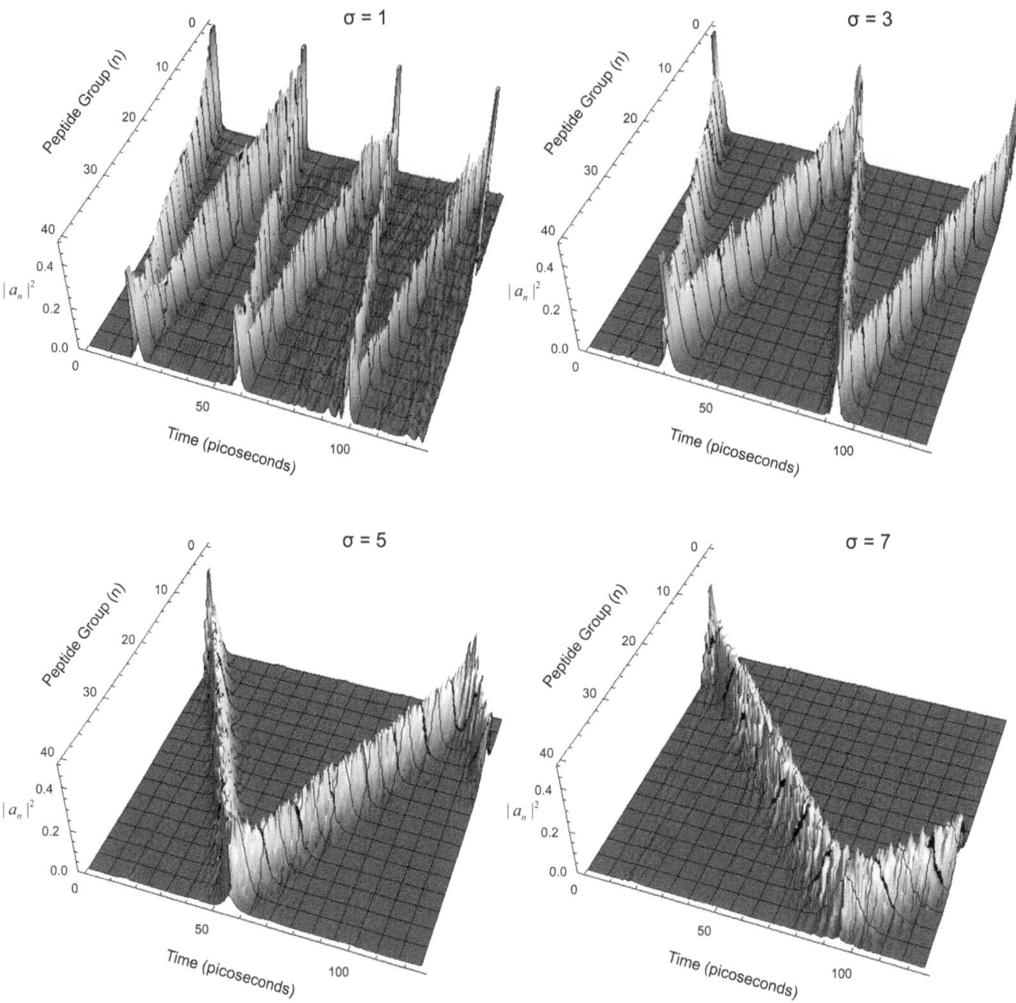

FIGURE 17.4 Davydov soliton dynamics visualized through amide I probability distribution $|a_n|^2$ for the symmetric Hamiltonian $\xi = 1$ at suprathreshold value $\chi = 35$ pN for different initial Gaussian step distributions σ of amide I energy over 1, 3, 5, or 7 peptide groups starting from the N-end of an α-helix spine composed of 40 peptide groups during a period of 125 ps. Solitons with larger initial spread have lower velocities.

17.8.4 Expectation Values for the Peptide Group Position and Momentum Operators

The phonon Hamiltonian for the lattice of hydrogen bonds (−C=O ··· H-N−) in the protein α-helix spine can be brought into the familiar form [191]

$$\hat{H}_{\text{ph}} = \sum_q \hbar\omega_q \left(\hat{b}_q^\dagger \hat{b}_q + \frac{1}{2} \right) \tag{17.29}$$

with the use of the phonon creation and annihilation operators, $\hat{b}_q^\dagger$ and $\hat{b}_q$, through which the momentum operator $\hat{p}_n$ for longitudinal displacement of the nth peptide group is expressed as

$$\hat{p}_n = \sum_q \sqrt{\frac{M\hbar\omega_q}{2N}} e^{\imath q l n} \imath \left(\hat{b}_{-q}^\dagger - \hat{b}_q \right) \tag{17.30}$$

and the position operator $\hat{u}_n$ for longitudinal displacement of the nth peptide group as

$$\hat{u}_n = \sum_q \sqrt{\frac{\hbar}{2NM\omega_q}} e^{\imath qln} \left(\hat{b}_{-q}^{\dagger} + \hat{b}_q\right) \tag{17.31}$$

where N is the number of peptide groups, $l = 0.45$ nm is the lattice spacing (the distance between peptide groups) and the phonon dispersion relation holds

$$\omega_q = 2\sqrt{\frac{w}{M}} \left|\sin\left(\frac{ql}{2}\right)\right| \tag{17.32}$$

From the zero expectation values for the phonon creation and annihilation operators with the vacuum

$$\langle 0_{\text{ph}} | \hat{b}_q^{\dagger} | 0_{\text{ph}} \rangle = \langle 0_{\text{ph}} | \hat{b}_q | 0_{\text{ph}} \rangle = 0 \tag{17.33}$$

together with equations 17.30 and 17.31, it can be seen that

$$\langle 0_{\text{ph}} | \hat{u}_n | 0_{\text{ph}} \rangle = \langle 0_{\text{ph}} | \hat{p}_n | 0_{\text{ph}} \rangle = 0 \tag{17.34}$$

Then, taking into consideration the quantum commutation relations

$$[\hat{u}_n, \hat{p}_n] = \imath\hbar; \qquad [\hat{p}_n, \hat{u}_n] = -\imath\hbar \tag{17.35}$$

and the fact that position and momentum operators with different indices commute, we can use equations 17.7 and 17.8 to obtain the expectation value of the position operator $\hat{u}_n$ as follows:

$$\begin{aligned}
\langle D_2 | \hat{u}_n | D_2 \rangle &= \langle 0_{\text{ph}} \big| e^{\frac{\imath}{\hbar}\sum_j (b_j\hat{p}_j - c_j\hat{u}_j)} \hat{u}_n e^{-\frac{\imath}{\hbar}\sum_j (b_j\hat{p}_j - c_j\hat{u}_j)} \big| 0_{\text{ph}} \rangle \\
&= \langle 0_{\text{ph}} \big| e^{\frac{\imath}{\hbar}(b_n\hat{p}_n - c_n\hat{u}_n)} e^{\frac{\imath}{\hbar}\sum_{j\neq n} (b_j\hat{p}_j - c_j\hat{u}_j)} \hat{u}_n e^{-\frac{\imath}{\hbar}\sum_{j\neq n} (b_j\hat{p}_j - c_j\hat{u}_j)} e^{-\frac{\imath}{\hbar}(b_n\hat{p}_n - c_n\hat{u}_n)} \big| 0_{\text{ph}} \rangle \\
&= \langle 0_{\text{ph}} \big| e^{\frac{\imath}{\hbar}(b_n\hat{p}_n - c_n\hat{u}_n)} e^{\frac{\imath}{\hbar}\sum_{j\neq n} (b_j\hat{p}_j - c_j\hat{u}_j)} e^{-\frac{\imath}{\hbar}\sum_{j\neq n} (b_j\hat{p}_j - c_j\hat{u}_j)} \hat{u}_n e^{-\frac{\imath}{\hbar}(b_n\hat{p}_n - c_n\hat{u}_n)} \big| 0_{\text{ph}} \rangle \\
&= \langle 0_{\text{ph}} \big| e^{\frac{\imath}{\hbar}(b_n\hat{p}_n - c_n\hat{u}_n)} \hat{u}_n e^{-\frac{\imath}{\hbar}(b_n\hat{p}_n - c_n\hat{u}_n)} \big| 0_{\text{ph}} \rangle \\
&= \left\langle 0_{\text{ph}} \left| \hat{u}_n + \left[\frac{\imath}{\hbar}(b_n\hat{p}_n - c_n\hat{u}_n), \hat{u}_n\right] \right| 0_{\text{ph}} \right\rangle \\
&= \langle 0_{\text{ph}} | \hat{u}_n | 0_{\text{ph}} \rangle + \langle 0_{\text{ph}} | b_n | 0_{\text{ph}} \rangle \\
&= b_n
\end{aligned}$$

The calculation of the expectation value of the momentum operator $\hat{p}_n$ is analogous

$$\begin{aligned}
\langle D_2 | \hat{p}_n | D_2 \rangle &= \langle 0_{\text{ph}} \big| e^{\frac{\imath}{\hbar}\sum_j (b_j\hat{p}_j - c_j\hat{u}_j)} \hat{p}_n e^{-\frac{\imath}{\hbar}\sum_j (b_j\hat{p}_j - c_j\hat{u}_j)} \big| 0_{\text{ph}} \rangle \\
&= \langle 0_{\text{ph}} \big| e^{\frac{\imath}{\hbar}(b_n\hat{p}_n - c_n\hat{u}_n)} e^{\frac{\imath}{\hbar}\sum_{j\neq n} (b_j\hat{p}_j - c_j\hat{u}_j)} \hat{p}_n e^{-\frac{\imath}{\hbar}\sum_{j\neq n} (b_j\hat{p}_j - c_j\hat{u}_j)} e^{-\frac{\imath}{\hbar}(b_n\hat{p}_n - c_n\hat{u}_n)} \big| 0_{\text{ph}} \rangle \\
&= \langle 0_{\text{ph}} \big| e^{\frac{\imath}{\hbar}(b_n\hat{p}_n - c_n\hat{u}_n)} e^{\frac{\imath}{\hbar}\sum_{j\neq n} (b_j\hat{p}_j - c_j\hat{u}_j)} e^{-\frac{\imath}{\hbar}\sum_{j\neq n} (b_j\hat{p}_j - c_j\hat{u}_j)} \hat{p}_n e^{-\frac{\imath}{\hbar}(b_n\hat{p}_n - c_n\hat{u}_n)} \big| 0_{\text{ph}} \rangle \\
&= \langle 0_{\text{ph}} \big| e^{\frac{\imath}{\hbar}(b_n\hat{p}_n - c_n\hat{u}_n)} \hat{p}_n e^{-\frac{\imath}{\hbar}(b_n\hat{p}_n - c_n\hat{u}_n)} \big| 0_{\text{ph}} \rangle \\
&= \left\langle 0_{\text{ph}} \left| \hat{p}_n + \left[\frac{\imath}{\hbar}(b_n\hat{p}_n - c_n\hat{u}_n), \hat{p}_n\right] \right| 0_{\text{ph}} \right\rangle \\
&= \langle 0_{\text{ph}} | \hat{p}_n | 0_{\text{ph}} \rangle + \langle 0_{\text{ph}} | c_n | 0_{\text{ph}} \rangle \\
&= c_n
\end{aligned}$$

17.8.5 Polaron Drift and the Rachet Effect

In [185, 186], Brizhik and co-workers studied the ratchet effect of a polaron drift in a temperature-dependent setting. Both in classical and quantum terms, the rachet effect is induced by energy dissipation and symmetry breaking within the system. Specifically, a polaron propagating in a spatially asymmetric discrete lattice encounters a Peierls–Nabarro barrier, which serves as a rachet potential, for as long as the electron field intensity and period are sufficiently large. The amplitude of the polaron drift has a non-monotonic dependence on temperature, hence leading to a degree of stochastic resonance, which in turn supports the prevailing drift, since when the field amplitude is too small, the polaron does not gain sufficient energy to cross the energy barrier that separates it from the upcoming lattice site. In the presence of thermal excitations, the polaron gains extra energy from a thermal bath to overcome the barrier (cf. [10]). Accordingly, [185, 186](cf. [175]) incorporate into their model two extra terms, a Langevin factor $F(t)$ describing the thermal excitations and a viscous liquid coefficient Γ.

REFERENCES

1. G. Finkelstein. *Emil du Bois-Reymond: Neuroscience, Self, and Society in Nineteenth-Century Germany*. MIT Press, Cambridge, Massachusetts, 2013.
2. B. Katz. *The Release of Neural Transmitter Substances*. Liverpool University Press, Liverpool, 1969.
3. T. C. Südhof. Neurotransmitter release: the last millisecond in the life of a synaptic vesicle. *Neuron*, 80(3):675–690, 2013.
4. R. S. Zucker. Exocytosis: a molecular and physiological perspective. *Neuron*, 17(6):1049–1055, 1996.
5. T. C. Südhof and J. Rizo. Synaptic vesicle exocytosis. *Cold Spring Harbor Perspectives in Biology*, 3(12):a005637, 2011.
6. T. C. Südhof. A molecular machine for neurotransmitter release: synaptotagmin and beyond. *Nature Medicine*, 19(10):1227–1231, 2013.
7. F. Beck and J. C. Eccles. Quantum aspects of brain activity and the role of consciousness. *Proceedings of the National Academy of Sciences*, 89(23):11357–11361, 1992.
8. J. C. Eccles. Do mental events cause neural events analogously to the probability fields of quantum mechanics? *Proceedings of the Royal Society of London B*, 227(1249):411–428, 1986.
9. D. D. Georgiev and J. F. Glazebrook. SNARE proteins as molecular masters of interneuronal communication. *Biomedical Reviews*, 21:17–23, 2010.
10. D. D. Georgiev and J. F. Glazebrook. Quasiparticle tunneling in neurotransmitter release. In W. A. Goddard III, D. Brenner, S. E. Lyshevski, and G. J. Iafrate, editors, *Handbook of Nanoscience, Engineering, and Technology, Third Edition*, Electrical Engineering Handbook, chapter 30, pages 983–1016. CRC Press, Boca Raton, 2012.
11. D. D. Georgiev and J. F. Glazebrook. Quantum interactive dualism: From Beck and Eccles tunneling model of exocytosis to molecular biology of snare zipping. *Biomedical Reviews*, 25:15–24, 2014.
12. D. D. Georgiev and J. F. Glazebrook. The quantum physics of synaptic communication via the SNARE protein complex. *Progress in Biophysics and Molecular Biology*, 135:16–29, 2018.
13. C. B. Anfinsen. Principles that govern the folding of protein chains. *Science*, 181(4096):223, 1973.
14. V. N. Uversky. Intrinsic disorder in proteins associated with neurodegenerative diseases. In J. Ovádi and F. Orosz, editors, *Protein Folding and Misfolding: Neurodegenerative Diseases*, Focus on Structural Biology, pages 21–75. Springer, Dordrecht, 2009.
15. P. E. Wright and H. J. Dyson. Linking folding and binding. *Current Opinion in Structural Biology*, 19(1):31–38, 2009.
16. A. S. Davydov. The theory of contraction of proteins under their excitation. *Journal of Theoretical Biology*, 38(3):559–569, 1973.
17. A. S. Davydov. Solitons and energy transfer along protein molecules. *Journal of Theoretical Biology*, 66(2):379–387, 1977.
18. A. S. Davydov. The role of proteins in electron transport at large distances. *Physica Status Solidi (b)*, 90(2):457–464, 1978.
19. A. C. Scott. Dynamics of Davydov solitons. *Physical Review A*, 26(1):578–595, 1982.

20. A. C. Scott. Davydov's soliton. *Physics Reports*, 217(1):1–67, 1992.
21. E. R. Chapman. Synaptotagmin: a Ca^{2+} sensor that triggers exocytosis? *Nature Reviews Molecular Cell Biology*, 3(7):498–508, 2002.
22. P. I. Hanson, J. E. Heuser, and R. Jahn. Neurotransmitter release—four years of SNARE complexes. *Current Opinion in Neurobiology*, 7(3):310–315, 1997.
23. C. F. Stevens and Y. Wang. Facilitation and depression at single central synapses. *Neuron*, 14(4):795–802, 1995.
24. C. F. Stevens and J. F. Wesseling. Activity-dependent modulation of the rate at which synaptic vesicles become available to undergo exocytosis. *Neuron*, 21(2):415–424, 1998.
25. Z.-W. Wang. *Molecular Mechanisms of Neurotransmitter Release*. Contemporary Neuroscience. Humana Press, Totowa, New Jersey, 2008.
26. D. D. Georgiev. *Quantum Information and Consciousness: A Gentle Introduction*. CRC Press, Boca Raton, 2017.
27. B. W. Connors and M. A. Long. Electrical synapses in the mammalian brain. *Annual Review of Neuroscience*, 27:393–418, 2004.
28. J. C. Eccles. The synapse: from electrical to chemical transmission. *Annual Review of Neuroscience*, 5:325–339, 1982.
29. C. Beaulieu, Z. Kisvarday, P. Somogyi, M. Cynader, and A. Cowey. Quantitative distribution of gaba-immunopositive and-immunonegative neurons and synapses in the monkey striate cortex (area 17). *Cerebral Cortex*, 2(4):295–309, 1992.
30. M. B. Dalva, A. C. McClelland, and M. S. Kayser. Cell adhesion molecules: signalling functions at the synapse. *Nature Reviews Neuroscience*, 8(3):206–220, 2007.
31. C. Dean, F. G. Scholl, J. Choih, S. DeMaria, J. Berger, E. Isacoff, and P. Scheiffele. Neurexin mediates the assembly of presynaptic terminals. *Nature Neuroscience*, 6(7):708–716, 2003.
32. C. Dean and T. Dresbach. Neuroligins and neurexins: linking cell adhesion, synapse formation and cognitive function. *Trends in Neurosciences*, 29(1):21–29, 2006.
33. D. D. Georgiev and J. F. Glazebrook. Subneuronal processing of information by solitary waves and stochastic processes. In S. E. Lyshevski, editor, *Nano and Molecular Electronics Handbook*, Nano and Microengineering Series, chapter 17, pages 1–41. CRC Press, Boca Raton, 2007.
34. T. C. Südhof. Neuroligins and neurexins link synaptic function to cognitive disease. *Nature*, 455(7215):903–911, 2008.
35. R. Jahn and R. H. Scheller. SNAREs—engines for membrane fusion. *Nature Reviews Molecular Cell Biology*, 7(9):631–643, 2006.
36. R. Jahn and D. Fasshauer. Molecular machines governing exocytosis of synaptic vesicles. *Nature*, 490(7419):201–207, 2012.
37. T. C. Südhof. The presynaptic active zone. *Neuron*, 75(1):11–25, 2012.
38. O.-H. Shin. Exocytosis and synaptic vesicle function. *Comprehensive Physiology*, 4(1):149–175, 2014.
39. J. Rizo and T. C. Südhof. The membrane fusion enigma: SNAREs, Sec1/Munc18 proteins, and their accomplices—guilty as charged? *Annual Review of Cell and Developmental Biology*, 28(1):279–308, 2012.
40. J. Rizo and J. Xu. The synaptic vesicle release machinery. *Annual Review of Biophysics*, 44(1):339–367, 2015.
41. T. C. Südhof and J. E. Rothman. Membrane fusion: grappling with SNARE and SM proteins. *Science*, 323(5913):474–477, 2009.
42. P. Zhou, Z. P. Pang, X. Yang, Y. Zhang, C. Rosenmund, T. Bacaj, and T. C. Südhof. Syntaxin-1 N-peptide and H_{abc}-domain perform distinct essential functions in synaptic vesicle fusion. *EMBO Journal*, 32(1):159–171, 2012.
43. Q. Zhou, P. Zhou, A. L. Wang, D. Wu, M. Zhao, T. C. Südhof, and A. T. Brunger. The primed SNARE–complexin–synaptotagmin complex for neuronal exocytosis. *Nature*, 548(7668):420–425, 2017.
44. S. Wang, U. B. Choi, J. Gong, X. Yang, Y. Li, A. L. Wang, X. Yang, A. T. Brunger, and C. Ma. Conformational change of syntaxin linker region induced by Munc13s initiates SNARE complex formation in synaptic exocytosis. *EMBO Journal*, 36(6):816–829, 2017.
45. Y. Yu, S. Chen, X. Mo, J. Gong, C. Li, and X. Yang. Accessory and central α-helices of complexin selectively activate Ca^{2+} triggering of synaptic exocytosis. *Frontiers in Molecular Neuroscience*, 11:61, 2018.

46. H. Sakamoto, T. Ariyoshi, N. Kimpara, K. Sugao, I. Taiko, K. Takikawa, D. Asanuma, S. Namiki, and K. Hirose. Synaptic weight set by Munc13-1 supramolecular assemblies. *Nature Neuroscience*, 21(1):41–49, 2018.
47. T. Weber, B. V. Zemelman, J. A. McNew, B. Westermann, M. Gmachl, F. Parlati, T. H. Söllner, and J. E. Rothman. SNAREpins: minimal machinery for membrane fusion. *Cell*, 92(6):759–772, 1998.
48. Y. A. Chen and R. H. Scheller. SNARE-mediated membrane fusion. *Nature Reviews Molecular Cell Biology*, 2(2):98–106, 2001.
49. J. Han, K. Pluhackova, and R. A. Böckmann. The multifaceted role of SNARE proteins in membrane fusion. *Frontiers in Physiology*, 8:5, 2017.
50. H. Mostafavi, S. Thiyagarajan, B. S. Stratton, E. Karatekin, J. M. Warner, J. E. Rothman, and B. O'Shaughnessy. Entropic forces drive self-organization and membrane fusion by SNARE proteins. *Proceedings of the National Academy of Sciences*, 114(21):5455–5460, 2017.
51. D. Ungar and F. M. Hughson. SNARE protein structure and function. *Annual Review of Cell and Developmental Biology*, 19(1):493–517, 2003.
52. S. R. Lane and Y. Liu. Characterization of the palmitoylation domain of SNAP-25. *Journal of Neurochemistry*, 69(5):1864–1869, 1997.
53. M. Veit, T. H. Söllner, and J. E. Rothman. Multiple palmitoylation of synaptotagmin and the t-SNARE SNAP-25. *FEBS Letters*, 385(1):119–123, 1996.
54. H. J. Risselada and H. Grubmüller. How SNARE molecules mediate membrane fusion: recent insights from molecular simulations. *Current Opinion in Structural Biology*, 22(2):187–196, 2012.
55. Q. Zhou, Y. Lai, T. Bacaj, M. Zhao, A. Y. Lyubimov, M. Uervirojnangkoorn, O. B. Zeldin, A. S. Brewster, N. K. Sauter, A. E. Cohen, S. M. Soltis, R. Alonso-Mori, M. Chollet, H. T. Lemke, R. A. Pfuetzner, U. B. Choi, W. I. Weis, J. Diao, T. C. Südhof, and A. T. Brunger. Architecture of the synaptotagmin-SNARE machinery for neuronal exocytosis. *Nature*, 525(7567):62–67, 2015.
56. P. Zhou, T. Bacaj, X. Yang, Z. P. Pang, and T. C. Südhof. Lipid-anchored SNAREs lacking transmembrane regions fully support membrane fusion during neurotransmitter release. *Neuron*, 80(2):470–483, 2013.
57. M. T. Palfreyman and E. M. Jorgensen. Roles of SNARE proteins in synaptic vesicle fusion. In Z.-W. Wang, editor, *Molecular Mechanisms of Neurotransmitter Release*, pages 35–59. Humana Press, Totowa, New Jersey, 2008.
58. R. B. Sutton, D. Fasshauer, R. Jahn, and A. T. Brunger. Crystal structure of a SNARE complex involved in synaptic exocytosis at 2.4 Å resolution. *Nature*, 395(6700):347–353, 1998.
59. M. E. Graham, P. Washbourne, M. C. Wilson, and R. D. Burgoyne. SNAP-25 with mutations in the zero layer supports normal membrane fusion kinetics. *Journal of Cell Science*, 114(24):4397–4405, 2001.
60. S.-Y. Hua and M. P. Charlton. Activity-dependent changes in partial VAMP complexes during neurotransmitter release. *Nature Neuroscience*, 2(12):1078–1083, 1999.
61. A. Yersin, H. Hirling, P. Steiner, S. Magnin, R. Regazzi, B. Hüni, P. Huguenot, P. De Los Rios, G. Dietler, S. Catsicas, and S. Kasas. Interactions between synaptic vesicle fusion proteins explored by atomic force microscopy. *Proceedings of the National Academy of Sciences*, 100(15):8736–8741, 2003.
62. M. C. Chicka, E. Hui, H. Liu, and E. R. Chapman. Synaptotagmin arrests the SNARE complex before triggering fast, efficient membrane fusion in response to Ca^{2+}. *Nature Structural & Molecular Biology*, 15(8):827–835, 2008.
63. J. T. Littleton, M. Stern, M. Perin, and H. J. Bellen. Calcium dependence of neurotransmitter release and rate of spontaneous vesicle fusions are altered in Drosophila synaptotagmin mutants. *Proceedings of the National Academy of Sciences*, 91(23):10888–10892, 1994.
64. K. C. Martin, Y. Hu, B. A. Armitage, S. A. Siegelbaum, E. R. Kandel, and B. K. Kaang. Evidence for synaptotagmin as an inhibitory clamp on synaptic vesicle release in Aplysia neurons. *Proceedings of the National Academy of Sciences*, 92(24):11307–11311, 1995.
65. F. A. Dodge and R. Rahamimoff. Co-operative action of calcium ions in transmitter release at the neuromuscular junction. *Journal of Physiology*, 193(2):419–432, 1967.
66. C. F. Stevens and J. M. Sullivan. The synaptotagmin C2A domain is part of the calcium sensor controlling fast synaptic transmission. *Neuron*, 39(2):299–308, 2003.
67. T. C. Südhof. Calcium control of neurotransmitter release. *Cold Spring Harbor Perspectives in Biology*, 4(1):a011353, 2012.

68. E. R. Chapman and A. F. Davis. Direct interaction of a Ca^{2+}-binding loop of synaptotagmin with lipid bilayers. *Journal of Biological Chemistry*, 273(22):13995–14001, 1998.
69. U. B. Choi, M. Zhao, Y. Zhang, Y. Lai, and A. T. Brunger. Complexin induces a conformational change at the membrane-proximal C-terminal end of the snare complex. *eLife*, 5:e16886, 2016.
70. C. G. Giraudo, W. S. Eng, T. J. Melia, and J. E. Rothman. A clamping mechanism involved in SNARE-dependent exocytosis. *Science*, 313(5787):676–680, 2006.
71. J. Rizo and C. Rosenmund. Synaptic vesicle fusion. *Nature Structural & Molecular Biology*, 15(7):665–674, 2008.
72. J. Iyer, C. J. Wahlmark, G. A. Kuser-Ahnert, and F. Kawasaki. Molecular mechanisms of complexin fusion clamp function in synaptic exocytosis revealed in a new Drosophila mutant. *Molecular and Cellular Neuroscience*, 56:244–254, 2013.
73. J. Xu, K. D. Brewer, R. Perez-Castillejos, and J. Rizo. Subtle interplay between synaptotagmin and complexin binding to the SNARE complex. *Journal of Molecular Biology*, 425(18):3461–3475, 2013.
74. S. S. Krishnakumar, D. T. Radoff, D. Kümmel, C. G. Giraudo, F. Li, L. Khandan, S. W. Baguley, J. Coleman, K. M. Reinisch, F. Pincet, and J. E. Rothman. A conformational switch in complexin is required for synaptotagmin to trigger synaptic fusion. *Nature Structural & Molecular Biology*, 18:934–940, 2011.
75. Y. Lai, J. Diao, Y. Liu, Y. Ishitsuka, Z. Su, K. Schulten, T. Ha, and Y.-K. Shin. Fusion pore formation and expansion induced by Ca^{2+} and synaptotagmin 1. *Proceedings of the National Academy of Sciences*, 110(4):1333–1338, 2013.
76. A. Vasin, D. Volfson, J. T. Littleton, and M. Bykhovskaia. Interaction of the complexin accessory helix with synaptobrevin regulates spontaneous fusion. *Biophysical Journal*, 111(9):1954–1964, 2016.
77. R. T. Wragg, D. Snead, Y. Dong, T. F. Ramlall, I. Menon, J. Bai, D. Eliezer, and J. S. Dittman. Synaptic vesicles position complexin to block spontaneous fusion. *Neuron*, 77(2):323–334, 2013.
78. Y. Lai, J. Diao, D. J. Cipriano, Y. Zhang, R. A. Pfuetzner, M. S. Padolina, and A. T. Brunger. Complexin inhibits spontaneous release and synchronizes Ca^{2+}-triggered synaptic vesicle fusion by distinct mechanisms. *eLife*, 3:e03756, 2014.
79. S. S. Krishnakumar, F. Li, J. Coleman, C. M. Schauder, D. Kümmel, F. Pincet, J. E. Rothman, and K. M. Reinisch. Re-visiting the trans insertion model for complexin clamping. *eLife*, 4:e04463, 2015.
80. M. Dhara, A. Yarzagaray, Y. Schwarz, S. Dutta, C. Grabner, P. K. Moghadam, A. Bost, C. Schirra, J. Rettig, K. Reim, N. Brose, R. Mohrmann, and D. Bruns. Complexin synchronizes primed vesicle exocytosis and regulates fusion pore dynamics. *Journal of Cell Biology*, 204(7):1123, 2014.
81. J. Diao, P. Grob, D. J. Cipriano, M. Kyoung, Y. Zhang, S. Shah, A. Nguyen, M. Padolina, A. Srivastava, M. Vrljic, A. Shah, E. Nogales, S. Chu, and A. T. Brunger. Synaptic proteins promote calcium-triggered fast transition from point contact to full fusion. *eLife*, 1:e00109, 2012.
82. J. Diao, D. J. Cipriano, M. Zhao, Y. Zhang, S. Shah, M. S. Padolina, R. A. Pfuetzner, and A. T. Brunger. Complexin-1 enhances the on-rate of vesicle docking via simultaneous SNARE and membrane interactions. *Journal of the American Chemical Society*, 135(41):15274–15277, 2013.
83. R. A. Jorquera, S. Huntwork-Rodriguez, Y. Akbergenova, R. W. Cho, and J. T. Littleton. Complexin controls spontaneous and evoked neurotransmitter release by regulating the timing and properties of synaptotagmin activity. *Journal of Neuroscience*, 32(50):18234, 2012.
84. Y. Lai, U. B. Choi, Y. Zhang, M. Zhao, R. A. Pfuetzner, A. L. Wang, J. Diao, and A. T. Brunger. N-terminal domain of complexin independently activates calcium-triggered fusion. *Proceedings of the National Academy of Sciences*, 113(32):E4698–E4707, 2016.
85. J. R. Schaub, X. Lu, B. Doneske, Y.-K. Shin, and J. A. McNew. Hemifusion arrest by complexin is relieved by Ca^{2+}–synaptotagmin I. *Nature Structural & Molecular Biology*, 13(8):748–750, 2006.
86. T. Trimbuch and C. Rosenmund. Should I stop or should I go? The role of complexin in neurotransmitter release. *Nature Reviews Neuroscience*, 17:118–125, 2016.
87. J. Kim, Y. Zhu, and Y.-K. Shin. Preincubation of t-SNAREs with complexin I increases content-mixing efficiency. *Biochemistry*, 55(26):3667–3673, 2016.
88. M. Bykhovskaia, A. Jagota, A. Gonzalez, A. Vasin, and J. T. Littleton. Interaction of the complexin accessory helix with the C-terminus of the SNARE complex: molecular-dynamics model of the fusion clamp. *Biophysical Journal*, 105(3):679–690, 2013.
89. K. Neuland and M. Frick. Vesicular control of fusion pore expansion. *Communicative & Integrative Biology*, 8(2):e1018496, 2015.

90. M. B. Jackson and E. R. Chapman. The fusion pores of Ca^{2+}-triggered exocytosis. *Nature Structural & Molecular Biology*, 15(7):684–689, 2008.
91. A. V. Pobbati, A. Stein, and D. Fasshauer. N- to C-terminal SNARE complex assembly promotes rapid membrane fusion. *Science*, 313(5787):673–676, 2006.
92. I. Fernandez, D. Araç, J. Ubach, S. H. Gerber, O. Shin, Y. Gao, R. G. W. Anderson, T. C. Südhof, and J. Rizo. Three-dimensional structure of the synaptotagmin 1 C2B-domain: synaptotagmin 1 as a phospholipid binding machine. *Neuron*, 32(6):1057–1069, 2001.
93. X. Han and M. B. Jackson. Electrostatic interactions between the syntaxin membrane anchor and neurotransmitter passing through the fusion pore. *Biophysical Journal*, 88(3):L20–L22, 2005.
94. Z.-H. Sheng, J. Rettig, T. Cook, and W. A. Catterall. Calcium-dependent interaction of N-type calcium channels with the synaptic core complex. *Nature*, 379(6564):451–454, 1996.
95. J. F. Ellena, B. Liang, M. Wiktor, A. Stein, D. S. Cafiso, R. Jahn, and L. K. Tamm. Dynamic structure of lipid-bound synaptobrevin suggests a nucleation-propagation mechanism for trans-SNARE complex formation. *Proceedings of the National Academy of Sciences*, 106(48):20306–20311, 2009.
96. K. C. Chou, G. M. Maggiora, G. Némethy, and H. A. Scheraga. Energetics of the structure of the four-α-helix bundle in proteins. *Proceedings of the National Academy of Sciences of the United States of America*, 85(12):4295–4299, 1988.
97. A. Stein, G. Weber, M. C. Wahl, and R. Jahn. Helical extension of the neuronal SNARE complex into the membrane. *Nature*, 460(7254):525–528, 2009.
98. M. Zhao, S. Wu, Q. Zhou, S. Vivona, D. J. Cipriano, Y. Cheng, and A. T. Brunger. Mechanistic insights into the recycling machine of the SNARE complex. *Nature*, 518:61, 2015.
99. C. N. Pace, H. Fu, K. Lee Fryar, J. Landua, S. R. Trevino, D. Schell, R. L. Thurlkill, S. Imura, J. M. Scholtz, K. Gajiwala, J. Sevcik, L. Urbanikova, J. K. Myers, K. Takano, E. J. Hebert, B. A. Shirley, and G. R. Grimsley. Contribution of hydrogen bonds to protein stability. *Protein Science*, 23(5):652–661, 2014.
100. N. S. Myshakina, Z. Ahmed, and S. A. Asher. Dependence of amide vibrations on hydrogen bonding. *Journal of Physical Chemistry B*, 112(38):11873–11877, 2008.
101. X.-Z. Li, B. Walker, and A. Michaelides. Quantum nature of the hydrogen bond. *Proceedings of the National Academy of Sciences*, 108(16):6369–6373, 2011.
102. O. Pusuluk, T. Farrow, C. Deliduman, K. Burnett, and V. Vedral. Proton tunnelling in hydrogen bonds and its implications in an induced-fit model of enzyme catalysis. *Proceedings of the Royal Society A*, 474(2218):20180037, 2018.
103. S. C. L. Kamerlin and Warshel A. At the dawn of the 21st century: Is dynamics the missing link for understanding enzyme catalysis? *Proteins*, 78(6):1339–1375, 2010.
104. M. J. Sutcliffe and N. S. Scrutton. Enzymology takes a quantum leap forward. *Philosophical Transactions of the Royal Society of London. Series A: Mathematical, Physical and Engineering Sciences*, 358(1766):367–386, 2000.
105. J. Basran, M. J. Sutcliffe, and N. S. Scrutton. Enzymatic H-transfer requires vibration-driven extreme tunneling. *Biochemistry*, 38(10):3218–3222, 1999.
106. J. Basran, S. Patel, M. J. Sutcliffe, and N. S. Scrutton. Importance of barrier shape in enzyme-catalyzed reactions: vibrationally assisted hydrogen tunneling in tryptophan tryptophylquinone-dependent amine dehydrogenases. *Journal of Biological Chemistry*, 276(9):6234–6242, 2001.
107. M. J. Sutcliffe and N. S. Scrutton. Enzyme catalysis: over-the-barrier or through-the-barrier? *Trends in Biochemical Sciences*, 25(9):405–408, 2000.
108. M. J. Knapp and J. P. Klinman. Environmentally coupled hydrogen tunneling. *European Journal of Biochemistry*, 269(13):3113–3121, 2002.
109. J. P. Klinman and A. Kohen. Hydrogen tunneling links protein dynamics to enzyme catalysis. *Annual Review of Biochemistry*, 82(1):471–496, 2013.
110. H.-H. Limbach, J. Hennig, and J. Stulz. IR-spectroscopic study of isotope effects on the NH/ND-stretching bands of meso-tetraphenylporphine and vibrational hydrogen tunneling. *Journal of Chemical Physics*, 78(9):5432–5436, 1983.
111. S. V. Yelamanchili, C. Reisinger, A. Becher, S. Sikorra, H. Bigalke, T. Binz, and G. Ahnert-Hilger. The C-terminal transmembrane region of synaptobrevin binds synaptophysin from adult synaptic vesicles. *European Journal of Cell Biology*, 84(4):467–475, 2005.
112. X. Lou and Y.-K. Shin. SNARE zippering. *Bioscience Reports*, 36(3):e00327, 2016.

113. L. Pauling, R. B. Corey, and H. R. Branson. The structure of proteins: two hydrogen-bonded helical configurations of the polypeptide chain. *Proceedings of the National Academy of Sciences*, 37(4):205–211, 1951.
114. A. S. Davydov. Solitons, bioenergetics, and the mechanism of muscle contraction. *International Journal of Quantum Chemistry*, 16(1):5–17, 1979.
115. P. S. Lomdahl, S. P. Layne, and I. J. Bigio. Solitons in biology. *Los Alamos Science*, 10:2–31, 1984.
116. L. Brizhik, A. Eremko, B. Piette, and W. Zakrzewski. Solitons in α-helical proteins. *Physical Review E*, 70(3):031914, 2004.
117. L. Cruzeiro. The Davydov/Scott model for energy storage and transport in proteins. *Journal of Biological Physics*, 35(1):43–55, 2009.
118. L. Cruzeiro. The VES hypothesis and protein misfolding. *Discrete and Continuous Dynamical Systems - Series S*, 4(5):1033–1046, 2011.
119. W. Förner and J. Ladik. Influence of heat bath and disorder on Davydov solitons. In P. L. Christiansen and A. C. Scott, editors, *Davydov's Soliton Revisited: Self-Trapping of Vibrational Energy in Protein*, pages 267–283. Springer, New York, 1990.
120. W. Förner. Davydov soliton dynamics in proteins: I. Initial states and exactly solvable special cases. *Journal of Molecular Modeling*, 2(5):70–102, 1996.
121. W. Förner. Davydov soliton dynamics in proteins: II. The general case. *Journal of Molecular Modeling*, 2(5):103–135, 1996.
122. W. Förner. Davydov soliton dynamics in proteins: III. Applications and calculation of vibrational spectra. *Journal of Molecular Modeling*, 3(2):78–116, 1997.
123. W. Förner. Davydov solitons in proteins. *International Journal of Quantum Chemistry*, 64(3):351–377, 1997.
124. P. Hamm and J. Edler. Nonlinear vibrational spectroscopy: a method to study vibrational self-trapping. In T. Dauxois, A. Litvak-Hinenzon, R. MacKay, and A. Spanoudaki, editors, *Energy Localisation and Transfer*, Advanced Series in Nonlinear Dynamics, pages 301–324. World Scientific, Singapore, 2004.
125. J. Edler, P. Hamm, and A. C. Scott. Femtosecond study of self-trapped vibrational excitons in crystalline acetanilide. *Physical Review Letters*, 88(6):067403, 2002.
126. A. F. Lawrence, J. C. McDaniel, B. C. Chang, B. M. Pierce, and R. R. Birge. Dynamics of the Davydov model in α-helical proteins: effects of the coupling parameter and temperature. *Physical Review A*, 33(2):1188–1201, 1986.
127. A. F. Lawrence, J. C. McDaniel, B. C. Chang, and R. R. Birge. The nature of phonons and solitary waves in α-helical proteins. *Biophysical Journal*, 51(5):785–793, 1987.
128. H. Fröhlich. Interaction of electrons with lattice vibrations. *Proceedings of the Royal Society of London. Series A. Mathematical and Physical Sciences*, 215(1122):291–298, 1952.
129. H. Fröhlich. Coherence in biology. In H. Fröhlich and F. Kremer, editors, *Coherent Excitations in Biological Systems*, Proceedings in Life Sciences, pages 1–5. Springer, Berlin, 1983.
130. H. Fröhlich. General theory of coherent excitations on biological systems. In W. R. Adey and A. F. Lawrence, editors, *Nonlinear Electrodynamics in Biological Systems*, pages 491–496. Plenum Press, New York, 1984.
131. H. Fröhlich. Coherent excitation in active biological systems. In F. Gutmann and H. Keyzer, editors, *Modern Bioelectrochemistry*, pages 241–261. Plenum Press, New York, 1986.
132. T. Holstein. Studies of polaron motion: Part I. The molecular-crystal model. *Annals of Physics*, 8(3):325–342, 1959.
133. T. Holstein. Studies of polaron motion: Part II. The "small" polaron. *Annals of Physics*, 8(3):343–389, 1959.
134. G. Careri, U. Buontempo, F. Galluzzi, A. C. Scott, E. Gratton, and E. Shyamsunder. Spectroscopic evidence for Davydov-like solitons in acetanilide. *Physical Review B*, 30(8):4689, 1984.
135. L. Cruzeiro-Hansson and S. Takeno. Davydov model: The quantum, mixed quantum-classical, and full classical systems. *Physical Review E*, 56(1):894–906, 1997.
136. A. Xie, L. van der Meer, W. Hoff, and R. H. Austin. Long-lived amide I vibrational modes in myoglobin. *Physical Review Letters*, 84(23):5435–5438, 2000.
137. R. H. Austin, A. Xie, L. van der Meer, M. Shinn, and G. Neil. Self-trapped states in proteins? *Journal of Physics: Condensed Matter*, 15(18):S1693–S1698, 2003.

138. A. Xie, B. Redlich, L. van der Meer, S. Vyawahare, and R. H. Austin. Hidden harmonic quantum states in proteins: did Davydov get the sign wrong? *Physica Scripta*, 2015(T165):014042, 2015.
139. J. Edler and P. Hamm. Self-trapping of the amide I band in a peptide model crystal. *Journal of Chemical Physics*, 117(5):2415–2424, 2002.
140. J. Edler, R. Pfister, V. Pouthier, C. Falvo, and P. Hamm. Direct observation of self-trapped vibrational states in α-helices. *Physical Review Letters*, 93(10):106405, 2004.
141. J. Edler and P. Hamm. Spectral response of crystalline acetanilide and n-methylacetamide: vibrational self-trapping in hydrogen-bonded crystals. *Physical Review B*, 69(21):214301, 2004.
142. E. Schwartz, P. Bodis, M. Koepf, J. J. L. M. Cornelissen, A. E. Rowan, S. Woutersen, and R. J. M. Nolte. Self-trapped vibrational states in synthetic β-sheet helices. *Chemical Communications*, 31:4675–4677, 2009.
143. C. R. Baiz, M. Reppert, and A. Tokmakoff. An introduction to protein 2D IR spectroscopy. In M. D. Fayer, editor, *Ultrafast Infrared Vibrational Spectroscopy*, pages 361–403. CRC Press, Boca Raton, 2013.
144. C. Fang, A. Senes, L. Cristian, W. F. DeGrado, and R. M. Hochstrasser. Amide vibrations are delocalized across the hydrophobic interface of a transmembrane helix dimer. *Proceedings of the National Academy of Sciences*, 103(45):16740–16745, 2006.
145. F. Li, D. Kümmel, J. Coleman, K. M. Reinisch, J. E. Rothman, and F. Pincet. A half-zippered SNARE complex represents a functional intermediate in membrane fusion. *Journal of the American Chemical Society*, 136(9):3456–3464, 2014.
146. V. Shahrezaei and K. R. Delaney. Consequences of molecular-level Ca^{2+} channel and synaptic vesicle colocalization for the Ca^{2+} microdomain and neurotransmitter exocytosis: a Monte Carlo study. *Biophysical Journal*, 87(4):2352–2364, 2004.
147. V. Shahrezaei and K. R. Delaney. Brevity of the Ca^{2+} microdomain and active zone geometry prevent Ca^{2+}-sensor saturation for neurotransmitter release. *Journal of Neurophysiology*, 94(3):1912–1919, 2005.
148. D. D. Georgiev and J. F. Glazebrook. Quantum tunneling of Davydov solitons through massive barriers. *Chaos, Solitons and Fractals*, 123:275–293, 2019.
149. S. Haroche. Quantum information in cavity quantum electrodynamics: logical gates, entanglement engineering and 'Schrödinger-cat states'. *Philosophical Transactions of the Royal Society of London. Series A: Mathematical, Physical and Engineering Sciences*, 361(1808):1339–1347, 2003.
150. H. Mabuchi and A. C. Doherty. Cavity quantum electrodynamics: coherence in context. *Science*, 298(5597):1372–1377, 2002.
151. Z.-X. Man, Y.-J. Xia, and R. Lo Franco. Cavity-based architecture to preserve quantum coherence and entanglement. *Scientific Reports*, 5:13843, 2015.
152. L. Cruzeiro. Protein folding. In M. Springborg, editor, *Chemical Modelling Applications and Theory*, volume 7 of *Specialist Periodical Reports*, pages 89–114. Royal Society of Chemistry, Cambridge, 2010.
153. L. Cruzeiro and L. Degrève. What is the shape of the distribution of protein conformations at equilibrium? *Journal of Biomolecular Structure and Dynamics*, 33(7):1539–1546, 2015.
154. T. J. Melia, T. Weber, J. A. McNew, L. E. Fisher, R. J. Johnston, F. Parlati, L. K. Mahal, T. H. Söllner, and J. E. Rothman. Regulation of membrane fusion by the membrane-proximal coil of the t-SNARE during zippering of SNAREpins. *Journal of Cell Biology*, 158(5):929–940, 2002.
155. M. Xue, K. Reim, X. Chen, H.-T. Chao, H. Deng, J. Rizo, N. Brose, and C. Rosenmund. Distinct domains of complexin I differentially regulate neurotransmitter release. *Nature Structural & Molecular Biology*, 14:949–958, 2007.
156. X. Zhang, A. A. Rebane, L. Ma, F. Li, J. Jiao, H. Qu, F. Pincet, J. E. Rothman, and Y. Zhang. Stability, folding dynamics, and long-range conformational transition of the synaptic t-SNARE complex. *Proceedings of the National Academy of Sciences*, 113(50):E8031–E8040, 2016.
157. S. Caspi and E. Ben-Jacob. Conformation changes and folding of proteins mediated by Davydov's soliton. *Physics Letters A*, 272(1-2):124–129, 2000.
158. Q. Fang, K. Berberian, L.-W. Gong, I. Hafez, J. B. Sørensen, and M. Lindau. The role of the C terminus of the SNARE protein SNAP-25 in fusion pore opening and a model for fusion pore mechanics. *Proceedings of the National Academy of Sciences*, 105(40):15388–15392, 2008.
159. H. J. Risselada, C. Kutzner, and H. Grubmüller. Caught in the act: visualization of SNARE-mediated fusion events in molecular detail. *ChemBioChem*, 12(7):1049–1055, 2011.

160. M. Lindau, B. A. Hall, A. Chetwynd, O. Beckstein, and M. S. P. Sansom. Coarse-grain simulations reveal movement of the synaptobrevin C-terminus in response to piconewton forces. *Biophysical Journal*, 103(5):959–969, 2012.
161. M. Dhara, A. Yarzagaray, M. Makke, B. Schindeldecker, Y. Schwarz, A. Shaaban, S. Sharma, R. A. Böckmann, M. Lindau, R. Mohrmann, and D. Bruns. v-SNARE transmembrane domains function as catalysts for vesicle fusion. *eLife*, 5:e17571, 2016.
162. W. Stelzer, B. C. Poschner, H. Stalz, A. J. Heck, and D. Langosch. Sequence-specific conformational flexibility of SNARE transmembrane helices probed by hydrogen/deuterium exchange. *Biophysical Journal*, 95(3):1326–1335, 2008.
163. L. Gammaitoni, P. Hänggi, P. Jung, and F. Marchesoni. Stochastic resonance. *Reviews of Modern Physics*, 70(1):223–287, 1998.
164. P. Hänggi. Stochastic resonance in biology. How noise can enhance detection of weak signals and help improve biological information processing. *ChemPhysChem*, 3(3):285–290, 2002.
165. G. Deco, E. T. Rolls, and R. Romo. Stochastic dynamics as a principle of brain function. *Progress in Neurobiology*, 88(1):1–16, 2009.
166. B. M. Kampa and G. J. Stuart. NMDA receptor kinetics are tuned for spike-timing dependent synaptic plasticity. *Physiology News*, 58:29–30, 2005.
167. M. Grifoni, L. Hartmann, S. Berchtold, and P. Hänggi. Quantum tunneling and stochastic resonance. *Physical Review E*, 53(6):5890–5898, 1996.
168. M. Grifoni and P. Hänggi. Coherent and incoherent quantum stochastic resonance. *Physical Review Letters*, 76(10):1611–1614, 1996.
169. V. Degtyar, I. M. Hafez, C. Bray, and R. S. Zucker. Dance of the SNAREs: assembly and rearrangements detected with FRET at neuronal synapses. *Journal of Neuroscience*, 33(13):5507–5523, 2013.
170. L. E. Dobrunz and C. F. Stevens. Heterogeneity of release probability, facilitation, and depletion at central synapses. *Neuron*, 18(6):995–1008, 1997.
171. T. C. Südhof. The synaptic vesicle cycle revisited. *Neuron*, 28(2):317–320, 2000.
172. L. E. Dobrunz. Release probability is regulated by the size of the readily releasable vesicle pool at excitatory synapses in hippocampus. *International Journal of Developmental Neuroscience*, 20(3):225–236, 2002.
173. T. Branco and K. Staras. The probability of neurotransmitter release: variability and feedback control at single synapses. *Nature Reviews Neuroscience*, 10(5):373–383, 2009.
174. L. D. Landau. Über Die Bewegung der Elektronen in Kristallgitter. *Physikalische Zeitschrift der Sowjetunion*, 3:644–645, 1933.
175. J. Luo and B. M. A. G. Piette. A generalised Davydov–Scott model for polarons in linear peptide chains. *European Physical Journal B*, 90(8):155, 2017.
176. A. S. Davydov. Quantum theory of muscular contraction. *Biophysics*, 19:684–691, 1974.
177. D. D. Georgiev and J. F. Glazebrook. On the quantum dynamics of Davydov solitons in protein α-helices. *Physica A: Statistical Mechanics and its Applications*, 517:257–269, 2019.
178. L. Cruzeiro. Influence of the nonlinearity and dipole strength on the amide I band of protein α-helices. *Journal of Chemical Physics*, 123(23):234909, 2005.
179. J. M. Hyman, D. W. McLaughlin, and A. C. Scott. On Davydov's alpha-helix solitons. *Physica D*, 3(1-2):23–44, 1981.
180. P. S. Lomdahl and W. C. Kerr. Do Davydov solitons exist at 300 k? *Physical Review Letters*, 55(11):1235–1238, 1985.
181. N. A. Nevskaya and Y. N. Chirgadze. Infrared spectra and resonance interactions of amide-I and II vibrations of α-helix. *Biopolymers*, 15(4):637–648, 1976.
182. Z. Sinkala. Soliton/exciton transport in proteins. *Journal of Theoretical Biology*, 241(4):919–927, 2006.
183. L. Cruzeiro-Hansson. Effect of long range and anharmonicity in the minimum energy states of the Davydov–Scott model. *Physics Letters A*, 249(5):465–473, 1998.
184. A. Biswas, A. Moran, D. Milovic, F. Majid, and K. C. Biswas. An exact solution for the modified nonlinear Schrödinger's equation for Davydov solitons in α-helix proteins. *Mathematical Biosciences*, 227(1):68–71, 2010.
185. L. S. Brizhik, A. A. Eremko, B. M. A. G. Piette, and W. J. Zakrzewski. Ratchet dynamics of large polarons in asymmetric diatomic molecular chains. *Journal of Physics: Condensed Matter*, 22(15):155105, 2010.

186. L. S. Brizhik, A. A. Eremko, B. M. A. G. Piette, and W. J. Zakrzewski. Thermal enhancement and stochastic resonance of polaron ratchets. *Physical Review E*, 89(6):062905, 2014.
187. M. Daniel and M. M. Latha. A generalized Davydov soliton model for energy transfer in alpha helical proteins. *Physica A*, 298(3-4):351–370, 2001.
188. M. Daniel and M. M. Latha. Soliton in alpha helical proteins with interspine coupling at higher order. *Physics Letters A*, 302(2-3):94–104, 2002.
189. X. Lü and F. Lin. Soliton excitations and shape-changing collisions in alpha helical proteins with interspine coupling at higher order. *Communications in Nonlinear Science and Numerical Simulation*, 32:241–261, 2016.
190. L. M. Ristovski, N. Nestorović, and G. S. Davidović. The unified theory of Davydov's and Fröhlich's models. *Zeitschrift für Physik B*, 88(2):145–157, 1992.
191. W. C. Kerr and P. S. Lomdahl. Quantum-mechanical derivation of the equations of motion for Davydov solitons. *Physical Review B*, 35(7):3629–3632, 1987.
192. A. S. Davydov. Solitons in quasi-one-dimensional molecular structures. *Soviet Physics Uspekhi*, 25(12):898–918, 1982.
193. J. Sun, B. Luo, and Y. Zhao. Dynamics of a one-dimensional Holstein polaron with the Davydov ansätze. *Physical Review B*, 82(1):014305, 2010.

18 Parametric Analysis of Plasmonic Nanostructures for Biomedical Applications

Kai Liu[1], *Innem V.A.K. Reddy*[1], *Viktor Sukhotskiy*[1], *Amir Mokhtare*[2], *Xiaozheng Xue*[2], *Edward P. Furlani*[1,2]

[1]Department of Electrical Engineering, The State University of New York at Buffalo, Buffalo, New York, USA

[2]Department of Chemical and Biological Engineering, The State University of New York at Buffalo, Buffalo, New York, USA

CONTENTS

18.1 INTRODUCTION

The interest in plasmonic nanostructures has grown steadily, as advances in particle synthesis, especially the synthesis of multifunctional inorganic nanoparticles, have enabled a proliferation of applications in fields such as nanophotonics, biomedicine, and analytical chemistry.[1,2] Many applications exploit the unique and highly tunable optical properties of plasmonic nanostructures, most notably those associated with the effects of localized surface plasmon resonance (LSPR). At plasmonic resonance, there is intense absorption and scattering of incident light and highly localized field enhancement. Moreover, the LSPR wavelength of a plasmonic nanostructure is highly dependent on its size, morphology, optical material properties, and the properties of the surrounding medium. This LSPR wavelength can be tuned within the ultraviolet to near-infrared (NIR) spectrum by manipulating these factors. A desired LSPR wavelength can be obtained, in principle, by controlling the dimensions and morphology of the structure during synthesis. The ability to tune the LSPR and associated behavior has proved useful for a broad range of applications involving cancer therapies,[3] Raman scattering,[4] fluorescent labeling,[5] nonlinear optical imaging,[6] and biosensing,[7] among others. For example, in selected published works,[8–10] plasmonic nanoparticles with advanced geometries (i.e., Au nanocages and nanostars) have been successfully synthesized with excellent controllability, and important biomedical applications have been demonstrated. Two emerging biomedical applications that directly exploit plasmon-enhanced optical effects are thermally modulated drug delivery and photothermal cancer therapy. Most plasmon-based photothermal applications in vivo utilize Au-based nanostructures with LSPR wavelengths in the NIR, i.e., 650–1300 nm. This is known as the NIR window as these are the optical wavelengths that have the deepest penetration into tissue.[11]

In this chapter, we use full-wave field analysis to study the optical properties of NIR absorbing nanostructures that are of interest for biomedical applications. The chapter is divided into two parts. In the first part, we investigate and compare the optical properties of particles with demonstrated efficacy for theranostic applications, namely core@Au-shell,[12] Au nanorod,[13] and Au nanocage structures.[14] We quantify field localization, field enhancement, and the photothermal transduction of these particles as a function of their morphology, dimensions, and orientation with respect to the incident field polarization. In the second part of the chapter, we study the optical behavior of self-assembled 1D chains of magnetic-plasmonic Fe_3O_4@Au core-shell particles. Specifically, we quantify the field enhancement and photothermal transduction as functions of the particle properties and chain length, i.e., number of particles in a chain.

18.2 OPTICAL ABSORPTION ANALYSIS OF PLASMONIC NANOSTRUCTURES

We used 3D full-wave computational models to study the NIR plasmonic behavior of the three distinct nanoparticles with demonstrated efficacy for theranostic applications: SiO_2@Au core-shell,[1] Au nanorod,[2] and Au nanocage structures[3] as shown in Figure 18.1. The core-shell particles

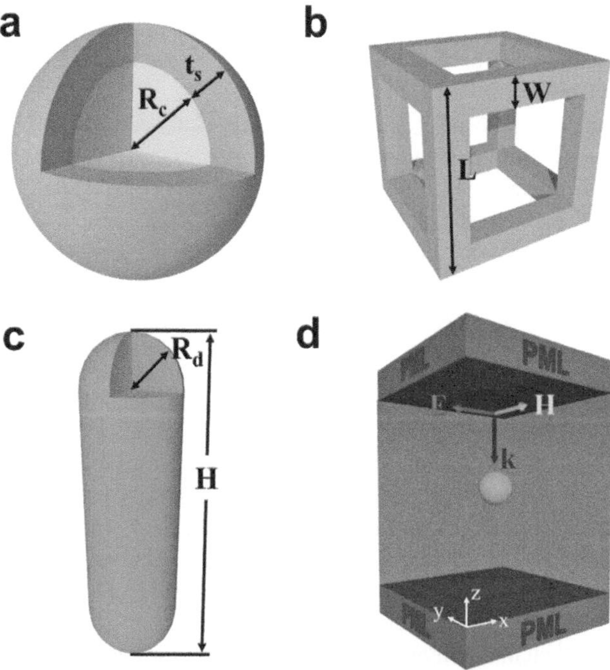

FIGURE 18.1 Plasmonic nanostructures and the computational model: (a) SiO_2@Au core-shell particles, (b) Au nanocages, and (c) Au nanorods. (d) CD showing the polarization and propagation direction of the incident field (adapted with permission from [51]).

consist of a silica (SiO_2) core with a radius R_c and a gold shell with a thickness t_s as shown in Figure 18.1a. The Au nanocages are cubic with 12 frame elements in the form of square Au nanowires, as shown in Figure 18.1b. The nanocage geometry is described by its length L, which defines the size of the cube, the width W that defines the cross-sectional area of the nanowire, and the aspect ratio $R = L/W$. In the literature, this structure is also referred to as a nanoframe.[3] The nanorod geometry, shown in Figure 18.1c, consists of a cylindrical body of radius R_d with hemispherical caps at either end. The total length of the nanorod is H, and the radius of the hemispherical cap, as well as the cylinder, is R_d. An example of the computational domain (CD) for this analysis is shown in Figure 18.1d. Here, a single core-shell particle is centered at the origin of the domain and immersed in a static carrier fluid, which we take to be H_2O. The computational model is described in detail in the following section.

These particles have attracted great attention because they can be synthesized in a controllable fashion using bottom-up chemical methods, which enables tuning of their optical properties. However, they also have drawbacks. Core-shell particles with an Au shell can have limited absorption in the NIR due to a relatively thin gold shell that is required to red-shift LSPR to that range. Nanorods have a solid metallic mass, but their absorption is a strong function of their orientation relative to the incident polarization, which results in less efficient heating for randomly oriented colloidal particles. Gold nanoframes are emerging as an alternative NIR nanomaterial for photothermal therapy[4] and drug delivery.[5] However, their NIR optical behavior (e.g., sensitivity of LSPR to spatial orientation with respect to the incident polarization) is not obvious and needs to be determined using complex 3D computational modeling. The optical properties of selected NIR nanomaterials, such as Au spheres, nanorods, nanotori, and nanoframes, along with an analysis of their use for photothermal applications, especially theranostic nanobubble generation is discussed in our previous work[6,7] as well as many other excellent references, many cited herein.[19–22] However, despite the growing interest and application of plasmonic particles, rational design is lacking in many cases and can be achieved using numerical multiphysics modeling. In our analysis below, we place more emphasis on optical absorption rather than scattering. The reason for this is twofold: first, the focus of our optical analysis is on photothermal and local-field–related applications in which the optical absorption is a more dominant factor that determines the efficiency of the system; second, the optical phenomena in the nanoparticles with relatively small dimensions are dominated by the optical absorption.[23]

18.2.1 Computational Domain

As noted above, we used 3D full-wave computational models to study the NIR plasmonic behaviors of the nanostructures (Figure 18.1a–c). Specifically, we used the finite-element–based radio frequency (RF) solver module in the commercial COMSOL multiphysics program (COMSOL Version 5.2, www.comsol.com). A typical CD is shown in Figure 18.1d. Here, a single core-shell particle is centered at the origin of the domain and immersed in H_2O. The particle is illuminated with a uniform downward-directed plane wave with the E field polarized along the x-axis. The height of the CD is 1000 nm, and perfectly matched layers (PMLs) (200 nm in height) are applied at the top and bottom of the domain to reduce backscattering from these boundaries. Perfect electric conductor (PEC) conditions are applied at the boundaries perpendicular to E, and perfect magnetic conductor (PMC) conditions are applied at the boundaries perpendicular to H. It is important to note that these symmetry boundary conditions (BCs) mimic the response of an infinite 2D array of coplanar identical nanoparticles with a center-to-center x and y lattice spacing equal to the spatial period $P = P_x = P_y$ of the CD. Thus, the field solution within the CD contains contributions from particles that exist outside the CD. The magnitude of these contributions, and hence their significance, depends on the lattice spacing, i.e., the spatial period of the CD. In our preliminary analysis, we used Mie theory to calibrate the CD, i.e., to determine a value of P that is large enough so that the field contributions from particles outside the CD are negligible, i.e., so that the analysis accurately reflects the optical response of a single-isolated colloidal particle.[24]

We used full-wave time-harmonic field theory for the analysis. A continuum electrodynamic analysis is justified, i.e., quantum effects can be neglected, since the nanostructures are several tens to hundreds of nanometers in size, much larger than the constituent atoms (e.g., ~0.2 nm for gold). As shown in Figure 18.1d, an incident plane wave is launched from the top surface of the domain and propagates downward toward the bottom PML. A source current is used to generate the field as described in the literature.[25–28] The incident light is linearly polarized at normal incidence with the E field along the x-axis and the H field along the y-axis. The time-harmonic E field within the domain satisfies the equation:

$$\nabla \times \left(\mu_r^{-1} \nabla \times E\right) - k_0^2 \left(\varepsilon_r - j\frac{\sigma}{\omega \varepsilon_0}\right) E = 0 \tag{18.1}$$

where μ_r, ϵ_r, and σ are the relative permeability, permittivity, and conductivity of the media, respectively. Throughout our computation analysis, it is assumed that the numerical models are structured with ideal geometric components without defects as well as the materials with ideal uniformity and homogeneity. However, the results we offered in this article can play a significant role in guiding the experimental practices and facilitating the understanding of clinical results, even when the defects exist in the fabricated or synthesized nanostructures. In the computational model, we compute the power absorbed by the particle Q_{abs} (W) and then use this to compute the cross-section $\sigma_{abs} = Q_{abs}/I_{laser}$, where I_{laser} (W/m^2) is the incident irradiance. To model the nanoparticles, we need expressions for optical constants $\varepsilon_r = (n - ik)^2$ of Au (ε_{Au}), SiO_2 (n_{SiO_2}), and the background medium (n_{H_2O}). Moreover, we need to consider the fact that the metallic materials (e.g., Au shell) can be thinner than the mean free path of free electrons (~42 nm). A dielectric function for gold that accounts for electron-surface scattering is expressed as follows[29]:

$$\varepsilon_{Au}(\omega, L_{eff}) = \varepsilon_{Au,bulk}(\omega) + \frac{\omega_p^2}{\omega^2 + i\omega \nu_f / l_\infty} - \frac{\omega_p^2}{\omega^2 + i\omega(\nu_f / l_\infty + A\nu_f / L_{eff})} \tag{18.2}$$

where $\varepsilon_{Au,bulk}$ is the bulk dielectric function of gold, ω is the angular frequency of incident light, $\omega_p = 0.93$ eV is the plasma frequency, $\nu_f = 1.4 \times 10^{15}$ nm/s is the Fermi velocity, $l_\infty = 42$ nm is the mean free path of the free electrons, A is a dimensionless parameter, usually assumed to be close to unity ($A = 1$), and $L_{eff} = t_s$ is the reduced effective mean free path of the free electrons. The bulk dielectric function is given by an analytical expression, equation 3, that is based on an experiment-fitted critical points model.[30–32] The detailed descriptions of parameters in equation 3 can be found in the literatures.[30–32] The material SiO_2 is assumed to be lossless, i.e., $k_{SiO_2} = 0$, with a dispersive index of refraction defined in equation 4.[29] The refractive index of the non-absorbing water surrounding is expressed in equation 5.[29] Moreover, all materials in our model have the permeability of $\mu_r = 1$.

$$\varepsilon_{Au,bulk}(\lambda) = \varepsilon_\infty - \frac{1}{\lambda_p^2\left(\frac{1}{\lambda^2} - i\frac{1}{\gamma_p \lambda}\right)} + \sum_{n=1,2} \frac{A_n}{\lambda_n}\left[\frac{e^{i\phi_n}}{\left(\frac{1}{\lambda_n} - \frac{1}{\lambda} + i\frac{1}{\gamma_n}\right)} + \frac{e^{-i\phi_n}}{\left(\frac{1}{\lambda_n} + \frac{1}{\lambda} - i\frac{1}{\gamma_n}\right)}\right] \tag{18.3}$$

$$n_{SiO_2}^2 = 1 + \frac{0.6961663\lambda^2}{\lambda^2 - (0.0684043)^2} + \frac{0.4079426\lambda^2}{\lambda^2 - (0.1162414)^2} + \frac{0.8974794\lambda^2}{\lambda^2 - (9.896161)^2} \tag{18.4}$$

$$n_{H_2O}^2 = 1 + \frac{0.5684027565\lambda^2}{\lambda^2 - 0.005101829712} + \frac{0.1726177391\lambda^2}{\lambda^2 - 0.01821153936} + \frac{0.02086189678\lambda^2}{\lambda^2 - 0.02620722293} + \frac{0.1130748644\lambda^2}{\lambda^2 - 10.69792721} \tag{18.5}$$

18.2.2 LSPR versus Particle Dimensions

We first study the LSPR tunability of the three nanostructures as a function of their dimensional parameters. The total particle volume is held fixed at $V_p = (50\text{ nm})^3$ for each particle. It is important to note that the fixed particle volume applies throughout this work, and hence, the volume fractions of the different colloids are identical. As mentioned earlier, we also assume that the colloids are sufficiently dilute so that interparticle photonic coupling is negligible. We begin by investigating the LSPR tunability of the SiO_2@Au structure. We calibrate and validate the 3D computational model for this structure using Mie theory. To this end, Figure 18.2a shows an analysis of the absorption spectrum of an SiO_2@Au particle as a function of the size of the CD (Figure 18.1d). Here, the length P that defines the square xy cross-section of the CD is systematically increased until the computed absorption spectrum equals that obtained using Mie theory. This occurs when $P = 2000$ nm, as seen in the inset of Figure 18.2a. This value of P is used throughout this work unless specified otherwise. This preliminary calibration is necessary because symmetry BCs are imposed on the lateral sides of the CD (i.e., transverse to the direction of propagation) to simplify the analysis. However, these BCs give rise to undesired interparticle coupling, which can contribute to the field solution and needs to

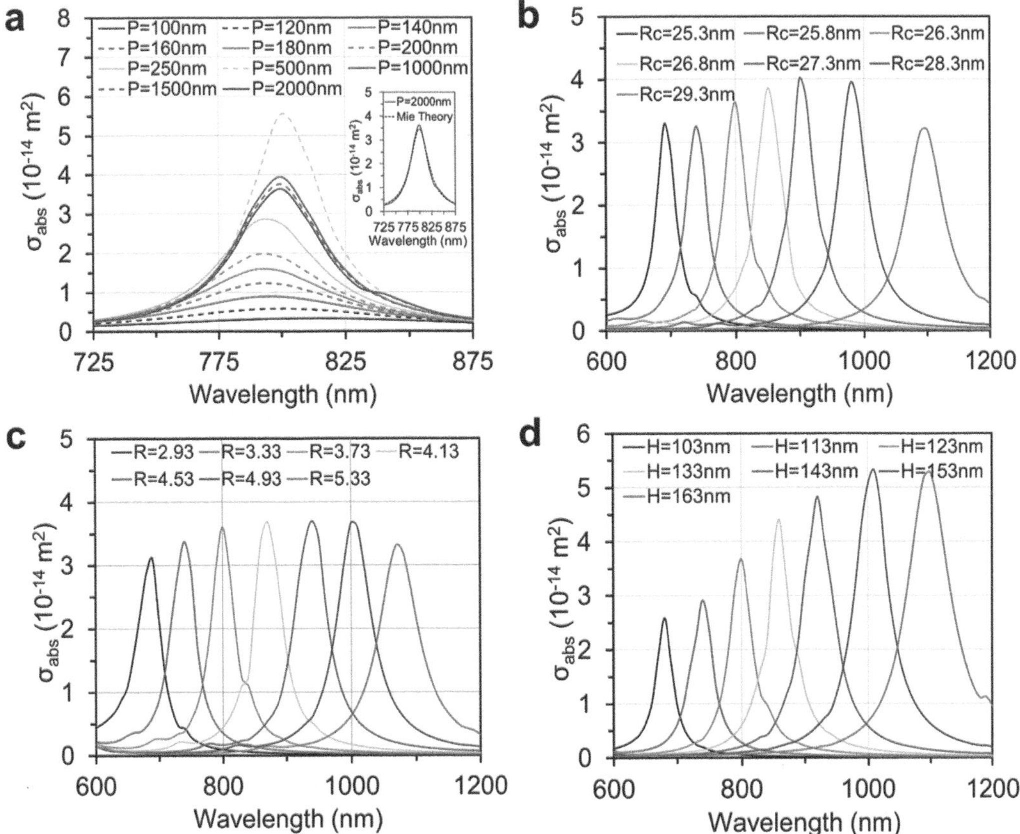

FIGURE 18.2 Absorption cross-section spectra σ_{abs} of particles versus variation of the domain period and dimensions: (a) σ_{abs} of the SiO_2@Au particle with $R_c = 27.3$ nm and $t_s = 3.7$ nm versus variation of the domain period P. The inset shows the results predicted by Mie theory. (b–d) σ_{abs} of three NIR colloids with the same volume V_p versus variation of dimensions: (b) SiO_2@Au core-shell particles, (c) Au nanoframes, and (d) Au nanorods (adapted with permission from Ref. [51]).

be minimized by choosing a sufficiently large spacing between the particles (i.e., sufficiently large P) as described in the previous section.

Once the computational model is calibrated for the SiO_2@Au particle, we study LSPR tunability wherein the SiO_2 core is enlarged and the shell thickness is reduced, i.e., R_c is systematically increased from 25.3 to 29.3 nm (t_s decreases from 5.7 to 1.7 nm). This produces a corresponding shift in the LSPR wavelength in the NIR from 690 to 1100 nm, as shown in Figure 18.2b. A similar analysis was performed for the nanocage. In this case, L is fixed at 50 nm and the aspect ratio $R = L/W$ is increased from 2.93 to 5.33 (i.e., W decreases from 17 to 9.38 nm), which produces a redshift in the LSPR wavelength from 690 nm to 1070 nm as shown in Figure 18.2c. Finally, for the Au nanorods, both the length H and radius R_d need to change in order to maintain a constant volume. As H increases from 103 to 163 nm, R_d decreases from 21.16 to 16.17 nm and the LSPR peak red-shifts from 680 to 1110 nm, as shown in Figure 18.2d.

This analysis shows that the three nanostructures are comparable with respect to their use for a prescribed NIR operating wavelength as the LSPR of each particle can be tuned to the NIR by controlling their dimensions during synthesis. This NIR tunability is especially attractive for laser-based theranostic applications, as discussed earlier. In the remaining sections, we perform an optical analysis of three specific structures that have an LSPR wavelength of 800 nm: an SiO_2@Au core-shell particle with $R_c = 27.3$ nm and $t_s = 3.7$ nm, a nanocage with $L = 50$ nm and $W = 13.4$ nm, and a nanorod with $H = 123$ nm and $R_d = 19$ nm.

18.2.3 LSPR versus Spatial Orientation

Next, we study the absorption of the nanostructures as a function of their orientation relative to the incident polarization. This is important because colloidal particles have random orientations that can impact their absorption. The SiO_2@Au particles are centrosymmetric, and therefore, their absorption cross-section is independent of orientation. However, nanocages and especially nanorods have less rotational symmetry, and their absorption changes with orientation. We use angles φ and θ to define the rotation of the particles relative to the x-axis and z-axis, respectively, as shown in Figure 18.3. In Figure 18.3a and b, the unit vector n is shown that is normal to the top area of the nanocage and along the principle axis of the nanorod, respectively. The angle φ lies in the xy plane and is measured from the x-axis to the projection of n onto xy plane, whereas $90° - \theta$ is the angle between n and the z-axis. We begin with an analysis of the nanocage as defined in the previous section ($L = 50$ nm and $W = 13.4$ nm) and compute σ_{abs} as a function of its orientation. The incident field is fixed at the LSPR wavelength of 800 nm, which is located within the NIR biological window and aligned with one of the most popularly used laser lines, i.e., 808 nm. As shown in Figure 18.3c, there is very little variation in σ_{abs} throughout the entire range of orientations. This is in sharp contrast to the Au nanorod ($H = 123$ nm and $R_d = 19$ nm), which exhibits a strong orientation-dependent absorption as shown in Figure 18.3d. Specifically, the amplitude of σ_{abs} decreases from its maximum to zero as the nanorod rotates away from its alignment with the field polarized along its long axis ($\varphi = 0°$, $\theta = 0°$). This orientation dependence of absorption can limit the use of nanorods for applications.

18.2.4 Field Enhancement

Another important feature of the plasmonic particles is their ability to generate highly localized enhanced fields at resonance. Such "hot spots" have been exploited for many applications, including fluorescent-based imaging, controlled drug delivery, nonlinear optics, surface-enhanced Raman scattering (SERS), and various biosensing modalities. In this section, we compare the field enhancement for particles that have the same volume and LSPR wavelength, i.e., the SiO_2@Au core-shell particle with $R_c = 27.3$ nm and $t_s = 3.7$ nm, and the nanocage with $L = 50$ nm and $W = 13.4$ nm. Since the Au nanorod is hardly an optimal choice for colloidal applications due to its undesirable

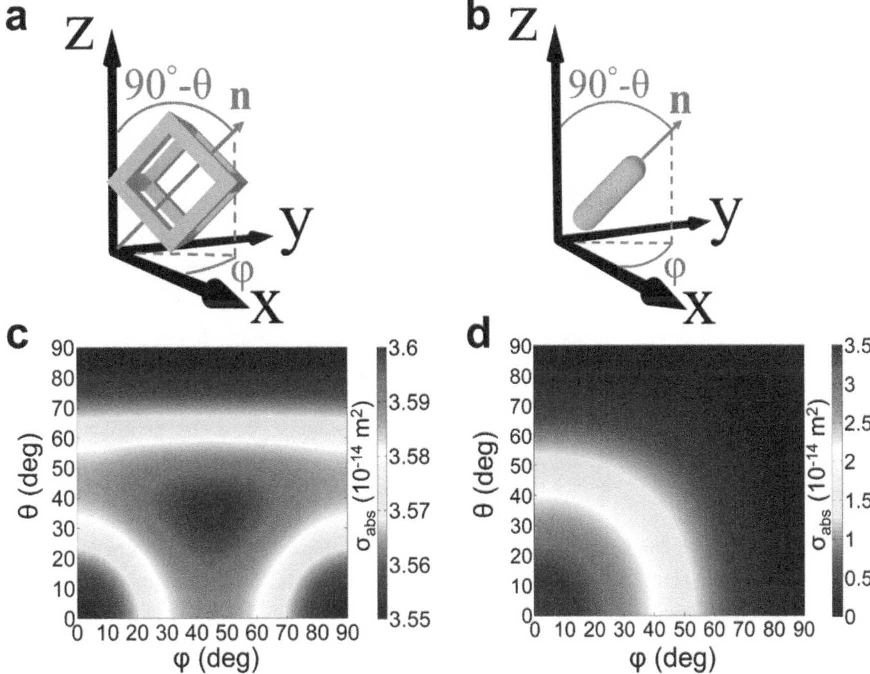

FIGURE 18.3 NIR absorption cross-section versus particle orientation: geometry and coordinates of spatial orientation of (a) nanocage and (b) nanorod; σ_{abs} versus spatial orientation (φ, θ) for (c) Au nanoframe, and (d) Au nanorod (adapted with permission from Ref. [51]).

sensitivity to the spatial orientation,[26] its spatial profiles of local field enhancement are moderately discussed.

In Figure 18.4, we plot the spatial profile of *E* field intensity enhancement ($|E|^2/|E_0|^2$) for the SiO_2@Au core-shell particle across four different cut planes. As shown in Figure 18.4a, two *yz*-planes (perpendicular to the field polarization) are labeled X1 and X2, where X1 is a symmetry plane through the center of the particle and X2 is parallel to X1 and tangential to the SiO_2 core. Figure 18.4b shows a weakly confined mode within the Au shell in X1, and Figure 18.4c shows a relatively strong mode distributed around the outer surface of Au shell in X2. The latter is due to the resonant dipolar moment of the Au core shell.

Figure 18.4d illustrates two *xz*-planes (parallel to the polarization), which are denoted *Y*1 and *Y*2, where *Y*1 cuts across the center of particle and *Y*2 is tangential to the SiO_2 core. Since *Y*1 is aligned with the polarization, the LSPR dipolar resonance gives rise to a strongly concentrated *E* field with an enhancement factor over 800 as depicted in Figure 18.4e. In contrast, the local field profile in Figure 18.4f also shows a similar dipolar resonance mode; however, the field intensity is much weaker, which is mainly due to the proximity of *Y*2-plane to the pole in *y* direction.

The *E* field enhancement profiles for the Au nanocage are shown in Figure 18.5. Four planes are chosen to render the field plots. As shown in Figure 18.5a, two planes, *X*1 and *X*2, are perpendicular to the polarization direction (*x*-axis). *X*1 overlaps the central symmetry plane, and *X*2 cuts the middle of the nanowires that form the edge of the structure. Figure 18.5b shows uniformly enhanced field intensity in *X*1 across the hollow interior of the nanocage.

This region can potentially be loaded with theranostic agents that can be modulated by the enhanced field. Figure 18.5c illustrates several strongly enhanced hot spots at the outer surface of edge nanowires, which are primarily due to the dipolar resonance in those nanowires as they are aligned parallel to the polarization. Two additional planes *Y*1 and *Y*2 are defined perpendicular to

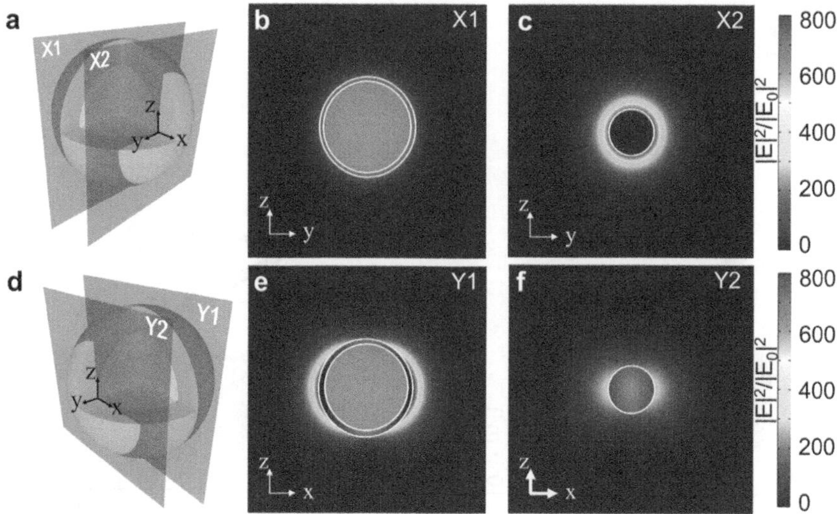

FIGURE 18.4 Local field enhancement of the SiO_2@Au core-shell particle (R_c = 27.3 nm and t_s = 3.7 nm) at the LSPR wavelength of 800 nm: (a), (d) illustrate four designated planes. (b), (c) and (e), (f) plot the profiles of LSPR-induced local field enhancement. The incidence is polarized along x direction (adapted with permission from Ref. [51]).

the y-axis as illustrated in Figure 18.5d. Strong field enhancement profiles can be observed in the hollow interior of the Au nanocage in Figure 18.5e and f. This localized field concentration in the interior of the nanocage is attributed to the strong mode coupling between adjacent Au nanowire frame elements.[33,34] The unique advantage of the nanocage over the core-shell particle is the abundance of coupled modes existing among the Au nanowires. The nanocage provides a larger number of hot spots on its surface that can be leveraged for theranostic applications. Moreover, the surface

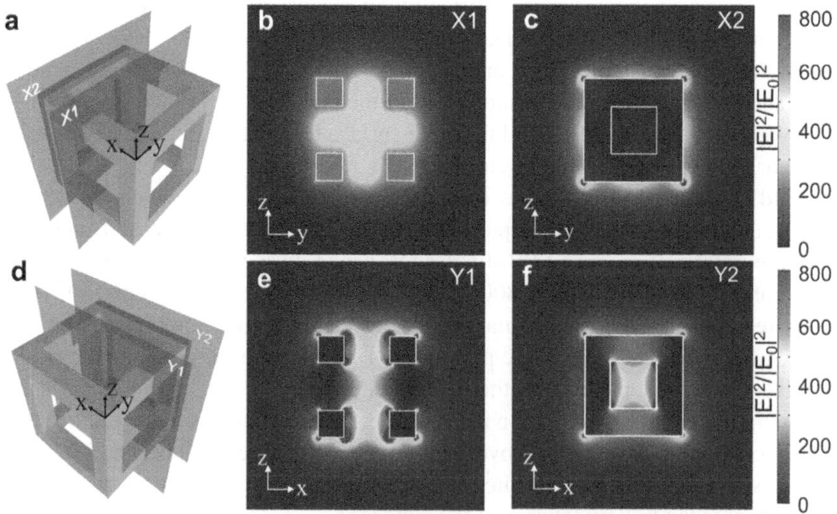

FIGURE 18.5 Local field enhancement profiles of the Au nanocage (L = 50 nm and W = 13.4 nm) at the LSPR wavelength of 800 nm: (a) and (d) are the conceptual schematics showing four designated planes. (b), (c), (e), and (f) show the profiles of LSPR-induced local field enhancement. The incident field is polarized along x direction (adapted with permission from Ref. [51]).

of nanocage can be functionalized with bio targeting agents to enable selective binding to a target biomaterial, e.g., cancer cells. Specifically, by manipulating thiolate-Au monolayer chemistry, excellent compatibility between Au surfaces and various molecules and ligands can be achieved.[35,36] During the functionalization process, fluorescent labels can be attached to the Au surface to enable spatial tracking and imaging.[15] The LSPR of the nanoparticles can be used to enhance fluorescent signal intensity,[37] i.e., to dramatically increase the signal intensity from surface-bound or encapsulated fluorescent molecules.[38,39] This enhanced fluorescence could enable high-resolution in vivo spatial imaging and tracking.

18.2.5 Photothermal Analysis

We used 3D computational thermodynamics to investigate the transient thermal behavior and heat transfer of the core-shell and nanocage structures.[40] For consistency, we analyzed particles with the same volume and dimensions as in the previous analysis, i.e., the SiO_2@Au particle with $R_c = 27.3$ nm and $t_s = 3.7$ nm, and the nanocage with $L = 50$ nm and $W = 13.4$ nm. Recall that these particles have comparable absorption cross-sections despite the substantial difference in their gold content, i.e., approximately a factor of 2. In particular, the SiO_2@Au particle contains much less gold (39,541 nm^3 or 0.764 fg) than the nanocage (69,238 nm^3 or 1.337 fg). The incident light induces currents within the gold that dissipate energy via Joule heating. The absorption cross-section depends on the magnitude and spatial extent of the LSPR-induced currents, which in turn depend on the amount and, importantly, the geometry (configuration) of the constituent gold. The current density and thermal power density (W nm^{-3}) distribution due to Joule heating within the two particles is shown in Figure 18.6a and b. The double-headed red arrows in these figures show the direction of polarization. The incident light propagates along the negative direction of z-axis. The small single-headed red arrows represent the current density in the constituent gold. It is important to note that the gold shell in the SiO_2@Au particle provides a continuous path for current to flow from one end of its induced dipole structure to the other. Consequently, it exhibits more uniform thermal dissipation over a relatively large volume of its Au shell. This is in contrast to the nanocage wherein only four nanowires, out of the 12 that form the nanoframe, carry significant current and produce heat. These four are aligned with the field polarization. The other eight nanowires in the frame are not aligned and have a negligible impact on optical absorption and heating when the particle has this orientation. Thus, even though the core-shell particle contains less gold than the nanocage, this difference is offset because a higher percentage of the gold in the former contributes to heating, which renders the overall photothermal conversion efficiency of the two particles comparable. An additional factor that can

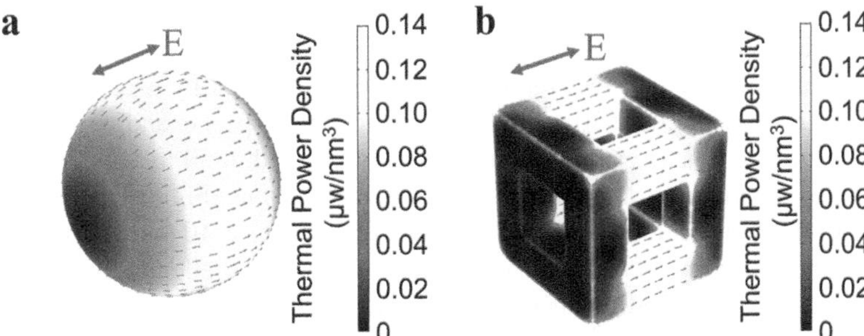

FIGURE 18.6 Photothermal profiles of SiO_2@Au core shell ($R_c = 27.3$ nm and $t_s = 3.7$ nm) and Au nanocage ($L = 50$ nm and $W = 13.4$ nm) particles with an LSPR wavelength of 800 nm: Spatial profiles of the thermal power density of (a) SiO_2@Au core-shell particle and (b) Au nanoframe. Small red arrows show the current density on the outer Au surfaces of two nanoparticles (adapted with permission from Ref. [52]).

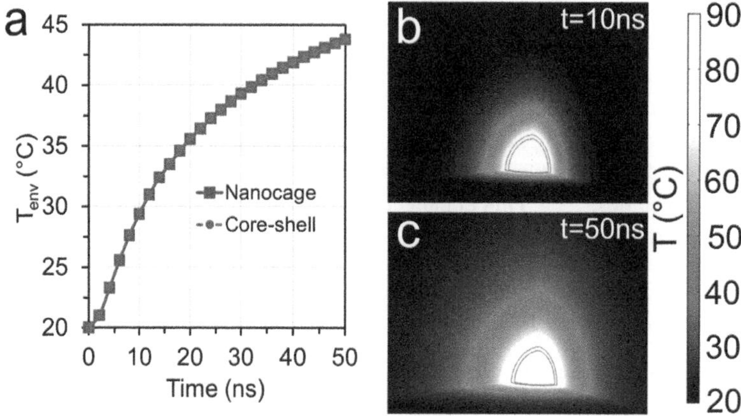

FIGURE 18.7 Transient temperature distribution of 62 nm SiO_2@Au nanoparticles (R_c = 27.3 nm, t_s = 3.7 nm): (a) Time-dependent average temperature in proximity to the particles: Temperature distribution due to the core-shell particle at (b) t = 10 ns and (c) 50 ns (adapted with permission from Ref. [52]).

influence absorption includes the special optical cavity configuration of the core-shell particle, which can result in a stronger LSPR resonance than in the nanocage with its open interior region.[41,42]

We perform 3D thermodynamic analysis of these structures using coupled heat transfer and RF modules of the COMSOL. The absorbed optical power density (heat generated per unit volume of gold) Q_h (W m^{-3}) is computed in the stationary (time harmonic) field solution of the RF model and acts as a heat source in the thermal model. The temperature T throughout the CD satisfies the equation:

$$\rho C_p \frac{\partial T}{\partial t} + \nabla \cdot (-k\nabla T) = Q$$

where Q (W m^{-3}) is the thermal energy generated per unit volume, i.e., $Q = Q_h$ inside the gold and $Q = 0$ in all other (loss less) materials. Here, ρ is the density, C_p is the specific heat capacity, and k is the thermal conductivity of the material. In many photothermal applications, higher concentrations of plasmonic nanoparticles are used, which gives rise to a cooperative interparticle heating that reduces the irradiance required to effect therapy, as discussed in the literature.[17,43] Specifically, the heat generated per unit volume in the carrier fluid increases with the particle concentration and this results in a more rapid rise in global fluid temperature for a given irradiance. The thermodynamic model is based on the same CD as the photonic analysis except that a thermal insulation condition is imposed on all boundaries. The initial temperature is set to 20°C.

Figure 18.7a shows essentially the same change in the environmental temperature T_{env} for the core-shell (blue curve) and nanocage (red curve). T_{env} is calculated by averaging the temperature over a spherical surface of radius of 70 nm that is centered with respect to and surrounds the particles. The results indicate that T_{env} rises quickly from 20 to 43°C within 50 ns which is consistent with typical laser pulse durations in photothermal therapy applications. Figure 18.7b and c shows the spatial profiles of the temperature for the SiO_2@Au core-shell at t = 10 and 50 ns, respectively. The results imply a rapid exchange of thermal energy between the core-shell particle and the surrounding H_2O environment, causing a continuous rise of environmental temperature during 50 ns.

18.3 OPTICAL PROPERTIES OF 1D PARTICLE CHAINS

Here, we consider the optical properties self-assembled colloidal magnetic-plasmonic Fe_3O_4@Au core-shell nanoparticles. The superparamagnetic core of these particles enables remote and adaptive magnetophoretic control of their position and arrangement using an external magnetic field and provides interparticle dipole–dipole coupling that drives the self-assembly process.[44–47] Moreover, when the field is withdrawn, the particles revert to their native unmagnetized state and redisperse in the carrier fluid, thereby enabling reconfigurable assembly. Magnetite (Fe_3O_4) is among the most commonly used materials for magnetic nanoparticles and is taken to be the particle core throughout this study. The particle's plasmonic shell provides unique optical and photothermal properties due to plasmonic effects as described above. The successful synthesis of uniform magnetic-plasmonic core-shell nanoparticles has been reported.[48,49] Similar to SiO_2@Au particles, such particles hold promise for a range of imaging, sensing, and theranostic applications, among others. However, the self-assembly of these particles has not been studied in any detail, and rational design to facilitate novel applications is lacking.

18.3.1 Computational Model

The CD for this analysis is shown in Figure 18.8. A single chain is centered at the origin of the domain and immersed in H_2O. The particle is illuminated with a uniform downward-directed plane wave with the *E* field polarized along the chain length, i.e., *x*-axis. Note that the visual orientation of the CD is different than in the previous section. As before, a surface current BC is employed as the excitation source. PMLs (200 nm in height) are applied at the top and bottom of the domain to reduce backscatter from these boundaries. PEC conditions are imposed on the boundaries perpendicular to *E*, and PMC conditions are imposed at the boundaries perpendicular to *H*. As before, it is important to note that these symmetry BCs mimic the response of an infinite 2D array of chains with a center-to-center *x* and *y* lattice spacing equal to the spatial period of the CD.[17,26,40,50,51] We chose a sufficiently large domain period so that interchain coupling is negligible; in effect, the model represents the response of a single isolated chain.

The dimensions of the Fe_3O_4@Au particles and the optical properties of Fe_3O_4 and Au can be found in the literature.[51,52] The carrier fluid has the optical properties of H_2O.

18.3.2 Finite and Infinite Chains

We investigate the spectral absorption cross-section of a 1D particle chain as a function the number of particles *N*. The chain length ranges from $N = 1$ to 12 particles and an infinite chain is also considered. The absorption spectra[52] (cross-section σ_{abs} vs. λ) for the different lengths (*N*) are compared with those of an infinite chain in Figure 18.8b. The results show that the absorption spectrum strongly red-shifts from that of a single particle ($N = 1$) and initially grows in amplitude as *N* increases to 6. As *N* increase further, the peak absorption decreases slightly and asymptotically approaches that of an infinite chain as the number of particles approaches ~10 (10 to 1 aspect ratio). This is intuitive and is clearly seen in Figure 18.8c, where the LSPR wavelength (triangular markers) of a finite

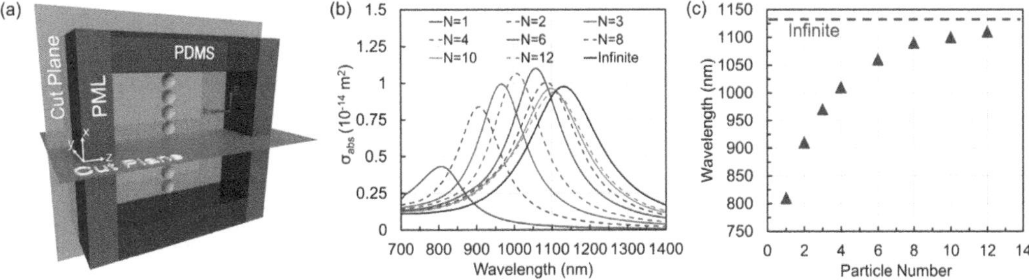

FIGURE 18.8 Photonic analysis of 1D particle chains: (a) CD for photonic analysis; (b) absorption cross-section σ_{abs} spectra versus number (*N*) of particles in a chain and for an infinite chain (solid black line); (c) LSPR wavelength of 1D chains versus *N*. The LSPR wavelength of an infinite chain is indicated by the dashed line (adapted with permission from Ref. [62]).

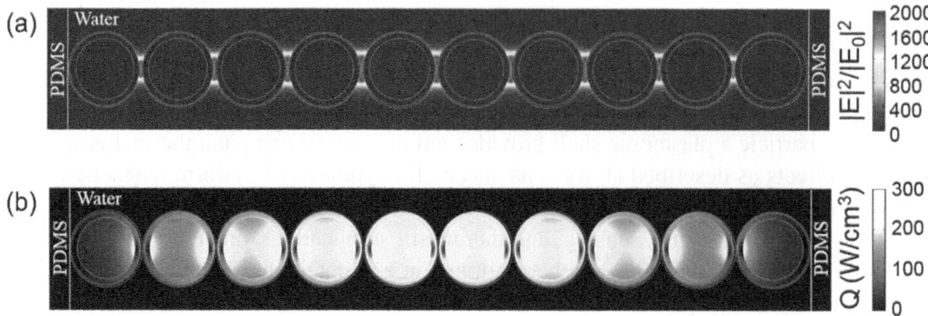

FIGURE 18.9 Field enhancement and photothermal transduction: (a) local enhancement $|E|^2/|E_0|^2$ and (b) thermal power density in a 1D chain of 10 polymer-coated Fe_3O_4@Au nanoparticles (adapted with permission from Ref. [62]).

chain rapidly approaches that of an infinite chain (dashed line). This result is consistent with previous LSPR studies of chains of solid plasmonic particles[53,54]; however, our results for Fe_3O_4@Au particle chains have not been reported before.

18.3.3 Field Enhancement

At plasmon resonance, the 1D chains exhibit localized field enhancement as well as strong absorption of light. Figure 18.9a shows the spatial profile of the local field enhancement for a 10-particle chain. Each particle has an ultrathin 1-nm-thick coating of surfactant. Thus, a nanogap exists between adjacent nanoparticles, and the analysis shows that the electric field is strongly concentrated in this deep-subwavelength volume.[55,56] Specifically, the local field enhancement factor $|E_{gap}|^2/|E_0|^2$ in the gap can exceed 2000. In fact, we observed a peak enhancement factor of 2×10^4 at the center of the nanogap. The local field enhancement factor is commonly treated as a standard figure of merit for many theranostic imaging applications, including fluorescent imaging and SERS. The enhancement factor of fluorescent imaging involving plasmonic nanostructures is found to be approximately proportional to the local field enhancement factor[57,58] $|E_{ex}|^2/|E_0|^2$, whereas the enhancement factor of Raman signals is approximately proportional to $|E_{ex}|^2/|E_0|^2 g|E_{stokes}|^2/|E_0|^2$, where $|E_{ex}|^2/|E_0|^2$ corresponds to field enhancement factor at excited laser wavelength and $|E_{stokes}|^2/|E_0|^2$ corresponds to laser wavelength specific to stokes mode.[58] The theoretical results, which are demonstrated in our research, can serve as an effective guidance for the theranostic imaging applications to enhance the predictability of experimental practices and shorten the development stage of the sample.[57] This field enhancement can also be exploited for biosensing or biomedical imaging applications, especially considering the fact that the strong near field region extends a sufficient distance to excite attached molecules.

For example, fluorophores attached to the surface of metallic nanoshell will have strong enhancement of fluorescent signals due to enhanced absorption. Additionally, a desirable reduction in radiative lifetime is also observed, resulting in substantial increases in the fluorescence quantum yield.[59] Similarly, Raman molecules and nonlinear materials can be attached to increase the intensities of Raman signals and two-photon luminescence signals. Therefore, the strong interparticle field enhancement is useful for molecular tracking and imaging as well as numerous other applications.

18.3.4 Photothermal Transduction

Steady-state analysis: We investigate the steady-state photothermal energy transduction for the 10-particle chain. Specifically, we compute the steady-state power density Q generated within the chain due to the conversion of photonic to thermal energy. Figure 18.9b shows the spatial profile

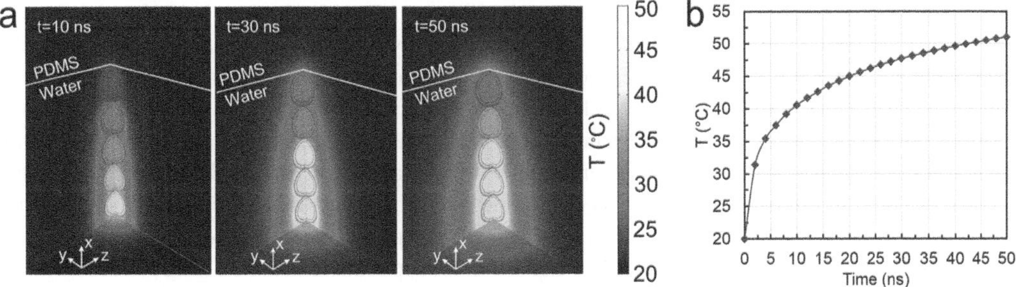

FIGURE 18.10 Photothermal transduction: (a) spatial profiles of temperature in a 1D chain with 10 Fe_3O_4@Au nanoparticles at t = 10 ns, t = 30 ns, and t = 50 ns. (b) Temperature at the center of the chain (adapted with permission from Ref. [62]).

of the local thermal energy generation. The strongest heating occurs throughout the center of the chain. This is due to a reduction in the depolarization field (along the length of the chain) in the central particles, which occurs because charges of opposite polarity are induced on the opposing surfaces of neighboring particles on either side of the gap between them. These paired charges of opposite polarity are balanced near the center of the chain due to symmetry (resulting in a substantial reduction of the depolarization field) and become progressively less balanced for particles farther from the center.[60] As a point of comparison, the absorption cross-section of the central particle in the 10-particle chain is $\sigma_{abs} = 1.5 \times 10^{-14}\,m^2$, which is more than three times greater than that of a single particle (i.e., $\sigma_{abs} = 0.4 \times 10^{-14}\,m^2$) at their respective LSPR wavelengths. Thus, the self-assembled structures can provide significant enhancement and nanoscale localization of photothermal transduction at plasmon resonance.

Time-dependent analysis: During photothermal therapy, a temperature rise to within a range of 40–49°C is required to ensure cell destruction.[61,62] In order to achieve this temperature, we choose an incident irradiance of 22.5 kW cm^{-2}, with the linear polarization along x-axis. This level of irradiance is appropriate for a single chain in a relatively large CD as described in this work. In photothermal applications, a higher concentration of 1D chains is used, which gives rise to a cooperative heating to reduce the irradiance required to effect therapy.[17] Figure 18.10a shows the half domain and spatial thermal profiles of a single chain of 10 particles at 10, 30, and 50 ns under the resonant optical excitation.

The peak temperature rises quickly from 20 to 50°C within 50 ns. This thermodynamic profile meets the requirement of photothermal cancer therapy based on a laser pulse with the duration of a few tens nanoseconds.[61,62] Figure 18.10b plots the trend of the temperature change at the center of the chain versus time. It shows that the temperature experiences a sharp increase at first, but the slope of the curve gradually decreases over time. The data predict that the temperature will reach a steady-state saturation value.

18.4 CONCLUSION

Colloidal nanoparticles with tunable plasmonic behavior are increasingly used to enable enhance existing theranostic applications, e.g., imaging, sensing, photothermal hyperthermia, therapeutic photothermal nanobubble generation, and optically activated drug delivery. Many such applications require operation within the NIR biological window. Accordingly, we have used 3D computational models to compare the NIR optical behavior of three distinct plasmonic nanostructures with demonstrated biomedical efficacy—the SiO_2@Au core-shell, Au nanocage, and Au nanorod—as well as self-assembled 1D chains of magnetic-plasmonic Fe_3O_4@Au core-shell particles. With regard to the individual nanostructures, we have found that while the LSPR of each of these can be readily tuned to the NIR biological window, there are significant differences in their behavior that impact their selection for a given application. Specifically, our analysis demonstrates the advantages of the core-shell and nanocage structures over the nanorods in terms of the absorption cross-section, insensitivity to spatial orientation, and local field enhancement. In general, particles with a higher degree of rotational symmetry provide more efficient photothermal transduction because their absorption cross-section is less sensitive to their orientation, which is random in a colloid. However, while the optical absorption efficiencies of the SiO_2@Au and nanocage structures are comparable, the latter has more abundant E-field hot spots along its surface and especially within its open interior that can be leveraged for imaging, plasmon-enhanced sensing, and optically controlled drug release. Finally, the computational approach applied here provides insight into fundamental mechanisms that govern the plasmonic behaviors of colloidal nanoparticles. It is useful for the rational design of plasmonic nanoparticles for a wide range of applications.

In our analysis of the 1D Fe_3O_4@Au particle chains, we have examined the absorption spectrum as a function of the chain length. We find that as the chain forms, the peak absorption red-shifts and initially intensifies as the length increases from a single to multiple particles (e.g., 6), but then decreases slightly and asymptotically approaches that of an infinite chain as the aspect ratio of the chain approaches 10 to 1. Moreover, the photothermal transduction of a finite chain is found to be focused within its central particles and greatly enhanced relative to that of an individual particle at the respective LSPR wavelengths. Self-assembled 1D chains of magnetic-plasmonic particles provide tunable, highly localized and extraordinary field enhancement that can be exploited for various imaging and sensing applications. The enhanced and focused photothermal transduction provided by such structures also holds promise for laser-induced hyperthermia and nanobubble-based cancer therapy.

Finally, computational models are clearly needed to explore the optical properties of plasmonic nanostructures. The numerical methods demonstrated here can be readily generalized to study such structures having multifunctional materials constituents and arbitrary shape. As such, this approach should prove useful for the rational design of nanostructured plasmonic media for a host of novel applications.

REFERENCES

1. Jain, P. K., Huang, X., El-Sayed, I. H. & El-Sayed, M. A. Noble Metals on the Nanoscale: Optical and Photothermal Properties and Some Applications in Imaging, Sensing, Biology, and Medicine. *Acc. Chem. Res.* **41**, 1578-1586 (2008).
2. Yang, X., Yang, M., Pang, B., Vara, M. & Xia, Y. Gold Nanomaterials at Work in Biomedicine. *Chem. Rev.* **115**, 10410-10488 (2015).
3. Lalisse, A., Tessier, G., Plain, J. & Baffou, G. Quantifying the Efficiency of Plasmonic Materials for Near-Field Enhancement and Photothermal Conversion. *J. Phys. Chem. C* **119**, 25518-25528 (2015).
4. Sivapalan, S. T. *et al.* Surface-Enhanced Raman Spectroscopy of Polyelectrolyte-Wrapped Gold Nanoparticles in Colloidal Suspension. *J. Phys. Chem. C* **117**, 10677-10682 (2013).
5. Oh, E. *et al.* Cellular Uptake and Fate of PEGylated Gold Nanoparticles Is Dependent on Both Cell-Penetration Peptides and Particle Size. *ACS Nano* **5**, 6434-6448 (2011).
6. Yuan, H. *et al.* In Vivo Particle Tracking and Photothermal Ablation Using Plasmon-Resonant Gold Nanostars. *Nanomedicine* **8**, 1355-1363 (2012).
7. Mayer, K. M. & Hafner, J. H. Localized Surface Plasmon Resonance Sensors. *Chem. Rev.* **111**, 3828-3857 (2011).
8. Pang, B., Yang, X. & Xia, Y. Putting Gold Nanocages to Work for Optical Imaging, Controlled Release and Cancer Theranostics. *Nanomedicine* **11**, 1715-1728 (2016).
9. Gilroy, K. D., Ruditskiy, A., Peng, H., Qin, D. & Xia, Y. Bimetallic Nanocrystals: Syntheses, Properties, and Applications. *Chem. Rev.*, doi: 10.1021/acs.chemrev.1026b00211 (2016).
10. Serrano-Montes, A. B. *et al.* Gold Nanostar-Coated Polystyrene Beads as Multifunctional Nanoprobes for SERS Bioimaging. *J. Phys. Chem. C*, doi: 10.1021/acs.jpcc.1026b02282 (2016).

11. Sun, Y. G. & Xia, Y. N. Alloying and Dealloying Processes Involved in the Preparation of Metal Nanoshells Through a Galvanic Replacement Reaction. *Nano Lett.* **3**, 1569-1572 (2003).
12. Byers, C. P. *et al.* From Tunable Core-shell Nanoparticles to Plasmonic Drawbridges: Active Control of Nanoparticle Optical Properties. *Sci. Adv.* **1**, e1500988 (2015).
13. Kuo, W. S. *et al.* Gold Nanorods in Photodynamic Therapy, as Hyperthermia Agents, and in Near-Infrared Optical Imaging. *Angew. Chem. Int. Ed.* **49**, 2711-2715 (2010).
14. Mahmoud, M. A. & El-Sayed, M. A. Gold Nanoframes: Very High Surface Plasmon Fields and Excellent Near-Infrared Sensors. *J. Am. Chem. Soc.* **132**, 12704-12710 (2010).
15. Skrabalak, S. E. *et al.* Gold Nanocages for Biomedical Applications. *Adv. Mater.* **19**, 3177-3184 (2007).
16. Yavuz, M. S. *et al.* Gold Nanocages Covered by Smart Polymers for Controlled Release with Near-Infrared Light. *Nat. Mater.* **8**, 935-939 (2009).
17. Karampelas, I. H., Liu, K., Alali, F. & Furlani, E. P. Plasmonic Nanoframes for Photothermal Energy Conversion. *J. Phys. Chem. C* **120**, 7256–7264 (2016).
18. Furlani, E. P., Karampelas, I. H. & Xie, Q. Analysis of Pulsed Laser Plasmon-Assisted Photothermal Heating and Bubble Generation at the Nanoscale. *Lab on a Chip* **12**, 3707-3719 (2012).
19. Baffou, G. & Quidant, R. Thermo-Plasmonics: Using Metallic Nanostructures as Nano-Sources of Heat. *Laser Photonics Rev.* **7**, 171-187 (2013).
20. Baffou, G., Polleux, J., Rigneault, H. & Monneret, S. Super-Heating and Micro-Bubble Generation around Plasmonic Nanoparticles under cw Illumination. *J. Phys. Chem. C* **118**, 4890-4898 (2014).
21. Lapotko, D. Pulsed Photothermal Heating of the Media During Bubble Generation Around Gold Nanoparticles. *Int. J. Heat Mass Trans.* **52**, 1540-1543 (2009).
22. Lukianova-Hleb, E. Y., Hanna, E. Y., Hafner, J. H. & Lapotko, D. O. Tunable Plasmonic Nanobubbles for Cell Theranostics. *Nanotechnology* **21** (2010).
23. Kelly, K. L., Coronado, E., Zhao, L. L. & Schatz, G. C. The Optical Properties of Metal Nanoparticles: The Influence of Size, Shape, and Dielectric Environment. *J. Phys. Chem. B* **107**, 668-677 (2003).
24. Henry, A. I. *et al.* Correlated Structure and Optical Property Studies of Plasmonic Nanoparticles. *J. Phys. Chem. C* **115**, 9291-9305 (2011).
25. Furlani, E. P. & Baev, A. Free-Space Excitation of Resonant Cavities Formed from Cloaking Metamaterial. *J. Mod. Optic.* **56**, 523-529 (2009).
26. Alali, F., Kim, Y. H., Baev, A. & Furlani, E. P. Plasmon-Enhanced Metasurfaces for Controlling Optical Polarization. *ACS Photonics* **1**, 507-515 (2014).
27. Furlani, E. P. & Baev, A. Optical Nanotrapping Using Cloaking Metamaterial. *Phys. Rev. E* **79**, 026607 (2009).
28. Furlani, E., Jee, H., Oh, H., Baev, A. & Prasad, P. Laser Writing of Multiscale Chiral Polymer Metamaterials. *Adv. OptoElectron.* **2012**, 861569 (2012).
29. Tuersun, P. & Han, X. Optical Absorption Analysis and Optimization of Gold Nanoshells. *Appl. Opt.* **52**, 1325-1329 (2013).
30. Etchegoin, P. G., Le Ru, E. C. & Meyer, M. An Analytic Model for the Optical Properties of Gold. *J. Chem. Phys.* **125**, 164705 (2006).
31. Etchegoin, P. G., Le Ru, E. C. & Meyer, M. Erratum: An Analytic Model for the Optical Properties of Gold. [*J. Chem. Phys.* 125, 164705 (2006)]. *J. Chem. Phys.* **127**, 189901 (2007).
32. Johnson, P. B. & Christy, R. W. Optical Constants of the Noble Metals. *Phys. Rev. B* 4370-4379 (1972).
33. Yang, S. *et al.* Feedback-Driven Self-Assembly of Symmetry-Breaking Optical Metamaterials in Solution. *Nat. Nanotechnol.* **9**, 1002-1006 (2014).
34. Funston, A. M., Novo, C., Davis, T. J. & Mulvaney, P. Plasmon Coupling of Gold Nanorods at Short Distances and in Different Geometries. *Nano Lett.* **9**, 1651-1658 (2009).
35. Chen, J. *et al.* Gold Nanocages: Bioconjugation and Their Potential Use as Optical Imaging Contrast Agents. *Nano Lett.* **5**, 473-477 (2005).
36. Xia, Y. N., Rogers, J. A., Paul, K. E. & Whitesides, G. M. Unconventional Methods for Fabricating and Patterning Nanostructures. *Chem. Rev.* **99**, 1823-1848 (1999).
37. Khatua, S. *et al.* Resonant Plasmonic Enhancement of Single-Molecule Fluorescence by Individual Gold Nanorods. *ACS Nano* **8**, 4440-4449 (2014).
38. Mackey, M. A., Saira, F., Mahmoud, M. A. & El-Sayed, M. A. Inducing Cancer Cell Death by Targeting Its Nucleus: Solid Gold Nanospheres Versus Hollow Gold Nanocages. *Bioconjug. Chem.* **24**, 897-906 (2013).

39. Xiong, W., Mazid, R., Yap, L. W., Li, X. & Cheng, W. Plasmonic Caged Gold Nanorods for Near-infrared Light Controlled Drug Delivery. *Nanoscale* **6**, 14388-14393 (2014).
40. Liu, K., Xue, X. Z. & Furlani, E. P. A Numerical Study of the Photothermal Behaviour of Near-Infrared Plasmonic Colloids. *RSC Adv.* **6**, 100670-100675 (2016).
41. Pu, Y., Grange, R., Hsieh, C. L. & Psaltis, D. Nonlinear Optical Properties of Core-Shell Nanocavities for Enhanced Second-Harmonic Generation. *Phys. Rev. Lett.* **104** (2010).
42. Penninkhof, J. J. *et al.* Optical Cavity Modes in Gold Shell Colloids. *J. Appl. Phys.* **103** (2008).
43. Tsai, M. F. *et al.* Au Nanorod Design as Light-Absorber in the First and Second Biological Near-Infrared Windows for in Vivo Photothermal Therapy. *Acs Nano* **7**, 5330-5342 (2013).
44. Henderson, J. R. & Crawford, T. M. Repeatability of Magnetic-Field Driven Self-Assembly of Magnetic Nanoparticles. *J. Appl. Phys.* **109** (2011).
45. Henderson, J., Shi, S., Cakmaktepe, S. & Crawford, T. M. Pattern Transfer Nanomanufacturing Using Magnetic Recording for Programmed Nanoparticle Assembly. *Nanotechnology* **23** (2012).
46. Yellen, B. B. & Friedman, G. Programmable Assembly of Colloidal Particles Using Magnetic Microwell Templates. *Langmuir* **20**, 2553-2559 (2004).
47. Yellen, B. B., Fridman, G. & Friedman, G. Ferrofluid Lithography. *Nanotechnology* **15**, S562-S565 (2004).
48. Levin, C. S. *et al.* Magnetic-Plasmonic Core-Shell Nanoparticles. *ACS Nano* **3**, 1379-1388 (2009).
49. Park, H. Y. *et al.* Fabrication of Magnetic Core @ Shell Fe Oxide @ Au Nanoparticles for Interfacial Bioactivity and Bio-separation. *Langmuir* **23**, 9050-9056 (2007).
50. Liu, K., Xue, X. Z., Sukhotskiy, V. & Furlani, E. P. Optical Fano Resonance in Self-Assembled Magnetic-Plasmonic Nanostructures. *J. Phys. Chem. C* **120**, 27555-27561 (2016).
51. Liu, K., Xue, X. Z. & Furlani, E. P. Theoretical Comparison of Optical Properties of Near-Infrared Colloidal Plasmonic Nanoparticles. *Sci. Rep.* **6**, 34189 (2016).
52. Xue, X. Z., Liu, K. & Furlani, E. P. Theoretical Study of the Self-Assembly and Optical Properties of 1D Chains of Magnetic-Plasmonic Nanoparticles. *J. Phys. Chem. C* **121**, 9489-9496 (2017).
53. Chen, T. H., Pourmand, M., Feizpour, A., Cushman, B. & Reinhard, B. M. Tailoring Plasmon Coupling in Self-Assembled One-Dimensional Au Nanoparticle Chains Through Simultaneous Control of Size and Gap Separation. *J. Phys. Chem. Lett.* **4**, 2147-2152 (2013).
54. Wei, Q. H., Su, K. H., Durant, S. & Zhang, X. Plasmon Resonance of Finite One-Dimensional Au Nanoparticle Chains. *Nano Lett.* **4**, 1067-1071 (2004).
55. Radziuk, D. & Moehwald, H. Prospects for Plasmonic Hot Spots in Single Molecule SERS Towards the Chemical Imaging of Live Cells. *Phys. Chem. Chem. Phys.* **17**, 21072-21093 (2015).
56. Zhou, X. *et al.* Selective Functionalization of the Nanogap of a Plasmonic Dimer. *ACS Photonics* **2**, 121-129 (2015).
57. Kawata, S., Inouye, Y. & Verma, P. Plasmonics for Near-Field Nano-Imaging and Superlensing. *Nat Photonics* **3**, 388-394 (2009).
58. Ye, J. *et al.* Plasmonic Nanoclusters: Near Field Properties of the Fano Resonance Interrogated with SERS. *Nano Lett.* **12**, 1660-1667 (2012).
59. Ayala-Orozco, C. *et al.* Fluorescence Enhancement of Molecules Inside a Gold Nanomatryoshka. *Nano Lett.* **14**, 2926-2933 (2014).
60. Willingham, B. A. & Link, S. Energy Transport in Metal Nanoparticle Chains Via Sub-Radiant Plasmon Modes. *Abstr. Pap. Am. Chem. S* **242** (2011).
61. Huang, X. & El-Sayed, M. A. Gold Nanoparticles: Optical Properties and Implementations in Cancer Diagnosis and Photothermal Therapy. *J. Adv. Res.* **1**, 13-28 (2010).
62. Liu, K., Mokhtare, A., Xue, X. Z. & Furlani, E. P. Theoretical Study of the Photothermal Behaviour of Self-Assembled Magnetic-Plasmonic Chain Structures. *Phys. Chem. Chem. Phys.* **19**, 31613-31620 (2017).

19 Nonlinear Electronic Heat Transport in Molecules

Kamil Walczak, Arthur Luniewski, and Rita Aghjayan
Queens College, City University of New York, Department of Physics, Flushing, New York, USA

CONTENTS

19.1 INTRODUCTION

The Fourier's physics for heat flow is based on two fundamental postulates. The *first Fourier's law* interrelates heat flux J with temperature difference ΔT in the ballistic heat transport regime [1], or its vector $\vec{J}$ with temperature gradient $\vec{\nabla} T$ in the diffusive heat transport regime [2]

$$J = K\Delta T, \quad \textbf{(ballisticity)} \tag{19.1}$$

$$\vec{J} = -\kappa A \vec{\nabla} T. \quad \textbf{(diffusivity)} \tag{19.2}$$

Here K is a size-dependent (extensive) quantity known as thermal conductance, κ is material-specific (intensive) quantity known as thermal conductivity, and A is the cross-sectional area associated with heat conduction process. Those transport laws are strictly related to the linear response regime and may be violated in the case of high-intensity heat fluxes, in the presence of time-dependent thermal effects, and because of confinement as well as quantization effects occurring at nanoscale. The *second Fourier's law* is a scaling rule for the length-dependent thermal conductance

$$K = \text{const}(L), \quad \textbf{(ballisticity)} \tag{19.3}$$

$$K = \kappa \frac{A}{L}. \quad \textbf{(diffusivity)} \tag{19.4}$$

Here L is the length of the conduction channel. It should be noted that the second Fourier's law may be broken in biological and mesoscopic systems, where certain scattering events, collective phenomena, as well as correlation effects become important. The experimental verifications of Fourier's law are usually based on Equations 19.3 and 19.4, where $K \sim L^{\beta-1}$ or $\kappa \sim L^{\beta}$ with $\beta = 0$ for diffusive heat transfer or $\beta = 1$ for ballistic heat flow [3–6].

In nanoscale systems, transport of energy carried by electrons turned out to be quantized [7–10] and may be a strongly nonlinear function of temperature difference. From quantum mechanical point of view, we have learnt how to treat nonlinear transport properties of molecular complexes and the associated noises in terms of Landauer conduction channels within the following set of equations [11,12]:

$$J = \frac{1}{2\pi\hbar}\sum_{n}\int_{-\infty}^{+\infty}(\varepsilon-\mu)P_n(\varepsilon)[f_L - f_R]\mathrm{d}\varepsilon, \tag{19.5}$$

$$S_{\mathrm{TH}} = \frac{1}{\pi\hbar}\sum_{n}\int_{-\infty}^{+\infty}(\varepsilon-\mu)^2P_n(\varepsilon)[f_L(1-f_L)+f_R(1-f_R)]\mathrm{d}\varepsilon, \tag{19.6}$$

$$S_{\mathrm{SN}} = \frac{1}{\pi\hbar}\sum_{n}\int_{-\infty}^{+\infty}(\varepsilon-\mu)^2P_n(\varepsilon)[1-P_n(\varepsilon)](f_L - f_R)^2\mathrm{d}\varepsilon. \tag{19.7}$$

The summation in Equations 19.5–19.7 is over each conduction channel, which contributes independently to electronic heat transport. Here, S_{TH} denotes equilibrium thermal noise, S_{SN} stands for non-equilibrium shot noise of quantum nature, $P_n(\varepsilon)$ is probability that an electron of energy ϵ will pass through nanoscale system, and f_L and f_R are Fermi-Dirac distribution functions that capture statistics of electrons in the left and right reservoirs, respectively. However, the proper definition of Landauer conduction channels remains an open question in nanoscale physics [13].

In this chapter, we use nonlinear heat transport theory applicable to molecular junctions composed of a single molecule coupled to two metallic reservoirs (thermal baths). Specifically, we focus our attention on three major aspects of nanoscale heat conduction. (1) Landauer conduction channels, which we define as discrete energy levels of a molecule characterized by delocalized orbitals and significant overlap with reservoir continuum of states (relatively strong reservoir–molecule coupling). (2) Molecular noises that may mask original signals (heat fluxes) but may also be analyzed, providing some useful information about individual heat carriers in the system under investigation, e.g., their statistical independence or correlation effects. Here we propose a new type of noise–signal relations, which will be tested numerically. (3) Nonlinear corrections to heat fluxes and the associated noises in the ballistic heat transport regime. It should be noted that Equations 19.5–19.7 become questionable when we are dealing with channel mixture effects in which electron passing through one energy level is scattered into a different one. It is particularly true in molecular systems, where the phase coherence of the electronic wave functions is maintained during heat propagation. This channel mixture and other interference effects call for transport formalism with globally defined transmission function (not as the sum of individual Landauer conduction channels), where all the interference and overlapping phenomena are included, while heat carriers from different channels do interfere and must remain indistinguishable.

19.2 ELECTRONIC HEAT FLUX

We analyze heat transport phenomena within a conceptually simple and transparent Landauer formalism, which relates transport properties of nanoscale system to its scattering properties, such as globally defined quantum-mechanical transmission probability function. Knowing the energy-dependent transmission function for electrons $P(\varepsilon)$, we can use the Landauer-type formula to calculate the so-called electronic heat flux [14,15]:

$$J = \int_{-\infty}^{+\infty} P(\varepsilon)F(\varepsilon, T)\mathrm{d}\varepsilon, \tag{19.8}$$

$$F(\varepsilon, T) = \frac{(\varepsilon - \mu)}{2\pi\hbar}[f_L - f_R]. \tag{19.9}$$

Here $f_L = f_L(\varepsilon, T + \Delta T)$ and $f_R = f_R(\varepsilon, T)$ are Fermi-Dirac occupation factors for the left (L) and right (R) thermal baths (heat reservoirs), T is absolute temperature, and ΔT is temperature difference between two reservoirs. For practical reasons, we expand Equation 19.9 into Taylor series with respect to temperature difference ΔT (it is usually a small parameter) to obtain expression for the heat flux in the form:

$$J = \sum_{m=1}^{\infty} K^{(m)}(\Delta T)^m, \tag{19.10}$$

where the particular transport coefficients are defined via the following integral:

$$K^{(m)} = \int_{-\infty}^{+\infty} P(\varepsilon)F_m(\varepsilon, T)\mathrm{d}\varepsilon. \tag{19.11}$$

The integral kernels in Equation 19.11 are temperature-dependent functions that for the first two terms take the form

$$F_1(y) = \frac{k_B y^2}{8\pi\hbar}\cos^{-2}\left(\frac{y}{2}\right), \tag{19.12}$$

$$F_2(y) = \frac{k_B y^2}{8\pi\hbar T}\cosh^{-2}\left(\frac{y}{2}\right)\left[\frac{y}{2}\tanh\left(\frac{y}{2}\right) - 1\right], \tag{19.13}$$

where variable y is defined as

$$y(\varepsilon, T) = \frac{\varepsilon - \mu}{k_B T}. \tag{19.14}$$

Obviously, we could expand Equation 19.9 to higher-order terms with respect to temperature difference ΔT (see Appendix A), but in this chapter, we restrict ourselves only to the most essential quadratic corrections. In Equation 19.10, the coefficient $K^{(1)}$ plays the role of a linear thermal conductance, while $K^{(2)}$ defines the first nonlinear coefficient in the modified Fourier's law.

To give an example, let us consider a special case of ballistic heat transport regime, where the heat is transferred by a single energy level without any scattering involved into the conduction process of electrons. In this case $P(\varepsilon) = 1$, and after performing integration in Equation 19.11, we obtain

$$J = \frac{\pi k_B^2 T}{6\hbar}\Delta T\left[1 + \frac{1}{2}\left\{\frac{\Delta T}{T}\right\}\right] \equiv K^{(1)}\Delta T + K^{(2)}(\Delta T)^2. \tag{19.15}$$

According to Equation 19.15, in the ballistic heat transport regime, when the scattering processes are negligible, each energy level contributes a universal quantum $\pi k_B^2/6\hbar$ to the reduced thermal conductance $K^{(1)}/T$. At strongly non-equilibrium conditions, when the temperature difference ΔT is significant in comparison to other energy parameters, there is additional universal correction $\pi k_B^2 \Delta T/12\hbar$ to the electronic thermal conductance $K^{(1)}$ (this is actually the only nonzero correction in the ballistic heat transport regime). Although the amount of energy transferred by individual electrons is not quantized, thermal conductance turned out to be a quantized quantity. Quantization

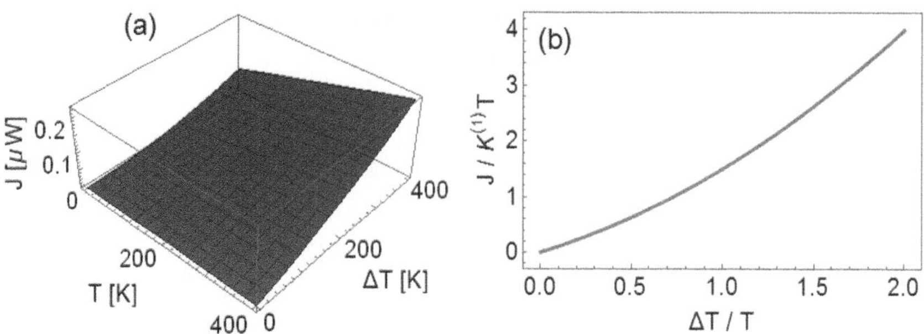

FIGURE 19.1 (a) 3D visualization of nonlinear (quadratic) ballistic Fourier's law; (b) 2D visualization of nonlinear (quadratic) ballistic Fourier's law.

of thermal conductance may also be understood as a consequence of Heisenberg uncertainty principle as combined with equipartition theorem from thermodynamics (see Appendix B). In Figure 19.1, we plot nonlinear (quadratic) ballistic Fourier's law, which is a linear function of temperature T and quandratic function of temperature difference ΔT. Interestingly, the quanta of thermal conductance and their nonlinear corrections associated with phonons [16–18] and photons [19,20] are expressed by exactly the same formulas as Equation 19.15.

19.3 NOISES AND NOISE–SIGNAL RELATIONS

In general, the electronic heat flux is fluctuating in time. The noise associated with nanoscale transport of conducting electrons is usually described by the symmetrized autocorrelation function of the form [21,22]:

$$S(t) = \frac{1}{2}\langle \delta J(t)\delta J(0) + \delta J(0)\delta J(t)\rangle. \tag{19.16}$$

Here the angular brackets indicate an ensemble average, while the variance (uncertainty) of the energy flux in the presence of a heat flow carried by individual electrons is defined as

$$\delta J(t) = J(t) - \langle J(t)\rangle. \tag{19.17}$$

In the asymptotic limit of infinite time ($t \to \infty$), the frequency-dependent power spectrum density related to heat flow carried by electrons is simply a Fourier transform of the autocorrelation function from Equation 19.16, namely

$$S(\omega) = \frac{1}{2\pi}\int_{-\infty}^{+\infty} S(t)\exp(-i\omega t)\mathrm{d}t. \tag{19.18}$$

Using the Landauer scattering formalism in the zero-frequency limit ($\omega \to 0$), the noise power of electronic heat flux fluctuations may be conveniently decomposed into two separate parts: the equilibrium (Johnson–Nyquist) thermal noise and non-equilibrium (Schottky) shot noise

$$S = S_{\mathrm{TH}} + S_{\mathrm{SN}}, \tag{19.19}$$

$$S_{\mathrm{TH}} = \int_{-\infty}^{+\infty} P(\varepsilon)\Phi(\varepsilon, T)\mathrm{d}\varepsilon, \tag{19.20}$$

$$S_{\mathrm{SN}} = \int_{-\infty}^{+\infty} P(\varepsilon)[1 - P(\varepsilon)]\Psi(\varepsilon, T)\mathrm{d}\varepsilon, \tag{19.21}$$

where the integral nuclei in Equations 19.20 and 19.21 are expressed via the following relations:

$$\Phi(\varepsilon, T) = \frac{(\varepsilon-\mu)^2}{\pi\hbar}[f_L(1-f_L)+f_R(1-f_R)], \tag{19.22}$$

$$\Psi(\varepsilon, T) = \frac{(\varepsilon-\mu)^2}{\pi\hbar}[f_L-f_R]^2. \tag{19.23}$$

Since the integral in Equation 19.21 is strictly semipositive, the non-equilibrium term can only enhance the noise above the level established by the equilibrium thermal fluctuations. The expansion of Equation 19.22 into the Taylor series with respect to temperature difference ΔT allows us to write down the following expression for the thermal noise:

$$S_{\mathrm{TH}} = \sum_{m=0}^{\infty} S_{\mathrm{TH}}^{(m)}(\Delta T)^m. \tag{19.24}$$

The particular noise coefficients in Equation 19.24 are defined via the following integral:

$$S_{\mathrm{TH}}^{(m)} = \int_{-\infty}^{+\infty} P(\varepsilon)\Phi_m(\varepsilon, T)d\varepsilon. \tag{19.25}$$

The integral kernels in Equation 19.25 are temperature-dependent functions that, for the first three terms, take the form:

$$\Phi_0(x) = \frac{(k_B T y)^2}{\pi\hbar}[\cosh(y)+1]^{-1}, \tag{19.26}$$

$$\Phi_1(x) = \frac{k_B^2 T y^3}{4\pi\hbar}\cosh^{-2}\left(\frac{y}{2}\right)\tanh\left(\frac{y}{2}\right), \tag{19.27}$$

$$\Phi_2(x) = \frac{k_B^2 y^3}{8\pi\hbar}\cosh^{-4}\left(\frac{y}{2}\right)\left[\frac{y}{2}[\cosh(y)-2]-\sinh(y)\right], \tag{19.28}$$

where the variable y is defined once again via Equation 19.14. Since Equation 19.24 is expressed via infinite series with the zeroth-order term, the equilibrium thermal noise is unavoidable even in the absence of temperature difference ΔT (or temperature gradient $\vec{\nabla}T$), which is fully consistent with our intuition.

As an example, let us consider a special case of ballistic heat transport regime, where the heat is carried by electrons passing through the single energy level without any scattering processes involved into the conduction process. In this case $P(\varepsilon) = 1$, and after performing integration in Equation 19.25, we obtain

$$S_{\mathrm{TH}} = \frac{2\pi k_B^3 T^3}{3\hbar}\left[1+\frac{3}{2}\left\{\frac{\Delta T}{T}\right\}+\frac{3}{2}\left\{\frac{\Delta T}{T}\right\}^2+\cdots\right] \equiv S_{\mathrm{TH}}^{(0)} + S_{\mathrm{TH}}^{(1)}\Delta T + S_{\mathrm{TH}}^{(2)}(\Delta T)^2 + \cdots \tag{19.29}$$

At thermal equilibrium, both reservoirs are kept at the same temperature ($\Delta T = 0$) and the average heat flux is zero $\langle J(t)\rangle = 0$, but the equilibrium thermal noise $S_{\mathrm{TH}}^{(0)}$ is directly proportional to the linear thermal conductance $K^{(1)}$ and may still be significant

$$S_{\mathrm{TH}}^{(0)} = 4k_B T^2 K^{(1)}. \tag{19.30}$$

Equation 19.30 is a simple manifestation of fluctuation–dissipation theorem (FDT) [23–26], where the noise component is related to the signal (transport coefficient). This relation is usually regarded as a universal equation, which is assumed to be independent of materials, chemical nature of interatomic connections, and the specific models used in simulations of transport characteristics associated with thermal devices. However, in our analysis related to ballistic heat transport regime, we can find two similar expressions for higher-order thermal noises

$$S_{\text{TH}}^{(1)} = 6k_BTK^{(1)}, \tag{19.31}$$

$$S_{\text{TH}}^{(2)} = 6k_BK^{(1)}. \tag{19.32}$$

Furthermore, we can also propose three analogous expressions for the considered thermal noise components by involving nonlinear correction to thermal conductance term $K^{(2)}$, namely

$$S_{\text{TH}}^{(0)} = 8k_BT^3K^{(2)}, \tag{19.33}$$

$$S_{\text{TH}}^{(1)} = 12k_BT^2K^{(2)}, \tag{19.34}$$

$$S_{\text{TH}}^{(2)} = 12k_BTK^{(2)}. \tag{19.35}$$

The obvious question is related to the universal character of all the noise–signal relations given by Equations 19.30–19.35. Further, calculating specific ratios of thermal noises related to different order terms, we can prove the following expressions:

$$S_{\text{TH}}^{(0)} = \frac{2}{3}TS_{\text{TH}}^{(1)} = \frac{2}{3}T^2S_{\text{TH}}^{(2)}. \tag{19.36}$$

The importance of Equation 19.36 stems from the fact that one can deduce the higher-order terms of equilibrium thermal noise on the basis of its zeroth-order component (which is usually easily measurable quantity at nearly equilibrium conditions). In Section 19.5, we provide numerical evidence that the FDT from Equation 19.30 is universally valid for molecular systems and arbitrary temperatures of heat reservoirs but is limited to small temperature difference between two thermal baths (as a consequence of linear response theory). Further, we show that all noise–signal relations from Equations 19.31–19.35 and noise–noise relations from Equation 36 are applicable to molecular systems in the transport regime when temperature difference between heat reservoirs is significant, while the average temperature of the system under consideration is relatively small (in comparison to room temperature).

In the next step, let us analyze non-equilibrium shot noise. The expansion of Equation 19.23 into the Taylor series with respect to temperature difference ΔT allows us to write down the following expression:

$$S_{\text{SN}} = \sum_{m=2}^{\infty} S_{\text{SN}}^{(m)}(\Delta T)^m. \tag{19.37}$$

Here, the particular noise coefficients are defined via the following integral transform:

$$S_{\text{SN}}^{(m)} = \int_{-\infty}^{+\infty} P(\varepsilon)[1 - P(\varepsilon)]\Psi_m(\varepsilon, T)d\varepsilon. \tag{19.38}$$

The integral kernels in Equation 19.38 are temperature-dependent functions. In the absence of temperature difference ($\Delta T = 0$), the non-equilibrium shot noise is identically equal to zero $S_{\text{SN}}^{(0)} = 0$. Further, we can show that the linear term with respect to the finite temperature difference ($\Delta T \neq 0$) also disappears $S_{SN}^{(1)} = 0$, and the first nonzero term is defined as a quadratic correction to the shot noise with the integral kernel of the form:

$$\Psi_2(x) = \frac{k_B^2y^4}{16\pi\hbar}\cosh^{-4}\left(\frac{y}{2}\right). \tag{19.39}$$

It should be noted that the non-equilibrium shot noise disappears for ideal ballistic conduction, for which $P(\varepsilon) = 1$, and reaches maximum value for $P(\varepsilon) = 1/2$. We can evaluate the maximum value of the non-equilibrium shot noise related to a single energy level by putting $P(\varepsilon) = 1/2$ into Equation 19.38 and after performing integration, we obtain

$$S^{(2)}_{\mathrm{SN,max}} \cong \frac{k_B^3 T}{4\pi\hbar}. \tag{19.40}$$

In this special case, we can come up with additional two equations relating the maximum value of non-equilibrium shot noise with transport coefficients

$$S^{(2)}_{\mathrm{SN,max}} = \frac{k_B}{2} K^{(1)}, \tag{19.41}$$

$$S^{(2)}_{\mathrm{SN,max}} = k_B T K^{(2)}. \tag{19.42}$$

Although Equations 19.41 and 19.42 look very similar to fluctuation–dissipation relations, their limited applicability is rather obvious, since they are valid only for specific energies ϵ for which electron has 50% chance to be transmitted $P(\varepsilon) = 1/2$ and 50% chance for being reflected $[1 - P(\varepsilon)] = 1/2$. Further, by comparing Equations 19.41 to 19.42 or equivalently Equations 19.30 to 19.33, or Equations 19.31 to 19.34, or even Equations 19.32 to 19.35, we obtain a formula that connects linear thermal conductance with its quadratic correction

$$K^{(1)} = 2TK^{(2)}. \tag{19.43}$$

In Section 19.5, we provide numerical evidence that the applicability of signal–signal relation from Equation 19.43 is limited to significantly large temperature difference between heat reservoirs, while the average temperature of the system under consideration is still relatively small.

19.4 ELECTRONIC TRANSMISSION FUNCTION

In order to use the perturbative transport theory developed in Section 19.3, and to test numerically all the signal–signal, signal–noise, and noise–noise relations introduced there, we need to define transmission function for the system composed of molecule coupled to two heat reservoirs, which are kept at different temperatures. The usual choice is the Caroli formula for the transmission probability expressed in terms of the effective propagators (Green's functions) [27]

$$T(\varepsilon) = \mathrm{Trace}[\Gamma_L G \Gamma_R G^+]. \tag{19.44}$$

Here, Γ_L and Γ_R are broadening functions of the contacts, while G is an effective propagator (Green's function) that captures the information about the electronic structure of the molecule as specifically modified by its connection to heat reservoirs. This propagator is a solution of the so-called Dyson equation

$$[\varepsilon I - H - \Sigma_L - \Sigma_R]G(\varepsilon) = I. \tag{19.45}$$

Here, ϵ is the energy of heat carrier, I is the unity matrix, H is Hamiltonian of molecular system, and Σ_L and Σ_R are self-energy terms related to broadening functions via the following expressions:

$$\Gamma_\alpha(\varepsilon) = i[\Sigma_\alpha - \Sigma_\alpha^+] = \tau_\alpha A(\varepsilon)\tau_\alpha^+, \tag{19.46}$$

Here, index $\alpha = L, R$ stands for the left and right reservoirs, $A(\varepsilon)$ is the energy-dependent surface spectral function, and τ_α is the matrix describing molecular connection to heat reservoirs, respectively. The self-energy terms can be expressed as follows:

$$\Sigma_\alpha(\varepsilon) = \tau_\alpha Q(\varepsilon)\tau_\alpha^+ = \Lambda_\alpha(\varepsilon) - \frac{i}{2}\Gamma_\alpha(\varepsilon). \tag{19.47}$$

Here, the surface propagator $Q(\varepsilon)$ includes the information about the electronic structure of thermal baths, while the real part of self-energy term $\Lambda_\alpha(\varepsilon)$ and the broadening function $\Gamma_\alpha(\varepsilon)$ are interrelated via the so-called Hilbert transform [28]. It should be noted that the real part of self-energy term describes the shift of discrete energy levels of the molecule itself, while its imaginary part is responsible for contact-induced broadening of those energy levels.

19.5 RESULTS AND DISCUSSION

So far, the presented formalism was very general, and many different levels of description may be used to study heat transfer at nanoscale. However, in this chapter, we limit ourselves to a very simple model applicable to molecular junction composed of benzene molecule coupled to two metallic (golden) reservoirs via thiol groups, as schematically shown in Figure 19.2. The organic molecule itself is treated within the tight-binding approaximation, where only delocalized pi-orbitals with hopping parameters $t = -2.7$ eV are included into our simplified Hamiltonian (all onsite energies have been made equal to zero) [29]. In the case of single-atom connection between molecule and individual heat reservoir, the coupling matrix τ_α is reduced to only one energy parameter τ. Both heat reservoirs are modeled via semi-elliptical density of states [30]

$$Q(\varepsilon) = \frac{1}{\gamma}\left[\frac{\varepsilon}{2\gamma} - i\sqrt{1 - \left(\frac{\varepsilon}{2\gamma}\right)^2}\right], \tag{19.48}$$

with $\gamma = 10$ eV to mimic wide-band metal and electrochemical potential or Fermi energy level $\mu = 0$ (this is the reference point on the energy scale). Here, 4γ is the energy bandwidth of the reservoir under consideration.

Figure 19.3 shows transmission function and its product with reflection function, because both of them are needed to perform calculations of transport characteristics and dynamical noise properties of nanoscale systems. We see that the most significant changes of mentioned functions are present at the vicinity of electrochemical potential (Fermi energy level) and therefore have direct impact on transport phenomena associated with electrons as heat carriers. In Figure 19.4, we plot linear thermal conductance and its nonlinear correction as functions of temperature. We noted a quadratic-to-linear transition in the temperature dependence of thermal conductance due to an increase of molecule–reservoir coupling in the range of intermediate temperatures. The nonlinear correction to thermal conductance starts at some finite value for $T \to 0$, increases to some maximum value, and then decreases linearly with temperature. From Figure 19.4, we see that the increase in molecule–reservoir coupling results in the shift of the maximum values of nonlinear corrections to thermal conductance (wide peaks) toward lower temperatures.

Figure 19.5 shows dynamical noise properties of heat fluxes flowing through molecular junction under consideration. From this picture, we see that both zeroth-order and first-order thermal noises gradually increase with temperature. Further, the second-order thermal noise increases with temperature up to some maximum value, and then decreases linearly when temperature is increased above certain critical value. Interestingly, the temperature at which the second-order thermal noise

FIGURE 19.2 Schematic representation of molecular junction under investigation.

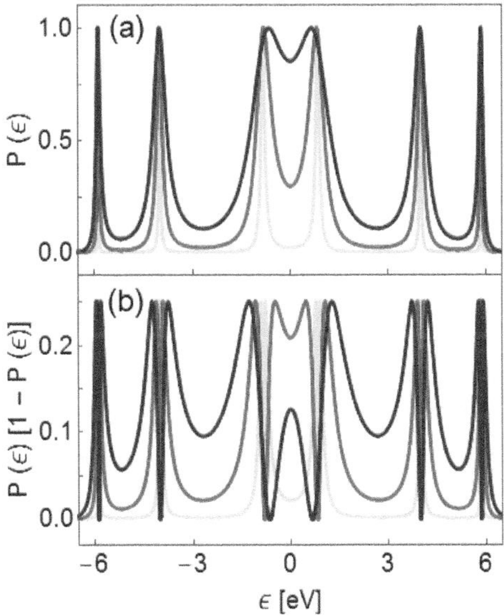

FIGURE 19.3 (a) Transmission function P and (b) its product with reflection function $P[1-P]$ plotted against energy of electron ϵ for different values of molecule–reservoir coupling strength (light grey: τ = 1 eV, dark grey: $\tau = 2$ eV, black: $\tau = 3$ eV).

reaches the maximum value only weakly depends on the molecule-reservoir coupling and for a set of parameters used in our model: $T_{\max} \approx 1900$ K. The general trend is the following: the larger the value of the molecule–reservoir coupling parameter, the greater are the values of all types of equilibrium thermal noises associated with nanoscale systems. Moreover, the second-order non-equilibrium shot noise strongly depends on the molecule–reservoir coupling and may be quite small even for large values of the molecule–reservoir coupling parameters, as clearly documented in Figure 19.5(d).

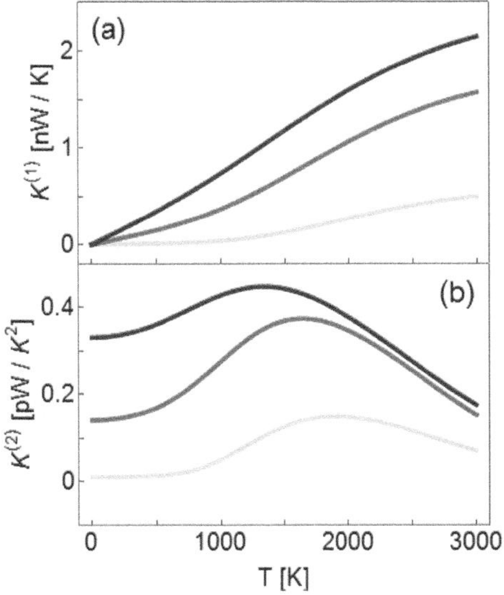

FIGURE 19.4 (a) Linear thermal conductance $K^{(1)}$ and (b) its nonlinear correction $K^{(2)}$ plotted against the temperature T for different values of molecule–reservoir coupling strength (light grey: τ = 1 eV, dark grey: $\tau = 2$ eV, black: $\tau = 3$ eV).

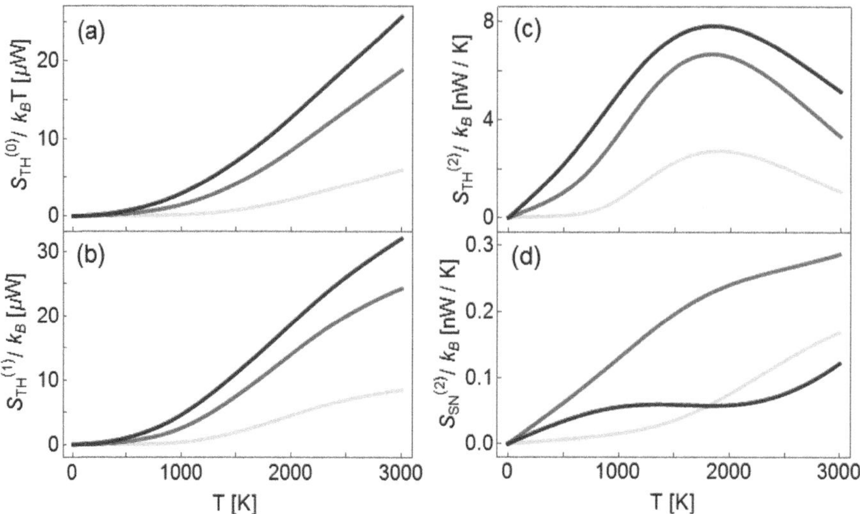

FIGURE 19.5 (a)–(c) Equilibrium thermal noise components $S_{\text{TH}}^{(0)}$, $S_{\text{TH}}^{(1)}$, and $S_{\text{TH}}^{(2)}$, and (d) the second-order non-equilibrium shot noise $S_{\text{SN}}^{(2)}$ plotted against the temperature T for different values of molecule–reservoir coupling strength (light grey: $\tau = 1$ eV, dark grey: $\tau = 2$ eV, black: $\tau = 3$ eV).

In statistical physics, the FDT is a powerful tool applicable to both classical as well as quantum systems, being a direct consequence of the linear response theory. The FDT is limited to weak non-equilibrium conditions in which the linear response of the system (thermal conductance) onto relatively small perturbation (temperature difference) is related to the statistical behavior of that system (zeroth-order thermal noise) kept at equilibrium conditions (constant temperature). The quantitative expression for FDT is given via Equation 19.30, and we checked numerically that it is valid for the molecular system under investigation for any temperature (results not shown). The importance of this particular formula stems from a simple fact that it is relatively easy to keep the system at constant temperature and conduct measurements at equilibrium thermodynamic conditions. Then we may use FDT to deduce the information about the linear reaction of the system onto applied temperature difference from a long-time observation of thermally activated fluctuations in that system (thermal noise).

However, it should also be noted that the FDT is violated in many nanoscale and mesoscopic systems, including glassy materials, proteins, and low-dimensional quantum systems [31–36]. At strongly non-equilibrium conditions with high-intensity heat fluxes, the linear conductance and the zeroth-order thermal noise are insufficient to describe nonlinear behavior of molecular junctions of practical importance [37–54]. The question of how to extract the information about nonlinear corrections to transport and noise coefficients from our knowledge of their linear terms immediately arises. Certainly, it should be possible, since quantum transport in molecular systems and the associated thermal noise are both related to the same quantum-mechanical transmission function. Based on the analysis performed in Section 19.3, we can rewrite the expressions for higher-order (nonlinear) terms as linear functions of lower-order quantities and the average temperature of the analyzed system, namely

$$K^{(2)} = \frac{K^{(1)}}{2T} = \frac{S_{\text{TH}}^{(0)}}{8k_BT^3}, \tag{19.49}$$

$$S_{\text{TH}}^{(1)} = 6k_BTK^{(1)} = \frac{3S_{\text{TH}}^{(0)}}{2T}, \tag{19.50}$$

$$S_{\text{TH}}^{(2)} = 6k_BK^{(1)} = \frac{3S_{\text{TH}}^{(0)}}{2T^2}. \tag{19.51}$$

Substituting from Equation 19.50 into 19.51, we obtain the noise–noise relation: $S^{(1)}_{\mathrm{TH}} = TS^{(2)}_{\mathrm{TH}}$. All those equations represent a completely new set of noise–signal relations that interconnect thermal noise components to the appropriate transport coefficients. Those quantitative relations are of great importance in non-equilibrium quantum thermodynamics, because nonlinear corrections can be calculated by using equilibrium thermal noise and linear transport coefficient, which are easily measurable quantities in comparison to any nonlinear terms. It should be noted that it is experimentally much easier to control the system at weak non-equilibrium situations than maintain strong non-equilibrium conditions for a long period of time (where the temperature difference plays the role of a driving force conjugate to the corresponding heat flux).

In Figure 19.6, we present numerical evidence that some of the noise–signal relations given by Equations 19.30–19.36 or by Equations 19.49–19.51 are valid when temperature difference between heat reservoirs is significant, while the average temperature of the system under consideration is relatively small (in comparison to room temperature). It should be noted that transport via atoms and molecules is ballistic or quantum in nature and the dissipation of energy during scattering

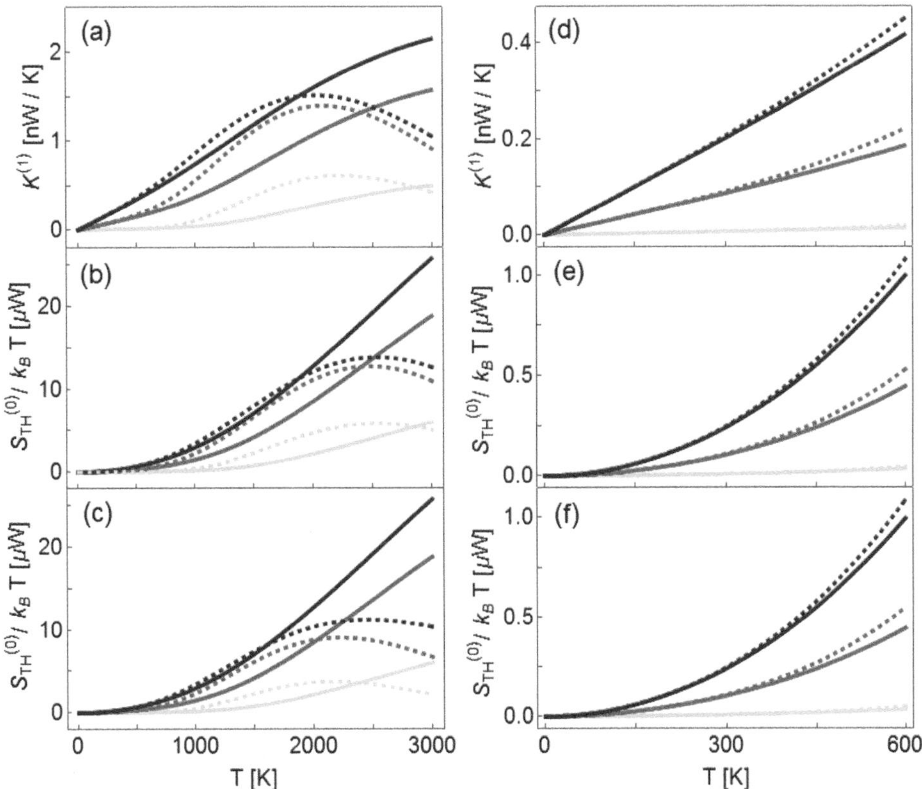

FIGURE 19.6 (a), (d) The exact solution for linear thermal conductance $K^{(1)}$ (solid line) and its approximated form $K^{(1)} = 2TK^{(2)}$ (dashed line) plotted against the average temperature of the molecular system under consideration. (b), (e) The exact solution for equilibrium thermal noise $S^{(0)}_{\mathrm{TH}}$ (solid line) and its approximated form $S^{(0)}_{\mathrm{TH}} = 8k_BT^3K^{(2)}$ (dashed line) plotted against the average temperature of the molecular system under consideration. (c), (f) The exact solution for equilibrium thermal noise $S^{(0)}_{\mathrm{TH}}$ (solid line) and its approximated form $S^{(0)}_{\mathrm{TH}} = 2T^2S^{(2)}_{\mathrm{TH}}/3$ (dashed line) plotted against the average temperature of the molecular system under consideration. All the results are presented for different values of molecule–reservoir coupling strength (green: $\tau = 1$ eV, red: $\tau = 2$ eV, blue: $\tau = 3$ eV).

phenomena of heat carriers takes place inside the macroscopic reservoirs. All the noise–signal relations, discussed in this chapter, couple statistical properties of a given system at equilibrium conditions (zeroth-order thermal noise) or during steady-state heat flow (higher-order noise terms) to relaxation processes described by exponential decay laws at non-equilibrium conditions (thermal conduction and its nonlinear corrections). In other words, there is the nonzero heat flux originated from thermally activated fluctuations inside particular reservoirs of molecular junctions, which are kept at constant temperature. This heat flux is exactly the same as generated by relatively large temperature difference (perturbation) between those reservoirs. Here, the characteristic relaxation times are identified with the reservoir–molecule escape rates, so they are closely related to the reservoir–molecule coupling parameters.

19.6 CONCLUDING REMARKS

Understanding transport of heat from the bottom-up perspective of quantum thermodynamics is of great scientific and technological importance. In particular, it is quite challenging to properly define and measure thermodynamic quantities for nanoscale and mesoscopic systems, ranging from simple molecules, quantum dots, single-electron devices, and trapped-ion systems to colloids, and biological settings. Thermodynamics becomes increasingly complicated as the system size decreases and fluctuations around average and steady-state quantities become significantly large. In those special cases, the probability of observing processes with entropy production opposite to that dictated by the second law of thermodynamics increases exponentially. Since the second law of thermodynamics is statistical in nature, there is nonzero probability that the entropy of an isolated nanoscale system may spontaneously decrease for a short period of time.

Although the questions of ergodicity, stochasticity, and thermalization are beyond the scope of this chapter, here we derived new noise–signal relations that are valid under the nearly equilibrium steady-state conditions.

Our original derivation is based on a scattering (Landauer) approach to nanoscale heat conduction processes carried by electrons (fermionic heat carriers). Since the signal (heat flux) and the associated noise (dynamical noise power) are equally important in nanoscale systems, both quantities are treated on equal footing. Specifically, we proved that the ballistic Fourier's law and the FDT are applicable to nanoscale systems under any thermal conditions as long as temperature difference between heat reservoirs is relatively small in comparison to other energies in the analyzed system (where linear response theory is still valid). Further, we derived a few asymptotic noise–signal relations with only limited applicability in non-equilibrium quantum thermodynamics. Such quantitative expressions allow one to calculate higher-order transport and noise terms on the basis of linear thermal conductance and equilibrium thermal noise. All those equations should be of great importance to scientists, who constantly deal with out-of-equilibrium open systems at molecular or mesoscopic scales [55].

It should be noted that practical realization of molecular junctions is always a compromise between the preservation of molecular electronic structure (resonant tunneling) and the formation of highly conductive connections (quasi-ballistic coupling). The simplicity of electronic structure of simple molecules (such as benzene) coupled to heat reservoirs may help in verification of existing models and validation of various approximations used to simplify calculations. In macroscopic systems, the more scattering we have in the system, the more noise we will measure by our apparatus. Following this logic, the ballistic heat conduction should be noiseless. However, noise in nanoscale systems is usually as important as signal itself, mainly due to quantum effects, such as Pauli exclusion principle, interference of electronic waves, and Coulomb blockade. Here noise may be viewed as discrepancy from the actual result or time-dependent fluctuations that mask the measured signal (heat flux). Furthermore, the detailed analysis of noise provides useful information about statistical independence of heat carriers under consideration [56]. The shot noise spectrum contains the information about the quantum correlations between fermionic heat carriers (electrons), which lead to the collective phenomenon of anti-bunching (see Appendix C).

Recently, the experimental techniques to measure shot noise were used in order to study transport properties of molecular complexes [57–62]. The shot noise spectrum, as the second moment of the heat flux, provides additional information about nanoscale transport, which is inaccessible via conventional measurements of thermal conductance. At nearly equilibrium conditions (low temperature and small bias voltage) and weak molecule–reservoir coupling regime, the measurements of shot noise allow one to estimate the whole spectrum of transmission function as well as the number of current-carrying conduction channels (number of modes). In this particular case, the thermal conductance of a coherent quantum system may be represented as a sum of independent conduction channels [58,59]. Combining shot noise and conductance measurements, it is also possible to study the effect of phonon activation on nanoscale transport in a view of the distribution of independent conduction channels, where the appearance of conductance enhancement may be a consequence of strong electron–phonon coupling or interchannel scattering [60]. Importantly, the formalism presented in this chapter is valid in the case of a strong molecule–reservoir coupling regime, where all the interference and hybridization phenomena are included.

19.7 APPENDIX A: HIGHER-ORDER CORRECTIONS

In this appendix, we provide the expressions for temperature-dependent integral kernels that may be used to calculate the third- and fourth-order corrections with respect to temperature difference ΔT. Those higher-order terms as related to electronic heat flux may be calculated by using Equation 19.11 with the following functions:

$$F_3(y) = \frac{k_B y^2}{16\pi\hbar T^2}\cosh^{-4}\left(\frac{y}{2}\right)\left[\left\{\frac{y^2}{6}+1\right\}\cosh(y) - y\sinh(y) - \frac{y^2}{3}+1\right], \tag{A1}$$

$$F_4(y) = \frac{k_B y^2}{16\pi\hbar T^3}\cosh^{-2}\left(\frac{y}{2}\right)\left[\left\{\frac{y^3}{12}+3y\right\}\tanh\left(\frac{y}{2}\right) - y^2 - 2 + \left\{\frac{3y^2}{2} - 4y^3\tanh\left(\frac{y}{2}\right)\right\}\cosh^{-2}\left(\frac{y}{2}\right)\right], \tag{A2}$$

where variable $y = (\varepsilon - \mu)/k_B T$ as in Equation 19.14. Further, the higher-order corrections to the equilibrium thermal noise associated with electronic heat flow may be calculated by using Equation 19.25 with the following functions:

$$\Phi_3(y) = \frac{k_B^2 y^3}{16\pi\hbar T}\cosh^{-5}\left(\frac{y}{2}\right)\left[\left\{1 - \frac{11y^2}{6}\right\}\sinh\left(\frac{y}{2}\right) + \left\{\frac{y^2}{6}+1\right\}\sinh\left(\frac{3y}{2}\right) + 3y\cosh\left(\frac{y}{2}\right) - y\cosh\left(\frac{3y}{2}\right)\right], \tag{A3}$$

$$\Phi_4(y) = \frac{k_B^2 y^3}{16\pi\hbar T^2}\cosh^{-4}\left(\frac{y}{2}\right)\left[\left\{\frac{y^3}{6}+6y\right\}\cosh^2\left(\frac{y}{2}\right) + \frac{5y^3}{4}\cosh^{-2}\left(\frac{y}{2}\right) - \{y^2+2\}\sinh(y) + 6y^2\tanh\left(\frac{y}{2}\right) - \frac{5y^3}{4} - 9y\right]. \tag{A4}$$

where variable $y = (\varepsilon - \mu)/k_B T$ as in Equation 19.14. Finally, the higher-order corrections to the non-equilibrium shot noise associated with electronic heat flow may be calculated by using Equation 19.38 with the following functions:

$$\Psi_3(y) = \frac{k_B^2 y^4}{8\pi\hbar T}\cosh^{-4}\left(\frac{y}{2}\right)\left[\frac{y}{2}\tanh\left(\frac{y}{2}\right) - 1\right], \tag{A5}$$

$$\Psi_4(y) = \frac{3k_B^2 y^4}{32\pi\hbar T^2}\cosh^{-6}\left(\frac{y}{2}\right)\left[\left\{\frac{7y^2}{36} + 1\right\}\cosh(y) - y\sinh(y) - \frac{11y^2}{36} + 1\right]. \tag{A6}$$

where variable $y = (\varepsilon - \mu)/k_B T$ as in Equation 19.14.

19.8 APPENDIX B: QUANTUM OF THERMAL CONDUCTANCE

In this appendix, we present a simple justification behind the concept of thermal conductance quantization by combining the equipartition theorem from thermodynamics and Heisenberg uncertainty principle from quantum theory. According to the equipartition theorem, the average energy associated with each degree of freedom of the system under investigation is $E = k_B T/2$, hence, the change of the energy during the process of heat transfer is simply

$$\Delta E = \frac{k_B}{2}\Delta T. \tag{B1}$$

The heat flux is defined as the amount of thermal energy transferred from one heat reservoir to the other via quantum system within certain time interval τ; therefore, $J = k_B T/2\tau$. From that relation, we can extract the following expression for the characteristic time:

$$\tau = \frac{k_B T}{2J}. \tag{B2}$$

By virtue of Heisenberg uncertainty principle for complementary variables of energy and time, we can write down the following inequality:

$$\Delta E \cdot \tau \geq \frac{\hbar}{2}. \tag{B3}$$

Substituting from Equations B1 and B2 into Heisenberg inequality from Equation B3, we obtain

$$\frac{k_B^2 T}{4J}\Delta T \geq \frac{\hbar}{2}. \tag{B4}$$

By taking into consideration, the formal definition of the linear thermal conductance as heat flux divided by temperature difference $K \equiv J/\Delta T$, Equation B4 may be rewritten in the form

$$K \leq \frac{k_B^2 T}{2\hbar}. \tag{B5}$$

From the above inequality, we see that the maximum value for the linear thermal conductance during the process of heat transport as carried by electrons via single energy level (or carried by acoustic phonons associated with individual vibrational mode) is very close to the quantum of thermal conductance $K^{(1)} = \pi k_B^2 T/6\hbar$ derived from the Landauer formula in Section 19.2 (the difference between two results is <5%).

19.9 APPENDIX C: ANALYSIS OF FANO FACTOR

In this appendix, we perform analysis of Fano factor for a purely ballistic transport regime. First of all, it should be noted that electrons are intrinsically noisy due to thermal phenomena and quantum effects. Thermal noise is a direct consequence of random motion of heat carriers because of nonzero temperature of the sample, while quantum (shot) noise is closely related to Heisenberg uncertainty principle and Pauli exclusion principle for fermionic heat carriers (like electrons). In Section 19.3, we showed that the stochastic nature of heat reservoir generates thermal and quantum fluctuations of heat fluxes that are not arbitrary, but these quantities are interrelated via the special-type equations, known as noise–signal relations. In particular, shot noise determines dynamic fluctuations of heat fluxes due to granularity of electrons (corpuscular property). Furthermore, the detailed analysis of noise provides information, which is unavailable via measurements of heat fluxes, about statistical independence of heat carriers under consideration. The steady-state heat flux is a measure of average energy transferred by individual carriers, while noise spectrum offers information about statistics of electrons and their quantum correlations leading to collective phenomenon of anti-bunching.

The Fano factor is defined as a dimensionless noise-to-signal ratio. In the case of linear response theory, the Fano factor may be defined as follows:

$$F \equiv \frac{S_{\mathrm{TH}}^{(0)}}{4k_B T^2 K^{(1)}} = 1. \tag{C1}$$

The statistical independence of electrons in the linear ballistic transport regime is reflected in the Poissonian value for Fano factor $F = 1$. Equation C1 is actually a simple demonstration of the FDT. However, in the case of nonlinear ballistic transport regime, the heat flux is expressed via Equation 19.15, and the Fano factor with up to the second-order noise terms may be defined as

$$F \equiv \frac{S_{\mathrm{TH}}^{(0)} + S_{\mathrm{TH}}^{(1)}\Delta T + S_{\mathrm{TH}}^{(2)}(\Delta T)^2 + S_{\mathrm{TH}}^{(2)}(\Delta T)^2}{4k_B T^2 (K^{(1)} + K^{(2)}\Delta T)} = \sum_{m=0}^{3} F^{(m)}. \tag{C2}$$

In the case of collisionless transport of electrons flowing via single energy level, we have $P(\varepsilon) = 1$, and the additive terms of the Fano factor from Equation C2 may be calculated analytically

$$F^{(0)} = \frac{S_{\mathrm{TH}}^{(0)}}{4k_B T^2 (K^{(1)} + K^{(2)}\Delta T)} = \frac{2}{2+x}, \tag{C3}$$

$$F^{(1)} = \frac{S_{\mathrm{TH}}^{(1)}\Delta T}{4k_B T^2 (K^{(1)} + K^{(2)}\Delta T)} = \frac{3x}{2+x}, \tag{C4}$$

$$F^{(2)} = \frac{S_{\mathrm{TH}}^{(2)}(\Delta T)^2}{4k_B T^2 (K^{(1)} + K^{(2)}\Delta T)} = \frac{3x^2}{2+x}. \tag{C5}$$

Here, the dimensionless variable $x = \Delta T/T$, and in this particular case, we obtain $S_{\mathrm{SN}}^{(2)} = 0$ (and consequently $F^{(3)} = 0$). If we add together Equations C3–C5, we obtain the temperature-dependent expression for the minimum value of the Fano factor

$$F_{\min} = \sum_{m=0}^{2} F^{(m)} = \frac{2 + 3x + 3x^2}{2+x}. \tag{C6}$$

Interestingly, thermal noise is a statistics-independent quantity and turned out to be exactly the same for both fermionic and bosonic heat carriers (like electrons and phonons) [63]. However, the non-equilibrium shot noise generated by electrons differs from that generated by acoustic phonons not only by its sign but also by its magnitude [64]. In general, the electronic shot noise is non-negative and usually increases the value of the Fano factor, while the phononic shot noise is non-positive and usually reduces the value of the Fano factor.

As mentioned in Section 19.3, the non-equilibrium shot noise is a non-negative quantity that reaches the maximum value for transmission function $P(\varepsilon) = 1/2$. In this specific situation, both heat flux and thermal noise are reduced by half, but their ratio still remains the same. As a consequence, the thermal noise components of the Fano factor will not change, while the shot noise component $F^{(3)}$ may be calculated as

$$F^{(3)} = \frac{S^{(2)}_{\mathrm{SN,max}}(\Delta T)^2}{2k_B T^2(K^{(1)} + K^{(2)}\Delta T)} = \frac{x^2/2}{2+x}. \tag{C7}$$

If we add together Equations C3–C5 and C7, we obtain the temperature-dependent expression for the maximum value of the Fano factor

$$F_{\max} = \sum_{m=0}^{3} F^{(m)} = \frac{2 + 3x + 7x^2/2}{2+x}. \tag{C8}$$

According to the spin-statistics theorem, and as a consequence of quantum correlation effects between electrons, the electronic wave function describing ensemble of fermionic particles must be antisymmetric with respect to exchange of two fermions. Moreover, the equilibrium distribution function of fermionic carriers takes the form of Fermi-Dirac statistics. However, at non-equilibrium conditions, electrons tend to avoid each other due to their anti-bunching property. Such tendency ultimately results in positive shot noise Fano factor $F^{(3)} > 0$, which increases super-Poissonian noise power spectral density above the level established by thermal fluctuations. It should be noted that the shot noise becomes dominant when the finite number of electrons is sufficiently small, so that uncertainties associated with Poissonian distribution become significant.

As shown in Figure 19.7, the Fano factor reaches its Poissonian limit $F = 1$ at equilibrium thermal conditions at which $\Delta T \to 0$. In this particular case, there are no correlations between electrons that are moving freely through the sample. However, at non-equilibrium conditions, when $\Delta T > 0$, we obtain positive value of shot noise Fano factor $F^{(3)} > 0$ within the super-Poissonian regime of the total noise power $F > 1$. Here, the enhancement of thermal super-Poissonian noise is due to the so-called anti-bunching effect of electrons. This anti-bunching property of fermionic heat carriers is a direct consequence of Pauli exclusion principle, which is a natural tendency of fermions to avoid the same quantum state (and consequently each other). Note that the non-equilibrium shot noise constitutes a relatively small contribution to the total noise power, which is entirely dominated by thermal fluctuations even in the case of large temperature difference between heat reservoirs.

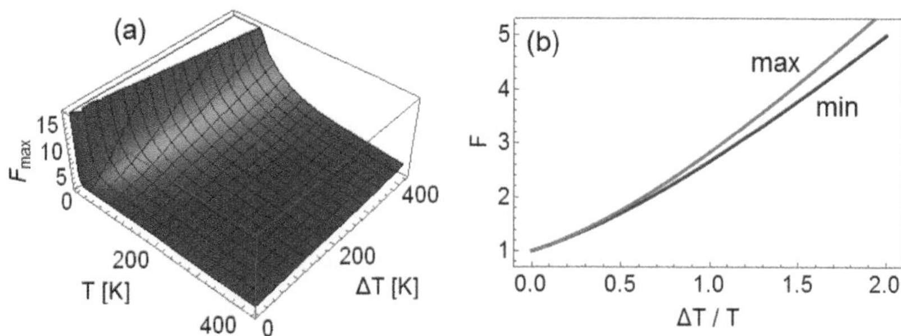

FIGURE 19.7 (a) 3D plot of maximum value of dimensionless Fano factor for ballistic transport of electrons as a function of absolute temperature T and temperature difference ΔT; (b) 2D plot of Fano factor F as a function of ratio $\Delta T/T$ (indicated both its minimum and maximum values).

REFERENCES

1. K. Walczak, *Nanoscale Heat Conduction: Modeling Prospects* (Taylor & Francis Group, 2016). Dekker Encyclopedia of Nanoscience and Nanotechnology, Third Edition.
2. J. Fourier, *Theorie Analytique de la Chaleur* (Didot, Paris 1822).
3. S. Maruyama, *Physica B* **323**, pp. 193–195 (2002).
4. B. Li and J. Wang, *Phys. Rev. Lett.* **91**, 044301 (2003).
5. C. W. Chang, D. Okawa, H. Garcia, A. Majumdar, and A. Zettl, *Phys. Rev. Lett.* **101**, 075903 (2008).
6. N. Yang, G. Zhang, and B. Li, *Nano Today* **5**, pp. 85–90 (2010).
7. L.W. Molenkamp, Th. Gravier, H. von Houten, O. J. A. Buijk, M. A. A. Mabesoone, and C.T. Foxon, *Phys. Rev. Lett.* **68**, pp. 3765–3768 (1992).
8. G. T. Guttman, E. Ben-Jacob, and D.J. Bergman, *Phys. Rev. B* **53**, pp. 15856–15862 (1996).
9. A. Greiner, L. Reggiani, T. Kuhn, and L. Varani, *Phys. Rev. Lett.* **78**, pp. 1114–1117 (1997).
10. J. T. Nicholls and O. Chiatti, *J. Phys.: Condens. Matter* **20**, 164210 (2008).
11. Y. M. Blanter and M. Büttiker, *Phys. Rep.* **336**, pp. 1–166 (2000).
12. C. Beenakker and C. Schönenberger, *Phys. Today* **56**, pp. 37–42 (2003).
13. G. A. Levin, W. A. Jones, K. Walczak, and K. L. Yerkes, *Phys. Rev. E* **85**, 031109 (2012).
14. R. Landauer, *IBM J. Res. Dev.* **1**, pp. 223–231 (1957).
15. R. Landauer, *Philos. Mag.* **21**, pp. 863–867 (1970).
16. L. G. C. Rego and G. Kirczenow, *Phys. Rev. Lett.* **81**, pp. 232–235 (1998).
17. K. Schwab, E. A. Henrlksen, J. M. Worlock, and M. L. Roukes, *Nature (London)* **404**, pp. 974–977 (2000).
18. T. Yamamoto, S. Watanabe, and K. Watanabe, *Phys. Rev. Lett.* **92**, 075502 (2004).
19. D. R. Schmidt, R. J. Schoelkopf, and A. N. Cleland, *Phys. Rev. Lett.* **93**, 045901 (2004).
20. M. Meschke, W. Guichard, and J. P. Pekola, *Nature* **444**, pp. 187–190 (2006).
21. F. Zhan, S. Denisov, and P. Hänggi, *Phys. Rev. B* **84**, 195117 (2011).
22. F. Zhan, S. Denisov, and P. Hänggi, *Phys. Stat. Sol.* B **250**, pp. 2355–2364 (2013).
23. H. B. Callen and T. A. Welton, *Phys. Rev.* **83**, pp. 34–40 (1951).
24. J. Weber, *Phys. Rev.* **101**, pp. 1620–1626 (1956).
25. R. Kubo, *Rep. Prog. Phys.* **29**, pp. 255–284 (1966).
26. B. U. Felderhof, *J. Math. Phys. A* **11**, pp. 921–927 (1978).
27. C. Caroli, R. Combescot, P. Nozieres, and D. Saint-James, *J. Phys. C: Solid State Phys.* **4**, pp. 916–929 (1971).
28. K. Walczak and K. L. Yerkes, *J. Appl. Phys.* **115**, 174308 (2014).
29. K. Walczak, *Phys. Stat. Sol. (b)* **241**, pp. 2555–2561 (2004).
30. K. Walczak, *Cent. Eur. J. Chem.* **2**, pp. 524–533 (2004).
31. A. Crisanti and F. Ritort, *J. Phys. A: Math. Gen.* **36**, pp. R181–R290 (2003).
32. A. Mitra and A. J. Millis, *Phys. Rev. B* **72**, 121102(R) (2005).
33. A. Mauri and D. Leporini, *Europhys. Lett.* **76**, pp. 1022–1028 (2006).
34. K. Hayashi and M. Takano, *Biophys. J.* **93**, pp. 895–901 (2007).
35. T. Kawasaki and H. Tanaka, *Phys. Rev. Lett.* **102**, 185701 (2009).
36. D. V. Averin and J. P. Pekola, *Phys. Rev. Lett.* **104**, 220601 (2010).
37. J. K. Gimzewski and C. Joachim, *Science* **283**, pp. 1683–1688 (1999).
38. C. Joachim, J. K. Gimzewski, M. A. Ratner, *Nature (London)* **408**, pp. 541–548 (2000).
39. J. C. Ellenbogen and C. Love, *Proc. IEEE* **88**, pp. 386–426 (2000).
40. C. Dekker and M. A. Ratner, *Phys. World* **14**, pp. 29–33 (2001).
41. R. L. Carroll and C. B. Gorman, *Angew. Chem. Int. Ed.* **41**, pp. 4378–4400 (2002).
42. A. Nitzan and M. A. Ratner, *Science* **300**, pp. 1384–1389 (2003).
43. J. R. Heath and M. A. Ratner, *Phys. Today* **56**, pp. 43–49 (2003).
44. J. M. Tour, *Molecular Electronics. Commercial Insights, Chemistry, Devices, Architecture and Programming* (World Scientific, London 2003).
45. C. Joachim and M. A. Ratner, *Proc. Natl. Acad. Sci. USA* **102**, pp. 8801–8808 (2005).
46. *Introducing Molecular Electronics, Lecture Notes in Physics*, edited by G. Cuniberti, G. Fagas, K. Richter (Springer-Verlag, Berlin Heidelberg 2005).
47. N. J. Tao, *Nat. Nanotechnol.* **1**, pp. 173–181 (2006).

48. S. E. Lyshevski, *Molecular Electronics, Circuits and Processing Platforms* (CRC Press, Boca Raton 2007).
49. J. R. Heath, *Annu. Rev. Matter. Res.* **39**, pp. 1–23 (2009).
50. R. L. McCreery and A. J. Bergren, *Adv. Mater.* **21**, pp. 1–20 (2009).
51. J. Del Nero, F. M. de Souza, and R. B. Capaz, *J. Comput. Theor. Nanosci.* **7**, pp. 1–14 (2010).
52. M. Tsutsui and M. Taniguchi, *Sensors* **12**, pp. 7259–7298 (2012).
53. S. E. Lyshevski, *Molecular and Biomolecular Processing: Solutions, Directions and Prospects, Handbook on Nano and Molecular Electronics*, edited by W. Goddard, D. Brenner, S. E. Lyshevski and G. Iafrate (CRC Press, Boca Raton 2012), pp. 125–177.
54. S. V. Aradhya and L. Venkataraman, *Nat. Nanotechnol.* **8**, 399–410 (2013).
55. J. Millen and A. Xuereb, *New J. Phys.* **18**, 011002 (2016).
56. M.B. Büttiker, *Phys. Rev. Lett.* **65**, pp. 2901–2904 (1990).
57. M. Kiguchi, O. Tal, S. Wohlthat, F. Pauly, M. Krieger, D. Djukic, J. C. Cuevas, and J. M. van Ruitenbeek, *Phys. Rev. Lett.* **101**, 046801 (2008).
58. R. Vardimon, M. Klionky, and O. Tal, *Phys. Rev. B* **88**, 161404(R) (2013).
59. M. Kumar, O. Tal, R. H. M. Smit, A. Smogunov, E. Tosatti, and J. M. van Ruitenbeek, *Phys. Rev. B* **88**, 245431 (2013).
60. R. Ben-Zvi, R. Vardimon, T. Yelin, and O. Tal, *ACS Nano* **7**, pp. 11147–11155 (2013).
61. R. Vardimon, T. Yelin, M. Klionsky, S. Sarkar, A. Biller, L. Kronik, and O. Tal, *Nano Lett.* **14**, pp. 2988–2993 (2014).
62. R. Vardimon, M. Matt, P. Nielaba, J. C. Cuevas and O. Tal, *Phys. Rev. B* **93**, 085439 (2016).
63. I.V. Krive, E.N. Bogachek, A.G. Scherbakov, and U. Landman, *Phys. Rev. B* **64**, 233304 (2001).
64. K. Walczak, *Nonlinear Transport and Noise Properties of Acoustic Phonons*. APS March Meeting 2016, Baltimore, MD.

20 Designing Gates to Realize a Full Adder Quantum Circuit in cQED Transmon Systems

Sahar Daraeizadeh[a,b], *Shavindra P. Premaratne*[b], *A. Y. Matsuura*[b], *Marek Perkowski*[a]

[a]Department of Electrical and Computer Engineering, Portland State University, Portland Oregon, USA
[b]Intel Labs, Intel Corporation, Hillsboro, Oregon, USA

CONTENTS

20.1 INTRODUCTION

Quantum computing has captured the imagination of many, from academic to corporate researchers, with the goal of harnessing the power of quantum mechanics to enable computations beyond the capabilities of traditional transistor-based classical computers. This technology has already inspired the naming of the current time period as the Noisy Intermediate Scale Quantum (NISQ)[1] era, referring to the state of the nascent quantum devices of 50–100 qubits that are being built today.

The fundamental element of computation for a quantum computer is the quantum bit (or qubit). In a quantum computer, a qubit is in a superposition of two states (state $|0\rangle$ and state $|1\rangle$). However, when the qubit is measured, its state collapses to either $|0\rangle$ or $|1\rangle$. Since a single qubit can represent two states, N qubits can exist in a superposition of 2^N states enabling a quantum computer to simultaneously operate on all 2^N states. Additionally, a group of qubits can become entangled resulting in a state which cannot be represented as a product of individual qubit states. Superposition and entanglement open the potential for exponential speed-up for certain classes of mathematical problems using special-purpose quantum algorithms.

While superposition and entanglement offer the potential of great computational power, it also makes simulation of quantum circuits using classical resources difficult, due to exponential overhead.[2] The best simulators for ideal qubits & ideal gates can only handle ~50 qubits. In this chapter, we consider ideal qubits & non-ideal gates, making simulation of quantum systems beyond 20 qubits nearly impossible due to memory limitations of current computational resources. This is unfortunate, as simulation is a fruitful practice in classical computing for architectural design. However, simulation of small qubit systems is still possible and can be used as an important tool for the architectural design of quantum computers and algorithms, especially for early NISQ devices.

The outline of the chapter is as follows: In Section 20.1, we introduce the concept of qubits. In Section 20.2, we provide some background on circuit quantum electrodynamics (cQED) systems. In Section 20.3, we describe a general methodology of simulating quantum systems dynamics, and design the required universal gate set to perform reversible logic quantum computation[3–6] using a simulator. In Section 20.4, we demonstrate how to realize a full adder quantum circuit using the universal gate set. Finally, given the constraints of the quantum system, we map this circuit to the physical layout of the quantum chip.

20.2 SUPERCONDUCTING QUBITS

20.2.1 Introduction to Qubits

A qubit is a two-level system that has two distinct energy levels referred to as the ground state and the excited state, $|0\rangle$ and $|1\rangle$, respectively. The state of a qubit can be in a superposition of the ground and excited states, which is a linear combination of $|0\rangle$ and $|1\rangle$. A general pure state of a qubit can be written as follows:

$$|\psi\rangle = \alpha|0\rangle + \beta|1\rangle, |\alpha|^2 + |\beta|^2 = 1 \tag{20.1}$$

where α and β are the probability amplitudes of being in states $|0\rangle$ and $|1\rangle$, respectively. The qubit state $|\psi\rangle$ is a vector in a complex vector space.[7]

There exist different physical realizations of qubits such as trapped ions, nitrogen-vacancy centers in diamond, quantum dots, nuclear magnetic resonance, neutral atoms in optical lattices, and superconducting devices. In 2017 and 2018, the International Roadmap for Devices and Systems (IRDS) emphasized the need for further development and scaling the number of qubits of superconducting quantum computing systems.[8,9] The IRDS reports indicate how Josephson junction-based superconducting devices are used for quantum computing applications.[10]

20.2.2 Introduction to Superconducting Quantum Computing with Transmons

The coherent control of quantum levels using superconducting devices was first demonstrated by Nakamura et al.[11] The quantum behavior observed in the charge qubit[12] used in the latter experiment had coherence times on the order of a few nanoseconds. Recent improvements in coherence times and coherent control of these devices have allowed the superconducting qubits to become a leading platform for realizing a universal quantum computer.[13,14] Transmons[15] and their variations,[16,17] have proliferated as the most popular choice of qubits for constructing superconducting quantum computing circuits. In this chapter, we focus on the use of transmons within the cQED architecture.[18,19]

20.2.3 A Quantum Harmonic Oscillator

The quantum harmonic oscillator (QHO) is a system that has been studied extensively in quantum physics and serves as a model for many physical systems. It is a quantum analog of the classical harmonic oscillator.[20] Given that an isolated QHO has equally spaced levels (perfect harmonicity), it is not possible to isolate two levels for use in quantum computing as qubit levels. Hence, it is necessary to identify other candidate systems with a certain degree of anharmonicity to facilitate quantum computation. Nevertheless, in superconducting quantum computation, resonators are coupled to anharmonic systems to serve as a qubit measurement proxy,[18,19] realize quantum memory elements,[21] and encode logical qubits.[22]

20.2.4 Transmon

The transmon's quantum-level structure is that of a weakly anharmonic oscillator to a good approximation. The relative anharmonicity is usually 3–5% of the characteristic system frequency.[23] The Hamiltonian of the transmon can be written as follows[15]:

$$\hat{H} = \left(\hat{b}^\dagger\hat{b} + \frac{1}{2}\right)\hbar\omega_q - \frac{E_C}{12}\left(\hat{b}^\dagger + \hat{b}\right)^4 - E_J \tag{20.2}$$

where $E_C = \frac{e^2}{2C_\Sigma}$ is the single-electron charging energy for the total capacitance C_Σ, and $E_J = \frac{\hbar I_c}{2e}$ is the Josephson energy for a junction having a critical current of I_c, $\omega_q = \hbar^{-1}\sqrt{8E_JE_C}$ is the characteristic frequency of the qubit transition, and $\hat{b}$ and $\hat{b}^\dagger$correspond to the annihilation and creation operators for the transmon, respectively. Using perturbation theory and considering E_C/E_J as the small parameter, the transmon energy spectrum $\{E_m\}$ can be approximated by the simplified expression:

$$E_m = \hbar\omega_m = \left(m + \frac{1}{2}\right)\sqrt{8E_JE_C} - \frac{E_C}{4}(2m^2 + 2m + 1) - E_J \tag{20.3}$$

with absolute anharmonicity approximately given by $-E_C$.[15,23] Note that the energy levels are spaced non-uniformly (i.e., $E_1 - E_0 \neq E_2 - E_1$).

A comparison of energy spectra for a transmon and a QHO is given in Figure 20.1, where both systems have a nominal fundamental frequency of 6 GHz. The transmon fundamental frequency is set by adjusting the Josephson junction parameter E_J relative to E_C. The ability to tune the transmon frequency gives great flexibility in designing physical quantum computing circuits. Generally, the fundamental frequencies of superconducting quantum devices are engineered in the range of 3–10 GHz in which compatible control electronics are readily available.

20.2.5 A Transmon Coupled to a Resonator

A two-level system or a qubit coupled to a QHO (or resonator) is described by the Jaynes–Cummings (JC) Hamiltonian.[24] If the qubit fundamental frequency is given by ω_q and the resonator frequency is given by ω_r, then the Hamiltonian can be written as follows:

$$H_{\text{JC}} = \hbar\omega_r\left(a^\dagger a + \frac{1}{2}\right) + \frac{1}{2}\hbar\omega_q\sigma_z + \hbar g\left(a\sigma^+ + a^\dagger\sigma^-\right). \tag{20.4}$$

Here, $\sigma_z = \begin{pmatrix} 1 & 0 \\ 0 & -1 \end{pmatrix}$ is the Pauli-Z matrix representing the qubit state, a and $a^\dagger$ are the lowering and raising operators for the resonator, respectively, and $\sigma^+ = \begin{pmatrix} 0 & 1 \\ 0 & 0 \end{pmatrix}$ and $\sigma^- = \begin{pmatrix} 0 & 0 \\ 1 & 0 \end{pmatrix}$ are the raising and lowering operators for the qubit, respectively. g is the effective coupling strength

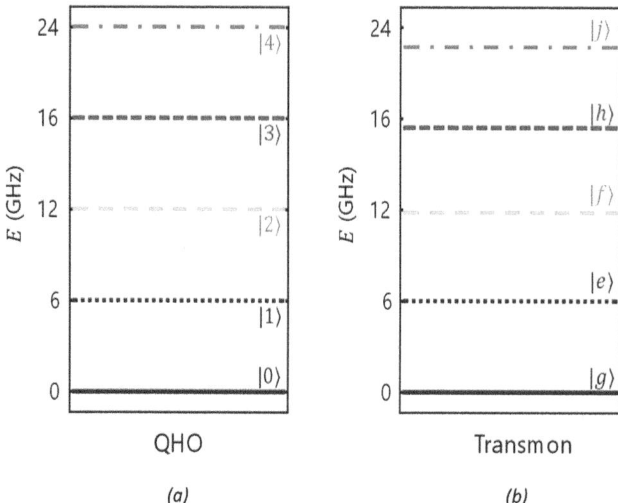

FIGURE 20.1 Comparison of the eigenenergies for (a) QHO and (b) transmon having identical fundamental frequency 6 GHz. The energy levels of the QHO are represented as numbers while the transmon energy levels are represented by letters. Here, the transmon has an absolute anharmonicity of 200 MHz. The higher energy levels of the transmon are slightly different from the QHO because of the anharmonicity. The ground state energy is set to 0 for both systems for ease in comparison.

between the resonator and the qubit, since g quantifies the rate at which the qubit and resonator would exchange quanta if they had identical characteristic frequencies. In expressing H_{JC}, the rotating wave approximation (RWA) has been used to eliminate the so-called counter-rotating terms that do not conserve the total number of photons.[25]

It is necessary to take the multi-level structure of the transmon into account when accurately describing a transmon coupled to a resonator. This is done by generalizing the JC Hamiltonian to include a multilevel system coupled to a resonator as follows:

$$\tilde{H}_{JC} = \hbar\omega_r\left(a^\dagger a + \frac{1}{2}\right) + \frac{1}{2}\hbar\sum_m \omega_m|m\rangle\langle m| + \hbar\sum_n g_{n,n+1}\left(a|n+1\rangle\langle n| + a^\dagger|n\rangle\langle n+1|\right) \quad (20.5)$$

where the kets represent the eigenstates of the transmon and $g_{n,n+1}$ represents the coupling strength between the transmon and the resonator within the n-excitation manifold.[15] This Hamiltonian is presented in the block diagonal form, but is not exactly solvable without further approximation. Applying the "dispersive approximation" where $|\omega_r - \omega_q| \gg g$, and truncating the transmon to its two lowest levels, the generalized JC Hamiltonian can be exactly solved to yield the following diagonal Hamiltonian:

$$H_{eff} = \hbar(\omega_r + \chi\sigma_z)a^\dagger a + \frac{1}{2}\hbar(\omega_q + \chi)\sigma_z \quad (20.6)$$

In Equation 20.6, the resonator and qubit frequencies are shifted, and χ is the dispersive shift that captures the effective interaction between the qubit levels of the transmon and the cavity.

20.2.6 Multiple Transmons Coupled to a Resonator

Following the methods developed in Section 20.2.5, it is straightforward to extend the system under study to multiple transmon qubits coupled to a single resonator. The resulting Hamiltonian is known as the Tavis–Cummings (TC) Hamiltonian,[26,27] and within the RWA, it is given by

$$H_{\mathrm{TC}} = \hbar\omega_r\left(a^\dagger a + \frac{1}{2}\right) + \sum_j \frac{1}{2}\hbar\omega_q^{(j)}\sigma_z^{(j)} + \hbar g_j\left(a\sigma_j^+ + a^\dagger\sigma_j^-\right) \tag{20.7}$$

where *j* is an index for enumerating qubits. As in the case of the JC Hamiltonian, the TC Hamiltonian can be generalized to multilevel systems, such as transmons, to obtain the following Hamiltonian:

$$\tilde{H}_{TC} = \hbar\omega_r\left(a^\dagger a + \frac{1}{2}\right) + \sum_j^{N_q}\left[\frac{1}{2}\hbar\sum_m \omega_m^{(j)}|m\rangle_{(j)}\langle m| + \hbar\sum_n g_{n,n+1}^{(j)}\left(a|n+1\rangle_{(j)}\langle n| + a^\dagger|n\rangle_{(j)}\langle n+1|\right)\right] \tag{20.8}$$

which enumerates the N_q different transmons with the index *j*, and the different energy levels of each transmon with the indices *m, n*. When simulating the Hamiltonian evolution over time, it is helpful to use approximations to reduce the computational overhead.

In particular, for the case of two transmons coupled to a single resonator (i.e., $N_q = 2$), using the dispersive approximation, subsequently using the two-level qubit approximation for the transmons, and assuming no resonator excitations, it is possible to obtain the following relatively simple Hamiltonian[28]:

$$H_{\mathit{eff}}^{2q} = \frac{1}{2}\hbar\omega_q^{(1)}\sigma_z^{(1)} + \frac{1}{2}\hbar\omega_q^{(2)}\sigma_z^{(2)} + \hbar\frac{g_1 g_2\left(\omega_q^{(1)} - \omega_r + \omega_q^{(2)} - \omega_r\right)}{2(\omega_q^{(1)} - \omega_r)(\omega_q^{(2)} - \omega_r)}\left(\sigma_{(1)}^-\sigma_{(2)}^+ + \sigma_{(1)}^+\sigma_{(2)}^-\right) \tag{20.9}$$

This effective two-qubit Hamiltonian captures the interaction between the qubits mediated through the cavity. The above Hamiltonian can be extended to any number of qubits in a straightforward fashion, simply by considering the physical connectivity of devices.

20.2.7 Single-Qubit Operations

Now that we explained the Hamiltonian for our specific system, we can describe how to perform operations on the qubits. The simplest and most straightforward operations are single qubit operations. This involves manipulation of the state of an individual qubit with no other qubits affected. In the case of transmon qubits, this is typically achieved by applying microwave pulse drives at the fundamental frequency of the qubit transition. If the ground and excited states of the transmon are represented by the kets $|g\rangle$ and $|e\rangle$, then a general pure state of the transmon is represented in the following form:

$$|\psi\rangle = \cos\frac{\theta}{2}|g\rangle + e^{i\phi}\sin\frac{\theta}{2}|e\rangle \tag{20.10}$$

Here θ is the angle measured from the *z*-axis of the Bloch sphere,[25] and ϕ is the phase measured from the *x*-axis on the *xy* plane of the Bloch sphere. By appropriately choosing the phase and amplitude of the microwave drive pulse, and by adjusting the duration of the microwave drive pulse, it is possible to realize arbitrary rotations around *x*, *y*, and *z* axes of the Bloch sphere and derive any desired pure state of a transmon qubit.

20.2.8 Two-Qubit Operations

Conventional single-qubit operations are always performed using microwave drives, which are pulsed to achieve the desired rotation. On the contrary, two-qubit operations are implemented in many different ways depending on the underlying type of transmon qubit, the nature of the coupling between qubits, and the characteristics of the gate desired.[29] Mainly, two classes of two-qubit operations are considered: operations for tunable qubits and operations for fixed-frequency (non-tunable) qubits. Non-tunable systems use specially constructed level spectra along with microwave drives to induce Stark shifts and thereby implement entangling gates. While the flux-tunable gates have faster gate times, the qubits are left vulnerable to lower coherence times. Non-tunable qubits generally have better coherence times but suffer from typically longer gate times.

A few examples for flux-tuning–based gates are the controlled-Phase (cPhase) gate,[28,30,31] the imaginary-SWAP (iSWAP) operation,[32] and the $\sqrt{\text{iSWAP}}$ gate.[33,34] In this chapter, we will describe the implementation of the two-qubit cPhase gate.[28]

The two-qubit state is described using Dirac notation as $|q_1 q_2\rangle$, where q_1, q_2 represent the state of the two transmon qubits. Then, the eigenstates in the computational subspace are given by $\{|00\rangle, |01\rangle, |10\rangle, |11\rangle\}$. For typical non-entangling operations, including single-qubit operations, the collection of the quantum phases for the different eigenstates follows the rule $\phi_{11} = \phi_{10} + \phi_{01}$. However, it is possible to utilize the noncomputational higher states of the transmons to impart a phase such that the latter rule is broken. When the characteristic frequency of the higher frequency qubit is lowered, the $|11\rangle$ level approaches an avoided energy level crossing with $|02\rangle$ due to cavity mediation. This allows the state $|11\rangle$ to collect the phase as follows:

$$\phi_{11} = \phi_{10} + \phi_{01} - \int \zeta(t)\mathrm{d}t \tag{20.11}$$

where $\zeta = \omega_{10} + \omega_{01} - \omega_{11}$ is the frequency shift of the doubly excited state with respect to the sum of the individual qubits being excited. If $\int \zeta(t)\mathrm{d}t = (2n + 1)\,\pi$ where n is an integer, this mechanism realizes a cPhase gate. The strong frequency detuning will result in the accumulation of single-qubit phases as well, which will need to be corrected to implement a true cPhase gate corresponding to the operator in matrix form

$$\text{cPhase} = \begin{pmatrix} 1 & 0 & 0 & 0 \\ 0 & 1 & 0 & 0 \\ 0 & 0 & 1 & 0 \\ 0 & 0 & 0 & -1 \end{pmatrix}. \tag{20.12}$$

20.3 QUANTUM GATE DESIGN USING A SIMULATOR

20.3.1 Quantum Simulation

Here, we simulate the dynamics of a quantum system in a decoherence-free subspace where we consider pure state simulation. If we want to consider noise, decoherence, and mixed states, a master equation simulation using a density matrix formalism is usually preferred.[25] The time evolution of the state of a quantum system is described by a time-dependent Schrödinger equation or a related equation (e.g., Liouville–von Neumann equation or Lindblad equation). In a decoherence-free subspace, knowing the initial state of the system and the Hamiltonian, the solution to the Schrödinger equation is as follows:

$$|\psi(t)\rangle = \hat{U}|\psi(t_0)\rangle \tag{20.13}$$

Here, the state $|\psi(t)\rangle$ represents the vector for the probability amplitudes of different eigenstates, and $\hat{U}$ is the time evolution operator. $\hat{U}(t, t_0)$ is a unitary transformation operator, which maps the initial state at time t_0, to the final state at time t. $\hat{U}(t, t0) = e^{-\frac{i\hat{H}t}{\hbar}}$, where $\hat{H}$ is the Hamiltonian operator that describes the energy of the system.

Any time-dependent Hamiltonian that can be decomposed to m local interactions can be written as a summation of m local Hamiltonians.[35,36]

$$\hat{H} = \hat{H}_1 + \hat{H}_2 + \cdots + \hat{H}_m \tag{20.14}$$

This Hamiltonian can be efficiently simulated using a universal quantum computer.[37] If we decompose the Hamiltonian to m local noncommuting Hamiltonians, we can estimate the term $e^{\frac{-i}{\hbar}\hat{H}(t)t}$ using Trotterization as below[35]:

$$e^{\frac{-i}{\hbar}\hat{H}t} \approx (e^{\frac{-i}{\hbar}\hat{H}_1\frac{t}{n}} e^{\frac{-i}{\hbar}\hat{H}_2\frac{t}{n}} \ldots e^{\frac{-i}{\hbar}\hat{H}_m\frac{t}{n}})^n \tag{20.15}$$

where $n \to \infty$, i.e., the accuracy of this estimation is increased by choosing very small Trotter steps (t/n). In this method, we approximate the time evolution operator as $\hat{U}(t, t_0) = \hat{U}(t, t_{n-1}) \ldots \hat{U}(t, t_1)\hat{U}(t, t_0), t_0 < t_1 < \cdots < t_{n-1} < t$, where at each time interval $(\frac{t}{n})$; the Hamiltonian is considered piecewise constant. The Trotterization methodology for quantum simulation can also be done in a classical computer for simulating the dynamics of a small quantum system, since simulation of quantum systems with many qubits using classical computational resources is limited by the memory and computational power.

The Hilbert space describing a single qubit can be specified by a 2×2 matrix. In general, the Hilbert space of an n qubit system can be described by a $2^n \times 2^n$ matrix. To solve the Schrödinger equation, we need to construct all the required operators in the Hilbert space of the system. Operators (including the Hamiltonian) can be represented as matrices and states as vectors. In order to solve the Schrodinger equation, one needs to calculate the exponentiation of a matrix at each Trotter step as described in Equation 20.15. There are many ways of calculating the exponentiation of a matrix; from approximation theory and series methods, ordinary differential equation methods, polynomial methods, to matrix decomposition methods, splitting methods, and Krylov space methods.[38] Depending on the matrix characteristics and programming environment and available linear algebra libraries, one may choose the method of calculating exponentiation of a matrix. We used MATLAB for our simulations and used the `expm()` function, which uses a scaling and squaring algorithm with a Pade approximation. For convenience, we set the Planck's constant $h = 1$ in our simulations $(\hbar = 1/2\pi)$.

During time evolution, we keep track of the state evolution of the system by storing the state of the system at each Trotter step. Suppose we define a small Trotter step δt, the state of the system at ith Trotter step $|\psi(t_i)\rangle$ can be calculated as follows:

$$|\psi(t_i)\rangle = \hat{U}_i|\psi(t_{i-1})\rangle \tag{20.16}$$

where $\hat{U}_i = e^{-i\hat{H}_i\delta t/\hbar}$ is the unitary transformation operator at time step t_i, $\hat{H}_i$ is the Hamiltonian operator at time step t_i, and $|\psi(t_{i-1})\rangle$ is the state at time step t_{i-1}. During evolution, we consider the full state space of the effective Hamiltonian to calculate $\hat{U}_i$. The dimension of the Hilbert space for the effective Hamiltonian is $3^4 \times 3^4$ since we have four transmons and consider three energy levels for each transmon, the state vector $|\psi(t_i)\rangle$ has a size of 3^4. Assuming that the system does not exit the computational subspace, we use a projection operator $\hat{P}$ to project 81×81 $\hat{U}_i$ to 16×16 $\hat{U}_i^{\text{comp}}$ in the computational subspace, when calculating the intermediate result states. The projected unitary operator $\hat{U}_i^{(\text{comp})}$ in the computational subspace is obtained as follows:

$$\hat{U}_i^{(\text{comp})} = \hat{P}\hat{U}_i\hat{P}^\dagger \tag{20.17}$$

In order to construct the projector $\hat{P}$, which is a 16×81 matrix, first we form a list of all the computational states of U_i. This list is $L = [1, 2, 4, 5, 10, 11, 13, 14, 28, 29, 31, 32, 37, 38, 40, 41]$ where the value of each element in the list L represents a computational state (a column/row) in $\hat{U}_i$. For $\hat{P}(j, k)$ matrix with rows $j = 1, 2, \ldots, 16$, and columns $k = 1, 2, \ldots, 81$, $\hat{P}(j, k) = 1 \quad \text{for} \quad k = L(j)$, otherwise, $\hat{P}(j, k) = 0 \quad \text{for} \quad k \neq L(j)$.

The state vector in the computational subspace is calculated at each time step as follows:

$$|\psi^{(\text{comp})}(t_i)\rangle = \hat{U}_i^{(\text{comp})}|\psi^{(\text{comp})}(t_{i-1})\rangle \tag{20.18}$$

We keep track of all the intermediate result states $|\psi^{(\text{comp})}(t_i)\rangle$ during the time and plot the behavior of the system using the stored states.

20.3.2 Designing a Two-Qubit Gate Set for Transmons in cQED

We need to be able to realize two-qubit entangling gates in order to perform universal quantum computation. Here, we describe the design of two-qubit cPhase, controlled-$\sqrt{\text{NOT}}$ (c$\sqrt{\text{NOT}}$), and c$\sqrt{\text{NOT}}^{\dagger}$ gates.

Redefining the ground-state energy in the TC Hamiltonian for multilevel transmons, the effective Hamiltonian for two transmons can be written as[39]:

$$\tilde{H}_{\text{TC}} = \hbar\omega_r\left(a^{\dagger}a\right) + \sum_{j}^{2}\left[\frac{1}{2}\hbar\sum_{m}\omega_m^{(j)}|m\rangle_{(j)}\langle m| + \hbar\sum_{n} g_{n,n+1}^{(j)}\left(a|n+1\rangle_{(j)}\langle n| + a^{\dagger}|n\rangle_{(j)}\langle n+1|\right)\right] \tag{20.19}$$

The last term in the above Hamiltonian describes the coupling between the transmons mediated through the resonator. It can be replaced by the direct coupling term as follows:

$$\tilde{H}_{\text{coupling}} = \hbar\sum_{j_1,j_2}\sqrt{j_1+1}\sqrt{j_2+1}J_{j_1,j_2}\left(|j_1,j_2+1\rangle\langle j_1+1,j_2| + |j_1+1,j_2\rangle\langle j_1,j_2+1|\right) \tag{20.20}$$

where J_{j_1,j_2} is the direct coupling between level j_1 from the first transmon and level j_2 from the second transmon. The coupling coefficients for different allowed energy levels of two transmons can be calculated as[39]:

$$J_{j_1,j_2} = \frac{g_1 g_2\left(\omega_q^{(1)} + \delta_1 j_1 - \omega_r + \omega_q^{(2)} + \delta_2 j_2 - \omega_r\right)}{2\left(\omega_q^{(1)} + \delta_1 j_1 - \omega_r\right)\left(\omega_q^{(2)} + \delta_2 j_2 - \omega_r\right)} \tag{20.21}$$

where δ_1 and δ_2 are the anharmonicity values associated with transmons 1 and 2, respectively. To realize a two-qubit controlled-phase gate, we need to consider the energy levels up to the second excitation manifold of each qubit. All the couplings except for those involving $|22\rangle$ (which results in a total of four system excitations) are considered. Then the effective Hamiltonian for two transmons can be represented by the following matrix. The order of the levels in the matrix is $\{|00\rangle, |01\rangle, |02\rangle, |10\rangle, |11\rangle, |12\rangle, |20\rangle, |21\rangle, |22\rangle\}$.

$$\hat{H} = \begin{bmatrix} 0 & 0 & 0 & 0 & 0 & 0 & 0 & 0 & 0 \\ 0 & \tilde{\omega}_1^{(2)} & 0 & J_{0,0} & 0 & 0 & 0 & 0 & 0 \\ 0 & 0 & \tilde{\omega}_2^{(2)} & 0 & \sqrt{2}J_{0,1} & 0 & 0 & 0 & 0 \\ 0 & J_{0,0} & 0 & \tilde{\omega}_1^{(1)} & 0 & 0 & 0 & 0 & 0 \\ 0 & 0 & \sqrt{2}J_{0,1} & 0 & \tilde{\omega}_1^{(1)} + \tilde{\omega}_1^{(2)} & 0 & \sqrt{2}J_{1,0} & 0 & 0 \\ 0 & 0 & 0 & 0 & 0 & \tilde{\omega}_1^{(1)} + \tilde{\omega}_2^{(2)} & 0 & 2J_{1,1} & 0 \\ 0 & 0 & 0 & 0 & \sqrt{2}J_{1,0} & 0 & \tilde{\omega}_2^{(1)} & 0 & 0 \\ 0 & 0 & 0 & 0 & 0 & 2J_{1,1} & 0 & \tilde{\omega}_2^{(1)} + \tilde{\omega}_1^{(2)} & 0 \\ 0 & 0 & 0 & 0 & 0 & 0 & 0 & 0 & \tilde{\omega}_2^{(1)} + \tilde{\omega}_2^{(2)} \end{bmatrix} \tag{20.22}$$

Here $\tilde{\omega}_{(j)}^{(k)}$ represents the angular frequency associated with the kth transmon at energy level j and is given[39] as follows:

$$\tilde{\omega}_j^{(k)} \equiv j\omega_q^{(k)} + \frac{\delta_k}{2}(j-1)j + \frac{jg_k^2}{\omega_q^{(k)} - \omega_r + (j-1)\delta_k} \tag{20.23}$$

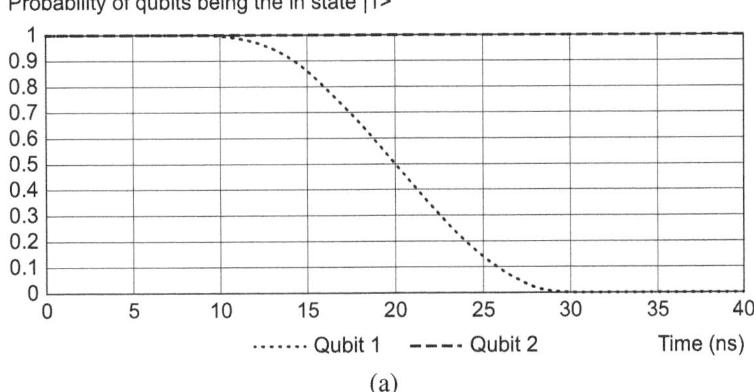

(a)

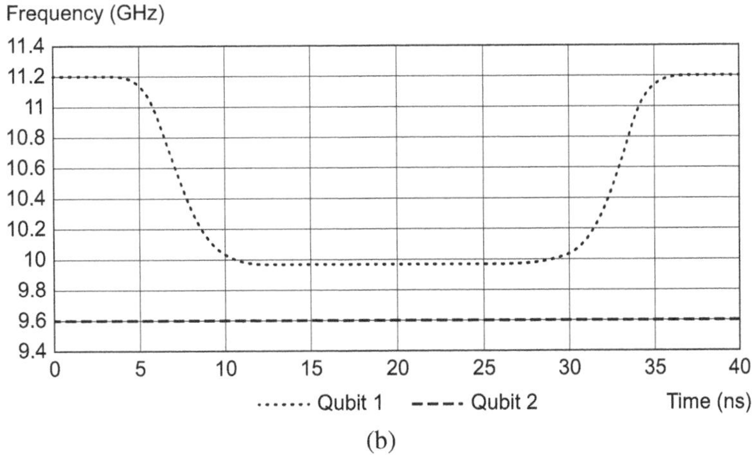

(b)

FIGURE 20.2 A controlled rotation around the Z-axis for 180°, sandwiched between two Hadamard operations on the first qubit, results in a CNOT(Q2, Q1) gate. (a) Probabilities of qubits Q1 and Q2, respectively, being in the state $|1\rangle$. (b) The required fundamental frequency (GHz) changes for qubits.

Since the couplings are small compared to the qubit transition frequencies, they can be considered as perturbations to the system. A Schrieffer–Wolff transformation is used to perform a full diagonalization of the Hamiltonian to second order in J.[39,40]

To design a two-qubit cPhase gate, the frequency difference between the two qubits is adjusted to reach the avoided level crossing region between levels $|11\rangle$ and $|02\rangle$, where couplings between levels introduce a phase shift as described in Section 20.2.8. When the desired phase is collected, the transmons are brought back to the original frequencies. The behavior of the system depends on the speed of the frequency change operation. The operations are performed in the adiabatic regime, where the frequency of qubits should change continuously with a smoothly shaped pulse when approaching the avoided crossing region. It should be adiabatic with respect to the timescale given by the level splitting.[28] Using our simulator, we can design the required pulses to achieve the controlled-rotation gates between two transmons.

We consider two transmons with fundamental frequencies of 11.2 and 9.6 GHz, with the resonator frequency fixed at 6.5 GHz. We assume both transmons have the same anharmonicity δ equal to −300 MHz, and the transmon–resonator coupling g is 200 MHz. The simulation result of a cPhase operation is shown in Figure 20.2. During simulation, U is calculated and subsequently single-qubit

Hadamard operations are directly applied before and after U at each time step. This composite gate then operates on the state at each time step to observe the evolution shown in Figure 20.2. The composite operation is necessary to ensure mapping of the phase change into an amplitude change.

20.3.3 Designing $c\sqrt{\text{NOT}}$ and $c\sqrt{\text{NOT}}^{\dagger}$ Gates

The design of quantum circuits like those from binary reversible gates starts from unitary and especially permutative matrices. A permutative matrix can be decomposed in many ways into a set of gates from a library. Example gates in such a library include unitary gates such as CNOT and CCNOT (Toffoli), $c\sqrt{\text{NOT}}$, $c\sqrt{\text{NOT}}^{\dagger}$, $cc\sqrt{\text{NOT}}$, and $cc\sqrt{\text{NOT}}^{\dagger}$. It was found that binary reversible quantum gates with a large number of inputs, or circuits such as arithmetic quantum circuits used in Grover's algorithm,[7] can be efficiently realized from unitary gates from the library described before. In some literature, $\sqrt{\text{NOT}}$ and $\sqrt{\text{NOT}}^{\dagger}$ are introduced by symbols V and $V^{\dagger}$, and $c\sqrt{\text{NOT}}$ and $c\sqrt{\text{NOT}}^{\dagger}$ are denoted as cV and $cV^{\dagger}$, respectively.[5] The unitary transformations of V and $V^{\dagger}$, operations are given as follows:

$$V = \sqrt{\text{NOT}} = \frac{1+i}{2}\begin{bmatrix} 1 & -i \\ -i & 1 \end{bmatrix} \tag{20.24}$$

$$V^{\dagger} = \sqrt{\text{NOT}}^{\dagger} = \frac{1-i}{2}\begin{bmatrix} 1 & i \\ i & 1 \end{bmatrix} \tag{20.25}$$

In gate design, we would like to set up the control parameters of the system such that the unitary transformation in the computational subspace is as close as possible to our target gate transformation matrix. However, certain gates are more natural to particular physical systems. For example, in transmons realizing a cPhase gate is more convenient than a native CNOT gate. Therefore, it is better to decompose gates such as CNOT to a set of native gates. It is proven in Barenco et al.[3] that for any 2×2 special unitary matrix U with the determinant equal to unity, the controlled-U operation can be decomposed as shown in Figure 20.3. In the quantum circuit shown in Figure 20.3, each qubit is represented by a line, and the sequence of gate operations on qubits are performed from left to right.

Thus, CNOT, $c\sqrt{\text{NOT}}$, and $c\sqrt{\text{NOT}}^{\dagger}$ gates can be realized using controlled-phase gates where the target qubit will be sandwiched between two Hadamard gates before and after the controlled-phase gate. A CNOT gate can be realized using a controlled-$R_z(\pi)$ gate since CNOT $= HcR_z(\pi)H$. As shown in Figure 20.4, this idea can be generalized, in particular, $c\sqrt{\text{NOT}} = H.\ cR_z(\frac{\pi}{2}).\ H$ and $c\sqrt{\text{NOT}}^{\dagger} = H.\ cR_z(\frac{3\pi}{2}).\ H$. Hence, by using $cR_z(\frac{\pi}{2})$ and $cR_z(\frac{3\pi}{2})$ gates, we can design $c\sqrt{\text{NOT}}$ and $c\sqrt{\text{NOT}}^{\dagger}$ gates.

The simulation results for $cR_z(\frac{\pi}{2})$ and $cR_z(\frac{3\pi}{2})$ gates for transmons are shown in Figures 20.5 and 20.6, respectively. We designed these gates by considering a fixed gate time of 40 ns. Once the two transmons reach the avoided crossing region, the wait time is adjusted to ensure a phase collection of 90° for the $cR_z(\frac{\pi}{2})$ gate and 270° for the $cR_z(\frac{3\pi}{2})$ gate.

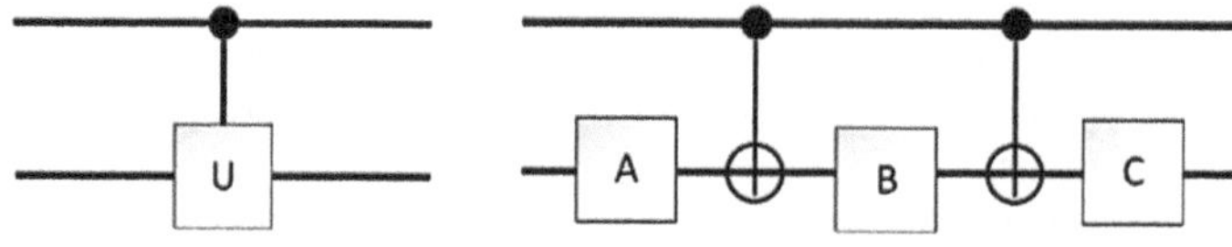

FIGURE 20.3 For any 2×2 special unitary matrix U, the controlled-U operation can be decomposed to a sequence of special unitary single qubit gates A, B, and C where $ABC = I$, and if the first qubit is in state $|1\rangle$, $A\sigma_x B\sigma_x C = U$ is applied to the second qubit.

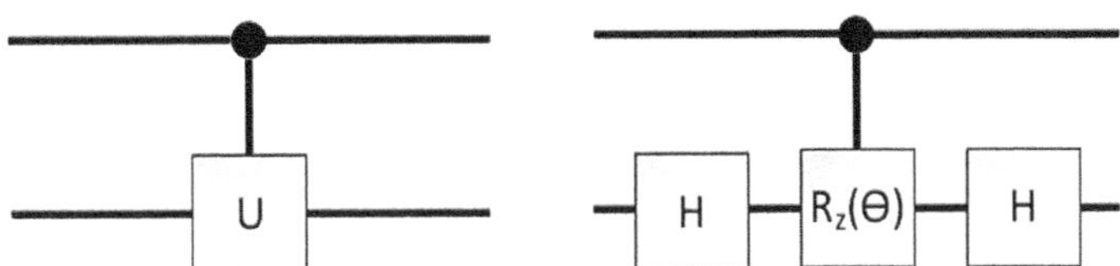

FIGURE 20.4 A controlled-unitary gate U is realized by decomposing it into a sequence of Hadamard and a controlled-rotation gate around the Z-axis since $U = HR_z(\theta)H$.

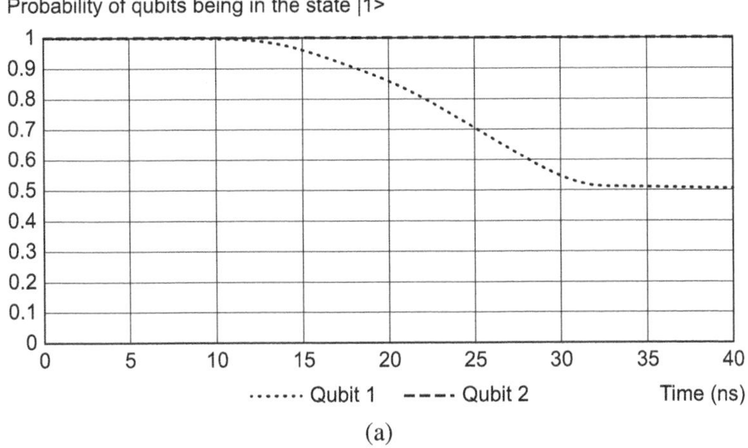

(a)

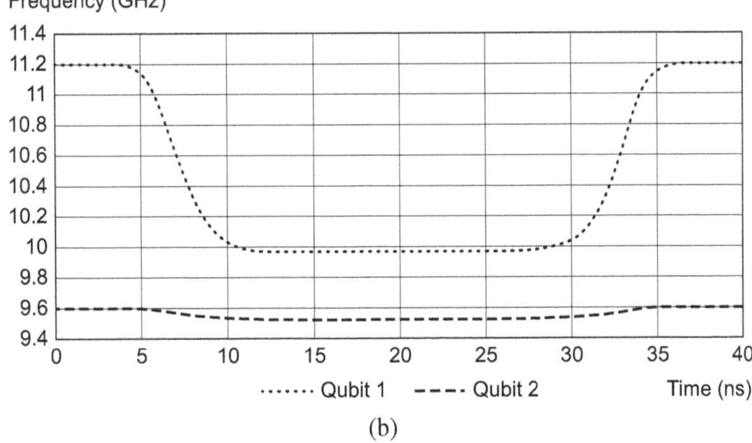

(b)

FIGURE 20.5 A controlled rotation around the Z-axis for 90° $cR_z(\frac{\pi}{2})$, sandwiched between two Hadamard operations on the first qubit, results in a c$\sqrt{\text{NOT}}$(Q2, Q1) gate. (a) Probabilities of qubits Q1 and Q2, respectively, being in the state $|1\rangle$. (b) The required fundamental frequency changes in GHz for qubits.

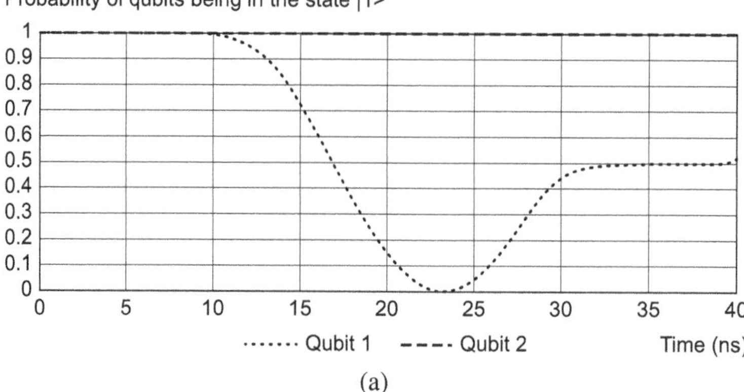

(a)

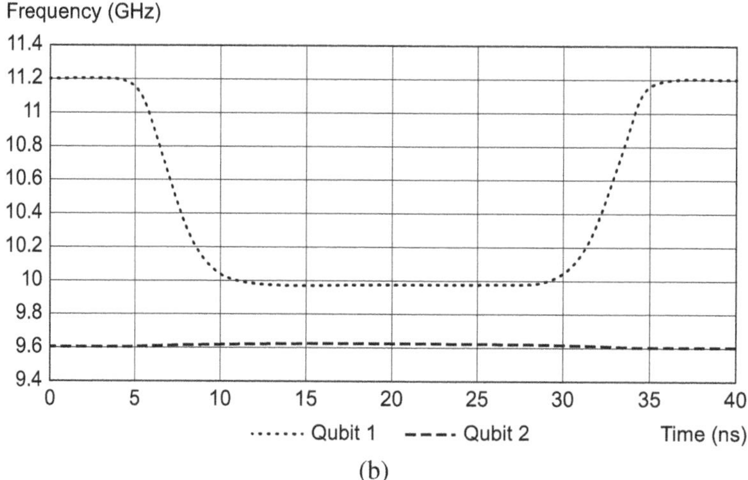

(b)

FIGURE 20.6 A controlled rotation around the Z-axis for 270° $cR_z(\frac{3\pi}{2})$, sandwiched between two Hadamard operations on the first qubit, results in a $c\sqrt{\text{NOT}}^{\dagger}$ (Q2, Q1) gate. (a) Probabilities of qubits Q1 and Q2, respectively, being in the state $|1\rangle$. (b) The required fundamental frequency changes in GHz for qubits.

20.4 REALIZING A FULL ADDER QUANTUM CIRCUIT

20.4.1 Quantum Logic Synthesis

Quantum logic synthesis consists of developing methods and algorithms to decompose larger matrices to elementary gates from predefined gate libraries in such a way that the quantum circuit has the smallest total gate cost, the maximum speed, and the reduced number of ancilla qubits. The implementation of operations such as Peres,[41] Toffoli, and Fredkin gates, as well as adders, multi-controlled AND gates, symmetric functions, and arithmetic functions based on controlled-$\sqrt[n]{\text{NOT}}$ have been proposed.[5,42–47] Here, we demonstrate a methodology to design small quantum circuits of this type in cQED transmon systems and how to map the circuit to the physical layout of the qubits. The principle of using only two qubit gates is represented in the full adder design shown in Figure 20.7.[47] In the figures, for simplicity, we use the notations of V and $V^{\dagger}$, for $\sqrt{\text{NOT}}$ and $\sqrt{\text{NOT}}^{\dagger}$, respectively.

In the design flow of the classical devices such as CMOS circuits in very large-scale integration (VLSI), the logic synthesis is a stage (abstraction layer) where a gate-level netlist is provided based

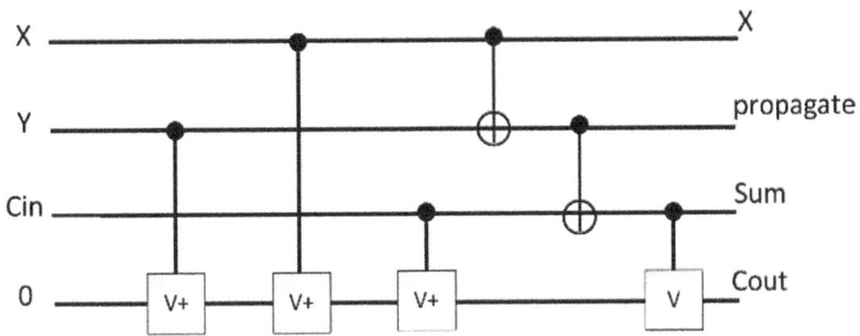

FIGURE 20.7 A full adder quantum circuit using cV ($c\sqrt{NOT}$), $cV^{\dagger}$ ($c\sqrt{NOT}^{\dagger}$), and CNOT gates.[47]

on register transfer level (RTL) design, and it happens before the physical design stage. Currently, the quantum logic synthesis is tightly bound to the research area of optimizing physical design (placement and routing) of the quantum circuit on the quantum chip layout. Therefore, in the next sections, we perform logic synthesis considering the quantum chip layout in a one- and two-dimensional space.

20.4.2 Quantum Circuit Implementation on a One-Dimensional Layout

To implement the quantum circuit in Figure 20.7, let us consider a chip consisting of four transmons, which are coupled through resonators.[48] Note that when operating a two-qubit cPhase gate between qubits, the frequency of either qubit must not cross any other fundamental frequency of nonparticipating transmons. This criterion must be ensured between all coupled transmons.

An example for coupling multiple transmons is shown in Figure 20.8, where four transmons are coupled in a linear array.

In our simulation for the full adder quantum circuit, we consider the fundamental frequencies of the transmons Q1–Q4 to be 8.4, 9.6, 11.2, and 12.6 GHz, respectively. Figure 20.9 shows the result of a simulation on four transmons.

In the NN architectures, in order to realize a two-qubit gate between two non-neighbor qubits, a SWAP operation is used to transfer information along the connected chain to yield the desired operation.

Notice in Figure 20.10 that the first $c\sqrt{NOT}^{\dagger}$ gate is operating between Q1 and Q3, which are not neighbors due to the current assignment. Therefore, a $c\sqrt{NOT}^{\dagger}$ (Q3, Q1) is physically realized as

$$c\sqrt{NOT}^{\dagger}(Q_3, Q_1) \equiv \text{SWAP}(Q_3, Q_2) \rightarrow c\sqrt{NOT}^{\dagger}(Q_2, Q_1) \rightarrow \text{SWAP}(Q_3, Q_2).$$

Similarly, $c\sqrt{NOT}^{\dagger}(Q_4, Q_1)$ is physically performed as

$$c\sqrt{NOT}^{\dagger}(Q_4, Q_1) \equiv \text{SWAP}(Q_4, Q_3) \rightarrow \text{SWAP}(Q_3, Q_2) \rightarrow c\sqrt{NOT}^{\dagger}(Q_2, Q_1) \rightarrow \text{SWAP}(Q_3, Q_2) \rightarrow \text{SWAP}(Q_4, Q_3)$$

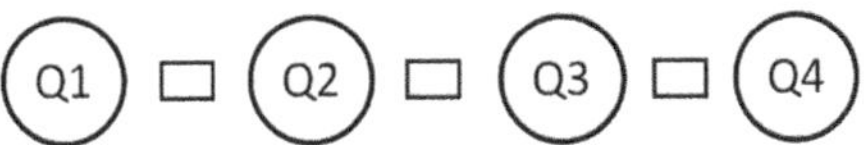

FIGURE 20.8 A one-dimensional nearest neighbor (NN) layout of transmons in cQED. The squares represent the coupling element, and circles represent the transmon qubits.

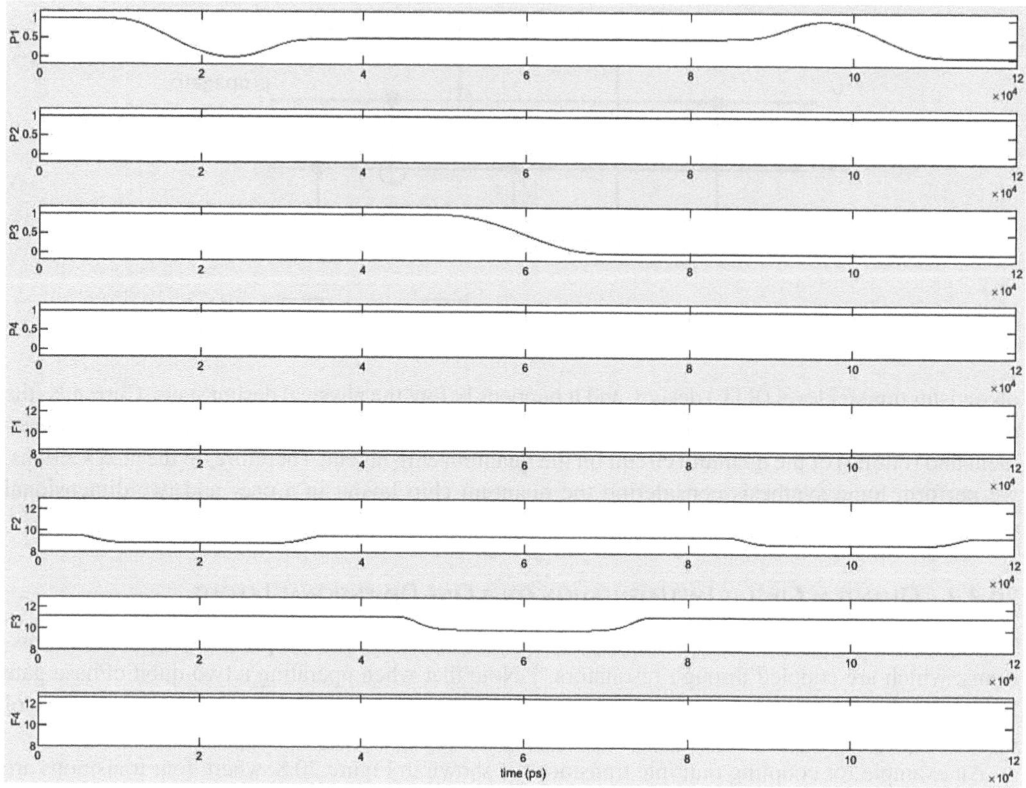

FIGURE 20.9 A $c\sqrt{\text{NOT}}^{\dagger}$ (Q2, Q1) gate is applied in the first and the third 40-ns time intervals, and a CNOT(Q2, Q3) gate is applied in the second 40-ns time interval. From top to bottom: P1–P4 represent the probability of qubits 1–4 being in state $|1\rangle$, respectively. F1–F4 show the applied fundamental frequencies (GHz) of qubits 1–4, respectively. The horizontal axes show the time in picoseconds.

This would require six extra SWAP gates to realize our full adder quantum circuit. However, it is possible to find a more optimal assignment of qubits considering connectivity of the logical quantum circuit. As input 0 has a direct interaction with all other three qubits, it is advantageous to assign it to either Q2 or Q3. It is also advantageous to assign X and C_{in} for the extremities, since they have no mutual gate. Using an efficient qubit assignment (see Figure 20.11), the number of SWAP gates is reduced from six to four.

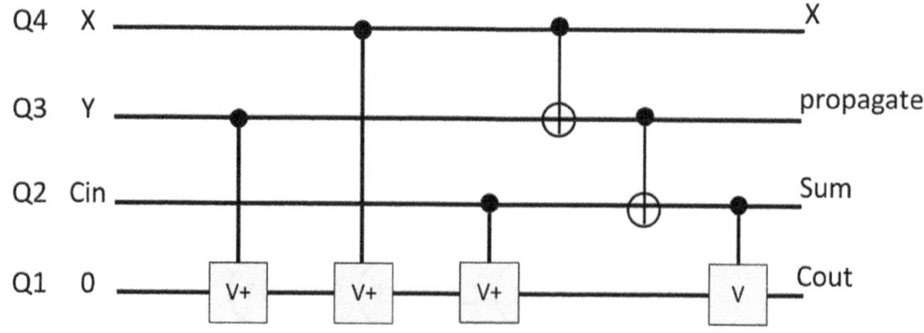

FIGURE 20.10 A placement option for realizing the full adder quantum circuit on the layout shown in Figure 20.8 is assigning 0, C_{in}, Y, and X to Q1, Q2, Q3, and Q4, respectively. The placement is simply assigning logical variables to qubits on the chip.

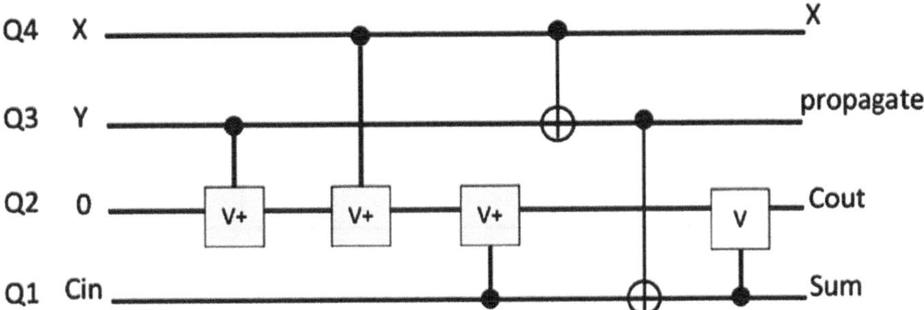

FIGURE 20.11 A placement option for realizing the full adder quantum circuit on the layout shown in Figure 20.8 is assigning C_{in}, 0, Y, and X to Q1, Q2, Q3, and Q4, respectively.

SWAP is not a native gate available in transmon systems. At a substantial operations cost, a SWAP gate can be decomposed to CNOT gates as shown in Figure 20.12.

The full adder circuit realized in terms of all NN interactions is shown in Figure 20.13.

The Quantum Assembly (QASM) code[49] for the full adder quantum circuit depicted in Figure 20.13 is as follows:

```
def    CV,1,'\sqrt{X}'
def    CVd,1,'{\sqrt{X}}^\dagger'
qubit q1
qubit q2,0
qubit q3
qubit q4
CVd    q3,q2
cnot   q4,q3
cnot   q3,q4
cnot   q4,q3
CVd    q3,q2
cnot   q4,q3
cnot   q3,q4
cnot   q4,q3
CVd    q1,q2
cnot   q4,q3
cnot   q2,q3
cnot   q3,q2
cnot   q2,q3
cnot   q2,q1
cnot   q2,q3
cnot   q3,q2
cnot   q2,q3
CV     q1,q2
```

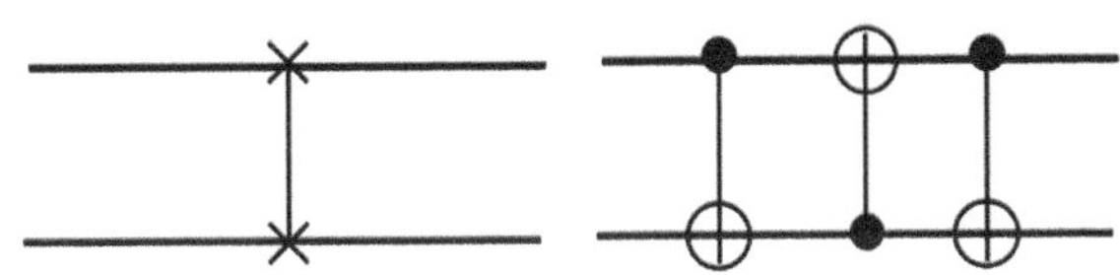

FIGURE 20.12 A SWAP gate can be realized using three CNOT gates.

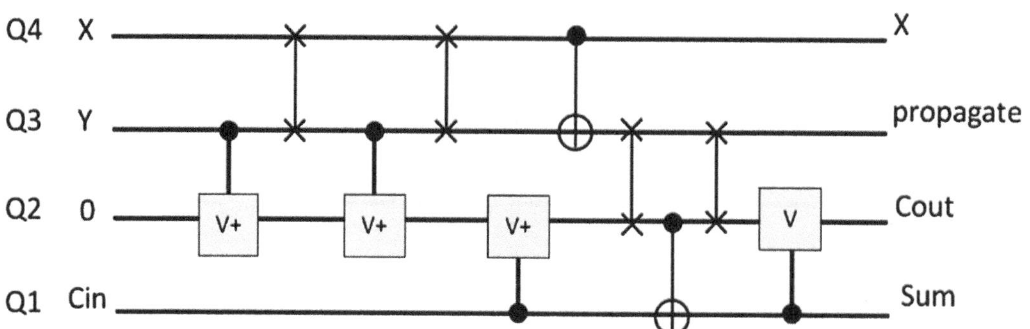

FIGURE 20.13 A full adder quantum circuit realization for transmons in the one-dimensional NN layout shown in Figure 20.8.

where the 12 extra CNOT gates (shown in bold) are from decomposing the SWAP operations. The total quantum cost (the number of two-qubit native gates)[6] of this circuit is 18.

This circuit can be simplified to reduce the number of SWAP gates and extra CNOT gates. In Figure 20.11, the first three $cV^{\dagger}$ ($c\sqrt{\text{NOT}}^{\dagger}$) gates commute with each other, and the order of applying them does not change the outcome of the quantum circuit (Figure 20.14).

Now we can change the placement of qubits with inputs X and Y such that Q1, Q2, Q3, and Q4 correspond to C_{in}, 0, X, and Y, respectively (see Figure 20.15).

In general, when the states of two qubits are swapped to realize a controlled-unitary operation, where only one of the qubits has a control role as shown in Figure 20.16, the same functionality can be realized using only four CNOT gates, as illustrated in Figure 20.17.

The pattern from Figure 20.16 is seen in the circuit of Figure 20.15 between positions 6 and 8. Therefore, we can reduce two more CNOT gates and realize the full adder circuit using only seven extra CNOT gates with a total cost of 13. The final full adder quantum circuit is shown in Figure 20.18.

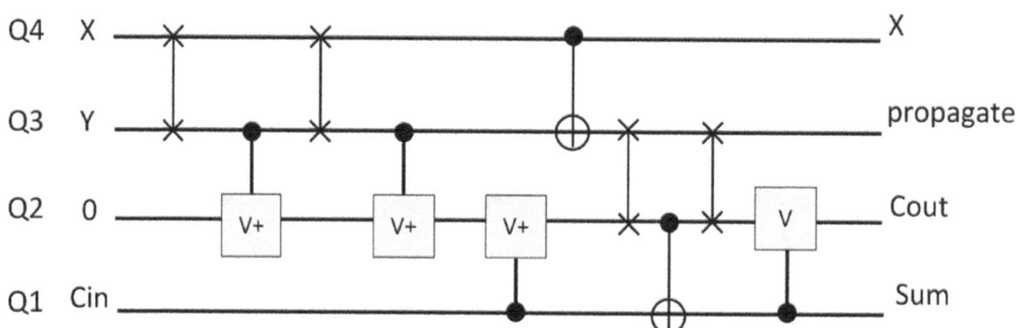

FIGURE 20.14 The full adder quantum circuit where the order of the first and the second $cV^{\dagger}$ ($c\sqrt{\text{NOT}}^{\dagger}$) gates are changed.

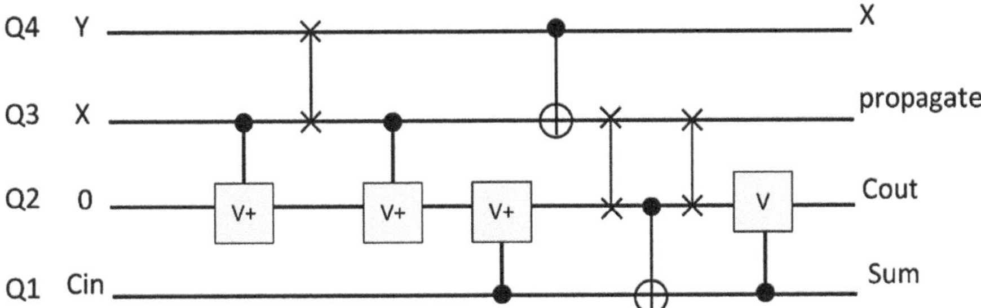

FIGURE 20.15 The full adder circuit where *X* and *Y* are assigned in reverse order (compared to Figure 20.14) to reduce one SWAP gate in the beginning of the circuit.

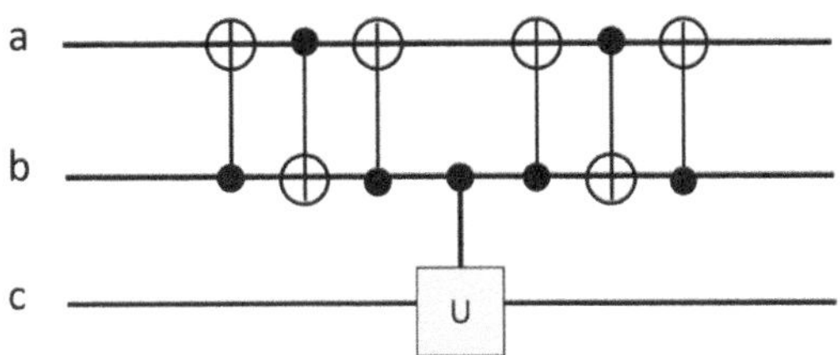

FIGURE 20.16 A general pattern where two qubits exchange their states to perform a controlled-unitary operation between two non-neighbor qubits.

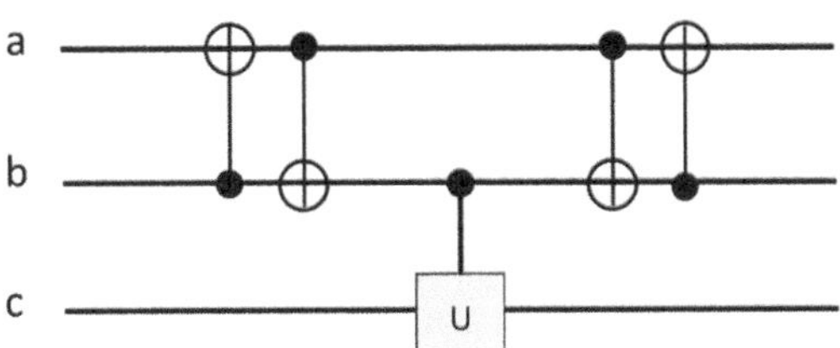

FIGURE 20.17 It is straightforward to show that any quantum subcircuit with the pattern shown in Figure 20.16 can be optimized to this quantum subcircuit with two fewer CNOT gates.

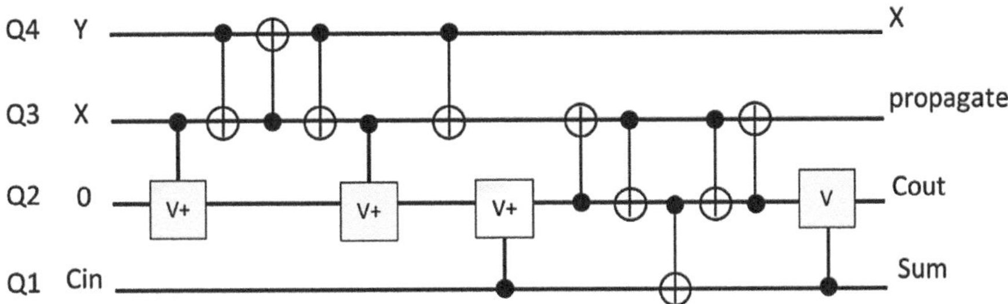

FIGURE 20.18 The full adder circuit following some logic synthesis and optimizing for the one-dimensional layout in Figure 20.8.

In Figure 20.18, note that the gates in positions 6 and 7 can be performed simultaneously to improve the latency of the circuit. The QASM code for the circuit shown in Figure 20.18 is as follows:

```
def    CV,1,'\sqrt{X}'
def    CVd,1,'{\sqrt{X}}^\dagger'
qubit q1
qubit q2,0
qubit q3
qubit q4
CVd    q3,q2
cnot   q4,q3
cnot   q3,q4
cnot   q4,q3
CVd    q3,q2
cnot   q4,q3
CVd    q1,q2
cnot   q2,q3
cnot   q3,q2
cnot   q2,q1
cnot   q3,q2
cnot   q2,q3
CV     q1,q2
```

The functionality of the quantum circuit shown in Figure 20.18 matches with the functionality of the circuit shown in Figure 20.13, and both are confirmed by simulation. This example illustrates how the abstract quantum logic synthesis, the one-dimensional logic synthesis, and the final placement to the actual layout of quantum chip are related. In the next section, we propose a two-dimensional qubit layout that requires no SWAP operation to implement a full adder quantum circuit.

20.4.3 Quantum Circuit Implementation on a Two-Dimensional Layout

Suppose we have a two-dimensional NN layout of transmons with resonator couplings as shown in Figure 20.19. In the two-dimensional layout, the frequencies of the resonators between each transmon pair are set differently so that they will not be resonant with each other. Since a resonator is shared only between a unique pair of qubits, turning on an interaction may cause only weak unwanted interactions between neighbor qubits that we can neglect to lowest order.

Here, if we place qubits as shown in Figure 20.20a–c, the quantum cost will be different. As it can be seen from the original full adder circuit in Figure 20.7, there is no interaction between X and C_{in}, while all other qubit pairs have direct interactions. Therefore, placement (a) is an optimal placement since the qubit pairs that are interacting directly are coupled directly in the layout as well, and there is no SWAP operation needed at all. In Figure 20.20, both (b) and (c) choices have equal cost and require two SWAP gates.

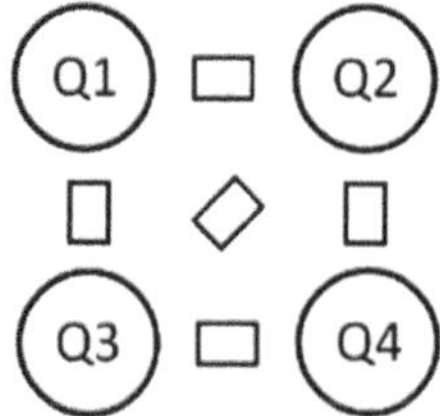

FIGURE 20.19 Two-dimensional layout for coupling four transmons Q1–Q4. The squares represent the resonators, and circles represent the transmon qubits.

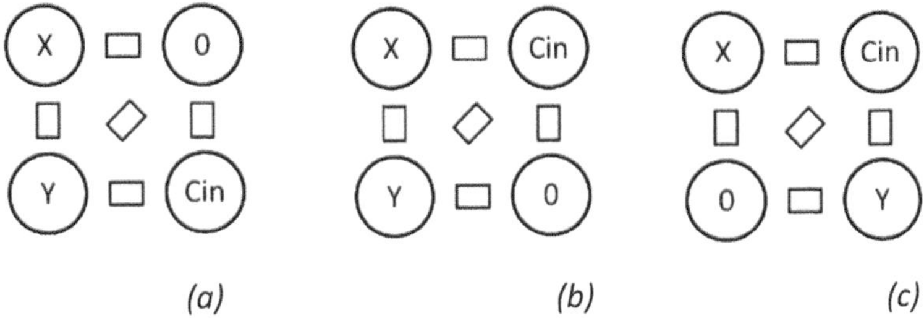

FIGURE 20.20 Some placement options to realize the circuit shown in Figure 20.7 on the layout shown in Figure 20.19.

If the optimal assignment shown in Figure 20.20a is chosen for the layout shown in Figure 20.19, the full adder circuit can be expressed without SWAP operations, resulting in a cost of 6. The QASM code for this placement is shown as follows:

```
def    CV,1,'\sqrt{X}'
def    CVd,1,'{\sqrt{X}}^\dagger'
qubit q1
qubit q2,0
qubit q3
qubit q4
CVd    q3,q2
CVd    q1,q2
CVd    q4,q2
cnot   q1,q3
cnot   q3,q4
CV     q4,q2
```

A more systematic way of finding the optimal physical design given the constraints of the physical system is by mapping the problem to a graph theory problem and using graph theory algorithms already being used in more mature fields such as VLSI physical design tools. For example, the physical layout of the qubits can be mapped to a *connectivity graph*, where vertices represent the qubits and edges represent the direct coupling between qubits. The connectivity graph for the layout in Figure 20.19 is given in Figure 20.21a. Similarly, the quantum circuit can be modeled as an *interaction graph* where vertices represent the qubits and edges represent the connection between qubits through the gates. The number of gates between two qubits can be specified as the weight on the edge between the two qubits. The interaction graph of the layout in Figure 20.7 is given in Figure 20.21b.

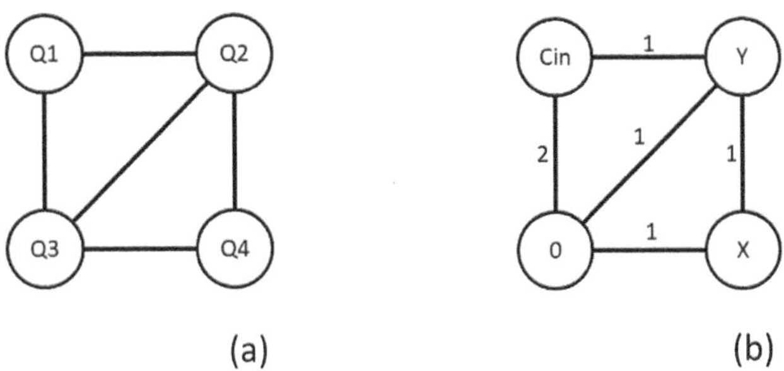

FIGURE 20.21 (a) The connectivity graph of the layout in Figure 20.19, (b) the interaction graph of the circuit in Figure 20.7.

Now the qubit assignment problem is mapped to a graph-embedding problem, where the interaction graph must be embedded into the connectivity graph such that the total distance between adjacent vertices in the interaction graph is minimized.[50] For this particular example, the optimal embedding is trivial and can be performed visually.

20.5 CONCLUSION

In this chapter, we show that Hamiltonian-level quantum simulators are very advantageous in gate design, logic synthesis, and physical design. We present a methodology to build a quantum simulator and use this simulator to design two-qubit controlled phase shift gates such as $cR_z(\pi)$, $cR_z(\frac{\pi}{2})$, and $cR_z(\frac{3\pi}{2})$. We used a full adder reversible logic circuit as an example to show constraints one may consider when implementing the circuit on an actual qubit chip. We showed how selecting different mappings to the qubit layout may affect the overall performance and cost of realizing a quantum logic circuit.

This study illustrates the need for computer-aided design (CAD) tools for physical design of quantum circuits with larger numbers of qubits, in order to optimize the placement and routing of qubits and to reduce the communication overhead introduced by SWAP gates. At some lower levels of abstraction, many CAD tools need to be optimized for specific quantum physical realizations. In modern CMOS technology design flow, some recent tools have developed an optimized combination of logic synthesis and physical layout design.[51] Similarly, a combined approach to logic synthesis and physical layout design based on regular structures was done for reversible logic, which is a model for permutative quantum circuits.[52] Perhaps, similar methodologies will be developed for quantum circuit design flow. As quantum devices mature and scale up, it will be increasingly important to create new quantum logic synthesis methods and physical design techniques that will be tailored to the specific constraints of emerging quantum devices.[50,53,54]

REFERENCES

1. J. Preskill (2018), Quantum computing in the NISQ era and beyond, *Quantum* 2, 79, doi: 10.22331/q-2018-08-06-79.
2. D. P. DiVincenzo (1997), "Topics in Quantum Computers", in Mesoscopic Electron Transport, ed. L. Kowenhove, G. Schon and L. Sohn.
3. A. Barenco, C. H. Bennett, C. H. Cleve, D. P. DiVincenzo, N. Margolus, N. Shor, T. Sleator, J. A. Smolin, and H. Weinfurter (1995), Elementary Gates for Quantum Computation, *Physical Review A*, 52, 3457–3467.
4. J. A. Smolin and D. P. DiVincenzo (1996), Five two-bit quantum gates are sufficient to implement the quantum Fredkin gate, Received 18 September 1995, American Physical Society, doi: 10.1103/PhysRevA.53.2855.
5. M. Szyprowski and P. Kerntopf (2013), Low Quantum Cost Realization of Generalized Peres and Toffoli Gates with Multiple-Control Signals, Proceedings of the 13th IEEE International Conference on Nanotechnology Beijing, China, August 5–8.
6. Soonchil Lee, Seong-Joo Lee, T. Kim, Jae-Seung Lee, J. Biamonte, and M. Perkowski. (2006), The cost of quantum gate primitives. *J. Multiple-Valued Logic Soft Comput.* 12. 561-573.
7. M. A. Nielsen and I. L. Chuang (2000), "Quantum Computation and Quantum Information."
8. International Roadmap for Devices and Systems (IRDS™) 2017 Edition, Beyond CMOS, https://irds.ieee.org/images/files/pdf/2017/2017IRDS_BC.pdf
9. https://irds.ieee.org/images/files/pdf/2017/1523037982875-IRDS_CryoE_HolmesDeBenedictis_v4b.pdf
10. G. Wendin, Quantum information processing with superconducting circuits: a review, Published 1 September 2017, 2017 IOP Publishing Ltd Reports on Progress in Physics, Volume 80, Number 10.
11. Y. Nakamura, Y. A. Pashkin, and J. S. Tsai, Coherent control of macroscopic quantum states in a single-Cooper-pair box. *Nature* 398, 786–788 (1999).
12. V. Bouchiat, D. Vion, P. Joyez, D. Esteve, and M. H. Devoret (1998), Quantum coherence with a single cooper pair. *Phys. Scr.* T76, 165.

13. X. Gu, A. F. Kockum, A. Miranowicz, Y. Liu, and F. Nori, Microwave photonics with superconducting quantum circuits. *Phys. Rep.* 718–719, 1–102 (2017).
14. M. H. Devoret and R. J. Schoelkopf (2013), Superconducting circuits for quantum information: an outlook. *Science* 339, 1169–1174.
15. J. Koch, T. M. Yu, J. Gambetta, A. A. Houck, D. I. Schuster, J. Majer, A. Blais, M. H. Devoret, S. M. Girvin, and R. J. Schoelkopf, Charge-insensitive qubit design derived from the Cooper pair box. *Phys. Rev. A* 76, 042319 (2007).
16. R. Barends, et al. (2013), Coherent Josephson qubit suitable for scalable quantum integrated circuits. *Phys. Rev. Lett.* 111, 080502.
17. H. Paik, D. I. Schuster, L. S. Bishop, G. Kirchmair, G. Catelani, A. P. Sears, B. R. Johnson, M. J. Reagor, L. Frunzio, L. Glazman, S. M. Girvin, M. H. Devoret, and R. J. Schoelkopf (2011), Observation of high coherence in Josephson junction qubits measured in a three-dimensional circuit QED architecture. *Phys. Rev. Lett.* 107, 240501.
18. A. Blais, R. S. Huang, A. Wallraff, S.M. Girvin, and R. J. Schoelkopf (2004), Cavity quantum electrodynamics for superconducting electrical circuits: an architecture for quantum computation. *Phys. Rev. A* 69, 062320.
19. A. Wallraff, D. I. Schuster, A. Blais, L. Frunzio, R.-S. Huang, J. Majer, S. Kumar, S. M. Girvin, and R. J. Schoelkopf, Strong coupling of a single photon to a superconducting qubit using circuit quantum electrodynamics. *Nature* 431, 162–167 (2004).
20. J. J. Sakurai and J. Napolitano (2017), Modern quantum mechanics. Cambridge University Press. doi:10.1017/9781108499996.
21. M. Reagor, W. Pfaff, Ch. Axline, R. W. Heeres, N. Ofek, K. Sliwa, E. Holland, Ch. Wang, J. Blumoff, K. Chou, M. J. Hatridge, L. Frunzio, M. H. Devoret, L. Jiang, and R. J. Schoelkopf (2016), Quantum memory with millisecond coherence in circuit QED. *Phys. Rev. B* 94, 014506.
22. M. Mirrahimi, Z. Leghtas, V. V. Albert, S. Touzard, R. J. Schoelkopf, L. Jiang, and M. H. Devoret (2014), Dynamically protected cat-qubits: a new paradigm for universal quantum computation. *New J. Phys.* 16.
23. L. S. Bishop (2010), Circuit quantum electrodynamics II. Yale University.
24. E. T. Jaynes and F. W. Cummings (1963), Comparison of quantum and semiclassical radiation theories with application to the beam maser. *Proc. IEEE* 51, 89–109.
25. D. A. Steck (2018), Quantum and atom optics (available online at http://steck.us/teaching [revision 0.12.3, 25 October 2018]).
26. M. Tavis and F. W. Cummings (1968), Exact solution for an N-molecule-radiation-field Hamiltonian. *Phys. Rev.* 170, 379–384.
27. J. M. Fink, R. Bianchetti, M. Baur, M. Goeppl, L. Steffen, S. Filipp, P. J. Leek, A. Blais, and A. Wallraff (2009), Dressed collective qubit states and the Tavis-Cummings model in circuit QED. *Phys. Rev. Lett.* 103, 083601.
28. L. DiCarlo, J. M. Chow, J. M. Gambetta, Lev S. Bishop, B. R. Johnson, D. I. Schuster, J. Majer, A. Blais, L. Frunzio, S. M. Girvin, and R. J. Schoelkopf (2009), Demonstration of two-qubit algorithms with a superconducting quantum processor. *Nature* 460, 240–244.
29. J. M. Gambetta, J. M. Chow, and M. Steffen (2017), Building logical qubits in a superconducting quantum computing system. *NPJ Quantum Inf.* 3, 2.
30. F. W. Strauch, Ph. R. Johnson, A. J. Dragt, C. J. Lobb, J. R. Anderson, and F. C. Wellstood (2003), Quantum logic gates for coupled superconducting phase qubits. *Phys. Rev. Lett.* 91, 167005.
31. R. Barends, J. Kelly, A. Megrant, A. Veitia, D. Sank, E. Jeffrey, T. C. White, J. Mutus, A. G. Fowler, B. Campbell, Y. Chen, Z. Chen, B. Chiaro, A. Dunsworth, C. Neill, P. O'Malley, P. Roushan, A. Vainsencher, J. Wenner, A. N. Korotkov, A. N. Cleland, and John M. Martinis, Superconducting quantum circuits at the surface code threshold for fault tolerance. *Nature* 508, 500–503 (2014).
32. R. McDermott, R. W. Simmonds, M. Steffen, K. B. Cooper, K. Cicak, K. D. Osborn, Seongshik Oh, D. P. Pappas, and John M. Martinis (2005), Simultaneous state measurement of coupled Josephson phase qubits. *Science* 307, 1299–1302.
33. R. C. Bialczak, M. Ansmann, M. Hofheinz, E. Lucero, M. Neeley, A. D. O'Connell, D. Sank, H. Wang, J. Wenner, M. Steffen, A. N. Cleland, and J. M. Martinis (2010), Quantum process tomography of a universal entangling gate implemented with Josephson phase qubits. *Nat. Phys.* 6, 409–413.

34. A. Dewes, F. R. Ong, V. Schmitt, R. Lauro, N. Boulant, P. Bertet, D. Vion, and D. Esteve (2012), Characterization of a two-transmon processor with individual single-shot qubit readout. *Phys. Rev. Lett.* 108, 057002.
35. M-H Yung, J. D. Whitfield, S.Boixo, D.G. Tempel, and A. Aspuru-Guzik, Introduction to quantum algorithms for physics and chemistry. *Adv. Chem. Phys.*, 21 March 2014, doi.: 10.1002/9781118742631.ch03.
36. S. Lloyd (1996), *Science* 273(5278), 1073–1078.
37. R. P. Feynman (1982), *Int. J. Theor. Phys.* 21(6-7), 467–488.
38. C. Moler and Ch. Van Loan, Nineteen dubious ways to compute the exponential of a matrix, twenty-five years later. *SIAM Rev.*, 45(1), 3–49, doi: 10.1137/S00361445024180.
39. S. Richer, Master thesis, Perturbative analysis of two-qubit gates on transmon qubits, September 2013.
40. J. M. Gambetta (2013). "Control of Superconducting Qubits". In: Lecture Notes of the 44th IFF Spring School. Ed. by David DiVincenzo. Schriften des Forschungszentrums Jülich.
41. A. Peres (1985), Reversible logic and quantum computers, *Phys. Rev. A*, 32, 6, pp. 3266–3276.
42. D. Maslov and G. W. Dueck (2003), Improved quantum cost for n-bit Toffoli gates, *IEEE Electronic Lett.*, 39, 25, pp. 1790–1791.
43. D. M. Miller, R. Wille, and Z. Sasanian (2011), Elementary quantum gate realizations for multiple-control Toffoli gates, Proceedings of 41st IEEE International Symposium on Multiple-Valued Logic, pp. 288–293.
44. D. M. Miller (2009), Lower cost quantum gate realizations of multiple-control Toffoli gates, IEEE Pacific Rim Conference on Communications, Computers and Signal Processing, pp. 308–313.
45. D. M. Miller and Z. Sasanian (2010), Lowering the Quantum Gate Cost of Reversible Circuits, 53rd IEEE International Midwest Symposium on Circuits and Systems, pp. 260–263.
46. C. Moraga and F. Z. Hadjam (2012), On Double Gates For Reversible Computing Circuits, Proceedings of 10th Workshop on Boolean Problems, pp. 131138.
47. W. N. N. Hung, X. Song, G. Yang, J. Yang, and M. Perkowski, Optimal synthesis of multiple output Boolean functions using a set of quantum gates by symbolic reachability analysis, *IEEE Trans. Comput. Aided Des. Integr. Circ. Syst.*, 25, 9, September 2006, doi: 10.1109/TCAD.2005.858352.
48. M. D. Reed, L. DiCarlo, S. E. Nigg, L. Sun, L. Frunzio, S. M. Girvin, and R. J. Schoelkopf (2012), Realization of three-qubit quantum error correction with superconducting circuits, *Nature* 482, 382-385, doi: 10.1038/nature10786.
49. I. Chuang. qasm2circ. http://www.media.mit.edu/quanta/qasm2circ/, 2005. accessed November 2016.
50. A. Shafaei, M. Saeedi, and M. Pedram, Qubit placement to minimize communication overhead in 2D quantum architectures, 2014 19th Asia and South Pacific Design Automation Conference (ASP-DAC), January 2014, doi: 10.1109/ASPDAC.2014.6742940.
51. T. Kutzschebauch and L. Stok, Congestion aware layout driven logic synthesis, IEEE/ACM International Conference on Computer Aided Design. ICCAD 2001. doi: 10.1109/ICCAD.2001.968621.
52. M. Perkowski, P. Kerntopf, A. Buller, M. Chrzanowska-Jeske, A. Mishchenko, X. Song, A. Al-Rabadi, L. Jozwiak, A. Coppola, and B. Massey (2012), Regular realization of symmetric functions using reversible logic, IEEE Proceedings Euromicro Symposium on Digital Systems Design, doi: 10.1109/DSD.2001.952289.
53. L. Amarú, P. Gaillardon, S. Mitra, and G. De Micheli, New logic synthesis as nanotechnology enabler, Proceedings of the IEEE (Volume: 103, Issue: 11, November 2015), doi: 10.1109/JPROC.2015.2460377.
54. N. Mohammadzadeh, M. Sedighi, and M. Saheb Zamani, Quantum physical synthesis: improving physical design by netlist modifications. doi: 10.1016/j.mejo.2010.02.005.

21 Microelectronics and Photonics in Quantum Processing and Communication

Sergey Edward Lyshevski and Ivan Puchades

Department of Electrical and Microelectronic Engineering, Rochester Institute of Technology, Rochester, New York, USA

CONTENTS

21.1 INTRODUCTION

Significant progress is accomplished in photonics, photoelectronics, and microelectronics. Fundamental, applied, and technological developments have led to high-performance computing and communication in physical and cyber-physical systems. The device physics of the majority of microelectronic and optoelectronic devices are based on principles of classical physics while minimizing adverse quantum phenomena. Research on new processing and communication paradigms to overcome fundamental and technology limits, quantum-enabled schemes, and devices are studied.

Quantum phenomena (photon absorption, photon emission, atomic electron transitions, spontaneous emission, etc.) are utilized by resonant-tunneling devices, quantum dots, single-photon avalanche diodes, as well as solid-state, inorganic, and organic lasers. These semiconductor microelectronic devices and quantum heterostructures exhibit *macroscopic* quantum *continuum* or coupled *microscopic–macroscopic* quantum *continuum*. Using microelectronic and photonic technologies, many sensing, communication, and processing devices are designed, fabricated, tested, characterized, and implemented.

Research and developments pertaining to quantum-attributed switching, computing, sensing, logic, and memory face: (1) enormous technological challenges, such as fabrication, synthesis, room-temperature detection, measurements, interfacing, and control; (2) theoretical challenges, such as high-fidelity modeling, analysis on operators, complementary quantum observables versus physical measurements, Heisenberg uncertainty principle, quantum statistics on distributions of states probabilities versus quantum statistical equilibrium, etc. Various quantum effects have been investigated, such as quantum tunneling, quantum conductivity, photon entanglement, photon–electron interactions, spin interactions, electron and spin tunneling, angular momentum coupling, energy splitting, single domain magnetics, and supermagnetism. Emerging devices should be supported by adequate integration, interfacing, interconnect, computing architecture, and processing calculi. There are debates on physics and practicality of room-temperature quantum-effect devices. Room-temperature quantum-ascertained sensing, memory, and information processing are exhibited by vertebrates, invertebrates, and prokaryotes. Biomimetics and biointerfacing cannot be accomplished unless baseline *computing* and *signaling* phenomena, pertained physics, system organization, and verified processing schemes will be comprehended. Despite long-standing unsolved problems, it is asserted that processing is accomplished by biomolecules and molecular scaffolds. The attempts to hypothesize on biophysics, biochemistry, mechanisms, and calculi of information processing using assumed postulates and doctrinaire conjectures lead to ambiguity.

Using the operator algebra, infinite-dimensional vector spaces and other concepts of quantum mechanics, modeling and analysis are performed. Quantum-probabilistic premises are applied to investigate and design solid-state microelectronic, optoelectronic, and photonic devices such as transistors, quantum cascade lasers, optical switches, and others. The sum of states of an assembly is in statistical equilibrium. The deterministic digital logic, arithmetic, and calculi are applied to design all digital circuits and processors. Measurable quantum transductions can be used to implement processing arithmetic for general-purpose computing. Physics-consistent quantum determinism and quantum statistical equilibrium are applied toward computing and communication schemes to ensure exceptional parallelism, processing speed, data rate, channel capacity, throughput, etc.

21.2 MOLECULAR AND BIOMOLECULAR PROCESSING

Enormous progress is achieved in microelectronics and photonics [1, 2]. *Microscopic* (molecular and macromolecular) and *macroscopic* (quantum dots, lasers, waveguides, etc.) devices exhibit quantum transductions [1–9]. There is a need to examine quantum state transitions and observable transductions in optoelectronic devices to research experimentally verifiable physics and biophysics of field propagation, photons emission, photon propagation, photon absorption, electron tunneling, etc. Processing tasks can be executed by physical devices that exhibit controllable quantum field- and photon-induced state transitions $s_i(t)$ with characterizable and measurable quantum transductions $\mathcal{T}_i$. We consider processing on directly measured processable variables $\boldsymbol{x}(t, \mathcal{T}) = [x_1, \ldots, x_k], \boldsymbol{x} \in \mathbb{R}$ that physically implement associated calculus pertained to arithmetic operations, logic, switching, etc. For a processing primitive $\mathcal{P}_i$, a tuple of initial, intermediary, and final physical variables $(\boldsymbol{x}_I, \boldsymbol{x}_R, \boldsymbol{x}_F)$ pertains to deterministic transductions $(\mathcal{T}_I, \mathcal{T}_R, \mathcal{T}_F)$, quantum statistical equilibrium transductions, or, quantum statistical ensembles (canonical ensembles) transductions $(\langle\mathcal{T}_I\rangle, \langle\mathcal{T}_R\rangle, \langle\mathcal{T}_F\rangle)$. We have

$$\mathcal{T}_I{:}\boldsymbol{x}_I \to \mathcal{T}_R{:}\boldsymbol{x}_R \to \mathcal{T}_F{:}\boldsymbol{x}_F, \quad \langle\mathcal{T}_I\rangle{:}\langle\boldsymbol{x}_I\rangle \to \langle\mathcal{T}_R\rangle{:}\langle\boldsymbol{x}_R\rangle \to \langle\mathcal{T}_F\rangle{:}\langle\boldsymbol{x}_F\rangle, \quad \boldsymbol{x} \in \mathbb{R}. \tag{21.1}$$

One may implement *distinguishable* and *computable* transforms $\mathcal{F}_i(\mathcal{T}, \boldsymbol{x})$ aiming to accomplish arithmetic operands $\mathcal{A}_i$ and switching functions $\boldsymbol{f}_i$ [5–7]. We examine quantum processing on:

1. Quantum-effect *utilizable* and *observable* transductions $\mathcal{T}(t, \boldsymbol{s})$ or $\langle\mathcal{T}(t, \boldsymbol{s})$ for *controllable* evolutions of *processable*, *detectable*, and *measurable* variables $\boldsymbol{x}(t, \mathcal{T})$.
2. *Achievable*, *distinguishable*, and *computable* transforms $\mathcal{F}(\mathcal{T}, \boldsymbol{x})$ pertained to computing logic, arithmetic, and design schemes.

Example 21.2.1: Processing and Communication in Living Organisms and Engineered Systems

The 375 to–1650 nm solid-state (GaN, AlGaN, InGaN, and GaAs) lasers, quantum well lasers, quantum cascade (GaAs/AlGaAs, GaInAs/AlInAs, and InGaAs/AlAsSb) lasers, vertical cavity surface emitting lasers, III–V diodes, and photodetectors are commercialized and used. Photons in the ultraviolet spectra are emitted by various eukaryotic cells and plants. The emission of biophotons, referred to as *bioluminescence, chemoluminescence*, and *dark luminescence*, is exhibited by bacteria, single-cell protists, mushrooms, fishes, squids, beetles, etc. Publications [10–19] provide experimental evidence of cellular electromagnetic radiation, photon emission, and photon absorption that may result in intra- and intercellular communication and processing governed by biophoton in 300 nm to the near-infrared spectrum. Physics of absorption of photons by retinal and other biomolecular receptors is of great importance. Photon detection by microbolometers, photodiodes, and other photodetectors was researched [1–9].

Visual systems of vertebrates and invertebrates sense, perceive, and process photons in near-infrared, visual, and near-ultraviolet spectra. They exhibit extraordinary resolution, sensitivity, selectivity, and image processing. Visual, somatosensory, and other systems are comprised of photoelectric-, photochemoelectric-, photochemomechanical-, and photothermo sensory receptors, as well as interfaced processing biomolecular scaffolds, which operate by emitting and absorbing photons and electromagnetic radiation in the following spectra: (1) near-infrared, 1500 to 750 nm, with *f* from 200 to 400 THz; (2) visible, 750 to 380 nm, 400 to 789.5 THz; (3) near- and middle-ultraviolet, 380 to 300 nm, 789.5 to 1000 THz. For example: (i) melanin absorbs ultraviolet radiation and transforms the energy into heat; (ii) photoisomerization of retinal in a visible spectrum yields phototransductions; (iii) chlorophyll absorbs photons in the visible spectrum causing photosynthesis; etc. The aforementioned transductions are described by Equation 21.1.

Despite unsolved problems and debates on photon-induced information processing in living organisms, optical computing and photonic signal processing are demonstrated in various analog and digital optoelectronic systems. The commercialized high-throughput optical communication platforms and free-space optics use high-sensitivity quadrature photodetectors, low-divergence tunable lasers, filters, ASICs and other components. In optical communication, the wavelength and frequency ranges are $\lambda \in [525\ 1550]$ nm, $f \in [194\ 571]$ THz. The IEEE 802.11 standard defines radiofrequency ranges for the Wi-Fi communication as 900 MHz, 2.4 GHz, 3.6 GHz, 4.9 GHz, 5 GHz, 5.9 GHz, and 60 GHz bands. The high-frequency wireless communication bands may reach hundreds of gigahertz. For the *W* band, $f\,|_{W\text{band}} \in [75\ 110]$ GHz. ■

Example 21.2.2

Photosensitive proteins (rhodopsin and photopsin) absorb photons. Photons are absorbed by the porphyrin ring of a chlorophyll molecule ensuring photosynthesis in plants, algae, photosynthetic bacteria, etc. The absorbed photon energy is $E = hf = h(c/\lambda)$, where h is the Plank constant, $h = 6.62 \times 10^{-34}$ J-s; f is the frequency, $f = c/\lambda$. The wavelengths $\lambda = 450$ nm and $\lambda = 700$ nm correspond to the maximum absorption. The resultant photon energies are $E_{\lambda=450\text{ nm}} = 4.42 \times 10^{-19}$ and $E_{\lambda=700\text{ nm}} = 2.84 \times 10^{-19}$ J. The de Broglie relation $\lambda = h/p$ yields the momentum p. The single-photon sensing, energy transductions, and energy conversion are accomplished at attojoules (1 aJ = 1×10^{-18} J).

Considering the particle in the infinite potential (Example 6.2), one finds the quantized energy eigenvalues E_n. For example, $E_n = \frac{h^2}{8ma^2}n^2 = \frac{\pi^2\hbar^2}{2ma^2}n^2$, $n = 1, 2, \ldots$. The ground state corresponds to the principal quantum number $n = 1$. For the harmonic oscillator, one has $E_n = \hbar\omega(n + \frac{1}{2})$, $n = 1, 2, \ldots$.

One observes the differences between the energy of photon E, as well as the energy of other *microscopic* systems, and, the quantized energy eigenvalues E_n, which are the mathematical operators. ■

Example 21.2.3

Bio, organic, and inorganic molecules, complexes, and compounds exhibit bioluminescence, chemiluminescence, electrochemiluminescence, and photoluminescence (fluorescence and phosphorescence). Fluorescence is exhibited by quinine $C_{20}H_{24}N_2O_2$ (contained in the bark of the cinchona tree), chlorophylls with different chemical structure, etc. Phosphorescence is exhibited by aromatic hydrocarbon naphthalene ($C_{10}H_8$), dopant-activated strontium aluminate $SrAl_2O_4$:Eu (europium is a dopant), zinc sulfide ZnS:Cu, etc.

Amino acids exhibit fluorescence, e.g., absorption and emission of photons characterized by the wavelength λ and energy E, which are measurable physical quantities. Fluorescence occurs as an atom, which is excited to a higher quantum state by absorbing a photon, returns to the lower energy level by emitting a photon. The excited singlet state S_1 returns (relaxes) to the ground singlet state S_0 by emitting a photon, $S_0 + hv \rightarrow S_1 \rightarrow S_0 + hv_1$. Phosphorescence involves the triple state transitions as $S_0 + hv \rightarrow S_1 \rightarrow S_2 \rightarrow S_0 + hv_1$. The aforementioned transductions can be experimentally examined on ensemble, as well as analyzed quantum mechanically with different level of fidelity applying assumptions to make the problem mathematically solvable. Modeling and simulation are carried out by applying concepts of theoretical quantum mechanics, such as solving Schrödinger and constitutive equations, using the Hilbert and Banach spaces, etc. Experimental studies for various fluorescent systems are performed implementing physical phenomena. Organic laser dyes (acridine, fluorecein, puronin, rhodamine, and others) and light-emitting diodes are used in communication, imaging, medicine, and biotechnology. ■

Example 21.2.4

Fluorescence of aromatic amino acids (tryptophan, tyrosine, and phenylalanine), absorption and emission wavelengths, lifetime, quantum yield, and other quantities are experimentally measured [11, 20, 21]. The fluorescence of a folded protein is superimposed by the fluorescence of amino acids. The intrinsic fluorescence of a folded protein depends on excitation and emission of tryptophan, which is affected by tyrosine and phenylalanine. Typically, the maximum absorption wavelength of tryptophan is 280 nm, while an emission peak varies from 300 to 350 nm depending on polarity, environment, and other factors. Fluorescence depends on the protein's conformational state. The tryptophan fluorescence is affected by the proximity of other amino acids. Nearby *protonated* groups, such as aspartic or glutamic acids, can cause quenching. There is an energy transfer between tryptophan and other fluorescent amino acids. The number of tryptophan molecules is less than other amino acids. The intrinsic tryptophan-induced fluorescence is applied to examine protein conformations using one or few tryptophans because a specific environment changes the emission spectra. The intensity, wavelength, and quantum yield of amino acid fluorescence are solvent dependent. The fluorescence spectrum shifts to a shorter wavelength, and the intensity of the fluorescence increases as the polarity of the solvent surrounding tryptophan decreases. The wavelength of tryptophan in the hydrophobic protein's core and on the protein's surface differs by 20 nm. Tyrosine significantly contributes to protein fluorescence. The tyrosine-induced fluorescence is quenched by nearby tryptophan due to energy transfer. The quantum yields of aromatic amino acids decrease if they are within a polypeptide chain. The quantum yield of tryptophan is higher compared with other amino acids. In the dimethyl sulfoxide organic solvent, the quantum yields of tryptophan, tyrosine, and phenylalanine are 0.67, 0.06, and 0.006, respectively. Due to enormous complexity, it is impossible to perform high-fidelity analysis by deriving the governing equations and solving them. ■

Example 21.2.5: Photoreception, Interfacing, and Processing in Living Organisms: Implications to Engineered Systems

In the human eye retina, there are 130 million rod photoreceptor cells and 5 million cone photoreceptor cells. Each rod cell contains hundreds of millions of light-sensitive rhodopsin molecules. A visual system exhibits photon absorption and detection (sensing), interfacing, data retrieval, and other processing tasks on billions of inputs in real time. Efforts to analyze analog *signaling* and

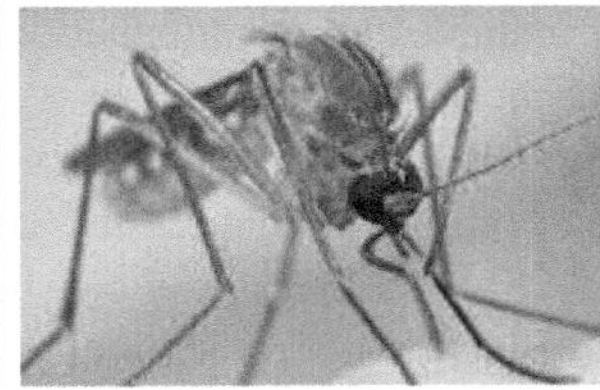

FIGURE 21.1 Insects perform cognition and learning, including information processing, consuming μW power. In fruit fly and honey bee, the number of neurons and synapses are 250,000 and 1×10^7 and 960,000 and 1×10^9, respectively.

computing have been conducted for many decades. Photosensitive proteins (rhodopsin and photopsin) absorb photons, and the *cis–trans* geometric isomerism within the Schiff base $R_2C{=}NR'$, $R' \neq H$ is considered in the literature. The so-called *optic* and *peripheral nerves* transmit signals. The *optic nerve* is composed of retinal ganglion cell axons and glial cells. Each human optic nerve contains of up to 2 million axons (*nerve fibers*) of the retinal ganglion cells of a single retina. To investigate *signaling* by neurofilament polymers in axons and microtubules (polymerized dimer of two globular proteins, alpha and beta tubulin), conductivity and other physical quantities are used. The microtubule length is from 100 nm to centimeters, and the inner and outer diameters are 12 and 24 nm. Microtubules are composed of polymerized arrays of protein subunits. Thirteen laterally aligned parallel protofilaments form a cavity in a microtubule. Alpha- and beta-tubulin monomers form dimers, which yield hexagonal lattices with different helical patterns. To some extent, one typifies interfacing by axons and microtubules by researching dielectric optical waveguides.

There are debates on baseline physics of *signaling, computing,* and processes in living organisms. The biophotoelectric sensing, communication, and processing on attojoules energy ($E_{\lambda=380\text{ nm}} = 5.23 \times 10^{-19}$ and $E_{\lambda=780\text{ nm}} = 2.55 \times 10^{-19}$ J) and high-fidelity spatiotemporal three-dimensional image constriction within 0.01 s are achieved on milliwatt power. High-resolution image processing and cognition are accomplished by primitive vertebrates. Insects, shown in Figure 21.1, consume <1 μW to perform information-processing tasks. The vertebrates' information processing cannot be accomplished by the most advanced processors with billions of transistors, device switching speed 1 THz, circuit speed 10 GHz, device switching energy 1×10^{-16} J, writing energy 5×10^{-15} J/bit, read/write time 1 ns, etc. Our aim is to investigate some aspects of quantum mechanics, device physics, and classical and quantum communication and processing. From engineering design and technology rationale, device physics and *microscopic–to–macroscopic* interfacing must be addressed. Inconsistent interfacing may significantly disturb and perturb transductions in quantum devices, preventing overall functionality. Living organisms substantiate quantum → classical and classical → quantum schemes. To ensure consistent analysis and design, there is a need to examine the implications of the Heisenberg uncertainty principle, research transductions, and investigate processing logic and algorithms. ■

21.3 WAVEGUIDES FUNDAMENTALS

To accomplish guided communication, optical fibers and optical waveguides are used. Depending on frequency, optical fibers and waveguides are designed using conductive, dielectric, and photonic media and metamaterials. Optical waveguides (planar, strip, rectangular, and circular) are characterized by operating wavelength, power, modes, losses, attenuation, etc. Waveguides, which guide photons and electromagnetic fields, are integrated with filters, crystals, lenses, lasers, photodetectors, and optoelectronic devices. Optical fibers operate at the hundreds of THz frequency, guiding light that propagates in 8 or 9 μm diameter core single-mode fibers or in 50, 62.5, or 100 μm diameter core multimode fibers with cladding (125 μm diameter), buffer, and jacket. The wavelength usually varies from 450 to 1660 nm, which corresponds to 670 and 180 THz. Due to internal reflection, the light is confined, and the attenuation varies from $0.4|_{1550\text{ nm}}$ and $0.5|_{1310\text{ nm}}$ (single-mode fibers) to 5 dB/km (multimode fibers).

Waveguides and materials: The core of optical fibers are polymers, silica, doped silica, fluorozirconate, fluoroaluminate, and chalcogenide glasses or crystalline materials such as sapphire [8, 9, 22-28]. The core glass can be doped with index-raising dopants such as germania GeO_2, phosphorus pentoxide P_2O_5, or alumina Al_2O_3. Cladding materials are glass or plastic. The refractive index of the cladding glass may be lowered by fluorine, boron oxide B_2O_3, or titania TiO_2. The III–V semiconductors (GaAs, InP, AlGaAs, InGaAsP, etc.), silica SiO_2 on silicon, silicon-on-insulator, and polymeric waveguides were designed, fabricated, and commercialized using materials with different refractive indexes, distinct bandgaps, and close lattice constants. The electro-optic electrically active lithium niobate $LiNbO_3$, $LiNbO_3$ with diffused Ti, lithium tantalite $LiTaO_3$, and barium titanate $BaTiO_3$ waveguides were implemented and commercialized in interfacing, networking, and communication applications.

We study planar and circular waveguides with nonlinear optical media (silicon nitride and polymers) and linear optical media (glass) as a core. Synthetic polymers, glass, and silicon dioxide can be used for cladding. Planar waveguides can be made on printed circuit boards (polyimide, polytetrafluoroethylene, or glass), graphite, graphene, III–V semiconductors, silicon, etc. Organic and inorganic polymers, silicon nitride Si_3N_4, doped silicon nitride, and other nonlinear optical materials are deposited on silicon using surface micromachining processes. The relative permittivity ε_r is frequency-, fabrication-, and temperature-dependent. For 1310 nm and 300K, the estimates are: $\varepsilon_r = 3.9$ for SiO_2; $\varepsilon_r = 3.9$ for Al_2O_3; $\varepsilon_r = 7.5$ for Si_3N_4; $\varepsilon_r = 22$ for Ta_2O_5; $\varepsilon_r = 80$ for TiO_2; $\varepsilon_r = 2000$ for $SrTiO_3$; and $\varepsilon_r = 25$ for ZrO_2. For polymers, ε_r varies from 3.2 to 3.6 for nonfluorinated aromatic polyimides and from 2.6 to 2.8 for fluorinated polyimide, poly(phenyl quinoxaline), poly(arylene ether oxazole), poly(arylene ether), polyquinoline, poly(norborene), poly(arylene ether), polynaphthalene, polystyrene, etc.

Governing equations: Maxwell's and constitutive equations for the field intensity vectors $(\boldsymbol{E}, \boldsymbol{H})$ and field density vectors $(\boldsymbol{D}, \boldsymbol{B})$ are

$$\nabla \times \boldsymbol{E} = -\frac{\partial \boldsymbol{B}}{\partial t}, \quad \nabla \times \boldsymbol{H} = \boldsymbol{J} + \frac{\partial \boldsymbol{D}}{\partial t}, \quad \nabla \cdot \boldsymbol{D} = \rho, \quad \nabla \cdot \boldsymbol{B} = 0, \quad \boldsymbol{D} = \varepsilon \boldsymbol{E}, \quad \boldsymbol{B} = \mu \boldsymbol{H}, \quad \boldsymbol{J} = \sigma \boldsymbol{E}, \tag{21.2}$$

where ε, μ, and σ are the electrical permittivity, permeability, and conductivity tensors.

In nonlinear media, ε depends on $\boldsymbol{E}$. Hence, $\boldsymbol{D} = \varepsilon(\boldsymbol{E})\boldsymbol{E}$, $D = \begin{bmatrix} \varepsilon_{xx}(\boldsymbol{E}) & \varepsilon_{xy}(\boldsymbol{E}) & \varepsilon_{xz}(\boldsymbol{E}) \\ \varepsilon_{yx}(\boldsymbol{E}) & \varepsilon_{yy}(\boldsymbol{E}) & \varepsilon_{yz}(\boldsymbol{E}) \\ \varepsilon_{zx}(\boldsymbol{E}) & \varepsilon_{zy}(\boldsymbol{E}) & \varepsilon_{zz}(\boldsymbol{E}) \end{bmatrix} \boldsymbol{E}$, $\varepsilon(\boldsymbol{E}) = \varepsilon_1 + \varepsilon_2(\boldsymbol{E}) + \varepsilon_3(\boldsymbol{E}^2) + \cdots$.

In a linear, homogeneous and isotropic media, the equations for the field intensity vectors $(\boldsymbol{E}, \boldsymbol{H})$ are

$$\nabla^2 \boldsymbol{E} = \mu_0 \sigma (\partial \boldsymbol{E}/\partial t) + \mu_0 \varepsilon (\partial^2 \boldsymbol{E}/\partial t^2), \qquad \nabla^2 \boldsymbol{H} = \mu_0 \sigma (\partial \boldsymbol{H}/\partial t) + \mu_0 \varepsilon (\partial^2 \boldsymbol{H}/\partial t^2). \tag{21.3}$$

The sinusoidal steady-state variations are described by three-dimensional vector phasors $\boldsymbol{E}$ and $\boldsymbol{H}$. In the frequency domain, the Maxwell equations are

$$\begin{aligned} &\nabla \times \boldsymbol{E} = j\omega\mu_0 \boldsymbol{H}, \nabla \cdot (\varepsilon \boldsymbol{E}) = 0, \\ &\nabla \times \boldsymbol{H} = -j\omega\varepsilon(\omega)\boldsymbol{E}, \nabla \cdot \boldsymbol{H} = 0, \varepsilon(\omega) = 1 + \xi(\omega). \end{aligned} \tag{21.4}$$

We have

$$\nabla^2 \boldsymbol{E} = j\omega\mu_0(\sigma + j\omega\varepsilon)\boldsymbol{E} = \gamma^2 \boldsymbol{E}, \nabla^2 \boldsymbol{H} = j\omega\mu_0(\sigma + j\omega\varepsilon)\boldsymbol{H} = \gamma^2 \boldsymbol{H}, \gamma^2 = j\omega\mu_0(\sigma + j\omega\varepsilon), \gamma = \alpha + j\beta. \tag{21.5}$$

where γ is the propagation constant; α and β are the frequency-dependent propagation attenuation and phase constants.

Example 21.3.1

Consider the uniform plane wave with the field vectors in the xy plane. The phasor Poynting vector propagates in the z direction. For $\boldsymbol{E} = E_x(z)\boldsymbol{a}_x$ and $\boldsymbol{H} = H_y(z)\boldsymbol{a}_y$, the solution of

$$\frac{d^2E_x}{dz^2} = \gamma^2 E_x,\ \frac{d^2H_y}{dz^2} = \gamma^2 H_y \text{ is } \frac{dE_x}{dz} = -j\omega\mu_0 H_y,\ \frac{dH_y}{dz} = -(\sigma + j\omega\varepsilon)E_x.$$

We have $E_x = E_x^+ + E_x^- = E_{x0}^+e^{-\gamma z} + E_{x0}^-e^{\gamma z}$ and $H_y = H_y^+ + H_y^- = H_{y0}^+e^{-\gamma z} + H_{y0}^-e^{\gamma z}$.

The forward- and backward-traveling waves attenuate exponentially due to the terms $e^{-\gamma z}$ and $e^{-\gamma z}$. The backward wave propagates in the $-z$ direction.

From $-\gamma E_{x0}^+e^{-\gamma z} + \gamma E_{x0}^-e^{\gamma z} = -j\omega\mu_0 H_{y0}^+e^{-\gamma z} + j\omega\mu_0 H_{y0}^-e^{\gamma z}$, one obtains

$$\frac{E_{x0}^+}{H_{y0}^+} = \frac{j\omega\mu_0}{\gamma} = \eta_c,\ \frac{E_{x0}^-}{H_{y0}^-} = -\frac{j\omega\mu_0}{\gamma} = -\eta_c.$$

The intrinsic impedance of the medium is $\eta_c = \frac{j\omega\mu_0}{\gamma} = \sqrt{\frac{j\omega\mu_0}{\sigma + j\omega\varepsilon}} = \sqrt{\frac{\mu_0}{\varepsilon}}\frac{1}{\sqrt{1 - j\frac{\sigma}{\omega\varepsilon}}}$, $\eta_c = \eta\angle\theta_\eta$.

For a lossless medium $\sigma = 0$ and $\gamma = j\omega\sqrt{\mu_0\varepsilon}$, $\eta_c = \eta = \sqrt{\frac{\mu_0}{\varepsilon}}$.

The phase velocity is $v_p = \frac{\omega}{\beta} = \frac{1}{\sqrt{\mu_0\varepsilon}}$.

The wavelength depends on media, and $\lambda = 2\pi/\beta = v_p/f$.

The power density Poynting vector is $\mathcal{P}(t, z) = \boldsymbol{E}(t, z) \times \boldsymbol{H}(t, z) = \text{Re}(\boldsymbol{E}(z)e^{j\omega t}) \times \text{Re}(\boldsymbol{H}(z)e^{j\omega t})$.

One finds $P(t, z) = \frac{E_{x0}^2}{\eta}e^{-2\alpha z}\cos(\omega t - \beta z)\cos(\omega t - \beta z - \theta_\eta)\boldsymbol{a}_z$ [W/m^2].

The time-averaging power density Poynting vector $P_{av}(z) = \frac{E_{x0}^2}{2\eta}e^{-2\alpha z}\cos\theta_\eta\boldsymbol{a}_z$ indicates that the power density attenuates as $e^{-2\alpha z}$, whereas E_x and H_y attenuate by $e^{-\alpha z}$. For the forward- and backward-traveling waves, we have

$$P_{av}^+(z) = \frac{(E_{x0}^+)^2}{2\eta}e^{-2\alpha z}\cos\theta_\eta\boldsymbol{a}_z \text{ and } P_{av}^-(z) = \frac{(E_{x0}^-)^2}{2\eta}e^{2\alpha z}\cos\theta_\eta\boldsymbol{a}_z \text{ [W/m}^2\text{]}.$$

In free space with (ε_0, μ_0), one obtains the intrinsic impedance $\eta_{c0} = 376.73\Omega$ and $v_{p0} = 3 \times 10^8$ m/s. Recall that $\varepsilon_0 = 8.854 \times 10^{-12}$ F/m and $\mu_0 = 4\pi \times 10^{-7}$ H/m. For $f = 10$ THz, one finds $\lambda_0 = 30$ μm.

For Teflon, $\sigma \approx 0$, $\varepsilon = \varepsilon_0\varepsilon_r$, $\varepsilon_r = 2.1$, and, $\mu = \mu_0\mu_r$, $\mu_r \approx 1$.

Thus, $\eta = 260$ Ω and $v_p = 2.07 \times 10^8$ m/s.

For lossy media, $\sigma \neq 0$.

The propagation constant $\gamma = \sqrt{j\omega\mu_0(\sigma + j\omega\varepsilon)} = j\omega\sqrt{\mu_0\varepsilon}\sqrt{1 - j\frac{\sigma}{\omega\varepsilon}} = \alpha + j\beta$ yields the frequency-dependent propagation attenuation and phase constants α and β. The electrical conductivity and permittivity tensors (σ, ε) vary as a function of frequency f. For conventional optical media, the ratio $\sigma/\omega\varepsilon$ is known. For polymers, silicon, Si_3N_4, SiO_2, and other media, (α, β) is a nonlinear function of f. ■

Example 21.3.2

For a hollow cylindrical waveguide, assume that the medium of propagation is linear, homogeneous, and isotropic, e.g., $\boldsymbol{D} = \varepsilon\boldsymbol{E}$ and $\boldsymbol{H} = \boldsymbol{B}/\mu$. Let the conductivity of the medium is $\sigma \approx 0$, and the charge density is $\rho = 0$. The $\boldsymbol{E}$ and $\boldsymbol{H}$ are mutually perpendicular. For a waveguide, comprised of a lossless dielectric $\sigma = 0$, characterized by (ε, μ), and, surrounded by a perfect conductor with $\sigma = \infty$, the electric and magnetic field intensities are obtained by solving the following wave equations

$$\nabla^2\boldsymbol{E} + \omega^2\mu_0\varepsilon\boldsymbol{E} = 0,\quad \nabla^2\boldsymbol{H} + \omega^2\mu_0\varepsilon\boldsymbol{H} = 0.$$

Hollow cylindrical waveguides do not support transverse electromagnetic (TEM) modes when electric and magnetic fields transverse to the direction of propagations. There exists high-order:

(i) transverse magnetic (TM) modes when a magnetic field is transverse to the line of axis, $H_z = 0$; (ii) transverse electric (TE) modes when an electric field is transverse to the line of axis, $E_z = 0$.

The propagation constant is $\gamma = \pm\sqrt{(\frac{1}{r_0}k_{nr}^{\text{TE TM}})^2 - \omega^2\mu_0\varepsilon} = \alpha + j\beta$, where $k_{nr}^{\text{TE TM}}$ is the rth positive root ($r = 1, 2, \ldots$) of the derivative of the nth-order Bessel function, which corresponds to TE and TM modes; r_0 is the inner radius.

The cutoff frequency at which $\gamma = 0$ is $f_c = \frac{v_0}{2\pi}\frac{k_{nr}^{\text{TE TM}}}{r_0} = \frac{1}{2\pi\sqrt{\mu_0\varepsilon}}\frac{k_{nr}^{\text{TE TM}}}{r_0}$.

The velocity in the medium is $v_0 = \frac{1}{\sqrt{\mu_0\varepsilon}}$.

The wavelength in z-direction and phase velocity are $\lambda = \frac{2\pi}{\beta} = \frac{\lambda_0}{\sqrt{1-(f_c/f)^2}}$ and $v_p = \frac{\omega}{\beta} = \frac{v_0}{\sqrt{1-(f_c/f)^2}}$, where λ_0 is the wavelength of an imaginary plane wave at the operating frequency f in an unbounded dielectric medium of the same type that fills guide characterized by μ and ε, $\lambda_0 = \frac{1}{f\sqrt{\mu_0\varepsilon}} = \frac{v_0}{f}$.

The group velocity is $v_g = v_0\sqrt{1-(f_c/f)^2}$.

The wavelength λ_{nr} and phase velocity $v_{p\,nr}$ of a TE_{nr} or TM_{nr} wave are $\lambda_{nr} = \frac{\lambda_0}{\sqrt{1-(f_{c\,nr}/f)^2}}$ and $v_{p\,nr} = \lambda_{nr}f = \frac{v_0}{\sqrt{1-(f_{c\,nr}/f)^2}}$.

One concludes that v_p of the TE and TM modes in the z-direction is $v_{p\,nr,z} = (\omega/\beta_{nr,z}) > c$ if $f > f_{c\,nr}$. Furthermore, $v_{p\,nr,z} = \infty$ for $f = f_{c\,nr}$. Signals cannot propagate faster than the speed of light c. The waves carry narrowband signals, and the group velocity is $v_{g\,nr} = \frac{d\omega}{d\beta_{nr,z}} = v_0\sqrt{1-(f_{c\,nr}/f)^2}$, $f > f_{c\,nr}$. Furthermore, $v_{p\,nr,z}v_{g\,nr,z} = v_0^2$.

For the operating modes, the operating frequency is $f > f_c$, $f = \omega/2\pi$, and γ is imaginary. The wave impedance η is always less than the intrinsic impedance of the dielectric medium, and η is purely resistive. For nonpropagating modes, $f < f_c$, and γ is real. The wave decays because all field components have the propagation attenuation factor $e^{-\gamma z} = e^{-\alpha z}$, and η is purely reactive. For the TE and TM modes, $\eta_{nr}^{\text{TE}} = \frac{\eta_0}{\sqrt{1-(\frac{\lambda_0 k_{nr}^{\text{TE}}}{2\pi r_0})^2}}$ and $\eta_{nr}^{\text{TM}} = \eta_0\sqrt{1-(\frac{\lambda_0 k_{nr}^{\text{TM}}}{2\pi r_0})}$, where η_0 is the plane-wave impedance of the lossless dielectric, $\eta_0 = \sqrt{\mu_0/\varepsilon}$, and, $\eta = |\boldsymbol{E}_T|/|\boldsymbol{H}_T|$.

The dominant mode has the lowest cutoff frequency, which is increasing with the TE_{01}, TE_{11}, TE_{21}, TM_{01}, TM_{11}, and higher-order modes. The time–average power transmitted in the $+z$-direction in near-lossless waveguide is found by using the complex Poynting vector over a transverse waveguide cross-section, and $P_z = \frac{1}{2}\text{Re}\int_s \boldsymbol{E}_T \times \boldsymbol{H}_T^* \cdot \boldsymbol{a}_z ds$. ■

Example 21.3.3

For a TE_{11} mode, the eigenvalue is $k_{11}^{\text{TE}} = 1.841$. For a circular fiber, let $r_0 = 4\ \mu\text{m}$ and $r_0 = 25\ \mu\text{m}$, $\mu_r = 1$, and $\varepsilon_r = 1$. From $f_c = \frac{v_0}{2\pi}\frac{k_{nr}^{\text{TE TM}}}{r_0} = \frac{1}{2\pi\sqrt{\mu_0\varepsilon}}\frac{k_{nr}^{\text{TE TM}}}{r_0}$, the cutoff frequencies $f_{c\text{TE}11}$ are 2.869×10^{13} and 3.514×10^{12} Hz if $r_0 = 4\ \mu\text{m}$ and $r_0 = 25\ \mu\text{m}$. For a TM_{01} mode, $k_{01}^{\text{TM}} = 2.405$. One finds $f_{c\,\text{TM}01} = 2.869 \times 10^{13}$ Hz and $f_{c\,\text{TM}01} = 4.591 \times 10^{12}$ Hz. ■

Example 21.3.4

For a TE_{11} mode, the time–average power transmission in a near-lossless waveguide is estimated by $P_{z\text{TE}11} = \frac{1}{4}\eta_0\pi r_0^2|H_{11}|^2(\frac{f}{f_{c\text{TE}11}})^2\sqrt{1-(\frac{f_{c\text{TE}11}}{f})^2}\frac{k_{11}^{\text{TE}2}}{k_{11}^{\text{TE}2}}J_1^2(k_{11}^{\text{TE}})$, where $|H_{11}|$ is the field amplitude; J_1 is the first-order Bessel function of the first kind. ■

Uniform plane waves: The power flow is analyzed by using a Poynting vector in the z-direction $\mathcal{P} = \boldsymbol{E} \times \boldsymbol{H}$. For orthogonal electromagnetic field vectors $\boldsymbol{E}$ and $\boldsymbol{H}$, the solution is found by using Faraday's and Ampere's laws in the interior of the guide. For $\sigma = 0$, one has

$$\nabla \times \boldsymbol{E} = -j\omega\mu_0\boldsymbol{H}, \quad \nabla \times \boldsymbol{H} = j\omega\varepsilon\boldsymbol{E}. \tag{21.6}$$

If $\sigma \neq 0$, for a wave traveling along the z-axis, we have

$$\frac{\partial^2 \boldsymbol{E}}{\partial z^2} = \varepsilon\mu_0 \frac{\partial^2 \boldsymbol{E}}{\partial t^2} + \sigma\mu_0 \frac{\partial \boldsymbol{E}}{\partial t}, \quad \frac{\partial^2 \boldsymbol{H}}{\partial z^2} = \varepsilon\mu_0 \frac{\partial^2 \boldsymbol{H}}{\partial t^2} + \sigma\mu_0 \frac{\partial \boldsymbol{H}}{\partial t}. \tag{21.7}$$

With $\nabla_t = j\omega$, letting $\boldsymbol{E}$ be parallel to the x-axis, one obtains

$$E_x = E_0 e^{j(\omega t - k_r z) - k_i z}, \quad H_y = H_0 e^{j(\omega t - k_r z - \theta) - k_i z}, \quad \theta = \arctan(k_i/k_r), \quad k_r = 2\pi/\lambda, \tag{21.8}$$

where k is the complex wave number, $k = k_r - jk_i$; k_i is the reciprocal of the distance over which the wave amplitude is attenuated by a factor e, $k_r > 0$ and $k_i > 0$. In vacuum, $k_r = 2\pi/\lambda_0$ and $k_i = 0$.

The $\boldsymbol{E}$ and $\boldsymbol{H}$ are in phase, and the electric and magnetic energy densities are equal, ${}^1\!/_2\varepsilon E^2 = {}^1\!/_2\mu_0 H^2$, while $\frac{E_0}{H_0} = \sqrt{\frac{\mu_0}{\varepsilon}}(1 + \frac{\sigma^2}{\omega^2\varepsilon^2})^{-1/4}$. For a lossy medium, $\sigma \neq 0$.

The forward- and backward-traveling waves are analyzed using the complex propagation constant $\gamma = \sqrt{j\omega\mu_0(\sigma + j\omega\varepsilon)} = \alpha + j\beta$. Having found $E_x = E_x^+ + E_x^- = E_m^+ e^{-\gamma z} + E_m^- e^{\gamma z}$, one concludes that the forward- and backward-traveling waves attenuate exponentially due to $e^{-\alpha z}$ and $e^{\alpha z}$ terms (the backward wave propagates in the $-z$-direction). For the forward-traveling wave, $E_x^+ = E_m^+ e^{-\alpha z}\cos(\omega t - \beta z + \theta^+)$ and $H_y^+ = \frac{1}{\eta} E_m^+ e^{-\alpha z}\cos(\omega t - \beta z + \theta^+ - \theta_\eta)$.

Example 21.3.5

We examine forward-traveling waves due to excitations. The cutoff frequencies $f_{c,\text{TE}nr}$ and $f_{c,\text{TM}nr}$ are found. For the operating modes, the operating frequency is $f > f_c$. Consider the frequency of excitation to be 2.29×10^{14} Hz (λ = 1310 nm). Let ε_r = 2.7 and σ = 1×10^{-4} S/m. From $\gamma = \sqrt{j\omega\mu_0(\sigma + j\omega\varepsilon)} = \alpha + j\beta$, for a planar wave, $\gamma = 1.15 \times 10^{-2} + j7.88 \times 10^6$, $\alpha = 1.15 \times 10^{-2}$ Np/m, and $\beta = 7.88 \times 10^6$ rad/m. The intrinsic impedance is $\eta_c = \sqrt{\frac{j\omega\mu_0}{\sigma + j\omega\varepsilon}}$, $\eta_c = 2.29 \times 10^2 + j3.34 \times 10^{-7}$. If the conductivity of walls is $\sigma_w = \infty$, the field is attenuated by a factor of 2 at a distance $d = \ln 2/\alpha = 60.5$ m. For silicon and glass, $\sigma = 1 \times 10^{-4}$ S/m and $\sigma = 1 \times 10^{-13}$ S/m. ■

Example 21.3.6: Analysis of planar waveguides

For a step-index waveguide, letting $\nabla\varepsilon = 0$, one finds homogeneous wave equations

$$\frac{\partial^2 E_z(x,y,z)}{\partial x^2} + \frac{\partial^2 E_z(x,y,z)}{\partial y^2} + (k_i^2 - \beta^2)E_z(x,y,z) = 0,$$

$$\frac{\partial^2 H_z(x,y,z)}{\partial x^2} + \frac{\partial^2 H_z(x,y,z)}{\partial y^2} + (k_i^2 - \beta^2)H_z(x,y,z) = 0,$$

where k_i is the constant for a region i, which has a constant index of refraction n_i, $k_i^2 = \omega^2\mu_0\varepsilon_i = n_i^2\frac{\omega^2}{c^2}$.

The resulting solutions $E \propto e^{\pm j\sqrt{k_i^2 - \beta^2}r}$ are sinusoidal in the core region ($\beta < k_{i\,\text{core}}$) and exponential in the cladding region ($\beta > k_{i\,\text{cladding}}$). Consider a linear and isotropic waveguide, illustrated in Figure 21.2a, $0 < x < a$, $0 < y < b$, $a > b$. The waveguide supports: (1) transverse electric waves (TE modes), $\boldsymbol{E} = (E_x, E_y, 0)$ and $\boldsymbol{H} = (H_x, H_y, H_z)$; and (2) transverse magnetic waves (TM modes), $\boldsymbol{E} = (E_x, E_y, E_z)$ and $\boldsymbol{H} = (H_x, H_y, 0)$. Assuming $\nabla\varepsilon = 0$, the governing Maxwell equations for the longitudinal components (E_z, H_z)

$$\frac{\partial E_z(x,y,z)}{\partial x^2} + \frac{\partial^2 E_z(x,y,z)}{\partial y^2} - k_c^2 E_z(x,y,z) = 0, \quad \frac{\partial^2 H_z(x,y,z)}{\partial x^2} + \frac{\partial^2 H_z(x,y,z)}{\partial y^2} - k_c^2 H_z(x,y,z) = 0$$

are solved using the boundary conditions. In general, (E_x, E_y, E_z) and (H_x, H_y, H_z) are coupled because $\nabla\varepsilon \neq 0$. Furthermore, the exhibited modes strongly depend on the geometry as well as on the index profile $n(x,y)$.

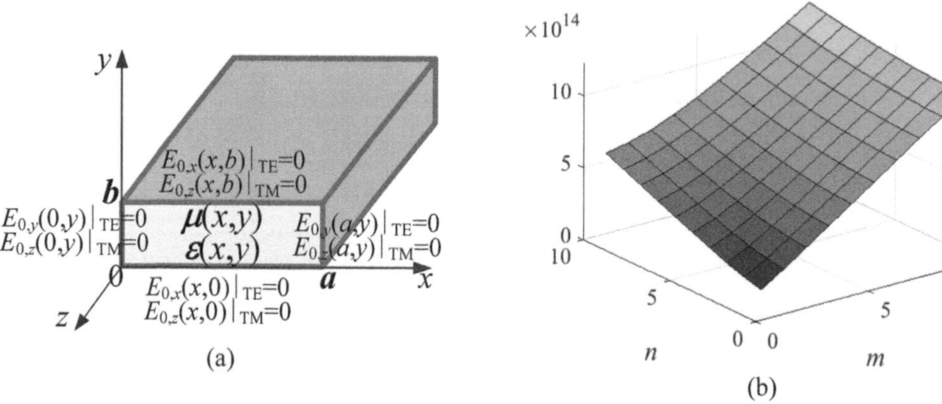

FIGURE 21.2 (a) Rectangular planar waveguide with two-dimensional transverse optical confinement, $0 < x < a$, $0 < y < b$ $(a > b)$. The boundary conditions for the TE and TM modes are documented; (b) Surface for cutoff frequency $f_{c,mn} = \frac{1}{2\pi\sqrt{\mu_0\varepsilon}}\sqrt{(\frac{m\pi}{a})^2 + (\frac{n\pi}{b})^2}$ for the TE_{mn} and TM_{mn} modes ($m = 1, \ldots, 10$, $n = 1, \ldots, 10$) if $\varepsilon_r = 7.5$, $a = 1000$ nm and $b = 500$ nm.

The wave number is $k = \omega/c$, $c = \frac{1}{\sqrt{\varepsilon_0\mu_0}}$. The cutoff wave number is $k_c^2 = k_x^2 + k_y^2 = (\frac{m\pi}{a})^2 + (\frac{n\pi}{b})^2$.

The phase constant is $\beta^2 = k^2 - k_c^2$, $\beta_{mn} = \sqrt{k^2 - k_c^2} = \sqrt{k^2 - (\frac{m\pi}{a})^2 - (\frac{n\pi}{b})^2}$.

For the TE modes, one has $\nabla\varepsilon \perp \boldsymbol{E}$, such that $\nabla\varepsilon \cdot \boldsymbol{E} = 0$. Each component of the electric field (E_x, E_y, E_z) for a TE mode satisfies a homogeneous scalar differential equation. The magnetic field components (H_x, H_y, H_z) for a TE mode are coupled. Applying the boundary conditions, solving

$$\frac{\partial^2 H_z(x,y,z)}{\partial x^2} + \frac{\partial^2 H_z(x,y,z)}{\partial y^2} - k_c^2 H_z(x,y,z) = 0,$$

$$E_z(x,y,z) = 0, m = 0, 1, 2, \ldots, n = 0, 1, 2, \ldots, (m,n) \neq (0,0)$$

we have

$$E_x(x,y,z) = j\frac{\omega\mu_0 n\pi}{k_c^2 b} A_{mn}\cos\left(\frac{m\pi}{a}x\right)\sin\left(\frac{n\pi}{b}y\right)e^{-j\beta z},$$

$$E_y(x,y,z) = -j\frac{\omega\mu_0 m\pi}{k_c^2 a} A_{mn}\sin\left(\frac{m\pi}{a}x\right)\cos\left(\frac{n\pi}{b}y\right)e^{-j\beta z}, E_z(x,y,z) = 0,$$

$$H_x(x,y,z) = j\frac{\beta m\pi}{k_c^2 a} A_{mn}\sin\left(\frac{m\pi}{a}x\right)\cos\left(\frac{n\pi}{b}y\right)e^{-j\beta z},$$

$$H_y(x,y,z) = j\frac{\beta n\pi}{k_c^2 b} A_{mn}\cos\left(\frac{m\pi}{a}x\right)\sin\left(\frac{n\pi}{b}y\right)e^{-j\beta z},$$

$$H_z(x,y,z) = A_{mn}\cos\left(\frac{m\pi}{a}x\right)\cos\left(\frac{n\pi}{b}y\right)e^{-j\beta z}.$$

For propagation of TE_{mn} modes, the inequality $[k^2 - (\frac{m\pi}{a})^2 - (\frac{n\pi}{b})^2] > 0$ should be met. The cutoff frequency for the TE_{mn} mode is $f_{c,mn} = \frac{c}{2\pi}\sqrt{(\frac{m\pi}{a})^2 + (\frac{n\pi}{b})^2}$ or $f_{c,mn} = \frac{1}{2\pi\sqrt{\mu_0\varepsilon}}\sqrt{(\frac{m\pi}{a})^2 + (\frac{n\pi}{b})^2}$.

If $\varepsilon_r = 1$, for the TE_{10} mode ($m = 1$, $n = 0$), the lowest cutoff frequency is $f_{c10} = c/2a$. If $a = 1000$ nm and $b = 500$ nm, one yields $f_{c,10} = 1.5 \times 10^{14}$ Hz and $\lambda = 2001$ nm. If $\varepsilon_r = 7.5$, for the TE_{10} mode, $f_{c,10} = 5.47 \times 10^{13}$ Hz and $\lambda = 5481$ nm.

The frequencies lower than f_c cannot propagate through the waveguide.

For the TM modes, one solves

$$\frac{\partial^2 E_z(x,y,z)}{\partial x^2} + \frac{\partial^2 E_z(x,y,z)}{\partial y^2} - k_c^2 E_z(x,y,z) = 0, H_z(x,y,z) = 0, m = 1, 2, \ldots, n = 1, 2, \ldots,$$

with the boundary conditions. We have

$$E_x(x,y,z) = -j\frac{\beta m\pi}{k_c^2 a}B_{mn}\cos\left(\frac{m\pi}{a}x\right)\sin\left(\frac{n\pi}{b}y\right)e^{-j\beta z},$$

$$E_y(x,y,z) = -j\frac{\beta n\pi}{k_c^2 b}B_{mn}\sin\left(\frac{m\pi}{a}x\right)\cos\left(\frac{n\pi}{b}y\right)e^{-j\beta z},$$

$$E_z(x,y,z) = B_{mn}\sin\left(\frac{m\pi}{a}x\right)\sin\left(\frac{n\pi}{b}y\right)e^{-j\beta z},$$

$$H_x(x,y,z) = j\frac{\omega\varepsilon n\pi}{k_c^2 b}B_{mn}\sin\left(\frac{m\pi}{a}x\right)\cos\left(\frac{n\pi}{b}y\right)e^{-j\beta z},$$

$$H_y(x,y,z) = -j\frac{\omega\varepsilon m\pi}{k_c^2 a}B_{mn}\cos\left(\frac{m\pi}{a}x\right)\sin\left(\frac{n\pi}{b}y\right)e^{-j\beta z}, H_z(x,y,z) = 0,$$

For propagation of TM_{mn} modes, the cutoff frequency is $f_{c,mn} = \frac{1}{2\pi\sqrt{\mu_0\varepsilon}}\sqrt{(\frac{m\pi}{a})^2 + (\frac{n\pi}{b})^2}$. In a rectangular waveguide with $\varepsilon_r = 7.5$, $a = 1000$ nm, and $b = 500$ nm, for the TM_{11} mode, one has $f_{c,11} = 1.22 \times 10^{14}$ Hz and $\lambda = 2451$ nm.

Figure 21.2b documents the surface for cutoff frequency $f_{c,mn}$ for the TE_{mn} and TM_{mn} modes ($m = 1, \ldots, 10$ and $n = 1, \ldots, 10$) in the 1000 × 500 nm waveguide, $\varepsilon_r = 7.5$. ■

Example 21.3.7: Surface Micromachined Dielectric Waveguides

The CMOS-compatible and silicon-compliant waveguide technologies are of importance in silicon photonics and optoelectronics. The waveguides under our consideration are buried-channel (rectangular, elliptic, and other geometries) planar waveguides. Our studies are applicable to diffused, ridge, rib, and strip-loaded, and rib waveguides. Surface micromachining (deposition, implantation, doping, etching, and lithography), of nanometers in thickness and width, optical media yields nonuniform heterogeneous core, cladding, and core-to-cladding interface. The fabrication and characterization of dielectric waveguides are reported in Section 21.7. We have spatial variations of $\sigma_i(x,y,z)$, $n_i(x,y,z)$, $\varepsilon_i(x,y,z)$, and $A_i(x,y,z)$. *Hybrid* modes with $f > f_c(E_z \neq 0$ and $H_z \neq 0)$, in addition to the TE and TM modes, are exhibited by the micromachined planar dielectric waveguides.

For dielectric materials, $\sigma \ll \omega\varepsilon$ for the operating frequency. Inhomogeneity of core and cladding results in distortions, losses, etc. The conductivity of the core and cladding are $\sigma_{core} \neq 0$ and $\sigma_{cladding} \neq \infty$. The waves in any propagating mode are attenuated, and the transmitted power decreases exponentially due to losses. The power losses α_i depend on modes, frequency, geometric heterogeneity, etc. These undesirable adverse features imply short waveguide length, use of error correction schemes, etc. It is impractical to take into account all effects and factors, phenomena, fabrication inconsistencies, etc. Under the aforementioned uncertainties, mathematical models with high-order tensors may not ensure adequateness and fidelity. One cannot find consistent analytic and numeric solutions to characterize the guided modes. The exploratory design may be accomplished by using the derived estimates, adequate materials, consistent fabrication processes, etc. The experimental characterization is needed to examine the *hybrid* modes. ■

Example 21.3.8: Guided Photon Propagation

Photon is the quantum of electromagnetic radiation, while the electromagnetic radiation is the *flux* of photons' quanta. Due to distortions, losses, and attenuation, quantum-deterministic and quantum-ensemble communication and processing may be achieved on a few or multiple photons. Consider photons, emitted by a 1 fW, 450 nm source such as laser or quantum dots. Thus, $E_{source} = 1 \times 10^{-15}$ J/s. The single-photon energy is $E_{\lambda=450\ nm} = 4.41 \times 10^{-19}$ J. The number of photons emitted is $N = E_{source}/E_{\lambda=450\ nm} = 2262$. The photodetectors are capable to detect the estimated number of photons at room temperature, despite distortions and losses. ■

Example 21.3.9

The number of photons with wavelength λ, emitted by a radiating device which produces power P during a period t, is $N = (Pt)/(hf) = (\lambda Pt)/(hc)$. The number of photons with $\lambda = 450$ nm, emitted by a radiating device that produces 1 μW during 10 ps, is 23. The energy of photons is 1.015×10^{-17} J. The Heisenberg uncertainty inequality $\sigma_E \sigma_t \geq {}^1/_2\hbar$ is guaranteed. We have $\Delta E \Delta t = 1.0151 \times 10^{-28} \geq {}^1/_2\hbar$, $\hbar = 1.0545718 \times 10^{-34}$ J-s. ■

Example 21.3.10: Nonlinear Optical Materials and Optical Phenomena in Waveguides

Optimized-by-specifications optical waveguides are designed using nonlinear optical media that exhibit nonlinear optical phenomena [29-37]. The mode- and wavelength-division-multiplexing [38-43], photon entanglement [44-46], and other communication- and processing-applicable effects can be implemented. Using surface micromachining, we fabricated optical waveguides with rectangular, trapezoidal, and ellipsoidal cross-sectional core and cladding geometries using different nonlinear optical structural and sacrificial materials, deposition, etching, and lithography [47]. Passive and active control schemes are examined for controlled photon propagation and transductions [29-43]. The refractive index n and thermo-optic coefficient depend on wavelength, core geometry, and dimensionality ($L \times W \times T$ for rectangular, $L \times \varnothing$ for circular, $L \times a \times b$ for ellipsoidal core), fabrication, media homogeneity, uniformity, etc. The controlled nonlinear electro-optical, magneto-optical, optical, and quantum effects in silicon nitride waveguides were demonstrated [22-28].

The refractive index can be changed. The difference in index of refraction Δn is due to the Kerr effect, $\Delta n = (n_{\parallel} - n_{\perp}) = \lambda \kappa E^2$, where κ is the media-dependent constant, and E is the electric field intensity. The thermo-optic coefficients dn/dT of silicon nitride and silicon dioxide are $(2.45 \pm 0.08) \times 10^{-5}$ K^{-1} and $(0.95 \pm 0.09) \times 10^{-5}$ K^{-1} at 300 K for $\lambda = 1550$ nm.

The index of refraction is controlled by changing the electric field intensity E. For nonlinear optical media

$$\boldsymbol{P}(\boldsymbol{E}) = \varepsilon_0(\chi^{(1)} \cdot \boldsymbol{E} + \chi^{(2)} \cdot \boldsymbol{EE} + \chi^{(3)} \cdot \boldsymbol{EEE} + \ldots) = \boldsymbol{P}^{(1)} + \boldsymbol{P}^{\mathrm{NL}},$$

where $\chi^{(n)}$ is the nth-order susceptibility tensor of rank $(n + 1)$ with 3^n components; $\chi^{(2)}$ is the third rank tensors with 27 elements; $\chi^{(3)}$ is the fourth rank tensor with 81 elements, and each element consists 48 terms.

For four-wave mixing in a $\chi^{(3)}$-medium with the idler frequency $f_i = (2f_p - f_s)$, the polarization is $P = \varepsilon_0(\chi E + \chi^{(3)}E^3)$. If a medium exhibits the third-order electric susceptibility optical Kerr effect, mixing a beam and signal with frequencies (f_p, f_s) yields the output tuple (f_p, f_s, f_i) with the idler frequency f_i.

One solves the Maxwell equation $\nabla \times \nabla \times \boldsymbol{E} + \frac{n^2}{c^2}\frac{\partial^2}{\partial t^2}\boldsymbol{E} = -\frac{1}{\varepsilon_0 c^2}\frac{\partial^2}{\partial t^2}\boldsymbol{P}^{NL}$.

Applying the slowly varying envelope approximation, one finds the spatiotemporal field magnitudes by solving [31]

$$\frac{\partial E_1}{\partial z} + \frac{1}{\nu_1}\frac{\partial E_1}{\partial t} = j2\pi\frac{\omega_1}{n_1 c}\left[\chi_{1234}^{NL} E_2 E_3^* E_4 e^{-i\Delta kz} + \sum_{j=1}^{4} \chi_{1j}^{NL} E_1 E_j E_j^*\right],$$

$$\frac{\partial E_2}{\partial z} + \frac{1}{\nu_2}\frac{\partial E_2}{\partial t} = j2\pi\frac{\omega_2}{n_2 c}\left[\chi_{1234}^{NL} E_1 E_3 E_4^* e^{i\Delta kz} + \sum_{j=1}^{4} \chi_{2j}^{NL} E_2 E_j E_j^*\right],$$

$$\frac{\partial E_3}{\partial z}+\frac{1}{\nu_3}\frac{\partial E_3}{\partial t}=j2\pi\frac{\omega_3}{n_3c}\left[\chi_{1234}^{NL}E_1^*E_2E_4e^{-i\Delta kz}+\sum_{j=1}^{4}\chi_{3j}^{NL}E_3E_jE_j^*\right],$$

$$\frac{\partial E_4}{\partial z}+\frac{1}{\nu_4}\frac{\partial E_4}{\partial t}=j2\pi\frac{\omega_4}{n_4c}\left[\chi_{1234}^{NL}E_1E_2^*E_3e^{i\Delta kz}+\sum_{j=1}^{4}\chi_{4j}^{NL}E_4E_jE_j^*\right],$$

where χ_{1234}^{NL} and χ_{ij}^{NL} are the nonlinear coefficients, $\chi_{1234}^{NL}=\hat{e}_1\hat{e}_2^*:\chi^{(3)}(-\omega_4,\omega_1,-\omega_2\omega_3):\hat{e}_3\hat{e}_4^*$ and $\chi_{ij}^{NL}=\hat{e}_i\hat{e}_i^*:\chi^{(3)}(-\omega_i,\omega_i,-\omega_j,\omega_j):\hat{e}_j\hat{e}_j^*$; e_1, e_2, e_3 and e_4 are the polarization vectors; and Δk is the wave vector mismatch, $\Delta k=(k_1-k_2+k_3-k_4)$.

The nonlinear Kerr coefficient is related to the third-order susceptibility $\chi^{(3)}$, $n_2=\frac{3\chi^{(3)}}{4c\varepsilon_0^2}$, where n_2 is the nonlinear refraction coefficient.

For Si_3N_4, n_2 varies from 1×10^{-19} m^2/W to 1×10^{-18} m^2/W depending on fabrication processes (PECVD, LPCVD, etc.), wavelength, dimensionality, core-cladding boundary, homogeneity, etc. Low-pressure chemical vapor deposition (LPCVD) at temperature (700–900°C) yields a stoichiometric homogenous Si_3N_4 with refractive index $n|_{\lambda=635\text{ nm}}=2.022-j0.000$ and $n|_{\lambda=1550\text{ nm}}=1.995-j0.000$. For silicon and SiO_2, if $\lambda=635$ nm, $n_{Si}=3.88-j0.019$ and $n_{SiO2}=1.456-j0.000$, respectively. The low temperature plasma-enhanced chemical vapor deposition (PECVD) yields silicon-rich Si_3N_4 with the refractive index n, which varies from 1.9 to 2.2 at $\lambda=1550$ nm. For nitrogen-rich Si_3N_4, $n<2$.

The intensity $\boldsymbol{I}$ is the time-averaged power through a unit area, $\boldsymbol{I}=\langle\boldsymbol{S}\rangle$. For a plane wave, propagating in the z-direction, $\boldsymbol{E}(t,\boldsymbol{r})=\boldsymbol{E}_0\cos(\omega t-\boldsymbol{k}\cdot\boldsymbol{r}+\phi)=\frac{1}{2}\tilde{\boldsymbol{E}}_0e^{j(\omega t-\boldsymbol{k}\cdot\boldsymbol{r})}+\text{c.c.}$

One finds $I=I_z=\frac{1}{2}c\varepsilon_0nE_0^2=\frac{1}{2}c\varepsilon_0n|\tilde{E}_0|^2$.

The intensity depends on the refractive index due to the third-order susceptibility. The effective refractive index is $n(\omega,I)=n_0(\omega)+n_2I, n_2(\omega)=\frac{3}{8n_0}\text{Re}(\chi^{(3)})$, where n_0 is the refractive index at the carrier frequency. The intensity dependence of n results in:

1. *Self-phase modulation (SPM)*: Pulses in a nonlinear optical medium induce a varying refractive index and produce a phase shift in the pulse with the magnitude $\phi_m=(n+n_2I)k_0L$ and $\phi_{spm}=n_2\,|\,E\,|^2\,k_0L$, where k_0 is the wavenumber, $k_0=2\pi/\lambda$, and L is the length.
2. *Cross-phase modulation (CPM)*: The optical fields exhibit nonlinear phaseshift in the presence of the others. For E_1 and E_2 with ω_1 and ω_2, $\phi_{cpm}=2n_2|E_2|^2k_0L$.
3. The total phase shift is the sum of the SPM and CPM terms, and $\phi_\Sigma=\phi_{spm}+\phi_{cpm}$.

We study the pulse propagation in the optical core, solving the partial differential equation for the pulse envelope $A(t,z)$

$$\frac{1}{2}j\delta\frac{\partial^2A(t,z)}{\partial t^2}+\frac{\partial A(t,z)}{\partial z}=-\frac{1}{2}\alpha A(t,z)+j\,bA(t,z)|A(t,z)|^2,$$

where δ, α, and b are the group velocity dispersion, loss, and effective scaling coefficients, $b=\omega n_2/(A_{eff}c)=2\pi n_2/(\lambda A_{eff})$; and A_{eff} is the effective area of the optical waveguide core. In the lossless media with $\alpha=0$, for the power P_0 and phase shift $\phi_\Sigma=P_0bz$, one finds a time-independent solution $A(z)=\sqrt{P_0}e^{jbP_0z}$. Silicon nitride exhibits relatively low losses and high refractive index, $n_{\lambda=300\text{ nm}}=2.18$ and $n_{\lambda=3000\text{ nm}}=1.95$. Therefore, Si_3N_4 is chosen as a nonlinear optical media for a core. Silicon dioxide ($n_{\lambda=160\text{ nm}}=1.55$ and $n_{\lambda=3000\text{ nm}}=1.40$) is chosen as the cladding material. It is possible to ensure the design-by-specification refractive index profile using dopants. Boron B and fluorine F dopants reduce the refractive index. For the 1000×500 nm silicon nitride waveguide with the Kerr coefficient $n_2=2.4\times10^{-19}$ m^2/W, the surfaces for the real and imaginary $A(P_0,z)$ are reported in Figure 21.3 for varying P_0 and z.

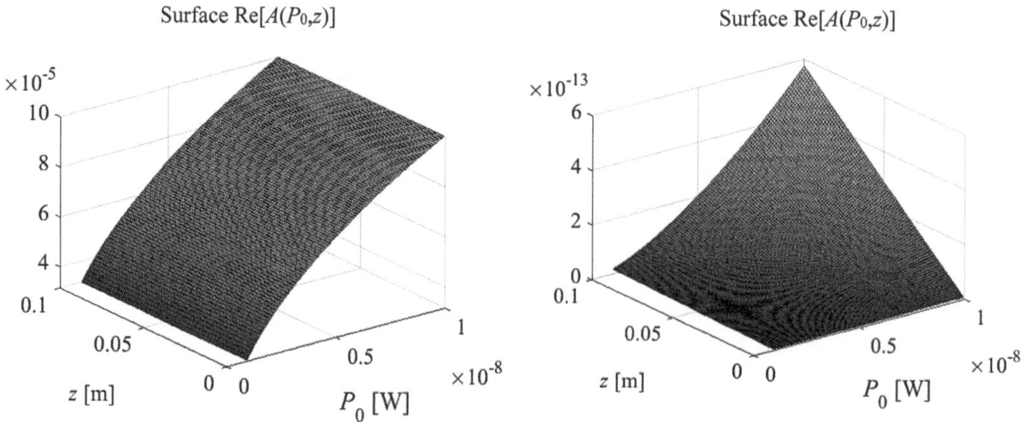

FIGURE 21.3 Surfaces for Re[$A(P_0,z)$] and Im[$A(P_0,z)$], $A(z) = \sqrt{P_0}e^{jbP_0z}$ when z and P_0 vary from 0.1 to 10 cm and 1 to 10 nW.

■

21.4 DESCRIPTIVE ANALYSIS OF PROCESSING AND COMMUNICATION: INFORMATION-THEORETIC ANALYSIS WITH APPLICATIONS

We focus on quantitative estimates and measures to analyze performance of communication and processing. The channel capacity, throughput, data rate, error estimates, data-processing capabilities, and other key features are investigated. Paradigms of theoretical information and computer sciences are applied.

Example 21.4.1

Consider an information source comprised from four symbols $\{a, b, c, d\}$. Let these symbols be random, and probabilities are $P\{a\} = 1/2, P\{b\} = 1/4, P(c\} = 1/8$ and $P\{d\} = 1/8$. The binary representation by means of *codewords* can be $a \to 00$, $b \to 01$, $c \to 10$, and $d \to 11$. Various coding schemes are used. The sample space can be (0,0), (0,1), (1,0), and (1,1). For $x \in \{a, b, c, d\}$, the estimate for the *information content* of x can be expressed as $I(x) \equiv \log_2 \frac{1}{p(x)} = -\log_2 p(x)$. ■

The *information content* of the information source is

$$\sum_x p(x)I(x) = -\sum_x p(x)\log_2 p(x). \tag{21.9}$$

The entropy measures the complexity of the set [48, 49]. The sets having larger entropies require more bits to represent them. For M symbols x_i, characterized by the probability density $p(x_i)$, the entropy is

$$H(X) = \sum_{i=1}^{M} p(x_i)\log_2 p(x_i), H \geq 0, i = 1, 2, \ldots, M-1, M. \tag{21.10}$$

The entropy of the information source $H(X)$ defines, or, equal to, the minimum number of digits per message required. The number of bits required by the Source Coding Theorem is positive, and $H \geq 0$. One concludes: (i) N independent and identically distributed random variables, each with entropy $H(X)$, can be compressed into more than $NH(X)$ bits with negligible risk of information loss as $N \to \infty$; and (b) if N random variables are compressed into fewer than $NH(X)$ bits, it is almost certain that information will be lost.

Example 21.4.2

Consider a random variable X whose realizations x are *letters* (symbols) in an *alphabet* $\mathcal{A}$. Let

$$X = \begin{cases} a \text{ with probability } \frac{1}{3} \\ b \text{ with probability } \frac{1}{6} \\ c \text{ with probability } \frac{1}{6} \\ d \text{ with probability } \frac{1}{12} \\ e \text{ with probability } \frac{1}{12} \\ f \text{ with probability } \frac{1}{12} \\ g \text{ with probability } \frac{1}{12} \end{cases}$$

. We have

$$H(X) = -\sum_{i=1}^{M} p(x_i)\log_2 p(x_i) = -\frac{1}{3}\log_2\frac{1}{3} - \frac{1}{6}\log_2\frac{1}{6} - \frac{1}{6}\log_2\frac{1}{6} - \frac{1}{12}\log_2\frac{1}{12} - \frac{1}{12}\log_2\frac{1}{12} - \frac{1}{12}\log_2\frac{1}{12} - \frac{1}{12}\log_2\frac{1}{12}$$

$$= 2.585 \text{ bit.}$$

■

Example 21.4.3

Consider a string of eight symbols (digits). Let each symbol be either binary (two-level) or quaternary. We have $2 \times 2 \times \ldots \times 2 \times 2 = 2^8 = 256$ and $4 \times 4 \times \ldots \times 4 \times 4 = 4^8 = 65{,}536$ different combinations (words).

Assume that each level is equally probable. One obtains the probabilities $p_i = (1/2)^8$ or $p_i = (1/4)^8$.

For the first and second cases, one has $H = (\frac{1}{2})^8 \log_2 \frac{1}{(\frac{1}{2})^8} = 0.0313$ bits and

$H = (\frac{1}{4})^8 \log_2 \frac{1}{(\frac{1}{4})^8} = 0.000244$ bits.

The maximum average information per symbol is of our interest. From $\log_2(1/p_i)$, one concludes that the maximum average information per symbol is 8 bits and 16 bits, $\log_2 \frac{1}{(\frac{1}{2})^8} = 8$ and $\log_2 \frac{1}{(\frac{1}{4})^8} = 16$.

The information rate is $R = r_s H_s$, where r_s is the symbol rate in symbols per second; H_s is the source entropy in bits per symbol. ■

Example 21.4.4

In communication systems, the transmitter T transmits a sequence of bits. One examines the probability that the receiver R receives data bits. Let the transmitter send "1" and "0" with probabilities $P[1] = p$ and $P[0] = 1 - p$. The binomial distribution $B(n,m)$ with parameters n and m characterizes the discrete probability distribution of a random variable. A sequence of N independent events (sequence of Boolean outcomes) is considered. The outcomes are 0 and 1, with the probabilities $P[1] = p$ and $P[0] = (1 - p)$. The outcomes form a symbolic sample space $S = \{0\ 1\}$. The random variable X is $X = \begin{cases} 1 \text{ with probability } p \\ 0 \text{ with probability } (1-p) \end{cases}$. The probability mass functions are $p_X[0] = P[X = 0] = (1 - p)$ and $p_X[1] = P[X = 1] = p$. That is, $p_X(x) = \begin{cases} p \\ 1-p \end{cases}$. The entropy (9) $H(X) = \sum_{x \in X} p_X(x)\log_2 p_X(x)$, as a function of p, is

$$H(X) = -\sum_{i=1}^{M} p(x_i)\log_2 p(x_i) = -p\log_2 p - (1-p)\log_2(1-p).$$

The Shannon entropy [49] is a quantitative measure of information. ■

Example 21.4.5

Processing is performed implementing hardware-consistent algorithmic operations and transforms, thereby implementing communication tasks (error correction, orthogonal frequency division multiplexing, quadrature amplitude modulation, etc.) and realizing switching functions [5-7, 50]. For m operations $x^m = x_1x_2\ldots x_{m-1}x_m$, for each x_i, there is a corresponding X_i. While examining processing schemes, assume that all X_i are independent and identical. Within the device-consistent software and algorithmic alphabet $\mathcal{A}$, distinguishable operations are $a_1, a_2, \ldots, a_A$. The number of occurrences of an operation a_i in x^j is $N(a_i, x^m)$. One has $p_{X^m}(x^m) = \prod_{i=1}^{m} p_X(x_i) = \prod_{i=1}^{A} p_X(a_i)^{N(a_i,x^m)}$ with $x^m \to \underbrace{a_1 \cdots a_1}_{N(a_1,x^m)} \underbrace{a_2 \cdots a_2}_{N(a_2,x^m)} \cdots \underbrace{a_A \cdots a_A}_{N(a_A,x^m)}$.

The information processing content is

$$I \equiv \log_2 \frac{1}{p_{X^m}(X^m)} = -\log_2 p_{X^m}(X^m) = -\log_2 \left(\prod_{i=1}^{A} p_X(a_i)^{N(a_i,\, X^m)}\right) = -\sum_{i=1}^{A} \log_2 p_X(a_i)^{N(a_i,\, X^m)}$$

$$= -\sum_{i=1}^{A} N(a_i, X^m)\log_2 p_X(a_i).$$

Hence, the processing implies the size for operations $\mathcal{A}^m = 2^{n\log\mathcal{A}}$ and the set size in the order of $2^{nH(X)}$. ■

Example 21.4.6

Consider a random variable X that is normally distributed with mean μ and variance σ^2. The corresponding probability density function (pdf) $p_X(x)$ is given by the Gaussian function $p_X(x) = \frac{1}{\sigma\sqrt{2\pi}} e^{-\frac{1}{2}\frac{(x-\mu)^2}{\sigma^2}}$, $-\infty < x < \infty$. Here, μ is the expectation (mean), $\mu = E[X] = \int_{-\infty}^{\infty} xf(x)dx$; σ is the standard deviation; and σ^2 is the variance, $\sigma^2 = E[(X-\mu)^2] = \int_{-\infty}^{\infty} (x-\mu)^2 f(x)dx$.

The normal distribution of X with a mean μ and variance σ^2 is denoted as $X \sim N(\mu, \sigma^2)$. If $\mu = 0$ and $\sigma^2 = 1$, one has the standard normal pdf $p_X(x) = \frac{1}{\sqrt{2\pi}} e^{-\frac{1}{2}x^2}$. If for X and Y, $\mu = 0$ ($\mu_X = 0$ and $\mu_Y = 0$) and $\sigma^2 = 1$($\sigma_X = 1$ and $\sigma_Y = 1$), the standard bivariate Gaussian pdf with the correlation ρ is $p_{X,Y}(x, y) = \frac{1}{2\pi\sqrt{1-\rho}} e^{-\frac{1}{2(1-c^2)}(x^2 - 2\rho xy + y^2)}$, $-1 < \rho < 1$, $-\infty < x < \infty$, $-\infty < y < \infty$.

For a multivariate normal distribution $\boldsymbol{X} \sim \mathcal{N}(\mu, \Sigma)$, $\boldsymbol{X} = (X_1, \ldots, X_n)$ with mean vector $\mu \in \mathbb{R}^n$ and covariance matrix $\Sigma \in \mathbb{R}^{n\times n}$, the pdf is $p(\boldsymbol{x}; \boldsymbol{\mu}, \boldsymbol{\Sigma}) = \frac{1}{\sqrt{2^n \pi^n |\boldsymbol{\Sigma}|}} e^{-\frac{1}{2}(\boldsymbol{X}-\boldsymbol{\mu})^T \Sigma^{-1}(x-\mu)}$.

For the Laplacian and Rayleigh distributions, the pdfs are $p_X(x) = \frac{1}{\sqrt{2\sigma^2}} e^{-\sqrt{\frac{2}{\sigma^2}}|x|}$ and $p_X(x) = \frac{x}{\sigma^2} e^{-\frac{1}{2}\frac{x^2}{\sigma^2}}$. ■

Example 21.4.7: Analog Signaling, Computing and Processing

In living organisms, analog *signaling* and *computing* are investigated [10-13]. The examples include, but are not limited to, action potentials in neurons, excitatory postsynaptic potential, inhibitory postsynaptic potential, electroencephalogram (electrical activity of the brain), electro-oculography (corneo-retinal potentials in the eye), electromyogram, magnetoencephalogram, etc. In *biometrics*, *brain–computer* interfacing, and other assertions, information-theoretic premises are used. Attempting to address various unsolved problems of neuroscience, bioengineering, and life sciences, continuous probability distributions, probabilistic models, and information estimates were investigated [5-7, 12, 51-54].

Various mechanical and electronic analog computing platforms were used for centuries, such as the Antikythera mechanism (100 BC), astronomical clock (1206), mechanical analog computer for fire control systems (1912), fully electronic analog computer for V-2 rockets (1942), the 1960s general-use analog computers with diodes and vacuum tubes, etc. The electromechanical switching devices for computing were demonstrated in the 1930s [55-57], the *Electronic Numerical Integrator and Computer* was revealed in 1943, and general-use analog computers and digital processors emerged in the 1950s and 1970s, respectively. The theoretical formalisms of quantum computing [58-63] and quantum communication [64-71] have been investigated since the 1950s, and these theoretical premises are still under study [72-76]. In many applications, analog operations and processing ensure superior performance compared with digital processing. For example, analog electronic filters and controllers are used. Furthermore, the majority of laws of physics, physical processes, devices, sensors, and systems are inherently continuous, including field-effect transistors, which are used in digital circuits and processors. Correspondingly, continuity formalism, analog processing, and continuous arithmetic (calculi) are topics of great importance and interest. Despite a very broad use of analog processing schemes (analog filters, analog multiplier and dividers, etc.) and pioneering publication in 1920 by Jan Łukasiewicz and 1936 by Garrett Birkhoff and John Von Neumann [58, 77], the general-purpose analog processing calculus is still under development [5-7, 78-82].

Examining analog computation and processing on continuous-time signals, a *differential entropy* can be applied. For a continuous-time random variable X, the *differential entropy* is

$$H(X) = -\int p_X(x)\log_2 p_X(x)dx, \int p_X(x)dx = 1 \text{ and } H(X_1, X_2, \ldots, X_{n-1}, X_n) = -\int p_X(\mathbf{x})\log_2 p_x(\mathbf{x})d\mathbf{x}$$

The relative entropy between pdfs $p_X(\mathbf{x})$ and $g_X(\mathbf{x})$ is $H_R(p_X||g_X) = \int p_X(\mathbf{x})\log_2 \frac{p_X(\mathbf{X})}{g_X(\mathbf{X})}d\mathbf{x}$.

For Cauchy, exponential, Laplace, Maxwell-Boltzmann, normal, and uniform distributions, using $p_X(x)$, we have the following pdfs and *differential entropies* $H(X)$:

(1) Cauchy distribution, $p_X(x) = \frac{a}{\pi}\frac{1}{a^2+x^2}, -\infty < x < \infty, a > 0, H(X) = ln(4\pi a)$;
(2) Exponential distribution, $p_X(x) = \frac{1}{a}e^{-\frac{x}{a}}, x > 0, a > 0, H(X) = 1 + \ln a$;
(3) Laplace distribution, $p_X(x) = \frac{1}{2a}e^{-\frac{|x-b|}{a}}, -\infty < x < \infty, a > 0, -\infty < b < \infty$, $H(X) = 1 + \ln(2a)$;
(4) Maxwell–Boltzmann distribution, $p_X(x) = 4\pi^{-\frac{1}{2}}a^{\frac{3}{2}}x^2e^{-ax^2}, x > 0, a > 0$, $H(X) = -0.023 + 0.5\ln(\pi/a)$;
(5) Normal distribution, $p_x(x) = \frac{1}{\sigma\sqrt{2\pi}}e^{-\frac{(x-\mu)^2}{2\sigma^2}}, -\infty < x < \infty, -\infty < \mu < \infty, \sigma > 0$, $H(X) = 0.5\ln(2\pi e\sigma^2)$ for $\mu = 0$;
(6) Uniform distribution, $p_X(x) = \frac{1}{b-a}, a \le x \le b, H(X) = \ln(b - a)$.

The *differential entropy* can be negative. The *differential entropy* of a Gaussian random variable with $p_X(x) = \frac{1}{\sigma\sqrt{2\pi}}e^{-\frac{x^2}{2\sigma^2}}$ is $H(X) = 0.5\ln(2\pi e\sigma^2)$. For a multivariate normal distribution $\mathbf{X} \sim \mathcal{N}(0, \mathbf{\Sigma})$ with covarience matrix $\mathbf{\Sigma} \in \mathbb{R}^{n\times n}$, one finds $H(X) = \frac{1}{2}\ln((2\pi e)^n|\Sigma|) = \frac{1}{2}\ln|2\pi e\Sigma|$. Thus, $H(X)$ can be positive, negative, or zero depending on the variance σ^2. The *differential entropy* depends on scaling, unit conversion, etc. If $Z = kX$, $H(Z) = H(X) + \log_2|k|$, where k is the scaling constant.

Continuous-time signals can be discretized. Let X_n denote a discretized continuous random variable with a sampling ΔT. We have, $\lim_{\Delta T\to 0} H(X_n) + \log_2\Delta T = H(X)$. The problems, however, are the inherent continuity of physical quantities, processing schemes, information carrying signals, information losses in analog to digital conversion, etc.

For memoryless channels, consider a source $[x_1, x_2, \ldots, x_{s-1}, x_s]$ and receiver $[y_1, y_2, \ldots, y_{r-1}, y_r]$ symbols. For random variables X and Y, which are statistically dependent, the conditional information content is analyzed using the marginal pdfs $p_X(x)$ and $p_Y(y)$ and the joint pdf $p_{X,Y}(x, y)$. The mutual information is $I(X; Y) = \int_Y\int_X p_{X,Y}(x, y)\log_2\frac{p_{X,Y}(x,y)}{p_X(x)p_Y(y)}dxdy$. Using the conditional probability $p_{X|Y}(x|y)$, one has $I(X | Y) \equiv -\log_2 p_{X|Y}(x | y)$. ■

Example 21.4.8: Estimates on Information Content, Information Losses, and Information Acquired

The conditional entropy $H(X \mid Y = y)$ of a random variable X, conditional on a particular realization y of Y, is the expected conditional information content, $H(X|Y=y) = -\sum_x p_{X|Y}(x|y)\log_2 p_{X|Y}(x|y)$. The expected conditional information content with respect to both X and Y is

$H(X|Y) = -\sum_{x\in X}\sum_{y\in Y} p_{X,Y}(x,y)\log_2 p_{X|Y}(x|y)$, $H(X \mid Y) \geq 0$.

The entropy $H(X)$ is greater than or equal to the positive-definite conditional entropy. That is, $H(X) \geq H(X \mid Y)$, $H(X) \geq 0$, $H(Y) \geq 0$, and $H(X \mid Y) \geq 0$. The conditional entropy $H(X \mid Y)$ corresponds to the average loss of information.

The joint entropy is the entropy of the joint random variable (X, Y),

$H(X, Y) = -\sum_{x,y} p_{X,Y}(x,y)\log_2 p_{X,Y}(x,y)$.

Furthermore, $H(X, Y) = H(X) + H(Y \mid X) = H(Y) + H(X \mid Y)$.

The mutual information $I(X; Y)$ between the stimulus and the response quantitatively defines the amount of information received on average. The mutual information $I(X; Y)$ provides a quantitative estimate on the dependence of two discrete random variables X and Y, and

$$I(X;Y) = I(Y;X) = H(X) - H(X \mid Y) = H(Y) - H(Y \mid X) = H(X) + H(Y) - H(X,Y),$$

$$I(X;Y) = \sum_{x\in X}\sum_{y\in Y} p_{X,Y}(x,y)\log_2 \frac{p_{X,Y}(x,y)}{p_X(x)p_Y(y)} \text{ [bits/symbol]},\ I(X;Y) \geq 0.$$

The positive-definite mutual information $I(X;Y) \geq 0$ provides the average amount of information received per symbol transmitted or processed. ■

Example 21.4.9: Information estimates on processing using the input–Output Mappings

For two jointly discrete random variables X and Y, the conditional probability mass function is

$$p_{Y|X}(y_j|x_i) = \frac{p_{X,Y}(x_i,y_j)}{p_X(x_i)},\ j = 1, 2, \ldots \text{ with } \sum_{j=1}^{\infty} p_{Y|X}(y_j|x_i) = \sum_{j=1}^{\infty} \frac{p_{X,Y}(x_i,y_j)}{p_X(x_i)} = 1.$$

The first equation gives the probability of the event $Y = y_j$ for $j = 1, 2, \ldots$ once $X = x_i$. For continuous random variables X and Y, the joint pdf $p_{X,Y}(x, y)$ is

$p_{X,Y}(x,y) = p_{Y|X}(y \mid x)p_X(x) = p_{X|Y}(x \mid y)p_Y(y)$.

The conditional pdfs are $p_{X|Y}(x|y) = \frac{p_{X,Y}(x,y)}{\int_{-\infty}^{\infty} p_{X,Y}(x,y)dx}$ and $p_{Y|X}(y|x) = \frac{p_{X,Y}(x,y)}{\int_{-\infty}^{\infty} p_{X,Y}(x,y)dy}$.

These pdfs are related as $p_{X|Y}(x|y) = \frac{p_{Y|X}(y|x)p_X(x)}{p_Y(y)}$. Furthermore, $p_{Y|X}(y|x) = \frac{p_{X,Y}(x,y)}{p_X(x)}$.

Two random variables X and Y possess zero bits of mutual information if and only if they are statistically independent. If X and Y are statistically independent, the joint pdf is $p_{X,Y}(x,y) = p_x(x)p_Y(y)$. Thus, $I(X; Y) = 0$. Any information on Y does not provide any information on X if they are statistically independent.

Two random variables X and Y possess $H(X)$ bits of mutual information if they are perfectly correlated as $Y = X$. ■

Example 21.4.10

Many data-processing schemes require the multiplication of random variables X_1 and X_2, $X = X_1X_2 \to Y$. Different algorithms and circuit-level solutions exist and have been commercialized. Fundamentally distinct continuous (analog multipliers and analog dividers), digital, and *hybrid* ICs are implemented. The commercialized low-power low-noise analog multipliers and dividers may ensure exceptional performance as compared to digital circuits. For continuous random variables X_1 and X_2, the pdf of $X = X_1X_2$ is $p_X(x) = \int_{-\infty}^{\infty} p_{X_1}\frac{1}{|x_1|}p_{X_2}(x/x_1)dx_1$. For $X_1 \sim \mathcal{N}(0,1)$ and $X_2 \sim \mathcal{N}(0,1)$, we have $p_X(x) = \mathcal{B}_\mathcal{K}(0, x\text{sgn}(x))/\pi$, $-\infty < x < \infty$, where $\mathcal{B}_\mathcal{K}$ is the modified Bessel function of the second kind, $\mathcal{B}_{\mathcal{K}} \to 0$ as $x \to \infty$ and $x \to -\infty$. One applies the aforementioned concepts and solutions to analyze analog processing. ■

Example 21.4.11

The data-processing inequality for three random variables X, Y, and Z, which form a Markov chain $X \to Y \to Z$, is $I(X;Y) \geq I(X;Z)$. Furthermore, $I(X;Y) \geq I(X;Y\,|\,Z) \geq 0$, where the mutual information of two random variables, conditioned to a third random variable, is $I(X;Y\,|\,Z) = \sum_{z\in Z}\sum_{y\in Y}\sum_{x\in X} p_{X,Y,Z}(x,y,z)\log_2 \frac{p_{X,Y|Z}(x,y|z)}{p_{X|Z}(x|z)p_{Y|Z}(y|z)}$. If random X and Y are conditionally independent through Z, the conditional mutual information is zero,

e.g., $I(X;Y\,|\,Z) = 0$ if $p_{X,Y|Z}(x,y\,|\,z) = p_{X|Z}(x\,|\,z)p_{Y|Z}(y\,|\,z)$. ■

Example 21.4.12: Relative Entropy and its Application

By using the relative entropy $D(p_{X1} \| p_{X2})$, one may evaluate equivalency, errors, failures, cybersecurity, malicious cyberattacks, vulnerability, risk factors, information risk, countermeasure effectiveness, and threats in communication and processing. The computing, coding, encryption, and other tasks can be analyzed and characterized. The relative entropy quantitatively measures the equivalence of pdfs and yields quantitative information estimates. For $p_{X1}(x)$ and $p_{X2}(x)$, the relative entropy is

$$D(p_{X_1} \| p_{X_2}) = \sum_x p_{X_1}(x)\log_2 \frac{p_{X_1}(x)}{p_{X_2}(x)}.$$

One yields $D(p_{X1} \| p_{X2}) \geq 0$ for any pdfs $p_{X1}(x)$ and $p_{X2}(x)$.

Furthermore, $I(X;Y) = D(p_{X,Y}(x,y) \| p_X(x)p_Y(y))$.

The relative entropy results in adequate quantitative estimates for various problems, starting form parallel computing to cybersecurity. ■

21.5 PERFORMANCE ESTIMATES OF COMMUNICATION AND PROCESSING

Processing and communication capabilities depend on device switching, throughput, data rate, bit error, latencies, noise power spectral density, channel capacity, and other factors. The processing mappings are

$$F : X \to Y \text{ and } F : X \times U(Y) \to Y, \tag{21.11}$$

where X, U, and Y are the input, feedback, and output.

Multiple-valued and multi-state algebra and switching theory [50], as well as Boolean algebra, are applicable in quantum-deterministic and quantum-ensemble processing. An n-variable r-valued function f is defined as a mapping from a finite set $\{0,1,\ldots,r-1\}^n$ into a finite set $\{0,1,\ldots,r-1\}$, and $f : \{0,1,\ldots,r-1\}^n \to \{0,1,\ldots,r-1\}$. A truth vector for an r-valued logic function f of n variables $(x_1,x_2,\ldots,x_{n-1},x_n)$ is $\boldsymbol{F} = [f(0), f(1), \ldots, f(r^n-1)]^T$.

The *achievable* digital signal processing capability C on n-variable, r-valued, multi-state inputs and outputs depends on the informational content of the finite alphabet $I(x_i) = -\log_2 p(x_i)$, channel capacity C, switching function f *realizability* $R(f)$ and *computability* $M(f)$, and data rate $D(f)$. We have $\boldsymbol{C} \cong I(x) \circ C \circ R(f) \circ M(f) \circ D(f)$. The processing energy for digital signal processing is $\mathcal{E} \cong \mathcal{E}_u \mathcal{N}_u$, where $\mathcal{E}_u$ and $\mathcal{N}_u$ is the unitary *switching energy* and required number of switching. Quantitative measures and estimates, which predefine or affect quintuple (I, C, R, M, D), are examined in this chapter.

Example 21.5.1

The ternary and quaternary logic functions are

f: $\{0, 1, 2\}^n \to \{0, 1, 2\}$ and f: $\{0, 1, 2, 3\}^n \to \{0, 1, 2, 3\}$.

The quaternary-input, binary-output logic function is f: $\{0, 1, 2, 3\}^n \to \{0, 1\}$.

The truth vector of a ternary *MIN* function for two variables is $\mathbf{F} = [0\ 0\ 0\ 0\ 1\ 1\ 0\ 1\ 2]^T$.

Various operations (MAX, MIN, SUM, and other), polynomial and word-level representations, fundamental expansions, and logics on multiple-valued functions are documented [50]. ■

Example 21.5.2

There are different line coding schemes, such as unipolar, polar, bipolar, and others. The *signal level* is the allowed signal values that represent the data. The pulse rate r_p is a number of pulses per second. The bit rate r_b is a number of bits per second, $r_b = r_p \times \log_2 L$, where L is the number of data levels, and $\log_2 L$ is the number of bits per level.

For unipolar (high "1" and low "0") and bipolar encoding, $L = 2$ and $L = 3$.

If the signal pulse duration is 10 ps and $L = 2$, the bit rate is $r_b = 1 \times 10^{11}$ bit/s. ■

Example 21.5.3

The effectiveness of the source code is estimated using the length of a word L, entropy $H(x)$, and the number of symbols in the encoding alphabet D. Using the average length of the code $\bar{L}$, one yields $\frac{\bar{L}_{\min}}{\bar{L}} = \frac{\bar{L}_{\min}}{\sum_{i=1}^{n} p(x_i) l_i}$, $\bar{L}_{\min} = \frac{H(x)}{\log_2 D}$. ■

Quantum transductions, transforms, and quantum processing: We consider the *distinguishable* and *computable* transforms $\mathcal{F}_i(\mathcal{T}, \mathbf{x})$, which are mappings on the *utilizable* initial, intermediate, and final *observable* transductions $(\mathcal{T}_I, \mathcal{T}_R, \mathcal{T}_F)$ or $(\langle\mathcal{T}_I\rangle, \langle\mathcal{T}_R\rangle, \langle\mathcal{T}_F\rangle)$ of variables $\mathbf{x}(t, \mathcal{T}) = [x_1, \ldots, x_k]^T$ in controlled physical *microscopic* systems. The devices accomplish the following irreversible and reversible transductions

$$\mathcal{T}_I : \mathbf{x}_I \to \mathcal{T}_R : \mathbf{x}_R \to \mathcal{T}_F : \mathbf{x}_F \text{ and } \mathcal{T}_I : \mathbf{x}_I \leftrightarrow \mathcal{T}_R : \mathbf{x}_R \leftrightarrow \mathcal{T}_F : \mathbf{x}_F, \mathbf{x} \in \mathbb{R}, \tag{21.12}$$

$$\langle\mathcal{T}_I\rangle : \langle\mathbf{x}_I\rangle \to \langle\mathcal{T}_R\rangle : \langle\mathbf{x}_R\rangle \to \langle\mathcal{T}_F\rangle : \langle\mathbf{x}_F\rangle \text{ and } \langle\mathcal{T}_I\rangle : \langle\mathbf{x}_I\rangle \leftrightarrow \langle\mathcal{T}_R\rangle : \langle\mathbf{x}_R\rangle \leftrightarrow \langle\mathcal{T}_F\rangle : \langle\mathbf{x}_F\rangle.$$

Consider a physical system with processing primitives $\mathcal{P}_1, \ldots, \mathcal{P}_k$. Each $\mathcal{P}_j$ exhibits transduction $\mathcal{T}_j(\mathbf{x})$ on *controllable* evolutions of *processable*, *detectable*, and *measurable* variables $\mathbf{x}_j$ yielding *distinguishable* and *computable* transforms $\mathcal{F}_j(\mathcal{T}, \mathbf{x})$. Using $\mathcal{F}_j(\mathcal{T}, \mathbf{x})$, consistent with device physics and the *admissible* arithmetic operand $\mathcal{A}_j$, one has $\mathcal{F}_\Sigma = \mathcal{F}_1 \circ \ldots \circ \mathcal{F}_k$.

The switching function on r-valued $\mathbf{x}_j$ is $f : \{0, \ldots, r-1\}^n \to {}^{\mathcal{F}}\{0, \ldots, r-1\}^m$ with a truth vector. Any f can be represented as

$$f = \mathcal{A}(\mathcal{F}, \mathcal{T}, \mathbf{x}). \tag{21.13}$$

The dynamic evolutions of variables $\mathbf{x}_{j,k} \rightsquigarrow^{\mathcal{Q}} \mathbf{x}_{j,k+1} = \mathcal{Q}_j(\mathbf{x}_{j,k})$ define the evolutions on $\mathcal{T}_{j,k} \rightsquigarrow^{\mathcal{Q}} \mathcal{T}_{j,k+1} = \mathcal{Q}_j(\mathcal{T}_{j,k})$ yielding the realization of physically realizable *computable* transforms $\mathcal{F}_j(\mathcal{T}, \mathbf{x})$, which must be consistent with $\mathcal{A}$. Quantum trunsductions ${}^{\mathcal{Q}}\mathcal{T}_j$ (quantum switching, molecular electron transition, and others) on $\mathbf{x}$ in physical systems can be controlled by using device-specific control principles.

The channel capacity C is found by maximizing the mutual information subject to the input probabilities. One has

$$C = \max_{p_X(\cdot)} I(X; Y) \qquad \text{[bit/symbol]}. \tag{21.14}$$

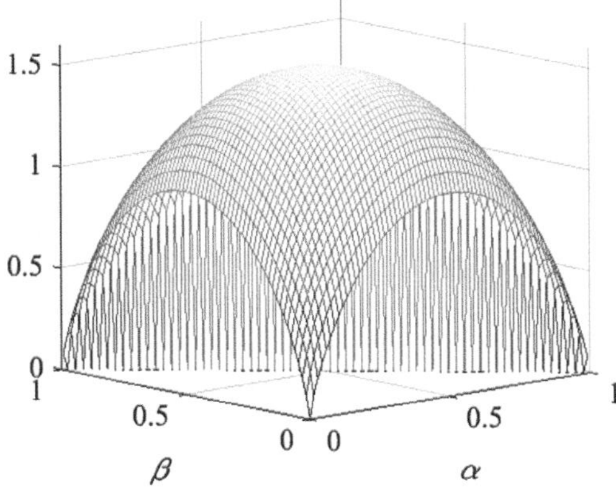

FIGURE 21.4 Entropy $H(x)$.

Example 21.5.4

In noiseless channels, $p(x_i, y_j) = 0$ and $p(x_i|y_j) = 0$ unless $i = j$. If $i = j$, $p(x_i|y_j) = 1$. Thus, $H(X|Y) = 0$. We have $I(X, Y) = H(X)$.

The entropy of a source is maximum if all source symbols are equally probable. If all source symbols are equally likely, one obtains $C = \sum_{i=1}^{n} \frac{1}{n}\log_2 n = \log_2 n$, or, $C = \log_2(1/p_i)$. ■

Example 21.5.5

Consider a ternary source x with probabilities $p(0) = \alpha$, $p(1) = \beta$ and $p(2) = 1 - \alpha - \beta$.

The entropy of the source is $H(x) = -\alpha\log_2\alpha - \beta\log_2\beta - (1 - \alpha - \beta)\log_2(1 - \alpha - \beta)$.

The three-dimensional plot is reported in Figure 21.4. ■

The analysis of mutual information yields the estimation of the channel capacity C, which depends on the conditional pdf $p_{Y|X}(y|x)$. The $p_{Y|X}(y|x)$ defines how the output changes with the input. It is very difficult to obtain distributions and to find pdfs. Optical communication, quantum communication, and interfacing can be viewed as a *point process channel*. The *instantaneous* data rate at which communication occurs cannot be lower than r_{min} and greater than r_{max}. The r_{max} depends on the transductions rates, latencies, noise, distortions, attenuation, etc. For a Poisson process, the channel capacity of the *point processes* for $r_{min} \leq r \leq r_{max}$ with the average data rate r_0 is [5-7]

$$C = r_{min}\left[e^{-1}\left(1 + \frac{r_{max} - r_{min}}{r_{min}}\right)^{\frac{1+r_{min}}{r_{max}-r_{min}}} - \left(1 + \frac{r_{min}}{r_{max} - r_{min}}\right)\ln\left(1 + \frac{r_{max} - r_{min}}{r_{min}}\right)\right], \quad (21.15)$$

which implies $C = \begin{cases} \frac{1}{\ln 2} r_{max}\left(e^{-1}\left(\frac{r_{max}}{r_{min}}\right)^{\frac{r_{max}}{r_{max}-r_{min}}} - \ln\left(\frac{r_{max}}{r_{min}}\right)^{\frac{r_{max}}{r_{max}-r_{min}}}\right) & \text{for } r_0 > \frac{1}{e} r_{min}\left(\frac{r_{max}}{r_{min}}\right)^{\frac{r_{max}}{r_{max}-r_{min}}} \\ \frac{1}{\ln 2}\left((r_0 - r_{min})\ln\left(\frac{r_{max}}{r_{min}}\right)^{\frac{r_{max}}{r_{max}-r_{min}}} - r_0\ln\left(\frac{r_0}{r_{min}}\right)\right) & \text{for } r_0 < \frac{1}{e} r_{min}\left(\frac{r_{max}}{r_{min}}\right)^{\frac{r_{max}}{r_{max}-r_{min}}} \end{cases}$

If the minimum rate is zero ($r_{min} = 0$), the channel capacity is $C = \begin{cases} \frac{1}{e\ln 2} r_{max} & \text{for } r_0 > \frac{1}{e} r_{max} \\ \frac{1}{\ln 2} r_0 \ln\left(\frac{r_{max}}{r_0}\right) & \text{for } r_0 < \frac{1}{e} r_{max} \end{cases}$.

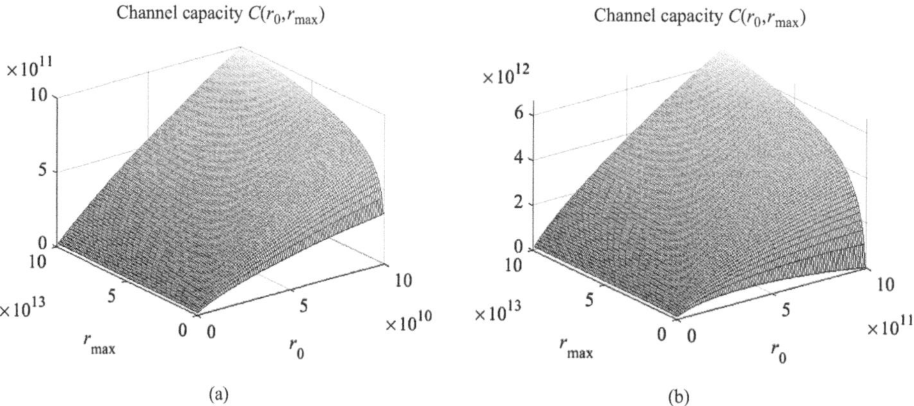

FIGURE 21.5 Channel capacity $C(r_0, r_{max})$ if $1 \times 10^{12} \le r_{max} \le 1 \times 10^{14}$ pulse/s and $r_{min} = 10$ pulse/s: $C_{max} = 9.97 \times 10^{11}$ bits/s if $1 \times 10^9 \le r_0 \le 1 \times 10^{11}$ pulse/s, and $C_{max} = 6.64 \times 10^{12}$ bits/s if $1 \times 10^{10} \le r_0 \le 1 \times 10^{12}$ pulse/s.

The maximum data rate r_{max} depends on many factors, such as the probability of the bit error p_b. For $r_{max}(p_b)$, one has

$$r_{max}(p_b) = \frac{C}{1 - H_2(p_b)}, H_2(p_b) = -[p_b \log_2 p_b + (1 - p_b)\log_2(1 - p_b)], \tag{21.16}$$

where p_b is the probability of a bit error; $H_2(p_b)$ is the binary entropy function.

Example 21.5.6

Quantum communication and processing schemes may support the data rate approaching 1×10^{14} pulse/s. From (21.15), the channel capacity $C(r_0, r_{max})$ is

$$C = \frac{1}{\ln 2}\left((r_0 - r_{min})\ln\left(\frac{r_{max}}{r_{min}}\right)^{\frac{r_{max}}{r_{max}-r_{min}}} - r_0 \ln\left(\frac{r_0}{r_{min}}\right)\right), r_{min} \le r \le r_{max} \text{ for } r_0 < \frac{1}{e} r_{min}\left(\frac{r_{max}}{r_{min}}\right)^{\frac{r_{max}}{r_{max}-r_{min}}}.$$

We compute $C(r_0, r_{max})$ if r_{max} varies from 1×10^{12} to 1×10^{14} pulse/s, and $r_{min} = 10$ pulse/s. Let the average rate r_0 vary as: (a) $1 \times 10^9 \le r_0 \le 1 \times 10^{11}$ pulse/s;
(b) $1 \times 10^{10} \le r_0 \le 1 \times 10^{12}$ pulse/s. Figure 21.5 reports the surfaces for $C(r_0, r_{max})$.

We find that for a single-channel communication, $C_{max} = 9.97 \times 10^{11}$ bits/s for $1 \times 10^9 \le r_0 \le 1 \times 10^{11}$ pulse/s, and $C_{max} = 6.64 \times 10^{12}$ bits/s for $1 \times 10^{10} \le r_0 \le 1 \times 10^{12}$ pulse/s. High channel capacity and throughput can be achieved. The probability of a bit error p_b depends on quantum transductions, transition probability $p_{a \to b}(t)$, and data rate r. We study achievable rates that depend on devices, computing arithmetic, communication schemes, etc. ■

Modulation in optical fiber links: The pulse-amplitude modulation (PAM) with N amplitude levels with no phase information is widely used in optical fiber links. Different amplitudes $(0, 1, \ldots, N - 1)$ are achieved by using a directly modulated laser or by an externally modulated semiconductor laser using an electro-absorption modulator. The coherent detection techniques are used to implement quadrature amplitude modulation (QAM), such as 64-QAM, 128-QAM, and 256-QAM. There are quadrature modulations, and optical carriers have a phase difference. For example, a 128-QAM signal consists of 128 constellation points. Dual-polarization QAMs are used. Although power-efficient PAM and QAM schemes ensure high data rate and high throughput, there are propagation, device, and algorithmic latencies.

Example 21.5.7: Analysis of Processing and Communication Parallelism

Parallelism, superpipelining, multithreading, and other schemes advance processing, communication, and networking capabilities. Multiprocessing, multichannel fiber optic, and multiple-channel

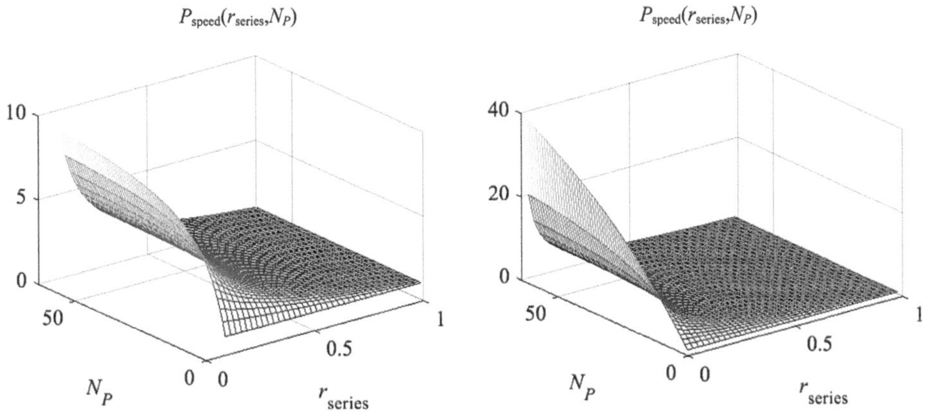

FIGURE 21.6 *Speed-up* estimate $P_{speed}(r_{series}, N_P)$: (a) $r_{series} \in [0.1\ 1]$ and $N_P \in [1\ 64]$, $P_{speed\,max} = 8.767$; (b) $r_{series} \in [0.01\ 1]$ and $N_P \in [1\ 64]$, $P_{speed\,max} = 39.26$.

optical communication solutions have been deployed and commercialized. Due to signal processing latencies, signal propagation delays, and device and circuit transients, we denote times needed to perform series and parallel tasks as t_{series} and $t_{parallel}$. The *speed-up* capabilities estimate P_{speed} is [5-7]

$$P_{speed} = \frac{1}{\frac{t_{series}}{t_{series}+t_{parallel}} + \frac{1}{N_P}\left(1 - \frac{t_{series}}{t_{series}+t_{parallel}}\right)} = \frac{1}{r_{series} + \frac{1}{N_P}(1 - r_{series})},$$

$$r_{series} = \frac{t_{series}}{t_{series} + t_{parallel}}, \quad r_{parallel} = 1 - r_{series} = \frac{t_{parallel}}{t_{series} + t_{parallel}},$$

where N_P is the number of processors or channels; r_{series} and $r_{parallel}$ are the series (not parallelizable) and parallelizable ratios.

The *speed-up* measure P_{speed} defines the degree of parallelism in processing and communication. The considered P_{speed} depends on r_{series} and $r_{parallel} = (1 - r_{series})$, $r_{series} > 0$, $r_{parallel} < 1$. Not parallelizable tasks are due to various latencies and delays, such as device transients, holds, protocols, algorithms, synchronization, flow control, propagation, transmission, and other delays. These latencies are associated with computing, memory, logic, coding, decoding, modulation, and other tasks accomplished by means of corresponding arithmetic and operations on state transductions.

Assume that from 90% to 99% of tasks and processes may be performed in parallel, while 10% to 1% are not parallelizable undertaking tasks. For not parallelizable tasks with $r_{series} = 0.1$ and $r_{series} = 0.01$, one finds $P_{speed\,max} = 8.767$ and $P_{speed\,max} = 39.26$ if $N_P = 64$. The surfaces for $P_{speed}(r_{series}, N_P)$ if $r_{series} \in [0.1\ 1]$ and $r_{series} \in [0.1\ 1]$ and $N_P \in [1\ 64]$ are illustrated in Figure 21.6. Performance and capabilities of data communication, interfacing, networking, and processing are significantly affected by the device and circuit switching frequencies, error rate, noise, distortions, etc. ■

The analog calculi and arithmetic, as well as Boolean algebra and digital logic, are reported in [5-7, 50, 56, 77-82]. These paradigms of computer sciences are supported by physical optoelectronic devices, which may consistently operate on quantum determinism and quantum statistical equilibrium phenomena. The aforementioned phenomena support computing and communication schemes.

21.6 QUANTUM MECHANICS AND *MICROSCOPIC* DEVICE SYSTEMS

To analyze, design, and evaluate nanoscaled transistors, lasers, quantum dots, and other devices that exhibit quantum phenomena, quantum mechanics is used. Quantum electrostatics and transport in transistors, photon emission and absorption in optoelectronic devices, as well as other quantum analyses are performed [1, 2, 7-9]. Monographs [8, 9] provide exceptional coverage for the aforementioned devices. Quantum-effect optoelectronic devices with some assumptions can be mathematically described by Schrödinger equations and mathematically characterized by a wave function $\Psi(\boldsymbol{r}, t) = \psi(\boldsymbol{r})\varphi(t)$ and other mathematical operators in abstract vector spaces. The time-dependent and time-independent Schrödinger equations, which yield the wave function, are

$$\hat{H}\Psi(\boldsymbol{r}, t) = i\hbar\frac{\partial\Psi(\boldsymbol{r}, t)}{\partial t}, \Psi(\boldsymbol{r}, t) = \psi(\boldsymbol{r})\varphi(t), \qquad (21.17)$$
$$\hat{H}\psi = E\psi,$$

where $\hat{H}$ is the *total* Hamiltonian operator, $\hat{H} = \hat{H}_0 + \hat{H}_E + \hat{H}_P$; $\hat{H}_0$ is the unperturbed Hamiltonian in the absence of external excitations, and, for an unperturbed *microscopic* system $\hat{H}_0\psi_n = E_n\psi_n$, $\hat{H}_0 = -\frac{\hbar^2}{2m}\nabla^2 + \hat{\Pi}$; $\hat{H}_E$ and $\hat{H}_P$ are the excitation and perturbation Hamiltonian operators; $\hat{\Pi}$ is the potential function.

The wave function is not detectable, not observable, and not measurable. Solving the Schrödinger equation, one finds the *allowed* quantized energy eigenvalues $E_1, E_2, \ldots, E_n$, as well as other quantities. For a *microscopic* system in quantum state n, the wave function is $\Psi_n = \psi_n e^{-iE_n/\hbar}$. In addition to E_n, the energy difference ΔE is calculated using the radiation absorbed or emitted during a transition between two states with E_1 and E_2, $f = (E_2 - E_1)/h = (E_2 - E_1)/(2\pi\hbar)$, $E_2 > E_1$.

Remark 21.6.1

The operators are used in (21.17). The operators for position $\boldsymbol{r}$, momentum $\boldsymbol{p}$, angular momentum $\boldsymbol{L}$, energy and kinetic energy Γ are $\hat{\boldsymbol{r}} \leftrightarrow \boldsymbol{r}$, $\hat{\boldsymbol{p}} \leftrightarrow -i\hbar\nabla$, $\hat{\boldsymbol{L}} \leftrightarrow -i\hbar(\boldsymbol{r} \times \nabla)$, $\hat{E} \leftrightarrow i\hbar\frac{\partial}{\partial t}$ and $\hat{\Gamma} \leftrightarrow -\frac{\hbar^2}{2m}\nabla^2$.

The classical total energy Hamiltonian H is fundamentally distinct compared with the Hamiltonian operator $\hat{H}$. The unperturbed Hamiltonians are $H_0 = \frac{1}{2m}p^2 + \Pi$ and $\hat{H}_0 = -\frac{\hbar^2}{2m}\nabla^2 + \hat{\Pi}$. ■

From $\Psi(\boldsymbol{r}, t) = \psi(\boldsymbol{r})\varphi(t)$, one obtains $i\hbar\frac{1}{\varphi}\frac{\partial\varphi}{\partial t} = E$ and $H\psi = E\psi$. From $\frac{\partial\varphi}{\partial t} = -i\frac{E}{\hbar}\varphi$, we have $\varphi(t) = e^{-iEt/\hbar}$.

The wave functions may mathematically describe and predict, with a particular degree of fidelity, statistical and transitions quantities pertaining to a mathematical description under underlying assumptions. Simplifications are used to derive and solve the governing Schrödinger and other equations. The general solution of the Schrödinger equation (21.17) is a linear combination of separable solutions with the associated allowed energies E_n. In particular,

$$\Psi(\boldsymbol{r}, t) = \sum_{n=1}^{\infty} c_n\Psi_n(\boldsymbol{r}, t) = \psi(\boldsymbol{r})\varphi(t) = \sum_{n=1}^{\infty} c_n\psi_n(\boldsymbol{r})e^{-i\frac{E_n}{\hbar}t} = \sum_{n=1}^{\infty} c_n\psi_n(\boldsymbol{r})e^{-i\omega_n t}, \sum_{n=1}^{\infty} |c_n|^2 = 1, \qquad (21.18)$$

where $|c_n|^2$ is the probability that a *microscopic* system is in a state with E_n, which implies that the probability that a measurement of the energy would yield E_n.

The solution of (21.17) gives the spatiotemporal evolution of $\Psi(\boldsymbol{r}, t)$ as the mixture of the stationary states. From $\langle H\rangle = \sum_{n=1}^{\infty} |c_n|^2 E_n$, one concludes that c_j represents the *weight* of ψ_j in $\Psi(\boldsymbol{r}, t)$. For a quantum-mechanically consistent observable C with an associated operator $\hat{C}$, the expectation value of C is

$$\langle C\rangle = \int \Psi^*(\boldsymbol{r}, t)\hat{C}\Psi(\boldsymbol{r}, t)dV. \qquad (21.19)$$

For any time-varying real-valued $w(\boldsymbol{r}, \boldsymbol{p}, t)$, one has $\langle w(\boldsymbol{r}, \boldsymbol{p}, t)\rangle = \int \Psi^*(\boldsymbol{r}, t)w(\boldsymbol{r}, -i\hbar\nabla, t)\Psi(\boldsymbol{r}, t)dV$, where $-i\hbar\nabla$ is the momentum operator. For specific C, the associated operator $\hat{C}$ is used.

The governing equation to describe the evolution of the expectation value of an observable is

$$\frac{d}{dt}\langle C\rangle = \left\langle \frac{\partial \hat{C}}{\partial t}\right\rangle - i\frac{1}{\hbar}\left\langle [\hat{C},\hat{H}]\right\rangle, [\hat{C},\hat{H}] = \hat{C}\hat{H} - \hat{H}\hat{C}. \quad (21.20)$$

For time-invariant $\hat{C}$, one has $\frac{d}{dt}\langle C\rangle = \frac{1}{i\hbar}\langle[\hat{C},\hat{H}]\rangle$. If $[\hat{C},\hat{H}] = 0$, the two operators commute.

While *microscopic* systems may be mathematically modeled by informative mathematical operators defined by a quadruple $(\hat{H}, \Psi, c_n, \hat{C})$, the processing can be achieved only on *processable* and *measurable observables*, e.g., on physical quantities pertained to physical devices. The pertained quantities on the quantum-mechanically consistent physical *observables* C (energy, momentum, and others) may be derived. For example,

$$\langle E\rangle = \sum_{n=1}^{\infty} |c_n|^2 E_n = \int \Psi^*(\mathbf{r},t)\hat{H}\Psi(\mathbf{r},t)dV = \langle \Psi|\hat{H}\Psi\rangle. \quad (21.21)$$

Example 21.6.1

Consider C to be x and p. The associated operators are $\hat{x} \leftrightarrow x$ and $\hat{p} \leftrightarrow -i\hbar\nabla$. From (21.20), one finds the quantum-mechanical analogy to the classical equations of motion

$$\frac{d}{dt}\langle x\rangle = \left\langle \frac{1}{m}p\right\rangle, \frac{d}{dt}\langle p\rangle = -\left\langle \frac{\partial \Pi(x)}{\partial x}\right\rangle,$$

where $\langle \frac{\partial \Pi(x)}{\partial x}\rangle \neq \frac{d}{d\langle x\rangle}\Pi(\langle x\rangle)$.

Furthermore, $\frac{d}{dt}\langle x\rangle = \frac{1}{m}\langle p\rangle = -\frac{i\hbar}{2m}\int (\Psi^*\frac{\partial\Psi}{\partial x} - \frac{\partial\Psi^*}{\partial x}\Psi)dx = \frac{i\hbar}{m}\int \Psi^*\frac{\partial\Psi}{\partial x}dx$. Similar as shown for $\langle x\rangle = \int \Psi^* x\Psi dx$ and $d\langle x\rangle/dt$, evolutions of other quantum-mechanical consistent quantities can be obtained. For example, the expectation value of the kinetic energy is $\langle \Gamma\rangle = -\frac{\hbar^2}{2m}\int \Psi^*\frac{\partial^2\Psi}{\partial x^2}dx$. ■

Example 21.6.2

Consider a *microscopic* particle that evolves in an infinite quantum well with $\Pi_{X=0} = \Pi_{X=a} = \infty$, $0 \le x \le a$. The solution of the Schrödinger equations is given, or approximated, by the wave function $\Psi(x,0) = A_1\sin(\pi x/a), 0 \le x \le a, A_1 \neq 0$ as shown in Figure 21.7. The given $\Psi(x,0)$ can be approximated by $\Psi(x,0) = Ax(a-x), 0 \le x \le a, A \neq 0$. Outside the well $\Psi = 0$.

Using normalization $\int_0^a \psi^*(x)\psi(x)dx = 1$, for $\Psi(x,0) = A_1\sin(\pi x/a)$, we have $\int_0^a [A_1\sin(\frac{1}{a}n\pi x)]^2 dx = 1$ yielding $A_1 = \sqrt{\frac{2}{a}}$.

Note that $\int_0^a [\sin(\frac{1}{a}\pi x)]^2 dx = \frac{a[\pi - \sin(\pi)]}{2\pi}$ and $\int_0^a x^2(a-x)^2 dx = \frac{1}{3}a^2x^3 - \frac{1}{3}ax^4 + \frac{1}{3}x^5$.

Using $c_n = \int_{-\infty}^{\infty} \psi_n^*(x)\Psi(x,0)dx$, one finds $c_n = \frac{4\sqrt{15}}{\pi^3 n^3}[1-\cos(\pi n)] = \begin{cases} 0, & \text{even } n \\ \frac{8\sqrt{15}}{\pi^3 n^3}, & \text{odd } n \end{cases}$.

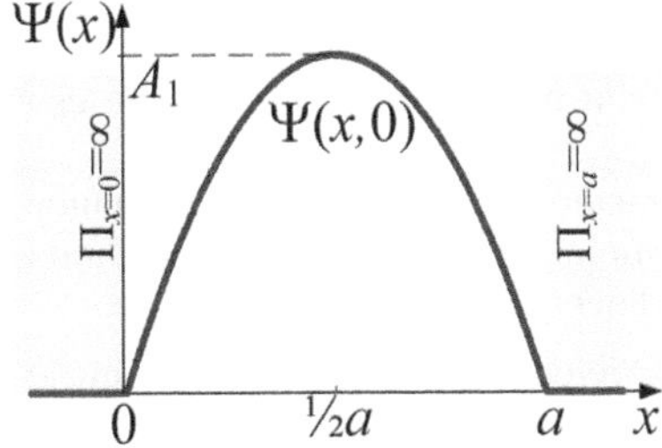

FIGURE 21.7 Wave function $\Psi(x,0) = A_1\sin(\pi x/a), 0 \le x \le a$.

For $\Psi(x,0) = Ax(a-x), 0 \le x \le a$, normalization may be carried out by applying $\int_0^a x(a-x)\sin(\frac{1}{a}\pi x)dx = \frac{4a^3}{\pi^3}$. One obtains $\Psi(x,t) = \sqrt{\frac{30}{a^5}}\frac{8}{\pi^3}\sum_{n=1,3,5}^{\infty}\frac{1}{n^3}\sin(\frac{\pi n}{a}x)e^{-i\frac{\pi^2\hbar n^2}{2ma^2}t}$, and the energy eigenvalues are $E_n = \frac{\pi^2\hbar^2}{2ma^2}n^2$. We found the probability that a system is at $E_1 = \frac{\pi^2\hbar^2}{2ma^2}(n=1)$ to be $|c_1|^2 = (8\sqrt{15}/\pi^3)^2 = 0.998555$.

There is an admixture of the excited states. From $\sum_{n=1}^{\infty}|c_n|^2 = \left(\frac{8\sqrt{15}}{\pi^3}\right)^2\sum_{n=1,3,5,\dots}^{\infty}\frac{1}{n^6} = 1$, the probability of being in all other excited states with $\forall n > 1$ is 1.445×10^{-3}.

The derived analytic results and $|c_n|^2$ yield

$$\Psi(x,t) = \sum_{n=1}^{\infty} c_n\psi_n(x)e^{-i\frac{E_n}{\hbar}t} \text{ with } \Psi(x,t)_{n=1} \approx \sqrt{\frac{30}{a^5}}\frac{8}{\pi^3}\sin\left(\frac{\pi}{a}x\right)e^{-i\frac{\pi^2\hbar}{2ma^2}t} = \sqrt{\frac{30}{a^5}}\frac{8}{\pi^3}\sin\left(\frac{\pi}{a}x\right)e^{-i\omega_1 t}.$$

The expectation value of the energy is

$\langle H\rangle = \sum_{n=1}^{\infty}|c_n|^2E_n = \sum_{n=1}^{\infty}|c_n|^2\frac{\pi^2\hbar^2}{2ma^2}n^2 = \frac{480\hbar^2}{\pi^4ma^2}\sum_{n=1,3,5,\dots}^{\infty}\frac{1}{n^4} = \frac{5\hbar^2}{ma^2}$.

One concludes that $\langle H\rangle$ is slightly greater than E_1, and, all $E_{n,\forall n\neq 1}$ are negligible as compared to E_1. ■

Example 21.6.3

Consider a *microscopic* particle, the evolution of which is described by a linear combination of two stationary (definite energy) states for which the time-independent wave function is $\Psi(\mathbf{r},0) = c_1\psi_1(\mathbf{r}) + c_2\psi_2(\mathbf{r})$.

The wave function is $\Psi(\mathbf{r},t) = c_1\psi_1(\mathbf{r})e^{-i\frac{E_1}{\hbar}t} + c_2\psi_2(\mathbf{r})e^{-i\frac{E_2}{\hbar}t}$, where E_1 and E_2 are the allowed energies associated with ψ_1 and ψ_2. We have

$$\Psi^*(\mathbf{r},t)\Psi(\mathbf{r},t) = \left(c_1^*\psi_1^*(\mathbf{r})e^{i\frac{E_1}{\hbar}} + c_2^*\psi_2^*(\mathbf{r})e^{i\frac{E_2}{\hbar}}\right)\left(c_1\psi_1(\mathbf{r})e^{-i\frac{E_1}{\hbar}} + c_2\psi_2(\mathbf{r})e^{-i\frac{E_2}{\hbar}}\right)$$
$$= c_1^*c_1\psi_1^*\psi_1 + c_2^*c_2\psi_2^*\psi_2 + c_1^*c_2\psi_1^*\psi_2e^{-i\frac{E_2-E_1}{\hbar}} + c_2^*c_1\psi_2^*\psi_1e^{i\frac{E_2-E_1}{\hbar}}.$$

Using the Euler identity $e^{-i\omega t} = (\cos\omega t - i\sin\omega t)$, we conclude that the probability density $P(\mathbf{r},t) = \Psi^*\Psi$ evolves with the frequency $f = (E_2 - E_1)/(2\pi\hbar)$.

Applying assumptions, let $\Psi^*\Psi = c_1^2\psi_1^2 + c_2^2\psi_2^2 + 2c_1c_2\psi_1\psi_2\cos(\frac{E_2-E_1}{\hbar}t)$. Hence, the probability density $P(\mathbf{r},t) = \Psi^*\Psi$ evolves (oscillates in time) with the angular frequency $\omega = (E_2 - E_1)/\hbar$. Thus, the *microscopic* system undergoes time- and spatially varying dynamic evolutions. For example, excited atoms emit radiations, and they return to the ground states. An electron undergoes a transition from an excited state with E_2 to a ground state with E_1, $(E_2 - E_1) > 0$. The probability density and charge distribution are evolving in time with the frequency

$f = (E_2 - E_1)/(2\pi\hbar) = (E_2 - E_1)/h$. ■

Example 21.6.4

Consider a quantum system with two stationary states with eigenstates $|1\rangle$ and $|2\rangle$. We have $\omega_{21} = (E_2 - E_1)/\hbar$.

At $t = 0$, the system is characterized by $|1\rangle$, and the excitation or perturbation H_E is applied. Using the time-dependent perturbation theory, $\langle 1|\hat{H}_E|1\rangle = 0$ and $\langle 2|\hat{H}_E|1\rangle = \hbar\omega_0$.

The time-dependent Schrödinger equation

$(\hat{H}_0 + \hat{H}_E)|\psi(t)\rangle = i\hbar\frac{\partial}{\partial t}|\psi(t)\rangle$ yields $|\psi(t)\rangle = c_1(t)e^{-i\frac{E_1}{\hbar}}|1\rangle + c_2(t)e^{-i\frac{E_2}{\hbar}t}|2\rangle$.

Having found $c_1(t) = \cos(\omega_0 t)$ and $c_2(t) = -i\sin(\omega_0 t)e^{i\omega_{21}t}$,

one obtains $|\psi(t)\rangle = e^{-i\frac{E_1}{\hbar}}[\cos(\omega_0 t)|1\rangle - i\sin(\omega_0 t)|2\rangle]$.

The transition probability is $P_{|1\rangle\to|2\rangle} = |\langle 2|\psi(t)\rangle|^2 = \sin^2(\omega_0 t)$.

The system is characterized by $\langle 2|$ when $P = 0$ as $\omega_0 t_{\langle 2|} = \pi(\pm {}^1/_2 + k), k \in N$, or, at times $t_{\langle 2|} = \pi(\pm {}^1/_2 + k)/\omega_0$. ■

Example 21.6.5: Hilbert Space and Wave Functions

A Hilbert space $\mathcal{H}$ is an abstract vector space with a finite or infinite number of dimensions. Using the bracket notations for the inner product of square-integrable functions $+q|g\rangle$, one may define $\Psi(x, t) = +x|\psi(t)\rangle$ in the Hilbert space by using the position eigenfunction $+x|$ of $\hat{x}$ with eigenvalue x, and eigenvector $|\psi(t)\rangle$. If the *microscopic* system admits a finite number of linearly independent states N, one has an N-dimensional vector space for $|\psi(t)\rangle$. Using the momentum eigenfunction $+p|$ of $\hat{p}$ with eigenvalue p, the momentum space wave functions is $\Phi(p, t) = +p|\psi(t)\rangle$. We have $\psi(x,t) = \sum_{n=1}^{\infty} c_n\psi_n(x)e^{-i\frac{E_n}{\hbar}t} = \frac{1}{\sqrt{2\pi\hbar}}\int_{-\infty}^{+\infty} e^{i\frac{1}{\hbar}px}\Phi(p,t)dp$ and $\Phi(p,t) = \frac{1}{\sqrt{2\pi\hbar}}\int_{-\infty}^{+\infty} e^{-i\frac{1}{\hbar}px}\psi(x,t)dx$. One may represent wave functions using different bases, spaces, expansions, etc. Although the wave functions can be defined in the abstract vector spaces, these spaces may not be a physically realizable "computing space."

An eigenvector $|\psi_n\rangle$ represents an eigenstate of the *microscopic* system in the Hilbert space of square-integrable functions. The Schrödinger *eigenfunction equations* in an abstract space are $\hat{H}|\psi(t)\rangle = i\hbar\frac{\partial}{\partial t}|\psi(t)\rangle$ and $\hat{H}|\psi_n\rangle = E|\psi_n\rangle$. Here, $|\psi(t)\rangle$ is the eigenvector in Hilbert space, which points in a direction that depends on time, $|\psi(t)\rangle e^{-i\frac{\hat{H}}{\hbar}t} = |\psi(0)\rangle\sum_{n=0}^{\infty}\frac{1}{n!}(-i\frac{\hat{H}}{\hbar}t)^n$, and $|\psi_n\rangle$ and E are the eigenvectors and eigenvalues of the Hamiltonian operator $\hat{H}$.

The general solution of the Schrödinger equation is a linear combination of separable solutions with the associated E_n. The solution
$\psi(\mathbf{r}, t) = \sum_{n=1}^{\infty} c_n\psi_n(\mathbf{r}, t) = \psi(\mathbf{r})\phi(t) = \sum_{n=1}^{\infty} c_n\psi_n(\mathbf{r})e^{-i\frac{E_n}{\hbar}t} = \sum_{n=1}^{\infty} c_n\psi_n(\mathbf{r})e^{-i\omega_n t}$
can be expressed as $|\psi(\mathbf{r}, t) = \sum_{n=1}^{\infty} c_n(t)\psi_n(\mathbf{r})$, or using the Dirac notation, as $|\psi\rangle = \sum_{n=1}^{\infty}|c_n\psi_n\rangle|$.
Applying the time-evolution operator $\hat{U} = e^{-i\frac{\hat{H}}{\hbar}t}$, we have $|\psi(t)\rangle = \hat{U}|\psi(0)\rangle$. ■

Example 21.6.6

Using Dirac's notations and orthonormal basis $B = \{|b_0\rangle\ldots, |b_{n-1}\rangle\} \in \mathcal{H}_n$, for a *pure* state, a "coherent superposition" is expressed as $|\gamma\rangle = \sum_{nC_n}|n\rangle$, where the complex *probability amplitudes* $c_n \in \mathbb{C}$ satisfy $|c_0|^2+|c_1|^2+\cdots+|c_{n-2}|^2+|c_{n-1}|^2 = 1$ or $\sum_n |c_n|^2 = 1$. One may expand an arbitrary *ket* $|\gamma\rangle$ in term of a complete set of orthonormal energy eigenstates $|n\rangle$ as

$$|\gamma\rangle = \sum_{nC_n}|n\rangle|\gamma\rangle = c_0|0\rangle + c_1|1\rangle + \cdots + c_{n-2}|n-2\rangle + c_{n-1}|n-1\rangle.$$

A set of $\{|0\rangle|1\rangle\ldots, |n-2\rangle|n-1\rangle\} \in \mathcal{H}_n$ forms a basis in an n-dimensional Hilbert space $\mathcal{H}_n$. The so-called computational basis states are $|\gamma\rangle \in \mathcal{H}_n$. For $|\gamma\rangle = c_0|0\rangle + c_1|1\rangle + c_2|2\rangle$, $|\gamma\rangle \in \mathcal{H}_3$, one may define $|0\rangle = (1\ 0\ 0)^T$, $|1\rangle = (0\ 1\ 0)^T$, $|2\rangle = (0\ 0\ 1)^T$.

A "state vector" $|\psi\rangle$ is said to be an eigenvector (also called *eigenket*, eigenstate, etc.) of an operator A if the use of A to $|\psi\rangle$ yields $A|\psi\rangle = a|\psi\rangle$, where a is a complex number, called an eigenvalue of A. ■

Example 21.6.7

A qubit *ket* $|\gamma\rangle$ represents eigenstates in a two-dimensional linear vector space $\mathcal{H}_2$ over the field of complex numbers with an inner product. For a qubit, $\{|0\rangle|1\rangle\} \in \mathcal{H}_2$, $|\gamma\rangle = c_0|0\rangle + c_1|1\rangle$, $(c_0, c_1) \in \mathbb{C}$. The probability amplitudes for $|0\rangle$ and $|1\rangle$ are $|c_0|^2$ and $|c_1|^2$, such that $|c_0|^2 + |c_1|^2 = 1$.

Using real $\nu, \kappa\rangle$ and ζ, a qubit *ket* can be visualized using the Bloch sphere as

$$|\gamma\rangle = e^{i\nu}[\cos({}^1/_2\zeta)|0\rangle + e^{i\kappa}\sin({}^1/_2\zeta)|1\rangle], c_0 = e^{i\nu}\cos({}^1/_2\zeta), c_1 = e^{i\nu}e^{i\kappa}\sin({}^1/_2\zeta).$$

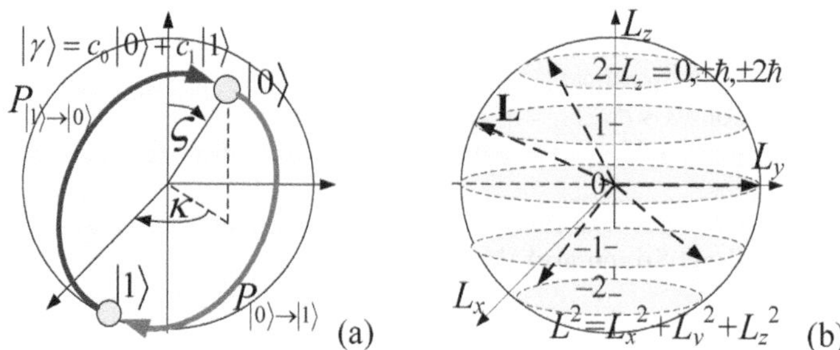

FIGURE 21.8 (a) A qubit *ket* $|\gamma\rangle$ on the Bloch sphere: schematic presentation of the transition probabilities $P_{|0\rangle\to|1\rangle}$ and $P_{|1\rangle\to|0\rangle}$ on stationary states with eigenstates $|0\rangle$ and $|1\rangle$; (b) Illustrative representation of the angular momentum eigenstates $|l\, m_l\rangle$ for five values of l. Vector $\boldsymbol{L}$, with length $\hbar\sqrt{l(l+1)}$, is "random" on one of the circular cones with a uniform probability. Here, $L_z = 0, \pm\hbar, \pm 2\hbar$, $\langle L_x\rangle = 0$ and $\langle L_x^2\rangle = {}^1/_2(6 - m_l^2)\hbar^2$. The operators L_x, L_y, and L_z are quantum-mechanically *incompatible observables.*

On a unit sphere, $|\gamma\rangle$ is defined by abstract coordinates κ and ζ, as illustrated in Figure 21.8a. Mathematically, a quibit *ket* $|\gamma\rangle$ is expressed as the energy eigenstates $|0\rangle$ and $|1\rangle$ on the Bloch sphere, and $|c_n|^2$ are the calculated probabilities that a system is in a state with E_n. Suggestions that quantum communication, data storage, and computing can be accomplished on a continuum or superposition of eigenstates and that a significant amount of information can be encoded by a qubit remain to be justified. It is unclear how to: (1) measure the superposition of eigenstates, and (2) simultaneously assess and manipulate mathematically narrative descriptive states, etc. Figure 21.8a schematically represents the transition probabilities $P_{|0\rangle\to|1\rangle}$ and $P_{|1\rangle\to|0\rangle}$. ■

Example 21.6.8

The angular momentum operator is $\boldsymbol{L} = -i\hbar(\boldsymbol{r}\times\nabla)$. The angular momentum operators $L_x = -i\hbar(y\frac{\partial}{\partial z} - z\frac{\partial}{\partial y})$, $L_y = -i\hbar(z\frac{\partial}{\partial x} - x\frac{\partial}{\partial z})$, and $L_z = -i\hbar(x\frac{\partial}{\partial y} - y\frac{\partial}{\partial x})$ do not commute, and they are *incompatible observables*.

One has $[L_x, L_y] = i\hbar L_z$, $[L_y, L_z] = i\hbar L_x$ and $[L_z, L_x] = i\hbar L_y$.

The Heisenberg uncertainty principle yields $\sigma_{L_x}^2\sigma_{L_y}^2 \geq \frac{1}{4}\hbar^2\langle L_z\rangle^2$ and $\sigma_{L_x}\sigma_{L_y} \geq \frac{1}{2}\hbar\,|\langle L_z\rangle|$.

Hence: (1) one cannot simultaneously yield L_x and L_y, or, other combinations; and (2) the square of the *total* angular momentum operator $L^2 = L_x^2 + L_y^2 + L_z^2$ does not commute with L_x, L_y and L_z, $[L^2, L_x] = 0$, $[L^2, L_y] = 0$ and $[L^2, L_z] = 0$. While a system has a finite angular momentum, theoretically one cannot obtain simultaneously eigenstates L_x and L_y or L_x and L^2.

The system states are labeled by using three quantum numbers (n,l,m_l). The allowed quantum numbers are $n = 1, 2, 3, \ldots, l \leq (n-1)$ and $m_l = -l, -l+1, \ldots, 0, \ldots, l-1, l$.

We use a tuple of quantum numbers (n,l,m_l), where:

1. The *principal quantum number n* determines the allowed energy.
2. The *orbital quantum number l* defines the *total* orbital angular momentum.
3. The *magnetic (azimuthal) quantum number* m_l gives the z-component of the angular momentum.

For one-electron atom, the eigenfunction equations are
$L_z\psi_{nlm_l} = m_l\hbar\psi_{nlm_l}$, $L^2\psi_{nlm_l} = l(l+1)\hbar^2\psi_{nlm_l}$.

Using the moment of inertia J, the eigenenergy is $E_l = l(l+1)\hbar^2/(2J)$. The energy E_l is $(2l+1)$-fold degenerate. For any value l, there are $(2l+1)$ eigenfunctions $(\psi_{l,l}, \ldots, \psi_{l,-l})$ with corresponding eigenenergies E_l. For a given l, there are $(2l+1)$ different values for m_l. These values can be visualized as rings of allowed m_l. Using the quantization relationships $L_z = m_l\hbar$ and $L^2 = l(l+1)\hbar^2$, Figure 21.8b documents the sphere to visualize angular momenta with the same length $\sqrt{l(l+1)}$ in the unit of $\hbar$ for the angular momentum eigenstates $|l\, m_l\rangle$. If $l = 2$, the allowed m_l are $m_l = -2, -1, 0, 1, 2$. In Figure 21.8b, the $\boldsymbol{L}$ is ambiguous because if L_x is defined, L_y and L_z are not defined. The eigenvalues of L_z are $m_l\hbar$. Despite the fact that L^2 and L_z were defined, the operators L_x, L_y, and L_z are *incompatible observables*. Theoretically, one may not perform any operations on (L_x, L_y, L_z).

The expectation values $\langle \mathcal{L}_z \rangle$ and $\langle \mathcal{L}^2 \rangle$ are

$$\langle \mathcal{L} \rangle = \int \psi^*_{nlm_l} L_z \psi_{nlm_l} dV = m_l\hbar \int \psi^*_{nlm_l} \psi_{nlm_l} dV = m_l\hbar,$$

$$\langle \mathcal{L}^2 \rangle = \int \psi^*_{nlm_l} L^2 \psi_{nlm_l} dV = l(l+1)\hbar^2 \int \psi^*_{nlm_l} \psi_{nlm_l} dV = l(l+1)\hbar^2.$$

The assumed measurements on the angular momentum of an electron in the (n, l, m_l) state of a one-electron atom will yield $L_z = m_l\hbar$ and $L^2 = l(l+1)\hbar^2$, while the relationships on the expectation values $\langle \mathcal{L}_z \rangle$ and $\langle \mathcal{L}^2 \rangle$ indicate that the values $m_l\hbar$ and $l(l+1)\hbar^2$ may be obtained using an immense number of measurement on average.

Operators L_x, L_y, and L_z can be expressed in the spherical coordinates, and the eigenvectors are spherical harmonics. Three commuting operators $(\hat{H}, L^2$, and $L_z)$ satisfy the following equations $\hat{H}\psi = E\psi$, $L_z\psi = m_l\hbar\psi$ and $L^2\psi = l(l+1)\hbar^2\psi$.

For a spin intrinsic angular momentum $\mathbf{S}$, which characterizes the motion about the center of mass, the aforementioned results hold. For the *incompatible observables*, one has the following commutation relationships:

$$[S_x, S_y] = i\hbar S_z,\ [S_y, S_z] = i\hbar S_x \ \text{ and } \ [S_z, S_x] = i\hbar S_y.$$

The eigenvectors satisfy $S^2|s\, m_l\rangle = s(s+1)\hbar^2|s\, m_l\rangle, S_z|s\, m_l\rangle = m_l\hbar|s\, m_l\rangle, S_\pm \equiv S_x \pm iS_y$ and $S \pm |sm_l\rangle = \sqrt{s(s+1) - m_l(m_l \pm 1)}\hbar|s(m_l \pm 1)\rangle$. ■

Example 21.6.9

Molecules have energy levels that are associated with the rotational motion of a molecule and with vibrational motion of the atoms relative to each other. Transitions between rotational and vibrational levels in molecules lead to molecular spectra. Using the vibrational and rotational quantum numbers m and l, one has E_m and E_l. The quantized vibrational energy levels are $E_m = (m + {}^1/_2)h\omega_v, m = 0, 1, 2, \ldots,$ where ω_v is the vibrational frequency.

Each electronic state E_e has its set of vibrational levels, while each vibrational level has its own set of rotational levels. The energies of all three modes are expressed as $\Delta E = \Delta E_e + (E_{m1} - E_{m2}) + (E_{l1} - E_{l2})$, where ΔE_e is the energy difference of two electronic states; $(E_{m1} - E_{m2})$ is the vibrational term, $(E_{m1} - E_{m2}) = (m_1 + {}^1/_2)h\omega_{v1} - (m_2 + {}^1/_2)h\omega_{v2}$; $(E_{l1} - E_{l2})$ is the rotational term, $(E_{l1} - E_{l2}) = (\hbar^2/2I_1)l_1(l_1+1) - (\hbar^2/2I_2)l_2(l_2+1)$; subscripts 1 and 2 correspond to the initial and final states.

One may examine electronic transitions. The vibrational selection rule for electronic dipole radiation is $\Delta m = \pm 1$ (harmonic oscillator) or $\Delta m = 2, 3, \ldots$ (anharmonic oscillator). This vibrational rule applies only if the electronic state does not change, e.g., for pure vibrational–rotational bands. If there are changes in the electronic state, the selection rule is given by the Frank–Codon principle. The rotational selection rule is $\Delta l = 0)\pm 1$ for electric dipole radiation. The characteristic time for an electronic transition is $\sim 1 \times 10^{-16}$ s, while the nuclear vibration time is $\sim 1 \times 10^{-13}$ s. ■

Example 21.6.10

An arbitrary *ket* $|\gamma\rangle$ can be mathematically represented using orthonormal eigenstates $|n\rangle$ as $|\gamma\rangle = \sum_n c_n|n\rangle = \sum_n |n\rangle\langle n|\gamma\rangle$. Alternatively, $|\gamma\rangle = \sum_{l=0}^{\infty} \sum_{m_l=-l}^{l} c_{lm_l}|l, m_l\rangle$. One many use many alternative expressions for *kets*. ■

Example 21.6.11

One implements arbitrary complexity switching and algebraic functions f by using logic gates, such as AND, NAND, OR, NOR, etc. The output of any two-input logic gates depends on the input states (x y) = (0 0), (0 1), (1 0) or (1 1). Here, 0 and 1 represent the *low* and *high* voltages, which are measurable *computable variables* consistent with device physics. The analog FETs physically implement binary and multiple-valued logic gates. *Three-state* gates are used in gate arrays to implement complex digital systems, such as memories, combinational circuits, processors, etc. Digital systems ensure specified input–output mappings.

Consider a system with two spin-$^1/_2$ particles (electrons). The wave function is $\psi(\boldsymbol{r}_1, \boldsymbol{r}_2)|s\ m_s\rangle$. It is impossible to simultaneously determine $\psi(\boldsymbol{r}_1, \boldsymbol{r}_2)$ and the spin states. The $|\gamma\rangle$ is expressed as a "superposition."

$$|\gamma\rangle = c_0|\downarrow\downarrow\rangle + c_1|\downarrow\uparrow\rangle + c_2|\uparrow\downarrow\rangle + c_3|\uparrow\uparrow\rangle.$$

The combined probability-dependent state $|s\ m_s\rangle$ with a *total* spin s and z-component m_s is in a linear combination and "superposition" of the composite states $|s_1\ m_{s1}\rangle|s_2\ m_{s2}\rangle$. One has $|s\ m_s\rangle = \sum_{m_s=m_{s1}+m_{s2}} C^{s_1 s_2 s}_{m_{s1} m_{s2} m_s}|s_1\ m_{s1}\rangle|s_2\ m_{s2}\rangle$, where $C^{s_1 s_2 s}_{m_{s1} m_{s2} m_s}$ are the Clebesh–Gordon coefficients.

Quantum-mechanical analysis may not assure: (i) practicality of computing or memories on qubits, (ii) possibility of processing on $|s\ m_s\rangle$, and (iii) parallelism and simultaneous manipulation of quantum states, etc. ■

Example 21.6.12

Using the quantum numbers (n, l, m_l), the wave function $\psi_{nlml}(r, \theta, \phi)$ for the hydrogen and hydrogen-like atoms was derived in the Coulomb and other potentials. The shell, characterized by n, contains n subshells, each corresponding to $l, l = 0, 1, 2, \ldots, n\!-\!1$. Each subshell contains $(2l + 1)$ distinct states corresponding to $m_l, m_l = -l, -l+1, \ldots, 0, \ldots, l-1, l$. The number of quantum states with a given E_n is n^2. Each electron has two possible spin states. The *total* degeneracy is $2n^2$. Thus, E_n corresponds to $2n^2$ different eigenfunctions. It is uncertain how processing tasks may not be accomplished assuming the use of *quantum indeterminacy* and *incompleteness* considering wave functions, *hidden variables*, state "superposition," etc. The energy-level splitting (fine and hyperfine structures, weak-field, and others) and various interactions may change the core assumptions and baseline results of mathematical descriptions and findings. ■

Example 21.6.13

Some quantum computing algorithms suggest using the density operator ρ, which provides a statistical mixture of *pure* states, $\rho = |\Psi\rangle\langle\Psi|$. An algorithmic transform for a system from $|\gamma_1\rangle$ with ρ_1 to a new $|\gamma_2\rangle$ with ρ_2 is $\rho_1 \to \rho_2 = R(\rho_1)$. By applying a unitary transform $U, UU^* = U^*U = I$, one has $R(\rho) = U\rho U^*$ and $\rho_1 \to^U \rho_2 = U\rho_1 U^*$. For a system in a *mixed* state, the wave function exists. The density operator $\rho = \frac{1}{N}\sum_{j=1}^{N}|\psi_j\rangle\langle\psi_j|$ yields the *ensemble averages*, e.g., the probability that a system chosen at random from the ensemble is found in the jth state. The normalization leads to $\rho^2 = \rho$ and $\langle\Psi|\Psi\rangle = \mathrm{tr}(\rho) = 1$. One has $\langle\Psi|\hat{C}|\Psi\rangle = \mathrm{tr}(\hat{C}\rho)$, where $\hat{C}$ is the *ensemble* observable operator, $\langle A\rangle = \frac{1}{N}\sum_{j=1}^{N}\langle\Psi_j|\hat{C}|\Psi_j\rangle = \mathrm{tr}(\rho\hat{C}), \mathrm{tr}\,\rho = 1$. Thus, one studies N states, each with distinct quantities. The unitary operations may be incomputable.

Using the eigenvalues λ_ρ of the source density matrix ρ, von Neumann's entropy $S(\rho_\mathcal{E})$ of a mixed state of non-orthogonal pure states is $S(\rho_E) = -\sum_{i=1}^{N}\lambda_{\rho i}\log_2\lambda_{\rho i} = -\mathrm{tr}(\rho\log_2\rho)$. The $S(\rho_\mathcal{E})$ may not provide a complete characterization of source because different sources of quantum information may have the same ρ. ■

Example 21.6.14

Consider a two-level *microscopic* system with two distinct unperturbed states, characterized by orthogonal ψ_a and ψ_b, $\langle\psi_a\,|\,\psi_b\rangle = \delta_{ab}$. These ψ_a and ψ_b are eigenstates of the unperturbed Hamiltonian $\hat{H}_0$. The time-independent Schrödinger equations are $\hat{H}_0\psi_a = E_a\psi_a$, $\hat{H}_0\psi_b = E_b\psi_b$.

The wave function is a linear combination of ψ_a and ψ_b, $\Psi(t) = c_a\psi_a e^{-i\frac{E_a}{\hbar}t} + c_b\psi_b e^{-i\frac{E_b}{\hbar}t}$, $|c_a|^2 + |c_b|^2 = 1$.

The probability amplitudes c_a and c_b define: (i) the probabilities that the *microscopic* system is in states ψ_a ψ_b; and (ii) the probabilities that a system is at the allowed E_a and E_b. For the perturbed system, the *total* Hamiltonian is $\hat{H}(t) = \hat{H}_0 + \hat{H}_E(t)$.

The time-dependent Schrödinger equation is solved. Denote $\hat{H}_{Eij} = \langle\psi_i|\hat{H}_E|\psi_j\rangle$, and assume that the diagonal matrix entities vanish, e.g., $\hat{H}_{Eii} = \hat{H}_{Ejj} = 0$. For a two-level *microscopic* system, the evolutions of $c_a(t)$ and $c_b(t)$ are

$$\frac{dc_a}{dt} = -i\frac{1}{\hbar}\hat{H}_{Eab}e^{-i\frac{E_b-E_a}{\hbar}t}c_b(t),\ \frac{dc_b}{dt} = -i\frac{1}{\hbar}\hat{H}_{Eab}e^{i\frac{E_b-E_a}{\hbar}t}c_a(t),\ |c_a(t)|^2 + |c_b(t)|^2 = 1.$$

For a sinusoidal excitation $\hat{H}_E(t,\mathbf{r}) = \mathcal{E}(\mathbf{r})\cos\omega t$, or, $\hat{H}_{Eab} = \mathcal{E}_{ab}\cos\omega t$, $\Pi_{ab} = \langle\psi_a|\mathcal{E}|\psi_b\rangle$.

Solving the differential equations for $c_a(t)$ and $c_b(t)$, one finds

$$c_b(t) = -\varepsilon_{ab}\frac{1}{2\hbar}\left[\frac{e^{i\left(\frac{E_b-E_a}{\hbar}+\omega\right)t}-1}{\frac{E_b-E_a}{\hbar}+\omega} + \frac{e^{i\left(\frac{E_b-E_a}{\hbar}-\omega\right)t}-1}{\frac{E_b-E_a}{\hbar}-\omega}\right].$$

The quantum transition probability defines the probability that a system, the quantum states of which is described by $\Psi(t)$, will evolve from a state with ψ_a to a state with ψ_b. The probability, as a function of time, evolves as

$$p_{a\to b}(t) = |c_b(t)|^2,$$

$$p_{a\to b}(t,\omega) = |\varepsilon_{ab}|^2\left|\frac{1}{2\hbar}\left[\frac{e^{i\left(\frac{E_b-E_a}{\hbar}+\omega\right)t}-1}{\frac{E_b-E_a}{\hbar}+\omega} + \frac{e^{i\left(\frac{E_b-E_a}{\hbar}-\omega\right)t}-1}{\frac{E_b-E_a}{\hbar}-\omega}\right]\right|^2$$

$$= |\varepsilon_{ab}|^2\left|\frac{1}{2\hbar}\left[\frac{e^{i\left(\frac{E_b-E_a}{\hbar}+\omega\right)t}-1}{\omega_0+\omega} + \frac{e^{i\left(\frac{E_b-E_a}{\hbar}-\omega\right)t}-1}{\omega_0-\omega}\right]\right|^2.$$

The quantum transition rate is $r = (dp/dt)$.

For a photon with wavelength $\lambda = 525$ nm, $E = 3.79 \times 10^{-19}$J. Let $\Delta E = (E_b - E_a) = 3.79 \times 10^{-19}$J. The *natural* transition frequency is $\omega_0 = (E_b - E_a)/\hbar$, which yields $\omega_0 = 3.589 \times 10^{15}$ rad/s, $f_0 = 5.71 \times 10^{14}$ Hz. Two-state switching is characterized by ΔE and $p_{a\to b}(t,\omega)$. The surface for the transition probability $p_{a\to b}(t,\omega)$ is shown in Figure 21.9a.

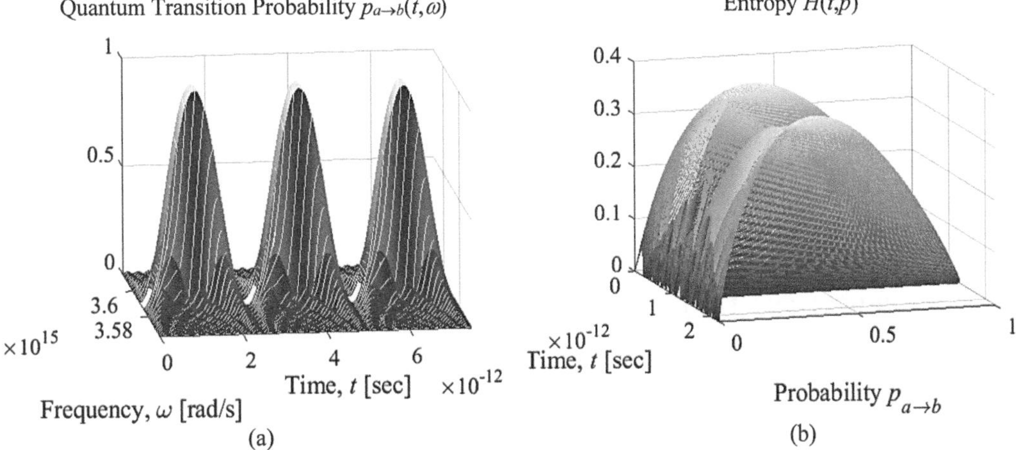

FIGURE 21.9 (a) Two-state *microscopic* device: quantum transition probability $p_{a\to b}(t,\omega)$; (b) surface for entropy $H(t,p)$.

The probability $p_{a\to b}(\omega)$ decreases as the difference $|\omega_0 - \omega|$ increases. To ensure a high transition probability, the system should be excited with ω relevant to the *natural* frequency ω_0.

A surface for the Shannon entropy, as a time-varying function $H(t, p) = -\sum_{i=1}^{N} P_i \ln p_i$, is illustrated in Figure 21.9b. The derived results demonstrate differences between quantum *microscopic* and *macroscopic* semiconductor devices. ■

Example 21.6.15

One may examine absorption, stimulated emission, spontaneous emission, etc. The expressions for $c_b(t)$ and transition probability $p_{a\to b}(t,\omega)$, $p_{a\to b}(t,\omega) = |c_b(t)|^2$ are found in Example 21.6.14. The *microscopic* system acquires energy $(E_b - E_a) = \hbar\omega_0$ by absorbing a photon. If the system in the upper state b, it can evolve to the lower state a, exhibiting stimulated emission (system absorbs an incident photon and emits photons) or spontaneous emission (system emits a photon without an external perturbation). Using the quantum transition probability $p_{b\to a}$, the transition rate is $r_{b\to a} = (dp_{b\to a}/dt)$.

The electromagnetic field energy density is $w = w_E + w_H = {}^1/_2\varepsilon E^2 + {}^1/_2\mu H^2$, and $w_E = w_H$. For the external electric field $E_0\cos(\omega t)$, one obtains $w = \varepsilon E_0^2\cos^2(\omega t)$. The averages of $\cos^2(\omega t)$ over a period is ${}^1/_2$. Thus, $w = {}^1/_2\varepsilon E_0^2$. For a monochromic wave, we have

$$r_{b\to a}(t,\omega) = \frac{dp_{b\to a}(t,\omega)}{dt},\; p_{b\to a}(t,\omega) = \frac{2w}{\varepsilon}\left|\frac{e\langle\psi_b|X|\psi_a\rangle}{2\hbar}\left[\frac{e^{i(\omega_0+\omega)t}-1}{\omega_0+\omega} + \frac{e^{i(\omega_0-\omega)t}-1}{\omega_0-\omega}\right]\right|^2.$$

For incoherent (independent) perturbations of electromagnetic waves with different modes characterized by different magnitudes, frequencies, attenuation, and other quantities, one expresses the energy density in the frequency range $d\omega$ as $\Omega(\omega)d\omega$.

We have $p_{b\to a}(t) = \frac{e^2}{2\varepsilon\hbar^2}\left|\langle\psi_b|x|\psi_a\rangle\right|^2 \int_0^\infty \Omega(\omega)\left|\left[\frac{e^{i(\omega_0+\omega)t}-1}{\omega_0+\omega} + \frac{e^{i(\omega_0-\omega)t}-1}{\omega_0-\omega}\right]\right|^2 d\omega.$ ■

Example 21.6.16: Quantum Mechanics and Probabilistic and Stochastic Computing

Using Boolean and multiple-valued algebra, efficient circuits and algorithms for data processing are developed and implemented in all existing general-purpose computing platforms. In many application, analog processing schemes are used. Quantum statistics on distributions of state probabilities, quantum statistical equilibrium, as well as physical transductions probabilities may be used applying the principles of stochastic and probabilistic computing. This concept was introduced in the 1950s [48, 83] and applied thereafter, attempting to analyze processing by neurons [10, 12, 48, 84]. As an example, consider a stochastic process $X = [x_1 x_2 \cdots x_k]$ where the x_i bits are randomly chosen to yield the probability p_x that $x_i = 1$. The adder uses n Boolean inputs and computes their sum as the output. For independent n Boolean inputs that have the same probability p of being 1, the output has the binomial distribution with the probability mass function $\binom{n}{k}p^k(1-p)^{n-k}$, where n is the number of trials $n \in \{0, 1, 2, \ldots\}$; p is the success probabilities of each trial, $p \in [0, 1]$; and k is the number of successes, $k \in \{0, 1, \ldots, n\}$. Let the data inputs of the multiplexer MUX be $(n+1)$ random $b_0, \ldots, b_n$. The computation scheme is $Y = \sum_{k=0}^{n} b_k\binom{n}{k}p^k(1-p)^{n-k}$. Stochastic and probabilistic general-purpose computing are found to be inadequate and impractical due to long computation time, algorithmic ambiguity, low accuracy, high error rate, etc. ■

21.7 FABRICATION AND CHARACTERIZATION OF OPTICAL WAVEGUIDES

The branched optical waveguides and controlling structures are fabricated on 6-in. p-type silicon wafers. As documented in Figure 21.10, after an RCA clean, a 750 nm SiO_2 layer is grown by a two-step thermal oxidation at 1000°C for 100 min. The alignment marks are patterned. A stoichiometric 500 nm Si_3N_4 layer is deposited using low-pressure chemical vapor deposition (LPCVD) using dichlorosilane ($SiCl_2H_2$) and ammonia (NH_3) at 810°C for 80 min. Fabrication of all structural

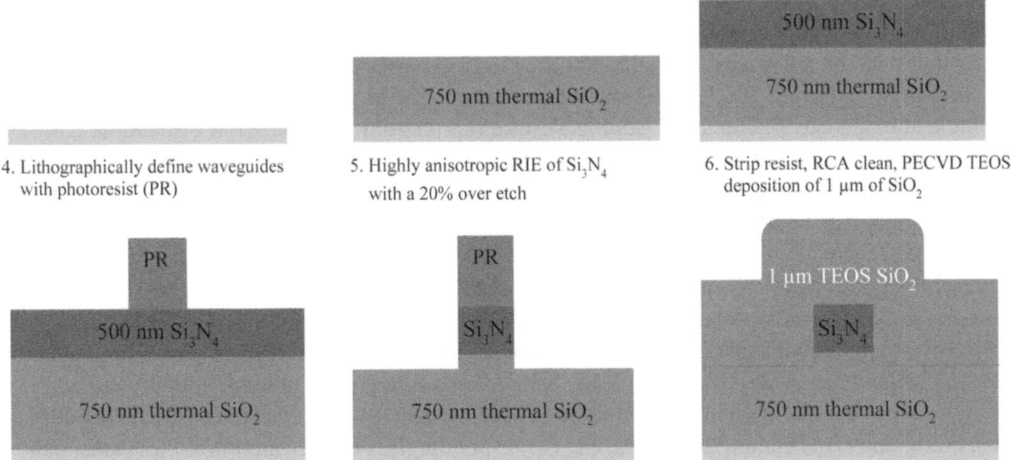

FIGURE 21.10 Simplified schematic of the process flow with the different materials and thickness used in the fabrication of Si_3N_4 optical waveguides. These optical waveguides guarantee the spectral transparency for photon wavelength from 475 nm to near-infrared spectrum with adequate propagation losses.

layers is performed to ensure low stress, adequate homogeneity, and uniformity by minimizing surface roughness. Different waveguide layouts are defined by photolithography. An i-line ASML PAS 5500 stepper ensures the 0.5 μm minimum feature size. We use AZ MiR701 photoresist (0.9 μm) and 150 mJ/cm^2 exposure dose. A dry reactive ion etch (RIE) process in a Trion etcher (CHF_3 and O_2 gases ratio is 12.5:1, 50 mTorr pressure, and 140 W power) ensures anisotropic Si_3N_4 structures. A 1-μm-thick SiO_2 is deposited using plasma-enhanced chemical vapor deposition (PECVD)–tetraethyl orthosilicate (TEOS) deposition, which yields silicon dioxide. An Applied Materials P5000 system ensures a consistent PECVD-TEOS deposition of SiO_2. The scanning electron microscopy (SEM) images of the fabricated optical waveguides are reported in Figure 21.11a and b. In addition to optical waveguides, openings in the TEOS cladding have been made by wet HF etch to deposit or implant colloidal quantum dots or dopants.

The fabricated optical waveguides can be integrated with photon sources, high-intensity electric field structures, optoelectronics, and controlling ASICs. Using additional steps, the electric filed radiating surfaces (evaporated or sputtered 20 nm chromium seed layer and 20 nm gold) can be

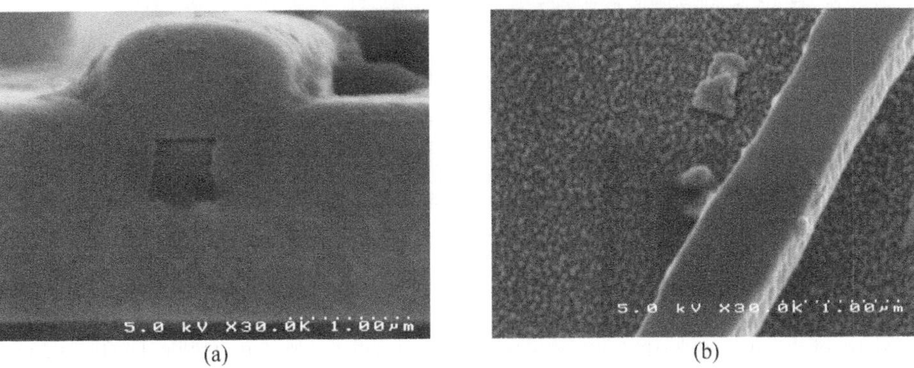

FIGURE 21.11 (a) SEM images of the cross-section 500 × 500 nm Si_3N_4 waveguide core with a SiO_2 cladding; (b) SEM image of the Si_3N_4 waveguide core on the wafer. The SiO_2 cladding has been removed.

deposited on Si, SiO_2, and Si_3N_4 insulators. We can vary the electric field intensity to control: (1) electro-optic effects in nonlinear optical media, which is the Si_3N_4 core; and (2) emission of photons by quantum dots or other active devices. A thermo-optic phase shifter can be fabricated. The power consumption, nonuniform heat transfer, and slow spatiotemporal thermodynamics are serious drawbacks that must be addresses. The silicon nitride optical waveguides losses are from 2 to 5 dB/cm for the wavelength from 450 to 1550 nm. Cross-sectional SEM images of waveguides allow one to examine homogeneity of core and cladding, surface morphology, etc. The SEM evaluation, reported in Figures 21.11 and 21.12, provides evidence that the waveguide core is fully and uniformly embedded within the cladding. For a Si_3N_4 with width a and thickness b, different a/b ratios can be achieved, and, a_{min} = 500 nm. The 1 μm PECVD-TEOS SiO_2 cladding is uniform and homogeneous. We achieved consistency of LPCVD, PECVD, RIE, wet etching, deposition, and other micromachining processes. While the homogeneity of SiO_2 and Si_3N_4 films up to 1 μm (b_{max} = 1μm) is ensured, the sidewalls of the Si_3N_4 waveguides are not ideally smooth. The angled side view of a Si_3N_4 waveguide with oxide cladding removed confirms the sidewall has significant roughness due to: (1) the RIE process [85]; and (2) the LPCVD and lithography processes. Improvements can be ensured by decreasing the thickness of the SiO_2 cladding on Si, reducing b, optimizing deposition, lithography, and etching processes, etc. Correspondingly, one may minimize the losses, reduce scattering, minimize dispersion, etc. The smooth sidewall surfaces can be achieved by optimizing LPCVD [86], improving the etch process [87, 88], and using a post-etch chemical treatment [89].

Various X- and Y-branched center- and side-split optical waveguides are designed and fabricated with 500 × 250 nm, 500 × 500 nm, 1000 × 500 nm, and other a/b Si_3N_4 cores and 1 μm SiO_2 cladding. As documented in Figures 21.11 and 21.12, we accomplished important milestones by fabricating and characterizing waveguides with nonlinear electro-optic media. Controlled schemes are investigated. Further high-fidelity analysis, characterization, and testing are needed.

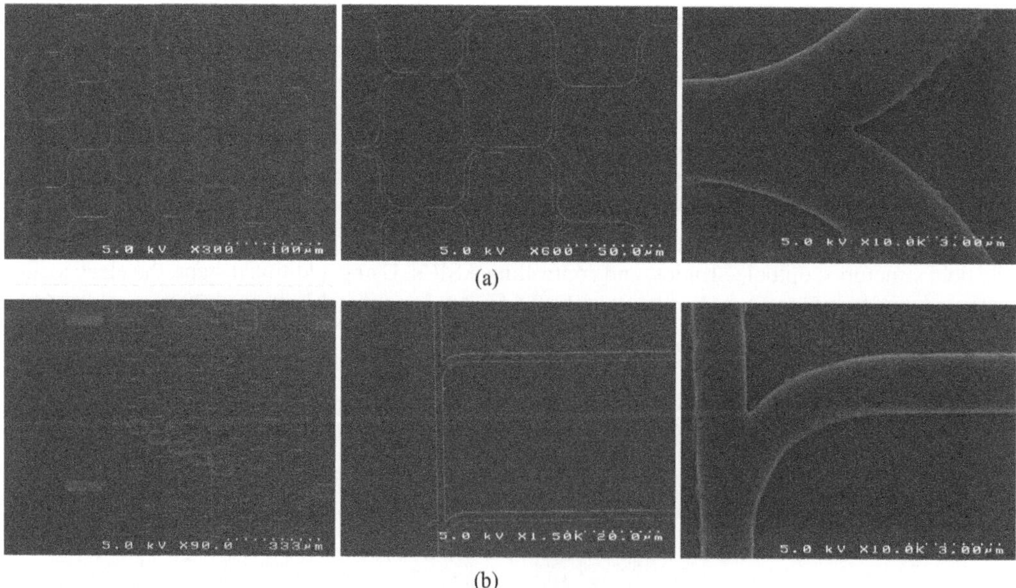

FIGURE 21.12 SEM images of optical waveguides for directed propagation, photon confinement, routing, and steering: (a) Top views of the SiO_2-cladded 1000 × 500 nm Si_3N_4 core in Y-branched center-split waveguides; (b) top views of the Si_3N_4 side-split waveguides with the specified curvature, profile, length, and openings (interfacing) to ensure control and integration capabilities. The width of the buried Si_3N_4 core is 1 μm, while the width of SiO_2 cladding is 3 μm.

Proof-of-concept physical devices and systems can be designed and implemented. Waveguides, microring, and microcavities exhibit nonlinear optical phenomena, *hybrid* modes, resonance, photon mixing, coupling, and other transductions that may implement formal mathematical operations. Control and room-temperature measurements schemes must support photon emission, absorption, detection, entanglements, mixing, etc. High-fidelity experimental studies are needed. The quantum-mechanical analysis can be accomplished as reported in Section 21.6 and in the illustrative examples below.

Example 21.7.1: Electromagnetic field quantization

An electromagnetic mode of N photons in a waveguide, having the same frequency f, has an energy $E = Nhf$. Using the conservation of energy principle, Dirac examined interactions between a charge and an electromagnetic field, as well as perturbations and induced transitions of the photon states. Quantization of an electromagnetic field results in the Bose statistics. A quantum state of the electromagnetic field is mathematically analyzed using a tensor product of the states for all electromagnetic modes. One has $|n_{\mathbf{k}0}\rangle \otimes |n_{\mathbf{k}1}\rangle\rangle \otimes \ldots \otimes |n_{\mathbf{k}_{n-1}}\rangle \otimes |n_{\mathbf{k}_n}\rangle$, where $|n_{\mathbf{k}_i}\rangle$ denotes the state in which n_{k_i} photons are in the mode $\mathbf{k}_i$. The creation of a new photon in mode $\mathbf{k}_i$, which is emitted due to an atomic transition, is denoted as $|n_{\mathbf{k}_i}\rangle \rightarrow |n_{\mathbf{k}_i}+1\rangle$. ■

Example 21.7.2: Photon transduction in waveguides, microrings, and microcavities

For a photon, the ground state energy is $E_0 = \frac{1}{2}\hbar\omega$, while the energy eigenvalues at any eigenstate ψ_n is $E_n = (n + \frac{1}{2})\hbar\omega$. Quantum-mechanically, the creation and annihilation operators $a^+(\mathbf{k})$ and $a^-(\mathbf{k})$ correspond to the creation of a photon with momentum $\hbar\mathbf{k}$ and helicity $\mathbf{J}\cdot\hat{\mathbf{k}} = \hbar$ and $\mathbf{J}\cdot\hat{\mathbf{k}} = -\hbar$, respectively.

The corresponding Schrödinger equation is

$$\hbar\omega(a^+a^- + {}^1/_2)\psi = E\psi,\ [a^-, a^+] = 1.$$

The time-evolution operator is $\hat{U} = e^{-\frac{1}{2}i\omega t}\sum_{k=0}^{\infty}\frac{1}{k!}(e^{-i\omega t} - 1)^k a^{+k}a^{-k}$.

Under many assumptions, the interacting Hamiltonian to study the quantum state transitions in the considered waveguides and physical interacting system (microrings and microcavities) with a *junction* is

$$\hat{H} = -i\hbar\left[c_a a^+a^- + c_b b^+b^- + c_{ab}a^+b^- + c_{ba}b^+a^-\right],$$

where c_i are the interaction operators.

For a photon in $[\ldots,|n_{-1}\rangle,|n\rangle,|n_1\rangle,\ldots]$, one has $a^+|\,n\rangle = \sqrt{n+1}|\,n+1\rangle$ and $a^-|\,n\rangle = \sqrt{n}|\,n-1\rangle$. A pair of photons interacts in $n(n-1)$ different modes with the probability that a pair annihilates is $p_{a-}n(n-1)$. The wave function evolves as

$\frac{\partial}{\partial t}|\,\psi\rangle = -p_{a+}\sum_i(a_i^+ - a_{i-1}^+)(a_i^- - a_{i-1}^-)|\,\psi\rangle + p_{a-}\sum_i(a_i^{-2} - a_i^{+2}a_i^{-2})|\psi\rangle$.

One may carry out theoretical analysis, which is extremely difficult to perform for the multiple photon interactions. ■

Practical quantum information sharing, steering, and processing schemes are of importance. The state transitions and transductions on single, multiple, and entangled photons may be supported by means of controlled photon absorption, emission, and mixing by electrostatically controlled quantum dots, lasers, waveguides, etc. The information can be encoded by controlled energy, varying charge, polarization, etc. Quantum determinisms on energy, polarization, frequencies, optical phases, and other measurable physical quantities can be used to implement communication and processing schemes by accomplishing binary and multiple-valued (ternary, quaternary, etc.) operations. The quantum determinism (Plank equation $E = hf$, Hertz photoelectric effect $E_K = h(f - f_0)$, Pauli exclusion principle, etc.) implies the use of real-valued and measurable physical quantiles. For the circularly polarized photons, a spin angular momentum $\pm\hbar$ is quantum-deterministic, with $J_z = +\hbar$ and $J_z = -\hbar$ for the left L and right R circular polarization. The Heisenberg uncertainty principle must be satisfied, which implies consideration of quantum-ensemble quantities. The waveguides may enable optoelectronic systems and processing schemes. While quantum mechanics fundamentals are of importance, consistent device- and system-level experimental studies are needed despite enormous challenges. Experimental substantiation may ensure a pathway to the technology justification, substantiation, and transfer.

In 1862, James Maxwell developed governing equations for the electromagnetic field. Philipp Lenard invented and demonstrated controlled electromagnetic radiation in vacuum tubes in 1888 and conducted photoelectric experiments in 1902. These discoveries resulted in the practical vacuum-tube triodes used in radio receivers since 1904. The principles of the field effect transistors were patented by Julius Lilienfeld in 1922. In 1947, the point-contact transistor was demonstrated by John Bardeen, Walter Brattain, and William Shockley. Morris Tanenbaum and Gordon Teal demonstrated a silicon transistor in 1954. It takes decades to demonstrate and commercialize new discoveries and technologies.

21.8 CONCLUSIONS

This chapter aims to accelerate fundamental research, motivate discoveries and innovations, enable engineering developments, and demonstrate technological feasibility of quantum processing. The *microscopic* and *macroscopic* optoelectronic devices exhibit quantum phenomena that can be used to perform computing, data storage and retrieval, logics, interfacing, and communication on measurable observables, which are due to controllable quantum transductions. Quantum-centric processing depends on pervasive understanding of devices and systems, fabrication, technology practicality, etc. Processing estimates, measures, and calculi were examined. Our results unify quantum mechanics, theoretical computer science, and engineering. The reported studies may contribute to analysis of information exchange and processing in living organisms, as well as advance engineering solutions. Significant theoretical, applied, experimental, and technological challenges remain. These challenges may be overcome by devising quantum-enabled processing schemes, devices, and systems. To substantiate our findings, we document applied, numerical, and experimental studies consistent with the *first principles* of physics, quantum mechanics, and microelectronic technologies. The illustrative examples justify and support our findings.

REFERENCES

1. *International Technology Roadmap for Semiconductors*, 2005, 2007 and 2009 Editions, Semiconductor Industry Association, Austin, Texas, USA.
2. *International Roadmap for Devices and Systems*, 2017 *Edition Beyond CMOS*, IEEE, 2018.
3. D. Dai, "Silicon nanophotonic integrated devices for on-chip multiplexing and switching," *J. Lightwave Technol.*, vol. 35, pp. 572–587, February 2017.
4. B. G. Lee et al., "FEC-free 60-Gb/s silicon photonic link using SiGe-driver ICs hybrid-integrated with photonics-enabled CMOS," *Proc. Optical Fiber Communication Conf.*, pp. 1–3, 2018.
5. S. E. Lyshevski, *Molecular Electronics, Circuits, and Processing Platform*, CRC Press, Boca Raton, FL, 2007.
6. S. E. Lyshevski, *Three-Dimensional Molecular Electronics and Integrated Circuits for Signal and Information Processing Platforms, Handbook on Nano and Molecular Electronics*, Ed. S. E. Lyshevski, CRC Press, Boca Raton, FL, pp. 6–1–6-102, 2007.
7. S. E. Lyshevski, *Molecular and Biomolecular Processing: Solutions, Directions and Prospects, Handbook of Nanoscience, Engineering and Technology*, Ed. W. Goddard, D. Brenner, S. E. Lyshevski and G. Iafrate, CRC Press, Boca Raton, FL, pp. 125–177, 2012.
8. S. M. Sze and K. K. Ng, *Physics of Semiconductor Devices*, John Wiley and Sons, NJ, 2007.
9. A. Yariv, *Quantum Electronics*, John Wiley and Sons, New York, 1988.
10. P. Dayan and L. F. Abbott, *Theoretical Neuroscience: Computational and Mathematical Modeling of Neural Systems*, MIT Press, Cambridge, MA, 2001.

11. J. J. Chang, J. Fisch and F. A. Popp, *Biophotons*, Kluwer Academic Publishers, Dordrecht-Boston-London, 1998.
12. *Computational Neuroscience: A Comprehensive Approach*, Ed. J. Feng, Chapman and Hall/CRC Press, Boca Raton, FL, 2003.
13. S. R. Hameroff and J. Tuszynski, *Search for Quantum and Classical Modes of Information Processing in Microtubules: Implications for "The Living Atate"*, Handbook on *Bioenergetic Organization in Living Systems*, Eds. F. Musumeci, M.-W. Ho, World Scientific, Singapore, 2003.
14. A. G. Gurwitsch, *Theory of the Biological Field*, Nauka, Moscow, 1944.
15. Y. Katoaka, Y. L. Cui, A. Yamagata, M. Niigaki, T. Hirohata, N. Oishi and Y. Watanabe, "Activity-dependent neural tissue oxidation emits intrinsic ultraweak photons," *Biochemical and Biophysical Research Communications*, vol. 285, pp. 1007–1011, 2001.
16. V. Kaznacheyev, "Electromagnetic bioinformation in intercellular interactions," *Psi. Research*, vol. 1, no. 1, pp. 47–76, 1982.
17. J. E. Morgan, "Circulation and axonal transport in the optic nerve", *Eye*, vol. 18, pp. 1089–1095, 2004.
18. W. D. McElroy and B. L. Strehler, "Bioluminescence," *Bacteriol. Rev.*, vol. 18, no. 3, pp. 177–194, 1954.
19. J. A. Tuszynski, A. Priel, J. A. Brown, H. F. Cantiello and J. M. Dixon, *Electronic and Ionic Conductivities of Microtubules and Actin Filaments, Their Consequences for Cell Signaling and Applications to Bioelectronics*, in Handbook on *Nano and Molecular Electronics*, Ed. S. E. Lyshevski, pp. 18-1 – 18.46, CRC Press, Boca Raton, FL, 2007.
20. K. Gupta and D. Panda, "Perturbation of microtubule polymerization by quercetin through tubulin binding: A novel mechanism of its antiproliferative activity," *Biochemistry*, vol. 41, no. 43, pp. 13029–13038, 2002.
21. E. Utreras, E. M. Jiménez-Mateos, E. Contreras-Vallejos, E. Tortosa, M. Pérez, S. Rojas, L. Saragoni, R. B. Maccioni, J. Avila and C. González-Billault, "Microtubule-associated protein 1B interaction with tubulin tyrosine ligase contributes to the control of microtubule tyrosination," *Developmental Neuroscience*, vol. 30, no. 1–3, pp. 200–210, 2008.
22. A. H. Hosseinnia et al., "High-quality silicon on silicon nitride integrated optical platform with an octave-spanning adiabatic interlayer coupler," *Opt. Express*, vol. 23, no. 23, pp. 30297–30307, 2015.
23. K. Ikeda et al., "Thermal and Kerr nonlinear properties of plasma-deposited silicon nitride/silicon dioxide waveguides," *Opt. Express*, vol. 16, no. 17, pp. 12987–12994, 2008.
24. M. Pantouvaki et al., "Active components for 50 Gb/s NRZ-OOK optical interconnects in a silicon photonics platform," *J. Lightwave Technol.*, vol. 35, no. 4, pp. 631–638, 2017.
25. M. Piels et al., "Low-loss silicon nitride AWG demultiplexer heterogeneously integrated with hybrid III–V/silicon photodetectors," *J. Lightwave Technol.*, vol. 32, no. 4, pp. 817–823, 2014.
26. W. D. Sacher et al., "Multilayer silicon nitride-on-silicon integrated photonic platforms and devices," *J. Lightwave Technol.*, vol. 33, no. 4, pp. 901–910, 2015.
27. W. D. Sacher et al., "Monolitithically integrated multilayer silicon nitride-on-silicon waveguide platforms for 3-D photonic circuits and devices," *Proc. IEEE*, vol. 106, no. 12. pp. 2232–2245, 2018.
28. W. D. Sacher et al., "Tri-layer silicon nitride-on-silicon photonic platform for ultra-low-loss crossings and interlayer transitions," *Opt. Express*, vol. 25, no. 25, pp. 30862–30875, 2017.
29. G. P. Agrawal, *Nonlinear Fiber Optics*, Academic Press, NY, 2012.
30. P. F. Antunes, A. N. Pinto and P. S. André, "Single-photon source by means of four-wave mixing inside a dispersion-shifted optical fiber," *J. Frontiers Optics*, FMJ3, 2006.
31. A. J. Almeida, N. A. Silva, N. J. Muga and A. N. Pinto, "Single-photon source using stimulated FWM in optical fibers for quantum communication," *Proc. SPIE*, 80013W, 2011.
32. N. Bloembergen, "Recent progress in four-wave mixing spectroscopy," *Laser Spectroscopy IV, Proc. Conf. Laser Spectroscopy*, Ed. H. Walther and K. W. Rothe, pp. 340–348, Springer, Berlin, 1979.
33. A. Dot et al., "Converting one photon into two via four-wave mixing in optical fibers," *Phys. Rev. A*, vol. 90, 043808, pp. 1–13, 2014.
34. E. Meyer-Scott et al., "Generating polarization-entangled photon pairs using cross-sliced birefringent fibers," *Optics Express*, vol. 21, no. 5, pp. 1–8, 2013.
35. C. Reimer et al., "Cross-polarized photon-pair generation and bi-chromatically pumped optical parametric oscillation on a chip," *Nature Communication*, 8236, 2015.
36. C. Reimer et al., "Integrated frequency comb source of heralded single photons," *Opt. Express*, vol. 22, no. 6, pp. 6535–6546, 2014.

37. K. Rottwitt et al., "Nonlinear optical signal processing and generation of quantum states using four-wave mixing," *Proc. Conf. Transparent Optical Networks*, pp. 1–4, 2018.
38. J. Van Erps et al., "High-resolution optical sampling of 640-Gb/s data using four-wave mixing in dispersion-engineered highly nonlinear As_2S_3 planar waveguides," *J. Lightwave Technology*, vol. 28, pp. 209–215, 2010.
39. D. F. Geraghty et al., "Wavelength conversion for WDM communication systems using four-wave mixing in semiconductor optical amplifiers," *J. Selected Topics in Quantum Electronics*, vol. 3, no. 5, pp. 1446–1155, 1997.
40. M. W. Maeda et al., "The effect of four-wave mixing in fibers on optical frequency-division multiplexed systems," *J. Lightwave Technology*, vol. 8, no. 9, pp. 1402–1408, 1990.
41. F. Ferreira et al., "Impact of inter-modular four-wave mixing on the performance of mode- and wavelength-division-multiplexing systems," *Proc. Conf. Transparent Optical Networks*, pp. 1–5, 2015.
42. Y. Fukuchi et al., "All-optical time division demultiplexing of 160 Gbit/s signal using cascaded secon-dorder nonlinear effect in quasi-phase matched LiNbO waveguide device," *Electron. Lett.*, vol. 39, pp. 789–790, 2003.
43. C. J. Krückel et al., "Linear and nonlinear characterization of low-stress high-confinement silicon-rich nitride waveguides," *Opt. Express*, vol. 23, pp. 25827–25837, 2015.
44. A. Agarwalet et al., "Entangled-pair transmission improvement using distributed phase-sensitive amplification," *Phys. Rev. X* 4, 041038, pp. 1–7, 2014.
45. N. Gregersen, P. Kaer and J. Mørk, "Modeling and design of high-efficiency single-photon sources," *IEEE Trans. J. Selected Topics in Quantum Electronics*, vol. 19, no. 5, 2013.
46. R. J. Young et al., "Quantum dot sources for single photon and entangled photon pair," *Proc. IEEE*, vol. 95, no. 9, pp. 1805–1814, 2007.
47. S. E. Lyshevski, *MEMS and NEMS: Systems, Devices, and Structures*, CRC Press, Boca Raton, FL, 2002.
48. J. von Neumann,*The Computer and the Brain*, Princeton Press, 1957.
49. C. E. Shannon, "A mathematical theory of communication," *The Bell System Technical Journal*, vol. 27, pp. 379–423, 623–656, 1948.
50. S. N. Yanushkevich, V. P. Shmerko and S. E. Lyshevski,*Computer Arithmetics for Nanoelectronics*, CRC Press, Boca Raton, FL, 2009.
51. S. E. Lyshevski et al., "Estimates and measures of data communication and processing in nanoscaled classical and quantum physical systems," *Proc. IEEE Nanotechnology Conf.*, pp. 1044–1047, 2014
52. S. E. Lyshevski, "Nano-, nanobio- and nanobiomedical-technologies: Enabling sensing, communication and processing paradigms," *Proc. Conf. Nanotechnology*, pp. 177–178, 2012.
53. S. E. Lyshevski, "Quantum molecular sensing, communication and processing by photons," *Proc. Conf. Nanotechnology*, pp. 476–481, 2012.
54. S. E. Lyshevski and L. Reznik, "Information-theoretic estimates of communication and processing in nanoscale and quantum optoelectronic systems," *Proc. Conf. Electronics and Nanotechnologies*, pp. 33–37, 2013.
55. A. Nakashima and M. Hanzava, "Expantion theorem and design of two-terminal relay networks", *Nippon Electrical Communication Engineering*, vol. 24, pp. 203–210, 1941.
56. C. E. Shannon, "A symbolic analysis of relay and switching circuits", *Transanctions of the American Institute of Electrical Engineering*, vol. 57, pp. 713–723, 1938.
57. V. I. Shestakov, "The algebra of two-terminal networks designed from two-terminal components: The algebra of A-schemata", *Avtomatika i Telemechanica*, vol. 2, no. 6, pp. 15–24, 1941.
58. G. Birkhoff and J. von Neumann, "The logic of quantum mechanics," *Annals of Mathematics*, vol. 37, no. 4, pp. 823–843, 1936.
59. R. P. Feynman, *Lectures on Computation*, Addison-Wesley, Reading, MA, 1986.
60. M. P. Frank and T. F. Kingght, "Ultimate theoretical models of nanocomputers," *Nanotechnology*, vol. 9, no. 3, pp. 162–176, 1998.
61. D. C. Marinescu and G. M. Marinescu, *Approaching Quantum Computing*, Prentice Hall, Upper Saddle River, NY, 2005.
62. M. Mosca, R. Jozsa, A. Steane and A. Ekert, "Quantum-enhanced information processing," *Transactions of The Royal Society*, vol. A 358, pp. 261–279, 2000.

63. A. Barenco, C. H. Bennett, R. Cleve, D. P. DiVincenzo, N. Margolus, P. Shor, T. Sleator, J. A. Smolin and H. Weinfurter, "Elementary gates for quantum computation," *Physical Review A*, vol. 52, no. 5. pp. 3457–3467, 1995.
64. A. J. Almeida et al., "Fiber-optical communication system using polarization encoding photons," *Proc. European Conf. Networks and Optical Communications*, pp. 127–132, 2010.
65. J. Breguet, A. Muller and N. Gisin, "Quantum cryptography with polarized photons in optical fibres: Experiment and practical limits," *Journal of Modern Optics*, vol. 41, pp. 2405–2412, 1994.
66. S. Dolinar et al., "On approaching the ultimate limits of photon-efficient and bandwidth-efficient optical communication," *Proc. Conf. Space Optical Systems and Applications*, pp. 269–278, 2011.
67. B. I. Erkmen et al., "On approaching the ultimate limits of communication using a photon-counting detector," *Proc. SPIE*, 824605, 2012.
68. A. N. Pinto et al., "Optical quantum communications: An experimental approach," *Proc. SPIE*, 80011M, 2011.
69. J. H. Shapiro, "The quantum theory of optical communication," *IEEE J. Selected Topics in Quantum Electronics*, vol. 15, no. 6, pp. 1547–1569, 2009.
70. H. Yuen and J. Shapiro, "Optical communication with two-photon coherent states – Part I: Quantum-state propagation and quantum-noise reduction," *IEEE Trans. Information Theory*, vol. 24, no. 6, pp. 657–668, 1978.
71. *Quantum Communication, Sensing and Measurement in Space*, Ed. B. I. Erkmen, J. H. Shapiro and K. Schwab, Keck Institute for Space Studies, 2012.
72. R. J. Bonneau, G. Ramseyer, T. Renz and C. Theim, "A mathematical architecture for molecular computing," *Proc. Applied Imagery Pattern Recognition Workshop*, pp. 80–86, 2007.
73. J. S. Clarke et al., "Quantum computing within the framework of advanced semiconductor manufacturing," *Proc. Electron Devices Meeting*, pp. 13.1.1–13.1.3, 2016.
74. A. Frisch, "IBM Q – Introduction into quantum computing (with live demo)," *Proc. Syctem-on-Chip Conf.*, pp. 1–2, 2017.
75. R. Hamerly et al., "Quantum vs. optical annealing: Benchmarking the OPO Ising machine and D-Wave," *Proc. Lasers and Electro-Optics Conf.*, pp. 1–2, 2018.
76. M. Steffen et al., "Progress, status, and prospects of superconducting qubits for quantum computing," *Proc. European Solid-State Devices Research Conf.*, pp. 17–20, 2016.
77. J. Łukasiewicz, "O logice trójwartościowej", *Ruch Filozoficzny*, vol. 5, pp. 170-171, 1920. English translation: "On three-valued logic", Ed. L. Borkowski, *Selected Works by Jan Łukasiewicz*, North-Holland, Amsterdam, pp. 87–88, 1970.
78. L. S. Hay, "Axiomatization of the infinite-valued predicate calculus", *Journal of Symbolic Logic*, vol. 28, pp. 77–86, 1963.
79. R. McNaughton, "A theorem about infinite-valued sentential logic", *Journal Symbolic Logic*, vol. 16, pp. 1–13, 1951.
80. D. Mundici, "Interpretation of AF C*-algebras in Łukasiewicz sentential calculus", *Journal of Functional Analysis*, vol. 65, pp. 15–63, 1986.
81. V. Marra and D. Mundici, "Consequence and complexity in infinite-valued logic: A survey", *Proc. IEEE Int. Symp. Multiple-Valued Logic*, pp. 104–114, 2002.
82. J. W. Mills and C. A. Daffinger, "CMOS VLSI Lukasiewicz logic arrays", *Proc. Int. Conf. Application Specific Array Processors*, pp. 469–480, 1990.
83. J. Von Neumann, *Lectures on Probabilistic Logics and the Synthesis of Reliable Organisms From Unreliable Components*, The Institute for Advanced Study, Princeton University, 1952.
84. M. Suri, D. Querlioz, O. Bichler, G. Palma, E. Vianello, D. Vuillaume, C. Gamrat, and B. DeSalvo, "Bio-inspired stochastic computing using binary CBRAM synapses," *IEEE Transactions on Electron Devices*, vol. 60, no. 7, pp. 2402–2409, 2013
85. S. Cheemalapati, M. Ladanov, J. Winskas and A. Pyayt, "Optimization of dry etching parameters for fabrication of polysilicon waveguides with smooth sidewall using a capacitively coupled plasma reactor," *Applied Optics*, vol. 53, pp. 5745–5749, 2014.
86. P. T. Lin, V. Singh, H. G.Lin, T. Tiwald, L. C. Kimerling and A. M. Agarwal, "Low-stress silicon nitride platform for mid-infrared broadband and monolithically integrated microphotonics." *Advanced Optical Materials*, vol. 1, pp. 732–739, 2013.

87. N. Marchack, M. Khater, J. Orcutt, J. Chang, S. Holmes, T. S. Kamlapurkar, W. Green and S. Engelmann, "Reducing line edge roughness in Si and SiN through plasma etch chemistry optimization for photonic waveguide applications," *Proc. SPIE* 10149, *Advanced Etch Technology for Nanopatterning* VI, 101490F, 2017.
88. A. Z. Şubramanian *et al.*, "Low-loss singlemode PECVD silicon nitride photonic wire waveguides for 532–900 nm wavelength window fabricated within a CMOS pilot line," *IEEE Photonics Journal*, vol. 5, no. 6, pp. 2202809–2202809, 2013.
89. M. Brunet, D. Aureau, P. Chantraine, F. Guillemot, A. Etcheberry, A. C. Gouget-Laemmel and F. Ozanam, " Etching and chemical control of the silicon nitride surface," *ACS Applied Materials and Interfaces*, vol. 9, no. 3, pp. 3075–3084, 2017.

22 A Knudsen Boundary Layer Model for Multiscale Micro and Nano Flows

Alexander S. Liberson

Mechanical Engineering Department, Rochester Institute of Technology, Rochester, New York, USA

CONTENTS

22.1 INTRODUCTION

When mechanical, thermal, or electrical device is decreased in size to a sufficiently small level, its characteristics become significantly different from their counterparts in a macroscale. Microscale effects become important, when the mean free path of the energy carrier becomes comparable to the characteristic length of the object. In such scale, the continuum approach based on a heuristic principle of continuity is no longer valid. Different models of a nonclassical mechanics and physics have been introduced using different names for the generalized continuum: Cosserat, gradient, nonlocal, nonsymmetric, microstructure, micropolar, couple stress, multipolar, micromorphic, multiscale, and others. A survey of different nonclassical theories can be found in Refs [1-6]. All these models include the higher order derivatives in mathematical formulations and additional empirical constants, whose experimental verification meets certain difficulties. The higher order gradients introduce additional degrees of freedom allowing one to continualize the discrete matter, representing better as a result the discrete microstructure. The higher order gradients provide also a regularization and smoothing of nonphysical singularities and discontinuities predicted by classical models of continuum.

It is agreeable to differentiate flow conditions in microchannels based on a rarefaction effect characterized by the Knudsen number. In the free molecular flow conditions ($Kn > 10$), the Boltzmann equations are solved using typical approaches as the lattice Boltzmann model, direct simulation Monte-Carlo method, and molecular dynamics. A comprehensive review of relating approaches could be found in Refs [7, 8]. Within the Knudsen number range $0.1 < \mathrm{Kn} < 10$, a rarefied gas can be considered as a continuum–transition flow, which is neither a continuous medium nor a free-molecular flow. To model these flows, a number of generalized hydrodynamics models have been proposed. The family of Burnett models replaces Boltzmann kinetic equation by series of differential equations in a reduced space with respect to the finite number of moments of kinetic function distribution. However, attempts at solving the Burnett equations have uncovered many physical and numerical difficulties produced by the model. The trace of evolution of Burnett models and description of the progress made by recent developments could be found in Ref. [9].

Within the range of $0.001 < \mathrm{Kn} < 0.1$, the micro fluid flow is often characterized as a slip flow [7, 8, 10]. Typical engineering analyses of fluid flows in microchannels and micromachined fluid systems are based on a simplified approach accounting for the slip flow at the boundary within the frame of classical Navier–Stokes model. However, slip boundary conditions are not the only intrinsic property of a microscale. Physical phenomena arousing in microscale are associated with the microstructure and microrotation of fluid molecules, which should be accounted for by the model to present an adequate description of microfluidics. The monographs [2, 4] provide a unified picture of relating mathematical theory based on a micropolar model. A number of publications indicate that flow analysis in microchannels, based on a gradient theory of micropolar fluids, gives better predictions of experimental results [11–13]. Additionally, we note that the presence of roughness may prevent flow from the slip at the solid boundary [14]. In this case, application of a simplified engineering microfluidic theory [10] completely ignores microstructure, making no difference to the application of the classical Navier-Stokes model to microscale. On the contrary, gradient models result in a correction for classical macro fluid predictions of a flow rate, average velocity, Darcy friction factor, and pressure distribution regardless of a roughness effect on boundary conditions at microscale.

The demarcation line between application to microflows of gradient models versus classical Navier–Stokes model, supplemented by slip conditions, is not clearly defined. In Refs [15, 16], the comparative results are presented for a one-dimensional case of a pressure-driven flow for the wide range of Knudsen numbers. The present work extends application of the developed gradient model to the two-dimensional case with a following comparison of both approaches. An asymptotic behavior of a two-dimensional flow through a microchannel is investigated via rigorous singular perturbation analysis with respect to the small parameter, proportional to the Knudsen number. This small parameter is involved in both partial differential equations and the Maxwell's slip boundary conditions. A singular perturbation technique reveals a thin boundary layer (Knudsen layer) region near the solid boundary as an inner solution, whereas the outer solution satisfies to the classical Navier–Stokes equations. The boundary layer solution is presented in an analytical form, whereas the outer region is modeled using analytical or numerical solutions for the canonical or noncanonical domains, respectively. Effect of a surface roughness on a slip boundary condition is discussed explaining controversial experimental measurements in microscale. Calculational results obtained compare favorably with the currently available experimental data.

22.2 MICROSCALE AVERAGING

Let $\partial\Omega$ be a microelement inside a volume Ω whose dimensions are significantly larger than those of $\partial\Omega$.

$\mathbf{X} = (x_1, x_2, \ldots, x_N)$ is the center of mass of the underformed element, $\Xi = (\zeta_1, \zeta_2, \ldots, \zeta_N)$ is the local coordinates in $\partial\Omega$, relative to the center of mass $\mathbf{X}$. We adopt the following procedure of averaging of the multidimensional function $\phi(\mathbf{X} + \Xi)$ across the microscale volume $\partial\Omega$

$$\bar{\phi}(\mathbf{X}) = \int_{\partial\Omega} \phi(\mathbf{X} + \Xi) W(\Xi) \mathrm{d}\Omega \tag{22.1}$$

where the weight function $W(\Xi)$ satisfies the normalization condition $\int_{\partial\Omega} W(\Xi)\mathrm{d}\Omega = 1$. The constant value of the weight function results in the simple averaging. Presenting weight as a multidimensional Dirac delta function, $W(\Xi) = \delta(\Xi)$, results in an equality of an averaged across microscale volume and local quantities, $\bar{\phi}(\mathbf{X}) = \phi(\mathbf{X})$, relating to the description of classical continuum.

Representing $\phi(\mathbf{X} + \Xi)$ as a multidimensional Taylor series, we arrive at (∇ is the Hamilton operator)

$$\phi(\mathbf{X} + \Xi) = \sum_{j=0}^{\infty} \frac{1}{j!}(\Xi \cdot \nabla)^j \phi(\mathbf{X}) \tag{22.2}$$

equivalent to the following scalar expression

$$\phi(x_1 + \zeta_1, \dots, x_N + \zeta_N) = \sum_{m_1}^{\infty} \sum_{m_2}^{\infty} \dots \sum_{m_N}^{\infty} \frac{\partial^{m_1 + \dots m_N} \phi(x_1, \dots, x_N)}{\partial x_1^{m_1} \dots \partial x_N^{m_N}} \prod_{j=1}^{N} \frac{1}{m_j!} \zeta_j^{m_j} \tag{22.3}$$

Integration according to Equation 22.1 yields

$$\bar{\phi}(x_1, \dots, x_N) = \phi(x_1, \dots, x_N) + \sum_{m_1}^{\infty} \sum_{m_2}^{\infty} \dots \sum_{m_N}^{\infty} C_{m_1 \dots m_N} \frac{\partial^{m_1 + \dots m_N} \phi(x_1, \dots, x_N)}{\partial x_1^{m_1} \dots \partial x_N^{m_N}} \tag{22.4}$$

where

$$C_{m_1 \dots m_N} = \int_{\partial\Omega} \prod_{j=1}^{N} \frac{1}{m_j!} \zeta_j^{m_j} W(\Xi)\mathrm{d}\Omega \tag{22.5}$$

Simple averaging ($W(\Xi) = 1$) boils down (Eq. 22.5) to the following analytical expression:

$$C_{m_1 \dots m_N} = \prod_{j=1}^{N} \frac{1}{(m_j + 1)!} l_j^{m_j} \tag{22.6}$$

in which l_j is the length scale parameter associated with the x_j direction. If all coordinates $\Xi = (\zeta_1, \zeta_2, \dots, \zeta_N)$ coincide with the symmetry axes inside microelement $\partial\Omega$, then odd coefficients (coefficients with odd number of indices) are equal to zero.

Consider spatial-temporal averaging of a one-dimensional time-dependent function $\phi(x, t)$ across the microscale segment $[-l, l]$ within the time interval $[0, T]$

$$\bar{\phi}(x, t) = \frac{1}{2lT} \int_0^T \int_{-l}^{l} \phi(x + \zeta, t - \tau)\mathrm{d}\zeta \mathrm{d}\tau \tag{22.7}$$

Using Taylor series expansion

$$\phi(x + \zeta, t - \tau) = \sum_{k=0}^{\infty} \sum_{n=0}^{\infty} \frac{(-1)^n \zeta^k \tau^n}{k!n!} \frac{\partial^{k+n} \phi(x, t)}{\partial x^k t^n} \tag{22.8}$$

we obtain

$$\bar{\phi}(x, t) = \sum_{n=0}^{\infty} \sum_{k=0,2,4,\dots}^{\infty} C_{kn} \frac{\partial^{k+n} \phi(x, t)}{\partial x^k t^n} \tag{22.9}$$

$$C_{kn} = \frac{(-1)^n l^k T^n}{(k + 1)!(n + 1)!} \tag{22.10}$$

The limited set of higher order derivatives is typically essential to describe the details of analyzed phenomena.

Preserving the lowest two terms in the truncated series (Eq. 22.9), we arrive at Equation 22.11

$$\bar{\phi}(x,t) = \left(1 - \frac{T}{2}\frac{\partial}{\partial t} + \frac{l^2}{6}\frac{\partial^2}{\partial x^2}\right)\phi(x,t) + O(T^2 + L^4) \tag{22.11}$$

which is asymptotically equivalent to

$$\phi(x,t) = \left(1 + \frac{T}{2}\frac{\partial}{\partial t} - \frac{l^2}{6}\frac{\partial^2}{\partial x^2}\right)\bar{\phi}(x,t) + O(T^2 + L^4) \tag{22.12}$$

In a multidimensional case characterized by the space vector coordinate **X**, the following continualization operators are introduced

$$\phi(\mathbf{X},t) = \left(1 + \frac{T}{2}\frac{\partial}{\partial t} - \frac{l^2}{6}\Delta\right)\bar{\phi}(\mathbf{X},t) + O(T^2 + L^4) \tag{22.13}$$

$$\frac{\partial}{\partial t}\phi(\mathbf{X},t) = \left(1 + \frac{T}{2}\frac{\partial}{\partial t} - \frac{l^2}{6}\Delta\right)\frac{\partial}{\partial t}\bar{\phi}(\mathbf{X},t) + O(T^2 + L^4) \tag{22.14}$$

$$\Delta\phi(\mathbf{X},t) = \left(1 + \frac{T}{2}\frac{\partial}{\partial t} - \frac{l^2}{6}\Delta\right)\Delta\bar{\phi}(\mathbf{X},t) + O(T^2 + L^4) \tag{22.15}$$

It was shown in [15] that the continualization procedure (Eq. 22.12), being applied to different physical phenomena, is unconditionally stable and preserve Galilean invariance.

22.3 FLUID FLOW IN MICROSCALE

The fluid flow with the negligible variation of density (Mach number < 0.3), when advective inertial forces are small compared with viscous forces, is governed by following system of Stokes equations [17]:

$$\boldsymbol{V}_t + \frac{1}{\rho}\nabla p = \nu\Delta\boldsymbol{V}, \quad \nabla\cdot\boldsymbol{V} = 0 \tag{22.16}$$

Applying multidimensional continualization procedure, we replace velocity vector and pressure gradient according to

$$\boldsymbol{V}_t \Rightarrow \left(I + \frac{\tau}{2}\frac{\partial}{\partial t} - \frac{l^2}{6}\Delta\right)\boldsymbol{V}_t \tag{22.17}$$

$$\nabla p \Rightarrow \left(I + \frac{\tau}{2}\frac{\partial}{\partial t} - \frac{l^2}{6}\Delta\right)\nabla p \tag{22.18}$$

$$\Delta\boldsymbol{V} \Rightarrow \left(I + \frac{\tau}{2}\frac{\partial}{\partial t} - \frac{l^2}{6}\Delta\right)\Delta\boldsymbol{V} \tag{22.19}$$

to obtain the generalized model for the fluid dynamics dependent on the microscale size effects

$$\frac{\partial}{\partial t}L(\boldsymbol{V},p) = \left(I - \frac{l^2}{6}\Delta\right)\left(\nu\Delta\boldsymbol{V} - \frac{\nabla p}{\rho}\right) \tag{22.20}$$

where I is the identity operator, and the differential operator L is defined as

$$L(\boldsymbol{V},p) = \boldsymbol{V} + \frac{\tau}{2}\left(\boldsymbol{V}_t + \frac{\nabla p}{\rho}\right) - \left(\frac{l^2}{6} + \frac{\nu\tau}{2}\right)\Delta\boldsymbol{V} \tag{22.21}$$

To specify correctly the boundary conditions, we apply the virtual work variational principle to Equation 22.20. Limiting the scope of analysis by the steady-state pressure-driven flow in a duct yields

$$\begin{aligned}&\iint\left(\Delta u-\frac{l^2}{6}\Delta\Delta u-\frac{p_x}{\mu}\right)\delta u\mathrm{d}A\\&=\oint\left[\left(\frac{\partial u}{\partial n}-\frac{l^2}{6}\frac{\partial\Delta u}{\partial n}\right)\delta u+\frac{l^2}{6}\Delta u\frac{\partial\delta u}{\partial n}\right]\mathrm{d}\Gamma-\delta\iint\left(\frac{(\nabla u)^2}{2}+\frac{l^2}{6}\frac{(\Delta u)^2}{2}\right)\mathrm{d}A\end{aligned} \tag{22.22}$$

The contour integral in Equation 22.22 specifies essential and natural boundary conditions imposed on the solution and its normal derivative.

$$\left(\frac{\partial u}{\partial n}-\frac{l^2}{6}\frac{\partial\Delta u}{\partial n}\right)\delta u=0,\quad \Delta u\frac{\partial\delta u}{\partial n}=0\quad on\ \partial\Omega \tag{22.23a}$$

Their weighted combination of essential conditions (Robin type of conditions, or the third type of conditions) is a correct one, and is a frequently applied alternative in different problems of mathematical physics. With application to microflows, the mixed boundary conditions represent the first-order velocity slip at a gas–solid interphase [7]

$$u-\sigma_p l\frac{\partial u}{\partial n}=0 \tag{22.23b}$$

The effect of roughness can change dramatically the slip coefficient σ_p, reducing it to zero if the roughness height is comparable to the mean free path [22, 23].

22.4 CONNECTION WITH MICROPOLAR FLUID DYNAMICS

One of the best-established theories of the fluid flow with microstructure is the theory of micro-continuum attributed to Eringen [2, 3]. The formulation of a micropolar theory acquires additional degrees of freedom—gyration—to determine the rotation of the microstructure observed, for instance, in molecular dynamics simulations [7]. As a result, the conservation laws of classical continuum are expanded with the additional conservation law of microinertia moments.

The nonsteady equations of the micropolar fluid theory are as follows [2-6, 11-12]:

$$\frac{\partial\rho}{\partial t}+\nabla\cdot(\rho\boldsymbol{V})=0 \tag{22.24}$$

$$\rho\frac{\partial\boldsymbol{V}}{\partial t}=(\lambda+2\mu+k)\nabla\nabla\cdot\boldsymbol{V}-(\mu+k)\nabla\times\nabla\times\boldsymbol{V}+k\nabla\times G-\nabla p \tag{22.25}$$

$$\rho j\frac{\partial\boldsymbol{G}}{\partial t}=(\alpha+\beta+\gamma)\nabla\nabla\cdot\boldsymbol{G}-\gamma\nabla\times\nabla\times\boldsymbol{G}+k\nabla\times\boldsymbol{V}-2k\boldsymbol{G} \tag{22.26}$$

For the case of fluid with the constant physical properties (ρ = constant), the constants λ, α, and β do not appear in the governing equations since the velocity and gyration vectors are solenoidal. Using vector identity [18] $\nabla\times\nabla\times\boldsymbol{V}=\nabla(\nabla\cdot(\boldsymbol{V}))-\Delta\boldsymbol{V}$, Equations 22.25 and 22.26 are further simplified as

$$\frac{\partial\boldsymbol{V}}{\partial t}-\nu\Delta\boldsymbol{V}-\frac{k}{\rho}\nabla\times\boldsymbol{G}+\frac{\nabla p}{\rho}=0 \tag{22.27}$$

$$j\frac{\partial\boldsymbol{G}}{\partial t}-\frac{\gamma}{\rho}\Delta\boldsymbol{G}+\frac{2k}{\rho}\boldsymbol{G}-\frac{k}{\rho}\nabla\times\boldsymbol{V}=0 \tag{22.28}$$

Since material constant k is smaller by order of magnitude in comparison with dynamic viscosity μ [19, 20], the coefficient $(\mu + k)/\rho$ in Equation 22.25 is replaced for simplicity by kinematic viscosity ν. To find connection of the generalized Navier–Stokes model (Eqs. 22.20 and 22.21) to the micropolar fluid mechanics, we need to eliminate the gyration vector $\boldsymbol{G}$ from Equations 22.27 and 22.28. Rewriting the system (Eqs. 22.25 and 22.26) in operators form, read

$$L_{11}\boldsymbol{V} + L_{12}\boldsymbol{G} = -\frac{\nabla p}{\rho} \tag{22.29}$$

$$L_{12}\boldsymbol{V} + L_{22}\boldsymbol{G} = 0 \tag{22.30}$$

where differential operators are

$$L_{11} = \frac{\partial(\cdot\cdot)}{\partial t} - \gamma\Delta(\cdot\cdot); \quad L_{12} = -\frac{k}{\rho}\nabla\times(\cdot\cdot); \quad L_{22} = j\frac{\partial(\cdot\cdot)}{\partial t} - \frac{\gamma}{\rho}\Delta(\cdot\cdot) + \frac{2k}{\rho}(\cdot\cdot); \tag{22.31}$$

Using operational version of Cramer rule, read

$$\left(L_{11}L_{22} - L_{12}^2\right)\boldsymbol{V} = -\frac{1}{\rho}L_{22}(\nabla p) \tag{22.32}$$

Substituting expressions for operators (Eq. 22.31) into Equation 22.30, one can present the micropolar fluid dynamics model in a form similar to the generalized Navier–Stokes model (Eqs. 22.20 and 22.21).

$$\frac{\partial}{\partial t}\bar{L}(\boldsymbol{V},p) = \left(I - \frac{\gamma}{2k}\Delta\right)\left(\nu\boldsymbol{V} - \frac{\nabla p}{\rho}\right) \tag{22.33}$$

$$\bar{L}(\boldsymbol{V},p) = \boldsymbol{V} + \frac{j}{2k}\left(\boldsymbol{V}_t + \frac{\nabla p}{\rho}\right) - \left(\frac{j\nu + \gamma}{2k}\right)\Delta\boldsymbol{V} \tag{22.34}$$

Equations of a micropolar theory in a form of Equations 22.33 and 22.34 coincide with the generalized Navier–Stokes model (Eqs. 22.20 and 22.21) if micropolar empirical material constants are specified in terms of a Knudsen number and a time characteristic constant as follows:

$$j = \tau\frac{k}{\rho}, \quad \frac{\gamma}{k} = \frac{l^2}{3} = \frac{Kn^2L^2}{3} \tag{22.35}$$

A simplified version of a micropolar theory (couple stress model) is presented by Stokes [6] neglecting all derivatives in the operator $\bar{L}$ (Eq. 22.34).

22.5 PRESSURE-DRIVEN STEADY FLOWS IN MICROCHANNELS

Consider pressure-driven steady-state isothermal flow in a cylindrical microchannel of an arbitrary shape. The governing equation [20] expresses the momentum balance, which is in a stationery condition that reads

$$\varepsilon^2\Delta v - v + \frac{p'}{\mu} = 0, \tag{22.36a}$$

$$\Delta u = v, \qquad \varepsilon = \frac{l}{\sqrt{6}} = Kn\frac{L}{\sqrt{6}} \tag{22.36b}$$

With application to the rectangular cross section (Figure 22.1), the low order boundary conditions are:

$$u_x|_{x=0} = 0, \qquad u_y|_{y=0} = 0 \tag{22.37a}$$

$$\left(u + \varepsilon_1\frac{\partial u}{\partial x}\right)|_{x=a} = 0, \qquad \left(u + \varepsilon_1\frac{\partial u}{\partial y}\right)|_{y=b} = 0, \qquad \varepsilon_1 = \sigma_p l = Kn * \sigma_p L \tag{22.37b}$$

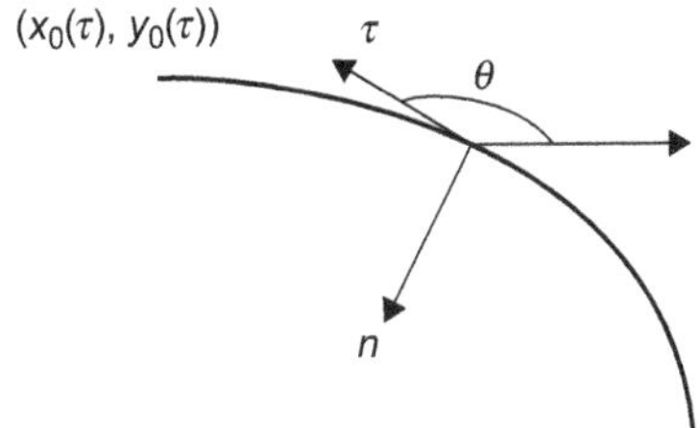

FIGURE 22.1 Natural coordinates (τ, n) in the vicinity to the boundary.

while the higher order boundary layer-related conditions read

$$v_x|_{x=0} = 0, \qquad v_y|_{y=0} = 0 \tag{22.38a}$$

$$v|_{x=a} = 0, \qquad v|_{y=b} = 0 \tag{22.38b}$$

Conditions (Eq. 22.37a) are the symmetry conditions, and conditions (Eq. 22.37b) are the first-order Maxwell slip boundary conditions at the liquid–solid interphase. The additional higher order conditions pertain to symmetry at $x = 0$ and $y = 0$, (Eq. 22.38a), and to the boundary layer conditions (Eq. 22.38b) at the solid–liquid interphase.

22.5.1 Asymptotic Expansions in Cartesian Coordinates

The difficulty of a multidimensional problem (Eq. 22.36 to Eq. 22.38) precludes from solving it exactly. However, the presence of a small parameter (squared Knudsen number) indicates existence of a thin boundary layer region, opening up a possibility to use a simplified model. The basic tool explored in this section is similar to the one used in Prandtl's boundary layer theory, where the flow around an airfoil is treated as inviscid far away from the wall, but viscous in its proximity. The present approach identifies the core flow by using classical Navier–Stokes model with the following boundary layer correction in a close vicinity to the wall.

A two-scale asymptotic expansion is defined as follows:

$$v(x, y) = \frac{p'}{\mu} + U\left(\frac{x-a}{\varepsilon}, y\right) + V\left(x, \frac{y-b}{\varepsilon}\right) \tag{22.39}$$

where the boundary layers coordinates are introduced $X = \frac{x-a}{\varepsilon}$, $Y = \frac{y-b}{\varepsilon}$ in the vicinities to the edges $x = a$ and $y = b$ accordingly. Introducing Equation 22.39 into Equation 22.36a, we obtain

$$\varepsilon^2 \Delta v - v + \frac{p'}{\mu} = \left(U_{XX} + \varepsilon^2 V_{YY}\right) + \left(\varepsilon^2 U_{XX} + V_{YY}\right) - \left(\frac{p'}{\mu} + U + V\right) + \frac{p'}{\mu} = 0 \tag{22.40}$$

The leading order equations for the boundary layers

$$U_{XX} - U = 0, \quad V_{YY} - V = 0 \tag{22.41}$$

identify solution exponentially decaying away from the walls, $U \sim e^{\frac{x-a}{\varepsilon}}$, $V \sim e^{\frac{y-b}{\varepsilon}}$. As an outcome, the leading order solution (Eq. 22.39) satisfying boundary conditions (Eq. 22.38) reads

$$v(x, y) = \frac{p'}{\mu}\left(1 - e^{\frac{x-a}{\varepsilon}} - e^{\frac{y-b}{\varepsilon}}\right) \tag{22.42}$$

The presented asymptotic solution (Eq. 22.42) is valid under the assumption that the boundary of a domain is a smooth curve. In case of a boundary, contained corner points, the solution (Eq. 22.42) should be supplemented by the corner boundary function [24], dying rapidly away from the point $x = a$, $y = b$. Considering geometry where the area relating to the corner boundary layer is negligible compared to the total area occupied by the edge boundary layer, the local corner effect can be neglected.

To find velocity distribution $u(x, y)$, consider the Poisson equation 22.36b. We are looking for the solution as an expansion by a complete set of eigenfunctions using the generalized double Fourier series

$$u(x,y) = \sum_{m=1}^{\infty}\sum_{n=1}^{\infty} A_{mn}\varphi_m(x)\psi_n(y) \tag{22.43}$$

where

$$\varphi_{m,xx} - \varphi_m = 0, \quad \varphi_{m,x}(0) = \varphi_m(a) + \varepsilon_1\varphi_{m,x}(a) = 0 \tag{22.44a}$$

$$\psi_{n,yy} - \psi_n = 0, \quad \psi_{n,y}(0) = \psi_n(b) + \varepsilon_1\psi_{n,y}(b) = 0 \tag{22.44b}$$

The solution of (44) yields $\varphi_m(x) = \cos(\lambda_m x)$, $\psi_n(y) = \cos\left(\lambda_n y\right)$, satisfying boundary conditions (Eq. 22.37), where the eigenvalues are the roots of the following transcendental equations:

$$\tan(\lambda_m a) = (\varepsilon_1\lambda_m)^{-1}, \qquad \tan(\lambda_n b) = (\varepsilon_1\lambda_n)^{-1} \tag{22.45}$$

To find unknown coefficients A_{mn}, we plug expansion (Eq. 22.43) into the Equation 22.36b, forming inner product with each of the elements of the set $\left\{\varphi_m(x)\psi_n(y)\right\}$

$$A_{mn} = -\frac{1}{\left(\lambda_m^2 + \lambda_n^2\right)}\int_0^a\int_0^b\left(1 - e^{\frac{x-a}{\varepsilon}} - e^{\frac{y-b}{\varepsilon}}\right)\varphi_m(x)\psi_n(y))dxdy \Big/ \int_0^a\int_0^b(\varphi_m(x)\psi_n(y))dxdy \tag{22.46}$$

Carrying out the leading order approximation terms leads to

$$A_{mn} = -\frac{4p'}{\mu\bar{a}\bar{b}}\frac{\sin(\lambda_m a)\sin(\lambda_n b)}{\lambda_m\lambda_n\left(\lambda_m^2 + \lambda_n^2\right)}\left[1 - \varepsilon^2\left(\lambda_m^2 + \lambda_n^2\right)\right] + O(\varepsilon^4) \tag{22.47}$$

$$\bar{a} = a + \frac{\sin(2\lambda_m a)}{2\lambda_m}, \quad \bar{b} = b + \frac{\sin(2\lambda_n b)}{2\lambda_n}$$

The presented solution (Eqs. 22.43 and 22.47) is a superposition of the classical Poiseuille solution, following from the above at $\varepsilon = \varepsilon_1 = 0$, and its correction, affected by the Knudsen boundary layer. We can simplify the procedure of finding the roots of transcendental equations (Eq. 22.45) by substituting (Eq. 22.45) with its asymptotic expansion in terms of a small parameter ε_1. When $\varepsilon_1 = 0$ (Eq. 22.45) reduces to $\tan(\lambda_m a) = \infty$, $\tan(\lambda_n b) = \infty$, whose roots are $\lambda_{m0}a = (m - 0.5)\pi$, $\lambda_{n0}a = (n - 0.5)\pi$. If ε_1 is small, but finite, we expect roots λ_m and λ_n to deviate slightly from λ_{m0} to λ_{n0}. In this case, we assume that the roots have an expansion $\lambda_m = \lambda_{m0} + \varepsilon_1\Delta\lambda_m + \ldots$, $\lambda_n = \lambda_{n0} + \varepsilon_1\Delta\lambda_n + \ldots$. By substituting the assumed expansion into the transcendental Equation 22.45 and neglecting all quadratic and higher order terms of ε_1, we obtain

$$\lambda_m = (1 - \varepsilon_1)\frac{(2m-1)\pi}{2a}, \quad \lambda_n = (1 - \varepsilon_1)\frac{(2n-1)\pi}{2b} \tag{22.48}$$

22.5.2 Asymptotic Expansions in Curvilinear Coordinates

The asymptotic solution of Equations 22.36a, 22.36b consists of regular and boundary layers. To construct the boundary layer in a domain of an arbitrary shape, we introduce a new local boundary—fitted coordinate system. This is an orthogonal system, consisting of an axis τ tangent to the boundary, and an axis n, oriented toward the center of the curvature. We assume that the boundary layer is sufficiently thin, so that the normal drawn through different points on the boundary do not intersect within the layer. Let the equation describing boundary have the parametric form $x = x_0(\tau), y = y_0(\tau), \theta = \text{atan}(y_{0,\tau}/x_{0,\tau}) = \theta(\tau)$ (Figure 22.1). The relationship between the global Cartesian and local curvilinear coordinates can be defined in the following way (Figure 22.1)

$$x(\tau, n) = x_0(\tau) - n * \sin(\theta(\tau)); \quad y(\tau, n) = y_0(\tau) + n * \cos(\theta(\tau)) \tag{22.49}$$

Equation 22.36b in an orthogonal curvilinear system of coordinates can be presented as [18]

$$\frac{\varepsilon^2}{h_n h_\tau}\left[\frac{\partial}{\partial n}\left(\frac{h_\tau}{h_n}\frac{\partial v}{\partial n}\right) + \frac{\partial}{\partial \tau}\left(\frac{h_n}{h_\tau}\frac{\partial v}{\partial \tau}\right)\right] - v + \frac{p'}{\mu} = 0 \tag{22.50}$$

where h_τ, h_n are the metric components in curvilinear coordinates (Lamé coefficients)

$$h_n = \sqrt{\left(\frac{\partial x}{\partial n}\right)^2 + \left(\frac{\partial y}{\partial n}\right)^2}, h_\tau = \sqrt{\left(\frac{\partial x}{\partial \tau}\right)^2 + \left(\frac{\partial y}{\partial \tau}\right)^2} \tag{22.51}$$

An asymptotic expansion for the solution of Equation 22.50 is defined as follows:

$$v(\tau, n) = \frac{p'}{\mu} + V\left(\tau, \frac{n}{\varepsilon}\right) \tag{22.52}$$

where the local stretched coordinate $\eta = \frac{n}{\varepsilon}$ is introduced in the vicinity to the boundary layer. The leading order equation for the boundary layer function $V(\tau, \eta)$ is obtained by substituting Equation 22.52 into Equation 22.50

$$V_{\eta\eta} - h_n^2 V = 0 \tag{22.53}$$

whose solution should exponentially decay leaving the boundary layer, $V \sim \exp(-h_n \frac{n}{\varepsilon})$. As a result, solution (Eq. 22.52) satisfying the Dirichlet boundary condition $v|_{\eta=0} = 0$ reads

$$v(\tau, n) = \frac{p'}{\mu}\left(1 - e^{\frac{-h_n n}{\varepsilon}}\right) \tag{22.54}$$

To find velocity distribution $u(x, y)$, the Poisson equation 22.36b with the right part presented by Equation 22.54 should be solved numerically in a relating two-dimensional domain.

22.6 RESULTS AND DISCUSSION

22.6.1 Plane Flow Between Parallel Plates

A one-dimensional model follows from equation 36 by setting all derivatives by x equal to zero. A normalized coordinate $\bar{y}=y/a$, so that $\bar{y} = 0$ corresponds to the axis of symmetry, and $\bar{y} = 1$ relates to the wall. A normalized velocity $\bar{u}$ and a normalized flow rate $\bar{Q}$ are defined as follows: $\bar{u} = u\mu/p'a^2$, $\bar{Q} = Q\mu/p'a^3$. In Figure 22.2 the normalized velocity distribution is plotted against the normalized coordinate $\bar{y}$ for $Kn = 0.2$.

Square markers indicate the classical Poiseuille solution, which does not depend on Knudsen number. Circles correspond to the gradient model satisfying the no-slip boundary conditions at the wall as a result of the roughness effect. Velocities predicted by the gradient model in this case are lower than the Navier–Stokes solution, which is attributed to the additional rotational degree of freedom assigned to the particle at a moderate level of a Knudsen number. The Maxwell's model assigning slip boundary conditions to the classical Navier–Stokes equations (diamond markers) predicts velocity distribution as a shift of the Navier–Stokes solution by the constant wall slip velocity.

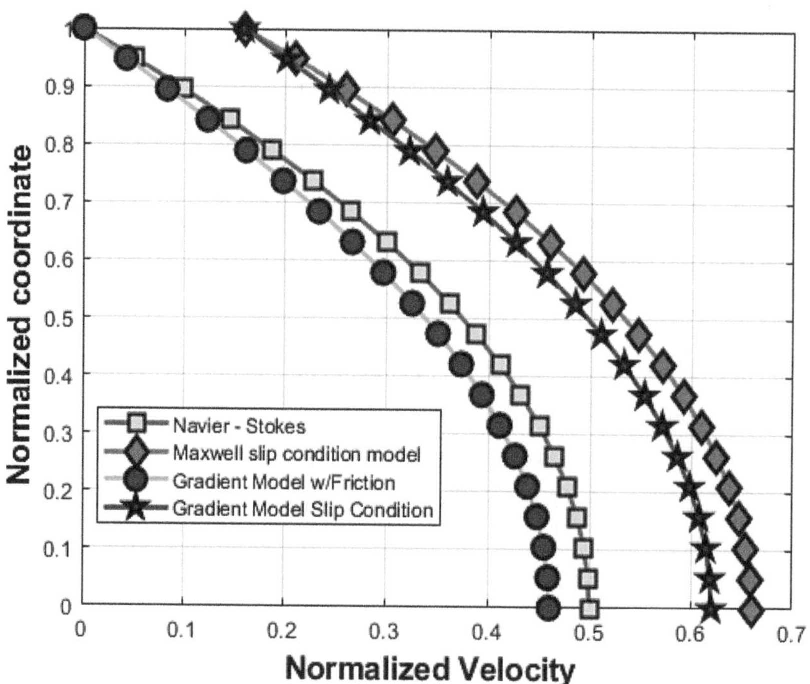

FIGURE 22.2 Comparison of velocity profile distributions for different models in microchannel.

Finally, the velocity distribution presented by pentagrams relates to the model accounting for the overall effect of particle rotations and the wall velocity slip. Since the effect of a velocity slip is dominant over the effect of spinning particles, velocity relating to the gradient model exceeds the Poiseuille velocity but with the lesser effect than the Maxwell's model.

In Figure 22.3, the normalized flow rate is plotted against the Knudsen number according to the four different models. Dash line indicates the Navier–Stokes solution independent on a Knudsen number. Diamonds relate to the gradient model satisfying the no-slip boundary conditions at the rough wall.

In this case, the flow rate prediction by the Poiseuille solution exceeds the one predicted by the gradient model, correlating with the velocity distributions for both models presented in Figure 22.1. Square markers designate the Maxwell's wall velocity slip model. Circles relate to the gradient model, accounting for both the slip boundary effect and the particle's rotation. The latter predicts flow rate, which exceeds the flow rate of a classical Navier–Stokes, but to a lesser extent than the Maxwell' model.

22.6.2 Flow in Rectangular Microchannels

Two-dimensional solutions in Cartesian coordinates were obtained using trigonometric expansion (43) with the coefficients calculated according to (47). It was sufficient to use 10 terms of the expansion (43) in each direction due to its fast convergence. The first 10 eigenvalues, $\lambda_m a$, obtained by solving the transcendental equation 45 for $Kn = 0.1$, are presented in Table 22.1.

The experimental data from [11] along with predictions from the Poiseuille model gradient models for the channel $3000 \times 600 \times 30\ \mu m^3$ ($L \times W \times H$) are presented in Figure 22.4. Square markers designate the testing data as the average data of three consecutive experimental runs. The gradient model at $Kn = 0.1$ (solid line) predicts the experimental data better than the classical Navier–Stokes theory (dash line).

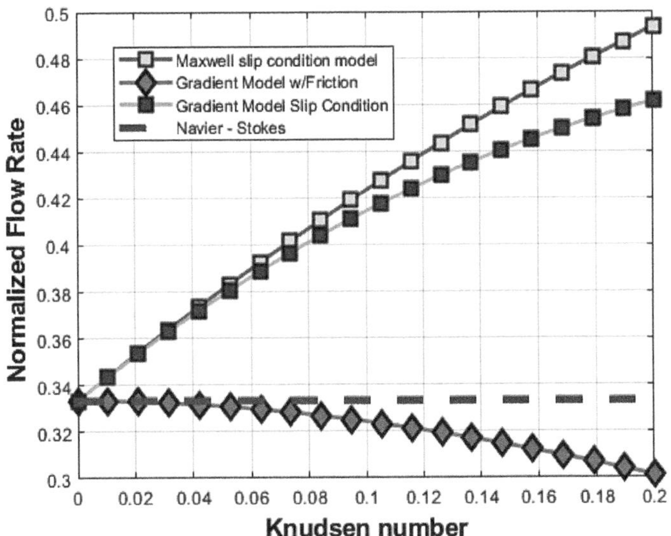

FIGURE 22.3 A normalized flow rate dependence on a Knudsen number for different models in microchannel.

TABLE 22.1
Eigenvalues of Equation 44 Calculated for $Kn = 0.1$

m	1	2	3	4	5	6	7	8	9	10
$\lambda_m a$	1.404	4.241	7.145	10.113	13.131	16.1835	19.257	22.348	25.449	28.558

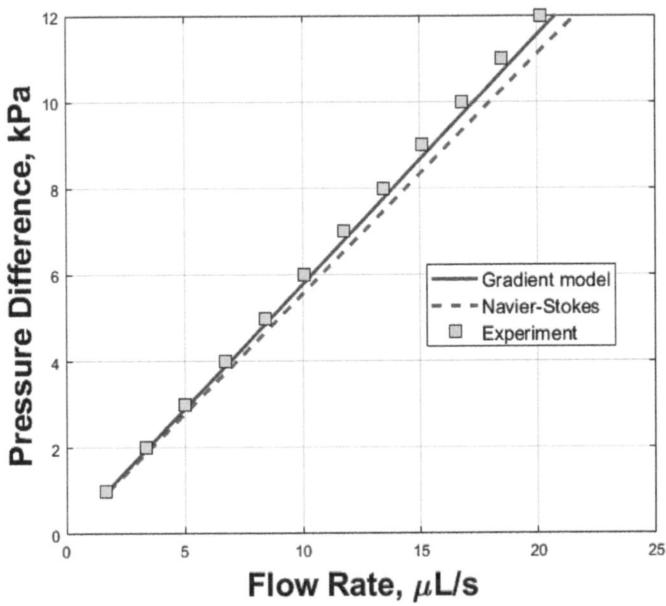

FIGURE 22.4 Comparison of the Navier–Stokes model, gradient model, and the experimental data for water flow in the microchannel 3000 × 600 × 30 μm^3 ($L \times W \times H$).

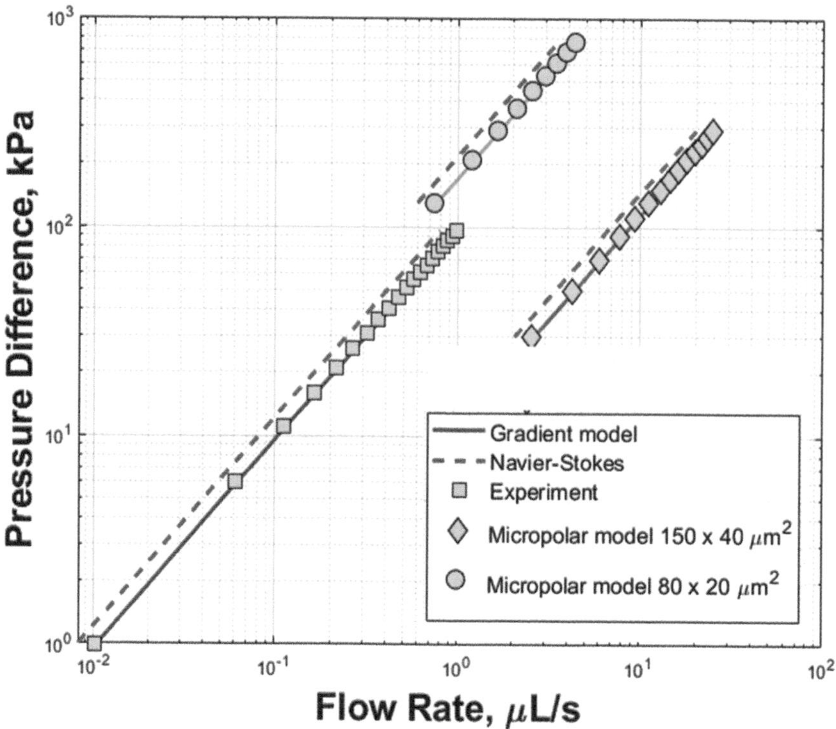

FIGURE 22.5 Comparison of the experimental data with the Navier–Stokes model, gradient model, and the micropolar model for water flow in three microchannels.

Figure 22.5 shows the experimental data from [11] compared to the predictions by the Navier–Stokes theory and micropolar and gradient models. The channel used to obtain experimental data, marked by squares, is $3000 \times 600 \times 30\ \mu m^3$. Microchannels used to compare to micropolar theory are $10{,}000 \times 60 \times 25.4\ \mu m^3$ and $11{,}700 \times 80 \times 20\ \mu m^3$. The gradient model provides a good prediction of experimental data over the wide range of applied pressure drop and a flow range.

22.7 CONCLUSION

A refined gradient model as an extension of classical fluid mechanics with a higher order of spatial and time derivatives of velocities, pressure, and acceleration is presented. Size effects, characterized by a mean free path or the size of a microstructural lattice, are captured by this model, making feasible its applications in micro- and nanoscales. In contrast to the number of gradient models, requiring a large quantity of phenomenological constants, the present model requires one additional constant, linked to the microscale characteristic length for the steady-state flow, and two additional parameters in a nonsteady counterpart. The derived model exactly matches the nonsteady linearized micropolar fluid flow model (small gyrations and velocity distributions), giving analytical interpretation of typical micropolar empirical coefficients in terms of a Knudsen number and a time characteristic constant.

An asymptotic behavior of a two-dimensional flow through a microchannel is investigated via rigorous singular perturbation analysis with respect to the small parameter, proportional to the Knudsen number. A singular perturbation technique reveals a thin boundary layer (Knudsen layer) region near the solid boundary described by the inner solution, whereas the outer solution relates to the classical Navier–Stokes model. For the boundary layer, solution is presented in an analytical form, while the outer region is modeled using analytical solution in case of a canonical domain or using a numerical solution for a noncanonical domain. Calculational results obtained compare favorably with the currently available experimental data.

REFERENCES

1. Askes H., Suiker A.S.J., Sluys L.J., A classification of higher order strain-gradient models-linear analysis. *Archive of Applied Mechanics*, 72 (2–3), (2002), 171-188.
2. Eringen A.S., *Micro Continuum Field Theories II: Fluent Media*, Springer Verlag, New York, 2001.
3. Eringen A.S., "Micropolar Theory of Liquid Crystals," In: J. F. Johnson and R. S. Porter, Eds., *Liquid Crystals and Ordered Fluids*, Vol. 3, Plenum Publishing, New York, 1978.
4. Lukaszewicz G., *Micropolar Fluid Theory and Applications, Modeling and Simulation in Science, Engineering and Technology*, Birkhauser, Boston, 1999.
5. Power P., *Bio-Fluid Mechanics, Advances in Fluid Mechanics*, W.I.T. Press, UK, 1995.
6. Stokes V.K., *Theories of Fluids with Microstructure*. Springer-Verlag, Berlin, 1984.
7. Karniadakis G., Beskok A., Alurtu N., "Microflows and Nanoflows." *Fundamentals and Simulation*, Springer, New York, 2000.
8. Song Y., Cheng D., Zhao L., *Microfluidics: Fundamentals, Devices and Applications: Fundamentals and Applications*, Wiley-VCH Verlag, Weinheim, Germany, 2018.
9. Agarwal R.K., Balakrishnan R., Beyond Navier-Stokes: Burnett equations for flows in the continuum – transition regime. *Physics of Fluids* 14, 1818, 2002.
10. Kandlikar S., Garimella S., Li D., Colin S., King M., *Heat Transfer and Fluid Flow in Minichannels and Microchannels*, Elsevier Ltd, Oxford, UK, 2014.
11. Papautsky I., Brazzle J., Ameel T., Frazier A.B., Laminar fluid behavior in microchannels using micropolar fluid theory. *Sensors Actuators*, 73, 101-108, 1999.
12. Papautsky I., Brazzle J., Ameel T., Frazier A.B., Microchannel fluid behavior using micropolar fluid theory. Proc. EMBS 97, Microelectromech. Sys., Chicago, 2285–2291, 1997.
13. Prokhorenko P., Migoun N.P., Stadthause M., *Theoretical Properties of Liquid Penetrant Testing*, DVS Verlag, Berlin, 1999.
14. Zhu, Y., Granick, S., Limits of the hydrodynamic no-slip boundary condition. *Physical Review Letters*, 88(10):106102.
15. Liberson A.S., Microscale: Wave Propagation Modeling Using a High-Order Gradient Continuum Theory. *Dekker Encyclopedia of Nanoscience and Nanotechnology*, Ed. E. Lyshevsky. Third Edition, Seven Volumes, CRC Press, Boca Raton, 2017.
16. Liberson A.S., Vahedein Y., A Variant of a Gradient Continuum Mechanics with Application to Flow in Microchannels. Proceedings of the 4th International Conference of Fluid Flow, Heat and Mass Transfer (FFHMT'17) Toronto, Canada, August 22–23, 2017 Paper No. 176.
17. Schlichting, H., *Boundary Layer Theory*, 9th edition, Springer - Verlag, Berlin Heidelberg, 2017.
18. Korn G.A., Korn T.A., *Mathematical Handbook for Scientists and Engineers*, McGraw-Hill, New York, 1968.
19. Kang C.K., Eringen A.S., The effect of microstructure on the rheological properties of blood, *Bulletin of Mathematical Biology* 38, 135-158, 1976.
20. Hazanka G.C., Phukan B., Effects of variable viscosity and thermal conductivity on MHD flow of micropolar fluid in a continuous moving flat plate. *International Journal of Computer Applications*, 122(8), 0975–8887, 2015.
21. Crnyaric-Zic, Mujacovic N., Numerical analysis of the solutions for 1D compressible viscous micropolar fluid flow with different boundary conditions. *Mathematics and Computers in Simulation*, 12645–62, 2016.
22. Richardson S., On the no-slip boundary conditions. *Journal of Fluid Mechanics* 58, 707, 1973.
23. Mo G., Rosenberger F. Molecular dynamics simulation of flow in a two dimensional channel with atomically rough walls. *Physical Review A*, 42, 4688-4692, 1990.
24. Han H., Kellogg R., Differentiability properties of solutions of the equation $-\varepsilon^2 \Delta u + ru = f(x, y)$ in a square. *SIAM Journal on Mathematical Analysis* 21(2), 394-408, 1990.

Index

Page numbers in bold indicate tables and italics indicate figures.

J

L

M

N

O

P

Q

R

S

T

U

V

W